Calculus

Fifth Edition

Frank Ayres, Jr., PhD

Formerly Professor and Head of the Department of Mathematics
Dickinson College

Elliott Mendelson, PhD

Professor of Mathematics
Queens College

Schaum's Outline Series

New York Chicago San Francisco Lisbon London
Madrid Mexico City Milan New Delhi San Juan
Seoul Singapore Sydney Toronto

The ***McGraw·Hill*** *Companies*

FRANK AYRES, Jr., PhD, was formerly Professor and Head of the Department at Dickinson College, Carlisle, Pennsylvania. He is the coauthor of *Schaum's Outline of Trigonometry* and *Schaum's Outline of College Mathematics*.

ELLIOTT MENDELSON, PhD, is Professor of Mathematics at Queens College. He is the author of *Schaum's Outline of Beginning Calculus*.

Schaum's Outline of CALCULUS

7 8 9 10 11 CUS CUS 1 5 4 3 2

MHID 0-07-150861-9
ISBN 978-0-07-150861-2

Sponsoring Editor: Charles Wall
Production Supervisor: Tama Harris McPhatter
Editing Supervisor: Maureen B. Walker
Interior Designer: Jane Tenenbaum
Project Manager: Madhu Bhardwaj

Library of Congress Cataloging-in-Publication Data is on file with the Library of Congress.

SCHAUM'S
outlines

Calculus

Preface

The purpose of this book is to help students understand and use the calculus. Everything has been aimed toward making this easier, especially for students with limited background in mathematics or for readers who have forgotten their earlier training in mathematics. The topics covered include all the material of standard courses in elementary and intermediate calculus. The direct and concise exposition typical of the Schaum Outline series has been amplified by a large number of examples, followed by many carefully solved problems. In choosing these problems, we have attempted to anticipate the difficulties that normally beset the beginner. In addition, each chapter concludes with a collection of supplementary exercises with answers.

This fifth edition has enlarged the number of solved problems and supplementary exercises. Moreover, we have made a great effort to go over ticklish points of algebra or geometry that are likely to confuse the student. The author believes that most of the mistakes that students make in a calculus course are not due to a deficient comprehension of the principles of calculus, but rather to their weakness in high-school algebra or geometry. Students are urged to continue the study of each chapter until they are confident about their mastery of the material. A good test of that accomplishment would be their ability to answer the supplementary problems.

The author would like to thank many people who have written to me with corrections and suggestions, in particular Danielle Cinq-Mars, Lawrence Collins, L.D. De Jonge, Konrad Duch, Stephanie Happ, Lindsey Oh, and Stephen B. Soffer. He is also grateful to his editor, Charles Wall, for all his patient help and guidance.

ELLIOTT MENDELSON

Contents

CHAPTER 1

Linear Coordinate Systems. Absolute Value. Inequalities

Linear Coordinate System

A linear coordinate system is a graphical representation of the real numbers as the points of a straight line. To each number corresponds one and only one point, and to each point corresponds one and only one number.

To set up a linear coordinate system on a given line: (1) select any point of the line as the *origin* and let that point correspond to the number 0; (2) choose a positive direction on the line and indicate that direction by an arrow; (3) choose a fixed distance as a unit of measure. If x is a positive number, find the point corresponding to x by moving a distance of x units from the origin in the positive direction. If x is negative, find the point corresponding to x by moving a distance of $-x$ units from the origin in the negative direction. (For example, if $x = -2$, then $-x = 2$ and the corresponding point lies 2 units from the origin in the negative direction.) See Fig. 1-1.

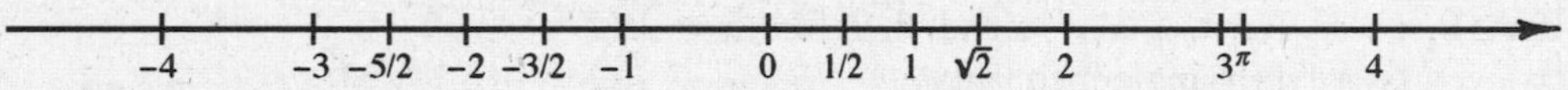

$-4 \quad -3 \quad -5/2 \quad -2 \quad -3/2 \quad -1 \quad 0 \quad 1/2 \quad 1 \quad \sqrt{2} \quad 2 \quad 3^{\pi} \quad 4$

Fig. 1-1

The number assigned to a point by a coordinate system is called the *coordinate* of that point. We often will talk as if there is no distinction between a point and its coordinate. Thus, we might refer to "the point 3" rather than to "the point with coordinate 3."

The absolute value $|x|$ of a number x is defined as follows:

$$|x| = \begin{cases} x & \text{if } x \text{ is zero or a positive number} \\ -x & \text{if } x \text{ is a negative number} \end{cases}$$

For example, $|4| = 4$, $|-3| = -(-3) = 3$, and $|0| = 0$. Notice that, if x is a negative number, then $-x$ is positive. Thus, $|x| \geq 0$ for all x.

The following properties hold for any numbers x and y.

(1.1) $|-x| = |x|$
When $x = 0$, $|-x| = |-0| = |0| = |x|$.
When $x > 0$, $-x < 0$ and $|-x| = -(-x) = x = |x|$.
When $x < 0$, $-x > 0$, and $|-x| = -x = |x|$.

(1.2) $|x - y| = |y - x|$
This follows from (**1.1**), since $y - x = -(x - y)$.

(1.3) $|x| = c$ implies that $x = \pm c$.
For example, if $|x| = 2$, then $x = \pm 2$. For the proof, assume $|x| = c$.
If $x \geq 0$, $x = |x| = c$. If $x < 0$, $-x = |x| = c$; then $x = -(-x) = -c$.

(1.4) $|x|^2 = x^2$
If $x \geq 0$, $|x| = x$ and $|x|^2 = x^2$. If $x \leq 0$, $|x| = -x$ and $|x|^2 = (-x)^2 = x^2$.

(1.5) $|xy| = |x| \cdot |y|$
By (**1.4**), $|xy|^2 = (xy)^2 = x^2y^2 = |x|^2|y|^2 = (|x| \cdot |y|)^2$. Since absolute values are nonnegative, taking square roots yields $|xy| = |x| \cdot |y|$.

(1.6) $\left|\frac{x}{y}\right| = \frac{|x|}{|y|}$ if $y \neq 0$

By (**1.5**), $|y|\left|\frac{x}{y}\right| = \left|y \cdot \frac{x}{y}\right| = |x|$. Divide by $|y|$.

(1.7) $|x| = |y|$ implies that $x = \pm y$

Assume $|x| = |y|$. If $y = 0$, $|x| = |0| = 0$ and (**1.3**) yields $x = 0$. If $y \neq 0$, then by (**1.6**),

$$\left|\frac{x}{y}\right| = \frac{|x|}{|y|} = 1$$

So, by (**1.3**), $x/y = \pm 1$. Hence, $x = \pm y$.

(1.8) Let $c \geq 0$. Then $|x| \leq c$ if and only if $-c \leq x \leq c$. See Fig. 1-2.

Assume $x \geq 0$. Then $|x| = x$. Also, since $c \geq 0$, $-c \leq 0 \leq x$. So, $|x| \leq c$ if and only if $-c \leq x \leq c$. Now assume $x < 0$. Then $|x| = -x$. Also, $x < 0 \leq c$. Moreover, $-x \leq c$ if and only if $-c \leq x$. (Multiplying or dividing an equality by a negative number reverses the inequality.) Hence, $|x| \leq c$ if and only if $-c \leq x \leq c$.

(1.9) Let $c \geq 0$. Then $|x| < c$ if and only if $-c < x < c$. See Fig. 1-2. The reasoning here is similar to that for (**1.8**).

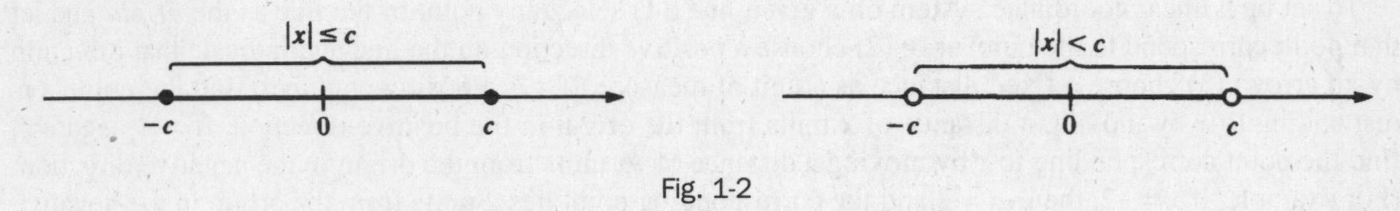

Fig. 1-2

(1.10) $-|x| \leq x \leq |x|$

If $x \geq 0$, $x = |x|$. If $x < 0$, $|x| = -x$ and, therefore, $x = -|x|$.

(1.11) $|x + y| \leq |x| + |y|$ (triangle inequality)

By (**1.8**), $-|x| \leq x \leq |x|$ and $-|y| \leq y \leq |y|$. Adding, we obtain $-(|x| + |y|) \leq x + y \leq |x| + |y|$. Then $|x + y| \leq |x| + |y|$ by (**1.8**). [In (**1.8**), replace c by $|x| + |y|$ and x by $x + y$.]

Let a coordinate system be given on a line. Let P_1 and P_2 be points on the line having coordinates x_1 and x_2. See Fig. 1-3. Then:

(1.12) $|x_1 - x_2| = \overline{P_1P_2}$ = distance between P_1 and P_2.

This is clear when $0 < x_1 < x_2$ and when $x_1 < x_2 < 0$. When $x_1 < 0 < x_2$, and if we denote the origin by O, then $\overline{P_1P_2} = \overline{P_1O} + \overline{OP_2} = (-x_1) + x_2 = x_2 - x_1 = |x_2 - x_1| = |x_1 - x_2|$.

As a special case of (**1.12**), when P_2 is the origin (and $x_2 = 0$):

(1.13) $|x_1|$ = distance between P_1 and the origin.

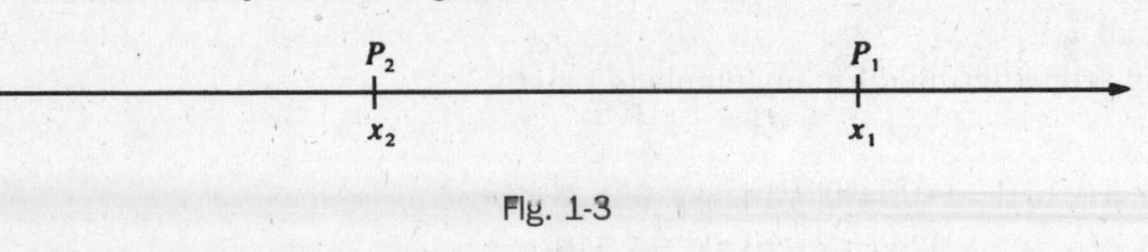

Fig. 1-3

Finite Intervals

Let $a < b$.

The *open interval* (a, b) is defined to be the set of all numbers between a and b, that is, the set of all x such that $a < x < b$. We shall use the term *open interval* and the notation (a, b) also for all the points between the points with coordinates a and b on a line. Notice that the open interval (a, b) does not contain the *endpoints* a and b. See Fig. 1-4.

The *closed interval* $[a, b]$ is defined to be the set of all numbers between a and b or equal to a or b, that is, the set of all x such that $a \leq x \leq b$. As in the case of open intervals, we extend the terminology and notation to points. Notice that the closed interval $[a, b]$ contains both endpoints a and b. See Fig. 1-4.

Open interval (a, b): $a < x < b$ **Closed interval** $[a, b]$: $a \le x \le b$

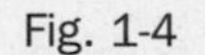

Fig. 1-4

By a *half-open interval* we mean an open interval (a, b) together with one of its endpoints. There are two such intervals: $[a, b)$ is the set of all x such that $a \le x < b$, and $(a, b]$ is the set of all x such that $a < x \le b$.

Infinite Intervals

Let (a, ∞) denote the set of all x such that $a < x$.
Let $[a, \infty)$ denote the set of all x such that $a \le x$.
Let $(-\infty, b)$ denote the set of all x such that $x < b$.
Let $(-\infty, b]$ denote the set of all x such that $x \le b$.

Inequalities

Any inequality, such as $2x - 3 > 0$ or $5 < 3x + 10 \le 16$, determines an interval. To solve an inequality means to determine the corresponding interval of numbers that satisfy the inequality.

EXAMPLE 1.1: Solve $2x - 3 > 0$.

$$2x - 3 > 0$$

$$2x > 3 \quad \text{(Adding 3)}$$

$$x > \tfrac{3}{2} \quad \text{(Dividing by 2}$$

Thus, the corresponding interval is $(\frac{3}{2}, \infty)$.

EXAMPLE 1.2: Solve $5 < 3x + 10 \le 16$.

$$5 < 3x + 10 \le 16$$

$$-5 < 3x \le 6 \quad \text{(Subtracting 10)}$$

$$-\tfrac{5}{3} < x \le 2 \quad \text{(Dividing by 3)}$$

Thus, the corresponding interval is $(-\frac{5}{3}, 2]$.

EXAMPLE 1.3: Solve $-2x + 3 < 7$.

$$-2x + 3 < 7$$

$$-2x < 4 \quad \text{(Subtracting 3)}$$

$$x > -2 \quad \text{(Dividing by } -2)$$

(Recall that dividing by a negative number reverses an inequality.) Thus, the corresponding interval is $(-2, \infty)$.

SOLVED PROBLEMS

1. Describe and diagram the following intervals, and write their interval notation, (a) $-3 < x < 5$; (*b*) $2 \le x \le 6$; (c) $-4 < x \le 0$; (d) $x > 5$; (e) $x \le 2$; (f) $3x - 4 \le 8$; (g) $1 < 5 - 3x < 11$.

(a) All numbers greater than -3 and less than 5; the interval notation is $(-3, 5)$:

−3 5

(b) All numbers equal to or greater than 2 and less than or equal to 6; [2, 6]:

(c) All numbers greater than –4 and less than or equal to 0; (–4, 0]:

(d) All numbers greater than 5; (5, ∞):

(e) All numbers less than or equal to 2; (–∞, 2]:

(f) $3x - 4 \le 8$ is equivalent to $3x \le 12$ and, therefore, to $x \le 4$. Thus, we get (–∞, 4]:

(g)

$$1 < 5 - 3x < 11$$

$$-4 < -3x < 6 \quad \text{(Subtracting 5)}$$

$$-2 < x < \tfrac{4}{3} \quad \text{(Dividing by – 3; note the reversal of inequalities)}$$

Thus, we obtain $(-2, \frac{4}{3})$:

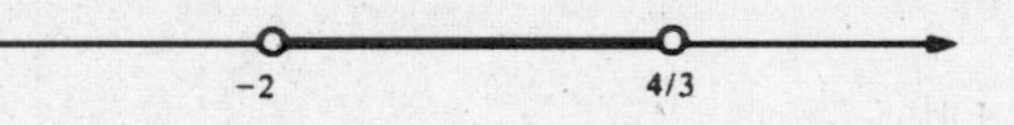

2. Describe and diagram the intervals determined by the following inequalities, (a) $|x| < 2$; (b) $|x| > 3$; (c) $|x - 3| < 1$; (d) $|x - 2| < \delta$ where $\delta > 0$; (e) $|x + 2| \le 3$; (f) $0 < |x - 4| < \delta$ where $\delta > 0$.

(a) By property (**1.9**), this is equivalent to $-2 < x < 2$, defining the open interval (–2, 2).

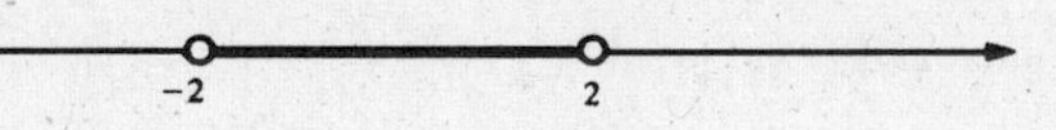

(b) By property (**1.8**), $|x| \le 3$ is equivalent to $-3 \le x \le 3$. Taking negations, $|x| > 3$ is equivalent to $x < -3$ or $x > 3$, which defines the union of the intervals (–∞, –3) and (3, ∞).

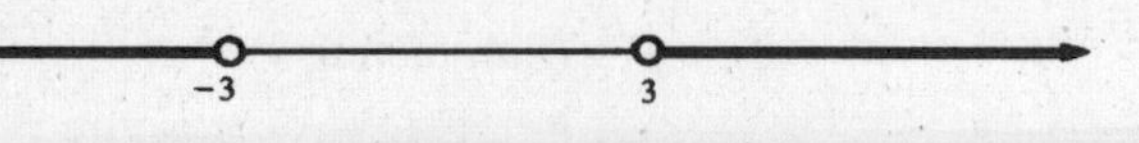

(c) By property (**1.12**), this says that the distance between x and 3 is less than 1, which is equivalent to $2 < x < 4$. This defines the open interval (2, 4).

We can also note that $|x - 3| < 1$ is equivalent to $-1 < x - 3 < 1$. Adding 3, we obtain $2 < x < 4$.

(d) This is equivalent to saying that the distance between x and 2 is less than δ, or that $2 - \delta < x < 2 + \delta$, which defines the open interval $(2 - \delta, 2 + \delta)$. This interval is called the *δ-neighborhood* of 2:

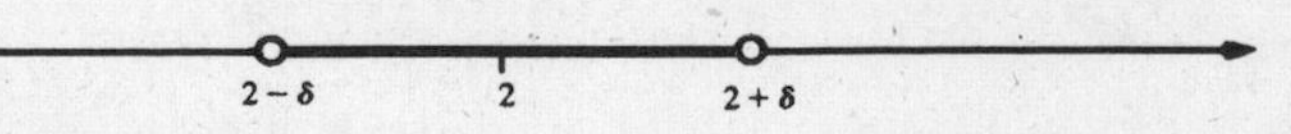

(e) $|x + 2| < 3$ is equivalent to $-3 < x + 2 < 3$. Subtracting 2, we obtain $-5 < x < 1$, which defines the open interval $(-5, 1)$:

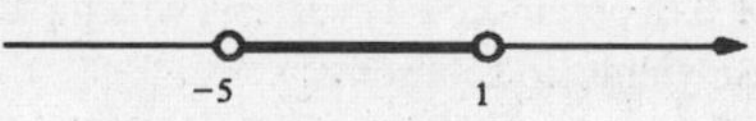

(f) The inequality $|x - 4| < \delta$ determines the interval $4 - \delta < x < 4 + \delta$. The additional condition $0 < |x - 4|$ tells us that $x \neq 4$. Thus, we get the union of the two intervals $(4 - \delta, 4)$ and $(4, 4 + \delta)$. The result is called the *deleted δ-neighborhood* of 4:

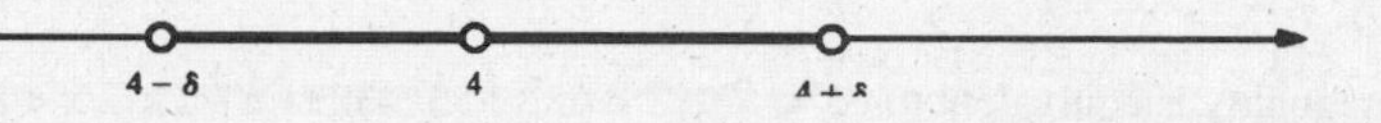

3. Describe and diagram the intervals determined by the following inequalities, (a) $|5 - x| \leq 3$; (b) $|2x - 3| < 5$; (c) $|1 - 4x| < \frac{1}{2}$.

(a) Since $|5 - x| = |x - 5|$, we have $|x - 5| \leq 3$, which is equivalent to $-3 \leq x - 5 \leq 3$. Adding 5, we get $2 \leq x \leq 8$, which defines the closed interval $[2, 8]$:

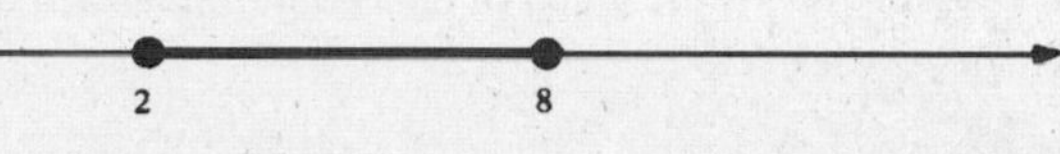

(b) $|2x - 3| < 5$ is equivalent to $-5 < 2x - 3 < 5$. Adding 3, we have $-2 < 2x < 8$; then dividing by 2 yields $-1 < x < 4$, which defines the open interval $(-1, 4)$:

(c) Since $|1 - 4x| = |4x - 1|$, we have $|4x - 1| < \frac{1}{2}$, which is equivalent to $-\frac{1}{2} < 4x - 1 < \frac{1}{2}$. Adding 1, we get $\frac{1}{2} < 4x < \frac{3}{2}$. Dividing by 4, we obtain $\frac{1}{8} < x < \frac{3}{8}$, which defines the open interval $(\frac{1}{8}, \frac{3}{8})$:

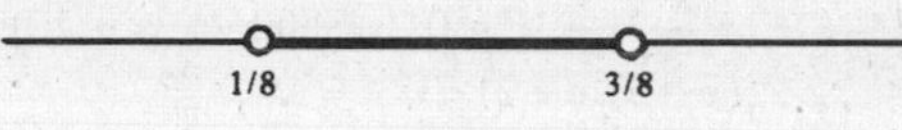

4. Solve the inequalities: (a) $18x - 3x^2 > 0$; (b) $(x + 3)(x - 2)(x - 4) < 0$; (c) $(x + 1)^2(x - 3) > 0$, and diagram the solutions.

(a) Set $18x - 3x^2 = 3x(6 - x) = 0$, obtaining $x = 0$ and $x = 6$. We need to determine the sign of $18x - 3x^2$ on each of the intervals $x < 0$, $0 < x < 6$, and $x > 6$, to determine where $18x - 3x^2 > 0$. Note that it is negative when $x < 0$ (since x is negative and $6 - x$ is positive). It becomes positive when we pass from left to right through 0 (since x changes sign but $6 - x$ remains positive), and it becomes negative when we pass through 6 (since x remains positive but $6 - x$ changes to negative). Hence, it is positive when and only when $0 < x < 6$.

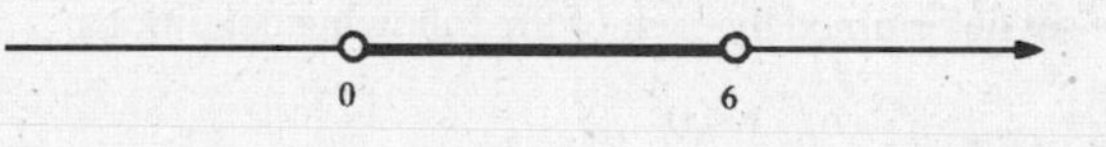

(b) The crucial points are $x = -3$, $x = 2$, and $x = 4$. Note that $(x + 3)(x - 2)(x - 4)$ is negative for $x < -3$ (since each of the factors is negative) and that it changes sign when we pass through each of the crucial points. Hence, it is negative for $x < -3$ and for $2 < x < 4$:

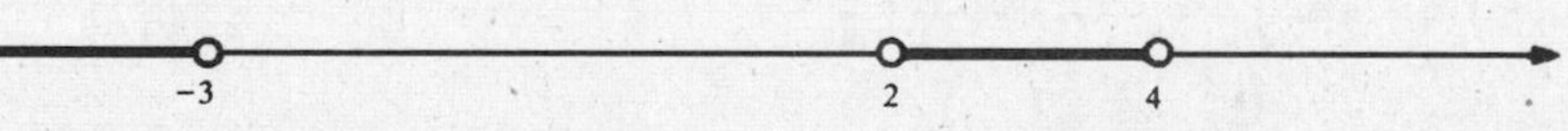

(c) Note that $(x + 1)$ is always positive (except at $x = -1$, where it is 0). Hence $(x + 1)^2 (x - 3) > 0$ when and only when $x - 3 > 0$, that is, for $x > 3$:

3

5. Solve $|3x - 7| = 8$.

By (**1.3**), $|3x - 7| = 8$ if and only if $3x - 7 = \pm 8$. Thus, we need to solve $3x - 7 = 8$ and $3x - 7 = -8$. Hence, we get $x = 5$ or $x = -\frac{1}{3}$.

6. Solve $\dfrac{2x+1}{x+3} > 3$.

Case 1: $x+3>0$. Multiply by $x+3$ to obtain $2x+1>3x+9$, which reduces to $-8>x$. However, since $x+3>0$, it must be that $x>-3$. Thus, this case yields no solutions.

Case 2: $x+3<0$. Multiply by $x+3$ to obtain $2x+1<3x+9$. (Note that the inequality is reversed, since we multiplied by a negative number.) This yields $-8<x$. Since $x+3<0$, we have $x<-3$. Thus, the only solutions are $-8<x<-3$.

7. Solve $\left|\dfrac{2}{x}-3\right|<5$.

The given inequality is equivalent to $-5 < \underline{2} - 3 < 5$. Add 3 to obtain $-2 < 2/x < 8$, and divide by 2 to get $-1 < 1/x < 4$.

Case 1: $x>0$. Multiply by x to get $-x<1<4x$. Then $x>\frac{1}{4}$ and $x>-1$; these two inequalities are equivalent to the single inequality $x>\frac{1}{4}$.

Case 2: $x<0$. Multiply by x to obtain $-x>1>4x$. (Note that the inequalities have been reversed, since we multiplied by the negative number x.) Then $x<\frac{1}{4}$ and $x<-1$. These two inequalities are equivalent to $x<-1$.

Thus, the solutions are $x>\frac{1}{4}$ or $x<-1$, the union of the two infinite intervals $(\frac{1}{4}, \infty)$ and $(-\infty, -1)$.

8. Solve $|2x-5| \geq 3$.

Let us first solve the negation $|2x-5|<3$. The latter is equivalent to $-3<2x-5<3$. Add 5 to obtain $2<2x<8$, and divide by 2 to obtain $1<x<4$. Since this is the solution of the negation, the original inequality has the solution $x \leq 1$ or $x \geq 4$.

9. Solve: $x^2 < 3x + 10$.

$$x^2 < 3x + 10$$
$$x^2 - 3x - 10 < 0 \quad \text{(Subtract } 3x+10\text{)}$$
$$(x-5)(x+2) < 0$$

The crucial numbers are -2 and 5. $(x-5)(x+2)>0$ when $x<-2$ (since both $x-5$ and $x+2$ are negative); it becomes negative as we pass through -2 (since $x+2$ changes sign); and then it becomes positive as we pass through 5 (since $x-5$ changes sign). Thus, the solutions are $-2<x<5$.

SUPPLEMENTARY PROBLEMS

10. Describe and diagram the set determined by each of the following conditions:

(a) $-5<x<0$
(b) $x \leq 0$
(c) $-2 \leq x < 3$
(d) $x \geq 1$
(e) $|x|<3$
(f) $|x| \geq 5$
(g) $|x-2|<\frac{1}{2}$
(h) $|x-3|>1$
(i) $0<|x-2|<1$
(j) $0<|x+3|<\frac{1}{4}$
(k) $|x-2| \geq 1$.

Ans. (e) $-3<x<3$; (f) $x \geq 5$ or $x \leq -5$; (g) $\frac{3}{2}<x<\frac{5}{2}$; (h) $x<2$ or $x>4$; (i) $x \neq 2$ and $1<x<3$; (j) $\frac{13}{4} < x < \frac{11}{4}$; (k) $x \geq 3$ or $x \leq 1$

11. Describe and diagram the set determined by each of the following conditions:

(a) $|3x-7|<2$
(b) $|4x-1| \geq 1$
(c) $\left|\dfrac{x}{3}-2\right| \leq 4$

(d) $\left|\frac{3}{x}-2\right| \le 4$

(e) $\left|2+\frac{1}{x}\right| > 1$

(f) $\left|\frac{4}{x}\right| < 3$

Ans. (a) $\frac{5}{3} < x < 3$; (b) $x \ge \frac{1}{2}$ or $x \le 0$; (c) $-6 \le x \le 18$; (d) $x \le -\frac{3}{2}$ or $x \ge \frac{1}{2}$; (e) $x > 0$ or $x < -1$ or $-\frac{1}{3} < x < 0$; (f) $x > \frac{4}{3}$ or $x < -\frac{4}{3}$

12. Describe and diagram the set determined by each of the following conditions:

(a) $x(x-5) < 0$
(b) $(x-2)(x-6) > 0$
(c) $(x+1)(x-2) < 0$
(d) $x(x-2)(x+3) > 0$
(e) $(x+2)(x+3)(x+4) < 0$
(f) $(x-1)(x+1)(x-2)(x+3) > 0$
(g) $(x-1)^2(x+4) > 0$
(h) $(x-3)(x+5)(x-4)^2 < 0$
(i) $(x-2)^3 > 0$
(j) $(x+1)^3 < 0$
(k) $(x-2)^3(x+1) < 0$
(l) $(x-1)^3(x+1)^4 < 0$
(m) $(3x-1)(2x+3) > 0$
(n) $(x-4)(2x-3) < 0$

Ans. (a) $0 < x < 5$; (b) $x > 6$ or $x < 2$; (c) $-1 < x < 2$; (d) $x > 2$ or $-3 < x < 0$; (e) $-3 < x < -2$ or $x < -4$; (f) $x > 2$ or $-1 < x < 1$ or $x < -3$; (g) $x > -4$ and $x \ne 1$; (h) $-5 < x < 3$; (i) $x > 2$; (j) $x < -1$; (k) $-1 < x < 2$; (l) $x < 1$ and $x \ne -1$; (m) $x > \frac{1}{3}$ or $x < -\frac{3}{2}$; (n) $\frac{3}{2} < x < 4$

13. Describe and diagram the set determined by each of the following conditions:

(a) $x^2 < 4$
(b) $x^2 \ge 9$
(c) $(x-2)^2 \le 16$
(d) $(2x+1)^2 > 1$
(e) $x^2 + 3x - 4 > 0$
(f) $x^2 + 6x + 8 \le 0$
(g) $x^2 < 5x + 14$
(h) $2x^2 > x + 6$
(i) $6x^2 + 13x < 5$
(j) $x^3 + 3x^2 > 10x$

Ans. (a) $-2 < x < 2$; (b) $x \ge 3$ or $x \le -3$; (c) $-2 \le x \le 6$; (d) $x > 0$ or $x < -1$; (e) $x > 1$ or $x < -4$; (f) $-4 \le x \le -2$; (g) $-2 < x < 7$; (h) $x > 2$ or $x < -\frac{3}{2}$; (i) $-\frac{5}{2} < x < \frac{1}{3}$; (j) $-5 < x < 0$ or $x > 2$

14. Solve: (a) $-4 < 2 - x < 7$ (b) $\frac{2x-1}{x} < 3$ (c) $\frac{x}{x+2} < 1$

(d) $\frac{3x-1}{2x+3} > 3$ (e) $\left|\frac{2x-1}{x}\right| > 2$ (f) $\left|\frac{x}{x+2}\right| \le 2$

Ans. (a) $-5 < x < 6$; (b) $x > 0$ or $x < -1$; (c) $x > -2$; (d) $-\frac{10}{3} < x < \frac{3}{2}$; (e) $x < 0$ or $0 < x < \frac{1}{4}$; (f) $x \le -4$ or $x \ge -1$

15. Solve:

(a) $|4x - 5| = 3$
(b) $|x + 6| = 2$
(c) $|3x - 4| = |2x + 1|$
(d) $|x + 1| = |x + 2|$
(e) $|x + 1| = 3x - 1$
(f) $|x + 1| < |3x - 1|$
(g) $|3x - 4| \geq |2x + 1|$

Ans. (a) $x = 2$ or $x = \frac{1}{2}$; (b) $x = -4$ or $x = -8$; (c) $x = 5$ or $x = \frac{3}{5}$; (d) $x = -\frac{3}{2}$; (e) $x = 1$; (f) $x > 1$ or $x < 0$; (g) $x \geq 5$ or $x \leq \frac{3}{5}$

16. Prove:

(a) $|x^2| = |x|^2$;
(b) $|x^n| = |x|^n$ for every integer n;
(c) $|x| = \sqrt{x^2}$;
(d) $|x - y| \leq |x| + |y|$;
(e) $|x - y| \geq ||x| - |y||$
[Hint: In (e), prove that $|x - y| \geq |x| - |y|$ and $|x - y| \geq |y| - |x|$.]

CHAPTER 2

Rectangular Coordinate Systems

Coordinate Axes

In any plane $\mathcal{P}$, choose a pair of perpendicular lines. Let one of the lines be horizontal. Then the other line must be vertical. The horizontal line is called the *x axis*, and the vertical line the *y axis*. (See Fig. 2-1.)

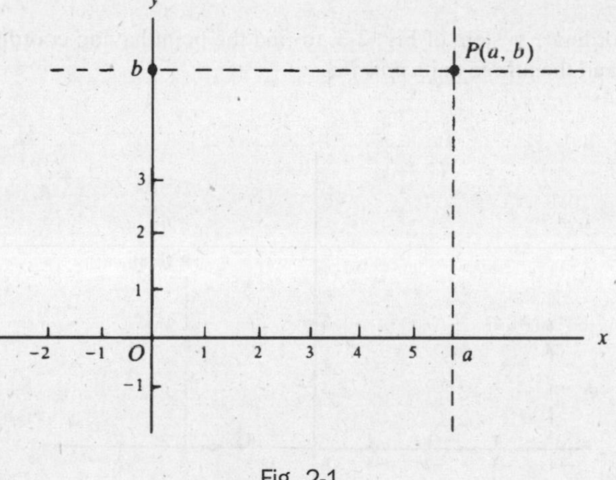

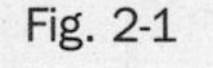

Fig. 2-1

Now choose linear coordinate systems on the x axis and the y axis satisfying the following conditions: The origin for each coordinate system is the point O at which the axes intersect. The x axis is directed from left to right, and the y axis from bottom to top. The part of the x axis with positive coordinates is called the *positive x axis*, and the part of the y axis with positive coordinates is called the *positive y axis*.

We shall establish a correspondence between the points of the plane $\mathcal{P}$ and pairs of real numbers.

Coordinates

Consider any point P of the plane (Fig. 2-1). The vertical line through P intersects the x axis at a unique point; let a be the coordinate of this point on the x axis. The number a is called the *x coordinate* of P (or the *abscissa* of P). The horizontal line through P intersects the y axis at a unique point; let b be the coordinate of this point on the y axis. The number b is called the *y coordinate* of P (or the *ordinate* of P). In this way, every point P has a unique pair (a, b) of real numbers associated with it. Conversely, every pair (a, b) of real numbers is associated with a unique point in the plane.

The coordinates of several points are shown in Fig. 2-2. For the sake of simplicity, we have limited them to integers.

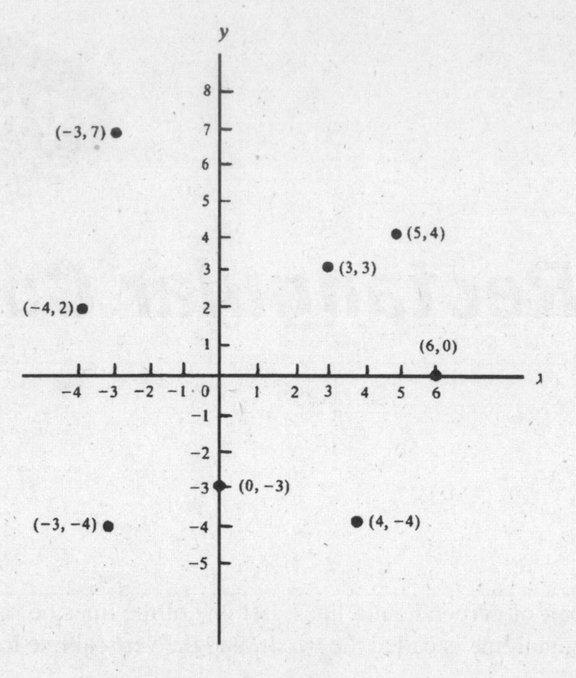

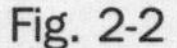

Fig. 2-2

EXAMPLE 2.1: In the coordinate system of Fig. 2-3, to find the point having coordinates (2, 3), start at the origin, move two units to the *right*, and then three units *upward*.

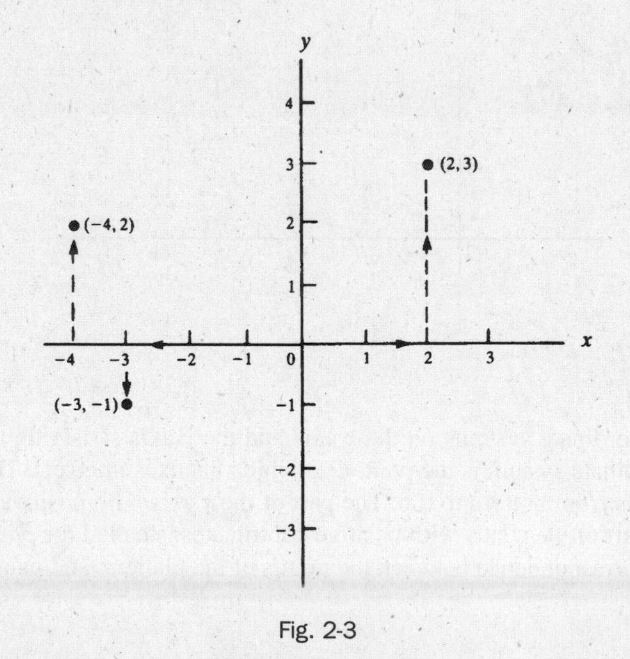

Fig. 2-3

To find the point with coordinates (–4, 2), start at the origin, move four units to the *left*, and then two units *upward*.

To find the point with coordinates (–3, –1), start at the origin, move three units to the *left*, and then one unit *downward*.

The order of these moves is not important. Hence, for example, the point (2, 3) can also be reached by starting at the origin, moving three units *upward*, and then two units to the *right*.

Quadrants

Assume that a coordinate system has been established in the plane $\mathcal{P}$. Then the whole plane $\mathcal{P}$, with the exception of the coordinate axes, can be divided into four equal parts, called *quadrants*. All points with both coordinates positive form the first quadrant, called quadrant I, in the upper-right-hand corner (see Fig. 2-4).

Quadrant II consists of all points with negative x coordinate and positive y coordinate. *Quadrants* III and IV are also shown in Fig. 2-4.

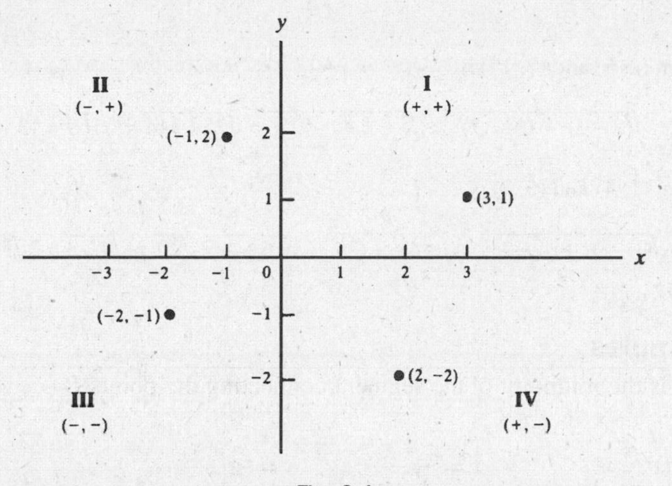

Fig. 2-4

The points on the x axis have coordinates of the form $(a, 0)$. The y axis consists of the points with coordinates of the form $(0, b)$.

Given a coordinate system, it is customary to refer to the point with coordinates (a, b) as "the point (a, b)." For example, one might say, "The point $(0, 1)$ lies on the y axis."

The Distance Formula

The distance $\overline{P_1P_2}$ between poinits P_1 and P_2 with coordinates (x_1, y_1) and (x_2, y_2) in a given coordinate system (see Fig. 2-5) is given by the following distance formula:

$$\overline{P_1P_2} = \sqrt{(x_1 - x_2)^2 + (y_1 - y_2)^2} \tag{2.1}$$

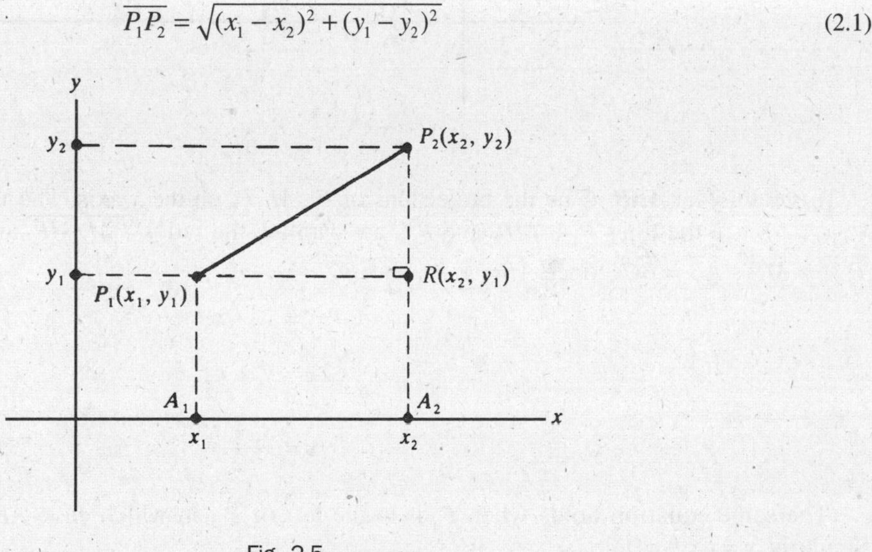

Fig. 2-5

To see this, let R be the point where the vertical line through P_2 intersects the horizontal line through P_1. The x coordinate of R is x_2, the same as that of P_2. The y coordinate of R is y_1, the same as that of P_1. By the Pythagorean theorem, $(\overline{P_1P_2})^2 = (\overline{P_1R})^2 + (\overline{P_2R})^2$. If A_1 and A_2 are the projections of P_1 and P_2 on the x axis, the segments P_1R and A_1A_2 are opposite sides of a rectangle, so that $\overline{P_1R} = \overline{A_1A_2}$. But $\overline{A_1A_2} = |x_1 - x_2|$ by property (**1.12**). So, $\overline{P_1R} = |x_1 - x_2|$. Similarly, $\overline{P_2R} = |y_1 - y_2|$. Hence, $(\overline{P_1P_2})^2 = |x_1 - x_2|^2 + |y_1 - y_2|^2 = (x_1 - x_2)^2 + (y_1 - y_2)^2$.

Taking square roots, we obtain the distance formula. (It can be checked that the formula also is valid when P_1 and P_2 lie on the same vertical or horizontal line.)

EXAMPLES:

(a) The distance between (2, 5) and (7, 17) is

$$\sqrt{(2-7)^2+(5-17)^2}=\sqrt{(-5)^2+(-12)^2}=\sqrt{25+144}=\sqrt{169}=13$$

(b) The distance between (1, 4) and (5, 2) is

$$\sqrt{(1-5)^2+(4-2)^2}=\sqrt{(-4)^2+(2)^2}=\sqrt{16+4}=\sqrt{20}=\sqrt{4}\sqrt{5}=2\sqrt{5}$$

The Midpoint Formulas

The point $M(x, y)$ that is the midpoint of the segment connecting the points $P_1(x_1, y_1)$ and $P_2(x_2, y_2)$ has the coordinates

$$x=\frac{x_1+x_2}{2} \qquad y=\frac{y_1+y_2}{2} \tag{2.2}$$

Thus, the coordinates of the midpoints are the averages of the coordinates of the endpoints. See Fig. 2-6.

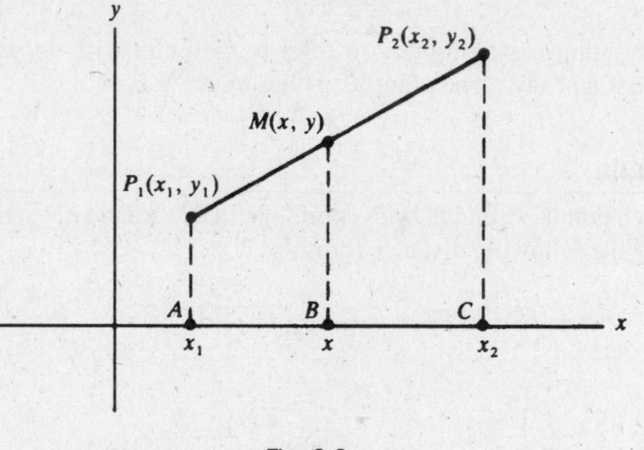

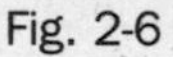
Fig. 2-6

To see this, let A, B, C be the projections of P_1, M, P_2 on the x axis. The x coordinates of A, B, C are x_1, x, x_2. Since the lines P_1A, MB, and P_2C are parallel, the ratios $\overline{P_1M}/\overline{MP_2}$ and $\overline{AB}/\overline{BC}$ are equal. Since $\overline{P_1M}=\overline{MP_2}$, $\overline{AB}=\overline{BC}$. Since $\overline{AB}=x-x_1$ and $\overline{BC}=x_2-x$,

$$x-x_1=x_2-x$$

$$2x=x_1+x_2$$

$$x=\frac{x_1+x_2}{2}$$

(The same equation holds when P_2 is to the left of P_1, in which case $\overline{AB}=x_1-x$ and $\overline{BC}=x-x_2$.) Similarly, $y=(y_1+y_2)/2$.

EXAMPLES:

(a) The midpoint of the segment connecting (2, 9) and (4, 3) is $\left(\frac{2+4}{2}, \frac{9+3}{2}\right)=(3, 6)$.

(b) The point halfway between (−5, 1) and (1, 4) is $\left(\frac{-5+1}{2}, \frac{1+4}{2}\right)=\left(-2, \frac{5}{2}\right)$.

Proofs of Geometric Theorems

Proofs of geometric theorems can often be given more easily by use of coordinates than by deductions from axioms and previously derived theorems. Proofs by means of coordinates are called *analytic*, in contrast to so-called *synthetic* proofs from axioms.

EXAMPLE 2.2: Let us prove analytically that the segment joining the midpoints of two sides of a triangle is one-half the length of the third side. Construct a coordinate system so that the third side AB lies on the positive x axis, A is the origin, and the third vertex C lies above the x axis, as in Fig. 2-7.

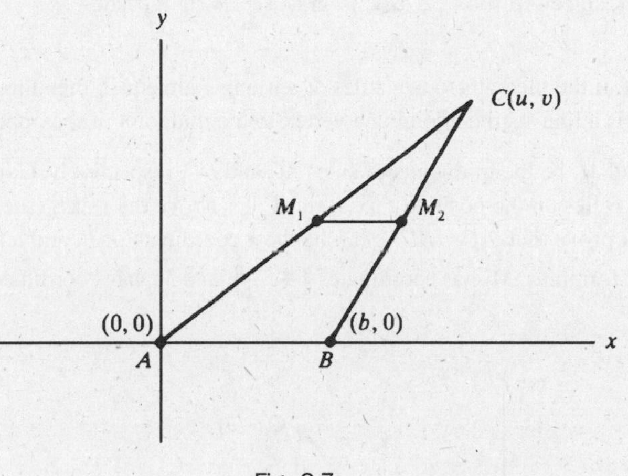

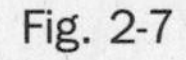

Fig. 2-7

Let b be the x coordinate of B. (In other words, let $b = \overline{AB}$.) Let C have coordinates (u, v). Let M_1 and M_2 be the midpoints of sides AC and BC, respectively. By the midpoint formulas (2.2), the coordinates of M_1 are $\left(\frac{u}{2}, \frac{v}{2}\right)$, and the coordinates of M_2 are $\left(\frac{u+b}{2}, \frac{v}{2}\right)$. By the distance formula (2.1),

$$\overline{M_1M_2} = \sqrt{\left(\frac{u}{2} - \frac{u+b}{2}\right)^2 + \left(\frac{v}{2} - \frac{v}{2}\right)^2} = \sqrt{\left(\frac{b}{2}\right)^2} = \frac{b}{2}$$

which is half the length of side AB.

SOLVED PROBLEMS

1. Show that the distance between a point $P(x, y)$ and the origin is $\sqrt{x^2 + y^2}$.

Since the origin has coordinates (0, 0), the distance formula yields $\sqrt{(x-0)^2 + (y-0)^2} = \sqrt{x^2 + y^2}$.

2. Is the triangle with vertices $A(1, 5)$, $B(4, 2)$, and $C(5, 6)$ isosceles?

$$\overline{AB} = \sqrt{(1-4)^2 + (5-2)^2} = \sqrt{(-3)^2 + (3)^2} = \sqrt{9+9} = \sqrt{18}$$

$$\overline{AC} = \sqrt{(1-5)^2 + (5-6)^2} = \sqrt{(-4)^2 + (-1)^2} = \sqrt{16+1} = \sqrt{17}$$

$$\overline{BC} = \sqrt{(4-5)^2 + (2-6)^2} = \sqrt{(-1)^2 + (-4)^2} = \sqrt{1+16} = \sqrt{17}$$

Since $\overline{AC} = \overline{BC}$, the triangle is isosceles.

3. Is the triangle with vertices $A(-5, 6)$, $B(2, 3)$, and $C(5, 10)$ a right triangle?

$$\overline{AB} = \sqrt{(-5-2)^2 + (6-3)^2} = \sqrt{(-7)^2 + (3)^2} = \sqrt{49+9} = \sqrt{58}$$

$$\overline{AC} = \sqrt{(-5-5)^2 + (6-10)^2} = \sqrt{(-10)^2 + (-4)^2}$$

$$= \sqrt{100+16} = \sqrt{116}$$

$$\overline{BC} = \sqrt{(2-5)^2 + (3-10)^2} = \sqrt{(-3)^2 + (-7)^2} = \sqrt{9+49} = \sqrt{58}$$

Since $\overline{AC}^2 = \overline{AB}^2 + \overline{BC}^2$, the converse of the Pythagorean theorem tells us that ΔABC is a right triangle, with right angle at B; in fact, since $\overline{AB} = \overline{BC}$, ΔABC is an isosceles right triangle.

4. Prove analytically that, if the medians to two sides of a triangle are equal, then those sides are equal. (Recall that a *median* of a triangle is a line segment joining a vertex to the midpoint of the opposite side.)

In ΔABC, let M_1 and M_2 be the midpoints of sides AC and BC, respectively. Construct a coordinate system so that A is the origin, B lies on the positive x axis, and C lies above the x axis (see Fig. 2-8). Assume that $\overline{AM_2} = \overline{BM_1}$. We must prove that $\overline{AC} = \overline{BC}$. Let b be the x coordinate of B, and let C have coordinates (u, v). Then, by the midpoint formulas, M_1 has coordinates $\left(\frac{u}{2}, \frac{v}{2}\right)$, and M_2 has coordinates $\left(\frac{u+b}{2}, \frac{v}{2}\right)$. Hence,

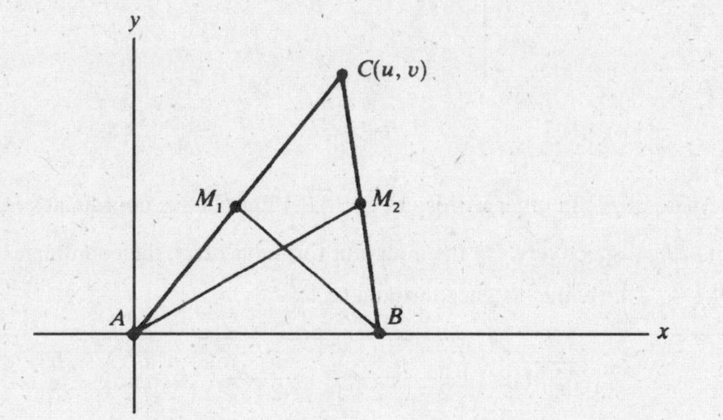

Fig. 2-8

$$\overline{AM_2} = \sqrt{\left(\frac{u+b}{2}\right)^2 + \left(\frac{v}{2}\right)^2} \quad \text{and} \quad \overline{BM_1} = \sqrt{\left(\frac{u}{2} - b\right)^2 + \left(\frac{v}{2}\right)^2}$$

Since $\overline{AM_2} = \overline{BM_1}$,

$$\left(\frac{u+b}{2}\right)^2 + \left(\frac{v}{2}\right)^2 = \left(\frac{u}{2} - b\right)^2 + \left(\frac{v}{2}\right)^2 = \left(\frac{u-2b}{2}\right)^2 + \left(\frac{v}{2}\right)^2$$

Hence, $\frac{(u+b)^2}{4} + \frac{v^2}{4} = \frac{(u-2b)^2}{4} + \frac{v^2}{4}$ and, therefore, $(u+b)^2 = (u-2b)^2$. So, $u + b = \pm(u - 2b)$. If $u + b = u - 2b$, then $b = -2b$, and therefore, $b = 0$, which is impossible, since $A \neq B$. Hence, $u + b = -(u - 2b) = -u + 2b$, whence $2u = b$. Now $\overline{BC} = \sqrt{(u-b)^2 + v^2} = \sqrt{(u-2u)^2 + v^2} = \sqrt{(-u)^2 + v^2} = \sqrt{u^2 + v^2}$, and $\overline{AC} = \sqrt{u^2 + v^2}$. Thus, $\overline{AC} = \overline{BC}$.

5. Find the coordinates (x, y) of the point Q on the line segment joining $P_1(1, 2)$ and $P_2(6, 7)$, such that Q divides the segment in the ratio $2:3$, that is, such that $\overline{P_1Q}/\overline{QP_2} = \frac{2}{3}$.

Let the projections of P_1, Q, and P_2 on the x axis be A_1, Q', and A_2, with x coordinates 1, x, and 6, respectively (see Fig. 2-9). Now $\overline{A_1Q'}/\overline{Q'A_2} = \overline{P_1Q}/\overline{QP_2} = \frac{2}{3}$. (When two lines are cut by three parallel lines, corresponding segments are in proportion.) But $\overline{A_1Q'} = x - 1$, and $\overline{Q'A_2} = 6 - x$. So $\frac{x-1}{6-x} = \frac{2}{3}$, and cross-multiplying yields $3x - 3 = 12 - 2x$. Hence $5x = 15$, whence $x = 3$. By similar reasoning, $\frac{y-2}{7-y} = \frac{2}{3}$, from which it follows that $y = 4$.

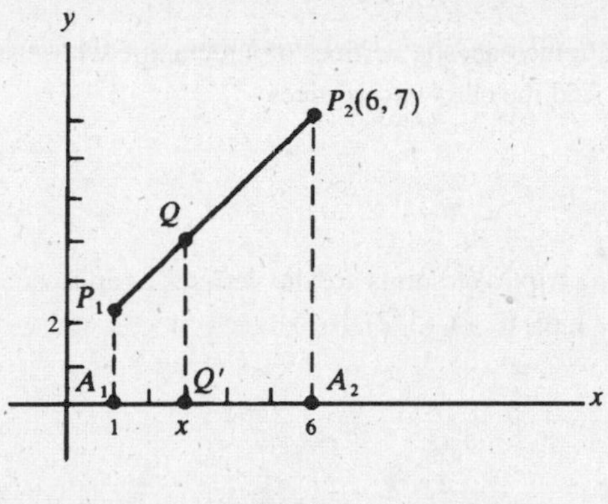

Fig. 2-9

SUPPLEMENTARY PROBLEMS

6. In Fig. 2-10, find the coordinates of points A, B, C, D, E, and F.

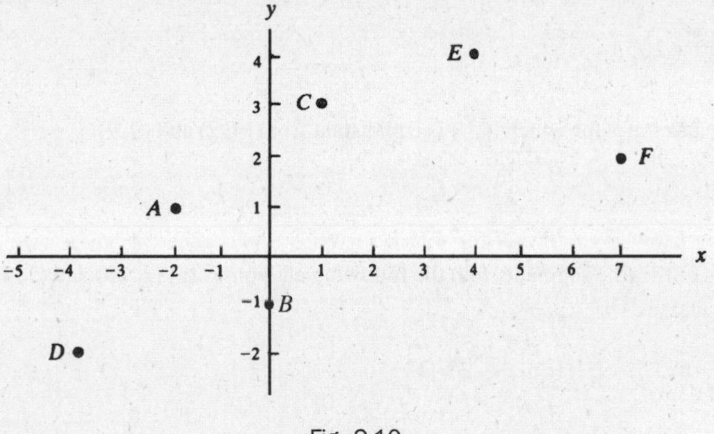

Fig. 2-10

Ans. (A) = (–2, 1); (B) = (0, –1); (C) = (1, 3); (D) = (–4, –2); (E) = (4, 4); (F) = (7, 2).

7. Draw a coordinate system and show the points having the following coordinates: (2, –3), (3, 3), (–1, 1), (2, –2), (0, 3), (3, 0), (–2, 3).

8. Find the distances between the following pairs of points:
(a) (3, 4) and (3, 6) (b) (2, 5) and (2, –2) (c) (3, 1) and (2, 1)
(d) (2, 3) and (5, 7) (e) (–2, 4) and (3, 0) (f) $\left(-2, \frac{1}{2}\right)$ and (4, –1)

Ans. (a) 2; (b) 7; (c) 1; (d) 5; (e) $\sqrt{41}$; (f) $\frac{3}{2}\sqrt{17}$

9. Draw the triangle with vertices $A(2, 5)$, $B(2, -5)$, and $C(-3, 5)$, and find its area.

Ans. Area = 25

10. If (2, 2), (2, –4), and (5, 2) are three vertices of a rectangle, find the fourth vertex.

Ans. (5, –4)

11. If the points (2, 4) and (–1, 3) are the opposite vertices of a rectangle whose sides are parallel to the coordinate axes (that is, the *x* and *y* axes), find the other two vertices.

Ans. (–1, 4) and (2, 3)

12. Determine whether the following triples of points are the vertices of an isosceles triangle: (a) (4, 3), (1, 4), (3, 10); (b) (–1, 1), (3, 3), (1, –1); (c) (2, 4), (5, 2), (6, 5).

Ans. (a) no; (b) yes; (c) no

13. Determine whether the following triples of points are the vertices of a right triangle. For those that are, find the area of the right triangle: (a) (10, 6), (3, 3), (6, –4); (b) (3, 1), (1, –2), (–3, –1); (c) (5, –2), (0, 3), (2, 4).

Ans. (a) yes, area = 29; (b) no; (c) yes, area $= \frac{15}{2}$

14. Find the perimeter of the triangle with vertices *A*(4, 9), *B*(–3, 2), and *C*(8, –5).

Ans. $7\sqrt{2} + \sqrt{170} + 2\sqrt{53}$

15. Find the value or values of *y* for which (6, *y*) is equidistant from (4, 2) and (9, 7).

Ans. 5

16. Find the midpoints of the line segments with the following endpoints: (a) (2, –3) and (7, 4); (b) $\left(\frac{5}{3}, 2\right)$ and (4, 1); (c) $(\sqrt{3}, 0)$ and (1, 4).

Ans. (a) $\left(\frac{9}{2}, \frac{1}{2}\right)$; (b) $\left(\frac{17}{6}, \frac{3}{2}\right)$; (c) $\left(\frac{1+\sqrt{3}}{2}, 2\right)$

17. Find the point (*x*, *y*) such that (2, 4) is the midpoint of the line segment connecting (*x*, *y*) and (1, 5).

Ans. (3, 3)

18. Determine the point that is equidistant from the points *A*(–1, 7), *B*(6, 6), and *C*(5, –1).

Ans. $\left(\frac{57}{25}, \frac{153}{50}\right)$

19. Prove analytically that the midpoint of the hypotenuse of a right triangle is equidistant from the three vertices.

20. Show analytically that the sum of the squares of the distance of any point *P* from two opposite vertices of a rectangle is equal to the sum of the squares of its distances from the other two vertices.

21. Prove analytically that the sum of the squares of the four sides of a parallelogram is equal to the sum of the squares of the diagonals.

22. Prove analytically that the sum of the squares of the medians of a triangle is equal to three-fourths the sum of the squares of the sides.

23. Prove analytically that the line segments joining the midpoints of opposite sides of a quadrilateral bisect each other.

24. Prove that the coordinates (x, y) of the point Q that divides the line segments from $P_1(x_1, y_1)$ to $P_2(x_2, y_2)$ in the ratio $r_1 : r_2$ are determined by the formulas

$$x = \frac{r_1 x_2 + r_2 x_1}{r_1 + r_2} \quad \text{and} \quad y = \frac{r_1 y_2 + r_2 y_1}{r_1 + r_2}$$

(*Hint*: Use the reasoning of Problem 5.)

25. Find the coordinates of the point Q on the segment P_1P_2 such that $\overline{P_1Q}/\overline{QP_2} = \frac{2}{7}$, if (a) $P_1 = (0, 0)$, $P_2 = (7, 9)$; (b) $P_1 = (-1, 0)$, $P_2 = (0, 7)$; (c) $P_1 = (-7, -2)$, $P_2 = (2, 7)$; (d) $P_1 = (1, 3)$, $P_2 = (4, 2)$.

Ans. (a) $\left(\frac{14}{9}, 2\right)$; (b) $\left(-\frac{7}{9}, \frac{14}{9}\right)$; (c) $\left(-5, \frac{28}{9}\right)$; (d) $\left(\frac{13}{9}, \frac{32}{9}\right)$

CHAPTER 3

Lines

The Steepness of a Line

The steepness of a line is measured by a number called the *slope* of the line. Let $\mathscr{L}$ be any line, and let $P_1(x_1, y_1)$ and $P_2(x_2, y_2)$ be two points of $\mathscr{L}$. The slope of $\mathscr{L}$ is defined to be the number $m = \frac{y_2 - y_1}{x_2 - x_1}$. The slope is the ratio of a change in the y coordinate to the corresponding change in the x coordinate. (See Fig. 3-1.)

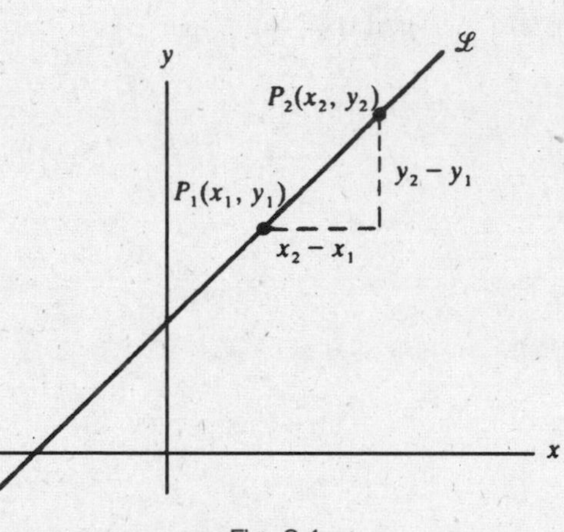

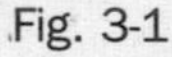
Fig. 3-1

For the definition of the slope to make sense, it is necessary to check that the number m is independent of the choice of the points P_1 and P_2. If we choose another pair $P_3(x_3, y_3)$ and $P_4(x_4, y_4)$, the same value of m must result. In Fig. 3-2, triangle P_3P_4T is similar to triangle P_1P_2Q. Hence,

$$\frac{\overline{QP_2}}{\overline{P_1Q}} = \frac{\overline{TP_4}}{\overline{P_3T}} \quad \text{or} \quad \frac{y_2 - y_1}{x_2 - x_1} = \frac{y_4 - y_3}{x_4 - x_3}$$

Therefore, P_1 and P_2 determine the same slope as P_3 and P_4.

EXAMPLE 3.1: The slope of the line joining the points (1, 2) and (4, 6) in Fig. 3-3 is $\frac{6-2}{4-1} = \frac{4}{3}$. Hence, as a point on the line moves 3 units to the right, it moves 4 units upwards. Moreover, the slope is not affected by the order in which the points are given: $\frac{2-6}{1-4} = \frac{-4}{-3} = \frac{4}{3}$. In general, $\frac{y_2 - y_1}{x_2 - x_1} = \frac{y_1 - y_2}{x_1 - x_2}$.

The Sign of the Slope

The sign of the slope has significance. Consider, for example, a line $\mathscr{L}$ that moves upward as it moves to the right, as in Fig. 3-4(*a*). Since $y_2 > y_1$ and $x_2 > x_1$, we have $m = \frac{y_2 - y_1}{x_2 - x_1} > 0$. *The slope of* $\mathscr{L}$ *is positive.*

Now consider a line $\mathscr{L}$ that moves downward as it moves to the right, as in Fig. 3-4(*b*). Here $y_2 < y_1$ while $x_2 > x_1$; hence, $m = \frac{y_2 - y_1}{x_2 - x_1} < 0$. *The slope of* $\mathscr{L}$ *is negative.*

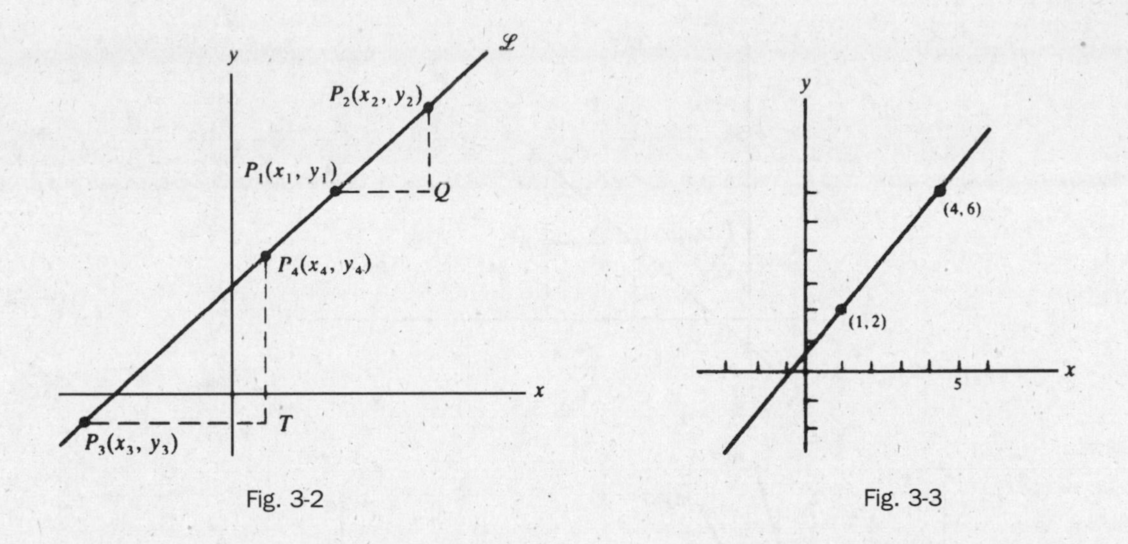

Fig. 3-2

Fig. 3-3

Now let the line $\mathscr{L}$ be horizontal, as in Fig. 3-4(*c*). Here $y_1 = y_2$, so that $y_2 - y_1 = 0$. In addition, $x_2 - x_1 \neq 0$. Hence, $m = \dfrac{0}{x_2 - x_1} = 0$. *The slope of* $\mathscr{L}$ *is zero.*

Line $\mathscr{L}$ is vertical in Fig. 3-4(*d*), where we see that $y_2 - y_1 > 0$ while $x_2 - x_1 = 0$. Thus, the expression $\dfrac{y_2 - y_1}{x_2 - x_1}$ is undefined. *The slope is not defined for a vertical line* $\mathscr{L}$. (Sometimes we describe this situation by saying that the slope of $\mathscr{L}$ is "infinite.")

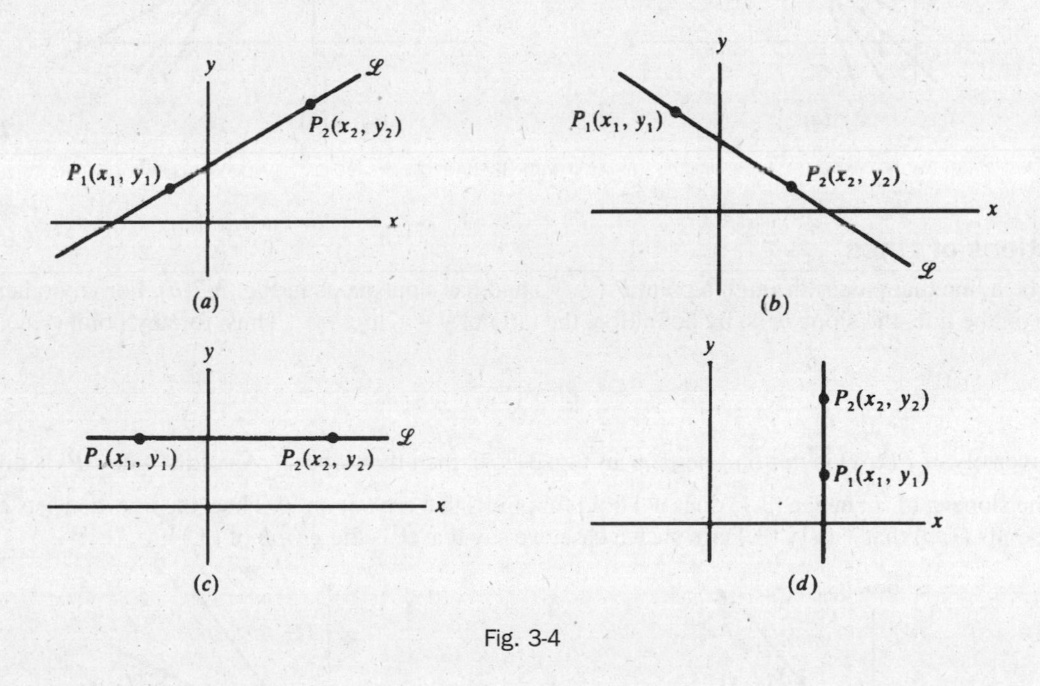

Fig. 3-4

Slope and Steepness

Consider any line $\mathscr{L}$ with positive slope, passing through a point $P_1(x_1, y_1)$; such a line is shown in Fig. 3-5. Choose the point $P_2(x_2, y_2)$ on $\mathscr{L}$ such that $x_2 - x_1 = 1$. Then the slope m of $\mathscr{L}$ is equal to the distance $\overline{AP_2}$. As the steepness of the line increases, $\overline{AP_2}$ increases without limit, as shown in Fig. 3-6(*a*). Thus, the slope of $\mathscr{L}$ increases without bound from 0 (when $\mathscr{L}$ is horizontal) to $+\infty$ (when the line is vertical). By a similar argument, using Fig. 3-6(*b*), we can show that as a negatively sloped line becomes steeper, the slope steadily decreases from 0 (when the line is horizontal) to $-\infty$ (when the line is vertical).

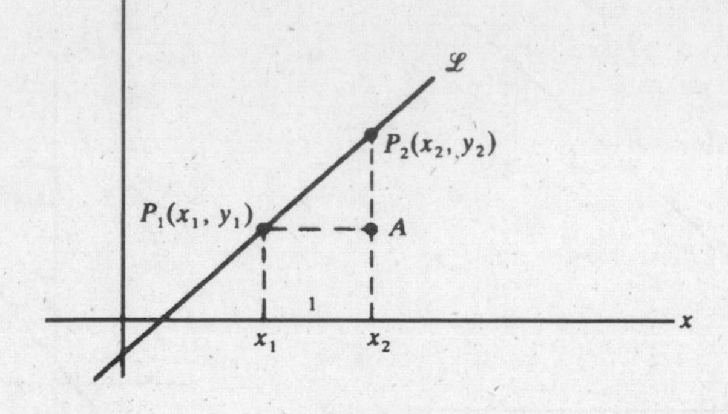

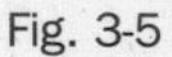

Fig. 3-5

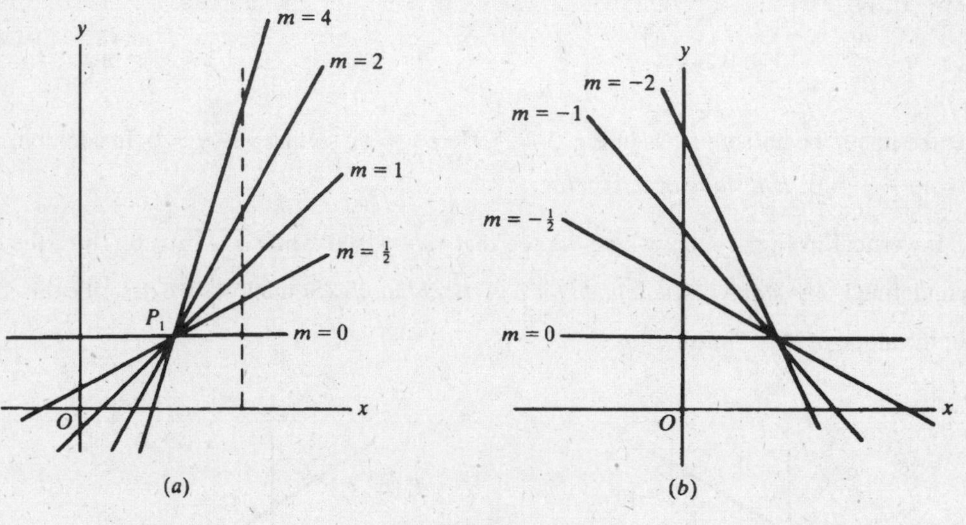

Fig. 3-6

Equations of Lines

Let $\mathscr{L}$ be a line that passes through a point $P_1(x_1, y_1)$ and has slope m, as in Fig. 3-7(*a*). For any other point $P(x, y)$ on the line, the slope m is, by definition, the ratio of $y - y_1$ to $x - x_1$. Thus, for any point (x, y) on $\mathscr{L}$,

$$m = \frac{y - y_1}{x - x_1} \tag{3.1}$$

Conversely, if $P(x, y)$ is *not* on line $\mathscr{L}$ as in Fig. 3-7(*b*), then the slope $\frac{y - y_1}{x - x_1}$ of the line PP_1 is different from the slope m of $\mathscr{L}$; hence (3.1) does not hold for points that are not on $\mathscr{L}$. Thus, the line consists of only those points (x, y) that satisfy (3.1). In such a case, we say that $\mathscr{L}$ is the *graph* of (3.1).

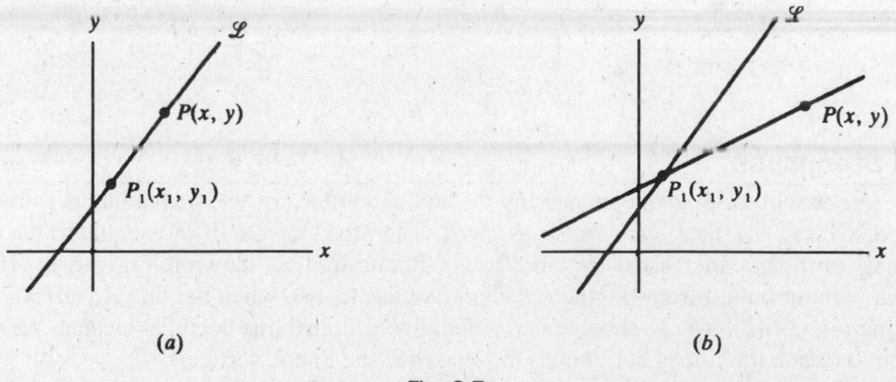

Fig. 3-7

A Point–Slope Equation

A point–slope equation of the line $\mathcal{L}$ is any equation of the form (3.1). If the slope m of $\mathcal{L}$ is known, then each point (x_1, y_1) of $\mathcal{L}$ yields a point–slope equation of $\mathcal{L}$. Hence, there is an infinite number of point–slope equations for $\mathcal{L}$. Equation (3.1) is equivalent to $y - y_1 = m(x - x_1)$.

EXAMPLE 3.2: (a) The line passing through the point (2, 5) with slope 3 has a point–slope equation $\frac{y-5}{x-2} = 3$. (b) Let $\mathcal{L}$ be the line through the points (3, −1) and (2, 3). Its slope is $m = \frac{3-(-1)}{2-3} = \frac{4}{-1} = -4$. Two point–slope equations of $\mathcal{L}$ are $\frac{y+1}{x-3} = -4$ and $\frac{y-3}{x-2} = -4$.

Slope–Intercept Equation

If we multiply (3.1) by $x - x_1$, we obtain the equation $y - y_1 = m(x - x_1)$, which can be reduced first to $y - y_1 = mx - mx_1$, and then to $y = mx + (y_1 - mx_1)$. Let b stand for the number $y_1 - mx_1$. Then the equation for line $\mathcal{L}$ becomes

$$y = mx + b \tag{3.2}$$

Equation (3.2) yields the value $y = b$ when $x = 0$, so the point $(0, b)$ lies on $\mathcal{L}$. Thus, b is the y coordinate of the intersection of $\mathcal{L}$ and the y axis, as shown in Fig. 3-8. The number b is called the *y intercept* of $\mathcal{L}$, and (3.2) is called the *slope–intercept equation* for $\mathcal{L}$.

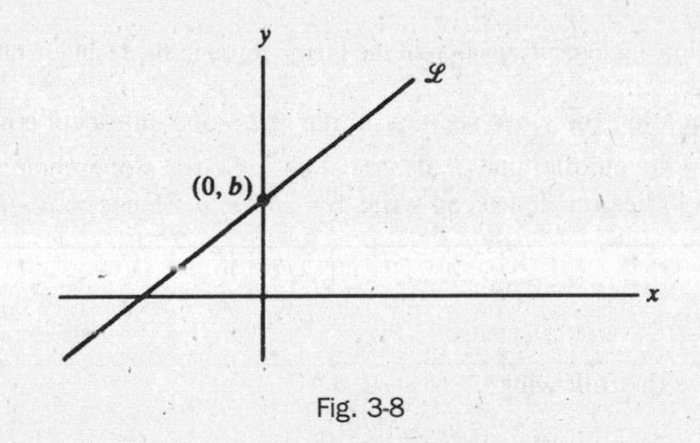

Fig. 3-8

EXAMPLE 3.3: The line through the points (2, 3) and (4, 9) has slope

$$m = \frac{9-3}{4-2} = \frac{6}{2} = 3$$

Its slope–intercept equation has the form $y = 3x + b$. Since the point (2, 3) lies on the line, (2, 3) must satisfy this equation. Substitution yields $3 = 3(2) + b$, from which we find $b = -3$. Thus, the slope–intercept equation is $y = 3x - 3$.

Another method for finding this equation is to write a point–slope equation of the line, say $\frac{y-3}{x-2} = 3$. Then multiplying by $x - 2$ and adding 3 yields $y = 3x - 3$.

Parallel Lines

Let $\mathcal{L}_1$ and $\mathcal{L}_2$ be parallel nonvertical lines, and let A_1 and A_2 be the points at which $\mathcal{L}_1$ and $\mathcal{L}_2$ intersect the y axis, as in Fig. 3-9(*a*). Further, let B_1 be one unit to the right of A_1, and B_2 one unit to the right of A_2. Let C_1 and C_2 be the intersections of the verticals through B_1 and B_2 with $\mathcal{L}_1$ and $\mathcal{L}_2$. Now, triangle $A_1B_1C_1$ is congruent to triangle $A_2B_2C_2$ (by the angle—side—angle congruence theorem). Hence, $\overline{B_1C_1} = \overline{B_2C_2}$ and

$$\text{Slope of } \mathcal{L}_1 = \frac{\overline{B_1C_1}}{1} = \frac{\overline{B_2C_2}}{1} = \text{slope of } \mathcal{L}_2$$

Thus, *parallel lines have equal slopes.*

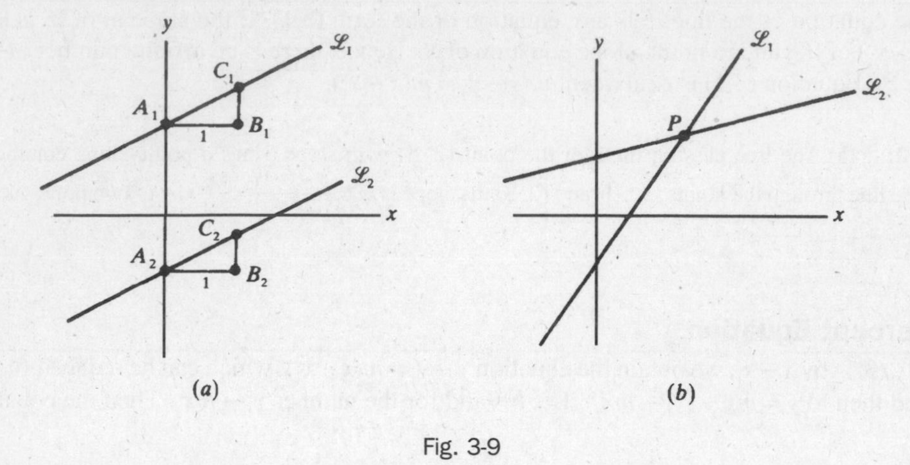

Fig. 3-9

Conversely, assume that two different lines $\mathcal{L}_1$ and $\mathcal{L}_2$ are not parallel, and let them meet at point P, as in Fig. 3-9(*b*). If $\mathcal{L}_1$ and $\mathcal{L}_2$ had the same slope, then they would have to be the same line. Hence, $\mathcal{L}_1$ and $\mathcal{L}_2$ have different slopes.

Theorem 3.1: Two distinct nonvertical lines are parallel if and only if their slopes are equal.

EXAMPLE 3.4: Find the slope–intercept equation of the line $\mathcal{L}$ through (4, 1) and parallel to the line $\mathcal{M}$ having the equation $4x - 2y = 5$.

By solving the latter equation for y, we see that $\mathcal{M}$ has the slope–intercept equation $y = 2x - \frac{5}{2}$. Hence, $\mathcal{M}$ has slope 2. The slope of the parallel line $\mathcal{L}$ also must be 2. So the slope–intercept equation of $\mathcal{L}$ has the form $y = 2x + b$. Since (4, 1) lies on $\mathcal{L}$, we can write $1 = 2(4) + b$. Hence, $b = -7$, and the slope–intercept equation of $\mathcal{L}$ is $y = 2x - 7$.

Perpendicular Lines

In Problem 5 we shall prove the following:

Theorem 3.2: Two nonvertical lines are perpendicular if and only if the product of their slopes is −1.

If m_1 and m_2 are the slopes of perpendicular lines, then $m_1 m_2 = -1$. This is equivalent to $m_2 = -\frac{1}{m_1}$; hence, *the slopes of perpendicular lines are negative reciprocals of each other.*

SOLVED PROBLEMS

1. Find the slope of the line having the equation $3x - 4y = 8$. Draw the line. Do the points (6, 2) and (12, 7) lie on the line?

   Solving the equation for y yields $y = \frac{3}{4}x - 2$. This is the slope–intercept equation; the slope is $\frac{3}{4}$ and the y intercept is −2.

   Substituting 0 for x shows that the line passes through the point (0, −2). To draw the line, we need another point. If we substitute 4 for x in the slope–intercept equation, we get $y = \frac{3}{4}(4) - 2 = 1$. So, (4, 1) also lies on the line, which is drawn in Fig. 3-10. (We could have found other points on the line by substituting numbers other than 4 for x.)

   To test whether (6, 2) is on the line, we substitute 6 for x and 2 for y in the original equation, $3x - 4y = 8$. The two sides turn out to be unequal; hence, (6, 2) is not on the line. The same procedure shows that (12, 7) lies on the line.

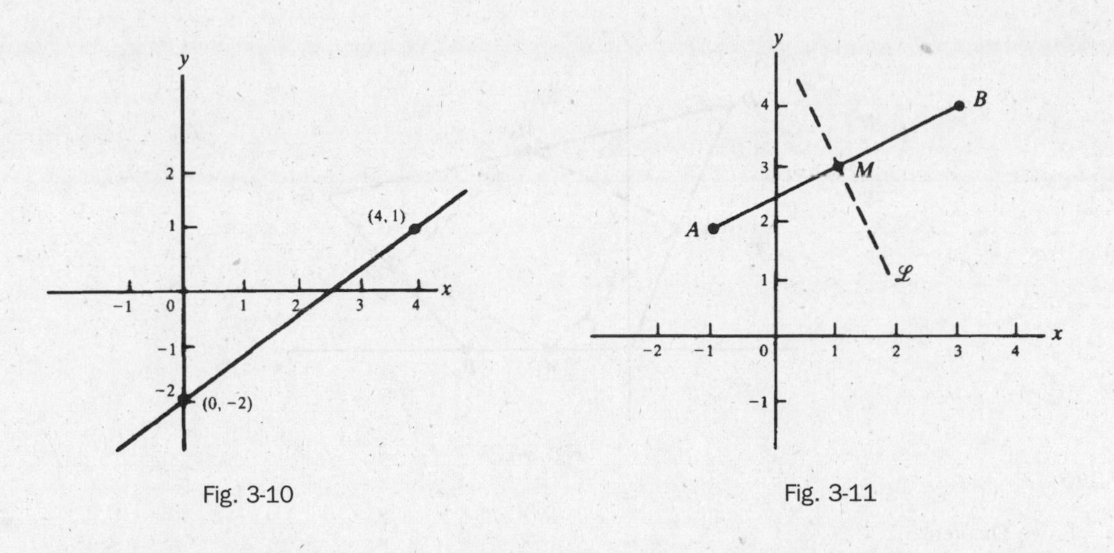

Fig. 3-10

Fig. 3-11

2. Let $\mathscr{L}$ be the perpendicular bisector of the line segment joining the points $A(-1, 2)$ and $B(3, 4)$, as shown in Fig. 3-11. Find an equation for $\mathscr{L}$.

$\mathscr{L}$ passes through the midpoint M of segment AB. By the midpoint formulas (2.2), the coordinates of M are (1, 3). The slope of the line through A and B is $\frac{4-2}{3-(-1)} = \frac{2}{4} = \frac{1}{2}$. Let m be the slope of $\mathscr{L}$. By Theorem 3.2, $\frac{1}{2}m = -1$, whence $m = -2$.

The slope–intercept equation for $\mathscr{L}$ has the form $y = -2x + b$. Since M (1, 3) lies on $\mathscr{L}$, we have $3 = -2(1) + b$. Hence, $b = 5$, and the slope–intercept equation of $\mathscr{L}$ is $y = -2x + 5$.

3. Determine whether the points $A(1, -1)$, $B(3, 2)$, and $C(7, 8)$ are collinear, that is, lie on the same line.

A, B, and C are collinear if and only if the line AB is identical with the line AC, which is equivalent to the slope of AB being equal to the slope of AC. The slopes of AB and AC are $\frac{2-(-1)}{3-1} = \frac{3}{2}$ and $\frac{8-(-1)}{7-1} = \frac{9}{6} = \frac{3}{2}$. Hence, A, B, and C are collinear.

4. Prove analytically that the figure obtained by joining the midpoints of consecutive sides of a quadrilateral is a parallelogram.

Locate a quadrilateral with consecutive vertices, A, B, C, and D on a coordinate system so that A is the origin, B lies on the positive x axis, and C and D lie above the x axis. (See Fig. 3-12.) Let b be the x coordinate of B, (u, v) the coordinates of C, and (x, y) the coordinates of D. Then, by the midpoint formula (2.2), the midpoints M_1, M_2, M_3, and M_4 of sides AB, BC, CD, and DA have coordinates $\left(\frac{b}{2}, 0\right)$, $\left(\frac{u+b}{2}, \frac{v}{2}\right)$, $\left(\frac{x+u}{2}, \frac{y+v}{2}\right)$, and $\left(\frac{x}{2}, \frac{y}{2}\right)$, respectively. We must show that $M_1M_2M_3M_4$ is a parallelogram. To do this, it suffices to prove that lines M_1M_2 and M_3M_4 are parallel and that lines M_2M_3 and M_1M_4 are parallel. Let us calculate the slopes of these lines:

$$\text{slope}(M_1M_2) = \frac{\frac{v}{2} - 0}{\frac{u+b}{2} - \frac{b}{2}} = \frac{\frac{v}{2}}{\frac{u}{2}} = \frac{v}{u} \qquad \text{slope}(M_3M_4) = \frac{\frac{y}{2} - \frac{y+v}{2}}{\frac{x}{2} - \frac{x+u}{2}} = \frac{-\frac{v}{2}}{-\frac{u}{2}} = \frac{v}{u}$$

$$\text{slope}(M_2M_3) = \frac{\frac{y+v}{2} - \frac{v}{2}}{\frac{x+u}{2} - \frac{u+b}{2}} = \frac{\frac{y}{2}}{\frac{x-b}{2}} = \frac{y}{x-b} \qquad \text{slope}(M_1M_4) = \frac{\frac{y}{2} - 0}{\frac{x}{2} - \frac{b}{2}} = \frac{y}{x-b}$$

Since $\text{slope}(M_1M_2) = \text{slope}(M_3M_4)$, M_1M_2 and M_3M_4 are parallel. Since $\text{slope}(M_2M_3) = \text{slope}(M_1M_4)$, M_2M_3 and M_1M_4 are parallel. Thus, $M_1M_2M_3M_4$ is a parallelogram.

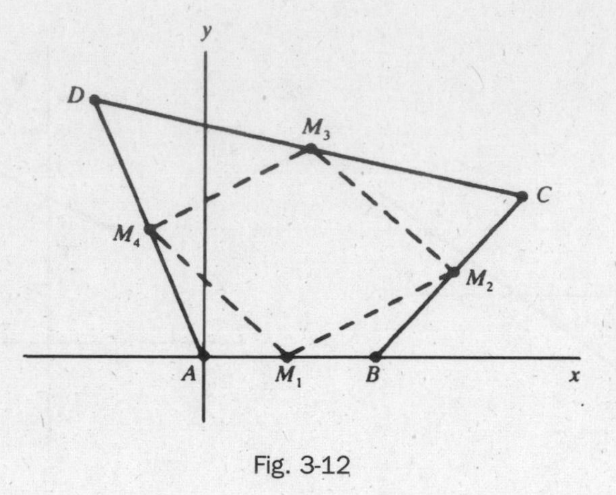

Fig. 3-12

5. Prove Theorem 3.2.

First we assume $\mathscr{L}_1$ and $\mathscr{L}_2$ are perpendicular nonvertical lines with slopes m_1 and m_2. We must show that $m_1m_2 = -1$. Let $\mathcal{M}_1$ and $\mathcal{M}_2$ be the lines through the origin O that are parallel to $\mathscr{L}_1$ and $\mathscr{L}_2$, as shown in Fig. 3-13(*a*). Then the slope of $\mathcal{M}_1$ is m_1, and the slope of $\mathcal{M}_2$ is m_2 (by Theorem 3.1). Moreover, $\mathcal{M}_1$ and $\mathcal{M}_2$ are perpendicular, since $\mathscr{L}_1$ and $\mathscr{L}_2$ are perpendicular.

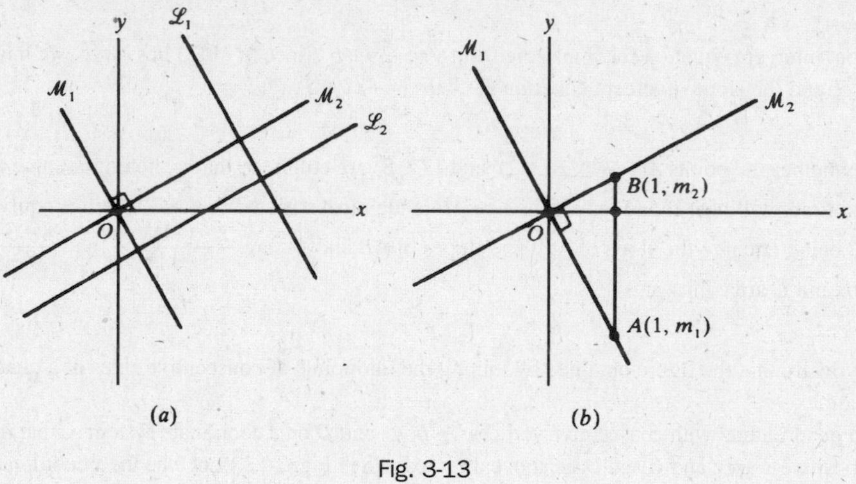

Fig. 3-13

Now let A be the point on $\mathcal{M}_1$ with x coordinate 1, and let B be the point on $\mathcal{M}_2$ with x coordinate 1, as in Fig. 3-13(*b*). The slope–intercept equation of $\mathcal{M}_1$ is $y = m_1x$; therefore, the y coordinate of A is m_1, since its x coordinate is 1. Similarly, the y coordinate of B is m_2. By the distance formula (2.1),

$$\overline{OB} = \sqrt{(1-0)^2 + (m_2 - 0)^2} = \sqrt{1 + m_2^2}$$

$$\overline{OA} = \sqrt{(1-0)^2 + (m_1 - 0)^2} = \sqrt{1 + m_1^2}$$

$$\overline{BA} = \sqrt{(1-1)^2 + (m_2 - m_1)^2} = \sqrt{(m_2 - m_1)^2}$$

Then by the Pythagorean theorem for right triangle BOA,

$$\overline{BA}^2 = \overline{OB}^2 + \overline{OA}^2$$

or

$$(m_2 - m_1)^2 = (1 + m_2^2) + (1 + m_1^2)$$

$$m_2^2 - 2m_2m_1 + m_1^2 = 2 + m_2^2 + m_1^2$$

$$m_2m_1 = -1$$

Now, conversely, we assume that $m_1m_2 = -1$, where m_1 and m_2 are the slopes of nonvertical lines $\mathcal{L}_1$ and $\mathcal{L}_2$. Then $\mathcal{L}_1$ is not parallel to $\mathcal{L}_2$. (Otherwise, by Theorem 3.1, $m_1 = m_2$ and, therefore, $m_1^2 = -1$, which contradicts the fact that the square of a real number is never negative.) We must show that $\mathcal{L}_1$ and $\mathcal{L}_2$ are perpendicular. Let P be the intersection of $\mathcal{L}_1$ and $\mathcal{L}_2$ (see Fig. 3-14). Let $\mathcal{L}_3$ be the line through P that is perpendicular to $\mathcal{L}_1$. If m_3 is the slope of $\mathcal{L}_3$, then, by the first part of the proof, $m_1m_3 - -1$ and, therefore, $m_1m_3 = m_1m_2$. Since $m_1m_3 = -1$, $m_1 \neq 0$; therefore, $m_3 = m_2$. Since $\mathcal{L}_2$ and $\mathcal{L}_3$ pass through the same point P and have the same slope, they must coincide. Since $\mathcal{L}_1$ and $\mathcal{L}_3$ are perpendicular, $\mathcal{L}_1$ and $\mathcal{L}_2$ are also perpendicular.

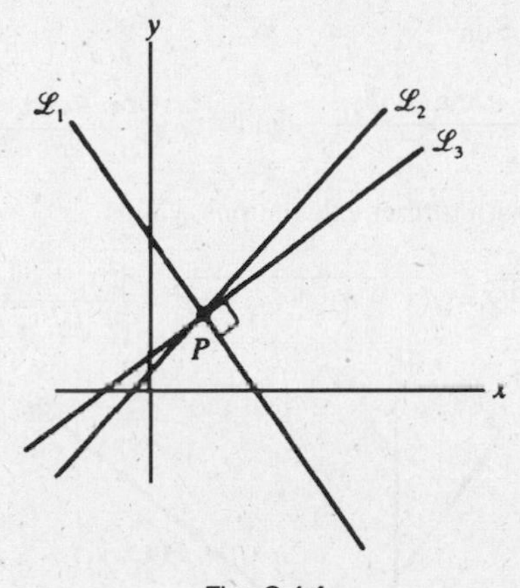

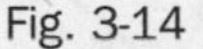
Fig. 3-14

6. Show that, if a and b are not both zero, then the equation $ax + by = c$ is the equation of a line and, conversely, every line has an equation of that form.

Assume $b \neq 0$. Then, if the equation $ax + by = c$ is solved for y, we obtain a slope–intercept equation $y = (-a/b)\,x + c/b$ of a line. If $b = 0$, then $a \neq 0$, and the equation $ax + by = c$ reduces to $ax = c$; this is equivalent to $x = c/a$, the equation of a vertical line.

Conversely, every nonvertical line has a slope–intercept equation $y = mx + b$, which is equivalent to $-mx + y = b$, an equation of the desired form. A vertical line has an equation of the form $x = c$, which is also an equation of the required form with $a = 1$ and $b = 0$.

7. Show that the line $y = x$ makes an angle of 45° with the positive x axis (that is, that angle BOA in Fig. 3-15 contains 45°).

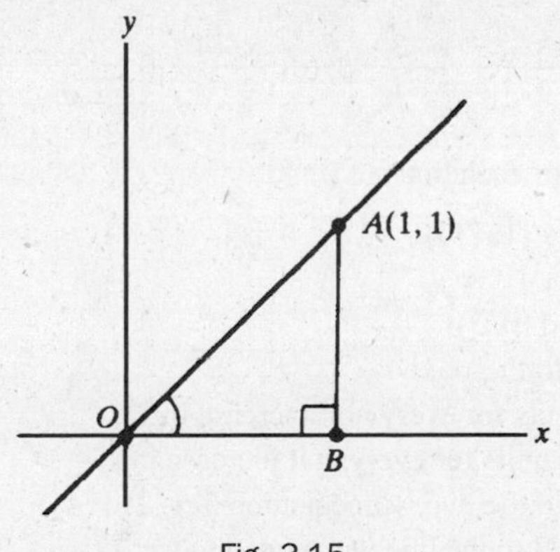

Fig. 3-15

Let A be the point on the line $y = x$ with coordinates (1, 1). Drop a perpendicular AB to the positive x axis. Then $\overline{AB} = 1$ and $\overline{OB} = 1$. Hence, angle OAB = angle BOA, since they are the base angles of isosceles triangle BOA. Since angle OBA is a right angle,

$$\text{Angle } OAB + \text{angle } BOA = 180° - \text{angle } OBA = 180° - 90° = 90°$$

Since angle BOA = angle OAB, they each contain 45°.

8. Show that the distance d from a point $P(x_1, y_1)$ to a line $\mathcal{L}$ with equation $ax + by = c$ is given by the formula $d = \dfrac{|ax + by - c|}{\sqrt{a^2 + b^2}}$.

Let $\mathcal{M}$ be the line through P that is perpendicular to $\mathcal{L}$. Then $\mathcal{M}$ intersects $\mathcal{L}$ at some point Q with coordinates (u, v), as in Fig. 3-16. Clearly, d is the length $\overline{PQ}$, so if we can find u and v, we can compute d with the distance formula. The slope of $\mathcal{L}$ is $-a/b$. Hence, by Theorem 3.2, the slope of $\mathcal{M}$ is b/a. Then a point–slope equation of $\mathcal{M}$ is $\dfrac{y - y_1}{x - x_1} = \dfrac{b}{a}$. Thus, u and v are the solutions of the pair of equations $au + bv = c$ and $\dfrac{v - y_1}{u - x_1} = \dfrac{b}{a}$. Tedious algebraic calculations yield the solution

$$u = \frac{ac + b^2x_1 + aby_1}{a^2 + b^2} \quad \text{and} \quad v = \frac{bc - abx_1 + a^2y_1}{a^2 + b^2}$$

The distance formula, together with further calculations, yields

$$d = \overline{PQ} = \sqrt{(x_1 - u)^2 + (y_1 - v)^2} = \frac{|ax_1 + by_1 - c|}{\sqrt{a^2 + b^2}}$$

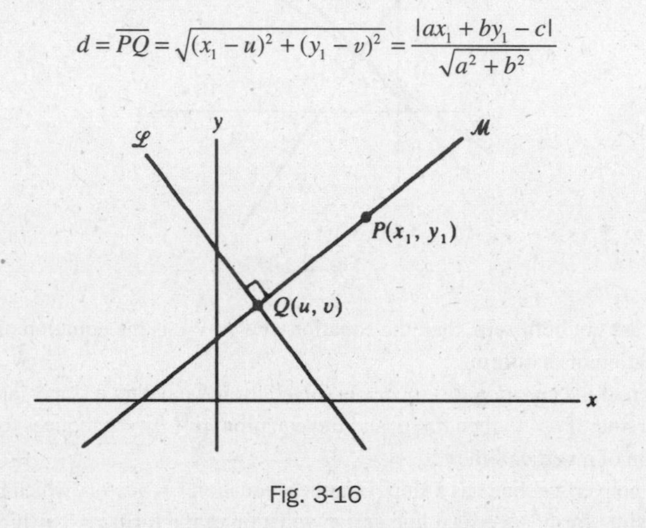

Fig. 3-16

SUPPLEMENTARY PROBLEMS

9. Find a point–slope equation for the line through each of the following pairs of points: (a) (3, 6) and (2, –4); (b) (8, 5) and (4, 0); (c) (1, 3) and the origin; (d) (2, 4) and (–2, 4).

Ans. (a) $\dfrac{y-6}{x-3} = 10$; (b) $\dfrac{y-5}{x-8} = \dfrac{5}{4}$; (c) $\dfrac{y-3}{x-1} = 3$; (d) $\dfrac{y-4}{x-2} = 0$

10. Find the slope–intercept equation of each line:

(a) Through the points (4, –2) and (1, 7)
(b) Having slope 3 and y intercept 4
(c) Through the points (–1, 0) and (0, 3)
(d) Through (2, –3) and parallel to the x axis
(e) Through (2, 3) and rising 4 units for every unit increase in x
(f) Through (–2, 2) and falling 2 units for every unit increase in x
(g) Through (3, –4) and parallel to the line with equation $5x - 2y = 4$
(h) Through the origin and parallel to the line with equation $y = 2$
(i) Through (–2, 5) and perpendicular to the line with equation $4x + 8y = 3$
(j) Through the origin and perpendicular to the line with equation $3x - 2y = 1$
(k) Through (2, 1) and perpendicular to the line with equation $x = 2$
(l) Through the origin and bisecting the angle between the positive x axis and the positive y axis

Ans. (a) $y = -3x + 10$; (b) $y = 3x + 3$; (c) $y = 3x + 3$; (d) $y = -3$; (e) $y = 4x - 5$; (f) $y = -2x - 2$; (g) $y = \frac{5}{2}x - \frac{23}{2}$; (h) $y = 0$; (i) $y = 2x + 9$; (j) $y = -\frac{2}{3}x$; (k) $y = 1$; (l) $y = x$

11. (a) Describe the lines having equations of the form $x = a$.
(b) Describe the lines having equations of the form $y = b$.
(c) Describe the line having the equation $y = -x$.

12. (a) Find the slopes and y intercepts of the lines that have the following equations: (i) $y = 3x - 2$; (ii) $2x - 5y = 3$; (iii) $y = 4x - 3$; (iv) $y = -3$; (v) $\frac{y}{2} + \frac{x}{3} = 1$.
(b) Find the coordinates of a point other than $(0, b)$ on each of the lines of part (a).

Ans. (a) (i) $m = 3, b = -2$; (ii) $m = \frac{2}{5}$ $b = -\frac{3}{5}$ (iii) $m = 4, b = -3$; (iv) $m = 0, b = -3$; (v) $m = -\frac{2}{3}$; $b = 2$;
(b) (i) (1, 1); (ii) (−6, −3); (iii) (1, 1); (iv) (1, −3); (v) (3, 0)

13. If the point $(3, k)$ lies on the line with slope $m = -2$ passing through the point (2, 5), find k.

Ans. $k = 3$

14. Does the point (3, −2) lie on the line through the points (8, 0) and (−7, −6)?

Ans. Yes

15. Use slopes to determine whether the points (7, −1), (10, 1), and (6, 7) are the vertices of a right triangle.

Ans. They are.

16. Use slopes to determine whether (8, 0), (−1, −2), (−2, 3), and (7, 5) are the vertices of a parallelogram.

Ans. They are.

17. Under what conditions are the points $(u, v + w)$, $(v, u + w)$, and $(w, u + v)$ collinear?

Ans. Always.

18. Determine k so that the points $A(7, 3)$, $B(-1, 0)$, and $C(k, -2)$ are the vertices of a right triangle with right angle at B.

Ans. $k = 1$

19. Determine whether the following pairs of lines are parallel, perpendicular, or neither:

(a) $y = 3x + 2$ and $y = 3x - 2$
(b) $y = 2x - 4$ and $y = 3x + 5$
(c) $3x - 2y = 5$ and $2x + 3y = 4$
(d) $6x + 3y = 1$ and $4x + 2y = 3$
(e) $x = 3$ and $y = -4$
(f) $5x + 4y = 1$ and $4x + 5y = 2$
(g) $x = -2$ and $x = 7$

Ans. (a) Parallel; (b) neither; (c) perpendicular; (d) parallel; (e) perpendicular; (f) neither; (g) parallel

20. Draw the line determined by the equation $2x + 5y = 10$. Determine whether the points (10, 2) and (12, 3) lie on this line.

21. For what values of k will the line $kx - 3y = 4k$ have the following properties: (a) have slope 1; (b) have y intercept 2; (c) pass through the point (2, 4); (d) be parallel to the line $2x - 4y = 1$; (e) be perpendicular to the line $x - 6y = 2$?

Ans. (a) $k = 3$; (b) $k = -\frac{3}{2}$; (c) $k = -6$; (d) $k = \frac{3}{2}$; (e) $k = -18$

22. Describe geometrically the families of lines (a) $y = mx - 3$ and (b) $y = 4x + b$, where m and b are any real numbers.

Ans. (a) Lines with y intercept -3; (b) lines with slope 4

23. In the triangle with vertices, $A(0, 0)$, $B(2, 0)$, and $C(3, 3)$, find equations for (a) the median from B to the midpoint of the opposite side; (b) the perpendicular bisector of side BC; and (c) the altitude from B to the opposite side.

Ans. (a) $y = -3x + 6$; (b) $x + 3y = 7$; (c) $y = -x + 2$

CHAPTER 4

Circles

Equations of Circles

For a point $P(x, y)$ to lie on the circle with center $C(a, b)$ and radius r, the distance $\overline{PC}$ must be equal to r (see Fig. 4-1). By the distance formula (2.1),

$$\overline{PC} = \sqrt{(x-a)^2 + (y-b)^2}$$

Thus, P lies on the circle if and only if

$$(x-a)^2 + (y-b)^2 = r^2 \tag{4.1}$$

Equation (4.1) is called the *standard equation* of the circle with center at (a, b) and radius r.

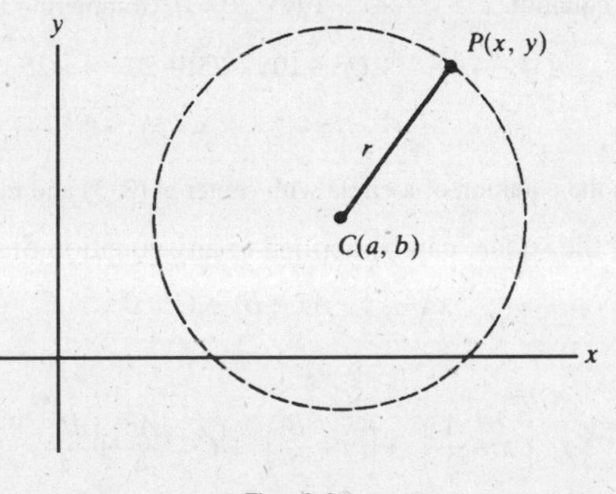

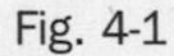

Fig. 4-1

EXAMPLE 4.1:

(a) The circle with center (3, 1) and radius 2 has the equation $(x-3)^2 + (y-1)^2 = 4$.
(b) The circle with center (2, −1) and radius 3 has the equation $(x-2)^2 + (y+1)^2 = 9$.
(c) What is the set of points satisfying the equation $(x-4)^2 + (y-5)^2 = 25$?

By (4.1), this is the equation of the circle with center at (4, 5) and radius 5. That circle is said to be the *graph* of the given equation, that is, the set of points satisfying the equation.

(d) The graph of the equation $(x+3)^2 + y^2 = 2$ is the circle with center at (−3, 0) and radius $\sqrt{2}$.

The Standard Equation of a Circle

The standard equation of a circle with center at the origin (0, 0) and radius r is

$$x^2 + y^2 = r^2 \tag{4.2}$$

For example, $x^2 + y^2 = 1$ is the equation of the circle with center at the origin and radius 1. The graph of $x^2 + y^2 = 5$ is the circle with center at the origin and radius $\sqrt{5}$.

The equation of a circle sometimes appears in a disguised form. For example, the equation

$$x^2 + y^2 + 8x - 6y + 21 = 0 \tag{4.3}$$

turns out to be equivalent to

$$(x+4)^2+(y-3)^2=4 \tag{4.4}$$

Equation (4.4) is the standard equation of a circle with center at (–4, 3) and radius 2.

Equation (4.4) is obtained from (4.3) by a process called *completing the square*. In general terms, the process involves finding the number that must be added to the sum $x^2 + Ax$ to obtain a square.

Here, we note that $\left(x+\frac{A}{2}\right)^2 = x^2 + Ax + \left(\frac{A}{2}\right)^2$. Thus, in general, *we must add* $\left(\frac{A}{2}\right)^2$ *to* $x^2 + Ax$ *to obtain the square* $\left(x+\frac{A}{2}\right)^2$. For example, to get a square from $x^2 + 8x$, we add $\left(\frac{8}{2}\right)^2$, that is, 16.

The result is $x^2 + 8x + 16$, which is $(x + 4)^2$. This is the process of completing the square.

Consider the original (4.3): $x^2 + y^2 + 8x - 6y + 21 = 0$. To complete the square in $x^2 + 8x$, we add 16. To complete the square in $y^2 - 6y$, we add $\left(-\frac{6}{2}\right)^2$, which is 9. Of course, since we added 16 and 9 to the left side of the equation, we must also add them to the right side, obtaining

$$(x^2+8x+16)+(y^2-6y+9)+21=16+9$$

This is equivalent to

$$(x+4)^2+(y-3)^2+21=25$$

and subtraction of 21 from both sides yields (4.4).

EXAMPLE 4.2: Consider the equation $x^2 + y^2 - 4x - 10y + 20 = 0$. Completing the square yields

$$(x^2-4x+4)+(y^2-10y+25)+20=4+25$$

$$(x-2)^2+(y-5)^2=9$$

Thus, the original equation is the equation of a circle with center at (2, 5) and radius 3.

The process of completing the square can be applied to any equation of the form

$$x^2+y^2+Ax+By+C=0 \tag{4.5}$$

to obtain

$$\left(x+\frac{A}{2}\right)^2+\left(y+\frac{B}{2}\right)^2+C=\frac{A^2}{4}+\frac{B^2}{4}$$

or

$$\left(x+\frac{A}{2}\right)^2+\left(y+\frac{B}{2}\right)^2=\frac{A^2+B^2-4C}{4} \tag{4.6}$$

There are three different cases, depending on whether $A^2 + B^2 - 4C$ is positive, zero, or negative.

Case 1: $A^2 + B^2 - 4C > 0$. In this case, (4.6) is the standard equation of a circle with center at $\left(-\frac{A}{2},-\frac{B}{2}\right)$ and radius $\frac{\sqrt{A^2+B^2-4C}}{2}$.

Case 2: $A^2 + B^2 - 4C = 0$. A sum of the squares of two quantities is zero when and only when each of the quantities is zero. Hence, (4.6) is equivalent to the conjunction of the equations $x + A/2 = 0$ and $y + B/2 = 0$ in this case, and the only solution of (4.6) is the point $(-A/2, -B/2)$. Hence, the graph of (4.5) is a single point, which may be considered a *degenerate circle* of radius 0.

Case 3: $A^2 + B^2 - 4C < 0$. A sum of two squares cannot be negative. So, in this case, (4.5) has no solution at all.

We can show that any circle has an equation of the form (4.5). Suppose its center is (a, b) and its radius is r; then its standard equation is

$$(x-a)^2+(y-b)^2=r^2$$

Expanding yields $x^2 - 2ax + a^2 + y^2 - 2by + b^2 = r^2$, or

$$x^2+y^2-2ax-2by+(a^2+b^2-r^2)=0$$

SOLVED PROBLEMS

1. Identify the graphs of (a) $2x^2 + 2y^2 - 4x + y + 1 = 0$; (b) $x^2 + y^2 - 4y + 7 = 0$; (c) $x^2 + y^2 - 6x - 2y + 10 = 0$.

(a) First divide by 2, obtaining $x^2 + y^2 - 2x + \frac{1}{2}y + \frac{1}{2} = 0$. Then complete the squares:

$$(x^2 - 2x + 1) + (y^2 + \tfrac{1}{2}y + \tfrac{1}{16}) + \tfrac{1}{2} = 1 + \tfrac{1}{16} = \tfrac{17}{16}$$

$$(x-1)^2 + (y + \tfrac{1}{4})^2 = \tfrac{17}{16} - \tfrac{1}{2} = \tfrac{17}{16} - \tfrac{8}{16} = \tfrac{9}{16}$$

Thus, the graph is the circle with center $(1, -\frac{1}{4})$ and radius $\frac{3}{4}$.

(b) Complete the square:

$$x^2 + (y-2)^2 + 7 = 4$$

$$x^2 + (y-2)^2 = -3$$

Because the right side is negative, there are no points in the graph.

(c) Complete the square:

$$(x-3)^2 + (y-1)^2 + 10 = 9 + 1$$

$$(x-3)^2 + (y-1)^2 = 0$$

The only solution is the point (3, 1).

2. Find the standard equation of the circle with center at $C(2, 3)$ and passing through the point $P(-1, 5)$.

The radius of the circle is the distance

$$\overline{CP} = \sqrt{(5-3)^2 + (-1-2)^2} = \sqrt{2^2 + (-3)^2} = \sqrt{4+9} = \sqrt{13}$$

So the standard equation is $(x-2)^2 + (y-3)^2 = 13$.

3. Find the standard equation of the circle passing through the points $P(3, 8)$, $Q(9, 6)$, and $R(13, -2)$.

First method: The circle has an equation of the form $x^2 + y^2 + Ax + By + C = 0$. Substitute the values of x and y at point P, to obtain $9 + 64 + 3A + 8B + C = 0$ or

$$3A + 8B + C = -73 \qquad (1)$$

A similar procedure for points Q and R yields the equations

$$9A + 6B + C = -117 \qquad (2)$$

$$13A - 2B + C = -173 \qquad (3)$$

Eliminate C from (1) and (2) by subtracting (2) from (1):

$$-6A + 2B = 44 \quad \text{or} \quad -3A + B = 22 \qquad (4)$$

Eliminate C from (1) and (3) by subtracting (3) from (1):

$$-10A + 10B = 100 \quad \text{or} \quad -A + B = 10 \qquad (5)$$

Eliminate B from (4) and (5) by subtracting (5) from (4), obtaining $A = -6$. Substitute this value in (5) to find that $B = 4$. Then solve for C in (1): $C = -87$.

Hence, the original equation for the circle is $x^2 + y^2 - 6x + 4y - 87 = 0$. Completing the squares then yields

$$(x-3)^2 + (y+2)^2 = 87 + 9 + 4 = 100$$

Thus, the circle has center $(3, -2)$ and radius 10.

Second method: The perpendicular bisector of any chord of a circle passes through the center of the circle. Hence, the perpendicular bisector $\mathcal{L}$ of chord PQ will intersect the perpendicular bisector $\mathcal{M}$ of chord QR at the center of the circle (see Fig. 4-2).

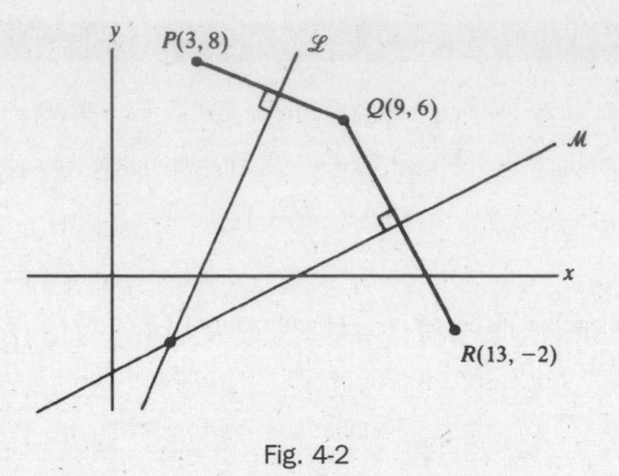

Fig. 4-2

The slope of line PQ is $-\frac{1}{3}$. So, by Theorem 3.2, the slope of $\mathscr{L}$ is 3. Also, $\mathscr{L}$ passes through the midpoint (6, 7) of segment PQ. Hence a point–slope equation of $\mathscr{L}$ is $\dfrac{y-7}{x-6}=3$, and therefore its slope–intercept equation is $y = 3x - 11$. Similarly, the slope of line QR is -2, and therefore the slope of $\mathscr{M}$ is $\frac{1}{2}$, Since $\mathscr{M}$ passes through the midpoint (11, 2) of segment QR, it has a point–slope equation $\dfrac{y-2}{x-11}=\dfrac{1}{2}$, which yields the slope–intercept equation $y=\frac{1}{2}x-\frac{7}{2}$. Hence, the coordinates of the center of the circle satisfy the two equations $y = 3x - 11$ and $y=\frac{1}{2}x-\frac{7}{2}$ and we may write

$$3x-11=\tfrac{1}{2}x-\tfrac{7}{2}$$

from which we find that $x = 3$. Therefore,

$$y = 3x - 11 = 3(3) - 11 = -2$$

So the center is at (3, −2). The radius is the distance between the center and the point (3, 8):

$$\sqrt{(-2-8)^2+(3-3)^2}=\sqrt{(-10)^2}=\sqrt{100}=10$$

Thus, the standard equation of the circle is $(x-3)^2+(y+2)^2=100$.

4. Find the center and radius of the circle that passes through $P(1, 1)$ and is tangent to the line $y = 2x - 3$ at the point $Q(3, 3)$. (See Fig. 4-3.)

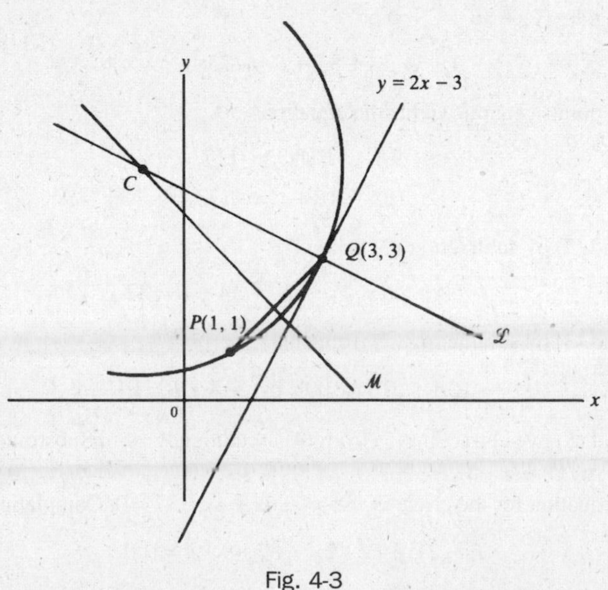

Fig. 4-3

The line $\mathscr{L}$ perpendicular to $y = 2x - 3$ at (3, 3) must pass through the center of the circle. By Theorem 3.2, the slope of $\mathscr{L}$ is $-\frac{1}{2}$. Therefore, the slope–intercept equation of $\mathscr{L}$ has the form $y=-\frac{1}{2}x+b$. Since (3, 3) is on $\mathscr{L}$, we have $3=-\frac{1}{2}(3)+b$; hence, $b=\frac{9}{2}$, and $\mathscr{L}$ has the equation $y=-\frac{1}{2}x+\frac{9}{2}$.

The perpendicular bisector $\mathcal{M}$ of chord PQ in Fig. 4-3 also passes through the center of the circle, so the intersection of $\mathcal{L}$ and $\mathcal{M}$ will be the center of the circle. The slope of PQ is 1. Hence, by Theorem 3.2, the slope of $\mathcal{M}$ is -1. So $\mathcal{M}$ has the slope–intercept equation $y = -x + b'$. Since the midpoint (2, 2) of chord PQ is a point on $\mathcal{M}$, we have $2 = -(2) + b'$; hence, $b' = 4$, and the equation of $\mathcal{M}$ is $y = -x + 4$. We must find the common solution of $y = -x + 4$ and $y = -\frac{1}{2}x + \frac{9}{2}$. Setting

$$-x + 4 = -\tfrac{1}{2}x + \tfrac{9}{2}$$

yields $x = -1$. Therefore, $y = -x + 4 = -(-1) + 4 = 5$, and the center C of the circle is $(-1, 5)$. The radius is the distance $\overline{PC} = \sqrt{(-1-3)^2 + (5-3)^2} = \sqrt{16+4} = \sqrt{20}$. The standard equation of the circle is then $(x+1)^2 + (y-5)^2 = 20$.

5. Find the standard equation of every circle that passes through the points $P(1, -1)$ and $Q(3, 1)$ and is tangent to the line $y = -3x$.

Let $C(c, d)$ be the center of one of the circles, and let A be the point of tangency (see Fig. 4-4). Then, because $\overline{CP} = \overline{CQ}$, we have

$$\overline{CP}^2 = \overline{CQ}^2 \quad \text{or} \quad (c-1)^2 + (d+1)^2 = (c-3)^2 + (d-1)^2$$

Expanding and simplifying, we obtain

$$c + d = 2 \tag{1}$$

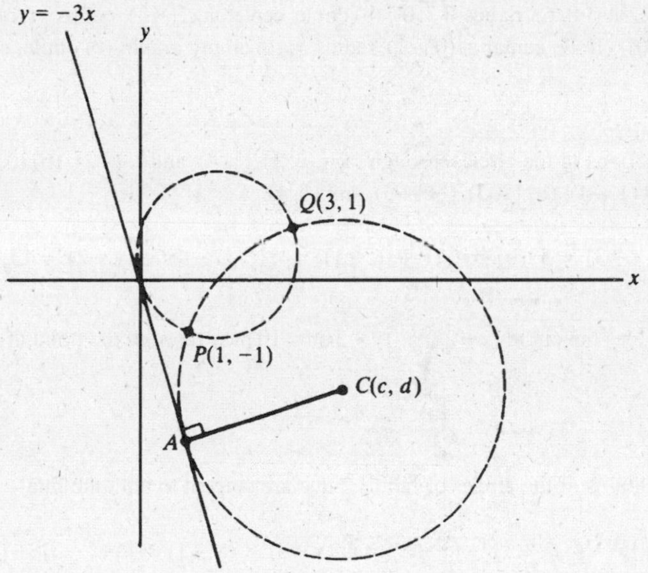

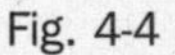

Fig. 4-4

In addition, $\overline{CP} = \overline{CA}$, and by the formula of Problem 8 in Chapter 3, $\overline{CA} = \dfrac{3c+d}{\sqrt{10}}$. Setting $\overline{CP}^2 = \overline{CA}^2$ thus yields $(c-1)^2 + (d+1)^2 = \dfrac{(3c+d)^2}{10}$. Substituting (1) in the right-hand side and multiplying by 10 then yields

$$10[(c-1)^2 + (d+1)^2] = (2c+2)^2 \quad \text{from which} \quad 3c^2 + 5d^2 - 14c + 10d + 8 = 0$$

By (1), we can replace d by $2 - c$, obtaining

$$2c^2 - 11c + 12 = 0 \quad \text{or} \quad (2c-3)(c-4) = 0$$

Hence, $c = \frac{3}{2}$ or $c = 4$. Then (1) gives us the two solutions $c = \frac{3}{2}$, $d = \frac{1}{2}$ and $c = 4$, $d = -2$. Since the radius $\overline{CA} = \dfrac{3c+d}{\sqrt{10}}$, these solutions produce radii of $\dfrac{\frac{10}{2}}{\sqrt{10}} = \dfrac{\sqrt{10}}{2}$ and $\dfrac{10}{\sqrt{10}} = \sqrt{10}$. Thus, there are two such circles, and their standard equations are

$$\left(x-\frac{3}{2}\right)^2+\left(y-\frac{1}{2}\right)^2=\frac{5}{2} \quad \text{and} \quad (x-4)^2+(y+2)^2=10$$

SUPPLEMENTARY PROBLEMS

6. Find the standard equations of the circles satisfying the following conditions:

(a) center at (3, 5) and radius 2
(b) center at (4, –1) and radius 1
(c) center at (5, 0) and radius $\sqrt{3}$
(d) center at (–2, –2) and radius $5\sqrt{2}$
(e) center at (–2, 3) and passing through (3, –2)
(f) center at (6, 1) and passing through the origin

Ans. (a) $(x-3)^2+(y-5)^2=4$; (b) $(x-4)^2+(y+1)^2=1$; (c) $(x-5)^2+y^2=3$; (d) $(x+2)^2+(y+2)^2=50$; (e) $(x+2)^2+(y-3)^2=50$; (f) $(x-6)^2+(y-1)^2=37$

7. Identify the graphs of the following equations:

(a) $x^2+y^2+16x-12y+10=0$
(b) $x^2+y^2-4x+5y+10=0$
(c) $x^2+y^2+x-y=0$
(d) $4x^2+4y^2+8y-3=0$
(e) $x^2+y^2-x-2y+3=0$
(f) $x^2+y^2+\sqrt{2}x-2=0$

Ans. (a) circle, center at (–8, 6), radius $3\sqrt{10}$; (b) circle, center at $(2,-\frac{5}{2})$, radius $\frac{1}{2}$; (c) circle, center at $(-\frac{1}{2},\frac{1}{2})$, radius $\sqrt{2}/2$; (d) circle, center at (0, –1), radius $\frac{7}{2}$; (e) empty graph; (f) circle, center at $(-\sqrt{2}/2,0)$, radius $\sqrt{5/2}$

8. Find the standard equations of the circles through (a) (–2, 1), (1, 4), and (–3, 2); (b) (0, 1), (2, 3), and $(1, 1+\sqrt{3})$; (c) (6, 1), (2, –5), and (1, –4); (d) (2, 3), (–6, –3), and (1, 4).

Ans. (a) $(x+1)^2+(y-3)^2=5$; (b) $(x-2)^2+(y-1)^2=4$; (c) $(x-4)^2+(y+2)^2=13$; (d) $(x+2)^2+y^2=25$

9. For what values of k does the circle $(x+2k)^2+(y-3k)^2=10$ pass through the point (1, 0)?

Ans. $k=\frac{9}{13}$ or $k=-1$

10. Find the standard equations of the circles of radius 2 that are tangent to both the lines $x=1$ and $y=3$.

Ans. $(x+1)^2+(y-1)^2=4$; $(x+1)^2+(y-5)^2=4$; $(x-3)^2+(y-1)^2=4$; $(x-3)^2+(y-5)^2=4$

11. Find the value of k so that $x^2+y^2+4x-6y+k=0$ is the equation of a circle of radius 5.

Ans. $k=-12$

12. Find the standard equation of the circle having as a diameter the segment joining (2, –3) and (6, 5).

Ans. $(x-4)^2+(y-1)^2=20$

13. Find the standard equation of every circle that passes through the origin, has radius 5, and is such that the y coordinate of its center is –4.

Ans. $(x-3)^2+(y+4)^2=25$ or $(x+3)^2+(y+4)^2=25$

14. Find the standard equation of the circle that passes through the points $(8, -5)$ and $(-1, 4)$ and has its center on the line $2x + 3y = 3$.

Ans. $(x - 3)^2 + (y + 1)^2 = 41$

15. Find the standard equation of the circle with center $(3, 5)$ that is tangent to the line $12x - 5y + 2 = 0$.

Ans. $(x - 3)^2 + (y - 5)^2 = 1$

16. Find the standard equation of the circle that passes through the point $(1, 3 + \sqrt{2})$ and is tangent to the line $x + y = 2$ at $(2, 0)$.

Ans. $(x - 5)^2 + (y - 3)^2 = 18$

17. Prove analytically that an angle inscribed in a semicircle is a right angle. (See Fig. 4-5.)

18. Find the length of a tangent from $(6, -2)$ to the circle $(x - 1)^2 + (y - 3)^2 = 1$. (See Fig. 4-6.)

Ans. 7

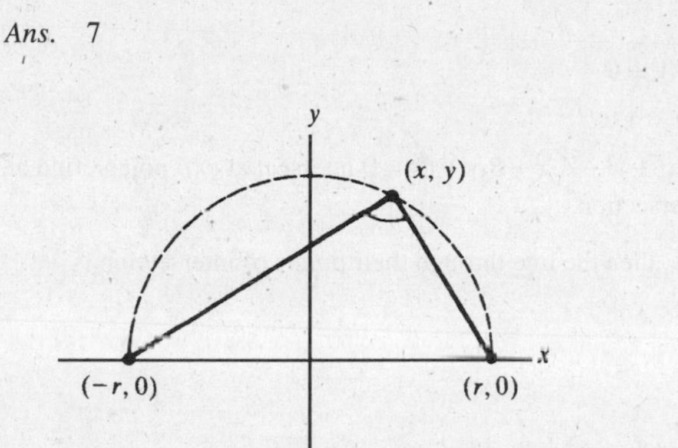

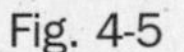
Fig. 4-5

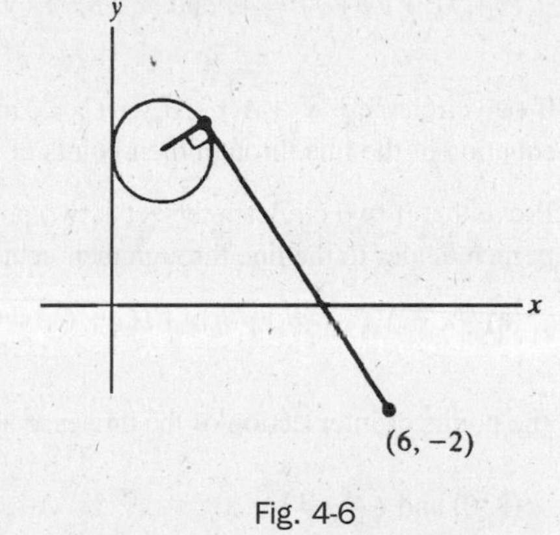

Fig. 4-6

19. Find the standard equations of the circles that pass through $(2, 3)$ and are tangent to both the lines $3x - 4y = -1$ and $4x + 3y = 7$.

Ans. $(x - 2)^2 + (y - 8)^2 = 25$ and $\left(x - \frac{6}{5}\right)^2 + \left(y - \frac{12}{5}\right)^2 = 1$

20. Find the standard equations of the circles that have their centers on the line $4x + 3y = 8$ and are tangent to both the lines $x + y = -2$ and $7x - y = -6$.

Ans. $(x - 1)^2 + y^2 = 2$ and $(x + 4)^2 + (y - 8)^2 = 18$

21. Find the standard equation of the circle that is concentric with the circle $x^2 + y^2 - 2x - 8y + 1 = 0$ and is tangent to the line $2x - y = 3$.

Ans. $(x - 1)^2 + (y - 4)^2 = 5$

22. Find the standard equations of the circles that have radius 10 and are tangent to the circle $x^2 + y^2 = 25$ at the point $(3, 4)$.

Ans. $(x - 9)^2 + (y - 12)^2 = 100$ and $(x + 3)^2 + (y + 4)^2 = 100$

23. Find the longest and shortest distances from the point (7, 12) to the circle $x^2 + y^2 + 2x + 6y - 15 = 0$.

Ans. 22 and 12

24. Let $\mathscr{C}_1$ and $\mathscr{C}_2$ be two intersecting circles determined by the equations $x^2 + y^2 + A_1x + B_1y + C_1 = 0$ and $x^2 + y^2 + A_2x + B_2y + C_2 = 0$. For any number $k \neq -1$, show that

$$x^2 + y^2 + A_1x + B_1y + C_1 + k(x^2 + y^2 + A_2x + B_2y + C_2) = 0$$

is the equation of a circle through the intersection points of $\mathscr{C}_1$ and $\mathscr{C}_2$. Show, conversely, that every such circle may be represented by such an equation for a suitable k.

25. Find the standard equation of the circle passing through the point (−3, 1) and containing the points of intersection of the circles $x^2 + y^2 + 5x = 1$ and $x^2 + y^2 + y = 7$.

Ans. (Use Problem 24.) $(x+1)^2 + \left(y + \frac{3}{10}\right)^2 = \frac{569}{100}$

26. Find the standard equations of the circles that have centers on the line $5x - 2y = -21$ and are tangent to both coordinate axes.

Ans. $(x + 7)^2 + (y + 7)^2 = 49$ and $(x + 3)^2 + (y - 3)^2 = 9$

27. (a) If two circles $x^2 + y^2 + A_1x + B_1y + C_1 = 0$ and $x^2 + y^2 + A_2x + B_2y + C_2 = 0$ intersect at two points, find an equation of the line through their points of intersection.

(b) Prove that if two circles intersect at two points, then the line through their points of intersection is perpendicular to the line through their centers.

Ans. (a) $(A_1 - A_2)x + (B_1 - B_2)y + (C_1 - C_2) = 0$

28. Find the points of intersection of the circles $x^2 + y^2 + 8y - 64 = 0$ and $x^2 + y^2 - 6x - 16 = 0$.

Ans. (8, 0) and $\left(\frac{24}{15}, \frac{24}{5}\right)$

29. Find the equations of the lines through (4, 10) and tangent to the circle $x^2 + y^2 - 4y - 36 = 0$.

Ans. $y = -3x + 22$ and $x - 3y + 26 = 0$

30. (GC) Use a graphing calculator to draw the circles in Problems 7(*d*), 10, 14, and 15. (*Note*: It may be necessary to solve for *y*.)

31. (GC) (a) Use a graphing calculator to shade the interior of the circle with center at the origin and radius 3. (b) Use a graphing calculator to shade the exterior of the circle $x^2 + (y - 2)^2 = 1$.

32. (GC) Use a graphing calculator to graph the following inequalities: (a) $(x - 1)^2 + y^2 < 4$; (b) $x^2 + y^2 - 6x - 8y > 0$.

CHAPTER 5

Equations and Their Graphs

The Graph of an Equation

The graph of an equation involving x and y as its only variables consists of all points (x, y) satisfying the equation.

EXAMPLE 5.1: (a) What is the graph of the equation $2x - y = 3$?

The equation is equivalent to $y = 2x - 3$, which we know is the slope–intercept equation of the line with slope 2 and y intercept -3.

(b) What is the graph of the equation $x^2 + y^2 - 2x + 4y - 4 = 0$?

Completing the square shows that the given equation is equivalent to the equation $(x - 1)^2 + (y + 2)^2 = 9$. Hence, its graph is the circle with center $(1, -2)$ and radius 3.

Parabolas

Consider the equation $y = x^2$. If we substitute a few values for x and calculate the associated values of y, we obtain the results tabulated in Fig. 5-1. We can plot the corresponding points, as shown in the figure. These points suggest the heavy curve, which belongs to a family of curves called *parabolas*. In particular, the graphs of equations of the form $y = cx^2$, where c is a nonzero constant, are parabolas, as are any other curves obtained from them by translations and rotations.

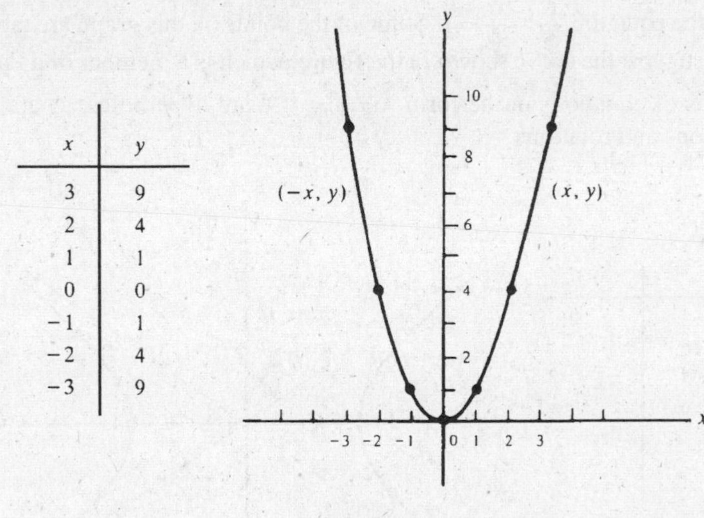

x	y
3	9
2	4
1	1
0	0
−1	1
−2	4
−3	9

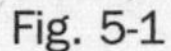

Fig. 5-1

In Fig. 5-1, we note that the graph of $y = x^2$ contains the origin (0, 0) but all its other points lie above the x axis, since x^2 is positive except when $x = 0$. When x is positive and increasing, y increases without bound. Hence, in the first quadrant, the graph moves up without bound as it moves right. Since $(-x)^2 = x^2$, it follows that, if any point (x, y) lies on the graph in the first quadrant, then the point $(-x, y)$ also lies on the graph in the second quadrant. Thus, the graph is symmetric with respect to the y axis. The y axis is called the *axis of symmetry* of this parabola.

Ellipses

To construct the graph of the equation $\frac{x^2}{9}+\frac{y^2}{4}=1$, we again compute a few values and plot the corresponding points, as shown in Fig. 5-2. The graph suggested by these points is also drawn in the figure; it is a member of a family of curves called *ellipses*. In particular, the graph of an equation of the form $\frac{x^2}{a^2}+\frac{y^2}{b^2}=1$ is an ellipse, as is any curve obtained from it by translation or rotation.

Note that, in contrast to parabolas, ellipses are bounded. In fact, if (x, y) is on the graph of $\frac{x^2}{9}+\frac{y^2}{4}=1$, then $\frac{x^2}{9}\leq\frac{x^2}{9}+\frac{y^2}{4}=1$, and, therefore, $x^2\leq 9$. Hence, $-3\leq x\leq 3$. So, the graph lies between the vertical lines $x=-3$ and $x=3$. Its rightmost point is $(3, 0)$, and its leftmost point is $(-3, 0)$. A similar argument shows that the graph lies between the horizontal lines $y=-2$ and $y=2$, and that its lowest point is $(0, -2)$ and its highest point is $(0, 2)$. In the first quadrant, as x increases from 0 to 3, y decreases from 2 to 0. If (x, y) is any point on the graph, then $(-x, y)$ also is on the graph. Hence, the graph is symmetric with respect to the y axis. Similarly, if (x, y) is on the graph, so is $(x, -y)$, and therefore the graph is symmetric with respect to the x axis.

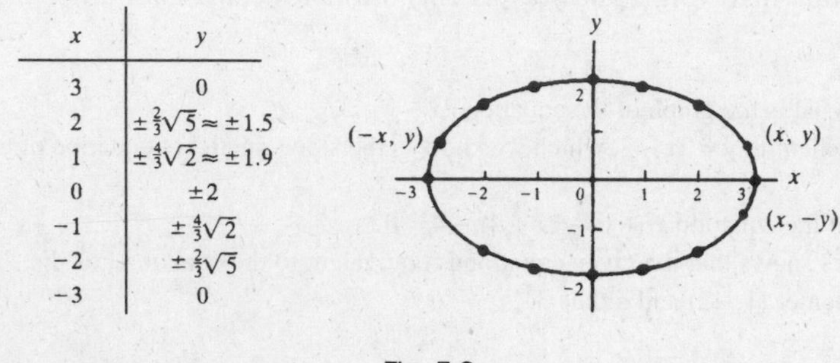

x	y
3	0
2	$\pm\frac{2}{3}\sqrt{5}\approx\pm1.5$
1	$\pm\frac{4}{3}\sqrt{2}\approx\pm1.9$
0	±2
-1	$\pm\frac{4}{3}\sqrt{2}$
-2	$\pm\frac{2}{3}\sqrt{5}$
-3	0

Fig. 5-2

When $a=b$, the ellipse $\frac{x^2}{a^2}+\frac{y^2}{b^2}=1$ is the circle with the equation $x^2+y^2=a^2$, that is, a circle with center at the origin and radius a. Thus, circles are special cases of ellipses.

Hyperbolas

Consider the graph of the equation $\frac{x^2}{9}-\frac{y^2}{4}=1$. Some of the points on this graph are tabulated and plotted in Fig. 5-3. These points suggest the curve shown in the figure, which is a member of a family of curves called *hyperbolas*. The graphs of equations of the form $\frac{x^2}{a^2}-\frac{y^2}{b^2}=1$ are hyperbolas, as are any curves obtained from them by translations and rotations.

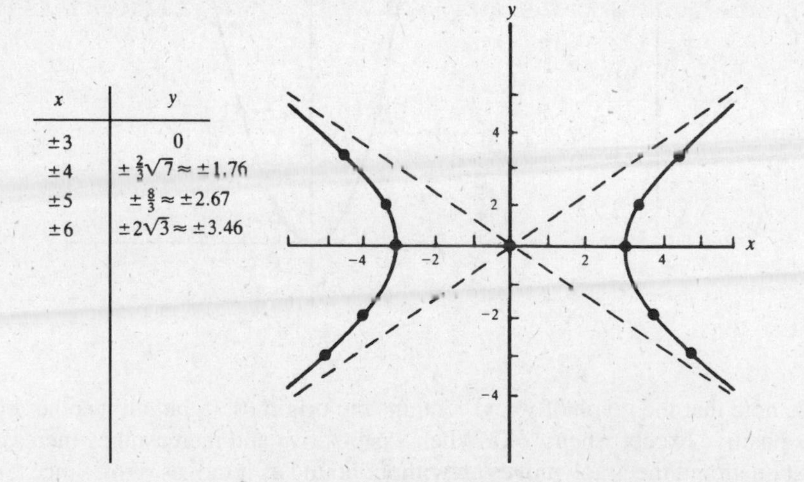

x	y
±3	0
±4	$\pm\frac{2}{3}\sqrt{7}\approx\pm1.76$
±5	$\pm\frac{8}{3}\approx\pm2.67$
±6	$\pm2\sqrt{3}\approx\pm3.46$

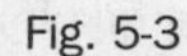

Fig. 5-3

Let us look at the hyperbola $\frac{x^2}{9}-\frac{y^2}{4}=1$ in more detail. Since $\frac{x^2}{9}=1+\frac{y^2}{4}\geq 1$, it follows that $x^2\geq 9$, and therefore, $|x|\geq 3$. Hence, there are no points on the graph between the vertical lines $x=-3$ and $x=3$. If (x, y)

is on the graph, so is $(-x, y)$; thus, the graph is symmetric with respect to the y axis. Similarly, the graph is symmetric with respect to the x axis. In the first quadrant, as x increases, y increases without bound.

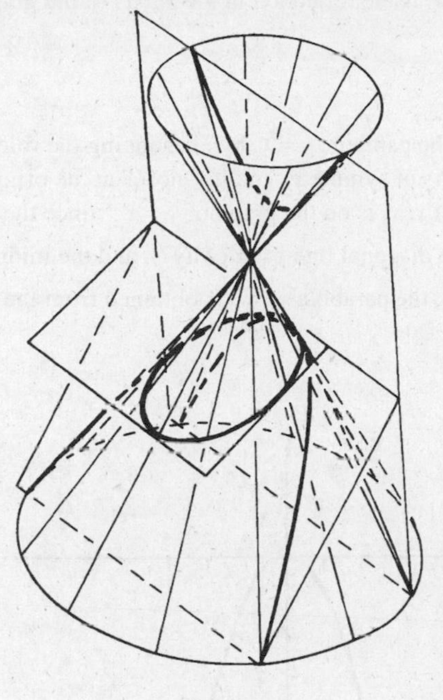

Fig. 5-4

Note the dashed lines in Fig. 5-3; they are the lines $y = \frac{2}{3}x$ and $y = -\frac{2}{3}x$, and they are called the *asymptotes* of the hyperbola: Points on the hyperbola get closer and closer to these asymptotes as they recede from the origin. In general, *the asymptotes of the hyperbola* $\frac{x^2}{a^2} - \frac{y^2}{b^2} = 1$ *are the lines* $y = \frac{b}{a}x$ *and* $y = -\frac{b}{a}x$.

Conic Sections

Parabolas, ellipses, and hyperbolas together make up a class of curves called *conic sections*. They can be defined geometrically as the intersections of planes with the surface of a right circular cone, as shown in Fig. 5-4.

SOLVED PROBLEMS

1. Sketch the graph of the *cubic curve* $y = x^3$.

The graph passes through the origin (0, 0). Also, for any point (x, y) on the graph, x and y have the same sign; hence, the graph lies in the first and third quadrants. In the first quadrant, as x increases, y increases without bound. Moreover, if (x, y) lies on the graph, then $(-x, -y)$ also lies on the graph. Since the origin is the midpoint of the segment connecting the points (x, y) and $(-x, -y)$, the graph is symmetric with respect to the origin. Some points on the graph are tabulated and shown in Fig. 5-5; these points suggest the heavy curve in the figure.

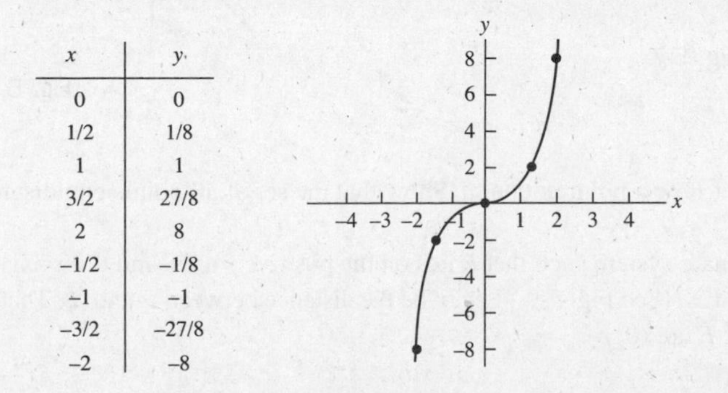

x	y
0	0
1/2	1/8
1	1
3/2	27/8
2	8
−1/2	−1/8
−1	−1
−3/2	−27/8
−2	−8

Fig. 5-5

2. Sketch the graph of the equation $y = -x^2$.

If (x, y) is on the graph of the parabola $y = x^2$ (Fig. 5-1), then $(x, -y)$ is on the graph of $y = -x^2$, and vice versa. Hence, the graph of $y = -x^2$ is the reflection in the x axis of the graph of $y = x^2$. The result is the parabola in Fig. 5-6.

3. Sketch the graph of $x = y^2$.

This graph is obtained from the parabola $y = x^2$ by exchanging the roles of x and y. The resulting curve is a parabola with the x axis as its axis of symmetry and its "nose" at the origin (see Fig. 5-7). A point (x, y) is on the graph of $x = y^2$ if and only if (y, x) is on the graph of $y = x^2$. Since the segment connecting the points (x, y) and (y, x) is perpendicular to the diagonal line $y = x$ (why?), and the midpoint $\left(\frac{x+y}{2}, \frac{x+y}{2}\right)$ of that segment is on the line $y = x$ (see Fig. 5-8), the parabola $x = y^2$ is obtained from the parabola $y = x^2$ by reflection in the line $y = x$.

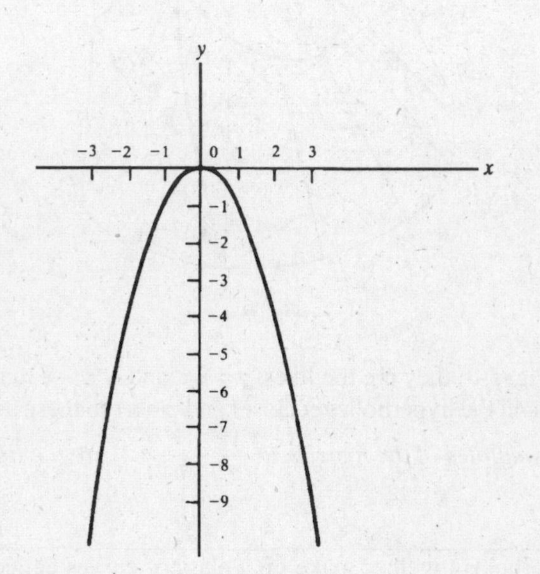

Fig. 5-6

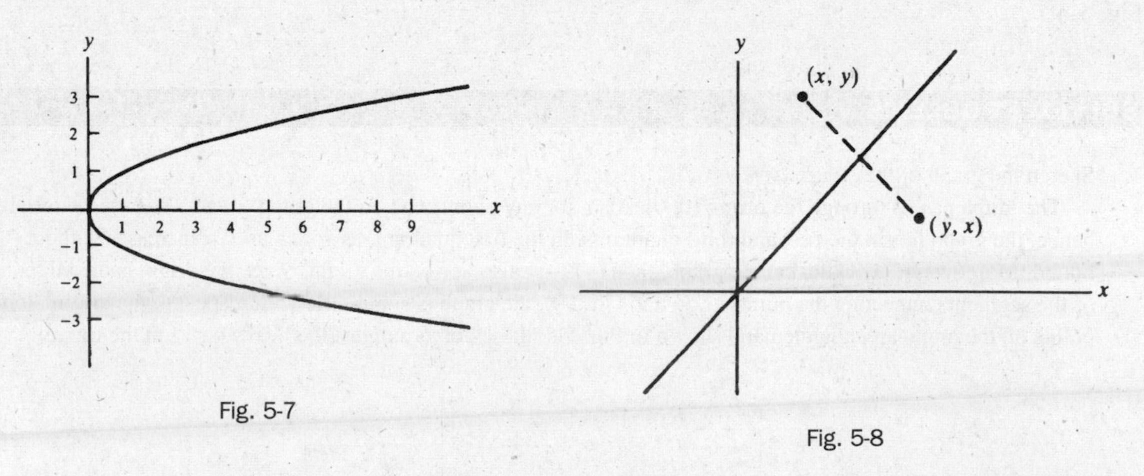

Fig. 5-7

Fig. 5-8

4. Let $\mathcal{L}$ be a line, and let F be a point not on $\mathcal{L}$. Show that the set of all points equidistant from F and $\mathcal{L}$ is a parabola.

Construct a coordinate system such that F lies on the positive y axis, and the x axis is parallel to $\mathcal{L}$ and halfway between F and $\mathcal{L}$. (See Fig. 5-9.) Let $2p$ be the distance between F and $\mathcal{L}$. Then $\mathcal{L}$ has the equation $y = -p$, and the coordinates of F are $(0, p)$.

Consider an arbitrary point $P(x, y)$. Its distance from $\mathscr{L}$ is $|y + p|$, and its distance from F is $\sqrt{x^2 + (y - p)^2}$. Thus, for the point to be equidistant from F and $\mathscr{L}$, we must have $|y + p| = \sqrt{x^2 + (y - p)^2}$. Squaring yields $(y + p)^2 = x^2 + (y - p)^2$, from which we find that $4py = x^2$. This is the equation of a parabola with the y axis as its axis of symmetry. The point F is called the *focus* of the parabola, and the line $\mathscr{L}$ is called its *directrix*. The chord AB through the focus and parallel to $\mathscr{L}$ is called the *latus rectum*. The "nose" of the parabola at (0, 0) is called its *vertex*.

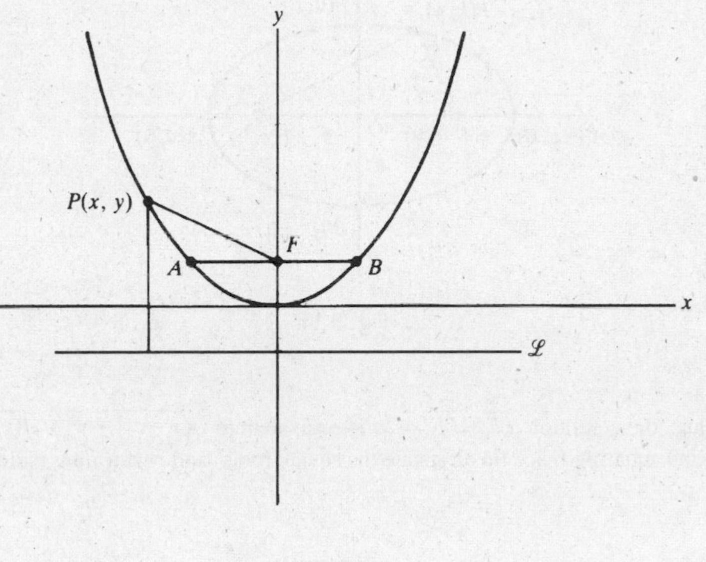

Fig. 5-9

5. Find the length of the latus rectum of a parabola $4py = x^2$.

The y coordinate of the endpoints A and B of the lactus rectum (see Fig. 5-9) is p. Hence, at these points, $4p^2 = x^2$ and, therefore, $x = \pm 2p$. Thus, the length AB of the latus rectum is $4p$.

6. Find the focus, directrix, and the length of the latus rectum of the parabola $y = \frac{1}{2}x^2$, and draw its graph.

The equation of the parabola can be written as $2y = x^2$. Hence, $4p = 2$ and $p = \frac{1}{2}$. Therefore, the focus is at $(0, \frac{1}{2})$, the equation of the directix is $y = -\frac{1}{2}$, and the length of the latus rectum is 2. The graph is shown in Fig. 5-10.

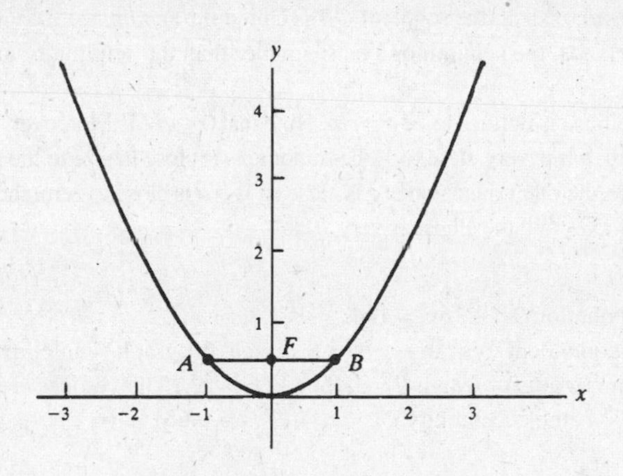

Fig. 5-10

7. Let F and F' be two distinct points at a distance $2c$ from each other. Show that the set of all points $P(x, y)$ such that $\overline{PF} + \overline{PF'} = 2a$, $a > c$ is an ellipse.

Construct a coordinate system such that the x axis passes through F and F', the origin is the midpoint of the segment FF', and F lies on the positive x axis. Then the coordinates of F and F' are $(c, 0)$ and $(-c, 0)$.

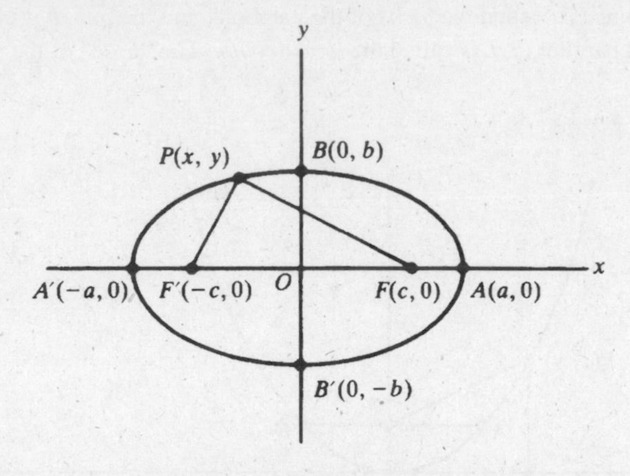

Fig. 5-11

(See Fig. 5-11.) Thus, the condition $\overline{PF} + \overline{PF'} = 2a$ is equivalent to $\sqrt{(x-c)^2 + y^2} + \sqrt{(x+c)^2 + y^2} = 2a$. After rearranging and squaring twice (to eliminate the square roots) and performing indicated operations, we obtain

$$(a^2 - c^2)x^2 + a^2y^2 = a^2(a^2 - c^2) \tag{1}$$

Since $a > c$, $a^2 - c^2 > 0$. Let $b = \sqrt{a^2 - c^2}$. Then (1) becomes $b^2x^2 + a^2y^2 = a^2b^2$, which we may rewrite as $\frac{x^2}{a^2} + \frac{y^2}{b^2} = 1$, the equation of an ellipse.

When $y = 0$, $x^2 = a^2$; hence, the ellipse intersects the x axis at the points $A'(-a, 0)$, and $A(a, 0)$, called the *vertices* of the ellipse (Fig. 5-11). The segment $A'A$ is called the *major axis*; the segment OA is called the *semimajor axis* and has length a. The origin is the *center* of the ellipse. F and F' are called the *foci* (each is a *focus*). When $x = 0$, $y^2 = b^2$. Hence, the ellipse intersects the y axis at the points $B'(0, -b)$ and $B(0, b)$. The segment $B'B$ is called the *minor axis*; the segment OB is called the *semiminor axis* and has length b. Note that $b = \sqrt{a^2 - c^2} < \sqrt{a^2} = a$. Hence, the semiminor axis is smaller than the semimajor axis. The basic relation among a, b, and c is $a^2 = b^2 + c^2$.

The *eccentricity* of an ellipse is defined to be $e = c/a$. Note that $0 < e < 1$. Moreover, $e = \sqrt{a^2 - b^2}/a = \sqrt{1 - (b/a)^2}$. Hence, when e is very small, b/a is very close to 1, the minor axis is close in size to the major axis, and the ellipse is close to being a circle. On the other hand, when e is close to 1, b/a is close to zero, the minor axis is very small in comparison with the major axis, and the ellipse is very "flat."

8. Identify the graph of the equation $9x^2 + 16y^2 = 144$.

The given equation is equivalent to $x^2/16 + y^2/9 = 1$. Hence, the graph is an ellipse with semimajor axis of length $a = 4$ and semiminor axis of length $b = 3$. (See Fig. 5-12.) The vertices are $(-4, 0)$ and $(4, 0)$. Since $c = \sqrt{a^2 - b^2} = \sqrt{16 - 9} = \sqrt{7}$, the eccentricity e is $c/a = \sqrt{7}/4 \approx 0.6614$.

9. Identify the graph of the equation $25x^2 + 4y^2 = 100$.

The given equation is equivalent to $x^2/4 + y^2/25 = 1$, an ellipse. Since the denominator under y^2 is larger than the denominator under x^2, the graph is an ellipse with the major axis on the y axis and the minor axis on the x axis (see Fig. 5-13). The vertices are at $(0, -5)$ and $(0, 5)$. Since $c = \sqrt{a^2 - b^2} = \sqrt{21}$, the eccentricity is $\sqrt{21}/5 \approx 0.9165$.

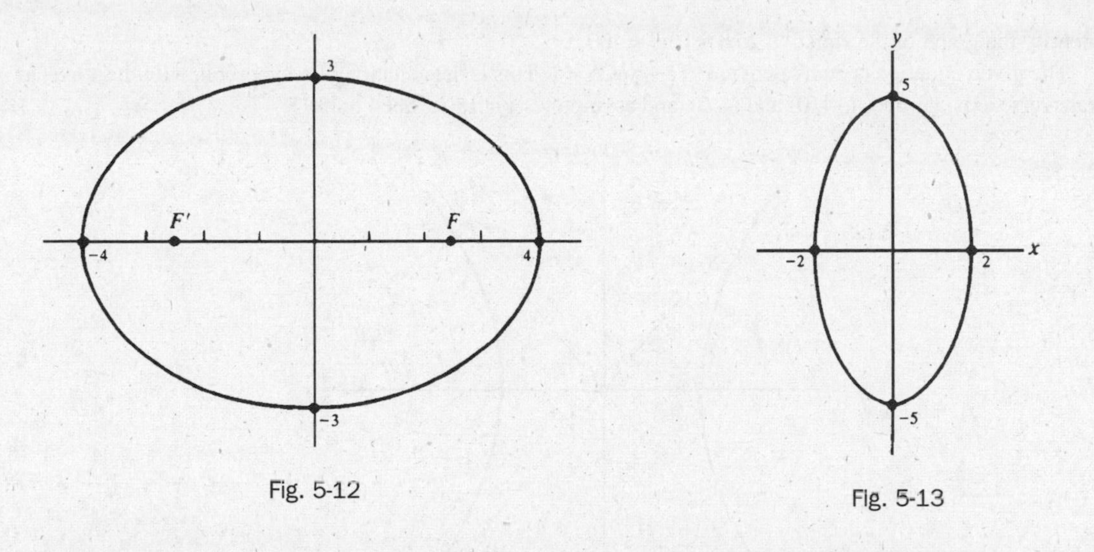

Fig. 5-12

Fig. 5-13

10. Let F and F' be distinct points at a distance of $2c$ from each other. Find the set of all points $P(x, y)$ such that $|\overline{PF} - \overline{PF'}| = 2a$, for $a < c$.

Choose a coordinate system such that the x axis passes through F and F', with the origin as the midpoint of the segment FF' and with F on the positive x axis (see Fig. 5-14). The coordinates of F and F' are $(c, 0)$ and $(-c, 0)$. Hence, the given condition is equivalent to $\sqrt{(x-c)^2 + y^2} - \sqrt{(x+c)^2 + y^2} = \pm 2a$. After manipulations required to eliminate the square roots, this yields

$$(c^2 - a^2)x^2 - a^2y^2 = a^2(c^2 - a^2) \tag{1}$$

Since $c > a$, $c^2 - a^2 > 0$. Let $b = \sqrt{c^2 - a^2}$. (Notice that $a^2 + b^2 = c^2$.) Then (1) becomes $b^2x^2 - a^2y^2 = a^2b^2$, which we rewrite as $\dfrac{x^2}{a^2} - \dfrac{y^2}{b^2} = 1$, the equation of a hyperbola.

When $y = 0$, $x = \pm a$. Hence, the hyperbola intersects the x axis at the points $A'(-a, 0)$ and $A(a, 0)$, which are called the *vertices* of the hyperbola. The asymptotes are $y = \pm \dfrac{b}{a}x$. The segment $A'A$ is called the *transverse axis*. The segment connecting the points $(0, -b)$ and $(0, b)$ is called the *conjugate axis*. The *center* of the hyperbola is the origin. The points F and F' are called the *foci*. The *eccentricity* is defined to be $e = \dfrac{c}{a} = \dfrac{\sqrt{a^2 + b^2}}{a} = \sqrt{1 + \left(\dfrac{b}{a}\right)^2}$. Since $c > a$, $e > 1$. When e is close to 1, b is very small relative to a, and the hyperbola has a very pointed "nose"; when e is very large, b is very large relative to a, and the hyperbola is very "flat."

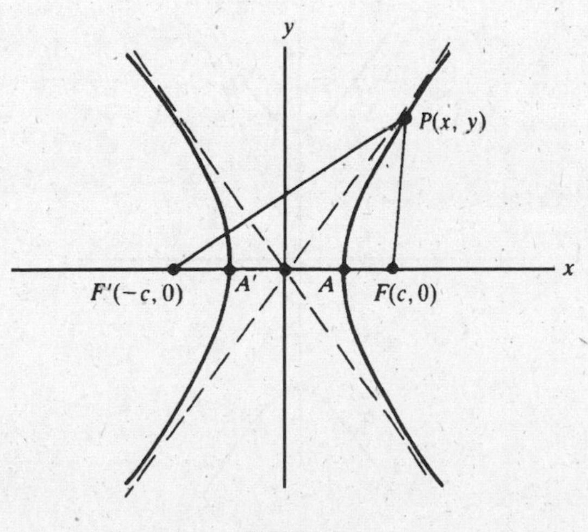

Fig. 5-14

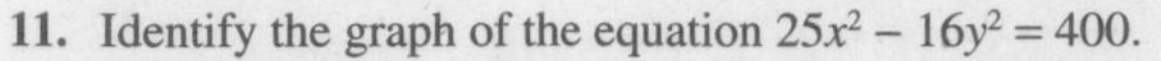

11. Identify the graph of the equation $25x^2 - 16y^2 = 400$.

The given equation is equivalent to $x^2/16 - y^2/25 = 1$. This is the equation of a hyperbola with the x axis as its transverse axis, vertices $(-4, 0)$ and $(4, 0)$, and asymptotes $y = \pm\frac{5}{4}x$. (See Fig. 5.15.)

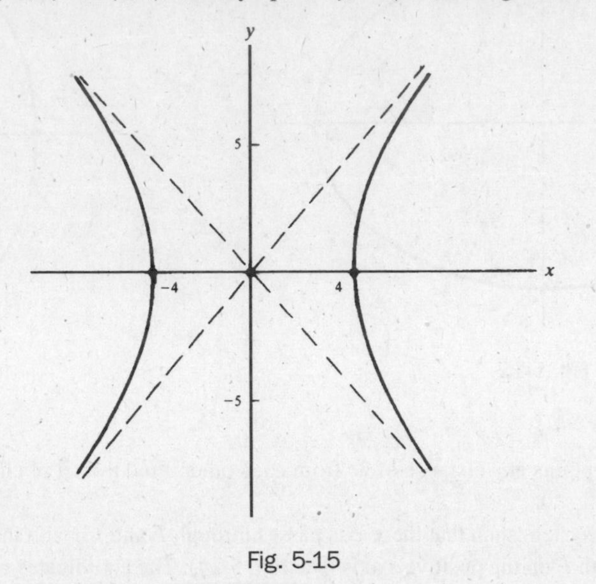

Fig. 5-15

12. Identify the graph of the equation $y^2 - 4x^2 = 4$.

The given equation is equivalent to $\frac{y^2}{4} - \frac{x^2}{1} = 1$. This is the equation of a hyperbola, with the roles of x and y interchanged. Thus, the transverse axis is the y axis, the conjugate axis is the x axis, and the vertices are $(0, -2)$ and $(0, 2)$. The asymptotes are $x = \pm\frac{1}{2}y$ or, equivalently, $y = \pm 2x$. (See Fig. 5-16.)

13. Identify the graph of the equation $y = (x - 1)^2$.

A point (u, v) is on the graph of $y = (x - 1)^2$ if and only if the point $(u - 1, v)$ is on the graph of $y = x^2$. Hence, the desired graph is obtained from the parabola $y = x^2$ by moving each point of the latter one unit to the right. (See Fig. 5-17.)

14. Identify the graph of the equation $\frac{(x-1)^2}{4} + \frac{(y-2)^2}{9} = 1$.

A point (u, v) is on the graph if and only if the point $(u - 1, v - 2)$ is on the graph of the equation $x^2/4 + y^2/9 = 1$. Hence, the desired graph is obtained by moving the ellipse $x^2/4 + y^2/9 = 1$ one unit to the right and two units upward. (See Fig. 5-18.) The center of the ellipse is at $(1, 2)$, the major axis is along the line $x = 1$, and the minor axis is along the line $y = 2$.

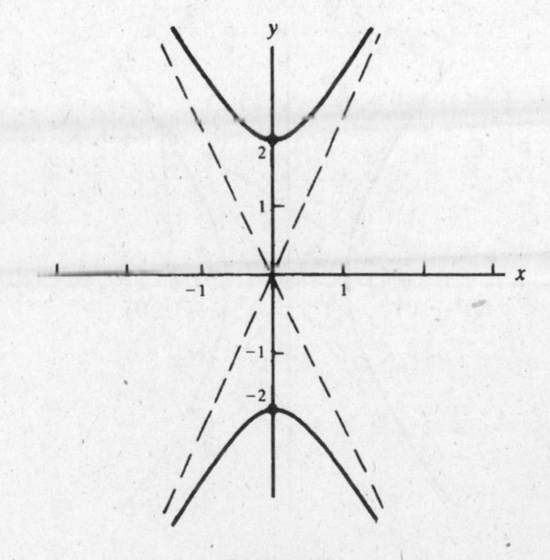

Fig. 5-16

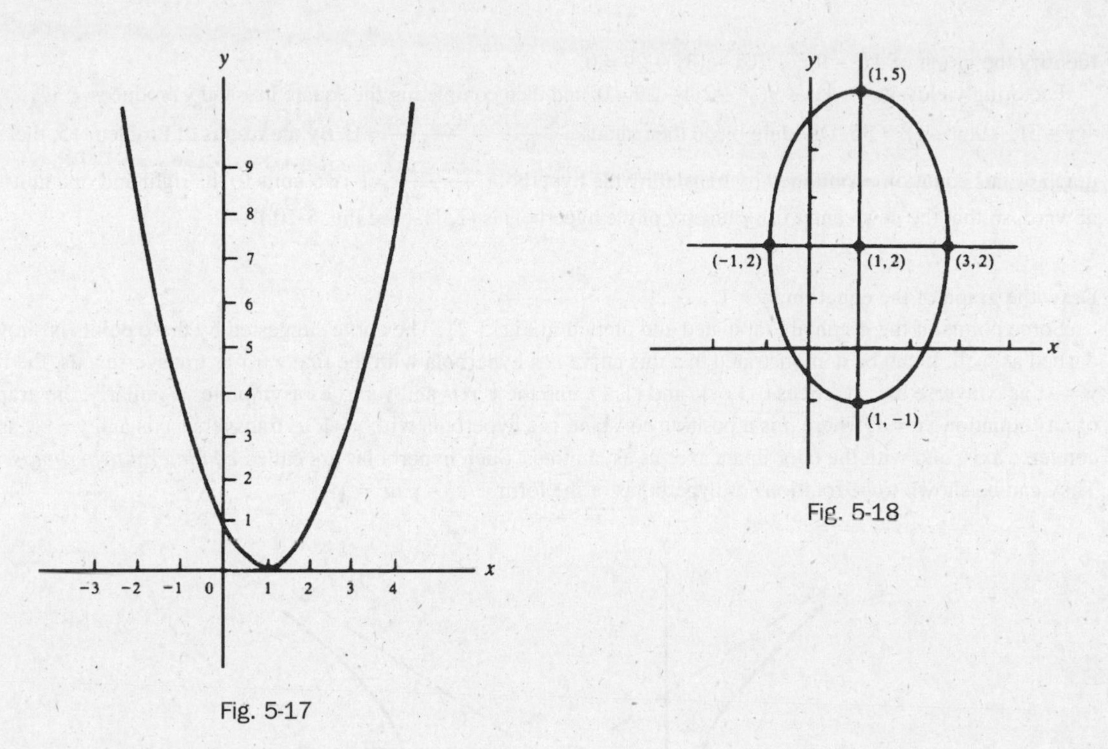

Fig. 5-17

Fig. 5-18

15. How is the graph of an equation $F(x - a, y - b) = 0$ related to the graph of the equation $F(x, y) = 0$?

A point (u, v) is on the graph of $F(x - a, y - b) = 0$ if and only if the point $(u - a, v - b)$ is on the graph of $F(x, y) = 0$. Hence, the graph of $F(x - a, y - b) = 0$ is obtained by moving each point of the graph of $F(x, y) = 0$ by a units to the right and b units upward. (If a is negative, we move the point $|a|$ units to the left. If b is negative, we move the point $|b|$ units downward.) Such a motion is called a *translation*.

16. Identify the graph of the equation $y = x^2 - 2x$.

Competing the square in x, we obtain $y + 1 = (x - 1)^2$. Based on the results of Problem 15, the graph is obtained by a translation of the parabola $y = x^2$ so that the new vertex is $(1, -1)$. [Notice that $y + 1$ is $y - (-1)$.] It is shown in Fig. 5-19.

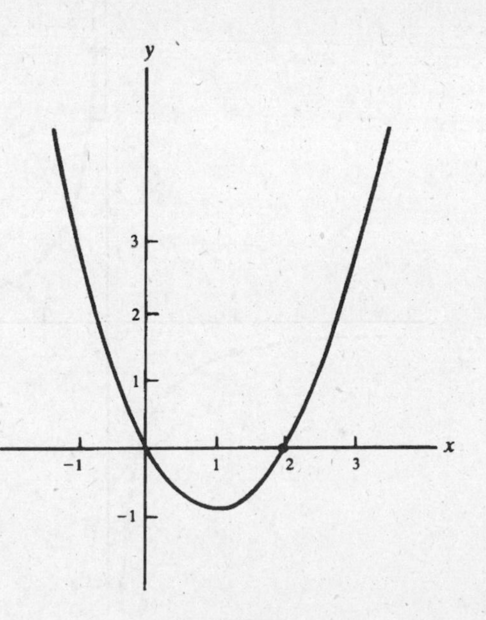

Fig. 5-19

17. Identify the graph of $4x^2 - 9y^2 - 16x + 18y - 29 = 0$.

Factoring yields $4(x^2 - 4x) - 9(y^2 - 2y) - 29 = 0$, and then completing the square in x and y produces $4(x-2)^2 - 9(y-1)^2 = 36$. Dividing by 36 then yields $\frac{(x-2)^2}{9} - \frac{(y-1)^2}{4} = 1$. By the results of Problem 15, the graph of this equation is obtained by translating the hyperbola $\frac{x^2}{9} - \frac{y^2}{4} = 1$ two units to the right and one unit upward, so that the new center of symmetry of the hyperbola is (2, 1). (See Fig. 5-20.)

18. Draw the graph of the equation $xy = 1$.

Some points of the graph are tabulated and plotted in Fig. 5-21. The curve suggested by these points is shown dashed as well. It can be demonstrated that this curve is a hyperbola with the line $y = x$ as transverse axis, the line $y = -x$ as converse axis, vertices $(-1, -1)$ and $(1, 1)$, and the x axis and y axis as asymptotes. Similarly, the graph of any equation $xy = d$, where d is a positive constant, is a hyperbola with $y = x$ as transverse axis and $y = -x$ as converse axis, and with the coordinate axes as asymptotes. Such hyperbolas are called *equilateral hyperbolas*. They can be shown to be rotations of hyperbolas of the form $x^2/a^2 - y^2/a^2 = 1$.

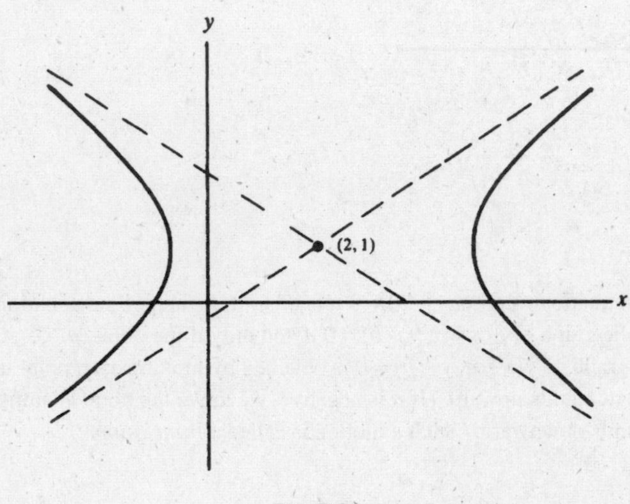

Fig. 5-20

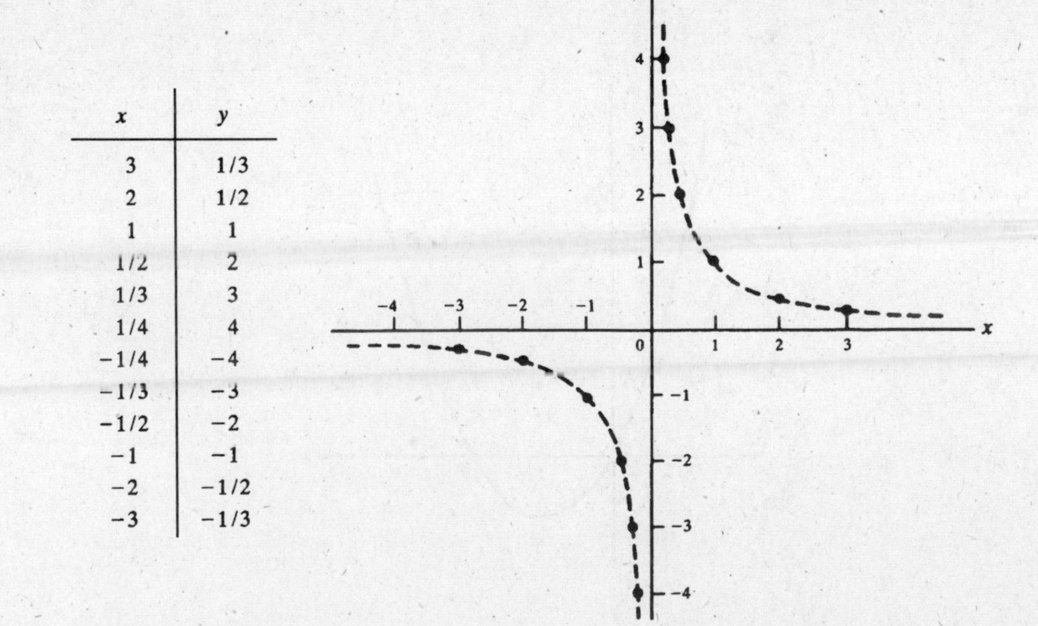

x	y
3	1/3
2	1/2
1	1
1/2	2
1/3	3
1/4	4
−1/4	−4
−1/3	−3
−1/2	−2
−1	−1
−2	−1/2
−3	−1/3

Fig. 5-21

SUPPLEMENTARY PROBLEMS

19. (a) On the same sheet of paper, draw the graphs of the following parabolas:

(i) $y = 2x^2$; (ii) $y = 3x^2$; (iii) $y = 4x^2$;
(iv) $y = \frac{1}{2}x^2$; (v) $y = \frac{1}{3}x^2$.

(b) (GC) Use a graphing calculator to check your answers to (a).

20. (a) On the same sheet of paper, draw the graphs of the following parabolas and indicate points of intersection:

(i) $y = x^2$; (ii) $y = -x^2$; (iii) $x = y^2$;
(iv) $x = -y^2$.

(b) (GC) Use a graphing calculator to check your answers to (a).

21. Draw the graphs of the following equations:

(a) $y = x^3 - 1$ (b) $y = (x - 2)^3$ (c) $y = (x + 1)^3 - 2$
(d) $y = -x^3$ (e) $y = -(x - 1)^3$ (f) $y = -(x - 1)^3 + 2$

22. (GC) Use a graphing calculator to answer Problem 21.

23. Identify and draw the graphs of the following equations:

(a) $y^2 - x^2 = 1$ (b) $25x^2 + 36y^2 = 900$ (c) $2x^2 - y^2 = 4$
(d) $xy = 4$ (e) $4x^2 + 4y^2 = 1$ (f) $8x = y^2$
(g) $10y = x^2$ (h) $4x^2 + 9y^2 = 16$ (i) $xy = -1$
(j) $3y^2 - x^2 = 9$

Ans. (a) hyperbola, y axis as transverse axis, vertices $(0, \pm 1)$, asymptotes $y = \pm x$; (b) ellipse, vertices $(\pm 6, 0)$ foci $(\pm\sqrt{11}, 0)$; (c) hyperbola, x axis as transverse axis, vertices $(\pm\sqrt{2}, 0)$, asymptotes $y = \pm x\sqrt{2}x$; (d) hyperbola, $y = x$ as transverse axis, vertices $(2, 2)$ and $(-2, -2)$, x and y axes as asymptotes; (e) circle, center $(0, 0)$, radius $\frac{1}{2}$; (f) parabola, vertex $(0, 0)$, focus $(2, 0)$, directrix $x = -2$; (g) parabola, vertex $(0, 0)$, focus $(0, \frac{5}{2})$, directrix $y = -\frac{5}{2}$; (h) ellipse, vertices $(\pm 2, 0)$, foci $(\pm\frac{2}{3}\sqrt{5}, 0)$; (i) hyperbola, $y = -x$ as transverse axis, vertices $(-1, 1)$ and $(1, -1)$, x and y axes as asymptotes; (j) hyperbola, y axis as transverse axis, vertices $(0, \pm\sqrt{3})$, asymptotes $y = \pm x\sqrt{3}x/3$

24. (GC) Use a graphing calculator to draw the graphs in Problem 23.

25. Identify and draw the graphs of the following equations:

(a) $4x^2 - 3y^2 + 8x + 12y - 4 = 0$ (b) $5x^2 + y^2 - 20x + 6y + 25 = 0$ (c) $x^2 - 6x - 4y + 5 = 0$
(d) $2x^2 + y^2 - 4x + 4y + 6 = 0$ (e) $3x^2 + 2y^2 + 12x - 4y + 15 = 0$ (f) $(x - 1)(y + 2) = 1$
(g) $xy - 3x - 2y + 5 = 0$ [*Hint*: Compare (f).] (h) $4x^2 + y^2 + 8x + 4y + 4 = 0$
(i) $2x^2 - 8x - y + 11 = 0$ (j) $25x^2 + 16y^2 - 100x - 32y - 284 = 0$

Ans. (a) empty graph; (b) ellipse, center at $(2, -3)$; (c) parabola, vertex at $(3, -1)$; (d) single point $(1, -2)$; (e) empty graph; (f) hyperbola, center at $(1, -2)$; (g) hyperbola, center at $(2, 3)$; (h) ellipse, center at $(-1, 2)$; (i) parabola, vertex at $(2, 3)$; (j) ellipse, center at $(2, 1)$

26. (GC) Use a graphing calculator to draw the graphs in Problem 25.

27. Find the focus, directrix, and length of the latus rectum of the following parabolas: (a) $10x^2 = 3y$; (b) $2y^2 = 3x$; (c) $4y = x^2 + 4x + 8$; (d) $8y = -x^2$.

Ans. (a) focus at $(0, \frac{3}{40})$, directrix $y = -\frac{3}{40}$, latus rectum $\frac{3}{10}$; (b) focus at $(\frac{3}{8}, 0)$, directrix $x = -\frac{3}{8}$, latus rectum $\frac{3}{2}$; (c) focus at $(-2, 2)$, directrix $y = 0$, latus rectum 4; (d) focus at $(0, -2)$, directrix $y = 2$, latus rectum 8

28. Find an equation for each parabola satisfying the following conditions:

(a) Focus at $(0, -3)$, directrix $y = 3$
(b) Focus at $(6, 0)$, directrix $x = 2$
(c) Focus at $(1, 4)$, directrix $y = 0$
(d) Vertex at $(1, 2)$, focus at $(1, 4)$
(e) Vertex at $(3, 0)$, directrix $x = 1$
(f) Vertex at the origin, y axis as axis of symmetry, contains the point $(3, 18)$
(g) Vertex at $(3, 5)$, axis of symmetry parallel to the y axis, contains the point $(5, 7)$
(h) Axis of symmetry parallel to the x axis, contains the points $(0, 1)$, $(3, 2)$, $(1, 3)$
(i) Latus rectum is the segment joining $(2, 4)$ and $(6, 4)$, contains the point $(8, 1)$
(j) Contains the points $(1, 10)$ and $(2, 4)$, axis of symmetry is vertical, vertex is on the line $4x - 3y = 6$

Ans. (a) $12y = -x^2$; (b) $8(x - 4) = y^2$; (c) $8(y - 2) = (x - 1)^2$; (d) $8(y - 2) = (x - 1)^2$; (e) $8(x - 3) = y^2$; (f) $y = 2x^2$; (g) $2(y - 5) = (x - 3)^2$; (h) $2\left(x - \frac{121}{40}\right) = -5\left(y - \frac{21}{10}\right)^2$; (i) $4(y - 5) = -(x - 4)^2$; (j) $y - 2 = 2(x - 3)^2$ or $y - \frac{2}{13} = 26\left(x - \frac{21}{13}\right)^2$

29. Find an equation for each ellipse satisfying the following conditions:

(a) Center at the origin, one focus at $(0, 5)$, length of semimajor axis is 13
(b) Center at the origin, major axis on the y axis, contains the points $(1, 2\sqrt{3})$ and $(\frac{1}{2}, \sqrt{15})$
(c) Center at $(2, 4)$, focus at $(7, 4)$, contains the point $(5, 8)$
(d) Center at $(0, 1)$, one vertex at $(6, 1)$, eccentricity $\frac{2}{3}$
(e) Foci at $\left(0, \pm\frac{4}{3}\right)$, contains $\left(\frac{4}{5}, 1\right)$
(f) Foci $(0, \pm 9)$, semiminor axis of length 12

Ans. (a) $\frac{x^2}{144} + \frac{y^2}{169} = 1$; (b) $\frac{x^2}{4} + \frac{y^2}{16} = 1$; (c) $\frac{(x-2)^2}{45} + \frac{(y-4)^2}{20} = 1$; (d) $\frac{x^2}{36} + \frac{(y-1)^2}{20} = 1$; (e) $x^2 + \frac{9y^2}{25} = 1$; (f) $\frac{x^2}{144} + \frac{y^2}{225} = 1$

30. Find an equation for each hyperbola satisfying the following conditions:

(a) Center at the origin, transverse axis the x axis, contains the points $(6, 4)$ and $(-3, 1)$
(b) Center at the origin, one vertex at $(3, 0)$, one asymptote is $y = \frac{2}{3}x$
(c) Has asymptotes $y = \pm\sqrt{2}x$, contains the point $(1, 2)$
(d) Center at the origin, one focus at $(4, 0)$, one vertex at $(3, 0)$

Ans. (a) $\frac{5x^2}{36} - \frac{y^2}{4} = 1$; (b) $\frac{x^2}{9} - \frac{y^2}{4} = 1$; (c) $\frac{y^2}{2} - x^2 = 1$; (d) $\frac{x^2}{9} - \frac{y^2}{7} = 1$

31. Find an equation of the hyperbola consisting of all points $P(x, y)$ such that $|\overline{PF} - \overline{PF'}| = 2\sqrt{2}$, where $F = (\sqrt{2}, \sqrt{2})$ and $F' = (-\sqrt{2}, -\sqrt{2})$.

Ans. $xy = 1$

32. (GC) Use a graphing calculator to draw the hyperbola $\frac{x^2}{9} - \frac{y^2}{4} = 1$ and its asymptotes $y = \pm\frac{2}{3}x$.

33. (GC) Use a graphing calculator to draw the ellipses $x^2 + 4y^2 = 1$ and $(x - 3)^2 + 4(y - 2)^2 = 1$. How is the latter graph obtained from the former one?

Functions

We say that a quantity y is a *function* of some other quantity x if the value of y is determined by the value of x. If f denotes the function, then we indicate the dependence of y on x by means of the formula $y = f(x)$. The letter x is called the *independent variable*, and the letter y is called the *dependent variable*. The independent variable is also called the *argument* of the function, and the dependent variable is called the *value* of the function.

For example, the area A of a square is a function of the length s of a side of the square, and that function can be expressed by the formula $A = s^2$. Here, s is the independent variable and A is the dependent variable.

The *domain* of a function is the set of numbers to which the function can be applied, that is, the set of numbers that are assigned to the independent variable. The *range* of a function is the set of numbers that the function associates with the numbers in the domain.

EXAMPLE 6.1: The formula $f(x) = x^2$ determines a function f that assigns to each real number x its square. The domain consists of all real numbers. The range can be seen to consist of all nonnegative real numbers. (In fact, each value x^2 is nonnegative. Conversely, if r is any nonnegative real number, then r appears as a value when the function is applied to $\sqrt{r}$, since $r = (\sqrt{r})^2$.)

EXAMPLE 6.2: Let g be the function defined by the formula $g(x) = x^2 - 4x + 2$ for all real numbers. Thus,

$$g(1) = (1)^2 - 4(1) + 2 = 1 - 4 + 2 = -1$$

and

$$g(-2) = (-2)^2 - 4(-2) + 2 = 4 + 8 + 2 = 14$$

Also, for any number a, $g(a + 1) = (a + 1)^2 - 4(a + 1) + 2 = a^2 + 2a + 1 - 4a - 4 + 2 = a^2 - 2a - 1$.

EXAMPLE 6.3: (a) Let the function $h(x) = 18x - 3x^2$ be defined for all real numbers x. Thus, the domain is the set of all real numbers. (b) Let the area A of a certain rectangle, one of whose sides has length x, be given by $A = 18x - 3x^2$. Both x and A must be positive. Now, by completing the square, we obtain

$$A = -3(x^2 - 6x) = -3[(x - 3)^2 - 9] = 27 - 3(x - 3)^2$$

Since $A > 0$, $3(x - 3)^2 < 27$, $(x - 3)^2 < 9$, $|x - 3| < 3$. Hence, $-3 < x - 3 < 3$, $0 < x < 6$. Thus, the function determining A has the open interval $(0, 6)$ as its domain. The graph of $A = 27 - 3(x - 3)^2$ is the parabola shown in Fig. 6-1. From the graph, we see that the range of the function is the half-open interval $(0, 27)$.

Notice that the function of part (b) is given by the same formula as the function of part (a), but the domain of the former is a proper subset of the domain of the latter.

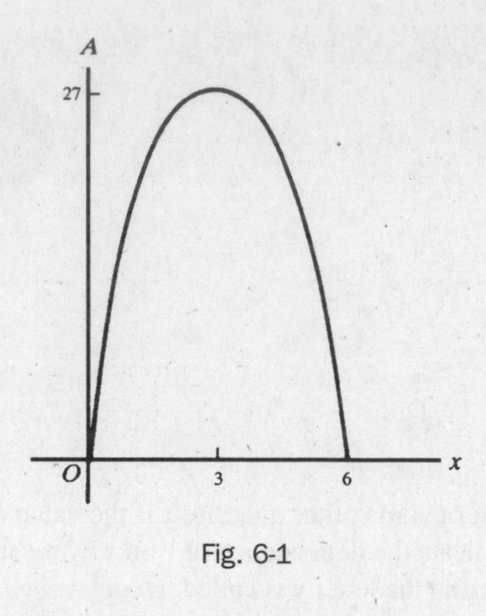

Fig. 6-1

The graph of a function f is defined to be the graph of the equation $y = f(x)$.

EXAMPLE 6.4: (a) Consider the function $f(x) = |x|$. Its graph is the graph of the equation $y = |x|$, and is indicated in Fig. 6-2. Notice that $f(x) = x$ when $x \geq 0$, whereas $f(x) = -x$ when $x \leq 0$. The domain of f consists of all real numbers. (*In general, if a function is given by means of a formula, then, if nothing is said to the contrary, we shall assume that the domain consists of all numbers for which the formula is defined.*) From the graph in Fig. 6-2, we see that the range of the function consists of all nonnegative real numbers. (*In general, the range of a function is the set of y coordinates of all points in the graph of the function.*) (b) The formula $g(x) = 2x + 3$ defines a function g. The graph of this function is the graph of the equation $y = 2x + 3$, which is the straight line with slope 2 and y intercept 3. The set of all real numbers is both the domain and range of g.

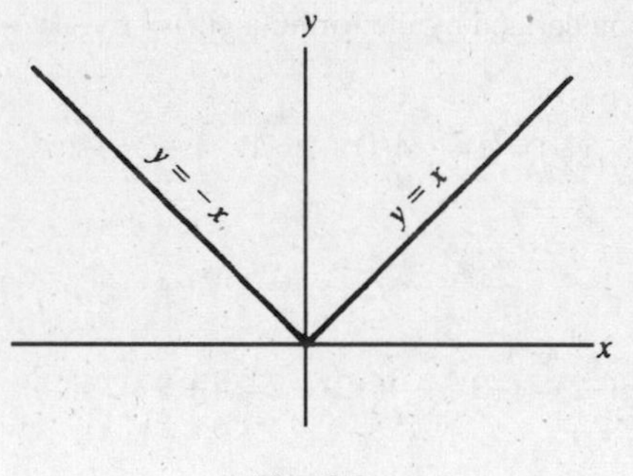

Fig. 6-2

EXAMPLE 6.5: Let a function g be defined as follows:

$$g(x) = \begin{cases} x^2 & \text{if } 2 \leq x \leq 4 \\ x+1 & \text{if } 1 \leq x < 2 \end{cases}$$

A function defined in this way is said to be *defined by cases*. Notice that the domain of g is the closed interval $[1, 4]$.

In a rigorous development of mathematics, a function f is defined to be a set of ordered pairs such that, if (x, y) and (x, z) are in the set f, then $y = z$. However, such a definition obscures the intuitive meaning of the notion of function.

SOLVED PROBLEMS

1. Given $f(x)=\dfrac{x-1}{x^2+2}$, find (a) $f(0)$; (b) $f(-1)$; (c) $f(2a)$; (d) $f(1/x)$; (e) $f(x+h)$.

(a) $f(0)=\dfrac{0-1}{0+2}=-\dfrac{1}{2}$ (b) $f(-1)=\dfrac{-1-1}{1+2}=-\dfrac{2}{3}$ (c) $f(2a)=\dfrac{2a-1}{4a^2+2}$

(d) $f(1/x)=\dfrac{1/x-1}{1/x^2+2}=\dfrac{x-x^2}{1+2x^2}$ (e) $f(x+h)=\dfrac{x+h-1}{(x+h)^2+2}=\dfrac{x+h-1}{x^2+2hx+h^2+2}$

2. If $f(x)=2^x$, show that (a) $f(x+3)-f(x-1)=\frac{15}{2}f(x)$ and (b) $\dfrac{f(x+3)}{f(x-1)}=f(4)$.

(a) $f(x+3)-f(x-1)=2^{x+3}-2^{x-1}=2^x(2^3-\frac{1}{2})=\frac{15}{2}f(x)$ (b) $\dfrac{f(x+3)}{f(x-1)}=\dfrac{2^{x+3}}{2^{x-1}}=2^4=f(4)$

3. Determine the domains of the functions

(a) $y=\sqrt{4-x^2}$ (b) $y=\sqrt{x^2-16}$ (c) $y=\dfrac{1}{x-2}$

(d) $y=\dfrac{1}{x^2-9}$ (e) $y=\dfrac{x}{x^2+4}$

(a) Since y must be real, $4-x^2\geq 0$, or $x^2\leq 4$. The domain is the interval $-2\leq x\leq 2$.
(b) Here, $x^2-16\geq 0$, or $x^2\geq 16$. The domain consists of the intervals $x\leq -4$ and $x\geq 4$.
(c) The function is defined for every value of x except 2.
(d) The function is defined for $x\neq\pm 3$.
(e) Since $x^2+4\neq 0$ for all x, the domain is the set of all real numbers.

4. Sketch the graph of the function defined as follows:

$f(x)=5$ when $0<x\leq 1$ $f(x)=10$ when $1<x\leq 2$

$f(x)=15$ when $2<x\leq 3$ $f(x)=20$ when $3<x\leq 4$ etc.

Determine the domain and range of the function.
The graph is shown in Fig. 6-3. The domain is the set of all positive real numbers, and the range is the set of integers, 5, 10, 15, 20, ...

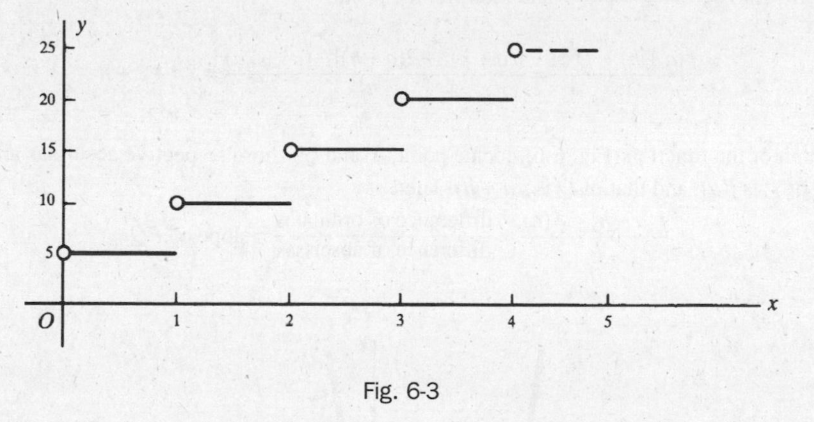

Fig. 6-3

5. A rectangular plot requires 2000 ft of fencing to enclose it. If one of its dimensions is x (in feet), express its area y (in square feet) as a function of x, and determine the domain of the function.

Since one dimension is x, the other is $\frac{1}{2}(2000-2x)=1000-x$. The area is then $y=x(1000-x)$, and the domain of this function is $0<x<1000$.

6. Express the length l of a chord of a circle of radius 8 as a function of its distance x from the center of the circle. Determine the domain of the function.

From Fig. 6-4 we see that $\frac{1}{2}l = \sqrt{64 - x^2}$, so that $l = 2\sqrt{64 - x^2}$. The domain is the interval $0 \le x < 8$.

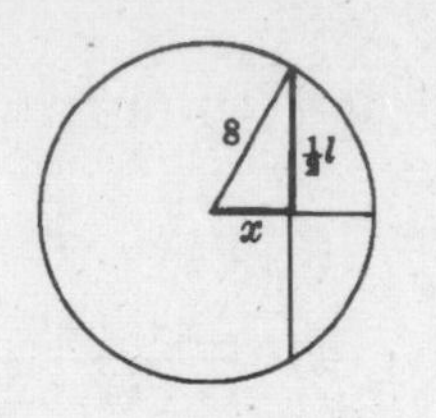

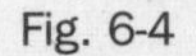
Fig. 6-4

7. From each corner of a square of tin, 12 inches on a side, small squares of side x (in inches) are removed, and the edges are turned up to form an open box (Fig. 6-5). Express the volume V of the box (in cubic inches) as a function of x, and determine the domain of the function.

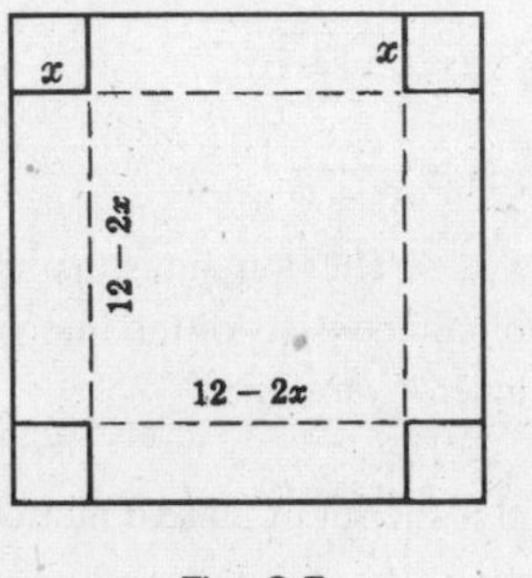

Fig. 6-5

The box has a square base of side $12 - 2x$ and a height of x. The volume of the box is then $V = x(12 - 2x)^2 = 4x(6 - x)^2$. The domain is the interval $0 < x < 6$.

As x increases over its domain, V increases for a time and then decreases thereafter. Thus, among such boxes that may be constructed, there is one of greatest volume, say M. To determine M, it is necessary to locate the precise value of x at which V ceases to increase. This problem will be studied in a later chapter.

8. If $f(x) = x^2 + 2x$, find $\dfrac{f(a+h) - f(a)}{h}$ and interpret the result.

$$\frac{f(a+h) - f(a)}{h} = \frac{[(a+h)^2 + 2(a+h)] - (a^2 + 2a)}{h} = 2a + 2 + h$$

On the graph of the function (Fig. 6-6), locate points P and Q whose respective abscissas are a and $a + h$. The ordinate of P is $f(a)$, and that of Q is $f(a + h)$. Then

$$\frac{f(a+h) - f(a)}{h} = \frac{\text{difference of ordinates}}{\text{difference of abscissas}} = \text{slope of } PQ$$

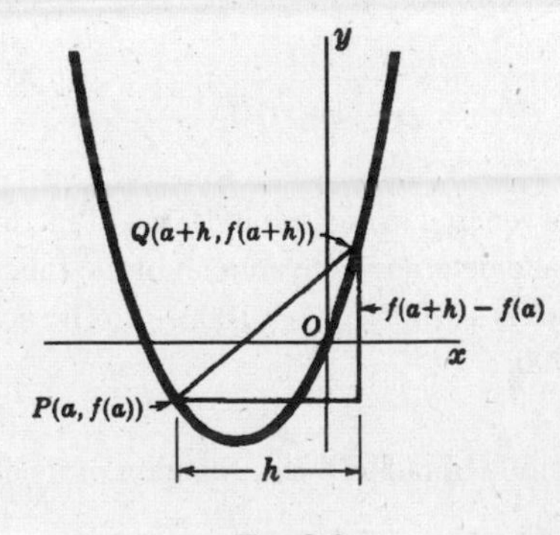

Fig. 6-6

9. Let $f(x) = x^2 - 2x + 3$. Evaluate (a) $f(3)$; (b) $f(-3)$; (c) $f(-x)$; (d) $f(x + 2)$; (e) $f(x - 2)$; (f) $f(x + h)$; (g) $f(x + h) - f(x)$; (h) $\dfrac{f(x+h)-f(x)}{h}$.

(a) $f(3) = 3^2 - 2(3) + 3 = 9 - 6 + 3 = 6$
(b) $f(-3) = (-3)^2 - 2(-3) + 3 = 9 + 6 + 3 = 18$
(c) $f(-x) = (-x)^2 - 2(-x) + 3 = x^2 + 2x + 3$
(d) $f(x + 2) = (x + 2)^2 - 2(x + 2) + 3 = x^2 + 4x + 4 - 2x - 4 + 3 = x^2 + 2x + 3$
(e) $f(x - 2) = (x - 2)^2 - 2(x - 2) + 3 = x^2 - 4x + 4 - 2x + 4 + 3 = x^2 - 6x + 11$
(f) $f(x + h) = (x + h)^2 - 2(x + h) + 3 = x^2 + 2hx + h^2 - 2x - 2h + 3 = x^2 + (2h - 2)x + (h^2 - 2h + 3)$
(g) $f(x + h) - f(x) - [x^2 + (2h - 2)x + (h^2 - 2h + 3)] - (x^2 - 2x + 3) = 2hx + h^2 - 2h = h(2x + h - 2)$
(h) $\dfrac{f(x+h)-f(x)}{h} = \dfrac{h(2x+h-2)}{h} = 2x+h-2$

10. Draw the graph of the function $f(x) = \sqrt{4-x^2}$, and find the domain and range of the function.

The graph of f is the graph of the equation $y = \sqrt{4-x^2}$. For points on this graph, $y^2 = 4 - x^2$; that is, $x^2 + y^2 = 4$. The graph of the last equation is the circle with center at the origin and radius 2. Since $y = \sqrt{4-x^2} \geq 0$, the desired graph is the upper half of that circle. Fig. 6-7 shows that the domain is the interval $-2 \leq x \leq 2$, and the range is the interval $0 \leq y \leq 2$.

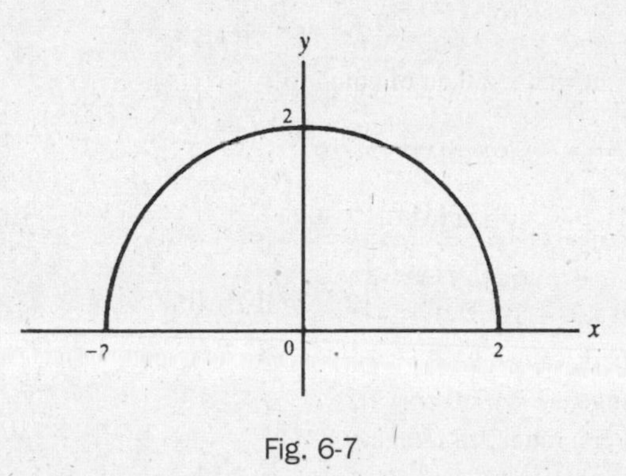

Fig. 6-7

SUPPLEMENTARY PROBLEMS

11. If $f(x) = x^2 - 4x + 6$, find (a) $f(0)$; (b) $f(3)$; (c) $f(-2)$. Show that $f(\frac{1}{2}) = f(\frac{7}{2})$ and $f(2 - h) = f(2 + h)$.

Ans. (a) –6; (b) 3; (c) 18

12. If $f(x) = \dfrac{x-1}{x+1}$, find (a) $f(0)$; (b) $f(1)$; (c) $f(-2)$. Show that $f\left(\dfrac{1}{x}\right) = -f(x)$ and $f\left(-\dfrac{1}{x}\right) = -\dfrac{1}{f(x)}$.

Ans. (a) –1; (b) 0; (c) 3

13. If $f(x) = x^2 - x$, show that $f(x + 1) = f(-x)$.

14. If $f(x) = 1/x$, show that $f(a) - f(b) = f\left(\dfrac{ab}{b-a}\right)$.

15. If $y = f(x) = \dfrac{5x+3}{4x-5}$, show that $x = f(y)$.

16. Determine the domain of each of the following functions:

(a) $y = x^2 + 4$ (b) $y = \sqrt{x^2 + 4}$ (c) $y = \sqrt{x^2 - 4}$ (d) $y = \dfrac{x}{x+3}$

(e) $y = \dfrac{2x}{(x-2)(x+1)}$ (f) $y = \dfrac{1}{\sqrt{9 - x^2}}$ (g) $y = \dfrac{x^2 - 1}{x^2 + 1}$ (h) $y = \sqrt{\dfrac{x}{2-x}}$

Ans. (a), (b), (g) all values of x; (c) $|x| \geq 2$; (d) $x \neq -3$; (e) $x \neq -1, 2$; (f) $-3 < x < 3$; (h) $0 \leq x < 2$

17. Compute $\dfrac{f(a+h) - f(a)}{h}$ in the following cases:

(a) $f(x) = \dfrac{1}{x-2}$ when $a \neq 2$ and $a + h \neq 2$

(b) $f(x) = \sqrt{x-4}$ when $a \geq 4$ and $a + h \geq 4$

(c) $f(x) = \dfrac{x}{x+1}$ when $a \neq -1$ and $a + h \neq -1$

Ans. (a) $\dfrac{-1}{(a-2)(a+h-2)}$; (b) $\dfrac{1}{\sqrt{a+h-4} + \sqrt{a-4}}$; (c) $\dfrac{1}{(a+1)(a+h+1)}$

18. Draw the graphs of the following functions, and find their domains and ranges:

(a) $f(x) = -x^2 + 1$ (b) $f(x) = \begin{cases} x - 1 & \text{if } 0 < x < 1 \\ 2x & \text{if } 1 \leq x \end{cases}$

(c) $f(x) = [x]$ = the greatest integer less than or equal to x

(d) $f(x) = \dfrac{x^2 - 4}{x - 2}$ (e) $f(x) = 5 - x^2$ (f) $f(x) = -4\sqrt{x}$

(g) $f(x) = |x - 3|$ (h) $f(x) = 4/x$ (i) $f(x) = |x|/x$

(j) $f(x) = x - |x|$ (k) $f(x) = \begin{cases} x & \text{if } x \geq 0 \\ 2 & \text{if } x < 0 \end{cases}$

Ans. (a) domain, all numbers; range, $y \leq 1$
(b) domain, $x > 0$; range, $-1 < y < 0$ or $y \geq 2$
(c) domain, all numbers; range, all integers
(d) domain, $x \neq 2$; range, $y \neq 4$
(e) domain, all numbers; range, $y \leq 5$
(f) domain, $x \geq 0$; range, $y \leq 0$
(g) domain, all numbers; range, $y \geq 0$
(h) domain, $x \neq 0$; range, $y \neq 0$
(i) domain, $x \neq 0$; range, $\{-1, 1\}$
(j) domain, all numbers; range, $y \leq 0$
(k) domain, all numbers; range, $y \geq 0$

19. (GC) Use a graphing calculator to verify your answers to Problem 18.

20. Evaluate the expression $\dfrac{f(x+h) - f(x)}{h}$ for the following functions f:

(a) $f(x) = 3x - x^2$ (b) $f(x) = \sqrt{2x}$
(c) $f(x) = 3x - 5$ (d) $f(x) = x^3 - 2$

Ans. (a) $3 - 2x - h$ (b) $\dfrac{2}{\sqrt{2(x+h)} + \sqrt{2x}}$ (c) 3 (d) $3x^2 + 3xh + h^2$

21. Find a formula for the function f whose graph consists of all points satisfying each of the following equations. (In plain language, solve each equation for y.)

(a) $x^5y + 4x - 2 = 0$ (b) $x = \dfrac{2+y}{2-y}$ (c) $4x^2 - 4xy + y^2 = 0$

Ans. (a) $f(x) = \dfrac{2-4x}{x^5}$; (b) $f(x) = \dfrac{2(x-1)}{x+1}$; (c) $f(x) = 2x$

22. Graph the following functions and find their domain and range:

(a) $f(x) = \begin{cases} x+2 & \text{if } -1 < x < 0 \\ x & \text{if } 0 \le x < 1 \end{cases}$ (b) $g(x) = \begin{cases} 2-x & \text{if } 0 < x < 2 \\ x-1 & \text{if } 3 \le x < 4 \end{cases}$ (c) $h(x) = \begin{cases} \dfrac{x^2-4}{x-2} & \text{if } x \ne 2 \\ 4 & \text{if } x = 2 \end{cases}$

Ans. (a) domain = (−1, 1], range = [0, 2)
(b) domain = union of (0, 2) and [3, 4), range = (0, 3)
(c) domain and range = set of all real numbers

23. (GC) Verify your answers to Problem 22 by means of a graphing calculator.

24. In each of the following cases, define a function that has the given set $\mathcal{D}$ as its domain and the given set $\mathcal{R}$ as its range: (a) $\mathcal{D} = (0, 2)$ and $\mathcal{R} = (1, 7)$; (b) $\mathcal{D} = (0, 1)$ and $\mathcal{R} = (1, \infty)$.

Ans. (a) One such function is $f(x) = 3x + 1$. (b) One such function is $f(x) = \dfrac{1}{1-x}$.

25. (a) Prove the vertical line test: A set of points in the *xy* plane is the graph of a function if and only if the set intersects every vertical line in at most one point.

(b) Determine whether each set of points in Fig. 6-8 is the graph of a function.

Ans. Only (b) is the graph of a function.

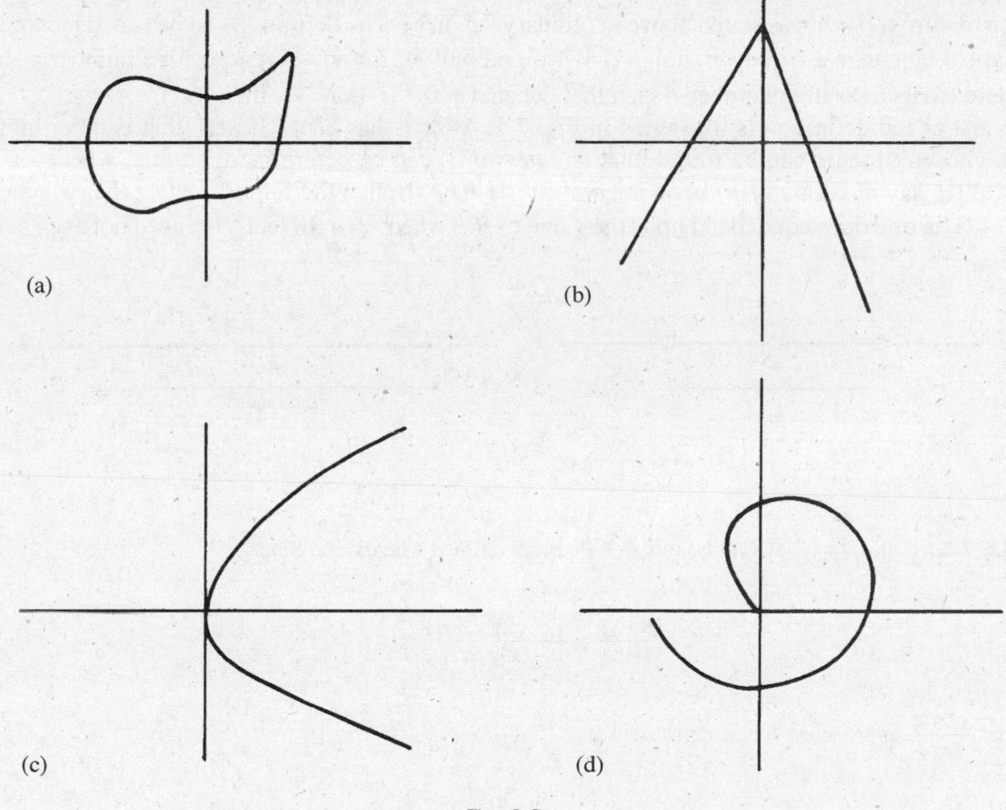

Fig. 6-8

Limits

Limit of a Function

If f is a function, then we say:

A is the limit of $f(x)$ as x approaches a

if the value of $f(x)$ gets arbitrarily close to A as x approaches a. This is written in mathematical notation as:

$$\lim_{x \to a} f(x) = A$$

For example, $\lim_{x \to 3} x^2 = 9$, since x^2 gets arbitrarily close to 9 as x approaches as close as one wishes to 3. The definition of $\lim_{x \to a} f(x) = A$ was stated above in ordinary language. The definition can be stated in more precise mathematical language as follows: $\lim_{x \to a} f(x) = A$ if and only if, for any given positive number $\in$, however small, there exists a positive number δ such that, whenever $0 < |x - a| < \delta$, then $|f(x) - A| < \in$.

The gist of the definition is illustrated in Fig. 7-1. After $\in$ has been chosen [that is, after interval (ii) has been chosen], then δ can be found [that is, interval (i) can be determined] so that, whenever $x \neq a$ is on interval (i), say at x_0, then $f(x)$ is on interval (ii), at $f(x_0)$. Notice the important fact that whether or not $\lim_{x \to a} f(x) = A$ is true does not depend upon the value of $f(x)$ when $x = a$. In fact, $f(x)$ need not even be defined when $x = a$.

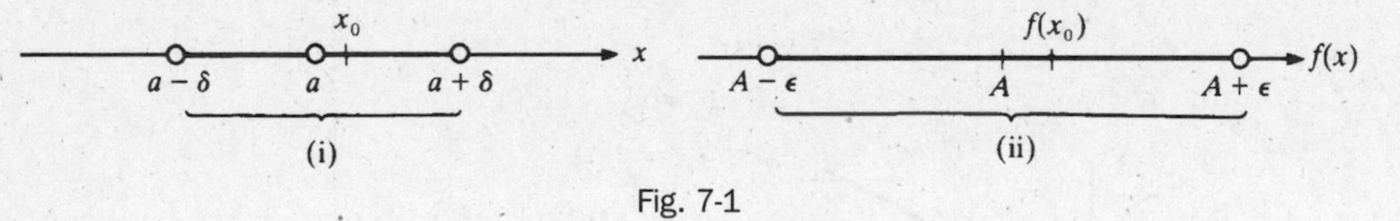

Fig. 7-1

EXAMPLE 7.1: $\lim_{x \to 2} \frac{x^2 - 4}{x - 2} = 4$, although $\frac{x^2 - 4}{x - 2}$ is not defined when $x = 2$. Since

$$\frac{x^2 - 4}{x - 2} = \frac{(x - 2)(x + 2)}{x - 2} = x + 2$$

we see that $\frac{x^2 - 4}{x - 2}$ approaches 4 as x approaches 2.

EXAMPLE 7.2: Let us use the precise definition to show that $\lim_{x \to 2} (4x - 5) = 3$. Let $\in > 0$ be chosen. We must produce some $\delta > 0$ such that, whenever $0 < |x - 2| < \delta$, then $|(4x - 5) - 3| < \in$.

First we note that $|(4x - 5) - 3| = |4x - 8| = 4|x - 2|$.

If we take δ to be $\in/4$, then, whenever $0 < |x - 2| < \delta$, $|(4x - 5) - 3| = 4|x - 2| < 4\delta = \in$.

Right and Left Limits

Next we want to talk about one-sided limits of $f(x)$ as x approaches a from the right-hand side or from the left-hand side. By $\lim_{x \to a^-} f(x) = A$ we mean that f is defined in some open interval (c, a) and $f(x)$ approaches A as x approaches a through values less than a, that is, as x approaches a *from the left*. Similarly, $\lim_{x \to a^+} f(x) = A$ means that f is defined in some open interval (a, d) and $f(x)$ approaches A as x approaches *a from the right*. If f is defined in an interval to the left of a and in an interval to the right of a, then the statement $\lim_{x \to a} f(x) = A$ is equivalent to the conjunction of the two statements $\lim_{x \to a^-} f(x) = A$ and $\lim_{x \to a^+} f(x) = A$. We shall see by examples below that the existence of the limit from the left does not imply the existence of the limit from the right, and conversely.

When a function is defined only on one side of a point a, then we shall identify $\lim_{x \to a} f(x)$ with the one-sided limit, if it exists. For example, if $f(x) = \sqrt{x}$, then f is defined only at and to the right of 0. Hence, since $\lim_{x \to 0^+} \sqrt{x} = 0$, we will also write $\lim_{x \to 0} \sqrt{x} = 0$. Of course, $\lim_{x \to 0^-} \sqrt{x}$ does not exist, since $\sqrt{x}$ is not defined when $x < 0$. This is an example where the existence of the limit from one side does not entail the existence of the limit from the other side. As another interesting example, consider the function $g(x) = \sqrt{1/x}$, which is defined only for $x > 0$. In this case, $\lim_{x \to 0^+} \sqrt{1/x}$ does not exist, since $1/x$ gets larger and larger without bound as x approaches 0 from the right. Therefore, $\lim_{x \to 0} \sqrt{1/x}$ does not exist.

EXAMPLE 7.3: The function $f(x) = \sqrt{9 - x^2}$ has the interval $-3 \le x \le 3$ as its domain. If a is any number on the interval $(-3, 3)$, then $\lim_{x \to a} \sqrt{9 - x^2}$ exists and is equal to $\sqrt{9 - a^2}$. Now consider $a = 3$. Let x approach 3 from the left; then $\lim_{x \to 3^-} \sqrt{9 - x^2} = 0$. For $x > 3$, $\sqrt{9 - x^2}$ is not defined, since $9 - x^2$ is negative. Hence, $\lim_{x \to 3} \sqrt{9 - x^2} = \lim_{x \to 3^-} \sqrt{9 - x^2} = 0$. Similarly, $\lim_{x \to -3} \sqrt{9 - x^2} = \lim_{x \to -3^+} \sqrt{9 - x^2} = 0$.

Theorems on Limits

The following theorems are intuitively clear. Proofs of some of them are given in Problem 11.

Theorem 7.1: If $f(x) = c$, a constant, then $\lim_{x \to a} f(x) = c$.

For the next five theorems, assume $\lim_{x \to a} f(x) = A$ and $\lim_{x \to a} g(x) = B$.

Theorem 7.2: $\lim_{x \to a} c \cdot f(x) = c \lim_{x \to a} f(x) = cA$.

Theorem 7.3: $\lim_{x \to a} [f(x) \pm g(x)] = \lim_{x \to a} f(x) \pm \lim_{x \to a} g(x) = A \pm B$.

Theorem 7.4: $\lim_{x \to a} [f(x)g(x)] = \lim_{x \to a} f(x) \cdot \lim_{x \to a} g(x) = A \cdot B$.

Theorem 7.5: $\lim_{x \to a} \left(\dfrac{f(x)}{g(x)} \right) = \dfrac{\lim_{x \to a} f(x)}{\lim_{x \to a} g(x)} = \dfrac{A}{B}$, if $B \neq 0$.

Theorem 7.6: $\lim_{x \to a} \sqrt[n]{f(x)} = \sqrt[n]{\lim_{x \to a} f(x)} = \sqrt[n]{A}$, if $\sqrt[n]{A}$ is defined.

Infinity

Let

$$\lim_{x \to a} f(x) = +\infty$$

mean that, as x approaches a, $f(x)$ eventually becomes greater than any preassigned positive number, however large. In such a case, we say that $f(x)$ approaches $+\infty$ as x approaches a. More precisely, $\lim_{x \to a} f(x) = +\infty$ if and only if, for any positive number M, there exists a positive number δ such that, whenever $0 < |x - a| < \delta$, then $f(x) > M$.

Similarly, let

$$\lim_{x\to a} f(x) = -\infty$$

mean that, as x approaches a, $f(x)$ eventually becomes less than any preassigned number. In that case, we say that $f(x)$ approaches $-\infty$ as x approaches a.

Let

$$\lim_{x\to a} f(x) = \infty$$

mean that, as x approaches a, $|f(x)|$ eventually becomes greater than any preassigned positive number. Hence, $\lim_{x\to a} f(x) = \infty$ if and only if $\lim_{x\to a} |f(x)| = +\infty$.

These definitions can be extended to one-sided limits in the obvious way.

EXAMPLE 7.4:

(a) $\lim_{x\to 0} \frac{1}{x^2} = +\infty$ (b) $\lim_{x\to 1} \frac{-1}{(x-1)^2} = -\infty$ (c) $\lim_{x\to 0} \frac{1}{x} = \infty$

EXAMPLE 7.5:

(a) $\lim_{x\to 0^+} \frac{1}{x} = +\infty$. As x approaches 0 from the right (that is, through positive numbers), $1/x$ is positive and eventually becomes larger than any preassigned number.

(b) $\lim_{x\to 0^-} \frac{1}{x} = -\infty$ since, as x approaches 0 from the left (that is, through negative numbers), $1/x$ is negative and eventually becomes smaller than any preassigned number.

The limit concepts already introduced can be extended in an obvious way to the case in which the variable approaches $+\infty$ or $-\infty$. For example,

$$\lim_{x\to +\infty} f(x) = A$$

means that $f(x)$ approaches A as $x \to +\infty$, or, in more precise terms, given any positive ϵ, there exists a number N such that, whenever $x > N$, then $|f(x) - A| < \epsilon$. Similar definitions can be given for the statements $\lim_{x\to -\infty} f(x) = A$, $\lim_{x\to +\infty} f(x) = +\infty$, $\lim_{x\to -\infty} f(x) = -\infty$, $\lim_{x\to a} f(x) = -\infty$, and $\lim_{x\to -\infty} f(x) = +\infty$.

EXAMPLE 7.6: $\lim_{x\to +\infty} \frac{1}{x} = 0$ and $\lim_{x\to +\infty} \left(2 + \frac{1}{x^2}\right) = 2$.

Caution: When $\lim_{x\to a} f(x) = \pm\infty$ and $\lim_{x\to a} g(x) = \pm\infty$, Theorems 7.3–7.5 do not make sense and cannot be used.

For example, $\lim_{x\to 0} \frac{1}{x^2} = +\infty$ and $\lim_{x\to 0} \frac{1}{x^4} = +\infty$, but

$$\lim_{x\to 0} \frac{1/x^2}{1/x^4} = \lim_{x\to 0} x^2 = 0$$

Note: We say that a limit, such as $\lim_{x\to a} f(x)$ or $\lim_{x\to +\infty} f(x)$ exists when the limit is a real number, but not when the limit is $+\infty$ or $-\infty$ or ∞. For example, since $\lim_{x\to 2} \frac{x^2-4}{x-2} = 4$, we say that $\lim_{x\to 2} \frac{x^2-4}{x-2}$ exists. However, although $\lim_{x\to 0} \frac{1}{x^2} = +\infty$, we do not say that $\lim_{x\to 0} \frac{1}{x^2}$ exists.

SOLVED PROBLEMS

1. Verify the following limit computations:

(a) $\lim_{x\to 2} 5x = 5 \lim_{x\to 2} x = 5 \cdot 2 = 10$

(b) $\lim_{x\to 2}(2x+3) = 2\lim_{x\to 2} x + \lim_{x\to 2} 3 = 2\cdot 2+3 = 7$

(c) $\lim_{x\to 2}(x^2-4x+1) = 4-8+1 = -3$

(d) $\lim_{x\to 3}\dfrac{x-2}{x+2} = \dfrac{\lim_{x\to 3}(x-2)}{\lim_{x\to 3}(x+2)} = \dfrac{1}{5}$

(e) $\lim_{x\to -2}\dfrac{x^2-4}{x^2+4} = \dfrac{4-4}{4+4} = 0$

(f) $\lim_{x\to 4}\sqrt{25-x^2} = \sqrt{\lim_{x\to 4}(25-x^2)} = \sqrt{9} = 3$

[*Note*: Do not assume from these problems that $\lim_{x\to a} f(x)$ is invariably $f(a)$.]

(g) $\lim_{x\to -5}\dfrac{x^2-25}{x+5} = \lim_{x\to -5}(x-5) = -10$

2. Verify the following limit computations:

(a) $\lim_{x\to 4}\dfrac{x-4}{x^2-x-12} = \lim_{x\to 4}\dfrac{x-4}{(x+3)(x-4)} = \lim_{x\to 4}\dfrac{1}{x+3} = \dfrac{1}{7}$

The division by $x-4$ before passing to the limit is valid since $x \neq 4$ as $x \to 4$; hence, $x-4$ is never zero.

(b) $\lim_{x\to 3}\dfrac{x^3-27}{x^2-9} = \lim_{x\to 3}\dfrac{(x-3)(x^2+3x+9)}{(x-3)(x+3)} = \lim_{x\to 3}\dfrac{x^2+3x+9}{x+3} = \dfrac{9}{2}$

(c) $\lim_{h\to 0}\dfrac{(x+h)^2-x^2}{h} = \lim_{h\to 0}\dfrac{x^2+2hx+h^2-x^2}{h} = \lim_{h\to 0}\dfrac{2hx+h^2}{h} = \lim_{h\to 0}(2x+h) = 2x$

Here, and again in Problems 4 and 5, h is a variable, so that it might be thought that we are dealing with functions of two variables. However, the fact that x is a variable plays no role in these problems; for the moment, x can be considered a constant.

(d) $\lim_{x\to 2}\dfrac{4-x^2}{3-\sqrt{x^2+5}} = \lim_{x\to 2}\dfrac{(4-x^2)(3+\sqrt{x^2+5})}{(3-\sqrt{x^2+5})(3+\sqrt{x^2+5})} = \lim_{x\to 2}\dfrac{(4-x^2)(3+\sqrt{x^2+5})}{4 \quad x^2} = \lim_{x\to 2}(3+\sqrt{x^2+5}) = 6$

(e) $\lim_{x\to 1}\dfrac{x^2+x-2}{(x-1)^2} = \lim_{x\to 1}\dfrac{(x-1)(x+2)}{(x-1)^2} = \lim_{x\to 1}\dfrac{x+2}{x-1} = \infty$; no limit exists.

3. In the following problems (a)–(c), you can interpret $\lim_{x\to\pm\infty}$ as either $\lim_{x\to+\infty}$ or $\lim_{x\to-\infty}$; it will not matter which. Verify the limit computations.

(a) $\lim_{x\to\pm\infty}\dfrac{3x-2}{9x+7} = \lim_{x\to\pm\infty}\dfrac{3-2/x}{9+7/x} = \dfrac{3-0}{9+0} = \dfrac{1}{3}$

(b) $\lim_{x\to\pm\infty}\dfrac{6x^2+2x+1}{5x^2-3x+4} = \lim_{x\to\pm\infty}\dfrac{6+2/x+1/x^2}{5-3/x+4/x^2} = \dfrac{6+0+0}{5-0+0} = \dfrac{6}{5}$

(c) $\lim_{x\to\pm\infty}\dfrac{x^2+x-2}{4x^3-1} = \lim_{x\to\pm\infty}\dfrac{1/x+1/x^2-2/x^3}{4-1/x^3} = \dfrac{0}{4} = 0$

(d) $\lim_{x\to-\infty}\dfrac{2x^3}{x^2+1} = \lim_{x\to-\infty}\dfrac{2x}{1+1/x^2} = -\infty$

(e) $\lim_{x\to+\infty}\dfrac{2x^3}{x^2+1} = \lim_{x\to+\infty}\dfrac{2x}{1+1/x^2} = +\infty$

(f) $\lim_{x\to+\infty}(x^5-7x^4-2x+5) = \lim_{x\to+\infty} x^5\left(1-\dfrac{7}{x}-\dfrac{2}{x^4}+\dfrac{5}{x^5}\right) = +\infty$ since

$\lim_{x\to+\infty}\left(1-\dfrac{7}{x}-\dfrac{2}{x^4}+\dfrac{5}{x^5}\right) = (1-0-0+0) = 1$ and $\lim_{x\to+\infty} x^5 = +\infty$

(g) $\lim_{x\to-\infty}(x^5-7x^4-2x+5) = \lim_{x\to-\infty} x^5\left(1-\dfrac{7}{x}-\dfrac{2}{x^4}+\dfrac{5}{x^5}\right) = -\infty$ since

$$\lim_{x\to-\infty}\left(1-\frac{7}{x}-\frac{2}{x^4}+\frac{5}{x^5}\right)=(1-0-0+0)=1 \text{ and } \lim_{x\to-\infty}x^5=-\infty$$

4. Given $f(x)=x^2-3x$, find $\lim_{h\to 0}\frac{f(x+h)-f(x)}{h}$.

Since $f(x)=x^2-3x$, we have $f(x+h)=(x+h)^2-3(x+h)$ and

$$\lim_{h\to 0}\frac{f(x+h)-f(x)}{h}=\lim_{h\to 0}\frac{(x^2+2hx+h^2-3x-3h)-(x^2-3x)}{h}=\lim_{h\to 0}\frac{2hx+h^2-3h}{h}$$
$$=\lim_{h\to 0}(2x+h-3)=2x-3.$$

5. Given $f(x)=\sqrt{5x+1}$, find $\lim_{h\to 0}\frac{f(x+h)-f(x)}{h}$ when $x>-\frac{1}{5}$.

$$\lim_{h\to 0}\frac{f(x+h)-f(x)}{h}=\lim_{h\to 0}\frac{\sqrt{5x+5h+1}-\sqrt{5x+1}}{h}$$
$$=\lim_{h\to 0}\frac{\sqrt{5x+5h+1}-\sqrt{5x+1}}{h}\,\frac{\sqrt{5x+5h+1}+\sqrt{5x+1}}{\sqrt{5x+5h+1}+\sqrt{5x+1}}$$
$$=\lim_{h\to 0}\frac{(5x+5h+1)-(5x+1)}{h(\sqrt{5x+5h+1}+\sqrt{5x+1})}$$
$$=\lim_{h\to 0}\frac{5}{\sqrt{5x+5h+1}+\sqrt{5x+1}}=\frac{5}{2\sqrt{5x+1}}$$

6. (a) In each of the following, (a) to (e), determine the points $x=a$ for which each denominator is zero. Then see what happens to y as $x\to a^-$ and as $x\to a^+$, and verify the given solutions.
(b) (GC) Check the answers in (*a*) with a graphing calculator.

(a) $y=f(x)=2/x$: The denominator is zero when $x=0$. As $x\to 0^-$, $y\to-\infty$; as $x\to 0^+$, $y\to+\infty$.

(b) $y=f(x)=\frac{x-1}{(x+3)(x-2)}$: The denominator is zero for $x=-3$ and $x=2$. As $x\to-3^-$, $y\to-\infty$; as $x\to-3^+$, $y\to+\infty$. As $x\to 2^-$, $y\to-\infty$; as $x\to 2^+$, $y\to+\infty$.

(c) $y=f(x)=\frac{x-3}{(x+2)(x-1)}$: The denominator is zero for $x=-2$ and $x=1$. As $x\to-2^-$, $y\to-\infty$; as $x\to-2^+$, $y\to+\infty$. As $x\to 1^-$, $y\to+\infty$; as $x\to 1^+$, $y\to-\infty$.

(d) $y=f(x)=\frac{(x+2)(x-1)}{(x-3)^2}$: The denominator is zero for $x=3$. As $x\to 3^-$, $y\to+\infty$; as $x\to 3^+$, $y\to+\infty$.

(e) $y=f(x)=\frac{(x+2)(1-x)}{x-3}$: The denominator is zero for $x=3$. As $x\to 3^-$, $y\to+\infty$; as $x\to 3^+$, $y\to-\infty$.

7. For each of the functions of Problem 6, determine what happens to y as $x\to-\infty$ and $x\to+\infty$.

(a) As $x\to\pm\infty$, $y=2/x\to 0$. When $x<0$, $y<0$. Hence, as $x\to-\infty$, $y\to 0^-$. Similarly, as $x\to+\infty$, $y\to 0^+$.

(b) Divide numerator and denominator of $\frac{x-1}{(x+3)(x-2)}$ by x^2 (the highest power of x in the denominator), obtaining

$$\frac{1/x-1/x^2}{(1+3/x)(1-2/x)}$$

Hence, as $x\to\pm\infty$,

$$y\to\frac{0-0}{(1+0)(1-0)}=\frac{0}{1}=0$$

As $x\to-\infty$, the factors $x-1$, $x+3$, and $x-2$ are negative, and, therefore, $y\to 0^-$. As $x\to+\infty$, those factors are positive, and, therefore, $y\to 0^+$.

(c) Similar to (*b*).

(d) $\dfrac{(x+2)(x-1)}{(x-3)^2} = \dfrac{x^2+x-2}{x^2-6x+9} = \dfrac{1+1/x-2/x^2}{1-6/x+9/x^2}$, after dividing numerator and denominator by x^2 (the highest power of x in the denominator). Hence, as $x \to \pm\infty$, $y \to \dfrac{1+0-0}{1-0+0} = \dfrac{1}{1} = 1$. The denominator $(x-3)^2$ is always nonnegative. As $x \to -\infty$, both $x+2$ and $x-1$ are negative and their product is positive; hence, $y \to 1^+$. As $x \to +\infty$, both $x+2$ and $x-1$ are positive, as is their product; hence, $y \to 1^+$.

(e) $\dfrac{(x+2)(1-x)}{x-3} = \dfrac{-x^2-x+2}{x-3} = \dfrac{-x-1+2/x}{1-3/x}$, after dividing numerator and denominator by x (the highest power of x in the denominator). As $x \to \pm\infty$, $2/x$ and $3/x$ approach 0, and $-x-1$ approaches $\pm\infty$. Thus, the denominator approaches 1 and the numerator approaches $\pm\infty$. As $x \to -\infty$, $x+2$ and $x-3$ are negative and $1-x$ is positive; so, $y \to +\infty$. As $x \to +\infty$, $x+2$ and $x-3$ are positive and $1-x$ is negative; so, $y \to -\infty$.

8. Examine the function of Problem 4 in Chapter 6 as $x \to a^-$ and as $x \to a^+$ when a is any positive integer.

Consider, as a typical case, $a = 2$. As $x \to 2^-$, $f(x) \to 10$. As $x \to 2^+$, $f(x) \to 15$. Thus, $\lim_{x\to 2} f(x)$ does not exist. In general, the limit fails to exist for all positive integers. (Note, however, that $\lim_{x\to 0} f(x) = \lim_{x\to 0^+} f(x) = 5$, since $f(x)$ is not defined for $x \le 0$.)

9. Use the precise definition to show that $\lim_{x\to 2}(x^2+3x) = 10$.

Let $\epsilon > 0$ be chosen. Note that $(x-2)^2 = x^2 - 4x + 4$, and so, $x^2 + 3x - 10 = (x-2)^2 + 7x - 14 = (x-2)^2 + 7(x-2)$. Hence $|(x^2+3x) - 10| = |(x-2)^2 + 7(x-2)| \le |x-2|^2 + 7|x-2|$. If we choose δ to be the minimum of 1 and $\epsilon/8$, then $\delta^2 \le \delta$, and, therefore, $0 < |x-2| < \delta$ implies $|(x^2+3x)-10| < \delta^2 + 7\delta \le \delta + 7\delta = 8\delta \le \epsilon$.

10. If $\lim_{x\to a} g(x) = B \neq 0$, prove that there exists a positive number δ such that $0 < |x-a| < \delta$ implies $|g(x)| > \dfrac{|B|}{2}$.

Letting $\epsilon = |B|/2$ we obtain a positive δ such that $0 < |x-a| < \delta$ implies $|g(x) - B| < |B|/2$. Now, if $0 < |x-a| < \delta$, then $|B| = |g(x) + (B - g(x))| \le |g(x)| + |B - g(x)| < |g(x)| + |B|/2$ and, therefore, $|B|/2 < |g(x)|$.

11. Assume (I) $\lim_{x\to a} f(x) = A$ and (II) $\lim_{x\to a} g(x) = B$. Prove:

(a) $\lim_{x\to a}[f(x)+g(x)] = A+B$ (b) $\lim_{x\to a} f(x)g(x) = AB$ (c) $\lim_{x\to a}\dfrac{f(x)}{g(x)} = \dfrac{A}{B}$ if $B \neq 0$

(a) Let $\epsilon > 0$ be chosen. Then $\epsilon/2 > 0$. By (I) and (II), there exist positive δ_1 and δ_2 such that $0 < |x-a| < \delta_1$ implies $|f(x) - A| < \epsilon/2$ and $0 < |x-a| < \delta_2$ implies $|g(x) - B| < \epsilon/2$. Let δ be the minimum of δ_1 and δ_2. Thus, for $0 < |x-a| < \delta$, $|f(x)-A| < \epsilon/2$ and $|g(x)-B| < \epsilon/2$. Therefore, for $0 < |x-a| < \delta$,

$$|(f(x)+g(x)-(A+B)| = |(f(x)-A)+(g(x)-B)|$$

$$\le |f(x)-A| + |g(x)-B| < \frac{\epsilon}{2} + \frac{\epsilon}{2} = \epsilon$$

(b) Let $\epsilon > 0$ be chosen. Choose ϵ^* to be the minimum of $\epsilon/3$ and 1 and $\epsilon/(3|B|)$ (if $B \neq 0$), and $\epsilon/(3|A|)$ (if $A \neq 0$). Note that $(\epsilon^*)^2 \le \epsilon^*$ since $\epsilon^* \le 1$ Moreover, $|B|\epsilon^* \le \epsilon/3$ and $|A|\epsilon^* \le \epsilon/3$. By (I) and (II), there exist positive δ_1 and δ_2 such that $0 < |x-a| < \delta_1$ implies $|f(x)-A| < \epsilon^*$ and $0 < |x-a| < \delta_2$ implies $|g(x)-B| < \epsilon^*$. Let δ be the minimum of δ_1 and δ_2. Now, for $0 < |x-a| < \delta$,

$$|f(x)g(x) - AB| = |(f(x)-A)(g(x)-B) + B(f(x)-A) + A(g(x)-B)|$$

$$\le |f(x)-A)(g(x)-B)| + |B(f(x)-A)| + |A(g(x)-B)|$$

$$= |f(x)-A||g(x)-B| + |B||f(x)-A| + |A||g(x)-B|$$

$$\le (\epsilon^*)^2 + |B|\epsilon^* + |A|\epsilon^* \le \epsilon^* + \frac{\epsilon}{3} + \frac{\epsilon}{3} \le \frac{\epsilon}{3} + \frac{\epsilon}{3} + \frac{\epsilon}{3} = \epsilon$$

(c) By part (b), it suffices to show that $\lim_{x\to a}\dfrac{1}{g(x)} = \dfrac{1}{B}$. Let $\epsilon > 0$ be chosen. Then $B^2\epsilon/2 > 0$. Hence, there exists a positive δ_1 such that $0 < |x-a| < \delta_1$ implies $|g(x) - B| < \dfrac{|B|^2\epsilon}{2}$.

By Problem 10, there exists a positive δ_2 such that $0 < |x - a| < \delta_2$ implies $|g(x)| > |B|/2$. Let δ be the minimum of δ_1 and δ_2. Then $0 < |x - a| < \delta$ implies that

$$\left|\frac{1}{g(x)} - \frac{1}{B}\right| = \frac{|B - g(x)|}{|B||g(x)|} < \frac{|B|^2 \in}{2} \cdot \frac{2}{|B|^2} = \in$$

12. Prove that, for any polynomial function

$$f(x) = a_n x^n + a_{n-1}x^{n-1} + \cdots + a_1 x + a_0, \qquad \lim_{x\to a} f(x) = f(a)$$

This follows from Theorems 7.1–7.4 and the obvious fact that $\lim_{x\to a} x = a$.

13. Prove the following generalizations of the results of Problem 3. Let $f(x) = a_n x^n + a_{n-1}x^{n-1} + \cdots + a_1 x + a_0$ and $g(x) = b_k x^k + b_{k-1}x^{k-1} + \cdots + b_1 x + b_0$ be two polynomials.

(a) $\lim_{x\to\pm\infty} \frac{f(x)}{g(x)} = \frac{a_n}{b_k}$ if $n = k$

(b) $\lim_{x\to\pm\infty} \frac{f(x)}{g(x)} = 0$ if $n < k$

(c) $\lim_{x\to+\infty} \frac{f(x)}{g(x)} = \pm\infty$ if $n > k$. (It is $+\infty$ if and only if a_n and b_k have the same sign.)

(d) $\lim_{x\to-\infty} \frac{f(x)}{g(x)} = \pm\infty$ if $n > k$. (The correct sign is the sign of $a_n b_k (-1)^{n-k}$.)

14. Prove (a) $\lim_{x\to 2^-} \frac{1}{(x-2)^3} = -\infty$; (b) $\lim_{x\to+\infty} \frac{x}{x+1} = 1$; (c) $\lim_{x\to+\infty} \frac{x^2}{x-1} = +\infty$.

(a) Let M be any negative number. Choose δ positive and equal to the minimum of 1 and $\frac{1}{|M|}$. Assume $x < 2$ and $0 < |x - 2| < \delta$. Then $|x-2|^3 < \delta^3 \le \delta \le \frac{1}{|M|}$. Hence, $\frac{1}{|x-2|^3} > |M| = -M$. But $(x - 2)^3 < 0$. Therefore, $\frac{1}{(x-2)^3} = -\frac{1}{|x-2|^3} < M$.

(b) Let $\in$ be any positive number, and let $M = 1/\in$. Assume $x > M$. Then

$$\left|\frac{x}{x+1} - 1\right| = \left|\frac{1}{x+1}\right| = \frac{1}{x+1} < \frac{1}{x} < \frac{1}{M} = \in$$

(c) Let M be any positive number. Assume $x > M + 1$. Then $\frac{x^2}{x-1} \ge \frac{x^2}{x} = x > M$.

15. Evaluate: (a) $\lim_{x\to 0^+} \frac{|x|}{x}$; (b) $\lim_{x\to 0^-} \frac{|x|}{x}$; (c) $\lim_{x\to 0} \frac{|x|}{x}$

(a) When $x > 0$, $|x| = x$. Hence, $\lim_{x\to 0^+} \frac{|x|}{x} = \lim_{x\to 0^+} 1 = 1$.

(b) When $x < 0$, $|x| = -x$. Hence, $\lim_{x\to 0^-} \frac{|x|}{x} = \lim_{x\to 0^-} -1 = -1$.

(c) $\lim_{x\to 0} \frac{|x|}{x}$ does not exist, since $\lim_{x\to 0^-} \frac{|x|}{x} \ne \lim_{x\to 0^+} \frac{|x|}{x}$.

SUPPLEMENTARY PROBLEMS

16. Evaluate the following limits:

(a) $\lim_{x\to 2} (x^2 - 4x)$

(b) $\lim_{x\to -1} (x^3 + 2x^2 - 3x - 4)$

(c) $\lim_{x\to 1} \frac{(3x-1)^2}{(x+1)^3}$

(d) $\lim_{x\to 0}\frac{3^x - 3^{-x}}{3^x + 3^{-x}}$

(e) $\lim_{x\to 2}\frac{x-1}{x^2-1}$

(f) $\lim_{x\to 2}\frac{x^2-4}{x^2-5x+6}$

(g) $\lim_{x\to -1}\frac{x^2+3x+2}{x^2+4x+3}$

(h) $\lim_{x\to 2}\frac{x-2}{x^2-4}$

(i) $\lim_{x\to 2}\frac{x-2}{\sqrt{x^2-4}}$

(j) $\lim_{x\to 2}\frac{\sqrt{x-2}}{x^2-4}$

(k) $\lim_{h\to 0}\frac{(x+h)^3-x^3}{h}$

(l) $\lim_{x\to 1}\frac{x-1}{\sqrt{x^2+3}-2}$

Ans. (a) –4; (b) 0; (c) $\frac{1}{2}$; (d) 0; (e) $\frac{1}{3}$; (f) –4; (g) $\frac{1}{2}$; (h) $\frac{1}{4}$; (i) 0; (j) ∞, no limit; (k) $3x^2$; (l) 2

17. Evalute the following limits:

(a) $\lim_{x\to +\infty}\frac{7x^9-4x^5+2x-13}{-3x^9+x^8-5x^2+2x}$

(b) $\lim_{x\to +\infty}\frac{14x^3-5x+27}{x^4+10}$

(c) $\lim_{x\to -\infty}\frac{2x^5+12x+5}{7x^3+6}$

(d) $\lim_{x\to +\infty}\frac{-2x^3+7}{5x^2-3x-4}$

(e) $\lim_{x\to +\infty}(3x^3-25x^2-12x-17)$

(f) $\lim_{x\to -\infty}(3x^3-25x^2-12x-17)$

(g) $\lim_{x\to -\infty}(3x^4-25x^3-8)$

Ans. (a) $-\frac{7}{3}$; (b) 0; (c) $+\infty$; (d) $-\infty$; (e) $+\infty$; (f) $-\infty$; (g) $+\infty$

18. Evaluate the following limits:

(a) $\lim_{x\to +\infty}\frac{2x+3}{4x-5}$

(b) $\lim_{x\to +\infty}\frac{2x^2+1}{6+x-3x^2}$

(c) $\lim_{x\to +\infty}\frac{x}{x^2+5}$

(d) $\lim_{x\to +\infty}\frac{x^2+5x+6}{x+1}$

(e) $\lim_{x\to +\infty}\frac{x+3}{x^2+5x+6}$

(f) $\lim_{x\to +\infty}\frac{3^x-3^{-x}}{3^x+3^{-x}}$

(g) $\lim_{x\to-\infty} \frac{3^x - 3^{-x}}{3^x + 3^{-x}}$

Ans. (a) $\frac{1}{2}$; (b) $-\frac{2}{3}$; (c) 0; (d) $+\infty$; (e) 0; (f) 1; (g) -1

19. Find $\lim_{h\to0} \frac{f(a+h)-f(a)}{h}$ for the functions *f* in Problems 11, 12, 13, 15, and 16 (*a, b, d, g*) of Chapter 6.

Ans. (11) $2a-4$; (12) $\frac{2}{(a+1)^2}$; (13) $2a-1$; (15) $-\frac{27}{(4a-5)^2}$; (16) (a) $2a$, (b) $\frac{a}{\sqrt{a^2+4}}$, (d) $\frac{3}{(a+3)^2}$, (g) $\frac{4a}{(a^2+1)^2}$

20. (GC) Investigate the behavior of

$$f(x) = \begin{cases} x & \text{if } x > 0 \\ x+1 & \text{if } x \le 0 \end{cases}$$

as $x \to 0$. Draw a graph and verify it with a graphing calculator.

Ans. $\lim_{x\to0^+} f(x) = 0$; $\lim_{x\to0^-} f(x) = 1$; $\lim_{x\to0} f(x)$ does not exist.

21. Use Theorem 7.4 and mathematical induction to prove $\lim_{x\to a} x^n = a^n$ for all positive integers *n*.

22. For $f(x) = 5x - 6$, find $\delta > 0$ such that, whenever $0 < |x-4| < \delta$, then $|f(x) - 14| < \epsilon$, when (a) $\epsilon = \frac{1}{2}$ and (b) $\epsilon = 0.001$.

Ans. (a) $\frac{1}{10}$; (b) 0.0002

23. Use the precise definition to prove: (a) $\lim_{x\to3} 5x = 15$; (b) $\lim_{x\to2} x^2 = 4$; (c) $\lim_{x\to2}(x^2 - 3x + 5) = 3$.

24. Use the precise definition to prove:

(a) $\lim_{x\to0} \frac{1}{x} = \infty$ (b) $\lim_{x\to1} \frac{x}{x-1} = \infty$ (c) $\lim_{x\to+\infty} \frac{x}{x-1} = 1$ (d) $\lim_{x\to-\infty} \frac{x^2}{x+1} = -\infty$

25. Let $f(x)$, $g(x)$, and $h(x)$ be such that (1) $f(x) \le g(x) \le h(x)$ for all values in certain intervals to the left and right of *a*, and (2) $\lim_{x\to a} f(x) = \lim_{x\to a} h(x) = A$. Prove $\lim_{x\to a} g(x) = A$.

(*Hint*: For $\epsilon > 0$, there exists $\delta > 0$ such that, whenever $0 < |x-a| < \delta$, then $|f(x) - A| < \epsilon$ and $|h(x) - A| < \epsilon$ and, therefore, $A - \epsilon < f(x) \le g(x) \le h(x) < A + \epsilon$.)

26. Prove: If $f(x) \le M$ for all *x* in an open interval containing *a* and if $\lim_{x\to a} f(x) = A$, then $A \le M$.

(*Hint*: Assume $A > M$. Choose $\epsilon = \frac{1}{2}(A - M)$ and derive a contradiction.)

27. (GC) Use a graphing calculator to confirm the limits found in Problems 1(*d, e, f*), 2(*a, b, d*), 16, and 18.

28. (a) Show that $\lim_{x\to+\infty}(x - \sqrt{x^2-1}) = 0$.

(*Hint*: Multiply and divide by $x + \sqrt{x^2-1}$.)

(b) Show that the hyperbola $\frac{x^2}{a^2} - \frac{y^2}{b^2} = 1$ gets arbitrarily close to the asymptote $y = \frac{b}{a}x$ as *x* approaches ∞.

29. (a) Find $\lim_{x\to0} \frac{\sqrt{x+3} - \sqrt{3}}{x}$.

(*Hint*: Multiply the numerator and denominator by $\sqrt{x+3} + \sqrt{3}$.)

(b) (GC) Use a graphing calculator to confirm the result of part (a).

30. Let $f(x) = \sqrt{x} - 1$ if $x > 4$ and $f(x) = x^2 - 4x + 1$ if $x < 4$. Find:

(a) $\lim_{x \to 4^+} f(x)$ (b) $\lim_{x \to 4^-} f(x)$ (c) $\lim_{x \to 4} f(x)$

Ans. (a) 1; (b) 1; (c) 1

31. Let $g(x) = 10x - 7$ if $x > 1$ and $g(x) = 3x + 2$ if $x < 1$. Find:

(a) $\lim_{x \to 1^+} g(x)$ (b) $\lim_{x \to 1^-} g(x)$ (c) $\lim_{x \to 1} g(x)$

Ans. (a) 3; (b) 5; (c) It does not exist.

CHAPTER 8

Continuity

Continuous Function

A function f is defined to be continuous at x_0 if the following three conditions hold:

(i) $f(x_0)$ is defined;

(ii) $\lim_{x \to x_0} f(x)$ exists;

(iii) $\lim_{x \to x_0} f(x) = f(x_0)$.

For example, $f(x) = x^2 + 1$ is continuous at 2, since $\lim_{x \to 2} f(x) = 5 = f(2)$. Condition (i) implies that a function can be continuous only at points of its domain. Thus, $f(x) = \sqrt{4 - x^2}$ is not continuous at 3 because $f(3)$ is not defined.

Let f be a function that is defined on an interval (a, x_0) to the left of x_0 and/or on an interval (x_0, b) to the right of x_0. We say that f is discontinuous at x_0 if f is not continuous at x_0, that is, if one or more of the conditions (i)–(iii) fails.

EXAMPLE 8.1:

(a) $f(x) = \dfrac{1}{x-2}$ is discontinuous at 2 because $f(2)$ is not defined and also because $\lim_{x \to 2} f(x)$ does not exist (since $\lim_{x \to 2} f(x) = \infty$). See Fig. 8-1.

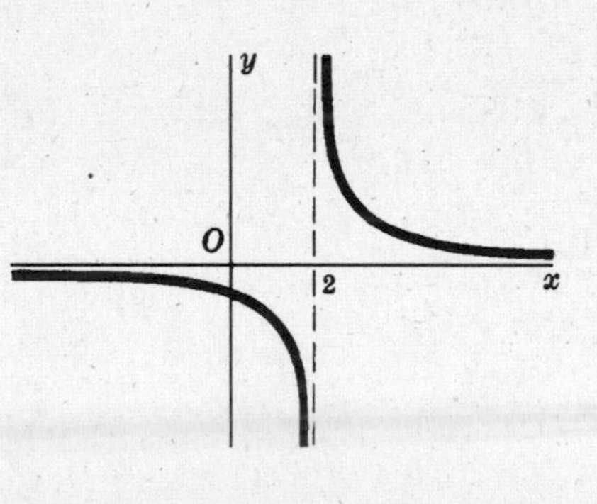

Fig. 8-1

(b) $f(x) = \dfrac{x^2 - 4}{x-2}$ is discontinuous at 2 because $f(2)$ is not defined. However, $\lim_{x \to 2} f(x) = \lim_{x \to 2} \dfrac{(x+2)(x-2)}{x-2} = \lim_{x \to 2}(x+2) = 4$ so that condition (ii) holds.

The discontinuity at 2 in Example 8.1(*b*) is said to be *removable* because, if we extended the function f by defining its value at $x = 2$ to be 4, then the extended function g would be continuous at 2. Note that $g(x) = x + 2$ for all x. The graphs of $f(x) = \dfrac{x^2 - 4}{x-2}$ and $g(x) = x + 2$ are identical except at $x = 2$, where the former has a "hole." (See Fig. 8-2.) Removing the discontinuity consists simply of filling the "hole."

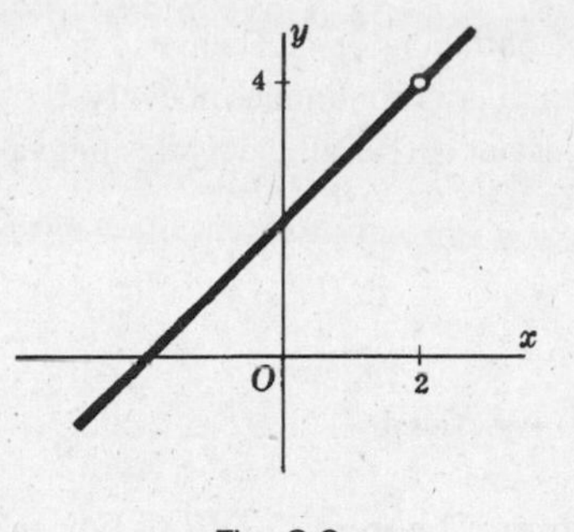

Fig. 8-2

The discontinuity at 2 in Example 8.1(*a*) is not removable. Redefining the value of f at 2 cannot change the fact that $\lim_{x\to 2}\frac{1}{x-2}$ does not exist.

We also call a discontinuity of a function f at x_0 *removable* when $f(x_0)$ is defined and changing the value of the function at x_0 produces a function that is continuous at x_0.

EXAMPLE 8.2: Define a function f as follows:

$$f(x)=\begin{cases}x^2 & \text{if } x\neq 2\\ 0 & \text{if } x=2\end{cases}$$

Here $\lim_{x\to 2} f(x)=4$, but $f(2)=0$. Hence, condition (iii) fails, so that f has a discontinuity at 2. But if we change the value of f at 2 to be 4, then we obtain a function h such that $h(x)=x^2$ for all x, and h is continuous at 2. Thus, the discontinuity of f at 2 was removable.

EXAMPLE 8.3: Let f be the function such that $f(x)=\frac{|x|}{x}$ for all $x\neq 0$. The graph of f is shown in Fig. 8-3. f is discontinuous at 0 because $f(0)$ is not defined. Moreover,

$$\lim_{x\to 0^+} f(x)=\lim_{x\to 0^+}\frac{x}{x}=1 \quad\text{and}\quad \lim_{x\to 0^-} f(x)=\lim_{x\to 0^-}\frac{-x}{x}=-1$$

Thus, $\lim_{x\to 0^-} f(x)\neq \lim_{x\to 0^+} f(x)$. Hence, the discontinuity of f at 0 is not removable.

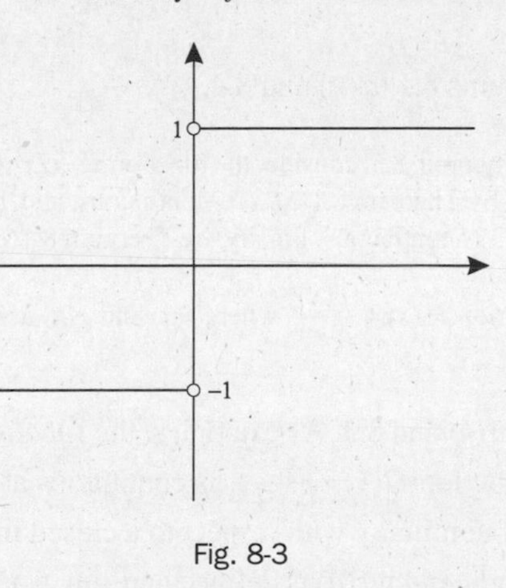

Fig. 8-3

The kind of discontinuity shown in Example 8.3 is called a *jump discontinuity*. In general, a function f has a jump discontinuity at x_0 if $\lim_{x\to x_0^-} f(x)$ and $\lim_{x\to x_0^+} f(x)$ both exist and $\lim_{x\to x_0^-} f(x)\neq \lim_{x\to x_0^+} f(x)$. Such a discontinuity is not removable.

EXAMPLE 8.4: The function of Problem 4 in Chapter 6 has a jump discontinuity at every positive integer.

Properties of limits lead to corresponding properties of continuity.

Theorem 8.1: Assume that f and g are continuous at x_0. Then:

(a) The constant function $h(x) = c$ for all x is continuous at every x_0.
(b) cf is continuous at x_0, for any constant c. (Recall that cf has the value $c \cdot f(x)$ for each argument x.)
(c) $f + g$ is continuous at x_0.
(d) $f - g$ is continuous at x_0.
(e) fg is continuous at x_0.
(f) f/g is continuous at x_0 if $g(x_0) \neq 0$.
(g) $\sqrt[n]{f}$ is continuous at x_0 if $\sqrt[n]{f(x_0)}$ is defined.

These results follow immediately from Theorems 7.1–7.6. For example, (*c*) holds because

$$\lim_{x \to x_0} (f(x) + g(x)) = \lim_{x \to x_0} f(x) + \lim_{x \to x_0} g(x) = f(x_0) + g(x_0)$$

Theorem 8.2: The identify function $I(x) = x$ is continuous at every x_0.

This follows from the fact that $\lim_{x \to x_0} x = x_0$.

We say that a function f is *continuous on a set A* if f is continuous at every point of A. Moreover, if we just say that *f is continuous*, we mean that f is continuous at every real number.

The original intuitive idea behind the notion of continuity was that the graph of a continuous function was supposed to be "continuous" in the intuitive sense that one could draw the graph without taking the pencil off the paper. Thus, the graph would not contain any "holes" or "jumps." However, it turns out that our precise definition of continuity goes well beyond that original intuitive notion; there are very complicated continuous functions that could certainly not be drawn on a piece of paper.

Theorem 8.3: Every polynomial function

$$f(x) = a_n x^n + a_{n-1} x^{n-1} + \cdots + a_1 x + a_0$$

is continuous.

This is a consequence of Theorems 8.1 (*a–e*) and 8.2.

EXAMPLE 8.5: As an instance of Theorem 8.3, consider the function $x^2 - 2x + 3$. Note that, by Theorem 8.2, the identity function x is continuous and, therefore, by Theorem 8.1(*e*), x^2 is continuous, and, by Theorem 8.1(*b*), $-2x$ is continuous. By Theorem 8.1(*a*), the constant function 3 is continuous. Finally, by Theorem 8.1(*c*), $x^2 - 2x + 3$ is continuous.

Theorem 8.4: Every *rational function* $H(x) = \frac{f(x)}{g(x)}$, where $f(x)$ and $g(x)$ are polynomial functions, is continuous on the set of all points at which $g(x) \neq 0$.

This follows from Theorems 8.1(*f*) and 8.3. As examples, the function $H(x) = \frac{x}{x^2 - 1}$ is continuous at all points except 1 and -1, and the function $G(x) = \frac{x - 7}{x^2 + 1}$ is continuous at all points (since $x^2 + 1$ is never 0).

We shall use a special notion of continuity with respect to a closed interval $[a, b]$. First of all, we say that a function f is *continuous on the right at a* if $f(a)$ is defined and $\lim_{x \to a^+} f(x)$ exists, and $\lim_{x \to a^+} f(x) = f(a)$. We say that f is *continuous on the left at b* if $f(b)$ is defined and $\lim_{x \to b^-} f(x)$ exists, and $\lim_{x \to b^-} f(x) = f(b)$.

Definition: f is *continuous* on $[a, b]$ if f is continuous at each point on the open interval (a, b), f is continuous on the right at a, and f is continuous on the left at b.

Note that whether f is continuous on $[a, b]$ does not depend on the values of f, if any, outside of $[a, b]$. Note also that every continuous function (that is, a function continuous at all real numbers) must be continuous on any closed interval. In particular, every polynomial function is continuous on any closed interval.

We want to discuss certain deep properties of continuous functions that we shall use but whose proofs are beyond the scope of this book.

Theorem 8.5 (Intermediate Value Theorem): If f is continuous on $[a, b]$ and $f(a) \neq f(b)$, then, for any number c between $f(a)$ and $f(b)$, there is at least one number x_0 in the open interval (a, b) for which $f(x_0) = c$.

Figure 8-4(a) is an illustration of Theorem 8.5. Fig. 8-5 shows that continuity throughout the interval is essential for the validity of the theorem. The following result is a special case of the Intermediate Value Theorem.

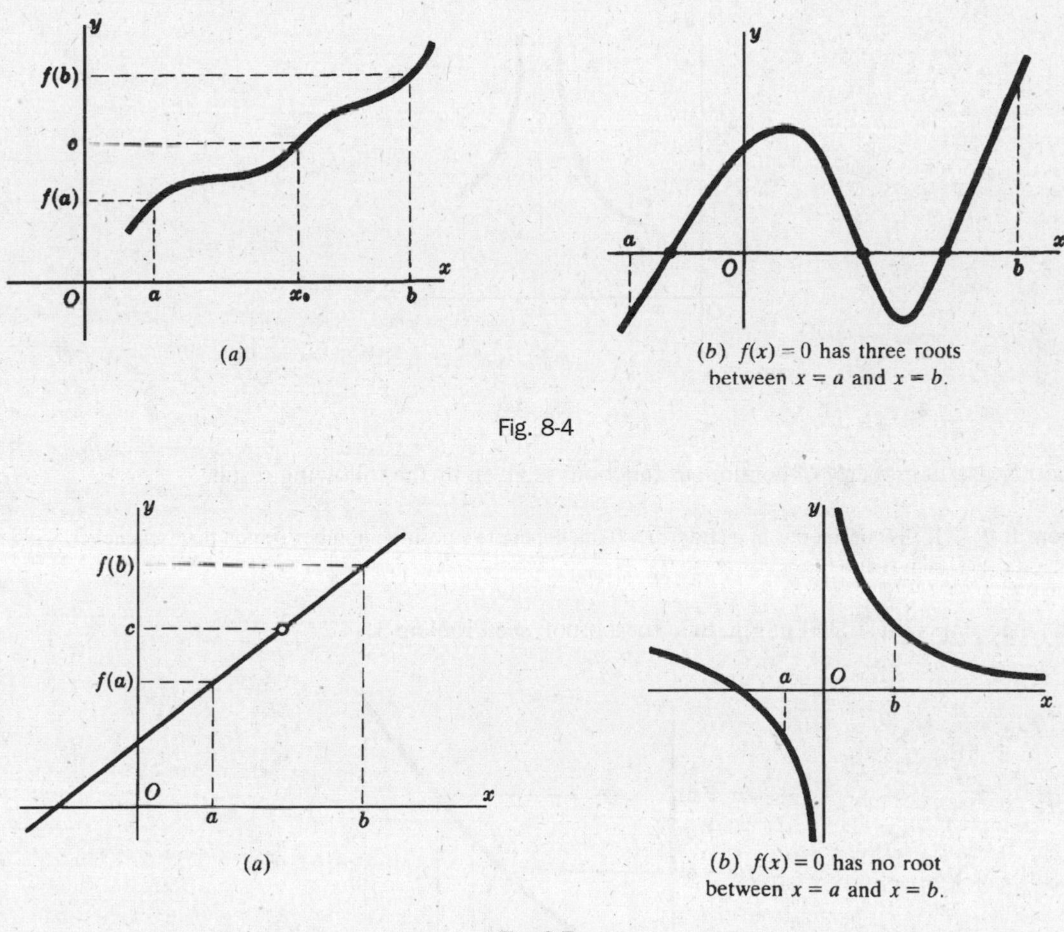

(a)

(b) $f(x) = 0$ has three roots between $x = a$ and $x = b$.

Fig. 8-4

(a)

(b) $f(x) = 0$ has no root between $x = a$ and $x = b$.

Fig. 8-5

Corollary 8.6: If f is continuous on $[a, b]$ and $f(a)$ and $f(b)$ have opposite signs, then the equation $f(x) = 0$ has at least one root in the open interval (a, b), and, therefore, the graph of f crosses the x-axis at least once between a and b. (See Fig. 8-4(b).)

Theorem 8.7 (Extreme Value Theorem): If f is continuous on $[a, b]$, then f takes on a least value m and a greatest value M on the interval.

As an illustration of the Extreme Value Theorem, look at Fig. 8-6(a), where the minimum value m occurs at $x = c$ and the maximum value M occurs at $x = d$. In this case, both c and d lie inside the interval. On the other hand, in Fig. 8-6(b), the minimum value m occurs at the endpoint $x = a$ and the maximum value M occurs inside the interval. To see that continuity is necessary for the Extreme Value Theorem to be true, consider the function whose graph is indicated in Fig. 8-6(c). There is a discontinuity at c inside the interval; the function has a minimum value at the left endpoint $x = a$ but the function has no maximum value.

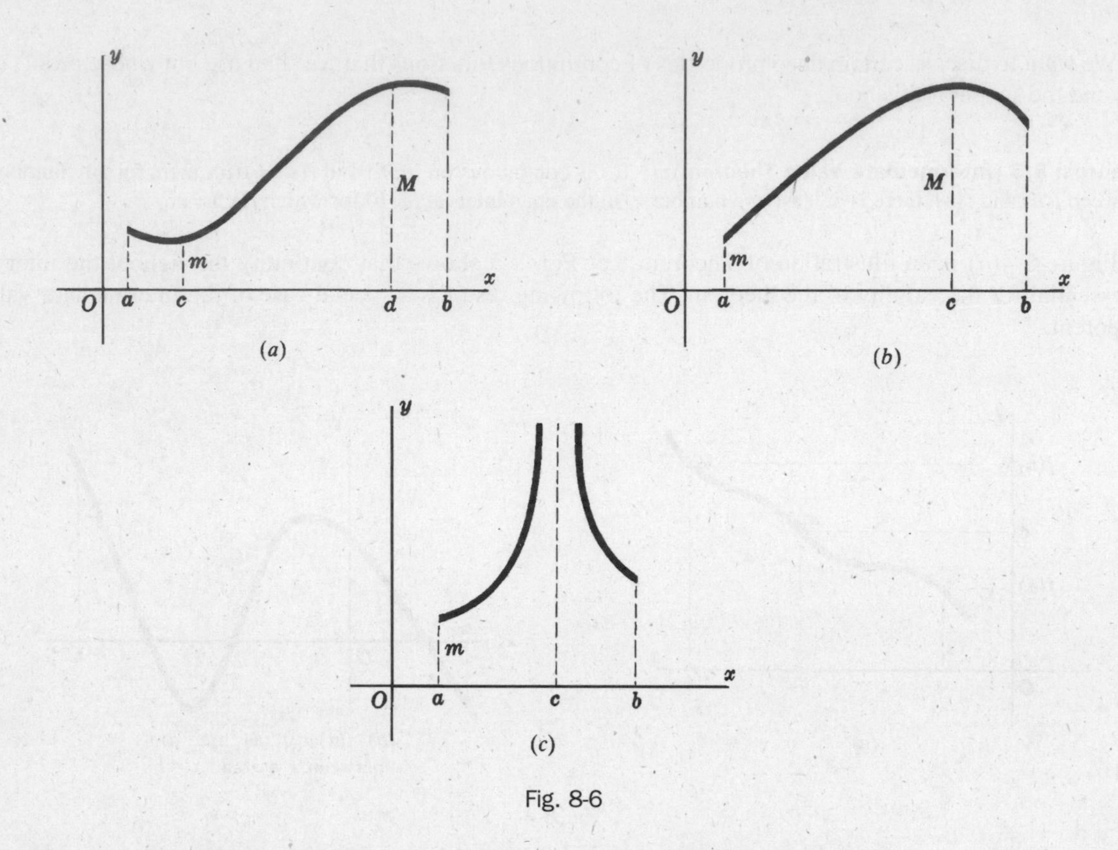

Fig. 8-6

Another useful property of continuous functions is given by the following result.

Theorem 8.8: If f is continuous at c and $f(c) > 0$, then there is a positive number δ such that, whenever $c - \delta < x < c + \delta$, then $f(x) > 0$.

This theorem is illustrated in Fig. 8-7. For a proof, see Problem 3.

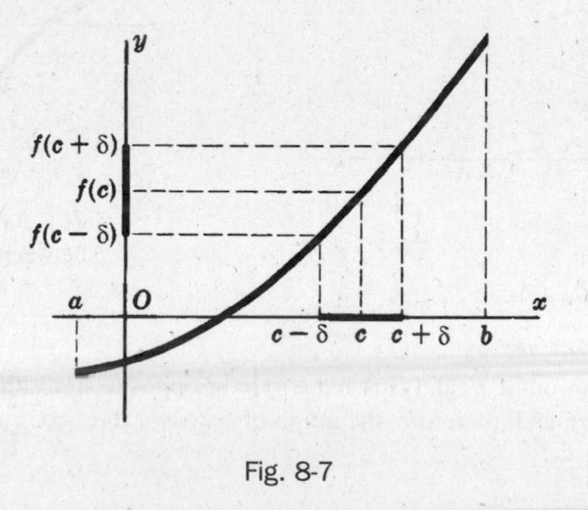

Fig. 8-7

SOLVED PROBLEMS

1. Find the discontinuities of the following functions. Determine whether they are removable. If not removable, determine whether they are jump discontinuities. (GC) Check your answers by showing the graph of the function on a graphing calculator.

(a) $f(x) = \frac{2}{x}$. Nonremovable discontinuity at $x = 0$.

(b) $f(x) = \frac{x-1}{(x+3)(x-2)}$. Nonremovable discontinuities at $x = -3$ and $x = 2$.

(c) $f(x) = \frac{(x+2)(x-1)}{(x-3)^2}$. Nonremovable discontinuity at $x = 3$.

(d) $f(x) = \frac{x^3 - 27}{x^2 - 9}$. Has a removable discontinuity at $x = 3$. (Note that $x^3 - 27 = (x-3)(x^2 + 3x + 9)$.) Also has a nonremovable discontinuity at $x = -3$.

(e) $f(x) = \frac{4 - x^2}{3 - \sqrt{x^2+5}}$. Has a removable discontinuity at $x = \pm 2$. Note that

$$\frac{4-x^2}{3-\sqrt{x^2+5}}\,\frac{3+\sqrt{x^2+5}}{3+\sqrt{x^2+5}} = 3 + \sqrt{x^2+5}.$$

(f) $f(x) = \frac{x^2 + x - 2}{(x-1)^2}$. Has a nonremovable discontinuity at $x = 1$.

(g) $f(x) = [x] =$ the greatest integer $\leq x$. Has a jump discontinuity at every integer.

(h) $f(x) = x - [x]$. Has a nonremovable discontinuity at every integer.

(i) $f(x) = 3x^3 - 7x^2 + 4x - 2$. A polynomial has no discontinuities.

(j) $f(x) = \begin{cases} 0 & \text{if } x = 0 \\ 2 & \text{if } x \neq 0 \end{cases}$ Removable discontinuity at $x = 0$.

(k) $f(x) = \begin{cases} x & \text{if } x \leq 0. \\ x^2 & \text{if } 0 < x < 1 \\ 2 - x & \text{if } x \geq 1. \end{cases}$ No discontinuities.

2. Show that the existence of $\lim_{h \to 0} \frac{f(a+h) - f(a)}{h}$ implies that f is continuous at $x = a$.

$$\lim_{h\to 0}(f(a+h) - f(a)) = \lim_{h\to 0}\left(\frac{f(a+h)-f(a)}{h}\cdot h\right) =$$

$$\lim_{h\to 0}\frac{f(a+h)-f(a)}{h}\cdot \lim_{h\to 0} h = \lim_{h\to 0}\frac{f(a+h)-f(a)}{h}\cdot 0 = 0$$

But

$$\lim_{h\to 0}(f(a+h) - f(a)) = \lim_{h\to 0} f(a+h) - \lim_{h\to 0} f(a) = \lim_{h\to 0} f(a+h) - f(a)$$

Hence, $\lim_{h\to 0} f(a+h) = f(a)$. Note that $\lim_{h\to 0} f(a+h) = \lim_{x\to a} f(x)$. So, $\lim_{x\to a} f(x) = f(a)$.

3. Prove Theorem 8.8.

By the continuity of f at c, $\lim_{x\to c} f(x) = f(c)$. If we let $\epsilon = f(c)/2 > 0$, then there exists a positive δ such that $0 < |x - c| < \delta$ implies that $|f(x) - f(c)| < f(c)/2$. The latter inequality also holds when $x = c$. Thus, $|x - c| < \delta$ implies $|f(x) - f(c)| < f(c)/2$. The latter implies $-f(c)/2 < f(x) - f(c) < f(c)/2$. Adding $f(c)$ to the left-hand inequality, we obtain $f(c)/2 < f(x)$.

SUPPLEMENTARY PROBLEMS

4. Determine the discontinuities of the following functions and state why the function fails to be continuous at those points. (GC) Check your answers by graphing the function on a graphing calculator.

(a) $f(x) = \frac{x^2 - 3x - 10}{x + 2}$

(b) $f(x) = \begin{cases} x + 3 & \text{if } x \geq 2 \\ x^2 + 1 & \text{if } x < 2 \end{cases}$

(c) $f(x) = |x| - x$

(d) $f(x) = \begin{cases} 4 - x & \text{if } x < 3 \\ x - 2 & \text{if } 0 < x < 3 \\ x - 1 & \text{if } x \leq 0 \end{cases}$

(e) $f(x) = \frac{x^4 - 1}{x^2 - 1}$

(f) $f(x) = \frac{x^3 + x^2 - 17x + 15}{x^2 + 2x - 15}$

(g) $f(x) = x^3 - 7x$ (h) $f(x) = \dfrac{x^2 - 4}{x^2 - 5x + 6}$

(i) $f(x) = \dfrac{x^2 + 3x + 2}{x^2 + 4x + 3}$ (j) $f(x) = \dfrac{x - 2}{x^2 - 4}$

(k) $f(x) = \dfrac{x - 1}{\sqrt{x^2 + 3} - 2}$

Ans. (a) Removable discontinuity at $x = -2$. (Note that $x^2 - 3x - 10 = (x + 2)(x - 5)$.)
(b, c, g) None.
(d) Jump discontinuity at $x = 0$.
(e) Removable discontinuities at $x = \pm 1$.
(f) Removable discontinuities at $x = 3$, $x = -5$. (Note that $x^2 + 2x - 5 = (x + 5)(x - 3)$ and $x^3 + x^2 - 17x + 15 = (x + 5)(x - 3)(x - 1)$.)
(h) Removable discontinuity at $x = 2$ and nonremovable discontinuity at $x = 3$.
(i) Removable discontinuity at $x = -1$ and nonremovable discontinuity at $x = -3$.
(j) Removable discontinuity at $x = 2$ and nonremovable discontinuity at $x = -2$.
(k) Removable discontinuity at $x = 1$ and nonremovable discontinuity at $x = -1$.

5. Show that $f(x) = \lfloor x \rfloor$ is continuous.

6. If Fig. 8-5(*a*) is the graph of $f(x) = \dfrac{x^2 - 4x - 21}{x - 7}$, show that there is a removable discontinuity at $x = 7$ and that $c = 10$ there.

7. Prove: If f is continuous on the interval $[a, b]$ and c is a number in (a, b) such that $f(c) < 0$, then there exists a positive number δ such that, whenever $c - \delta < x < c + \delta$, then $f(x) < 0$.

(*Hint*: Apply Theorem 8.8 to $-f$.)

8. Sketch the graphs of the following functions and determine whether they are continuous on the closed interval $[0, 1]$:

(a) $f(x) = \begin{cases} -1 & \text{if } x < 0 \\ 0 & \text{if } 0 \le x \le 1 \\ 1 & \text{if } x > 1 \end{cases}$ (b) $f(x) = \begin{cases} \dfrac{1}{x} & \text{if } x > 0 \\ 1 & \text{if } x \le 0 \end{cases}$

(c) $f(x) = \begin{cases} -x^2 & \text{if } x \le 0 \\ x^2 & \text{if } x > 0 \end{cases}$ (d) $f(x) = 1$ if $0 < x \le 1$

(e) $f(x) = \begin{cases} x & \text{if } x \le 0 \\ 0 & \text{if } 0 < x < 1 \\ x & \text{if } x \ge 1 \end{cases}$

Ans. (a) Yes. (b) No. Not continuous on the right at 0. (c) Yes. (d) No. Not defined at 0. (*e*) No. Not continuous on the left at 1.

CHAPTER 9

The Derivative

Delta Notation

Let f be a function. As usual, we let x stand for any argument of f, and we let y be the corresponding value of f. Thus, $y = f(x)$. Consider any number x_0 in the domain of f. Let Δx (read "delta x") represent a small change in the value of x, from x_0 to $x_0 + \Delta x$, and then let Δy (read "delta y") denote the corresponding change in the value of y. So, $\Delta y = f(x_0 + \Delta x) - f(x_0)$. Then the ratio

$$\frac{\Delta y}{\Delta x} = \frac{\text{change in } y}{\text{change in } x} = \frac{f(x_0 + \Delta x) - f(x_0)}{\Delta x}$$

is called the *average rate of change* of the function f on the interval between x_0 and $x_0 + \Delta x$.

EXAMPLE 9.1: Let $y = f(x) = x^2 + 2x$. Starting at $x_0 = 1$, change x to 1.5. Then $\Delta x = 0.5$. The corresponding change in y is $\Delta y = f(1.5) - f(1) = 5.25 - 3 = 2.25$. Hence, the average rate of change of y on the interval between $x = 1$ and $x = 1.5$ is $\frac{\Delta y}{\Delta x} = \frac{2.25}{0.5} = 4.5$.

The Derivative

If $y = f(x)$ and x_0 is in the domain of f, then by the *instantaneous rate of change* of f at x_0 we mean the limit of the average rate of change between x_0 and $x_0 + \Delta x$ as Δx approaches 0:

$$\lim_{\Delta x \to 0} \frac{\Delta y}{\Delta x} = \lim_{\Delta x \to 0} \frac{f(x_0 + \Delta x) - f(x_0)}{\Delta x}$$

provided that this limit exists. This limit is also called the *derivative* of f at x_0.

Notation for Derivatives

Let us consider the derivative of f at an arbitrary point x in its domain:

$$\lim_{\Delta x \to 0} \frac{\Delta y}{\Delta x} = \lim_{\Delta x \to 0} \frac{f(x + \Delta x) - f(x)}{\Delta x}$$

The value of the derivative is a function of x, and will be denoted by any of the following expressions:

$$D_x y = \frac{dy}{dx} = y' = f'(x) = \frac{d}{dx}y = \frac{d}{dx}f(x) = \lim_{\Delta x \to 0} \frac{\Delta y}{\Delta x}$$

The value $f'(a)$ of the derivative of f at a particular point a is sometimes denoted by $\left.\frac{dy}{dx}\right|_{x=a}$

Differentiability

A function is said to be *differentiable* at a point x_0 if the derivative of the function exists at that point. Problem 2 of Chapter 8 shows that differentiability implies continuity. That the converse is false is shown in Problem 11.

SOLVED PROBLEMS

1. Given $y = f(x) = x^2 + 5x - 8$, find Δy and $\Delta y/\Delta x$ as x changes (a) from $x_0 = 1$ to $x_1 = x_0 + \Delta x = 1.2$ and (b) from $x_0 = 1$ to $x_1 = 0.8$.

(a) $\Delta x = x_1 - x_0 = 1.2 - 1 = 0.2$ and $\Delta y = f(x_0 + \Delta x) - f(x_0) = f(1.2) - f(1) = -0.56 - (-2) = 1.44$.
So $\dfrac{\Delta y}{\Delta x} = \dfrac{1.44}{0.2} = 7.2$.

(b) $\Delta x = 0.8 - 1 = -0.2$ and $\Delta y = f(0.8) - f(1) = -3.36 - (-2) = -1.36$. So $\dfrac{\Delta y}{\Delta x} = \dfrac{-1.36}{-0.2} = 6.8$.

Geometrically, $\Delta y/\Delta x$ in (a) is the slope of the secant line joining the points $(1, -2)$ and $(1.2, -0.56)$ of the parabola $y = x^2 + 5x - 8$, and in (b) is the slope of the secant line joining the points $(0.8, -3.36)$ and $(1, -2)$ of the same parabola.

2. If a body (that is, a material object) starts out at rest and then falls a distance of s feet in t seconds, then physical laws imply that $s = 16t^2$. Find $\Delta s/\Delta t$ as t changes from t_0 to $t_0 + \Delta t$. Use the result to find $\Delta s/\Delta t$ as t changes: (a) from 3 to 3.5, (b) from 3 to 3.2, and (c) from 3 to 3.1.

$$\frac{\Delta s}{\Delta t} = \frac{16(t_0 + \Delta t)^2 - 16t_0^2}{\Delta t} = \frac{32t_0\Delta t + 16(\Delta t)^2}{\Delta t} = 32t_0 + 16\,\Delta t$$

(a) Here $t_0 = 3$, $\Delta t = 0.5$, and $\Delta s/\Delta t = 32(3) + 16(0.5) = 104$ ft/sec.
(b) Here $t_0 = 3$, $\Delta t = 0.2$, and $\Delta s/\Delta t = 32(3) + 16(0.2) = 99.2$ ft/sec.
(c) Here $t_0 = 3$, $\Delta t = 0.1$, and $\Delta s/\Delta t = 97.6$ ft/sec.

Since Δs is the displacement of the body from time $t = t_0$ to $t = t_0 + \Delta t$,

$$\frac{\Delta s}{\Delta t} = \frac{\text{displacement}}{\text{time}} = \text{average velocity of the body over the time interval}$$

3. Find dy/dx, given $y = x^3 - x^2 - 4$. Find also the value of dy/dx when (a) $x = 4$, (b) $x = 0$, (c) $x = -1$.

$$y + \Delta y = (x + \Delta x)^3 - (x + \Delta x)^2 - 4$$
$$= x^3 + 3x^2(\Delta x) + 3x(\Delta x)^2 + (\Delta x)^3 - x^2 - 2x(\Delta x) - (\Delta x)^2 - 4$$
$$\Delta y = (3x^2 - 2x)\Delta x + (3x - 1)(\Delta x)^2 + (\Delta x)^3$$
$$\frac{\Delta y}{\Delta x} = 3x^2 - 2x + (3x - 1)\Delta x + (\Delta x)^2$$
$$\frac{dy}{dx} = \lim_{\Delta x \to 0} [3x^2 - 2x + (3x - 1)\Delta x + (\Delta x)^2] = 3x^2 - 2x$$

(a) $\left.\dfrac{dy}{dx}\right|_{x=4} = 3(4)^2 - 2(4) = 40$; (b) $\left.\dfrac{dy}{dx}\right|_{x=0} = 3(0)^2 - 2(0) = 0$; (c) $\left.\dfrac{dy}{dx}\right|_{x=-1} = 3(-1)^2 - 2(-1) = 5$

4. Find the derivative of $y = f(x) = x^2 + 3x + 5$.

$$\Delta y = f(x+\Delta x) - f(x) = [(x+\Delta x)^2 + 3(x+\Delta x)+5)] - [x^2+3x+5]$$
$$= [x^2 + 2x\Delta x + (\Delta x)^2 + 3x + 3\Delta x + 5] - [x^2+3x+5] = 2x\Delta x + (\Delta x)^2 + 3\Delta x$$
$$= (2x + \Delta x + 3)\Delta x$$
$$\frac{\Delta y}{\Delta x} = 2x + \Delta x + 3$$

So, $\dfrac{dy}{dx} = \lim\limits_{\Delta x \to 0}(2x + \Delta x + 3) = 2x + 3$.

5. Find the derivative of $y = f(x) = \dfrac{1}{x-2}$ at $x = 1$ and $x = 3$.

$$\Delta y = f(x+\Delta x) - f(x) = \frac{1}{(x+\Delta x)-2} - \frac{1}{x-2} = \frac{(x-2)-(x+\Delta x-2)}{(x-2)(x+\Delta x-2)}$$
$$= \frac{-\Delta x}{(x-2)(x+\Delta x-2)}$$
$$\frac{\Delta y}{\Delta x} = \frac{-1}{(x-2)(x+\Delta x-2)}$$

So, $\dfrac{dy}{dx} = \lim\limits_{\Delta x \to 0}\dfrac{-1}{(x-2)(x+\Delta x-2)} = \dfrac{-1}{(x-2)^2}$.

At $x = 1$, $\dfrac{dy}{dx} = \dfrac{-1}{(1-2)^2} = -1$. At $x = 3$, $\dfrac{dy}{dx} = \dfrac{-1}{(3-2)^2} = -1$.

6. Find the derivative of $f(x) = \dfrac{2x-3}{3x+4}$.

$$f(x+\Delta x) = \frac{2(x+\Delta x)-3}{3(x+\Delta x)+4}$$
$$f(x+\Delta x) - f(x) = \frac{2x+2\Delta x-3}{3x+3\Delta x+4} - \frac{2x-3}{3x+4}$$
$$= \frac{(3x+4)[(2x-3)+2\Delta x] - (2x-3)[(3x+4)+3\Delta x]}{(3x+4)(3x+3\Delta x+4)}$$
$$= \frac{(6x+8-6x+9)\Delta x}{(3x+4)(3x+3\Delta x+4)} = \frac{17\Delta x}{(3x+4)(3x+3\Delta x+4)}$$
$$\frac{f(x+\Delta x)-f(x)}{\Delta x} = \frac{17}{(3x+4)(3x+3\Delta x+4)}$$
$$f'(x) = \lim_{\Delta x\to 0}\frac{17}{(3x+4)(3x+3\Delta x+4)} = \frac{17}{(3x+4)^2}$$

7. Find the derivative of $y = f(x) = \sqrt{2x+1}$.

$$y + \Delta y = (2x + 2\Delta x + 1)^{1/2}$$
$$\Delta y = (2x+2\Delta x+1)^{1/2} - (2x+1)^{1/2}$$
$$= [(2x+2\Delta x+1)^{1/2} - (2x+1)^{1/2}]\frac{(2x+2\Delta x+1)^{1/2} + (2x+1)^{1/2}}{(2x+2\Delta x+1)^{1/2} + (2x+1)^{1/2}}$$
$$= \frac{(2x+2\Delta x+1)-(2x+1)}{(2x+2\Delta x+1)^{1/2} + (2x+1)^{1/2}} = \frac{2\Delta x}{(2x+2\Delta x+1)^{1/2} + (2x+1)^{1/2}}$$
$$\frac{\Delta y}{\Delta x} = \frac{2}{(2x+2\Delta x+1)^{1/2} + (2x+1)^{1/2}}$$
$$\frac{dy}{dx} = \lim_{\Delta x\to 0}\frac{2}{(2x+2\Delta x+1)^{1/2} + (2x+1)^{1/2}} = \frac{1}{(2x+1)^{1/2}}$$

8. Find the derivative of $f(x) = x^{1/3}$. Examine $f'(0)$.

$$f(x+\Delta x) = (x+\Delta x)^{1/3}$$

$$f(x+\Delta x) - f(x) = (x+\Delta x)^{1/3} - x^{1/3}$$

$$= \frac{[(x+\Delta x)^{1/3} - x^{1/3}][(x+\Delta x)^{2/3} + x^{1/3}(x+\Delta x)^{1/3} + x^{2/3}]}{(x+\Delta x)^{2/3} + x^{1/3}(x+\Delta x)^{1/3} + x^{2/3}}$$

$$= \frac{x+\Delta x - x}{(x+\Delta x)^{2/3} + x^{1/3}(x+\Delta x)^{1/3} + x^{2/3}}$$

$$\frac{f(x+\Delta x) - f(x)}{\Delta x} = \frac{1}{(x+\Delta x)^{2/3} + x^{1/3}(x+\Delta x)^{1/3} + x^{2/3}}$$

$$f'(x) = \lim_{\Delta x \to 0} \frac{1}{(x+\Delta x)^{2/3} + x^{1/3}(x+\Delta x)^{1/3} + x^{2/3}} = \frac{1}{3x^{2/3}}$$

The derivative does not exist at $x = 0$ because the denominator is zero there. Note that the function f is continuous at $x = 0$.

9. Interpret dy/dx geometrically.

From Fig. 9-1 we see that $\Delta y/\Delta x$ is the slope of the secant line joining an arbitrary but fixed point $P(x, y)$ and a nearby point $Q(x + \Delta x, y + \Delta y)$ of the curve. As $\Delta x \to 0$, P remains fixed while Q moves along the curve toward P, and the line PQ revolves about P toward its limiting position, the tangent line PT moves to the curve at P. Thus, dy/dx gives the slope of the tangent line at P to the curve $y = f(x)$.

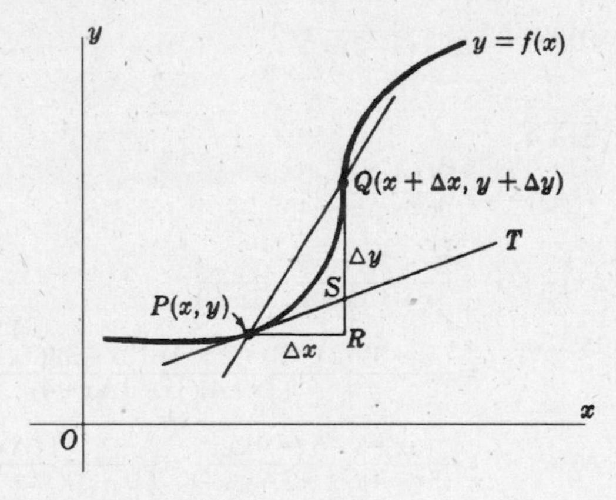

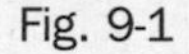

Fig. 9-1

For example, from Problem 3, the slope of the cubic $y = x^3 - x^2 - 4$ is $m = 40$ at the point $x = 4$; it is $m = 0$ at the point $x = 0$; and it is $m = 5$ at the point $x = -1$.

10. Find ds/dt for the function of Problem 2 and interpret the result.

$$\frac{\Delta s}{\Delta t} = 32t_0 + 16\Delta t. \quad \text{Hence,} \quad \frac{ds}{dt} = \lim_{\Delta t \to 0} (32t_0 + 16\Delta t) = 32t_0$$

As $\Delta t \to 0$, $\Delta s/\Delta t$ gives the average velocity of the body for shorter and shorter time intervals Δt. Then we can consider ds/dt to be the *instantaneous velocity* v of the body at time t_0.

For example, at $t = 3$, $v = 32(3) = 96$ ft/sec. In general, if an object is moving on a straight line, and its position on the line has coordinate s at time t, then its instantaneous velocity at time t is ds/dt. (See Chapter 19.)

11. Find $f'(x)$ when $f(x) = |x|$.

The function is continuous for all values of x. For $x < 0$, $f(x) = -x$ and

$$f'(x) = \lim_{\Delta x \to 0} \frac{-(x+\Delta x) - (-x)}{\Delta x} = \lim_{\Delta x \to 0} \frac{-\Delta x}{\Delta x} = \lim_{\Delta x \to 0} -1 = -1$$

Similarly, for $x > 0, f(x) = x$ and

$$f'(x) = \lim_{\Delta x \to 0} \frac{(x+\Delta x) - x}{\Delta x} = \lim_{\Delta x \to 0} \frac{\Delta x}{\Delta x} = \lim_{\Delta x \to 0} 1 = 1$$

At $x = 0, f(x) = 0$ and $\lim_{\Delta x \to 0} \frac{f(0+\Delta x) - f(0)}{\Delta x} = \lim_{\Delta x \to 0} \frac{|\Delta x|}{\Delta x}$.

As $\Delta x \to 0^-$, $\frac{|\Delta x|}{\Delta x} = \frac{-\Delta x}{\Delta x} = -1 \to -1$. But, as $\Delta x \to 0^+$, $\frac{|\Delta x|}{\Delta x} = \frac{\Delta x}{\Delta x} = 1 \to 1$. Hence, the derivative does not exist at $x = 0$.

Since the function is continuous at 0, this shows that continuity does not imply differentiability.

12. Compute $\epsilon = \frac{\Delta y}{\Delta x} - \frac{dy}{dx}$ for the function of (a) Problem 3 and (b) Problem 5. Verify that $\epsilon \to 0$ as $\Delta x \to 0$.

(a) $\epsilon = [3x^2 - 2x + (3x-1)\Delta x + (\Delta x)^2] - (3x^2 - 2x) = (3x - 1 + \Delta x)\,\Delta x$

(b) $\epsilon = \frac{1}{(x-2)(x+\Delta x-2)} - \frac{-1}{(x-2)^2} = \frac{-(x-2)+(x+\Delta x-2)}{(x-2)^2(x+\Delta x-2)} = \frac{1}{(x-2)^2(x+\Delta x-2)}\Delta x$

Both obviously go to zero as $\Delta x \to 0$.

13. Interpret $\Delta y = \frac{dy}{dx}\Delta x + \epsilon\,\Delta x$ of Problem 12 geometrically.

In Fig. 9-1, $\Delta y = RQ$ and $\frac{dy}{dx}\Delta x = PR \tan \angle TPR = RS$; thus, $\epsilon\,\Delta x = SQ$. For a change Δx in x from $P(x, y)$, Δy is the corresponding change in y *along the curve* while $\frac{dy}{dx}\Delta x$ is the corresponding change in y *along the tangent line PT*. Since their difference $\epsilon\,\Delta x$ is a multiple of $(\Delta x)^2$, it goes to zero faster than Δx, and $\frac{dy}{dx}\Delta x$ can be used as an approximation of Δy when $|\Delta x|$ is small.

SUPPLEMENTARY PROBLEMS

14. Find Δy and $\Delta y/\Delta x$, given

(a) $y = 2x - 3$ and x changes from 3.3 to 3.5.
(b) $y = x^2 + 4x$ and x changes from 0.7 to 0.85.
(c) $y = 2/x$ and x changes from 0.75 to 0.5.

Ans. (a) 0.4 and 2; (b) 0.8325 and 5.55; (c) $\frac{4}{3}$ and $-\frac{16}{3}$

15. Find Δy, given $y = x^2 - 3x + 5$, $x = 5$, and $\Delta x = -0.01$. What then is the value of y when $x = 4.99$?

Ans. $\Delta y = -0.0699$; $y = 14.9301$

16. Find the average velocity (see Problem 2), given: (a) $s = (3t^2 + 5)$ feet and t changes from 2 to 3 seconds. (b) $s = (2t^2 + 5t - 3)$ feet and t changes from 2 to 5 seconds.

Ans. (a) 15 ft/sec; (b) 19 ft/sec

17. Find the increase in the volume of a spherical balloon when its radius is increased (a) from r to $r + \Delta r$ inches; (b) from 2 to 3 inches. (Recall that volume $V = \frac{4}{3}\pi r^3$.)

Ans. (a) $\frac{4}{3}\pi[3r^2 + 3r\,\Delta r + (\Delta r)^2]\Delta r$ in^3; (b) $\frac{76}{3}\pi$ in^3

18. Find the derivative of each of the following:

(a) $y = 4x - 3$ (b) $y = 4 - 3x$ (c) $y = x^2 + 2x - 3$
(d) $y = 1/x^2$ (e) $y = (2x - 1)/(2x + 1)$ (f) $y = (1 + 2x)/(1 - 2x)$
(g) $y = \sqrt{x}$ (h) $y = 1/\sqrt{x}$ (i) $y = \sqrt{1+2x}$
(j) $y = 1/\sqrt{2+x}$

Ans. (a) 4; (b) −3; (c) $2(x + 1)$; (d) $-2/x^3$; (e) $\frac{4}{(2x+1)^2}$; (f) $\frac{4}{(1-2x)^2}$; (g) $\frac{1}{2\sqrt{x}}$; (h) $-\frac{1}{2x\sqrt{x}}$; (i) $\frac{1}{\sqrt{1+2x}}$; (j) $-\frac{1}{2(2+x)^{3/2}}$

19. Find the slope of the tangent line to the following curves at the point $x = 1$ (see Problem 9): (a) $y = 8 - 5x^2$; (b) $y = \frac{4}{x+1}$; (c) $\frac{2}{x+3}$.

Ans. (a) −10; (b) −1; (c) $-\frac{1}{8}$

20. (GC) Use a graphing calculator to verify your answers in Problem 19. (Graph the curve and the tangent line that you found.)

21. Find the coordinates of the vertex (that is, the turning point) of the parabola $y = x^2 - 4x + 1$ by making use of the fact that, at the vertex, the slope of the tangent line is zero. (See Problem 9.) (GC) Check your answer with a graphing calculator.

Ans. (2, −3)

22. Find the slope m of the tangent lines to the parabola $y = -x^2 + 5x - 6$ at its points of intersection with the x axis.

Ans. At $x = 2$, $m = 1$. At $x = 3$, $m = -1$.

23. When an object is moving on a straight line and its coordinate on that line is s at time t (where s is measured in feet and t in seconds), find the velocity at time $t = 2$ in the following cases:

(a) $s = t^2 + 3t$ (b) $s = t^3 - 3t^2$ (c) $s = \sqrt{t+2}$
(See Problem 10.)

Ans. (a) 7 ft/sec; (b) 0 ft/sec; (c) $\frac{1}{4}$ ft/sec

24. Show that the instantaneous rate of change of the volume V of a cube with respect to its edge x (measured in inches) is 12 in^3/in when $x = 2$ in.

CHAPTER 10

Rules for Differentiating Functions

Differentiation

Recall that a function f is said to be *differentiable* at x_0 if the derivative $f'(x_0)$ exists. A function is said to be differentiable on a set if the function is differentiable at every point of the set. If we say that a function is differentiable, we mean that it is differentiable at every real number. The process of finding the derivative of a function is called *differentiation*.

Theorem 10.1 (Differentiation Formulas): In the following formulas, it is assumed that u, v, and w are functions that are differentiable at x; c and m are assumed to be constants.

(1) $\frac{d}{dx}(c) = 0$ (The derivative of a constant function is zero.)

(2) $\frac{d}{dx}(x) = 1$ (The derivative of the identity function is 1.)

(3) $\frac{d}{dx}(cu) = c\frac{du}{dx}$

(4) $\frac{d}{dx}(u+v+\ldots) = \frac{du}{dx} + \frac{dv}{dx} + \ldots$ (Sum Rule)

(5) $\frac{d}{dx}(u-v) = \frac{du}{dx} - \frac{dv}{dx}$ (Difference Rule)

(6) $\frac{d}{dx}(uv) = u\frac{dv}{dx} + v\frac{du}{dx}$ (Product Rule)

(7) $\frac{d}{dx}\left(\frac{u}{v}\right) = \frac{v\frac{du}{dx} - u\frac{dv}{dx}}{v^2}$ provided that $v \neq 0$ (Quotient Rule)

(8) $\frac{d}{dx}\left(\frac{1}{x}\right) = -\frac{1}{x^2}$ provided that $x \neq 0$

(9) $\frac{d}{dx}(x^m) = mx^{m-1}$ (Power Rule)

Note that formula (8) is a special case of formula (9) when $m = -1$. For proofs, see Problems 1–4.

EXAMPLE 10.1: $D_x(x^3 + 7x + 5) = D_x(x^3) + D_x(7x) + D_x(5)$ (Sum Rule)

$= 3x^2 + 7D_x(x) + 0$ (Power Rule, formulas (3) and (1))

$= 3x^2 + 7$ (formula (2))

Every polynomial is differentiable, and its derivative can be computed by using the Sum Rule, Power Rule, and formulas (1) and (3).

Composite Functions. The Chain Rule

The *composite function* $f \circ g$ of functions g and f is defined as follows: $(f \circ g)(x) = f(g(x))$. The function g is applied first and then $f \cdot g$ is called the *inner function*, and f is called the *outer function*. $f \circ g$ is called the *composition* of g and f.

EXAMPLE 10.2: Let $f(x) = x^2$ and $g(x) = x + 1$. Then:

$$(f \circ g)(x) = f(g(x)) = f(x+1) = (x+1)^2 = x^2 + 2x + 1$$

$$(g \circ f)(x) = g(f(x)) = g(x^2) = x^2 + 1$$

Thus, in this case, $f \circ g \neq g \circ f$.

When f and g are differentiable, then so is their composition $f \circ g$. There are two procedures for finding the derivative of $f \circ g$. The first method is to compute an explicit formula for $f(g(x))$ and differentiate.

EXAMPLE 10.3: If $f(x) = x^2 + 3$ and $g(x) = 2x + 1$, then

$$y = f(g(x)) = f(2x+1) = (2x+1)^2 + 3 = 4x^2 + 4x + 4 \quad \text{and} \quad \frac{dy}{dx} = 8x + 4$$

Thus, $D_x(f \circ g) = 8x + 4$.

The second method of computing the derivative of a composite function is based on the following rule.

Chain Rule

$$D_x(f(g(x)) = f'(g(x)) \cdot g'(x)$$

Thus, the derivative of $f \circ g$ is the product of the derivative of the outer function f (evaluated at $g(x)$) and the derivative of the inner function (evaluated at x). It is assumed that g is differentiable at x and that f is differentiable at $g(x)$.

EXAMPLE 10.4: In Example 10.3, $f'(x) = 2x$ and $g'(x) = 2$. Hence, by the Chain Rule,

$$D_x(f(g(x)) = f'(g(x)) \cdot g'(x) = 2g(x) \cdot 2 = 4g(x) = 4(2x+1) = 8x + 4$$

Alternative Formulation of the Chain Rule

Let $u = g(x)$ and $y = f(u)$. Then the composite function of g and f is $y = f(u) = f(g(x))$, and we have the formula:

$$\frac{dy}{dx} = \frac{dy}{du}\frac{du}{dx} \qquad \text{(Chain Rule)}$$

EXAMPLE 10.5: Let $y = u^3$ and $u = 4x^2 - 2x + 5$. Then the composite function $y = (4x^2 - 2x + 5)^3$ has the derivative

$$\frac{dy}{dx} = \frac{dy}{du}\frac{du}{dx} = 3u^2(8x-2) = 3(4x^2 - 2x + 5)^2(8x-2)$$

Warning. In the Alternative Formulation of the Chain Rule, $\frac{dy}{dx} = \frac{dy}{du}\frac{du}{dx}$, the y on the left denotes the composite function of x, whereas the y on the right denotes the original function of u. Likewise, the two occurrences of u have different meanings. This notational confusion is made up for by the simplicity of the alternative formulation.

Inverse Functions

Two functions f and g such that $g(f(x)) = x$ and $f(g(y)) = y$ are said to be *inverse functions*. Inverse functions reverse the effect of each other. Given an equation $y = f(x)$, we can find a formula for the inverse of f by solving the equation for x in terms of y.

EXAMPLE 10.6:

(a) Let $f(x) = x + 1$. Solving the equation $y = x + 1$ for x, we obtain $x = y - 1$. Then the inverse g of f is given by the formula $g(y) = y - 1$. Note that g reverses the effect of f and f reverses the effect of g.

(b) Let $f(x) = -x$. Solving $y = -x$ for x, we obtain $x = -y$. Hence, $g(y) = -y$ is the inverse of f. In this case, the inverse of f is the same function as f.

(c) Let $f(x) = \sqrt{x}$. f is defined only for nonnegative numbers, and its range is the set of nonnegative numbers. Solving $y = \sqrt{x}$ for x, we get $x = y^2$, so that $g(y) = y^2$. Note that, since g is the inverse of f, g is only defined for nonnegative numbers, since the values of f are the nonnegative numbers. (Since $y = f(g(y))$, then, if we allowed g to be defined for negative numbers, we would have $-1 = f(g(-1)) = f(1) = 1$, a contradiction.)

(d) The inverse of $f(x) = 2x - 1$ is the function $g(y) = \frac{y+1}{2}$.

Notation

The inverse of f will be denoted f^{-1}.

Do not confuse this with the exponential notation for raising a number to the power -1. The context will usually tell us which meaning is intended.

Not every function has an inverse function. For example, the function $f(x) = x^2$ does not possess an inverse. Since $f(1) = 1 = f(-1)$, an inverse function g would have to satisfy $g(1) = 1$ and $g(1) = -1$, which is impossible. (However, if we restricted the function $f(x) = x^2$ to the domain $x \geq 0$, then the function $g(y) = \sqrt{y}$ would be an inverse function of f.)

The condition that a function f must satisfy in order to have an inverse is that f is *one-to-one*, that is, for any x_1 and x_2, if $x_1 \neq x_2$, then $f(x_1) \neq f(x_2)$. Equivalently, f is one-to-one if and only if, for any x_1 and x_2, if $f(x_1) = f(x_2)$, then $x_1 = x_2$.

EXAMPLE 10.7: Let us show that the function $f(x) = 3x + 2$ is one-to-one. Assume $f(x_1) = f(x_2)$. Then $3x_1 + 2 = 3x_2 + 2$, $3x_1 = 3x_2$, $x_1 = x_2$. Hence, f is one-to-one. To find the inverse, solve $y = 3x + 2$ for x, obtaining $x = \frac{y-2}{3}$. Thus, $f^{-1}(y) = \frac{y-2}{3}$. (In general, if we can solve $y = f(x)$ for x in terms of y, then we know that f is one-to-one.)

Theorem 10.2 (Differentiation Formula for Inverse Functions): Let f be one-to-one and continuous on an interval (a, b). Then:

(a) The range of f is an interval I (possibly infinite) and f is either increasing or decreasing. Moreover, f^{-1} is continuous on I.

(b) If f is differentiable at x_0 and $f'(x_0) \neq 0$, then f^{-1} is differentiable at $y_0 = f(x_0)$ and $(f^{-1})'(y_0) = \frac{1}{f'(x_0)}$.

The latter equation is sometimes written

$$\frac{dx}{dy} = \frac{1}{\frac{dy}{dx}}$$

where $x = f^{-1}(y)$.

For the proof, see Problem 69.

EXAMPLE 10.8:

(a) Let $y = f(x) = x^2$ for $x > 0$. Then $x = f^{-1}(y) = \sqrt{y}$. Since $\frac{dy}{dx} = 2x$, $\frac{dx}{dy} = \frac{1}{2x} = \frac{1}{2\sqrt{y}}$. Thus, $D_y(\sqrt{y}) = \frac{1}{2\sqrt{y}}$. (Note that this is a special case of Theorem 8.1(9) when $m = \frac{1}{2}$.)

(b) Let $y = f(x) = x^3$ for all x. Then $x = f^{-1}(y) = \sqrt[3]{y} = y^{1/3}$ for all y. Since $\frac{dy}{dx} = 3x^2$, $\frac{dx}{dy} = \frac{1}{3x^2} = \frac{1}{3y^{2/3}}$. This holds for all $y \neq 0$. (Note that $f^{-1}(0) = 0$ and $f'(0) = 3(0)^2 = 0$.)

Higher Derivatives

If $y = f(x)$ is differentiable, its derivative y' is also called the *first derivative* of f. If y' is differentiable, its derivative is called the *second derivative* of f. If this second derivative is differentiable, then its derivative is called the *third derivative* of f, and so on.

Notation

First derivative: $y', \quad f'(x), \quad \frac{dy}{dx}, \quad D_x y$

Second derivative: $y'', \quad f''(x), \quad \frac{d^2y}{dx^2}, \quad D_x^2 y$

Third derivative: $y''', \quad f'''(x), \quad \frac{d^3y}{dx^3}, \quad D_x^3 y$

n^{th} derivative: $y^{(n)}, \quad f^{(n)}, \quad \frac{d^ny}{dx^n}, \quad D_x^n y$

SOLVED PROBLEMS

1. Prove Theorem 10.1, (1)–(3): (1) $\frac{d}{dx}(c) = 0$; (2) $\frac{d}{dx}(x) = 1$; (3) $\frac{d}{dx}(cu) = c\frac{du}{dx}$.

Remember that $\frac{d}{dx}f(x) = \lim_{\Delta x \to 0} \frac{f(x+\Delta x) - f(x)}{\Delta x}$.

(1) $\frac{d}{dx}c = \lim_{\Delta x \to 0} \frac{c - c}{\Delta x} = \lim_{\Delta x \to 0} 0 = 0$

(2) $\frac{d}{dx}(x) = \lim_{\Delta x \to 0} \frac{(x+\Delta x) - x}{\Delta x} = \lim_{\Delta x \to 0} \frac{\Delta x}{\Delta x} = \lim_{\Delta x \to 0} 1 = 1$

(3) $\frac{d}{dx}(cu) = \lim_{\Delta x \to 0} \frac{cu(x+\Delta x) - cu(x)}{\Delta x} = \lim_{\Delta x \to 0} c\frac{u(x+\Delta x) - u(x)}{\Delta x}$

$$= c \lim_{\Delta x \to 0} \frac{u(x+\Delta x) - u(x)}{\Delta x} = c\frac{du}{dx}$$

2. Prove Theorem 10.1, (4), (6), (7):

(4) $\frac{d}{dx}(u + v + \cdots) = \frac{du}{dx} + \frac{dv}{dx} + \cdots$

(6) $\frac{d}{dx}(uv) = u\frac{dv}{dx} + v\frac{du}{dx}$

(7) $\frac{d}{dx}\left(\frac{u}{v}\right) = \frac{v\frac{du}{dx} - u\frac{dv}{dx}}{v^2}$ provided that $v \neq 0$

(4) It suffice to prove this for just two summands, u and v. Let $f(x) = u + v$. Then

$$\frac{f(x+\Delta x) - f(x)}{\Delta x} = \frac{u(x+\Delta x) + v(x+\Delta x) - u(x) - v(x)}{\Delta x}$$

$$= \frac{u(x+\Delta x) - u(x)}{\Delta x} + \frac{v(x+\Delta x) - v(x)}{\Delta x}$$

Taking the limit as $\Delta x \to 0$ yields $\frac{d}{dx}(u + v) = \frac{du}{dx} + \frac{dv}{dx}$.

(6) Let $f(x) = uv$. Then

$$\begin{aligned}\frac{f(x+\Delta x)-f(x)}{\Delta x} &= \frac{u(x+\Delta x)v(x+\Delta x)-u(x)v(x)}{\Delta x}\\ &= \frac{[u(x+\Delta x)v(x+\Delta x)-v(x)u(x+\Delta x)]+[v(x)u(x+\Delta x)-u(x)v(x)]}{\Delta x}\\ &= u(x+\Delta x)\frac{v(x+\Delta x)-v(x)}{\Delta x}+v(x)\frac{u(x+\Delta x)-u(x)}{\Delta x}\end{aligned}$$

Taking the limit as $\Delta x \to 0$ yields

$$\frac{d}{dx}(uv) = u(x)\frac{d}{dx}v(x)+v(x)\frac{d}{dx}u(x) = u\frac{dv}{dx}+v\frac{du}{dx}$$

Note that $\lim_{\Delta x\to 0} u(x+\Delta x) = u(x)$ because the differentiability of u implies its continuity.

(7) Set $f(x) = \frac{u}{v} = \frac{u(x)}{v(x)}$, then

$$\begin{aligned}\frac{f(x+\Delta x)-f(x)}{\Delta x} &= \frac{\frac{u(x+\Delta x)}{v(x+\Delta x)}-\frac{u(x)}{v(x)}}{\Delta x} = \frac{u(x+\Delta x)v(x)-u(x)v(x+\Delta x)}{\Delta x\{v(x)v(x+\Delta x)\}}\\ &= \frac{[u(x+\Delta x)v(x)-u(x)v(x)]-[u(x)v(x+\Delta x)-u(x)v(x)]}{\Delta x[v(x)v(x+\Delta x)]}\\ &= \frac{v(x)\frac{u(x+\Delta x)-u(x)}{\Delta x}-u(x)\frac{v(x+\Delta x)-v(x)}{\Delta x}}{v(x)v(x+\Delta x)}\end{aligned}$$

and for $\Delta x \to 0$, $\frac{d}{dx}f(x) = \frac{d}{dx}\left(\frac{u}{v}\right) = \frac{v(x)\frac{d}{dx}u(x)-u(x)\frac{d}{dx}v(x)}{[v(x)]^2} = \frac{v\frac{du}{dx}-u\frac{dv}{dx}}{v^2}$

3. Prove Theorem 10.1(9): $D_x(x^m) = mx^{m-1}$, when m is a nonnegative integer.
Use mathematical induction. When $m = 0$,

$$D_x(x^m) = D_x(x^0) = D_x(1) = 0 = 0\cdot x^{-1} = mx^{m-1}$$

Assume the formula is true for m. Then, by the Product Rule,

$$\begin{aligned}D_x(x^{m+1}) &= D_x(x^m\cdot x) = x^m D_x(x)+xD_x(x^m) = x^m\cdot 1 + x\cdot mx^{m-1}\\ &= x^m + mx^m = (m+1)x^m\end{aligned}$$

Thus, the formula holds for $m + 1$.

4. Prove Theorem 10.1(9): $D_x(x^m) = mx^{m-1}$, when m is a negative integer.
Let $m = -k$, where k is a positive integer. Then, by the Quotient Rule and Problem 3,

$$\begin{aligned}D_x(x^m) &= D_x(x^{-k}) = D_x\left(\frac{1}{x^k}\right)\\ &= \frac{x^k D_x(1)-1\cdot D_x(x^k)}{(x^k)^2} = \frac{x^k\cdot 0 - kx^{k-1}}{x^{2k}}\\ &= -k\frac{x^{k-1}}{x^{2k}} = -kx^{-k-1} = mx^{m-1}\end{aligned}$$

5. Differentiate $y = 4 + 2x - 3x^2 - 5x^3 - 8x^4 + 9x^5$.

$$\frac{dy}{dx} = 0 + 2(1) - 3(2x) - 5(3x^2) - 8(4x^3) + 9(5x^4) = 2 - 6x - 15x^2 - 32x^3 + 45x^4$$

6. Differentiate $y = \frac{1}{x} + \frac{3}{x^2} + \frac{2}{x^3} = x^{-1} + 3x^{-2} + 2x^{-3}$.

$$\frac{dy}{dx} = -x^{-2} + 3(-2x^{-3}) + 2(-3x^{-4}) = -x^{-2} - 6x^{-3} - 6x^{-4} = -\frac{1}{x^2} - \frac{6}{x^3} - \frac{6}{x^4}$$

7. Differentiate $y = 2x^{1/2} + 6x^{1/3} - 2x^{3/2}$.

$$\frac{dy}{dx} = 2\left(\frac{1}{2}x^{-1/2}\right) + 6\left(\frac{1}{3}x^{-2/3}\right) - 2\left(\frac{3}{2}x^{1/2}\right) = x^{-1/2} + 2x^{-2/3} - 3x^{1/2} = \frac{1}{x^{1/2}} + \frac{2}{x^{2/3}} - 3x^{1/2}$$

8. Differentiate $y = \frac{2}{x^{1/2}} + \frac{6}{x^{1/3}} - \frac{2}{x^{3/2}} - \frac{4}{x^{3/4}} = 2x^{-1/2} + 6x^{-1/3} - 2x^{-3/2} - 4x^{-3/4}$.

$$\frac{dy}{dx} = 2\left(-\frac{1}{2}x^{-3/2}\right) + 6\left(-\frac{1}{3}x^{-4/3}\right) - 2\left(-\frac{3}{2}x^{-5/2}\right) - 4\left(-\frac{3}{4}x^{-7/4}\right)$$

$$= -x^{-3/2} - 2x^{-4/3} + 3x^{-5/2} + 3x^{-7/4} = -\frac{1}{x^{3/2}} - \frac{2}{x^{4/3}} + \frac{3}{x^{5/2}} + \frac{3}{x^{7/4}}$$

9. Differentiate $y = \sqrt[3]{3x^2} - \frac{1}{\sqrt{5x}} = (3x^2)^{1/3} - (5x)^{-1/2}$.

$$\frac{dy}{dx} = \frac{1}{3}(3x^2)^{-2/3}(6x) - \left(-\frac{1}{2}\right)(5x)^{-3/2}(5) = \frac{2x}{(9x^4)^{1/3}} + \frac{5}{2(5x)(5x)^{1/2}} = \frac{2}{\sqrt[3]{9x}} + \frac{1}{2x\sqrt{5x}}$$

10. Prove the Power Chain Rule: $D_x(y^m) = my^{m-1}D_x y$.
This is simply the Chain Rule, where the outer function is $f(x) = x^m$ and the inner function is y.

11. Differentiate $s = (t^2 - 3)^4$.
By the Power Chain Rule, $\frac{ds}{dt} = 4(t^2 - 3)^3(2t) = 8t(t^2 - 3)^3$.

12. Differentiate (a) $z = \frac{3}{(a^2 - y^2)^2} = 3(a^2 - y^2)^{-2}$; (b) $f(x) = \sqrt{x^2 + 6x + 3} = (x^2 + 6x + 3)^{1/2}$.

(a) $\frac{dz}{dy} = 3(-2)(a^2 - y^2)^{-3}\frac{d}{dy}(a^2 - y^2) = 3(-2)(a^2 - y^2)^{-3}(-2y) = \frac{12y}{(a^2 - y^2)^3}$

(b) $f'(x) = \frac{1}{2}(x^2 + 6x + 3)^{-1/2}\frac{d}{dx}(x^2 + 6x + 3) = \frac{1}{2}(x^2 + 6x + 3)^{-1/2}(2x + 6) = \frac{x + 3}{\sqrt{x^2 + 6x + 3}}$

13. Differentiate $y = (x^2 + 4)^2(2x^3 - 1)^3$.
Use the Product Rule and the Power Chain Rule:

$$y' = (x^2 + 4)^2\frac{d}{dx}(2x^3 - 1)^3 + (2x^3 - 1)^3\frac{d}{dx}(x^2 + 4)^2$$

$$= (x^2 + 4)^2(3)(2x^3 - 1)^2\frac{d}{dx}(2x^3 - 1) + (2x^3 - 1)^3(2)(x^2 + 4)\frac{d}{dx}(x^2 + 4)$$

$$= (x^2 + 4)^2(3)(2x^3 - 1)^2(6x^2) + (2x^3 - 1)^3(2)(x^2 + 4)(2x)$$

$$= 2x(x^2 + 4)(2x^3 - 1)^2(13x^3 + 36x - 2)$$

14. Differentiate $y = \frac{3-2x}{3+2x}$.

Use the Quotient Rule:

$$y' = \frac{(3+2x)\frac{d}{dx}(3-2x) - (3-2x)\frac{d}{dx}(3+2x)}{(3+2x)^2} = \frac{(3+2x)(-2)-(3-2x)(2)}{(3+2x)^2} = \frac{-12}{(3+2x)^2}$$

15. Differentiate $y = \frac{x^2}{\sqrt{4-x^2}} = \frac{x^2}{(4-x^2)^{1/2}}$.

$$\frac{dy}{dx} = \frac{(4-x^2)^{1/2}\frac{d}{dx}(x^2) - x^2\frac{d}{dx}(4-x^2)^{1/2}}{4-x^2} = \frac{(4-x^2)^{1/2}(2x)-(x^2)(\frac{1}{2})(4-x^2)^{-1/2}(-2x)}{4-x^2}$$

$$= \frac{(4-x^2)^{1/2}(2x)+x^3(4-x^2)^{-1/2}}{4-x^2}\frac{(4-x^2)^{1/2}}{(4-x^2)^{1/2}} = \frac{2x(4-x^2)+x^3}{(4-x^2)^{3/2}} = \frac{8x-x^3}{(4-x^2)^{3/2}}$$

16. Find $\frac{dy}{dx}$, given $x = y\sqrt{1-y^2}$.

By the Product Rule,

$$\frac{dx}{dy} = y\cdot\frac{1}{2}(1-y^2)^{-1/2}(-2y)+(1-y^2)^{1/2} = \frac{1-2y^2}{\sqrt{1-y^2}}$$

By Theorem 10.2,

$$\frac{dy}{dx} = \frac{1}{dx/dy} = \frac{\sqrt{1-y^2}}{1-2y^2}$$

17. Find the slope of the tangent line to the curve $x = y^2 - 4y$ at the points where the curve crosses the y axis.

The intersection points are (0, 0) and (0, 4). We have $\frac{dx}{dy} = 2y-4$ and so $\frac{dy}{dx} = \frac{1}{dx/dy} = \frac{1}{2y-4}$. At (0, 0) the slope is $-\frac{1}{4}$, and at (0, 4) the slope is $\frac{1}{4}$.

18. Derive the Chain Rule: $D_x(f(g(x)) = f'(g(x))\cdot g'(x))$.

Let $H = f\circ g$. Let $y = g(x)$ and $K = g(x+h)-g(x)$. Also, let $F(t) = \frac{f(y+t)-f(y)}{t} - f'(y)$ for $t \neq 0$. Since $\lim_{t\to 0} F(t) = 0$, let $F(0) = 0$. Then $f(y+t)-f(y) = t(F(t)+f'(y))$ for all t. When $t = K$,

$$f(y+K)-f(y) = K(F(K)+f'(y))$$

$$f(g(x+h))-f(g(x)) = K(F(K)+f'(y))$$

Hence,

$$\frac{H(x+h)-H(x)}{h} = \frac{K}{h}(F(K)+f'(y))$$

Now,

$$\lim_{h\to 0}\frac{K}{h} = \lim_{h\to 0}\frac{g(x+h)-g(x)}{h} = g'(x)$$

Since $\lim_{h\to 0} K = 0$, $\lim_{h\to 0} F(K) = 0$. Hence,

$$H'(x) = f'(y)g'(x) = f'(g(x))g'(x).$$

19. Find $\frac{dy}{dx}$, given $y = \frac{u^2-1}{u^2+1}$ and $u = \sqrt[3]{x^2+2}$.

$$\frac{dy}{du} = \frac{4u}{(u^2+1)^2} \quad \text{and} \quad \frac{du}{dx} = \frac{2x}{3(x^2+2)^{2/3}} = \frac{2x}{3u^2}$$

Then
$$\frac{dy}{dx} = \frac{dy}{du}\frac{du}{dx} = \frac{4u}{(u^2+1)^2}\frac{2x}{3u^2} = \frac{8x}{3u(u^2+1)^2}$$

20. A point moves along the curve $y = x^3 - 3x + 5$ so that $x = \frac{1}{2}\sqrt{t} + 3$, where t is time. At what rate is y changing when $t = 4$?

We must find the value of dy/dt when $t = 4$. First, $dy/dx = 3(x^2 - 1)$ and $dx/dt = 1/(4\sqrt{t})$. Hence,

$$\frac{dy}{dt} = \frac{dy}{dx}\frac{dx}{dt} = \frac{3(x^2-1)}{4\sqrt{t}}$$

When $t = 4$, $x = \frac{1}{2}\sqrt{4} + 3 = 4$, and $\frac{dy}{dt} = \frac{3(16-1)}{4(2)} = \frac{45}{8}$ units per unit of time.

21. A point moves in the plane according to equations $x = t^2 + 2t$ and $y = 2t^3 - 6t$. Find dy/dx when $t = 0, 2$, and 5.

Since the first equation may be solved for t and this result substituted for t in the second equation, y is a function of x. We have $dy/dt = 6t^2 - 6$. Since $dx/dt = 2t + 2$, Theorem 8.2 gives us $dt/dx = 1/(2t + 2)$. Then

$$\frac{dy}{dx} = \frac{dy}{dt}\frac{dt}{dx} = 6(t^2-1)\frac{1}{2(t+1)} = 3(t-1).$$

The required values of dy/dx are -3 at $t = 0$, 3 at $t = 2$, and 12 at $t = 5$.

22. If $y = x^2 - 4x$ and $x = \sqrt{2t^2+1}$, find dy/dt when $t = \sqrt{2}$.

$$\frac{dy}{dx} = 2(x-2) \quad \text{and} \quad \frac{dx}{dt} = \frac{2t}{(2t^2+1)^{1/2}}$$

So
$$\frac{dy}{dt} = \frac{dy}{dx}\frac{dx}{dt} = \frac{4t(x-2)}{(2t^2+1)^{1/2}}$$

When $t = \sqrt{2}$, $x = \sqrt{5}$ and $\frac{dy}{dt} = \frac{4\sqrt{2}(\sqrt{5}-2)}{\sqrt{5}} = \frac{4\sqrt{2}}{5}(5 - 2\sqrt{5})$.

23. Show that the function $f(x) = x^3 + 3x^2 - 8x + 2$ has derivatives of all orders and find them.

$f'(x) = 3x^2 + 6x - 8$, $f''(x) = 6x + 6$, $f'''(x) = 6$, and all derivatives of higher order are zero.

24. Investigate the successive derivatives of $f(x) = x^{4/3}$ at $x = 0$.

$$f'(x) = \tfrac{4}{3}x^{1/3} \quad \text{and} \quad f'(0) = 0$$

$$f''(x) = \tfrac{4}{9}x^{-2/3} = \frac{4}{9x^{2/3}} \quad \text{and} \quad f''(0) \text{ does not exist}$$

$f^{(n)}(0)$ does not exist for $n \geq 2$.

25. If $f(x) = \frac{2}{1-x} = 2(1-x)^{-1}$, find a formula for $f^{(n)}(x)$.

$$f'(x) = 2(-1)(1-x)^{-2}(-1) = 2(1-x)^{-2} = 2(1!)(1-x)^{-2}$$

$$f''(x) = 2(1!)(-2)(1-x)^{-3}(-1) = 2(2!)(1-x)^{-3}$$

$$f'''(x) = 2(2!)(-3)(1-x)^{-4}(-1) = 2(3!)(1-x)^{-4}$$

which suggest $f^{(n)}(x) = 2(n!)(1-x)^{-(n+1)}$. This result may be established by mathematical induction by showing that if $f^{(k)}(x) = 2(k!)(1-x)^{-(k+1)}$, then

$$f^{(k+1)}(x) = -2(k!)(k+1)(1-x)^{-(k+2)}(-1) = 2[(k+1)!](1-x)^{-(k+2)}$$

SUPPLEMENTARY PROBLEMS

26. Prove Theorem 10.1 (5): $D_x(u - v) = D_xu - D_xv$.

Ans. $D_x(u - v) = D_x(u + (-v)) = D_xu + D_x(-v) = D_xu + D_x((-1)v) = D_xu + (-1)D_xv = D_xu - D_xv$ by Theorem 8.1(4, 3)

In Problems 27 to 45, find the derivative.

27. $y = x^5 + 5x^4 - 10x^2 + 6$ *Ans.* $\dfrac{dy}{dx} = 5x(x^3 + 4x^2 - 4)$

28. $y = 3x^{1/2} - x^{3/2} + 2x^{-1/2}$ *Ans.* $\dfrac{dy}{dx} = \dfrac{3}{2\sqrt{x}} - \frac{3}{2}\sqrt{x} - 1/x^{3/2}$

29. $y = \dfrac{1}{2x^2} + \dfrac{4}{\sqrt{x}} = \dfrac{1}{2}x^{-2} + 4x^{-1/2}$ *Ans.* $\dfrac{dy}{dx} = -\dfrac{1}{x^3} - \dfrac{2}{x^{3/2}}$

30. $y = \sqrt{2x} + 2\sqrt{x}$ *Ans.* $y' = (1 + \sqrt{2})/\sqrt{2x}$

31. $f(t) = \dfrac{2}{\sqrt{t}} + \dfrac{6}{\sqrt[3]{t}}$ *Ans.* $f'(t) = -\dfrac{t^{1/2} + 2t^{2/3}}{t^2}$

32. $y = (1 - 5x)^6$ *Ans.* $y' = -30(1 - 5x)^5$

33. $f(x) = (3x - x^3 + 1)^4$ *Ans.* $f'(x) = 12(1 - x^2)(3x - x^3 + 1)^3$

34. $y = (3 + 4x - x^2)^{1/2}$ *Ans.* $y' = (2 - x)/y$

35. $\theta = \dfrac{3r + 2}{2r + 3}$ *Ans.* $\dfrac{d\theta}{dr} = \dfrac{5}{(2r + 3)^2}$

36. $y = \left(\dfrac{x}{1 + x}\right)^5$ *Ans.* $y' = \dfrac{5x^4}{(1 + x)^6}$

37. $y = 2x^2\sqrt{2 - x}$ *Ans.* $y' = \dfrac{x(8 - 5x)}{\sqrt{2 - x}}$

38. $f(x) = x\sqrt{3 - 2x^2}$ *Ans.* $f'(x) = \dfrac{3 - 4x^2}{\sqrt{3 - 2x^2}}$

39. $y = (x - 1)\sqrt{x^2 - 2x + 2}$ *Ans.* $\dfrac{dy}{dx} = \dfrac{2x^2 - 4x + 3}{\sqrt{x^2 - 2x + 2}}$

40. $z = \dfrac{w}{\sqrt{1 - 4w^2}}$ *Ans.* $\dfrac{dz}{dw} = \dfrac{1}{(1 - 4w^2)^{3/2}}$

41. $y = \sqrt{1 + \sqrt{x}}$ *Ans.* $y' = \dfrac{1}{4\sqrt{x}\sqrt{1 + \sqrt{x}}}$

42. $f(x) = \sqrt{\dfrac{x - 1}{x + 1}}$ *Ans.* $f'(x) = \dfrac{1}{(x + 1)\sqrt{x^2 - 1}}$

43. $y = (x^2 + 3)^4(2x^3 - 5)^3$ *Ans.* $y' = 2x(x^2 + 3)^3(2x^3 - 5)^2(17x^3 + 27x - 20)$

44. $s = \dfrac{t^2 + 2}{3 - t^2}$ *Ans.* $\dfrac{ds}{dt} = \dfrac{10t}{(3 - t^2)^2}$

45. $y = \left(\dfrac{x^2 - 1}{2x^3 + 1}\right)^4$ *Ans.* $y' = \dfrac{8x(1 + 3x - x^3)(x^2 - 1)^3}{(2x^3 + 1)^5}$

46. For each of the following, compute dy/dx by two different methods and check that the results are the same: (a) $x = (1 + 2y)^3$ (b) $x = \dfrac{1}{2+y}$.

In Problems 47 to 50, use the Chain Rule to find $\dfrac{dy}{dx}$.

47. $y = \dfrac{u-1}{u+1}$, $u = \sqrt{x}$ *Ans.* $\dfrac{dy}{dx} = \dfrac{1}{\sqrt{x}(1+\sqrt{x})^2}$

48. $y = u^3 + 4$, $u = x^2 + 2x$ *Ans.* $\dfrac{dy}{dx} = 6x^2(x+2)^2(x+1)$

49. $y = \sqrt{1+u}, u = \sqrt{x}$ *Ans.* See Problem 42.

50. $y = \sqrt{u}$, $u = v(3 - 2v)$, $v = x^2$ *Ans.* See Problem 39.

$\left(\textit{Hint: } \dfrac{dy}{dx} = \dfrac{dy}{du}\dfrac{du}{dv}\dfrac{dv}{dx}.\right)$

In Problems 51 to 54, find the indicated derivative.

51. $y = 3x^4 - 2x^2 + x - 5$; y''' *Ans.* $y''' = 72x$

52. $y = \dfrac{1}{\sqrt{x}}$; $y^{(4)}$ *Ans.* $y^{(4)} = \dfrac{105}{16x^{9/2}}$

53. $f(x) = \sqrt{2 - 3x^2}$; $f''(x)$ *Ans.* $f''(x) = -\dfrac{6}{(2-3x^2)^{3/2}}$

54. $y = \dfrac{x}{\sqrt{x-1}}$; y'' $y'' = \dfrac{4-x}{4(x-1)^{5/2}}$

In Problems 55 and 56, find a formula for the *n*th derivative.

55. $y = \dfrac{1}{x^2}$ *Ans.* $y^{(n)} = \dfrac{(-1)^n[(n+1)!]}{x^{n+2}}$

56. $f(x) = \dfrac{1}{3x+2}$ *Ans.* $f^{(n)}(x) = (-1)^n \dfrac{3^n(n!)}{(3x+2)^{n+1}}$

57. If $y = f(u)$ and $u = g(x)$, show that

(a) $\dfrac{d^2y}{dx^2} = \dfrac{dy}{du}\cdot\dfrac{d^2u}{dx^2} + \dfrac{d^2y}{du^2}\left(\dfrac{du}{dx}\right)^2$ (b) $\dfrac{d^3y}{dx^3} = \dfrac{dy}{du}\cdot\dfrac{d^3u}{dx^3} + 3\dfrac{d^2y}{du^2}\cdot\dfrac{d^2u}{dx^2}\cdot\dfrac{du}{dx} + \dfrac{d^3y}{du^3}\left(\dfrac{du}{dx}\right)^3$

58. From $\dfrac{dx}{dy} = \dfrac{1}{y'}$, derive $\dfrac{d^2x}{dy^2} = -\dfrac{y''}{(y')^3}$ and $\dfrac{d^3x}{dy^3} = \dfrac{3(y'')^2 - y'y'''}{(y')^5}$.

In Problems 59 to 64, determine whether the given function has an inverse; if it does, find a formula for the inverse f^{-1} and calculate its derivative.

59. $f(x) = 1/x$ *Ans.* $x = f^{-1}(y) = 1/y$; $dx/dy = -x^2 = -1/y^2$

60. $f(x) = \frac{1}{3}x + 4$ *Ans.* $x = f^{-1}(y) = 3y - 12$; $dx/dy = 3$.

61. $f(x) = \sqrt{x-5}$ *Ans.* $x = f^{-1}(y) = y^2 + 5$; $dx/dy = 2y = 2\sqrt{x-5}$

62. $f(x) = x^2 + 2$ *Ans.* no inverse function

63. $f(x) = x^3$ *Ans.* $x = f^{-1}(y) = \sqrt[3]{y};\ \dfrac{dx}{dy} = \dfrac{1}{3x^2} = \dfrac{1}{3}y^{-2/3}$

64. $f(x) = \dfrac{2x-1}{x+2}$ *Ans.* $x = f^{-1}(y) = -\dfrac{2y+1}{y-2};\ \dfrac{dx}{dy} = \dfrac{5}{(y-2)^2}$

65. Find the points at which the function $f(x) = |x + 2|$ is differentiable.

Ans. All points except $x = -2$

66. (GC) Use a graphing calculator to draw the graph of the parabola $y = x^2 - 2x$ and the curve $y = |x^2 - 2x|$. Find all points of discontinuity of the latter curve.

Ans. $x = 0$ and $x = 2$

67. Find a formula for the *n*th derivative of the following functions: (a) $f(x) = \dfrac{x}{x+2}$; (b) $f(x) = \sqrt{x}$.

Ans. (a) $f^{(n)}(x) = (-1)^{n+1}\dfrac{2n!}{(x+2)^{n+1}}$

(b) $f^{(n)}(x) = (-1)^{n+1}\dfrac{3\cdot 5\cdot 7\cdot \cdots \cdot(2n-3)}{2n}x^{-(2n-1)/2}$

68. Find the second derivatives of the following functions:

(a) $f(x) = 2x - 7$ (b) $f(x) = 3x^2 + 5x - 10$

(c) $f(x) = \dfrac{1}{x+4}$ (d) $f(x) = \sqrt{7-x}$

Ans. (a) 0; (b) 6; (c) $\dfrac{2}{(x+4)^3}$; (d) $-\dfrac{1}{4}\dfrac{1}{(7-x)^{3/2}}$

69. Prove Theorem 10.2.

Ans. *Hints*: (a) Use the intermediate value theorem to show that the range is an interval. That f is increasing or decreasing follows by an argument that uses the extreme value and intermediate value theorems. The continuity of f^{-1} is then derived easily.

(b) $\dfrac{f^{-1}(y) - f^{-1}(y_0)}{y - y_0} = \dfrac{1}{\dfrac{f(f^{-1}(y)) - f(f^{-1}(y_0))}{f^{-1}(y) - f^{-1}(y_0)}} = \dfrac{1}{\dfrac{f(x) - f(x_0)}{x - x_0}}$

By the continuity of f^{-1}, as $y \to y_0$, $x \to x_0$, and we get $(f^{-1})'(y_0) = \dfrac{1}{f'(x_0)}$.

CHAPTER 11

Implicit Differentiation

Implicit Functions

An equation $f(x, y) = 0$ defines y *implicitly* as a function of x. The domain of that implicitly defined function consists of those x for which there is a unique y such that $f(x, y) = 0$.

EXAMPLE 11.1:

(a) The equation $xy + x - 2y - 1 = 0$ can be solved for y, yielding $y = \frac{1-x}{x-2}$. This function is defined for $x \neq 2$.

(b) The equation $4x^2 + 9y^2 - 36 = 0$ does not determine a unique function y. If we solve the equation for y, we obtain $y = \pm\frac{2}{3}\sqrt{9-x^2}$. We shall think of the equation as implicitly defining two functions, $y = \frac{2}{3}\sqrt{9-x^2}$ and $y = -\frac{2}{3}\sqrt{9-x^2}$. Each of these functions is defined for $|x| \leq 3$. The ellipse determined by the original equation is the union of the graphs of the two functions.

If y is a function implicitly defined by an equation $f(x, y) = 0$, the derivative y' can be found in two different ways:

1. Solve the equation for y and calculate y' directly. Except for very simple equations, this method is usually impossible or impractical.
2. Thinking of y as a function of x, differentiate both sides of the original equation $f(x, y) = 0$ and solve the resulting equation for y'. This differentiation process is known as *implicit differentiation.*

EXAMPLE 11.2:

(a) Find y', given $xy + x - 2y - 1 = 0$. By implicit differentiation, $xy' + y\,D_x(x) - 2y' - D_x(1) = D_x(0)$. Thus, $xy' + y - 2y' = 0$. Solve for y': $y' = \frac{1+y}{2-x}$. In this case, Example 11.1(*a*) shows that we can replace y by $\frac{1-x}{x-2}$ and find y' in terms of x alone. We see that it would have been just as easy to differentiate $y = \frac{1-x}{x-2}$ by the Quotient Rule. However, in most cases, we cannot solve for y or for y' in terms of x alone.

(b) Given $4x^2 + 9y^2 - 36 = 0$, find y' when $x = \sqrt{5}$. By implicit differentiation, $4D_x(x^2) + 9D_x(y^2) - D_x(36) = D_x(0)$. Thus, $4(2x) + 9(2yy') = 0$. (Note that $D_x(y^2) = 2yy'$ by the Power Chain Rule.) Solving for y', we get $y' = -4x/9y$. When $x = \sqrt{5}$, $y = \pm\frac{4}{3}$. For the function y corresponding to the upper arc of the ellipse (see Example 11.1(*b*)), $y = -\frac{4}{3}$ and $y' = -\sqrt{5}/3$. For the function y corresponding to the lower arc of the ellipse, $y = -\frac{4}{3}$ and $y' = -\sqrt{5}/3$.

Derivatives of Higher Order

Derivatives of higher order may be obtained by implicit differentiation or by a combination of direct and implicit differentiation.

EXAMPLE 11.3: In Example 11.2(*a*), $y' = \frac{1+y}{2-x}$. Then

$$y'' = D_x(y') = D_x\left(\frac{1+y}{2-x}\right) = \frac{(2-x)y' - (1+y)(-1)}{(2-x)^2}$$

$$= \frac{(2-x)y' + 1 + y}{(2-x)^2} = \frac{(2-x)\left(\frac{1+y}{2-x}\right) + 1 + y}{(2-x)^2} = \frac{2+2y}{(2-x)^2}$$

EXAMPLE 11.4: Find the value of y'' at the point $(-1, 1)$ of the curve $x^2y + 3y - 4 = 0$.

We differentiate implicitly with respect to x twice. First, $x^2y' + 2xy + 3y' = 0$, and then $x^2y'' + 2xy' + 2xy' + 2y + 3y'' = 0$. We could solve the first equation for y'' and then solve the second equation for y''. However, since we only wish to evaluate y'' at the particular point $(-1, 1)$, we substitute $x = -1$, $y = 1$ in the first equation to find $y' = \frac{1}{2}$ and then substitute $x = -1$, $y = 1$, $y' = \frac{1}{2}$ in the second equation to get $y'' - 1 - 1 + 2 + 3y' = 0$, from which we obtain $y'' = 0$. Notice that this method avoids messy algebraic calculations.

SOLVED PROBLEMS

1. Find y', given $x^2y - xy^2 + x^2 + y^2 = 0$.

$$D_x(x^2y) - D_x(xy^2) + D_x(x^2) + D_x(y^2) = 0$$

$$x^2y' + yD_x(x^2) - xD_x(y^2) - y^2D_x(x) + 2x + 2yy' = 0$$

$$x^2y' + 2xy - x(2yy') - y^2 + 2x + 2yy' = 0$$

$$(x^2 - 2xy + 2y)y' + 2xy - y^2 + 2x = 0$$

$$y' = \frac{y^2 - 2xy - 2x}{x^2 - 2xy + 2y}$$

2. If $x^2 - xy + y^2 = 3$, find y' and y''.

$$D_x(x^2) - D_x(xy) + D_x(y^2) = 0$$

$$2x - xy' - y + 2yy' = 0$$

Hence, $y' = \dfrac{2x - y}{x - 2y}$. Then,

$$y'' = \frac{(x-2y)D_x(2x-y) - (2x-y)D_x(x-2y)}{(x-2y)^2}$$

$$= \frac{(x-2y)(2-y') - (2x-y)(1-2y')}{(x-2y)^2}$$

$$= \frac{2x - xy' - 4y + 2yy' - 2x + 4xy' + y - 2yy'}{(x-2y)^2} = \frac{3xy' - 3y}{(x-2y)^2}$$

$$= \frac{3x\left(\dfrac{2x-y}{x-2y}\right) - 3y}{(x-2y)^2} = \frac{3x(2x-y) - 3y(x-2y)}{(x-2y)^3} = \frac{6(x^2 - xy + y^2)}{(x-2y)^3}$$

$$= \frac{18}{(x-2y)^3}$$

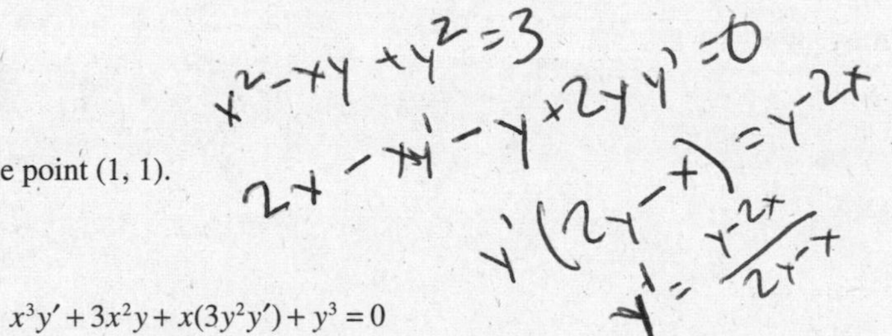

3. Given $x^3y + xy^3 = 2$, find y' and y'' at the point $(1, 1)$.

By implicit differentiation twice,

$$x^3y' + 3x^2y + x(3y^2y') + y^3 = 0$$

and
$$x^3y'' + 3x^2y' + 3x^2y' + 6xy + 3xy^2y'' + y'[6xyy' + 3y^2] + 3y^2y' = 0$$

Substituting $x = 1$, $y = 1$ in the first equation yields $y' = -1$. Then substituting $x = 1$, $y = 1$, $y' = -1$ in the second equation yields $y'' = 0$.

SUPPLEMENTARY PROBLEMS

4. Find y'', given: (a) $x + xy + y = 2$; (b) $x^3 - 3xy + y^3 = 1$.

Ans. (a) $y'' = \dfrac{2(1+y)}{(1+x)^2}$; (b) $y'' = -\dfrac{4xy}{(y^2-x)^3}$

5. Find y', y'', and y''' at: (a) the point (2, 1) on $x^2 - y^2 - x = 1$; (b) the point (1, 1) on $x^3 + 3x^2y - 6xy^2 + 2y^3 = 0$.

Ans. (a) $\frac{3}{2}, -\frac{5}{4}, \frac{45}{8}$; (b) 1, 0, 0

6. Find the slope of the tangent line at a point (x_0, y_0) of: (a) $b^2x^2 + a^2y^2 = a^2b^2$; (b) $b^2x^2 - a^2y^2 = a^2b^2$; (c) $x^3 + y^3 - 6x^2y = 0$.

Ans. (a) $-\dfrac{b^2x_0}{a^2y_0}$; (b) $\dfrac{b^2x_0}{a^2y_0}$; (c) $\dfrac{4x_0y_0 - x_0^2}{y_0^2 - 2x_0^2}$

7. Prove that the lines tangent to the curves $5y - 2x + y^3 - x^2y = 0$ and $2y + 5x + x^4 - x^3y^2 = 0$ at the origin intersect at right angles.

8. (a) The total surface area of a closed rectangular box whose base is a square with side y and whose height is x is given by $S = 2y^2 + 4xy$. If S is constant, find dy/dx without solving for y.

(b) The total surface area of a right circular cylinder of radius r and height h is given by $S = 2\pi r^2 + 2\pi rh$. If S is constant, find dr/dh.

Ans. (a) $-\dfrac{y}{x+y}$; (b) $-\dfrac{r}{2r+h}$

9. For the circle $x^2 + y^2 = r^2$, show that $\left|\dfrac{y''}{[1+(y')^2]^{3/2}}\right| = \dfrac{1}{r}$.

10. Given $S = \pi x(x + 2y)$ and $V = \pi x^2y$, show that $dS/dx = 2\pi(x - y)$ when V is a constant, and $dV/dx = -\pi x(x - y)$ when S is a constant.

11. Derive the formula $D_x(x^m) = mx^{m-1}$ of Theorem 10.1(9) when $m = p/q$, where p and q are nonzero integers. You may assume that $x^{p/q}$ is differentiable. (*Hint*: Let $y = x^{p/q}$. Then $y^q = x^p$. Now use implicit differentiation.)

12. (GC) Use implicit differentation to find an equation of the tangent line to $\sqrt{x} + \sqrt{y} = 4$ at (4, 4), and verify your answer on a graphing calculator.

Ans. $y = -x + 8$

CHAPTER 12

Tangent and Normal Lines

An example of a graph of a continuous function f is shown in Fig. 12-1(*a*). If P is a point of the graph having abscissa x, then the coordinates of P are $(x, f(x))$. Let Q be a nearby point having abscissa $x + \Delta x$. Then the coordinates of Q are $(x + \Delta x, f(x + \Delta x))$. The line PQ has slope $\dfrac{f(x+\Delta x)-f(x)}{\Delta x}$. As Q approaches P along the graph, the lines PQ get closer and closer to the tangent line $\mathscr{T}$ to the graph at P. (See Fig. 12-1 (*b*).) Hence, the slope of PQ approaches the slope of the tangent line. Thus, the slope of the tangent line is $\lim\limits_{\Delta x \to 0} \dfrac{f(x+\Delta x)-f(x)}{\Delta x}$, which is the derivative $f'(x)$.

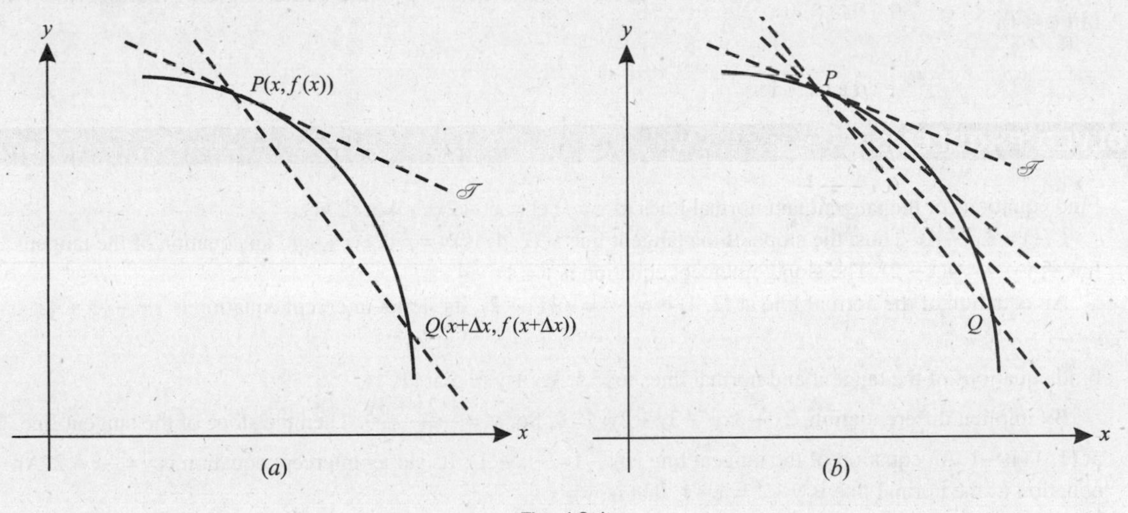

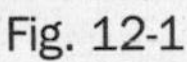
Fig. 12-1

If the slope m of the tangent line at a point of the curve $y = f(x)$ is zero, then the curve has a horizontal tangent line at that point, as at points A, C, and E of Fig. 12-2. In general, if the derivative of f is m at a point (x_0, y_0), then the point–slope equation of the tangent line is $y - y_0 = m(x - x_0)$. If f is continuous at x_0, but $\lim\limits_{x \to x_0} f'(x) = \infty$, then the curve has a vertical tangent line at x_0, as at points B and D of Fig. 12-2.

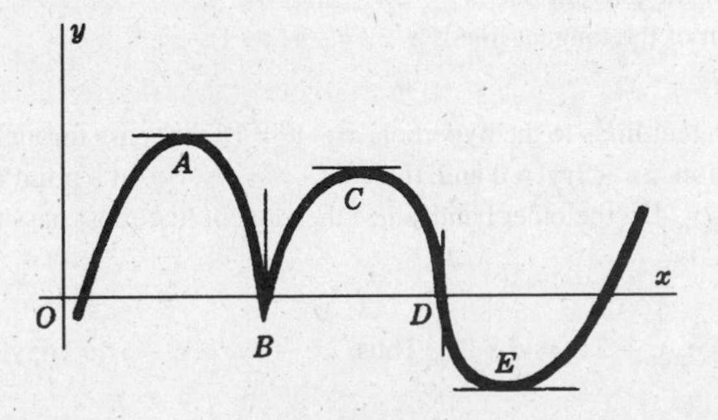

Fig. 12-2

The *normal line* to a curve at one of its points (x_0, y_0) is the line that passes through the point and is perpendicular to the tangent line at that point. Recall that a perpendicular to a line with nonzero slope m has slope $-1/m$. Hence, if $m \neq 0$ is the slope of the tangent line, then $y - y_0 = -(1/m)(x - x_0)$ is a point–slope equation of the normal line. If the tangent line is horizontal, then the normal line is vertical and has equation $x = x_0$. If the tangent line is vertical, then the normal line is horizontal and has equation $y = y_0$.

The Angles of Intersection

The angles of intersection of two curves are defined as the angles between the tangent lines to the curves at their point of intersection.

To determine the angles of intersection of the two curves:

1. Solve the equations of the curves simultaneously to find the points of intersection.
2. Find the slopes m_1 and m_2 of the tangent lines to the two curves at each point of intersection.
3. If $m_1 = m_2$, the angle of intersection is 0°, and if $m_1 = -1/m_2$, the angle of intersection is 90°; otherwise, the angle of intersection ϕ can be found from the formula

$$\tan \phi = \frac{m_1 - m_2}{1 + m_1 m_2}$$

ϕ is the acute angle of intersection when $\tan \phi > 0$, and $180° - \phi$ is the acute angle of intersection when $\tan \phi < 0$.

SOLVED PROBLEMS

1. Find equations of the tangent and normal lines to $y = f(x) = x^3 - 2x^2 + 4$ at (2, 4).

$f'(x) = 3x^2 - 4x$. Thus, the slope of the tangent line at (2, 4) is $m = f'(2) = 4$, and an equation of the tangent line is $y - 4 = 4(x - 2)$. The slope–intercept equation is $y = 4x - 4$.

An equation of the normal line at (2, 4) is $y - 4 = -\frac{1}{4}(x - 2)$. Its slope–intercept equation is $y = -\frac{1}{4}x + \frac{9}{2}$.

2. Find equations of the tangent and normal lines to $x^2 + 3xy + y^2 = 5$ at (1, 1).

By implicit differentiation, $2x + 3xy' + 3y + 2yy' = 0$. So, $y' = -\frac{2x + 3y}{3x + 2y}$. Then the slope of the tangent line at (1, 1) is -1. An equation of the tangent line is $y - 1 = -(x - 1)$. Its slope–intercept equation is $y = -x + 2$. An equation of the normal line is $y - 1 = x - 1$, that is, $y = x$.

3. Find the equations of the tangent lines with slope $m = -\frac{2}{9}$ to the ellipse $4x^2 + 9y^2 = 40$.

By implicit differentiation, $y' = -4x/9y$. So, at a point of tangency (x_0, y_0), $m = -4x_0/9y_0 = -\frac{2}{9}$. Then $y_0 = 2x_0$. Since the point is on the ellipse, $4x_0^2 + 9y_0^2 = 40$. So, $4x_0^2 + 9(2x_0)^2 = 40$. Therefore, $x_0^2 = 1$, and $x_0 = \pm 1$. The required points are (1, 2) and (−1, −2).

At (1, 2), an equation of the tangent line is $y - 2 = -\frac{2}{9}(x - 1)$.

At (−1, −2), an equation of the tangent line is $y + 2 = -\frac{2}{9}(x + 1)$.

4. Find an equation of the tangent lines to the hyperbola $x^2 - y^2 = 16$ that pass through the point (2, −2).

By implicit differentiation, $2x - 2yy' = 0$ and, therefore, $y' = x/y$. So, at a point of tangency (x_0, y_0), the slope of the tangent line must be x_0/y_0. On the other hand, since the tangent line must pass through (x_0, y_0) and (2, −2), the slope is $\frac{y_0 + 2}{x_0 - 2}$.

Thus, $\frac{x_0}{y_0} = \frac{y_0 + 2}{x_0 - 2}$. Hence, $x_0^2 - 2x_0 = y_0^2 + 2y_0$. Thus, $2x_0 + 2y_0 = x_0^2 - y_0^2 = 16$, yielding $x_0 + y_0 = 8$, and, therefore, $y_0 = 8 - x_0$.

If we substitute $8 - x_0$ for y_0 in $x_0^2 - y_0^2 = 16$ and solve for x_0, we get $x_0 = 5$. Then $y_0 = 3$. Hence, an equation of the tangent line is $y - 3 = \frac{5}{3}(x - 5)$.

5. Find the points of tangency of horizontal and vertical tangent lines to the curve $x^2 - xy + y^2 = 27$.

By implicit differentiation, $2x - xy' - y + 2yy' = 0$, whence $y' = \frac{y-2x}{2y-x}$.

For horizontal tangent lines, the slope must be zero. So, the numerator $y - 2x$ of y' must be zero, yielding $y = 2x$. Substituting $2x$ for y in the equation of the curve, we get $x^2 = 9$. Hence, the points of tangency are (3, 6) and (−3, −6).

For vertical tangent lines, the slope must be infinite. So, the denominator $2y - x$ of y' must be zero, yielding $x = 2y$. Replacing x in the equation of the curve, we get $y^2 = 9$. Hence, the points of tangency are (6, 3) and (−6, −3).

6. Find equations of the vertical lines that meet the curves (a) $y = x^3 + 2x^2 - 4x + 5$ and (b) $3y = 2x^3 + 9x^2 - 3x - 3$ in points at which the tangent lines to the two curves are parallel.

Let $x = x_0$ be such a line. The tangent lines at x_0 have slopes:
For (a): $y' = 3x^2 + 4x - 4$; at x_0, $m_1 = 3x_0^2 + 4x_0 - 4$
For (b): $3y' = 6x^2 + 18x - 3$; at x_0, $m_2 = 2x_0^2 + 6x_0 - 1$

Since $m_1 = m_2$, $3x_0^2 + 4x_0 - 4 = 2x_0^2 + 6x_0 - 1$. Then $x_0^2 - 2x_0 - 3 = 0$, $(x_0 - 3)(x_0 + 1) = 0$. Hence, $x_0 = 3$ or $x_0 = -1$. Thus, the vertical lines are $x = 3$ and $x = -1$.

7. (a) Show that the slope–intercept equation of the tangent line of slope $m \neq 0$ to the parabola $y^2 = 4px$ is $y = mx + p/m$.
(b) Show that an equation of the tangent line to the ellipse $b^2x^2 + a^2y^2 = a^2b^2$ at the point $P_0(x_0, y_0)$ on the ellipse is $b^2x_0x + a^2y_0y = a^2b^2$.

(a) $y' = 2p/y$. Let $P_0(x_0, y_0)$ be the point of tangency. Then $y_0^2 = 4px_0$ and $m = 2p/y_0$. Hence, $y_0 = 2p/m$ and $x_0 = \frac{1}{4}y_0^2/p = p/m^2$. The equation of the tangent line is then $y - 2p/m = m(x - p/m^2)$, which reduces to $y = mx + p/m$.

(b) $y' = -\frac{b^2x}{a^2y}$. At P_0, $m = -\frac{b^2x_0}{a^2y_0}$. An equation of the tangent line is $y - y_0 = -\frac{b^2x_0}{a^2y_0}(x - x_0)$, which reduces to $b^2x_0x + a^2y_0y = b^2x_0^2 + a^2y_0^2 = a^2b^2$ (since (x_0, y_0) satisfies the equation of the ellipse).

8. Show that at a point $P_0(x_0, y_0)$ on the hyperbola $b^2x^2 - a^2y^2 = a^2b^2$, the tangent line bisects the angle included between the focal radii of P_0.

At P_0 the slope of the tangent to the hyperbola is b^2x_0/a^2y_0 and the slopes of the focal radii P_0F' and P_0F (see Fig. 12-3) are $y_0/(x_0 + c)$ and $y_0/(x_0 - c)$, respectively. Now

$$\tan\alpha = \frac{\frac{b^2x_0}{a^2y_0} - \frac{y_0}{x_0+c}}{1 + \frac{b^2x_0}{a^2y_0}\cdot\frac{y_0}{x_0+c}} = \frac{(b^2x_0^2 - a^2y_0^2) + b^2cx_0}{(a^2+b^2)x_0y_0 + a^2cy_0} = \frac{a^2b^2 + b^2cx_0}{c^2x_0y_0 + a^2cy_0} = \frac{b^2(a^2 + cx_0)}{cy_0(a^2 + cx_0)} = \frac{b^2}{cy_0}$$

since $b^2x_0^2 - a^2y_0^2 = a^2b^2$ and $a^2 + b^2 = c^2$, and

$$\tan\beta = \frac{\frac{y_0}{x_{0-c}} - \frac{b^2x_0}{a^2y_0}}{1 + \frac{b^2x_0}{a^2x_0}\cdot\frac{y_0}{x_0+c}} = \frac{b^2cx_0 - (b^2x_0^2 - a^2y_0^2)}{(a^2+b^2)x_0y_0 - a^2cy_0} = \frac{b^2cx_0 - a^2b^2}{c^2x_0y_0 - a^2cy_0} = \frac{b^2}{cy_0}$$

Hence, $\alpha = \beta$ because $\tan\alpha = \tan\beta$.

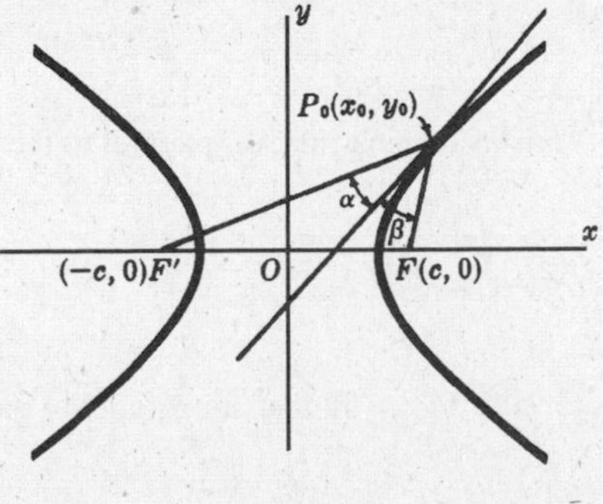

Fig. 12-3

9. One of the points of intersection of the curves (a) $y^2 = 4x$ and (b) $2x^2 = 12 - 5y$ is (1, 2). Find the acute angle of intersection of the curves at that point.

For (a), $y' = 2/y$. For (b), $y' = -4x/5$. Hence, at (1, 2), $m_1 = 1$ and $m_2 = -\frac{4}{5}$. So,

$$\tan\phi = \frac{m_1 - m_2}{1 + m_1 m_2} = \frac{1 + \frac{4}{5}}{1 - \frac{4}{5}} = 9$$

Then $\phi \approx 83° \, 40'$ is the acute angle of intersection.

10. Find the angles of intersection of the curves (a) $2x^2 + y^2 = 20$ and (b) $4y^2 - x^2 = 8$.

Solving simultaneously, we obtain $y^2 = 4$, $y = \pm 2$. Then the points of intersection are $(\pm 2\sqrt{2}, \, 2)$ and $(\pm 2\sqrt{2}, \, -2)$. For (a), $y' = -2x/y$, and for (b), $y' = x/4y$. At the point $(2\sqrt{2}, \, 2)$, $m_1 = -2\sqrt{2}$ and $m_2 = \frac{1}{4}\sqrt{2}$. Since $m_1 m_2 = -1$, the angle of intersection is 90° (that is, the curves are *orthogonal*). By symmetry, the curves are orthogonal at each of their points of intersection.

11. A cable of a certain suspension bridge is attached to supporting pillars 250 ft apart. If it hangs in the form of a parabola with the lowest point 50 ft below the point of suspension, find the angle between the cable and the pillar.

Take the origin at the vertex of the parabola, as in Fig. 12-4. The equation of the parabola is $y = \frac{2}{625}x^2$ and $y' = 4x/625$.

At (125, 50), $m = 4(125)/625 = 0.8000$ and $\theta = 38°40'$. Hence, the required angle is $\phi = 90° - \theta = 51° \, 20'$.

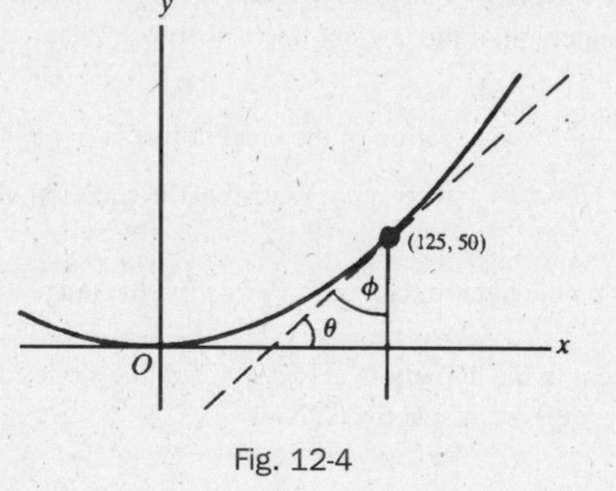

Fig. 12-4

SUPPLEMENTARY PROBLEMS

12. Examine $x^2 + 4xy + 16y^2 = 27$ for horizontal and vertical tangent lines.

Ans. Horizontal tangents at $(3, \, -\frac{3}{2})$ and $(-3, \, \frac{3}{2})$. Vertical tangents at $(6, \, -\frac{3}{4})$ and $(-6, \, -\frac{3}{4})$.

13. Find equations of the tangent and normal lines to $x^2 - y^2 = 7$ at the point (4, −3).

Ans. $4x + 3y = 7$ and $3x - 4y = 24$

14. At what points on the curve $y = x^3 + 5$ is its tangent line: (a) parallel to the line $12x - y = 17$; (b) perpendicular to the line $x + 3y = 2$?

Ans. (a) (2, 13), (−2, −3); (b) (1, 6), (−1, 4)

15. Find equations of the tangent lines to $9x^2 + 16y^2 = 52$ that are parallel to the line $9x - 8y = 1$.

Ans. $9x - 8y = \pm 26$

16. Find equations of the tangent lines to the hyperbola $xy = 1$ that pass through the point $(-1, 1)$.

Ans. $y = (2\sqrt{2} - 3)x + 2\sqrt{2} - 2$; $y = -(2\sqrt{2} + 3)x - 2\sqrt{2} - 2$

17. For the parabola $y^2 = 4px$, show that an equation of the tangent line at one of its points $P(x_0, y_0)$ is $y_0 y = 2p(x + x_0)$.

18. For the ellipse $b^2x^2 + a^2y^2 = a^2b^2$, show that the equations of its tangent lines of slope m are $y = mx \pm \sqrt{a^2m^2 + b^2}$

19. For the hyperbola $b^2x^2 - a^2y^2 = a^2b^2$, show that (a) an equation of the tangent line at one of its points $P(x_0, y_0)$ is $b^2x_0x - a^2y_0y = a^2b^2$; and (b) the equations of its tangent lines of slope m are $y = mx \pm \sqrt{a^2m^2 - b^2}$.

20. Show that the normal line to a parabola at one of its points P bisects the angle included between the focal radius of P and the line through P parallel to the axis of the parabola.

21. Prove: Any tangent line to a parabola, except at the vertex, intersects the directrix and the latus rectum (produced if necessary) in points equidistant from the focus.

22. Prove: The chord joining the points of contact of the tangent lines to a parabola from any point on its directrix passes through the focus.

23. Prove: The normal line to an ellipse at any of its points P bisects the angle included between the focal radii of P.

24. Prove: (a) The sum of the intercepts on the coordinate axes of any tangent line to $\sqrt{x} + \sqrt{y} = \sqrt{a}$ is a constant. (b) The sum of the squares of the intercepts on the coordinate axes of any tangent line to $x^{2/3} + y^{2/3} = a^{2/3}$ is a constant.

25. Find the acute angles of intersection of the circles $x^2 - 4x + y^2 = 0$ and $x^2 + y^2 = 8$.

Ans. $45°$

26. Show that the curves $y = x^3 + 2$ and $y = 2x^2 + 2$ have a common tangent line at the point $(0, 2)$ and intersect at the point $(2, 10)$ at an angle ϕ such that $\tan\phi = \frac{4}{97}$.

27. Show that the ellipse $4x^2 + 9y^2 = 45$ and the hyperbola $x^2 - 4y^2 = 5$ are orthogonal (that is, intersect at a right angle).

28. Find equations of the tangent and normal lines to the parabola $y = 4x^2$ at the point $(-1, 4)$.

Ans. $y + 8x + 4 = 0$; $8y - x - 33 = 0$

29. At what points on the curve $y = 2x^3 + 13x^2 + 5x + 9$ does its tangent line pass through the origin?

Ans. $x = -3, \ -1, \ \frac{3}{4}$

CHAPTER 13

Law of the Mean. Increasing and Decreasing Functions

Relative Maximum and Minimum

A function f is said to have a *relative maximum* at x_0 if $f(x_0) \geq f(x)$ for all x in some open interval containing x_0 (and for which $f(x)$ is defined). In other words, the value of f at x_0 is greater than or equal to all values of f at nearby points. Similarly, f is said to have a *relative minimum* at x_0 if $f(x_0) \leq f(x)$ for all x in some open interval containing x_0 (and for which $f(x)$ is defined). In other words, the value of f at x_0 is less than or equal to all values of f at nearby points. By a *relative extremum* of f we mean either a relative maximum or a relative minimum of f.

Theorem 13.1: If f has a relative extremum at a point x_0 at which $f'(x_0)$ is defined, then $f'(x_0) = 0$.

Thus, if f is differentiable at a point at which it has a relative extremum, then the graph of f has a horizontal tangent line at that point. In Fig. 13-1, there are horizontal tangent lines at the points A and B where f attains a relative maximum value and a relative minimum value, respectively. See Problem 5 for a proof of Theorem 13.1.

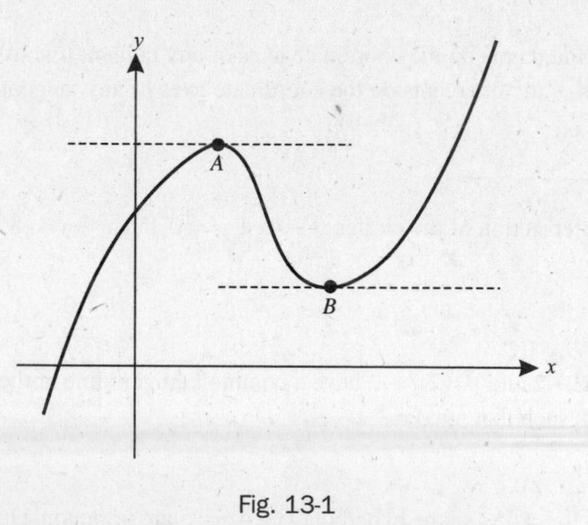

Fig. 13-1

Theorem 13.2 (Rolle's Theorem): Let f be continuous on the closed interval $[a, b]$ and differentiable on the open interval (a, b). Assume that $f(a) = f(b) = 0$. Then $f'(x_0) = 0$ for at least one point x_0 in (a, b).

This means that, if the graph of a continuous function intersects the x axis at $x = a$ and $x = b$, and the function is differentiable between a and b, then there is at least one point on the graph between a and b where the tangent line is horizontal. See Fig. 13-2, where there is one such point. For a proof of Rolle's Theorem, see Problem 6.

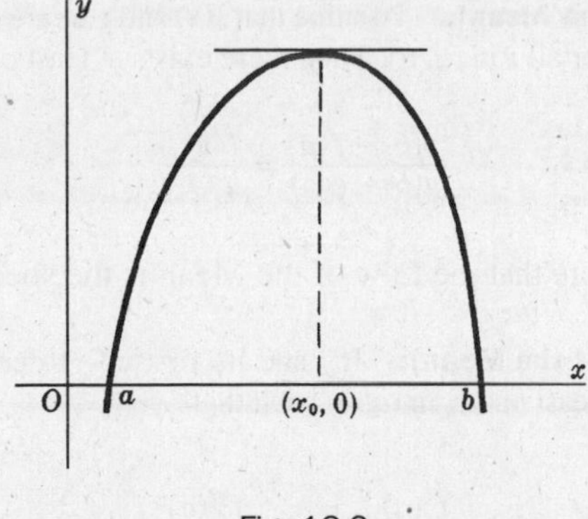

Fig. 13-2

Corollary 13.3 (Generalized Rolle's Theorem): Let g be continuous on the closed interval $[a, b]$ and differentiable on the open interval (a, b). Assume that $g(a) = g(b)$. Then $g'(x_0) = 0$ for at least one point x_0 in (a, b).

See Fig. 13-3 for an example in which there is exactly one such point. Note that Corollary 13.3 follows from Rolle's Theorem if we let $f(x) = g(x) - g(a)$.

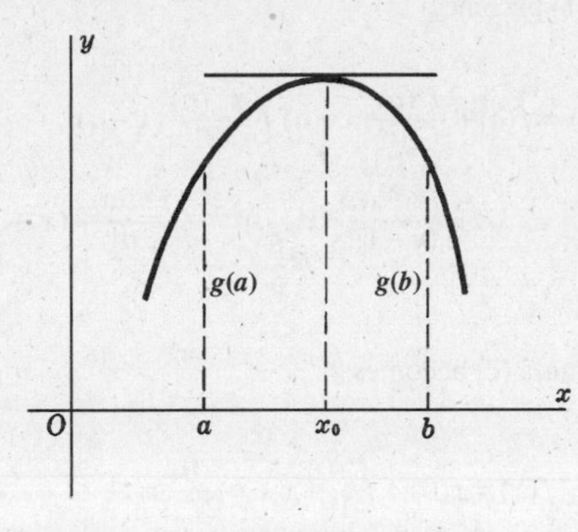

Fig. 13-3

Theorem 13.4 (Law of the Mean)†: Let f be continuous on the closed interval $[a, b]$ and differentiable on the open interval (a, b). Then there is at least one point x_0 in (a, b) for which

$$\frac{f(b) - f(a)}{b - a} = f'(x_0)$$

See Fig. 13-4. For a proof, see Problem 7. Geometrically speaking, the conclusion says that there is some point inside the interval where the slope $f'(x_0)$ of the tangent line is equal to the slope $(f(b) - f(a))/(b - a)$ of the line P_1P_2 connecting the points $(a, f(a))$ and $(b, f(b))$ of the graph. At such a point, the tangent line is parallel to P_1P_2, since their slopes are equal.

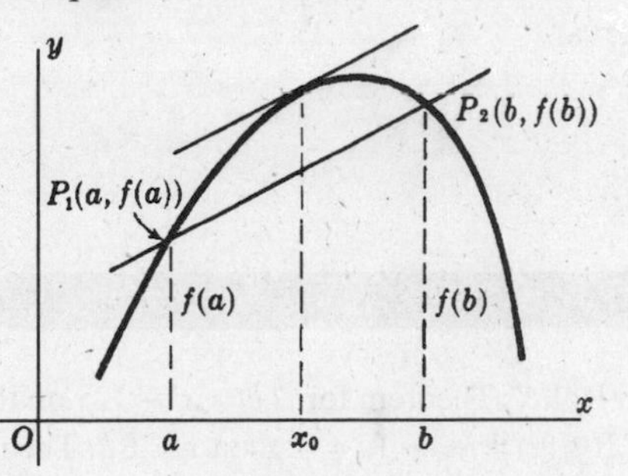

Fig. 13-4

† The Law of the Mean is also called the Mean-Value Theorem for Derivatives.

Theorem 13.5 (Extended Law of the Mean): Assume that $f(x)$ and $g(x)$ are continuous on $[a, b]$, and differentiable on (a, b). Assume also that $g'(x) \neq 0$ for all x in (a, b). Then there exists at least one point x_0 in (a, b) for which

$$\frac{f(b)-f(a)}{g(b)-g(a)} = \frac{f'(x_0)}{g'(x_0)}$$

For a proof, see Problem 13. Note that the Law of the Mean is the special case when $g(x) = x$.

Theorem 13.6 (Higher-Order Law of the Mean): If f and its first $n-1$ derivatives are continuous on $[a, b]$ and $f^{(n)}(x)$ exists on (a, b), then there is at least one x_0 in (a, b) such that

$$f(b) = f(a) + \frac{f'(a)}{1!}(b-a) + \frac{f''(a)}{2!}(b-a)^2 + \cdots + \frac{f^{(n-1)}(a)}{(n-1)!}(b-a)^{n-1} + \frac{f^{(n)}(x_0)}{n!}(b-a)^n \tag{1}$$

(For a proof, see Problem 14.)

When b is replaced by x, formula (1) becomes

$$f(x) = f(a) + \frac{f'(a)}{1!}(x-a) + \frac{f''(a)}{2!}(x-a)^2 + \cdots + \frac{f^{(n-1)}(a)}{(n-1)!}(x-a)^{n-1} + \frac{f^{(n)}(x_0)}{n!}(x-a)^n \tag{2}$$

for some x_0 between a and x.

In the special case when $a = 0$, formula (2) becomes

$$f(x) = f(0) + \frac{f'(0)}{1!}x + \frac{f''(0)}{2!}x^2 + \cdots + \frac{f^{(n-1)}(0)}{(n-1)!}x^{n-1} + \frac{f^{(n)}(x_0)}{n!}x^n \tag{3}$$

for some x_0 between 0 and x.

Increasing and Decreasing Functions

A function f is said to be *increasing* on an interval if $u < v$ implies $f(u) < f(v)$ for all u and v in the interval. Similarly, f is said to be *decreasing* on an interval if $u < v$ implies $f(u) > f(v)$ for all u and v in the interval.

Theorem 13.7: (a) If f' is positive on an interval, then f is increasing on that interval. (b) If f' is negative on an interval, then f is decreasing on that interval.

For a proof, see Problem 9.

SOLVED PROBLEMS

1. Find the value of x_0 prescribed in Rolle's Theorem for $f(x) = x^3 - 12x$ on the interval $0 \leq x \leq 2\sqrt{3}$.

Note that $f(0) = f(2\sqrt{3}) = 0$. If $f'(x) = 3x^2 - 12 = 0$, then $x = \pm 2$. Then $x_0 = 2$ is the prescribed value.

2. Does Rolle's Theorem apply to the functions (a) $f(x) = \dfrac{x^2-4x}{x-2}$, and (b) $f(x) = \dfrac{x^2-4x}{x+2}$ on the interval (0, 4)?

(a) $f(x) = 0$ when $x = 0$ or $x = 4$. Since f has a discontinuity at $x = 2$, a point on $[0, 4]$, the theorem does not apply.

(b) $f(x) = 0$ when $x = 0$ or $x = 4$. f has a discontinuity at $x = -2$, a point not on $[0, 4]$. In addition, $f'(x) = (x^2 + 4x - 8)/(x + 2)^2$ exists everywhere except at $x = -2$. So, the theorem applies and $x_0 = 2(\sqrt{3} - 1)$, the positive root of $x^2 + 4x - 8 = 0$.

3. Find the value of x_0 prescribed by the law of the mean when $f(x) = 3x^2 + 4x - 3$ and $a = 1$, $b = 3$.
$f(a) = f(1) = 4$, $f(b) = f(3) = 36$, $f'(x_0) = 6x_0 + 4$, and $b - a = 2$. So, $6x_0 + 4 = \frac{36-4}{2} = 16$. Then $x_0 = 2$.

4. Find a value x_0 prescribed by the extended law of the mean when $f(x) = 3x + 2$ and $g(x) = x^2 + 1$, on $[1, 4]$.
We have to find x_0 so that

$$\frac{f(b)-f(a)}{g(b)-g(a)} = \frac{f(4)-f(1)}{g(4)-g(1)} = \frac{14-5}{17-2} = \frac{3}{5} = \frac{f'(x_0)}{g'(x_0)} = \frac{3}{2x_0}$$

Then $x_0 = \frac{5}{2}$.

5. Prove Theorem 13.1: If f has a relative extremum at a point x_0 at which $f'(x_0)$ is defined, then $f'(x_0) = 0$.

Consider the case of a relative maximum. Since f has a relative maximum at x_0, then, for sufficiently small $|\Delta x|$, $f(x_0 + \Delta x) < f(x_0)$, and so $f(x_0 + \Delta x) - f(x_0) < 0$. Thus, when $\Delta x < 0$, $\frac{f(x_0+\Delta x)-f(x_0)}{\Delta x} > 0$. So,

$$\begin{aligned} f'(x_0) &= \lim_{\Delta x \to 0} \frac{f(x_0+\Delta x)-f(x_0)}{\Delta x} \\ &= \lim_{\Delta x \to 0^-} \frac{f(x_0+\Delta x)-f(x_0)}{\Delta x} \geq 0 \end{aligned}$$

When $\Delta x > 0$, $\frac{f(x_0+\Delta x)-f(x_0)}{\Delta x} < 0$. Hence,

$$\begin{aligned} f'(x_0) &= \lim_{\Delta x \to 0} \frac{f(x_0+\Delta x)-f(x_0)}{\Delta x}. \\ &= \lim_{\Delta x \to 0^+} \frac{f(x_0+\Delta x)-f(x_0)}{\Delta x} \leq 0 \end{aligned}$$

Since $f'(x_0) \geq 0$ and $f'(x_0) \leq 0$, it follows that $f'(x_0) = 0$.

6. Prove Rolle's Theorem (Theorem 13.2): If f is continuous on the closed interval $[a, b]$ and differentiable on the open interval (a, b), and if $f(a) = f(b) = 0$, then $f'(x_0) = 0$ for some point x_0 in (a, b).
If $f(x) = 0$ throughout $[a, b]$, then $f'(x) = 0$ for all x in (a, b). On the other hand, if $f(x)$ is positive (negative) somewhere in (a, b), then, by the Extreme Value Theorem (Theorem 8.7), f has a maximum (minimum) value at some point x_0 on $[a, b]$. That maximum (minimum) value must be positive (negative), and, therefore, x_0 lies on (a, b), since $f(a) = f(b) = 0$. Hence, f has a relative maximum (minimum) at x_0. By Theorem 13.1, $f'(x_0) = 0$.

7. Prove the Law of the Mean (Theorem 13.4): Let f be continuous on the closed interval $[a, b]$ and differentiable on the open interval (a, b). Then there is at least one point x_0 in (a, b) for which $(f(b) - f(a))/(b - a) = f'(x_0)$.

Let $F(x) = f(x) - f(a) - \frac{f(b)-f(a)}{b-a}(x-a)$.

Then $F(a) = 0 = F(b)$. So, Rolle's Theorem applies to F on $[a, b]$. Hence, for some x_0 in (a, b), $F'(x_0) = 0$.

But $F'(x) = f'(x) - \frac{f(b)-f(a)}{b-a}$. Thus, $f'(x_0) - \frac{f(b)-f(a)}{b-a} = 0$.

8. Show that, if g is increasing on an interval, then $-g$ is decreasing on that interval.
Assume $u < v$. Then $g(u) < g(v)$. Hence, $-g(u) > -g(v)$.

9. Prove Theorem 13.7: (a) If f' is positive on an interval, then f is increasing on that interval, (b) If f' is negative on an interval, then f is decreasing on that interval.

(a) Let a and b be any two points on the interval with $a < b$. By the Law of the Mean, $(f(b) - f(a))/(b - a) = f'(x_0)$ for some point x_0 in (a, b). Since x_0 is in the interval, $f'(x_0) > 0$. Thus, $(f(b) - f(a))/(b - a) > 0$. But, $a < b$ and, therefore, $b - a > 0$. Hence, $f(b) - f(a) > 0$. So, $f(a) < f(b)$.

(b) Let $g = -f$. So, g' is positive on the interval. By part (a), g is increasing on the interval. So, f is decreasing on the interval.

10. Show that $f(x) = x^5 + 20x - 6$ is an increasing function for all values of x.

$f'(x) = 5x^4 + 20 > 0$ for all x. Hence, by Theorem 13.7(a), f is increasing everywhere.

11. Show that $f(x) = 1 - x^3 - x^7$ is a decreasing function for all values of x.

$f'(x) = -3x^2 - 7x^6 < 0$ for all $x \neq 0$. Hence, by Theorem 13.7(b), f is decreasing on any interval not containing 0. Note that, if $x < 0$, $f(x) > 1 = f(0)$, and, if $x > 0$, $f(0) = 1 > f(x)$. So, f is decreasing for all real numbers.

12. Show that $f(x) = 4x^3 + x - 3 = 0$ has exactly one real solution.

$f(0) = -3$ and $f(1) = 2$. So, the intermediate value theorem tells us that $f(x) = 0$ has a solution in (0, 1). Since $f'(x) = 12x^2 + 1 > 0$, f is an increasing function. Therefore, there cannot be two values of x for which $f(x) = 0$.

13. Prove the Extended Law of the Mean (Theorem 13.5): If $f(x)$ and $g(x)$ are continuous on $[a, b]$, and differentiable on (a, b), and $g'(x) \neq 0$ for all x in (a, b), then there exists at least one point x_0 in (a, b) for which $\dfrac{f(b)-f(a)}{g(b)-g(a)} = \dfrac{f'(x_0)}{g'(x_0)}$.

Suppose that $g(b) = g(a)$. Then, by the generalized Rolle's Theorem, $g'(x) = 0$ for some x in (a, b), contradicting our hypothesis. Hence, $g(b) \neq g(a)$.

Let $F(x) = f(x) - f(b) - \dfrac{f(b)-f(a)}{g(b)-g(a)}(g(x) - g(b))$.

Then
$$F(a) = 0 = F(b) \quad \text{and} \quad F'(x) = f'(x) - \frac{f(b)-f(a)}{g(b)-g(a)}g'(x)$$

By Rolle's Theorem, there exists x_0 in (a, b) for which $f'(x_0) - \dfrac{f(b)-f(a)}{g(b)-g(a)}g'(x_0) = 0$.

14. Prove the Higher-Order Law of the Mean (Theorem 13.6): If f and its first $n - 1$ derivatives are continuous on $[a, b]$ and $f^{(n)}(x)$ exists on (a, b), then there is at least one x_0 in (a, b) such that

$$f(b) = f(a) + \frac{f'(a)}{1!}(b-a) + \frac{f''(a)}{2!}(b-a)^2 + \cdots + \frac{f^{(n-1)}(a)}{(n-1)!}(b-a)^{(n-1)} + \frac{f^{(n)}(x_0)}{n!}(b-a)^n \tag{1}$$

Let a constant K be defined by

$$f(b) = f(a) + \frac{f'(a)}{1!}(b-a) + \frac{f''(a)}{2!}(b-a)^2 + \cdots + \frac{f^{(n-1)}(a)}{(n-1)!}(b-a)^{(n-1)} + K(b-a)^n \tag{2}$$

and consider

$$F(x) = f(x) - f(b) + \frac{f'(x)}{1!}(b-x) + \frac{f''(x)}{2!}(b-x)^2 + \cdots + \frac{f^{(n-1)}(x)}{(n-1)!}(b-x)^{n-1} + K(b-x)^n$$

Now $F(a) = 0$ by (2), and $F(b) = 0$. By Rolle's Theorem, there exists x_0 in (a, b) such that

$$F'(x_0) = f'(x_0) + [f''(x_0)(b - x_0) - f'(x_0)] + \left[\frac{f'''(x_0)}{2!}(b - x_0)^2 - f''(x_0)(b - x_0)\right]$$

$$+ \cdots + \left[\frac{f^{(n)}(x_0)}{(n-1)!}(b - x_0)^{n-1} - \frac{f^{(n-1)}(x_0)}{(n-2)!}(b - x_0)^{n-2}\right] - Kn(b - x_0)^{n-1}$$

$$= \frac{f^{(n)}(x_0)}{(n-1)!}(b - x_0)^{n-1} - Kn(b - x_0)^{n-1} = 0$$

Then $K = \dfrac{f^{(n)}(x_0)}{n!}$, and (2) becomes (*1*).

15. If $f'(x) = 0$ for all x on (a, b), then f is constant on (a, b).

Let u and v be any two points in (a, b) with $u < v$. By the Law of the Mean, there exists x_0 in (u, v) for which $\dfrac{f(v) - f(u)}{v - u} = f'(x_0)$. By hypothesis, $f'(x_0) = 0$. Hence, $f(v) - f(u) = 0$, and, therefore, $f(v) = f(u)$.

SUPPLEMENTARY PROBLEMS

16. If $f(x) = x^2 - 4x + 3$ on $[1, 3]$, find a value prescribed by Rolle's Theorem.

Ans. $x_0 = 2$

17. Find a value prescribed by the Law of the Mean, given:

(a) $y = x^3$ on $[0, 6]$ *Ans.* $x_0 = 2\sqrt{3}$

(b) $y = ax^2 + bx + c$ on $[x_1, x_2]$ *Ans.* $x_0 = \frac{1}{2}(x_1 + x_2)$

18. If $f'(x) = g'(x)$ for all x in (a, b), prove that there exists a constant K such that $f(x) = g(x) + K$ for all x in (a, b). (*Hint*: $D_x(f(x) - g(x)) = 0$ in (a, b). By Problem 15, there is a constant K such that $f(x) - g(x) = K$ in (a, b).)

19. Find a value x_0 precribed by the extended law of the mean when $f(x) = x^2 + 2x - 3$, $g(x) = x^2 - 4x + 6$ on the interval $[0, 1]$.

Ans. $\frac{1}{2}$

20. Show that $x^3 + px + q = 0$ has: (a) one real root if $p > 0$, and (b) three real roots if $4p^3 + 27q^2 < 0$.

21. Show that $f(x) = \dfrac{ax + b}{cx + d}$ has neither a relative maximum nor a relative minimum. (*Hint*: Use Theorem 13.1.)

22. Show that $f(x) = 5x^3 + 11x - 20 = 0$ has exactly one real solution.

23. (a) Where are the following functions (i)–(vii) increasing and where are they decreasing? Sketch the graphs.
(b) (GC) Check your answers to (*a*) by means of a graphing calculator.

(i) $f(x) = 3x + 5$ *Ans.* Increasing everywhere

(ii) $f(x) = -7x + 20$ *Ans.* Decreasing everywhere

(iii) $f(x) = x^2 + 6x - 11$ *Ans.* Decreasing on $(-\infty, -3)$, increasing on $(-3, +\infty)$

(iv) $f(x) = 5 + 8x - x^2$ *Ans.* Increasing on $(-\infty, 4)$, decreasing on $(4, +\infty)$

(v) $f(x) = \sqrt{4 - x^2}$ *Ans.* Increasing on $(-2, 0)$, decreasing on $(0, 2)$

(vi) $f(x) = |x - 2| + 3$ *Ans.* Decreasing on $(-\infty, 2)$, increasing on $(2, +\infty)$

(vii) $f(x) = \dfrac{x}{x^2 - 4}$ *Ans.* Decreasing on $(-\infty, -2)$, $(-2, 2)$, $(2, +\infty)$; never increasing

24. (GC) Use a graphing calculator to estimate the intervals on which $f(x) = x^5 + 2x^3 - 6x + 1$ is increasing, and the intervals on which it is decreasing.

25. For the following functions, determine whether Rolle's Theorem is applicable. If it is, find the prescribed values.

(a) $f(x) = x^{3/4} - 2$ on $[-3, 3]$ *Ans.* No. Not differentiable at $x = 0$.

(b) $f(x) = |x^2 - 4|$ on $[0, 8]$ *Ans.* No. Not differentiable at $x = 2$.

(c) $f(x) = |x^2 - 4|$ on $[0, 1]$ *Ans.* No. $f(0) \neq f(1)$

(d) $f(x) = \dfrac{x^2 - 3x - 4}{x - 5}$ on $[-1, 4]$ *Ans.* Yes. $x_0 = 5 - \sqrt{6}$

CHAPTER 14

Maximum and Minimum Values

Critical Numbers

A number x_0 in the domain of f such that either $f'(x_0) = 0$ or $f'(x_0)$ is not defined is called a *critical number* of f.

Recall (Theorem 13.1) that, if f has a relative extremum at x_0 and $f'(x_0)$ is defined, then $f'(x_0) = 0$ and, therefore, x_0 is a critical number of f. Observe, however, that the condition that $f'(x_0) = 0$ does not guarantee that f has a relative extremum at x_0. For example, if $f(x) = x^3$, then $f'(x) = 3x^2$, and therefore, 0 is a critical number of f; but f has neither a relative maximum nor a relative minimum at 0. (See Fig. 5-5).

EXAMPLE 14.1:

(a) Let $f(x) = 7x^2 - 3x + 5$. Then $f'(x) = 14x - 3$. Set $f'(x) = 0$ and solve. The only critical number of f is $\frac{3}{14}$.
(b) Let $f(x) = x^3 - 2x^2 + x + 1$. Then $f'(x) = 3x^2 - 4x + 1$. Solving $f'(x) = 0$, we find that the critical numbers are 1 and $\frac{1}{3}$.
(c) Let $f(x) = x^{2/3}$. Then $f'(x) = \frac{2}{3}x^{-1/3} = \frac{2}{3x^{1/3}}$. Since $f'(0)$ is not defined, 0 is the only critical number of f.

We shall find some conditions under which we can conclude that a function f has a relative maximum or a relative minimum at a given critical number.

Second Derivative Test for Relative Extrema

Assume that $f'(x_0) = 0$ and that $f''(x_0)$ exists. Then:

(i) if $f''(x_0) < 0$, then f has a relative maximum at x_0;
(ii) if $f''(x_0) > 0$, then f has a relative minimum at x_0;
(iii) if $f''(x_0) = 0$, we do not know what is happening at x_0.

A proof is given in Problem 9. To see that (iii) holds, consider the three functions $f(x) = x^4$, $g(x) = -x^4$, and $h(x) = x^3$. Since $f'(x) = 4x^3$, $g'(x) = -4x^3$, and $h'(x) = 3x^2$, 0 is a critical number of all three functions. Since $f''(x) = 12x^2$, $g''(x) = -12x^2$, and $h''(x) = 6x$, the second derivative of all three functions is 0 at 0. However, f has a relative minimum at 0, g has a relative maximum at 0, and h has neither a relative maximum nor a relative minimum at 0.

EXAMPLE 14.2:

(a) Consider the function $f(x) = 7x^2 - 3x + 5$ of Example 1(*a*). The only critical number was $\frac{3}{14}$. Since $f''(x) = 14$, $f''(\frac{3}{14}) = 14 > 0$. So, the second derivative test tells us that f has a relative minimum at $\frac{3}{14}$.
(b) Consider the function $f(x) = x^3 - 2x^2 + x + 1$ of Example 1(*b*). Note that $f''(x) = 6x - 4$. At the critical numbers 1 and $\frac{1}{3}$, $f''(1) = 2 > 0$ and $f''(\frac{1}{3}) = -2 < 0$. Hence f has a relative minimum at 1 and a relative maximum at $\frac{1}{3}$.
(c) In Example 1(*c*), $f(x) = x^{2/3}$ and $f'(x) = \frac{2}{3}x^{-1/3}$. The only critical number is 0, where f' is not defined. Hence, $f''(0)$ is not defined and the second derivative test is not applicable.

If the second derivative test is not usable or convenient, either because the second derivative is 0, or does not exist, or is difficult to compute, then the following test can be applied. Recall that $f'(x)$ is the slope of the tangent line to the graph of f at x.

First Derivative Test

Assume $f'(x_0) = 0$.

Case {+, −}

If f' is positive in an open interval immediately to the left of x_0, and negative in an open interval immediately to the right of x_0, then f has a relative maximum at x_0. (See Fig. 14-1(*a*).)

Case {−, +}

If f' is negative in an open interval immediately to the left of x_0, and positive in an open interval immediately to the right of x_0, then f has a relative minimum at x_0. (See Fig. 14-1(*b*).)

Cases {+, +} and {−, −}

If f' has the same sign in open intervals immediately to the left and to the right of x_0, then f has neither a relative maximum nor a relative minimum at x_0. (See Fig. 14-1(*c, d*).)

For a proof of the first derivative test, see Problem 8.

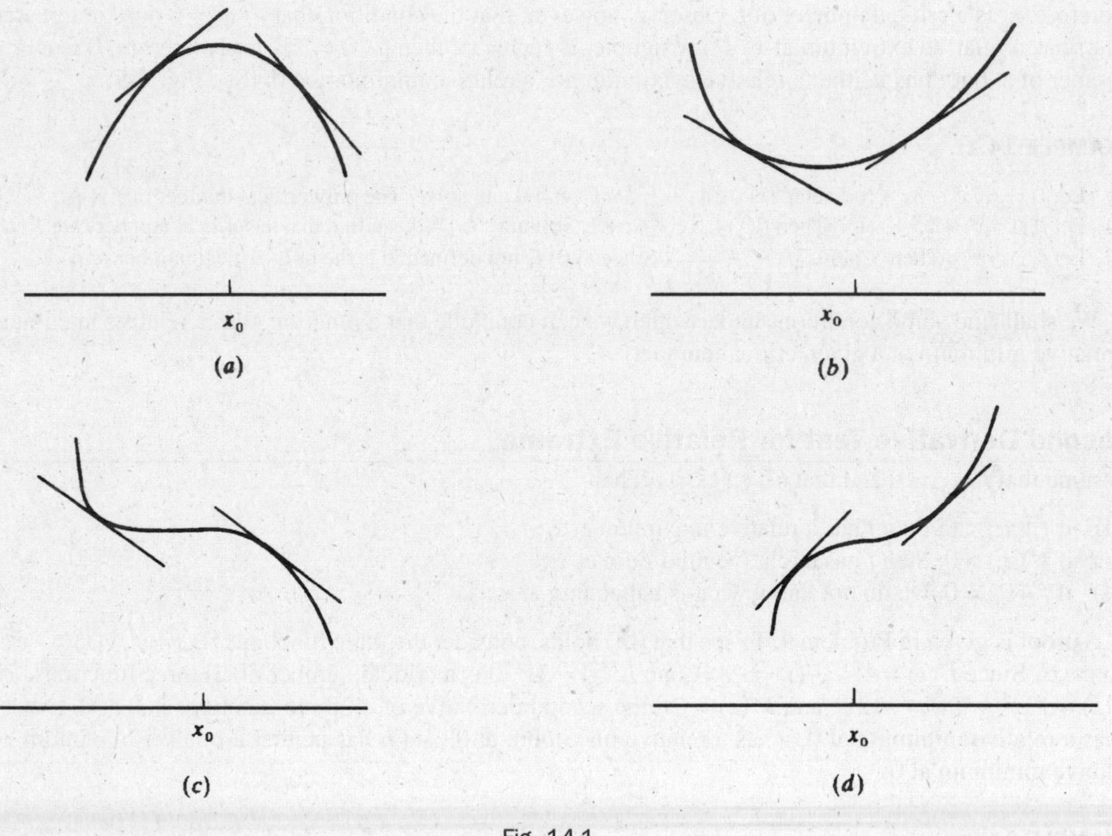

Fig. 14-1

EXAMPLE 14.3: Consider the three functions $f(x) = x^4$, $g(x) = -x^4$, and $h(x) = x^3$ discussed above. At their critical number 0, the second derivative test was not applicable because the second derivative was 0. Let us try the first derivative test.

(a) $f'(x) = 4x^3$. To the left of 0, $x < 0$, and so, $f'(x) < 0$. To the right of 0, $x > 0$, and so, $f'(x) > 0$. Thus, we have the case {−, +} and f must have a relative minimum at 0.

(b) $g'(x) = -4x^3$. To the left of 0, $x < 0$, and so, $g'(x) > 0$. To the right of 0, $x > 0$, and so, $g'(x) < 0$. Thus, we have the case {+, −} and g must have a relative maximum at 0.

(c) $h'(x) = 3x^2$. $h'(x) > 0$ on both sides of 0. Thus, we have the case {+, +} and h has neither a relative maximum nor a relative minimum at 0. There is an *inflection point* at $x = 0$.

These results can be verified by looking at the graphs of the functions.

Absolute Maximum and Minimum

An *absolute maximum* of a function f on a set S occurs at x_0 in S if $f(x) \le f(x_0)$ for all x in S. An *absolute minimum* of a function f on a set S occurs at x_0 in S if $f(x) \ge f(x_0)$ for all x in S.

Tabular Method for Finding the Absolute Maximum and Minimum

Let f be continuous on $[a, b]$ and differentiable on (a, b). By the Extreme Value Theorem, we know that f has an absolute maximum and minimum on $[a, b]$. Here is a tabular method for determining what they are and where they occur. (See Fig. 14-2.)

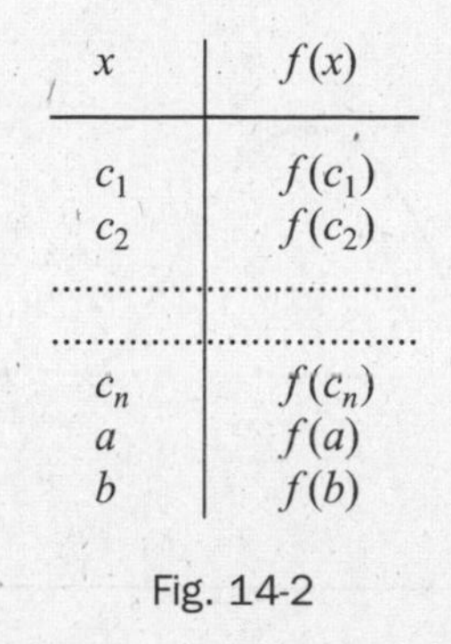

x	$f(x)$
c_1	$f(c_1)$
c_2	$f(c_2)$
⋯	⋯
c_n	$f(c_n)$
a	$f(a)$
b	$f(b)$

Fig. 14-2

First, find the critical numbers (if any) $c_1, c_2, \ldots$ of f in (a, b). Second, list these numbers in a table, along with the endpoints a and b of the interval. Third, calculate the value of f for all the numbers in the table. Then:

1. The largest of these values is the absolute maximum of f on $[a, b]$.
2. The smallest of these values is the absolute minimum of f on $[a, b]$.

EXAMPLE 14.4: Let us find the absolute maximum and minimum of $f(x) = x^3 - x^2 - x + 2$ on $[0, 2]$.

$f'(x) = 3x^2 - 2x - 1 = (3x + 1)(x - 1)$. Hence, the critical numbers are $x = -\frac{1}{3}$ and $x = 1$. The only critical number in $[0, 2]$ is 1. From the table in Fig. 14-3, we see that the maximum value of f on $[0, 2]$ is 4, which is attained at the right endpoint 2, and the minimum value is 1, attained at 1.

x	$f(x)$
1	1
0	2
2	4

Fig. 14-3

Let us see why the method works. By the Extreme Value Theorem, f achieves maximum and minimum values on the closed interval $[a, b]$. If either of those values occurs at an endpoint, that value will appear in the table and, since it is actually a maximum or minimum, it will show up as the largest or smallest value. If the maximum or minimum is assumed at a point x_0 inside the interval, f has a relative maximum or minimum at x_0 and, therefore, by Theorem 13.1, $f'(x_0) = 0$. Thus, x_0 will be a critical number and will be listed in the table, so that the corresponding maximum or minimum value $f(x_0)$ will be the largest or smallest value in the right-hand column.

Theorem 14.1: Assume that f is a continuous function defined on an interval J. The interval J can be a finite or infinite interval. If f has a unique relative extremum within J, then that relative extremum is also an absolute extremum on J.

To see why this is so, look at Fig. 14-4, where f is assumed to have a unique extremum, a relative maximum at c. Consider any other number d in J. The graph moves downward on both sides of c. So, if $f(d)$

were greater than $f(c)$, then, by the Extreme Value Theorem for the closed interval with endpoints c and d, f would have an absolute minimum at some point u between c and d. (u could not be equal to c or d.) Then f would have a relative minimum at u, contradicting our hypothesis that f has a relative extremum only at c. We can extend this argument to the case where f has a relative minimum at c by applying the result we have just obtained to $-f$.

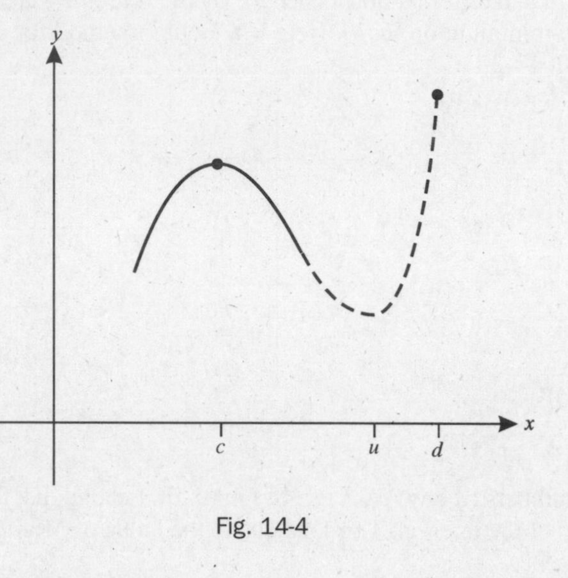

Fig. 14-4

SOLVED PROBLEMS

1. Locate the absolute maximum or minimum of the following functions on their domains:

(a) $y = -x^2$; (b) $y = (x-3)^2$; (c) $y = \sqrt{25-4x^2}$; (d) $y = \sqrt{x-4}$.

(a) $y = -x^2$ has an absolute maximum (namely, 0) when $x = 0$, since $y < 0$ when $x \neq 0$. It has no relative minimum, since its range is $(-\infty, 0)$. The graph is a parabola opening downward, with vertex at (0, 0).

(b) $y = (x-3)^2$ has an absolute minimum, 0, when $x = 3$, since $y > 0$ when $x \neq 3$. It has no absolute maximum, since its range is $(0, +\infty)$. The graph is a parabola opening upward, with vertex at (3, 0).

(c) $y = \sqrt{25-4x^2}$ has 5 as its absolute maximum, when $x = 0$, since $25 - 4x^2 < 25$ when $x \neq 0$. It has 0 as its absolute minimum, when $x = \frac{5}{2}$. The graph is the upper half of an ellipse.

(d) $y = \sqrt{x-4}$ has 0 as its absolute minimum when $x = 4$. It has no absolute maximum. Its graph is the upper half of a parabola with vertex at (4, 0) and the x axis as its axis of symmetry.

2. Let $f(x) = \frac{1}{3}x^3 + \frac{1}{2}x^2 - 6x + 8$. Find: (a) the critical numbers of f; (b) the points at which f has a relative maximum or minimum; (c) the intervals on which f is increasing or decreasing.

(a) $f'(x) = x^2 + x - 6 = (x+3)(x-2)$. Solving $f'(x) = 0$ yields the critical numbers -3 and 2.

(b) $f''(x) = 2x + 1$. Thus, $f''(-3) = -5 < 0$ and $f''(2) = 5$. Hence, by the second derivative test, f has a relative maximum at $x = -3$, where $f(-3) = \frac{43}{2}$. By the second derivative test, f has a relative minimum at $x = 2$, where $f(2) = \frac{2}{3}$.

(c) Look at $f'(x) = (x+3)(x-2)$. When $x > 2$, $f'(x) > 0$. For $-3 < x < 2$, $f'(x) < 0$. For $x < -3$, $f'(x) > 0$. Thus, by Theorem 13.7, f is increasing for $x < -3$ and $2 < x$, and decreasing for $-3 < x < 2$.

A sketch of part of the graph of f is shown in Fig. 14-5. Note that f has neither absolute maximum nor absolute minimum.

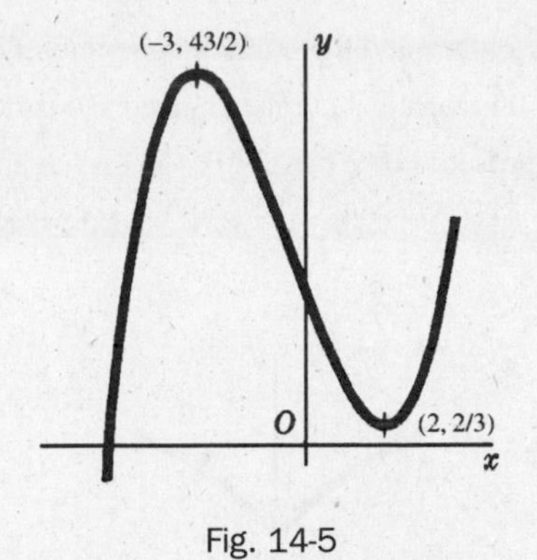

Fig. 14-5

3. Let $f(x) = x^4 + 2x^3 - 3x^2 - 4x + 4$. Find: (a) the critical numbers of f; (b) the points at which f has a relative extremum; (c) the intervals on which f is increasing or decreasing.

(a) $f'(x) = 4x^3 + 6x^2 - 6x - 4$. It is clear that $x = 1$ is a zero of $f'(x)$. Dividing $f'(x)$ by $x - 1$ yields $4x^2 + 10x + 4$, which factors into $2(2x^2 + 5x + 2) = 2(2x + 1)(x + 2)$. Thus, $f'(x) = 2(x - 1)(2x + 1)(x + 2)$, and the critical numbers are 1, $-\frac{1}{2}$, and -2.

(b) $f''(x) = 12x^2 + 12x - 6 = 6(2x^2 + 2x - 1)$. Using the second derivative test, we find: (i) at $x = 1, f''(1) = 18 > 0$, and there is a relative minimum; (ii) at $x = -\frac{1}{2}$, $f''(-\frac{1}{2}) = -9 < 0$, so that there is a relative maximum; (iii) at $x = -2, f''(-2) = 18 > 0$, so that there is a relative minimum.

(c) $f'(x) > 0$ when $x > 1$, $f'(x) < 0$ when $-\frac{1}{2} < x < 1$, $f'(x) > 0$ when $-2 < x < -\frac{1}{2}$, and $f'(x) < 0$ when $x < -2$. Hence, f is increasing when $x > 1$ or $-2 < x < -\frac{1}{2}$, and decreasing when $-\frac{1}{2} < x < 1$ or $x < -2$.

The graph is sketched in Fig. 14-6.

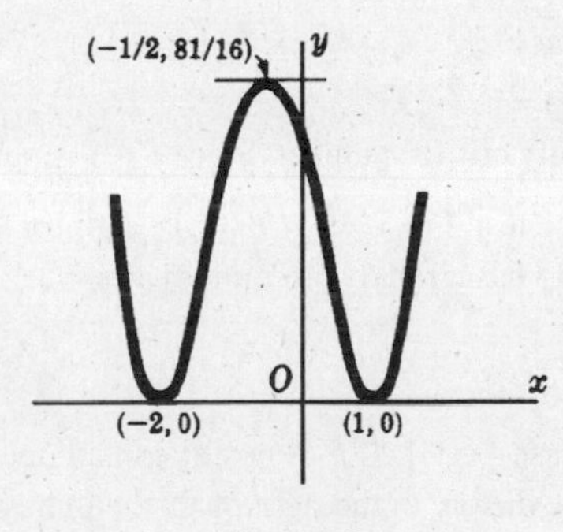

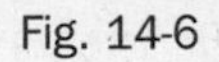

Fig. 14-6

4. Examine $f(x) = \dfrac{1}{x-2}$ for relative extrema, and find the intervals on which f is increasing or decreasing.

$f(x) = (x - 2)^{-1}$, so that $f'(x) = -(x-2)^{-2} = -\dfrac{1}{(x-2)^2}$. Thus, f' is never 0, and the only number where f' is not defined is the number 2, which is not in the domain of f. Hence, f has no critical numbers. So, f has no relative extrema. Note that $f'(x) < 0$ for $x \neq 2$. Hence, f is decreasing for $x < 2$ and for $x > 2$. There is a nonremovable discontinuity at $x = 2$. The graph is shown in Fig. 14-7.

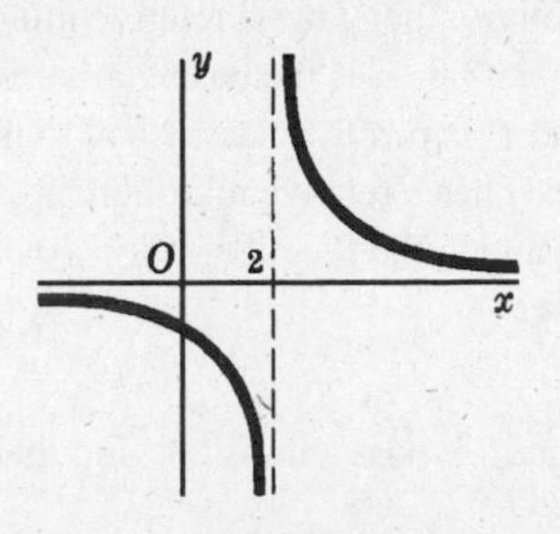

Fig. 14-7

5. Locate the relative extrema of $f(x) = 2 + x^{2/3}$ and the intervals on which f is increasing or decreasing.

$f'(x) = \frac{2}{3}x^{-1/3} = \frac{2}{3x^{1/3}}$. Then $x = 0$ is a critical number, since $f'(0)$ is not defined (but 0 is in the domain of f). Note that $f'(x)$ approaches ∞ as x approaches 0. When $x < 0$, $f'(x)$ is negative and, therefore, f is decreasing. When $x > 0$, $f'(x)$ is positive and, therefore, f is increasing. The graph is sketched in Fig. 14-8. f has an absolute minimum at $x = 0$.

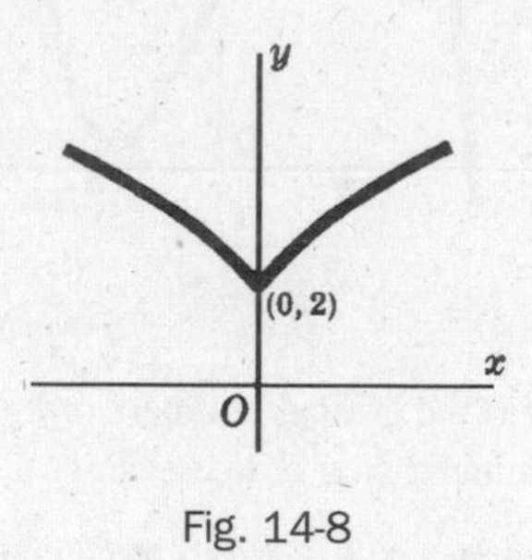

Fig. 14-8

6. Use the second derivative test to examine the relative extrema of the following functions: (a) $f(x) = x(12 - 2x)^2$; (b) $f(x) = x^2 + \frac{250}{x}$.

(a) $f'(x) = x(2)(12 - 2x)(-2) + (12 - 2x)^2 = (12 - 2x)(12 - 6x) = 12(x - 6)(x - 2)$. So, 6 and 2 are the critical numbers. $f''(x) = 12(2x - 8) = 24(x - 4)$. So, $f''(6) = 48 > 0$, and $f''(2) = -48 < 0$. Hence, f has a relative minimum at $x = 6$ and a relative maximum at $x = 2$.

(b) $f'(x) = 2x - \frac{250}{x^2} = 2\left(\frac{x^3 - 125}{x^2}\right)$. So, the only critical number is 5 (where $x^3 - 125 = 0$). $f''(x) = 2 + 500/x^3$. Since $f''(5) = 6 > 0$, f has a relative minimum at $x = 5$.

7. Determine the relative extrema of $f(x) = (x - 2)^{2/3}$.

$f'(x) = \frac{2}{3(x-2)^{1/3}}$. So, 2 is the only critical number. Since $f'(2)$ is not defined, $f''(2)$ will be undefined. Hence, we shall try the first derivative test. For $x < 2$, $f'(x) < 0$, and, for $x > 2$, $f'(x) > 0$. Thus, we have the case $\{-, +\}$ of the first derivative test, and f has a relative minimum at $x = 2$.

8. Prove the first derivative test.

Assume $f'(x_0) = 0$. Consider the case $\{+, -\}$: If f' is positive in an open interval immediately to the left of x_0, and negative in an open interval immediately to the right of x_0, then f has a relative maximum at x_0. To see this, notice that, by Theorem 13.8, since f' is positive in an open interval immediately to the left of x_0, f is increasing in that interval, and, since f' is negative in an open interval immediately to the right of x_0, f is decreasing in that interval. Hence, f has a relative maximum at x_0. The case $\{-, +\}$ follows from the case $\{+, -\}$ applied to $-f$. In the case $\{+, +\}$, f will be increasing in an interval around x_0, and, in the case $\{-, -\}$, f will be decreasing in an interval around x_0. So, in both cases, f has neither a relative maximum nor minimum at x_0.

9. Prove the second derivative test: If $f(x)$ is differentiable on an open interval containing a critical value x_0 of f, and $f''(x_0)$ exists and $f''(x_0)$ is positive (negative), then f has a relative minimum (maximum) at x_0.

Assume $f''(x_0) > 0$. Then, by Theorem 13.8, f' is increasing at x_0. Since $f'(x_0) = 0$, this implies that f' is negative nearby and to the left of x_0, and f' is positive nearby and to the right of x_0. Thus, we have the case $\{-, +\}$ of the first derivative test and, therefore, f has a relative minimum at x_0. In the opposite situation, where $f''(x_0) < 0$, the result we have just proved is applicable to the function $g(x) = -f(x)$. Then g has a relative minimum at x_0 and, therefore, f has a relative maximum at x_0.

10. Among those positive real numbers u and v whose sum is 50, find that choice of u and v that makes their product P as large as possible.

$P = u(50 - u)$. Here, u is any positive number less than 50. But we also can allow u to be 0 or 50, since, in those cases, $P = 0$, which will certainly not be the largest possible value. So, P is a continuous function $u(50 - u)$, defined on [0, 50]. $P = 50u - u^2$ is also differentiable everywhere, and $dP/du = 50 - 2u$. Setting $dP/du = 0$ yields a unique critical number $u = 25$. By the tabular method (Fig. 14-9), we see that the maximum value of P is 625, when $u = 25$ (and, therefore, $v = 50 - u = 25$).

u	P
25	625
0	0
50	0

Fig. 14-9

11. Divide the number 120 into two parts such that the product P of one part and the square of the other is a maximum.

Let x be one part and $120 - x$ the other part. Then $P = (120 - x)x^2$ and $0 \le x \le 120$. Since $dP/dx = 3x(80 - x)$, the critical numbers are 0 and 80. Using the tabular method, we find $P(0) = 0$, $P(80) = 256{,}000$ and $P(120) = 0$. So, the maximum value occurs when $x = 80$, and the required parts are 80 and 40.

12. A sheet of paper for a poster is to be 18 ft² in area. The margins at the top and bottom are to be 9 inches, and the margins at the sides 6 inches. What should be the dimensions of the sheet to maximize the printed area?

Let x be one dimension, measured in feet. Then $18/x$ is the other dimension. (See Fig. 14-10.) The only restriction on x is that $x > 0$. The printed area in square feet is $A = (x-1)\left(\frac{18}{x} - \frac{3}{2}\right)$, and $\frac{dA}{dx} = \frac{18}{x^2} - \frac{3}{2}$.

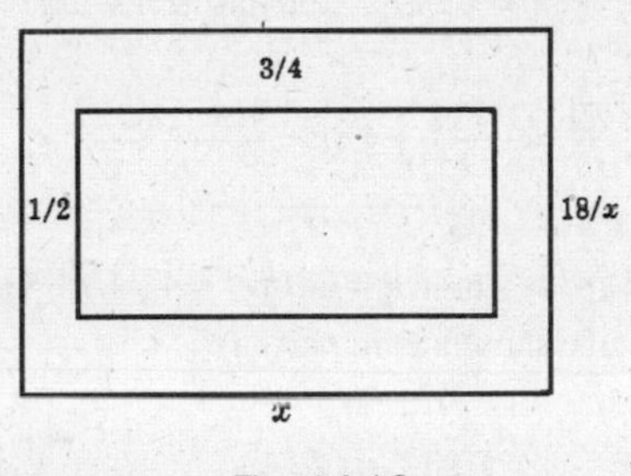

Fig. 14-10

Solving $dA/dx = 0$ yields the critical number $x = 2\sqrt{3}$. Since $d^2A/dx^2 = -36/x^3$ is negative when $x = 2\sqrt{3}$, the second derivative test tells us that A has a relative maximum at $x = 2\sqrt{3}$. Since $2\sqrt{3}$ is the only critical number in the interval $(0, +\infty)$, Theorem 14.1 tells us that A has an absolute maximum at $x = 2\sqrt{3}$. Thus, one side is $2\sqrt{3}$ ft and the other side is $18/(2\sqrt{3}) = 3\sqrt{3}$ ft.

13. At 9 A.M., ship B is 65 miles due east of another ship A. Ship B is then sailing due west at 10 mi/h, and A is sailing due south at 15 mi/h. If they continue on their respective courses, when will they be nearest one another, and how near? (See Fig. 14–11.)

Let A_0 and B_0 be the positions of the ships at 9 A.M., and A_t and B_t their positions t hours later. The distance covered in t hours by A is $15t$ miles; by B, $10t$ miles. The distance D between the ships is determined by $D^2 = (15t)^2 + (65 - 10t)^2$. Then

$$2D\frac{dD}{dt} = 2(15t)(15) + 2(65 - 10t)(-10); \text{ hence, } \frac{dD}{dt} = \frac{325t - 650}{D}$$

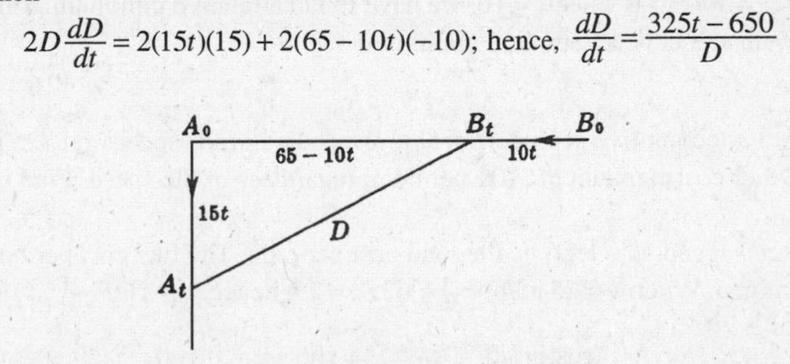

Fig. 14-11

Solving $dD/dt = 0$, yields the critical number $t = 2$. Since $D > 0$ and $325t - 650$ is negative to the left of 2 and positive to the right of 2, the case (–, +) of the first derivative test tells us that $t = 2$ yields a relative minimum for D. Since $t = 2$ is the only critical number, Theorem 14.1 implies that there is an absolute minimum at $t = 2$.

Setting $t = 2$ in $D^2 = (15t)^2 + (65 - 10t)^2$ yields $D = 15\sqrt{13}$ miles. Hence, the ships are nearest at 11 A.M., at which time they are $15\sqrt{13}$ miles apart.

14. A cylindrical container with circular base is to hold 64 in³. Find its dimensions so that the amount (surface area) of metal required is a minimum when the container is (a) an open can and (b) a closed can.

Let r and h be, respectively, the radius of the base and the height in inches, A the amount of metal, and V the volume of the container.

(a) Here $V = \pi r^2 h = 64$, and $A = 2\pi rh + \pi r^2$. To express A as a function of one variable, we solve for h in the first relation (because it is easier) and substitute in the second, obtaining

$$A = 2\pi r \frac{64}{\pi r^2} + \pi r^2 = \frac{128}{r} + \pi r^2 \quad \text{and} \quad \frac{dA}{dr} = -\frac{128}{r^2} + 2\pi r = \frac{2(\pi r^3 - 64)}{r^2}$$

and the critical number is $r = 4/\sqrt[3]{\pi}$. Then $h = 64/\pi r^2 = 4/\sqrt[3]{\pi}$. Thus, $r = h = 4/\sqrt[3]{\pi}$ in.

Now $dA/dr > 0$ to the right of the critical number, and $dA/dr < 0$ to the left of the critical number. So, by the first derivative test, we have a relative minimum. Since there is no other critical number, that relative minimum is an absolute minimum.

(b) Here again $V = \pi r^2 h = 64$, but $A = 2\pi rh + 2\pi r^2 = 2\pi r(64/\pi r^2) + 2\pi r^2 = 128/r + 2\pi r^2$. Hence,

$$\frac{dA}{dr} = -\frac{128}{r^2} + 4\pi r = \frac{4(\pi r^3 - 32)}{r^2}$$

and the critical number is $r = 2\sqrt[3]{4/\pi}$. Then $h = 64/\pi r^2 = 4\sqrt[3]{4/\pi}$. Thus, $h = 2r = 4\sqrt[3]{4/\pi}$ in. That we have found an absolute minimum can be shown as in part (a).

15. The total cost of producing x radio sets per day is $\$(\frac{1}{4}x^2 + 35x + 25)$, and the price per set at which they may be sold is $\$(50 - \frac{1}{2}x)$.

(a) What should be the daily output to obtain a maximum total profit?
(b) Show that the cost of producing a set is a relative minimum at that output.

(a) The profit on the sale of x sets per day is $P = x(50 - \frac{1}{2}x) - (\frac{1}{4}x^2 + 35x + 25)$. Then $dP/dx = 15 - 3x/2$; solving $dP/dx = 0$ gives the critical number $x = 10$.

Since $d^2P/dx^2 = -\frac{3}{2} < 0$, the second derivative test shows that we have found a relative maximum. Since $x = 10$ is the only critical number, the relative maximum is an absolute maximum. Thus, the daily output that maximizes profit is 10 sets per day.

(b) The cost of producing a set is $C = \dfrac{\frac{1}{4}x^2 + 35x + 25}{x} = \dfrac{1}{4}x + 35 + \dfrac{25}{x}$. Then $\dfrac{dC}{dx} = \dfrac{1}{4} - \dfrac{25}{x^2}$; solving $dC/dx = 0$ gives the critical number $x = 10$.

Since $d^2C/dx^2 = 50/x^3 > 0$ when $x = 10$, we have found a relative minimum. Since there is only one critical number, this must be an absolute minimum.

16. The cost of fuel to run a locomotive is proportional to the square of the speed and \$25 per hour for a speed of 25 miles per hour. Other costs amount to \$100 per hour, regardless of the speed. Find the speed that minimizes the cost per mile.

Let v be the required speed, and let C be the total cost per mile. The fuel cost per hour is kv^2, where k is a constant to be determined. When $v = 25$ mi/h, $kv^2 = 625k = 25$; hence, $k = 1/25$.

$$C = \frac{\text{cost in \$/h}}{\text{speed in mi/h}} = \frac{v^2/25 + 100}{v} = \frac{v}{25} + \frac{100}{v}$$

Then $$\frac{dC}{dv} = \frac{1}{25} - \frac{100}{v^2} = \frac{(v-50)(v+50)}{25v^2}$$

Since $v > 0$, the only relevant critical number is $v = 50$. Since $d^2C/dv^2 = 200/v^3 > 0$ when $v = 50$, the second derivative test tells us that C has a relative minimum at $v = 50$. Since $v = 50$ is the only critical number in $(0, +\infty)$, Theorem 14.1 tells us that C has an absolute minimum at $v = 50$. Thus, the most economical speed is 50 mi/h.

17. A man in a rowboat at P in Fig. 14-12, 5 miles from the nearest point A on a straight shore, wishes to reach a point B, 6 miles from A along the shore, in the shortest time. Where should he land if he can row 2 mi/h and walk 4 mi/h?

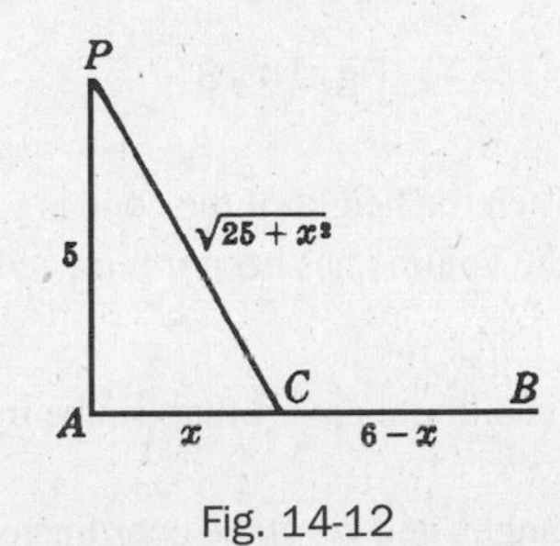

Fig. 14-12

Let C be the point between A and B at which the man lands, and let $AC = x$. The distance rowed is $PC = \sqrt{25+x^2}$, and the rowing time required is $t_1 = \frac{\text{distance}}{\text{speed}} = \frac{\sqrt{25+x^2}}{2}$. The distance walked is $CB = 6 - x$, and the walking time required is $t_2 = (6-x)/4$. Hence, the total time required is

$$t = t_1 + t_2 = \frac{\sqrt{25+x^2}}{2} + \frac{6-x}{4} \qquad \text{Then } \frac{dt}{dx} = \frac{x}{2\sqrt{25+x^2}} - \frac{1}{4} = \frac{2x - \sqrt{25+x^2}}{4\sqrt{25+x^2}}.$$

The critical number obtained from setting $2x - \sqrt{25+x^2} = 0$ is $x = \frac{5}{3}\sqrt{3} \sim 2.89$. Thus, he should land at a point about 2.89 miles from A toward B. (How do we know that this point yields the *shortest* time?)

18. A given rectangular area is to be fenced off in a field that lies along a straight river. If no fencing is needed along the river, show that the least amount of fencing will be required when the length of the field is twice its width.

Let x be the length of the field, and y its width. The area of the field is $A = xy$. The fencing required is $F = x + 2y$, and $dF/dx = 1 + 2\, dy/dx$. When $dF/dx = 0$, $dy/dx = -\frac{1}{2}$.

Also, $dA/dx = 0 = y + x\, dy/dx$. Then $y - \frac{1}{2}x = 0$, and $x = 2y$ as required.

To see that F has been minimized, note that $dy/dx = -y^2/A$ and

$$\frac{d^2F}{dx^2} = 2\frac{d^2y}{dx^2} = 2\left(-2\frac{y}{A}\frac{dy}{dx}\right) = -4\frac{y}{A}\left(-\frac{1}{2}\right) = 2\frac{y}{A} > 0 \quad \text{when} \quad \frac{dy}{dx} = -\frac{1}{2}$$

Now use the second derivative test and the uniqueness of the critical number.

19. Find the dimensions of the right circular cone of minimum volume V that can be circumscribed about a sphere of radius 8 inches.

Let x be the radius of the base of the cone, and $y + 8$ the height of the cone. (See Fig. 14-13.) From the similar right triangles ABC and AED, we have

$$\frac{x}{8} = \frac{y+8}{\sqrt{y^2-64}} \quad \text{and therefore} \quad x^2 = \frac{64(y+8)^2}{y^2-64}.$$

Also, $$V = \frac{\pi x^2(y+8)}{3} = \frac{64\pi(y+8)^2}{3(y-8)}. \qquad \text{So,} \quad \frac{dV}{dy} = \frac{64\pi(y+8)(y-24)}{3(y-8)^2}.$$

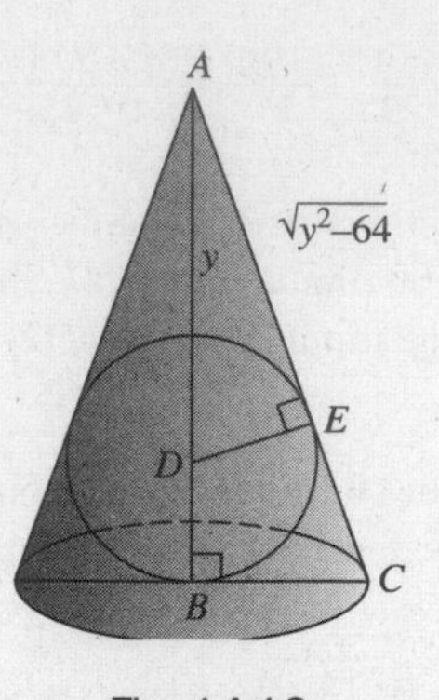

Fig. 14-13

The relevant critical number is $y = 24$. Then the height of the cone is $y + 8 = 32$ inches, and the radius of the base is $8\sqrt{2}$ inches. (How do we know that the volume has been minimized?)

20. Find the dimensions of the rectangle of maximum area A that can be inscribed in the portion of the parabola $y^2 = 4px$ intercepted by the line $x = a$.

Let $PBB'P'$ in Fig. 14-14 be the rectangle, and (x, y) the coordinates of P. Then

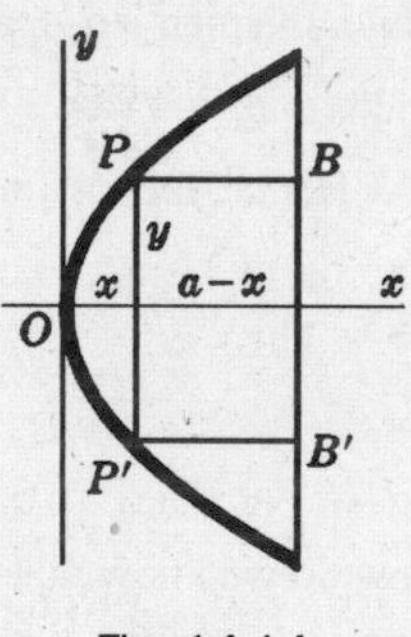

Fig. 14-14

$$A = 2y(a - x) = 2y\left(a - \frac{y^2}{4p}\right) = 2ay - \frac{y^3}{2p} \qquad \text{and} \qquad \frac{dA}{dy} = 2a - \frac{3y^2}{2p}$$

Solving $dA/dy = 0$ yields the critical number $y = \sqrt{4ap/3}$. The dimensions of the rectangle are $2y = \frac{4}{3}\sqrt{3ap}$ and $a - x = a - (y^2/4p) = 2a/3$.

Since $d^2A/dy^2 = -3y/p < 0$, the second derivative test and the uniqueness of the critical number ensure that we have found the maximum area.

21. Find the height of the right circular cylinder of maximum volume V that can be inscribed in a sphere of radius R. (See Fig. 14-15.)

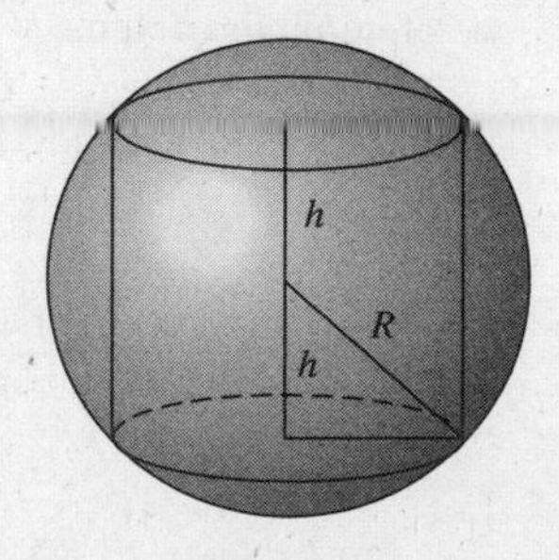

Fig. 14-15

Let r be the radius of the base, and $2h$ the height of the cylinder. From the geometry, $V = 2\pi r^2 h$ and $r^2 + h^2 = R^2$. Then

$$\frac{dV}{dr} = 2\pi\left(r^2\frac{dh}{dr} + 2rh\right) \quad \text{and} \quad 2r + 2h\frac{dh}{dr} = 0$$

From the last relation, $\frac{dh}{dr} = -\frac{r}{h}$, so $\frac{dV}{dr} = 2\pi\left(-\frac{r^3}{h} + 2rh\right)$. When V is a maximum, $\frac{dV}{dr} = 0$, from which $r^2 = 2h^2$.

Then $R^2 = r^2 + h^2 = 2h^2 + h^2$, so that $h = R/\sqrt{3}$ and the height of the cylinder is $2h = 2R/\sqrt{3}$. The second-derivative test can be used to verify that we have found a maximum value of V.

22. A wall of a building is to be braced by a beam that must pass over a parallel wall 10 ft high and 8 ft from the building. Find the length L of the shortest beam that can be used.

See Fig. 14-16. Let x be the distance from the foot of the beam to the foot of the parallel wall, and let y be the distance (in feet) from the ground to the top of the beam. Then $L = \sqrt{(x+8)^2 + y^2}$.

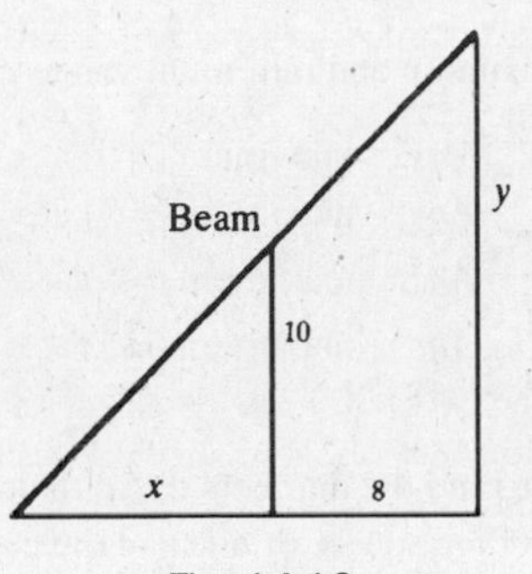

Fig. 14-16

Also, from similar triangles, $\frac{y}{10} = \frac{x+8}{x}$ and, therefore, $y = \frac{10(x+8)}{x}$. Hence,

$$L = \sqrt{(x+8)^2 + \frac{100(x+8)^2}{x^2}} = \frac{x+8}{x}\sqrt{x^2+100}$$

$$\frac{dL}{dx} = \frac{x[(x^2+100)^{1/2} + x(x+8)(x^2+100)^{-1/2}] - (x+8)(x^2+100)^{1/2}}{x^2} = \frac{x^3 - 800}{x^2\sqrt{x^2+100}}$$

The relevant critical number is $x = 2\sqrt[3]{100}$. The length of the shortest beam is

$$\frac{2\sqrt[3]{100} + 8}{2\sqrt[3]{100}}\sqrt{4\sqrt[3]{10{,}000} + 100} = (\sqrt[3]{100} + 4)^{3/2} \text{ ft}$$

The first derivative test and Theorem 14.1 guarantee that we really have found the shortest length.

SUPPLEMENTARY PROBLEMS

23. Examine each of the following for relative maximum and minimum values, using the first derivative test.

(a) $f(x) = x^2 + 2x - 3$ *Ans.* $x = -1$ yields relative minimum -4

(b) $f(x) = 3 + 2x - x^2$ *Ans.* $x = 1$ yields relative maximum 4

(c) $f(x) = x^3 + 2x^2 - 4x - 8$ *Ans.* $x = \frac{2}{3}$ yields relative minimum $-\frac{256}{27}$; $x = -2$ yields relative maximum 0

(d) $f(x) = x^3 - 6x^2 + 9x - 8$ *Ans.* $x = 1$ yields relative maximum -4; $x = 3$ yields relative minimum -8

(e) $f(x) = (2 - x)^3$ *Ans.* neither relative maximum nor relative minimum

(f) $f(x) = (x^2 - 4)^2$ *Ans.* $x = 0$ yields relative maximum 16; $x = \pm 2$ yields relative minimum 0

(g) $f(x) = (x - 4)^4(x + 3)^3$ *Ans.* $x = 0$ yields relative maximum 6912; $x = 4$ yields relative minimum 0; $x = -3$ yields neither

(h) $f(x) = x^3 + 48/x$ *Ans.* $x = -2$ yields relative maximum -32; $x = 2$ yields relative minimum 32

(i) $f(x) = (x - 1)^{1/3}(x + 2)^{2/3}$ *Ans.* $x = -2$ yields relative maximum 0; $x = 0$ yields relative minimum $-\sqrt[3]{4}$; $x = 1$ yields neither

24. Examine the functions of Problem 23 ($a - f$) for relative maximum and minimum values, using the second derivative test.

25. Show that $y = (a_1 - x)^2 + (a_2 - x)^2 + \cdots + (a_n - x)^2$ has an absolute minimum when $x = \dfrac{a_1 + a_2 + \cdots + a_n}{n}$.

26. Examine the following for absolute maximum and minimum values on the given interval.

(a) $y = -x^2$ on $-2 < x < 2$ *Ans.* maximum $(= 0)$ at $x = 0$

(b) $y = (x - 3)^2$ on $0 \le x \le 4$ *Ans.* maximum $(= 9)$ at $x = 0$; minimum $(= 0)$ at $x = 3$

(c) $y = \sqrt{25 - 4x^2}$ on $-2 \le x \le 2$ *Ans.* maximum $(= 5)$ at $x = 0$; minimum $(= 3)$ at $x = \pm 2$

(d) $y = \sqrt{x - 4}$ on $4 \le x \le 29$ *Ans.* maximum $(= 5)$ at $x = 29$; minimum $(= 0)$ at $x = 4$

27. The sum of two positive numbers is 20. Find the numbers if: (a) their product is a maximum; (b) the sum of their squares is a minimum; (c) the product of the square of one and the cube of the other is a maximum.

Ans. (a) 10, 10; (b) 10, 10; (c) 8, 12

28. The product of two positive numbers is 16. Find the numbers if: (a) their sum is least; (b) the sum of one and the square of the other is least.

Ans. (a) 4, 4; (b) 8, 2

29. An open rectangular box with square ends is to be built to hold 6400 ft³ at a cost of \$0.75/ft² for the base and \$0.25/ft² for the sides. Find the most economical dimensions.

Ans. $20 \times 20 \times 16$

30. A wall 8 ft high is $3\frac{3}{8}$ ft from a house. Find the shortest ladder that will reach from the ground to the house when leaning over the wall.

Ans. $15\frac{5}{8}$ ft

31. A company offers the following schedule of charges: \$30 per thousand for orders of 50,000 or less, with the charge decreased by $37\frac{1}{2}$¢ for each thousand above 50,000. Find the order size that makes the company's receipts a maximum.

Ans. 65,000

32. Find an equation of the line through the point (3, 4) that cuts from the first quadrant a triangle of minimum area.

Ans. $4x + 3y - 24 = 0$

33. At what point in the first quadrant on the parabola $y = 4 - x^2$ does the tangent line, together with the coordinate axes, determine a triangle of minimum area?

Ans. $(2\sqrt{3}/3, 8/3)$

34. Find the minimum distance from the point (4, 2) to the parabola $y^2 = 8x$.

Ans. $2\sqrt{2}$

35. (a) Examine $2x^2 - 4xy + 3y^2 - 8x + 8y - 1 = 0$ for maximum and minimum values of *y*. (b) (GC) Check your answer to (a) on a graphing calculator.

Ans. (a) Maximum at (5, 3); (b) minimum at (–1, –3)

36. (GC) Find the absolute maximum and minimum of $f(x) = x^5 - 3x^2 - 8x - 3$ on [–1, 2] to three-decimal-place accuracy.

Ans. Maximum 1.191 at $x = -0.866$; minimum -14.786 at $x = 1.338$

37. An electric current, when flowing in a circular coil of radius *r*, exerts a force $F = \dfrac{kx}{(x^2 + r^2)^{5/2}}$ on a small magnet located at a distance *x* above the center of the coil. Show that *F* is greatest when $x = \frac{1}{2}r$.

38. The work done by a voltaic cell of constant electromotive force *E* and constant internal resistance *r* in passing a steady current through an external resistance *R* is proportional to $E^2R/(r + R)^2$. Show that the work done is greatest when $R = r$.

39. A tangent line is drawn to the ellipse $\frac{x^2}{25} + \frac{y^2}{16} - 1$ so that the part intercepted by the coordinate axes is a minimum. Show that its length is 9.

40. A rectangle is inscribed in the ellipse $\frac{x^2}{400} + \frac{y^2}{225} = 1$ with its sides parallel to the axes of the ellipse. Find the dimensions of the rectangle of (a) maximum area and (b) maximum perimeter that can be so inscribed.

Ans. (a) $20\sqrt{2} \times 15\sqrt{2}$; (b) 32×18

41. Find the radius *R* of the right circular cone of maximum volume that can be inscribed in a sphere of radius *r*. (Recall that the volume of a right circular cone of radius *R* and height *h* is $\frac{1}{3}\pi R^2h$.)

Ans. $R = \frac{2}{3}r\sqrt{2}$

42. A right circular cylinder is inscribed in a right circular cone of radius *r*. Find the radius *R* of the cylinder if: (a) its volume is a maximum; (b) its lateral area is a maximum. (Recall that the volume of a right circular cylinder of radius *R* and height *h* is πR^2h, and its lateral area is $2\pi Rh$.)

Ans. (a) $R = \frac{2}{3}r$; (b) $R = \frac{1}{2}r$

43. Show that a conical tent of given volume will require the least amount of material when its height *h* is $\sqrt{2}$ times the radius *r* of the base. [Note first that the surface area $A = \pi(r^2 + h^2)$.]

44. Show that the equilateral triangle of altitude $3r$ is the isosceles triangle of least area circumscribing a circle of radius *r*.

45. Determine the dimensions of the right circular cylinder of maximum lateral surface area that can be inscribed in a sphere of radius 8.

Ans. $h = 2r = 8\sqrt{2}$

46. Investigate the possibility of inscribing a right circular cylinder of maximum total area (including its top and bottom) in a right circular cone of radius r and height h.

Ans. If $h > 2r$, radius of cylinder $= \frac{1}{2}\left(\frac{hr}{h-r}\right)$

CHAPTER 15

Curve Sketching. Concavity. Symmetry

Concavity

From an intuitive standpoint, an arc of a curve is said to be *concave upward* if it has the shape of a cup (see Fig. 15-1(*a*)) and is said to be *concave downward* if it has the shape of a cap (see Fig. 15-1(*b*)). Note that a more precise definition is available. An arc is concave upward if, for each x_0, the arc lies above the tangent line at x_0 in some open interval around x_0. Similarly, an arc is concave downward if, for each x_0, the arc lies below the tangent line at x_0 in some open interval around x_0. Most curves are combinations of concave upward and concave downward. For example, in Fig. 15-1(*c*), the curve is concave downward from *A* to *B* and from *C* to *D*, but concave upward from *B* to *C*.

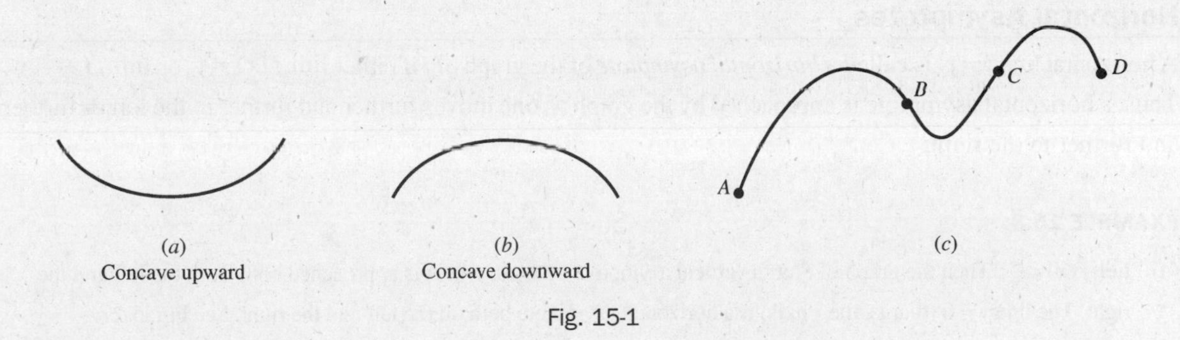

Fig. 15-1

The second derivative of *f* tells us about the concavity of the graph of *f*.

Theorem 15.1:

(a) If $f''(x) > 0$ for x in (a, b), then the graph of *f* is concave upward for $a < x < b$.
(b) If $f''(x) < 0$ for x in (a, b), then the graph of *f* is concave downward for $a < x < b$.

For the proof, see Problem 17.

EXAMPLE 15.1:

(a) Let $f(x) = x^2$. Then $f'(x) = 2x$, $f''(x) = 2$. Since $f''(x) > 0$ for all x, the graph of *f* is always concave upward. This was to be expected, since the graph is a parabola that opens upward.
(b) Let $f(x) = y = \sqrt{1-x^2}$. Then $y^2 = 1 - x^2$, $x^2 + y^2 = 1$. So, the graph is the upper half of the unit circle with the center at the origin. By implicit differentiation, we obtain $x + yy' = 0$ and then $1 + yy'' + (y')^2 = 0$. So, $y'' = -[1 + (y')^2]/y$. Since $y > 0$ (except at $x = 1$), $y'' < 0$. Hence, the graph is always concave downward, which is what we would expect.

Points of Inflection

A point of inflection on a curve $y = f(x)$ is a point at which the concavity changes, that is, the curve is concave upward on one side and concave downward on the other side of the point. So, if y'' exists in an open interval containing x_0, then $y'' < 0$ on one side of x_0 and $y'' > 0$ on the other side of x_0. Therefore, if y'' is continuous at x_0, then $y'' = 0$ at x_0. Thus, we have:

Theorem 15.2: If the graph of f has an inflection point at x_0 and f'' exists in an open interval containing x_0 and f'' is continuous at x_0, then $f''(x_0) = 0$.

EXAMPLE 15.2:

(a) Let $f(x) = x^3$. Then $f'(x) = 3x^2$, $f''(x) = 6x$. Thus, $f''(x) < 0$ for $x < 0$, and $f''(x) > 0$ for $x < 0$. Hence, the graph of f has an inflection point at $x = 0$. (See Fig. 5-5.) Note that $f''(0) = 0$, as predicted by Theorem 15.2.

(b) Let $f(x) = x^4$. Then $f'(x) = 4x^3$, and $f''(x) = 12x^2$. Solving $f''(x) = 0$ yields $x = 0$. However, the graph of f does not have an inflection point at $x = 0$. It is concave upward everywhere. This example shows that $f''(x_0) = 0$ does not necessarily imply that there is an inflection point at x_0.

(c) Let $f(x) = \frac{1}{3}x^3 + \frac{1}{2}x^2 - 6x + 8$. Solving $f''(x) = 2x + 1 = 0$, we find that the graph has an inflection point at $(-\frac{1}{2}, \frac{133}{12})$. Note that this is actually an inflection point, since $f''(x) < 0$ for $x < -\frac{1}{2}$ and $f''(x) > 0$ for $x > -\frac{1}{2}$. See Fig. 14-5.

Vertical Asymptotes

A vertical line $x = x_0$ such that $f(x)$ approaches $+\infty$ or $-\infty$ as x approaches x_0 either from the left or the right is called a *vertical asymptote* of the graph of f. If $f(x)$ has the form $g(x)/h(x)$, where g and h are continuous functions, then the graph of f has a vertical asymptote $x = x_0$ for every x_0 such that $h(x_0) = 0$ (and $g(x_0) \neq 0$).

Horizontal Asymptotes

A horizontal line $y = y_0$ is called a *horizontal asymptote* of the graph of f if either $\lim_{x\to-\infty} f(x) = y_0$ or $\lim_{x\to+\infty} f(x) = y_0$. Thus, a horizontal asymptote is approached by the graph as one moves further and further to the left or further and further to the right.

EXAMPLE 15.3:

(a) Let $f(x) = \frac{1}{x}$. Then the graph of f has a vertical asymptote at $x = 0$, which is approached both from the left and the right. The line $y = 0$ (that is, the x axis) is a horizontal asymptote both on the left and the right. See Fig. 5-21.

(b) Let $f(x) = \frac{1}{x-2}$. Then $x = 2$ is a vertical asymptote of the graph of f, which is approached both from the left and the right. The line $y = 0$ is a horizontal asymptote, which is approached both on the left and the right. See Fig. 14-7.

(c) Let $f(x) = \frac{x-2}{(x-1)(x+3)}$. Then the graph of f has vertical asymptotes at $x = 1$ and $x = -3$. The line $y = 0$ is a horizontal asymptote, which is approached both on the left and the right.

(d) Let $f(x) = \frac{x+4}{x-3}$. Then the graph of f has a vertical asymptote at $x = 3$, which is approached both from the left and the right. The line $y = 1$ is a horizontal asymptote, which is approached both on the left and the right.

Symmetry

We say that two points P and Q are *symmetric with respect to a line l* if l is the perpendicular bisector of the line segment connecting P and Q. [See Fig. 15-2(*a*).]

We say that two points P and Q are *symmetric with respect to a point B* if B is the midpoint of the segment connecting P and Q.

A curve is said to be symmetric with respect to a line l (respectively, point B) if, for any point P on the curve, there is another point Q on the curve such that P and Q are symmetric with respect to l (respectively, B). [See Fig. 15-2(*b*, *c*).]

If a curve is symmetric with respect to a line l, then l is called an *axis of symmetry* of l. For example, any line through the center of a circle is an axis of symmetry of that circle.

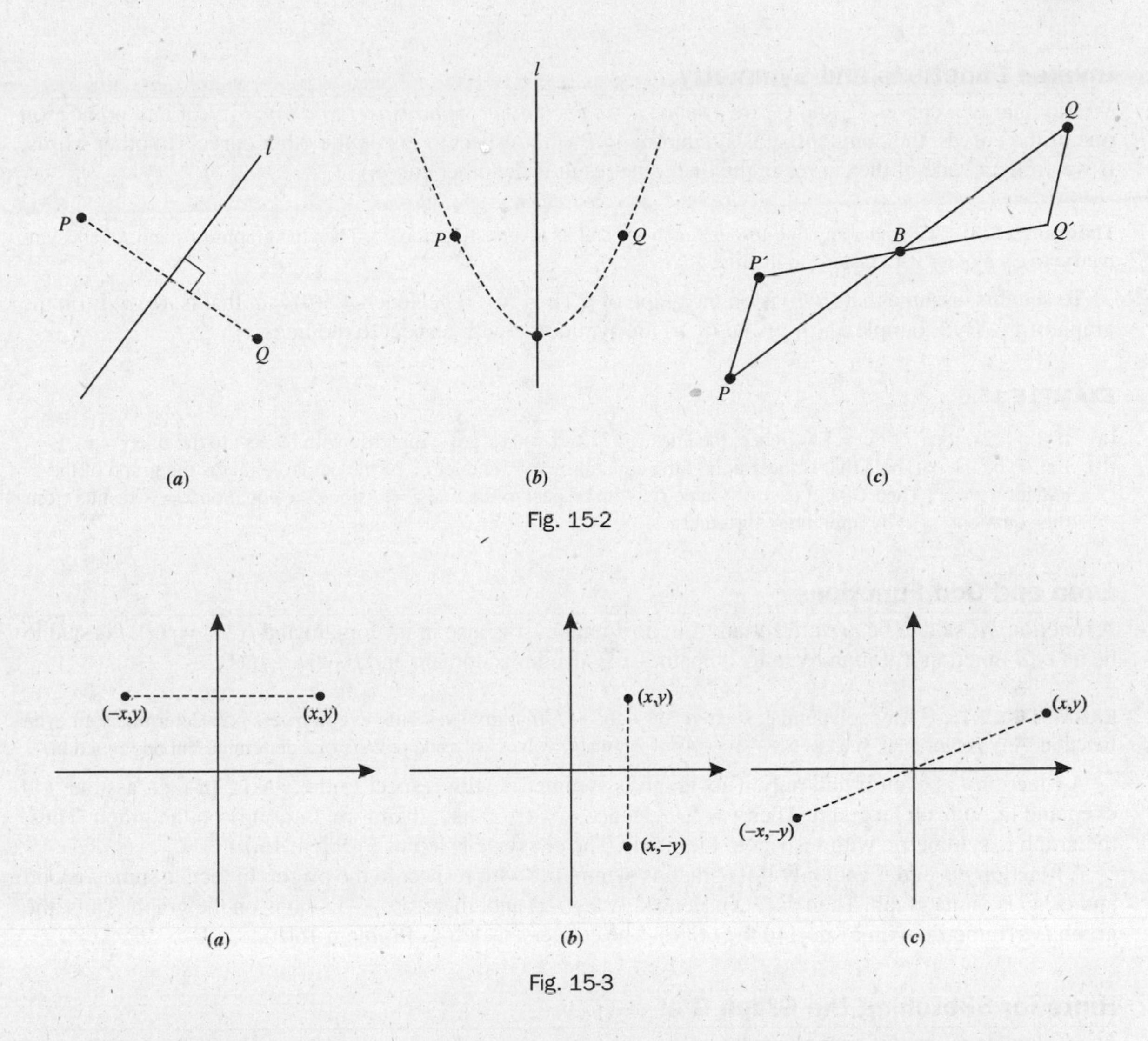

Fig. 15-2

Fig. 15-3

Points (x, y) and $(-x, y)$ are symmetric with respect to the y axis, and points (x, y) and $(x, -y)$ are symmetric with respect to the x axis. Points (x, y) and $(-x, -y)$ are symmetric with respect to the origin. See Fig. 15-3(*a-c*).

Consider the graph of an equation $F(x, y) = 0$. Then:

(i) The graph is symmetric with respect to the y axis if and only if $F(x, y) = 0$ implies $F(-x, y) = 0$.
(ii) The graph is symmetric with respect to the x axis if and only if $F(x, y) = 0$ implies $F(x, -y) = 0$.
(iii) The graph is symmetric with respect to the origin if and only if $F(x, y) = 0$ implies $F(-x, -y) = 0$.

EXAMPLE 15.4

(a) The parabola $y = x^2$ is symmetric with respect to the y axis.
(b) The parabola $x = y^2$ is symmetric with respect to the x axis.
(c) A circle $x^2 + y^2 = r^2$, an ellipse $\frac{x^2}{a^2} + \frac{y^2}{b^2} = 1$, and a hyperbola $\frac{x^2}{a^2} - \frac{y^2}{b^2} = 1$ are symmetric with respect to the y axis, the x axis, and the origin.

EXAMPLE 15.5: A point $P(a, b)$ is symmetric to the point $Q(b, a)$ with respect to the line $y = x$. To see this, note first that the line PQ has slope -1. Since the line $y = x$ has slope 1, the line PQ is perpendicular to the line $y = x$. In addition, the midpoint of the segment connecting P and Q is $\left(\frac{a+b}{2}, \frac{b+a}{2}\right)$, which is on the line $y = x$. Hence, the line $y = x$ is the perpendicular bisector of that segment.

Inverse Functions and Symmetry

We say that two curves C_1 and C_2 are *symmetric to each other with respect to a line l* if, for any point P on one of the curves, the point Q that is symmetric to P with respect to l is on the other curve. (In other words, if we "reflect" one of the curves in the line l, the result is the other curve.)

Theorem 15.3: Consider any one-to-one function f and its inverse function f^{-1}. Then the graphs of f and f^{-1} are symmetric to each other with respect to the line $y = x$.

To see this, assume that (a, b) is on the graph of f. Then $f(a) = b$. Hence, $f^{-1}(b) = a$, that is, (b, a) is on the graph of f^{-1}. By Example 5, (a, b) and (b, a) are symmetric with respect to the line $y = x$.

EXAMPLE 15.6:

(a) If $f(x) = 2x$, then $f^{-1}(x) = \frac{1}{2}x$. Hence, the lines $y = 2x$ and $y = \frac{1}{2}x$ are symmetric with respect to the line $y = x$.
(b) Let C_1 be the parabola that is the graph of the equation $y = x^2$, and let C_2 be the parabola that is the graph of the equation $x = y^2$. Then C_1 and C_2 are symmetric with respect to the line $y = x$, since the equation $x = y^2$ results from the equation $y = x^2$ by interchanging x and y.

Even and Odd Functions

A function f is said to be *even* if, for any x in its domain, $-x$ is also in its domain and $f(-x) = f(x)$. f is said to be an *odd* function if, for any x in its domain, $-x$ is also in its domain and $f(-x) = -f(x)$.

EXAMPLE 15.7: Any polynomial, such as $3x^6 - 8x^4 + 7$, that involves only even powers of x determines an even function. Any polynomial, such as $5x^9 + 2x^5 - 4x^3 + 3x$, that involves only odd powers of x determines an odd function.

A function f is even if and only if its graph is symmetric with respect to the y axis. In fact, assume f is even and (x, y) is on its graph. Then $y = f(x)$. Hence, $y = f(-x)$ and, therefore, $(-x, y)$ is on the graph. Thus, the graph is symmetric with respect to the y axis. The converse is left as Problem 16(*a*).

A function f is odd if and only if its graph is symmetric with respect to the origin. In fact, assume f is odd and (x, y) is on its graph. Then $y = f(x)$. Hence, $-y = f(-x)$ and, therefore, $(-x, -y)$ is on the graph. Thus, the graph is symmetric with respect to the origin. The converse is left as Problem 16(*b*).

Hints for Sketching the Graph G of $y = f(x)$

1. Calculate y', and, if convenient, y''.
2. Use y' to find any critical numbers (where $y' = 0$, or y' is undefined and y is defined). Determine whether these critical numbers yield relative maxima or minima by using the second derivative test or the first derivative test.
3. Use y' to determine the intervals on which y is increasing (when $y' > 0$) or decreasing (when $y' < 0$).
4. Use y'' to determine where G is concave upward (when $y'' > 0$) or concave downward (when $y'' < 0$). Check points where $y'' = 0$ to determine whether they are inflection points (if $y'' > 0$ on one side and $y'' < 0$ on the other side of the point).
5. Look for vertical asymptotes. If $y = g(x)/h(x)$, there is a vertical asymptote $x = x_0$ if $h(x_0) = 0$ and $g(x_0) \neq 0$.
6. Look for horizontal asymptotes. If $\lim_{x\to+\infty} f(x) = y_0$, then $y = y_0$ is a horizontal asymptote on the right. If $\lim_{x\to-\infty} f(x) = y_0$. then $y = y_0$ is a horizontal asymptote on the left.
7. Determine the behavior "at infinity." If $\lim_{x\to+\infty} f(x) = +\infty$ (respectively, $-\infty$), then the curve moves upward (respectively, downward) without bound to the right. Similarly, if $\lim_{x\to-\infty} f(x) = +\infty$ (respectively, $-\infty$), then the curve moves upward (respectively, downward) without bound to the left.
8. Find the y intercepts (where the curve cuts the y axis, that is, where $x = 0$) and the x intercepts (where the curve cuts the x axis, that is, where $y = 0$).
9. Indicate any corner points, where y' approaches one value from the left and another value from the right. An example is the origin on the graph of $y = |x|$.
10. Indicate any cusps, where y' approaches $+\infty$ from both sides or y' approaches $-\infty$ from both sides. An example is the origin on the graph of $y = \sqrt{|x|}$.
11. Find any oblique asymptotes $y = mx + b$ such that $\lim_{x\to+\infty}(f(x) - (mx + b)) = 0$ or $\lim_{x\to-\infty}(f(x) - (mx + b)) = 0$. An oblique asymptote is an asymptote that is neither vertical nor horizontal.

SOLVED PROBLEMS

1. Examine $y = 3x^4 - 10x^3 - 12x^2 + 12x - 7$ for concavity and points of inflection.

We have

$$y' = 12x^3 - 30x^2 - 24x + 12$$

$$y'' = 36x^2 - 60x - 24 = 12(3x+1)(x-2)$$

Set $y'' = 0$ and solve to obtain the possible points of inflection $x = -\frac{1}{3}$ and 2. Then:

When $x < -\frac{1}{3}$.	$y'' = +$, and the arc is concave upward.
When $-\frac{1}{3} < x < 2$.	$y'' = -$, and the arc is concave downward.
When $x > 2$.	$y'' = +$, and the arc is concave upward.

The points of inflection are $(-\frac{1}{3}, -\frac{322}{27})$ and $(2, -63)$, since y'' changes sign at $x = -\frac{1}{3}$ and $x = 2$. See Fig. 15-4.

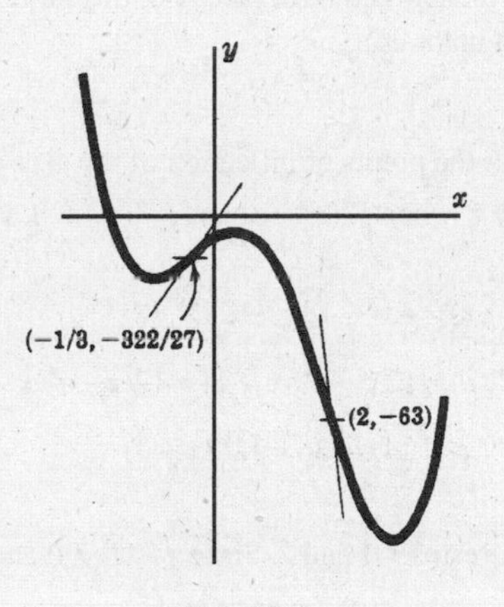

Fig. 15-4

2. Examine $y = x^4 - 6x + 2$ for concavity and points of inflection, and sketch the graph.

We have $y'' = 12x^2$. By Theorem 15.2, the possible point of inflection is at $x = 0$. On the intervals $x < 0$ and $x > 0$, y'' is positive, and the arcs on both sides of $x = 0$ are concave upward. The point (0, 2) is not a point of inflection. Setting $y' = 4x^3 - 6 = 0$, we find the critical number $x = \sqrt[3]{3/2}$. At this point, $y'' = 12x^2 > 0$ and we have a relative minimum by the second derivative test. Since there is only one critical number, there is an absolute minimum at this point (where $x \sim 1.45$ and $y \sim -3.15$). See Fig. 15-5.

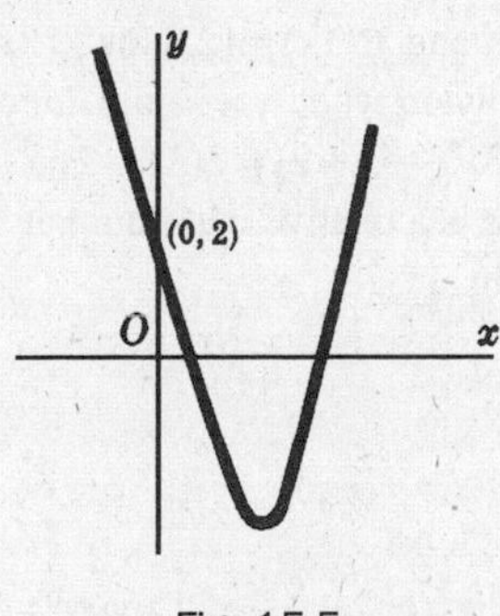

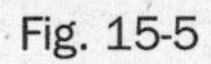
Fig. 15-5

3. Examine $y = 3x + (x+2)^{3/5}$ for concavity and points of inflection, and sketch the graph.

$y' = 3 + \dfrac{3}{5(x+2)^{2/5}}$ and $y'' = \dfrac{-6}{25(x+2)^{7/5}}$. The possible point of inflection is at $x = -2$. When $x > -2$, y'' is negative and the arc is concave downward. When $x < -2$, y'' is positive and the arc is concave upward. Hence, there is an inflection point at $x = -2$, where $y = -6$. (See Fig. 15-6.) Since $y' > 0$ (except at $x = -2$), y is an increasing function, and there are no relative extrema.

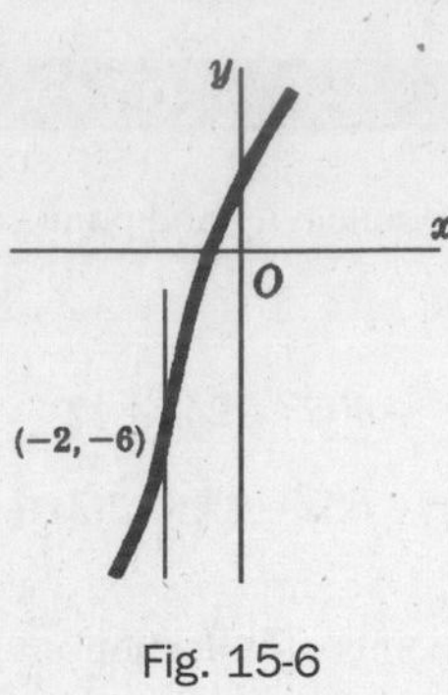

Fig. 15-6

4. If $f''(x_0) = 0$ and $f'''(x_0) \neq 0$, then there is an inflection point at x_0.

Since $f'''(x_0) = 0$, $f'''(x_0)$ is either positive or negative. Hence, f'' is either increasing or decreasing at x_0. Since $f''(x_0) = 0$, f'' has opposite signs to the left and right of x_0. So, the curve will have opposite concavity on the two sides of x_0, and there is an inflection point at x_0.

5. Find equations of the tangent lines at the points of inflection of $y = f(x) = x^4 - 6x^3 + 12x^2 - 8x$.

A point of inflection exists at $x = x_0$ when $f''(x_0) = 0$ and $f'''(x_0) \neq 0$. Here,

$$f'(x) = 4x^3 - 18x^2 + 24x - 8$$

$$f''(x) = 12x^2 - 36x + 24 = 12(x-1)(x-2)$$

$$f'''(x) = 24x - 36 = 12(2x-3)$$

The possible points of inflection are at $x = 1$ and 2. Since $f'''(1) \neq 0$ and $f'''(2) \neq 0$, the points $(1, -1)$ and $(2, 0)$ are points of inflection.

At $(1, -1)$, the slope of the tangent line is $m = f'(1) = 2$, and its equation is

$$y = y_1 = m(x - x_1) \quad \text{or} \quad y + 1 = 2(x-1) \quad \text{or} \quad y = 2x - 3$$

At $(2, 0)$, the slope is $f'(2) = 0$, and the equation of the tangent line is $y = 0$.

6. Sketch the graph of $y = f(x) = 2x^3 - 5x^2 + 4x - 7$.

$f'(x) = 6x^2 - 10x + 4$, $f''(x) = 12x - 10$, and $f'''(x) = 12$. Now, $12x - 10 > 0$ when $x > \frac{5}{6}$ and $12x - 10 < 0$ when $x < \frac{5}{6}$. Hence, the graph of f is concave upward when $x > \frac{5}{6}$, and concave downward when $x < \frac{5}{6}$. Thus, there is an inflection point at $x = \frac{5}{6}$. Since $f''(x) = 2(3x^2 - 5x + 2) = 2(3x-2)(x-1)$, the critical numbers are $x = \frac{2}{3}$ and $x = 1$. Since $f''(\frac{2}{3}) = -2 < 0$ and $f''(1) = 2$, there is a relative maximum at $x = \frac{2}{3}$ (where $y = -\frac{161}{27} \sim -5.96$) and a relative minimum at $x = 1$ (where $y = -6$). See Fig. 15-7.

7 Sketch the graph of $y = f(x) = \dfrac{x^2}{x-2}$.

$$y = \frac{x^2 - 4 + 4}{x-2} = \frac{x^2 - 4}{x-2} + \frac{4}{x-2} = x + 2 + \frac{4}{x-2}. \text{ Then } y' = 1 - \frac{4}{(x-2)^2} \text{ and } y'' = \frac{8}{(x-2)^3}.$$

Solving $y' = 0$, we obtain the critical numbers $x = 4$ and $x = 0$. Since $f''(4) = 1 > 0$ and $f''(0) = -1 < 0$, there is a relative minimum at $x = 4$ (where $y = 8$) and a relative maximum at $x = 0$ (where $y = 0$). Since y'' is never 0, there are no inflection points. The line $x = 2$ is a vertical asymptote. The line $y = x + 2$ is an oblique asymptote on both sides, since, on the curve, $y - (x + 2) = \dfrac{4}{x-2} \to 0$ as $x \to \pm\infty$. See Fig. 15-8.

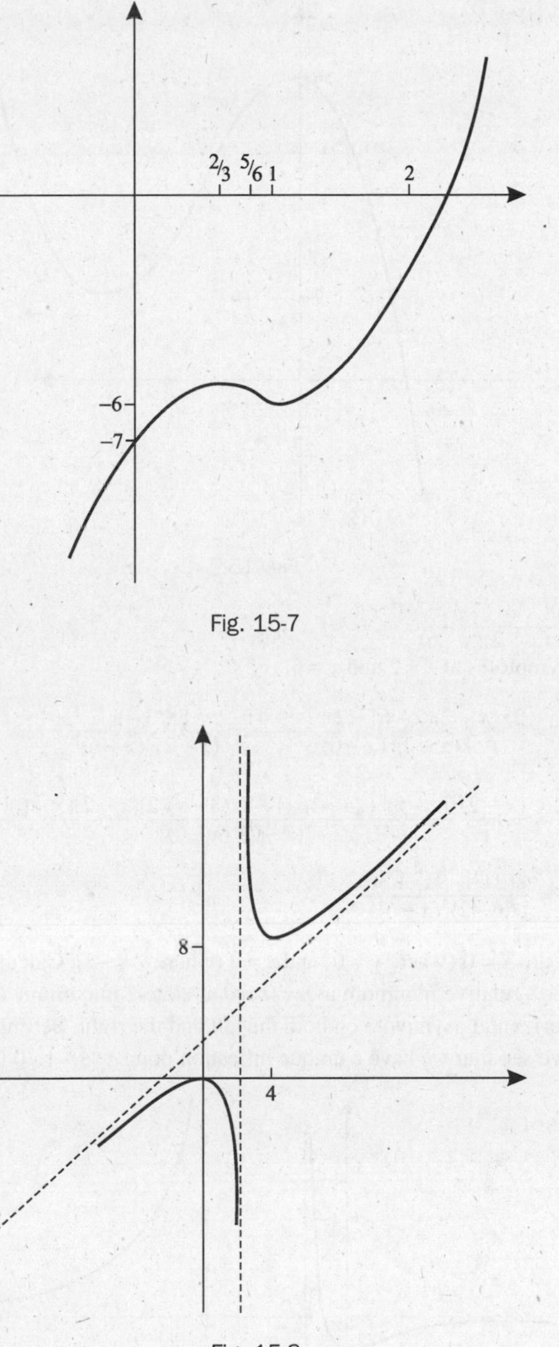

Fig. 15-7

Fig. 15-8

8. Sketch the graph of $g(x) = 2x^3 - 9x^2 + 36$.

$g'(x) = 6x^2 - 18x = 6x(x - 3)$ and $g''(x) = 12x - 18 = 6(2x - 3)$. So, the critical numbers are $x = 0$ (where $y = 36$) and $x = 3$ (where $y = 9$). Since $g''(0) = -18 < 0$ and $g''(3) = 18 > 0$, there is a relative maximum at $x = 0$ and a relative minimum at $x = 3$. Setting $g''(x) = 0$ yields $x = \frac{3}{2}$, where there is an inflection point, since $g''(x) = 6(2x - 3)$ changes sign at $x = \frac{3}{2}$.

$g(x) \to +\infty$ as $x \to +\infty$, and $g(x) \to -\infty$ as $x \to -\infty$. Since $g(-1) = 29$ and $g(-2) = -16$, the intermediate value theorem implies that there is a zero x_0 of g between -1 and -2. (A graphing calculator shows $x_0 \sim -1.70$.) That is the only zero because g is increasing up to the point (0, 36), decreasing from (0, 36) to (3, 9), and then increasing from (3, 9). See Fig. 15-9.

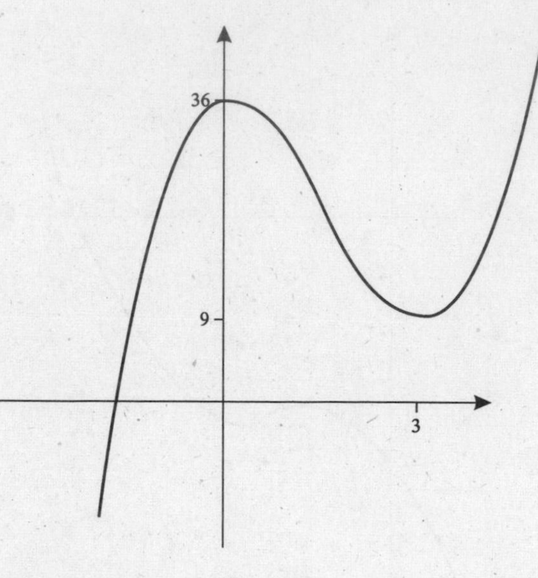

Fig. 15-9

9. Sketch the graph of $y = \dfrac{x^2}{(x-2)(x-6)}$.

There are vertical asymptotes at $x = 2$ and $x = 6$.

$$y' = \frac{2x(x-2)(x-6) - 2x^2(x-4)}{(x-2)^2(x-6)^2} = \frac{8x(3-x)}{(x-2)^2(x-6)^2}$$

$$y'' = \frac{(x-2)^2(x-6)^2(24-16x) - 8x(3-x)(2)(x-2)(x-6)(2x-8)}{(x-2)^4(x-6)^4}$$

$$= \frac{8(2x^3 - 9x^2 + 36)}{(x-2)^3(x-6)^3}$$

The critical numbers are $x = 0$ (where $y = 0$) and $x = 3$ (where $y = -3$). Calculation shows that $y''(0) > 0$ and $y''(3) < 0$. Hence, there is a relative minimum at $x = 0$ and a relative maximum at $x = 3$. Since $y \to 1$ when $x \to \pm\infty$, the line $y = 1$ is a horizontal asymptote on both the left and the right. Setting $y'' = 0$ yields $g(x) = 2x^3 - 9x^2 + 36 = 0$. By Problem 8, we see that we have a unique inflection point $x_0 \sim -1.70$ (where $y \sim 0.10$). See Fig. 15-10.

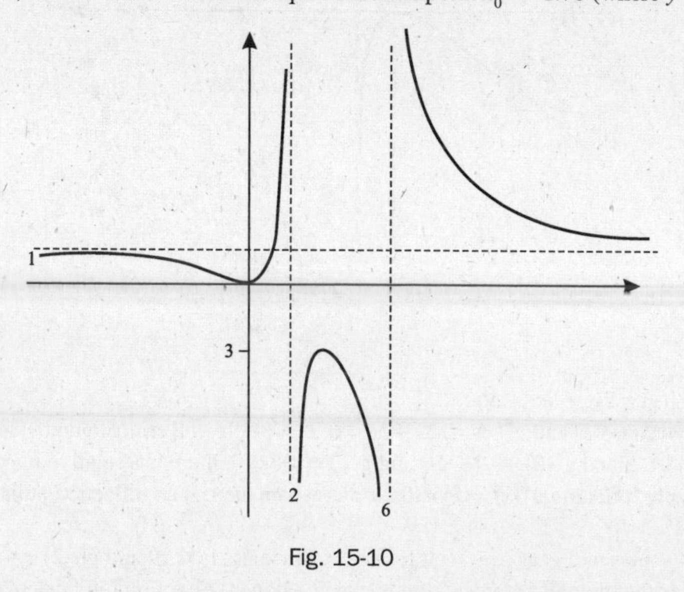

Fig. 15-10

10. Sketch the graph of $y^2(x^2 - 4) = x^4$.

$y^2 = \dfrac{x^4}{x^2-4}$. Then $y = \pm\dfrac{x^2}{\sqrt{x^2-4}}$. The curve exists only for $x^2 > 4$, that is, for $x > 2$ or $x < -2$, plus the isolated point (0, 0).

The curve is symmetric with respect to both coordinate axes and the origin. So, from now on, we shall consider only the first quadrant. Then

$$y' = \frac{x^3 - 8x}{(x^2-4)^{3/2}} \quad \text{and} \quad y'' = \frac{4x^2+32}{(x^2-4)^{5/2}}$$

The only critical number is $2\sqrt{2}$ (where $y = 4$). Since $y'' > 0$, the graph is concave upward and there is a relative minimum at $(2\sqrt{2}, 4)$. The lines $x = 2$ and $x = -2$ are vertical asymptotes. The rest of the graph in the other quadrants is obtained by reflection in the axes and origin. Note that there is also an oblique asymptote $y = x$, since $y^2 - x^2 = x^4/(x^2-4) - x^2 = 4/(x^2-4) \to 0$ as $x \to \pm\infty$. By symmetry, $y = -x$ is also an asymptote. See Fig. 15-11.

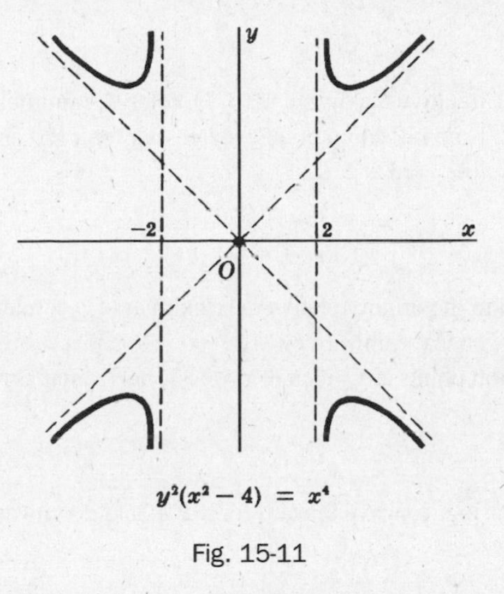

Fig. 15-11

SUPPLEMENTARY PROBLEMS

11. Examine the functions of Problem 23(*a–f*) of Chapter 14.

Ans. (a) No inflection point, concave upward everywhere
(b) No inflection point, concave downward everywhere
(c) Inflection point at $x = -\frac{2}{3}$, concave upward for $x > -\frac{2}{3}$, concave downward for $x < -\frac{2}{3}$
(d) Inflection point at $x = 2$, concave upward for $x > 2$, concave downward for $x < 2$
(e) Inflection point at $x = 2$, concave downward for $x > 2$, concave upward for $x < 2$
(f) Inflection point at $x = \pm\frac{2\sqrt{3}}{3}$, concave upward for $x > \frac{2\sqrt{3}}{3}$ and $x < -\frac{2\sqrt{3}}{3}$, concave downward for $-\frac{2\sqrt{3}}{3} < x < \frac{2\sqrt{3}}{3}$

12. Prove: If $f(x) = ax^3 + bx^2 + cx + d$ has two critical numbers, their average is the abscissa at the point of inflection. If there is just one critical number, it is the abscissa at the point of inflection.

13. Discuss and sketch the graphs of the following equations:

(a) $xy = (x^2 - 9)^2$

Ans. Symmetric with respect to the origin, vertical asymptote $x = 0$, relative minimum at (3, 0), relative maximum at (−3, 0), no inflection points, concave upward for $x > 0$

(b) $y = \frac{x^4}{1 - x^2}$

Ans. Symmetric with respect to the y axis, vertical asymptotes $x = \pm 1$, relative minimum at (0, 0), relative maxima at $(\pm\sqrt{2}, -4)$, no inflection points, concave upward for $|x| < 2$

(c) $y = x^2 + \frac{2}{x}$

Ans. Vertical asymptote $x = 0$, relative minimum at (1, 3), inflection point at $(-\sqrt[3]{2}, 0)$, concave upward for $x < -\sqrt[3]{2}$ and $x > 0$

(d) $y^3 = 6x^2 - x^3$

Ans. Relative maximum at $(4, 2\sqrt[3]{4})$, relative minimum at (0, 0), where there is a "cusp," inflection point (6, 0), concave upward for $x > 6$, oblique asymptote $y = -x + 2$ to the left and the right

(e) $y = 1 + \frac{x^2}{x-1}$

Ans. Vertical asymptote $x = 1$, relative maximum at (0, 1), relative minimum at (2, 5), concave upward for $x > 1$ and downward for $x < 1$, no inflection points, increasing for $x < 0$ and $x > 2$, decreasing for $0 < x < 1$ and $1 < x < 2$, oblique asymptote $y = x + 2$

(f) $y = \frac{x}{x^2 + 1}$

Ans. Symmetric with respect to the origin, relative maximum at $(1, \frac{1}{2})$, relative minimum at $(-1, -\frac{1}{2})$, increasing on $-1 < x < 1$, concave upward on $-\sqrt{3} < x < 0$ and $x > \sqrt{3}$, concave downward on $x < -\sqrt{3}$ and $0 < x < \sqrt{3}$, inflection points at $x = 0$ and $x = \pm\sqrt{3}$, horizontal asymptote $y = 0$ on both sides

(g) $y = x\sqrt{x-1}$

Ans. Defined for $x \geq 1$, increasing, concave upward for $x > \frac{4}{3}$, and downward for $x < \frac{4}{3}$, inflection point $(\frac{4}{3}, \frac{4}{9}\sqrt{3})$

(h) $y = x\sqrt[3]{2-x}$

Ans. Relative maximum at $x = \frac{3}{2}$, increasing for $x < \frac{3}{2}$, concave downward for $x < 3$, inflection point (3, −3)

(i) $y = \frac{x+1}{x^2}$

Ans. Vertical asymptote $x = 0$, horizontal asymptote $y = 0$ on both sides, relative minimum $(-2, -\frac{1}{4})$, increasing for $-2 < x < 0$, concave upward for $-3 < x < 0$ and $x > 0$, inflection point at $(-3, -\frac{2}{9})$, $y \to +\infty$ as $x \to 0$

14. Show that any function $F(x)$ that is defined for all x may be expressed in one and only one way as the sum of an even and an odd function. [*Hint*: Let $E(x) = \frac{1}{2}(F(x) + F(-x))$.]

15. Find an equation of the new curve C_1 that is obtained when the graph of the curve C with the equation $x^2 - 3xy + 2y^2 = 1$ is reflected in: (a) the x axis; (b) the y axis; (c) the origin.

Ans. (a) $x^2 + 3xy + 2y^2 = 1$; (b) same as (a); (c) C itself

16. (a) If the graph of f is symmetric with respect to the y axis, show that f is even. (b) If the graph of f is symmetric with respect to the origin, then show that f is odd. [*Hint*: For (a), if x is in the domain of f, $(x, f(x))$ is on the graph and, therefore, $(-x, f(x))$ is on the graph. Thus, $f(-x) = f(x)$.]

17. Prove Theorem 15.1: (a) If $f''(x) > 0$ for x in (a, b), then the graph of f is concave upward for $a < x < b$. (b) If $f''(x) < 0$ for x in (a, b), then the graph of f is concave downward for $a < x < b$.

[For (*a*), let x_0 belong to (a, b). Since $f''(x_0) > 0$, f' is increasing in some open interval I containing x_0. Assume x in I and $x > x_0$. By the law of the mean, $f(x) - f(x_0) = f'(x^*)(x - x_0)$ for some x^* with $x_0 < x^* < x$. Since f' is increasing, $f'(x_0) < f'(x^*)$. Then $f(x) = f'(x^*)(x - x_0) + f(x_0) > f'(x_0)(x - x_0) + f(x_0)$. But $y = f'(x_0)(x - x_0) + f(x_0)$ is an equation of the tangent line at x_0. A similar argument works when $x < x_0$. Thus, the curve lies above the tangent line and, therefore, is concave upward.]

18. (GC) Use a graphing calculator to draw the graph of $f(x) = x^3 - 3x^2 + 4x - 2$. Show analytically that f is increasing and that there is an inflection point at $(-1, 3)$. Use the calculator to draw the graph of f^{-1} and $y = x$, and observe that the graphs of f and f^{-1} are symmetric with respect to $y = x$.

19. (GC) Try to sketch the graph of $y = \dfrac{x^2}{x^3 - 3x^2 + 5}$ by standard methods and then use a graphing calculator for additional information (such as the location of any vertical asymptotes).

CHAPTER 16

Review of Trigonometry

Angle Measure

The traditional unit for measuring angles is the degree. 360 degrees make up a complete rotation. However, it turns out that a different unit, the radian, is more useful in calculus. Consider a circle of radius 1 and with center at a point C. (See Fig. 16-1.) Let CA and CB be two radii for which the arc $\widehat{AB}$ of the circle has length 1. Then one *radian* is taken to be the measure of the central angle ACB.

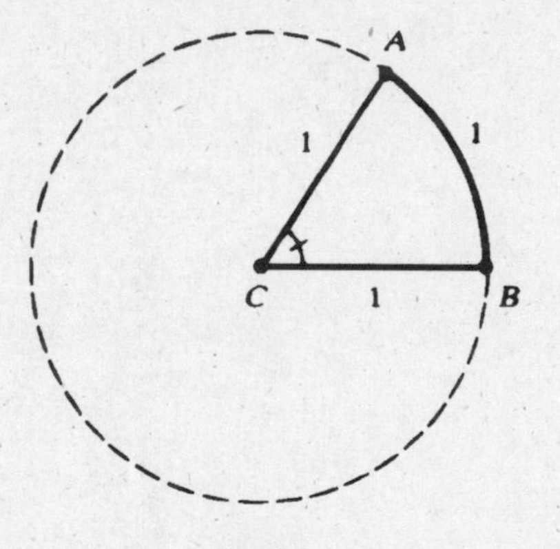

Fig. 16-1

If u is the number of degrees in angle ACB, then the ratio of u to 360° is equal to the ratio of $\widehat{AB}$ to the circumference 2π. Since $\widehat{AB} = 1$, $u/360 = 1/2\pi$ and, therefore, $u = 180/\pi$. So,

$$1 \text{ radian} = \frac{180}{\pi} \text{ degrees} \tag{1}$$

If π is approximated as 3.14, then 1 radian is about 57.3 degrees. Multiplying equation (1) by $\pi/180$, we obtain:

$$1 \text{ degree} = \frac{\pi}{180} \text{ radians} \tag{2}$$

The table in Fig. 16-2 shows the radian measure of some important degree measures.

Now take any circle of radius r with center O. (See Fig. 16-3.) Let $\angle DOE$ contain θ radians and let s be the length of arc DE. The ratio of θ to the number 2π of radians in a complete rotation is equal to the ratio of s to the entire circumference $2\pi r$. So, $\theta/2\pi = s/2\pi r$. Therefore,

$$s = r\theta \tag{3}$$

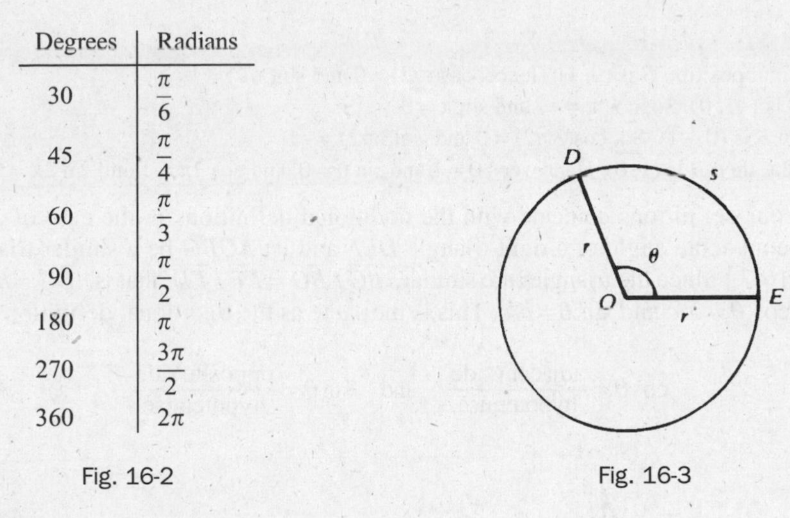

Degrees	Radians
30	$\frac{\pi}{6}$
45	$\frac{\pi}{4}$
60	$\frac{\pi}{3}$
90	$\frac{\pi}{2}$
180	π
270	$\frac{3\pi}{2}$
360	2π

Fig. 16-2

Fig. 16-3

Directed Angles

If an angle is thought of as being generated by a rotation, then its measure will be counted as positive if the rotation is counterclockwise and negative if the rotation is clockwise. See, for example, angles of $\pi/2$ radians and $-\pi/2$ radians in Fig. 16-4. We shall also allow angles of more than one complete rotation. For example, Fig. 16-5 shows a counterclockwise angle generated by a complete rotation plus another quarter of a complete rotation, yielding an angle of $2\pi + \pi/2 = 5\pi/2$ radians, and an angle of 3π radians generated by one and a half turns in the counterclockwise direction.

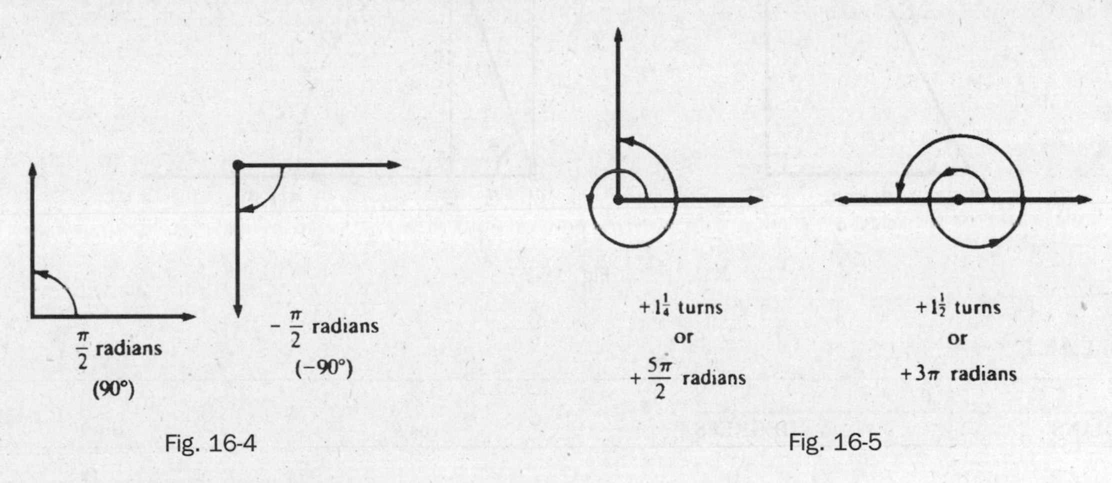

Fig. 16-4

Fig. 16-5

Sine and Cosine Functions

Consider a coordinate system with origin at O and point A at (1, 0). Rotate the arrow OA through an angle of θ degrees to a new position OB. Then (see Fig. 16-6):

1. $\cos\theta$ is defined to be the x coordinate of the point B.
2. $\sin\theta$ is defined to be the y coordinate of the point B.

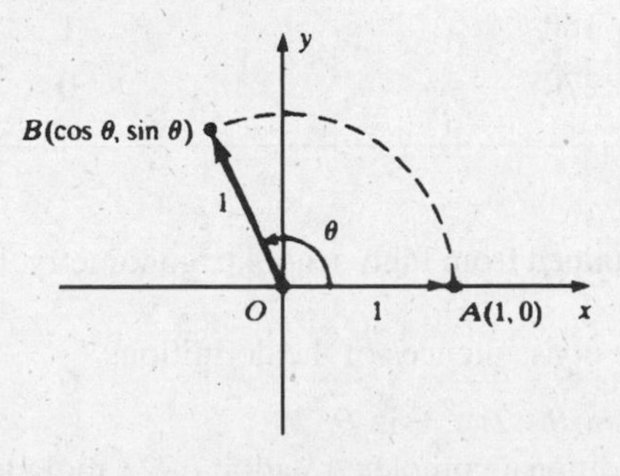

Fig. 16-6

EXAMPLE 16.1:

(a) If $\theta = \pi/2$, the final position B is (0, 1). Hence, $\cos(\pi/2) = 0$ and $\sin(\pi/2) = 1$.
(b) If $\theta = \pi$, then B is $(-1, 0)$. So, $\cos \pi = -1$ and $\sin \pi = 0$.
(c) If $\theta = 3\pi/2$, then B is $(0, -1)$. So, $\cos(3\pi/2) = 0$ and $\sin(3\pi/2) = -1$.
(d) If $\theta = 0$ or $\theta = 2\pi$, then B is (1, 0). Hence, $\cos 0 = 1$ and $\sin 0 = 0$, and $\cos 2\pi = 1$ and $\sin 2\pi = 0$.

Let us see that our definitions coincide with the traditional definitions in the case of an acute angle of a triangle. Let θ be an acute angle of a right triangle DEF and let ΔOBG be a similar triangle with hypotenuse 1. (See Fig. 16-7.) Since the triangles are similar, $\overline{BG}/\overline{BO} = \overline{EF}/\overline{ED}$, that is, $\overline{BG} = b/c$, and, likewise, $\overline{OG} = a/c$ Hence, $\cos \theta = a/c$ and $\sin \theta = b/c$. This is the same as the traditional definitions:

$$\cos\theta = \frac{\text{adjacent side}}{\text{hypotenuse}} \quad \text{and} \quad \sin\theta = \frac{\text{opposite side}}{\text{hypotenuse}}$$

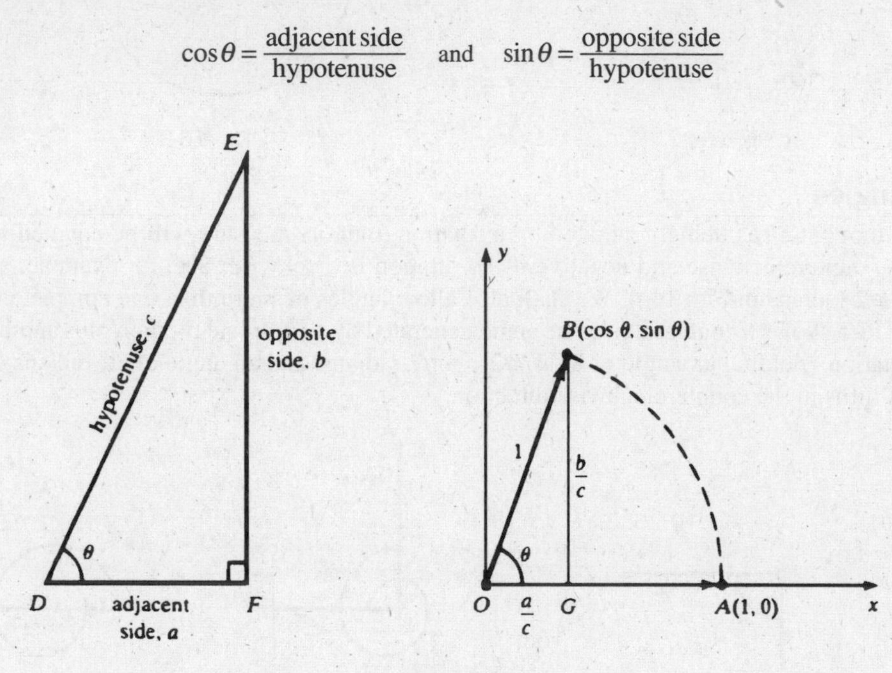

Fig. 16-7

TABLE 16.1

θ			
RADIANS	**DEGREES**	**cos θ**	**sin θ**
0	0	1	0
$\pi/6$	30	$\sqrt{3}/2$	1/2
$\pi/4$	45	$\sqrt{2}/2$	$\sqrt{2}/2$
$\pi/3$	60	1/2	$\sqrt{3}/2$
$\pi/2$	90	0	1
π	180	−1	0
$3\pi/2$	270	0	−1

We now can use the values obtained from high-school trigonometry. [See Problem 22(*a–c*).] Table 16-1 lists the most useful values.

Let us first collect some simple consequences of the definitions.

(16.1) $\cos(\theta + 2\pi) = \cos \theta$ and $\sin(\theta + 2\pi) = \sin \theta$

This holds because an additional complete rotation of 2π radians brings us back to the same point.

(16.2) $\cos(-\theta) = \cos \theta$ and $\sin(-\theta) = -\sin \theta$ (see Fig. 16-8)

(16.3) $\sin^2 \theta + \cos^2 \theta = 1$ [In accordance with tradiational notation, $\sin^2 \theta$ and $\cos^2 \theta$ stand for $(\sin \theta)^2$ and $(\cos \theta)^2$.]

In Fig. 16-6, $1 = \overline{OB} = \sqrt{\cos^2\theta + \sin^2\theta}$ by Problem 1 of Chapter 2. (16.3) implies $\sin^2 \theta = 1 - \cos^2 \theta$ and $\cos^2 \theta = 1 - \sin^2 \theta$.

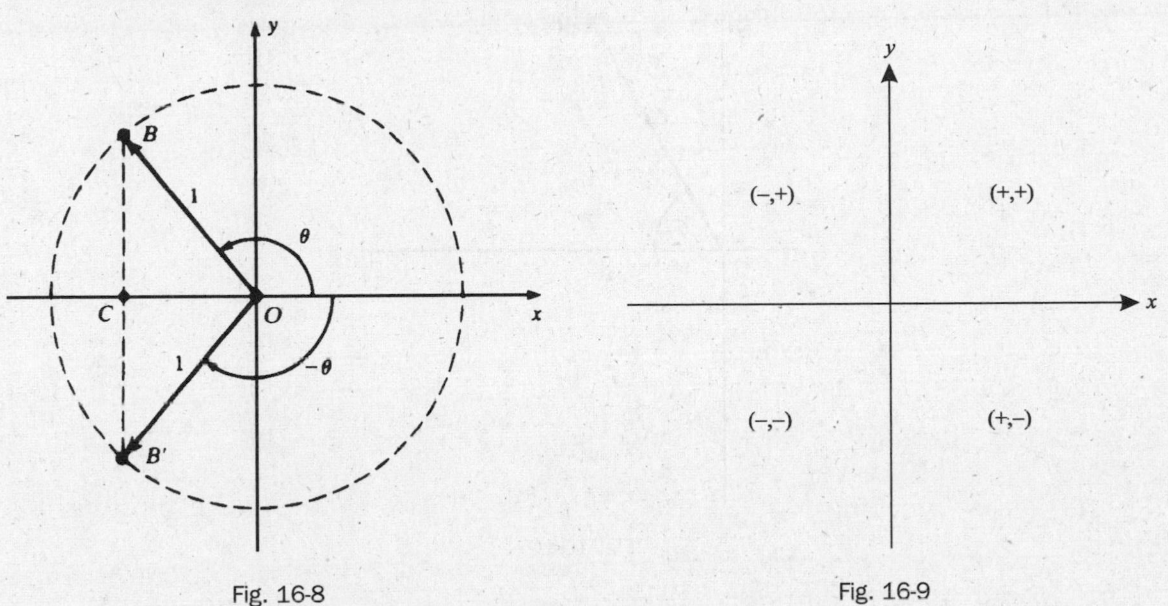

Fig. 16-8 Fig. 16-9

(16.4) In the four quadrants, the sine and cosine have the signs shown in Fig. 16-9.

(16.5) For any point $A(x, y)$ different from the origin O, let r be its distance from the origin, and let θ be the radian measure of the angle from the positive x axis to the arrow OA. (See Fig. 16-10.) The pair (r, θ) are called *polar coordinates* of A. Then $x = r \cos \theta$ and $y = r \sin \theta$. (See Problem 8.)

For the derivation of more complicated formulas, we shall depend on the following result.

(16.6) $\cos(u - v) = \cos u \cos v + \sin u \sin v$

For the proof, see Problem 11.

(16.7) $\cos(u + v) = \cos u \cos v - \sin u \sin v$

Replace v by $-v$ in (16.6) and use (16.2).

(16.8) $\cos(\pi/2 - v) = \sin v$ and $\sin(\pi/2 - v) = \cos v$

Replace u by $\pi/2$ in (16.6) and use $\cos(\pi/2) = 0$ and $\sin(\pi/2) = 1$. This yields $\cos(\pi/2 - v) = \sin v$. In this formula, replace v by $(\pi/2 - v)$ to obtain $\cos v = \sin(\pi/2 - v)$.

(16.9) $\sin(u + v) = \sin u \cos v + \cos u \sin v$

By (16.6) and (16.8),

$$\sin(u+v) = \cos[(\pi/2) - (u+v)] = \cos[(\pi/2 - u) - v]$$

$$= \cos(\pi/2 - u)\cos v + \sin(\pi/2 - u)\sin v = \sin u \cos v + \cos u \sin v$$

(16.10) $\sin(u - v) = \sin u \cos v - \cos u \sin v$

Replace v by $-v$ in (16.9) and use (16.2).

(16.11) $\cos 2u = \cos^2 u - \sin^2 u = 2\cos^2 u - 1 = 1 - 2\sin^2 u$

Replace v by u in (16.7) to get $\cos 2u = \cos^2 u - \sin^2 u$. Use $\sin^2 u = 1 - \cos^2 u$ and $\cos^2 u = 1 - \sin^2 u$ to obtain the other two forms.

(16.12) $\sin 2u = 2 \sin u \cos u$

Replace v by u in (16.9).

(16.13) $\cos^2\left(\frac{u}{2}\right) = \frac{1+\cos u}{2}$

$$\cos u = \cos\left(2 \cdot \frac{u}{2}\right) = 2\cos^2\left(\frac{u}{2}\right) - 1$$

by (16.11). Now solve for $\cos^2\left(\frac{u}{2}\right)$.

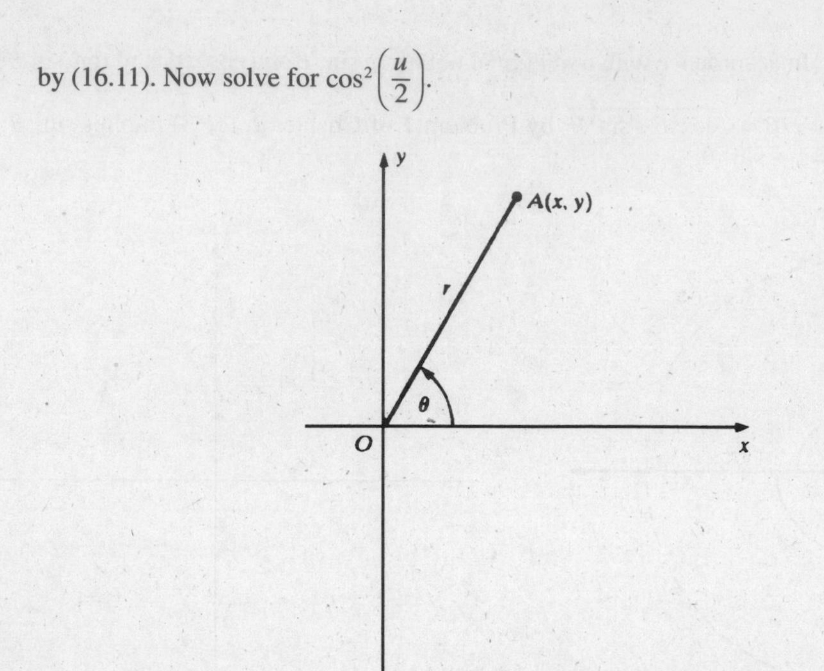

Fig. 16-10

(16.14) $\sin^2\left(\frac{u}{2}\right)=\frac{1-\cos u}{2}$

By (16.3) and (16.13),

$$\sin^2\left(\frac{u}{2}\right)=1-\cos^2\left(\frac{u}{2}\right)=1-\frac{1+\cos u}{2}=\frac{1-\cos u}{2}$$

(16.15) (a) (Law of Cosines). In any triangle ΔABC (see Fig. 16-11),

$$c^2=a^2+b^2-2ab\cos\theta$$

For a proof, see Problem 11(a).

(b) (Law of Sines)

$$\frac{\sin A}{a}=\frac{\sin B}{b}=\frac{\sin C}{c}$$

where $\sin A$ is $\sin(\angle BAC)$, and similarly for $\sin B$ and $\sin C$.

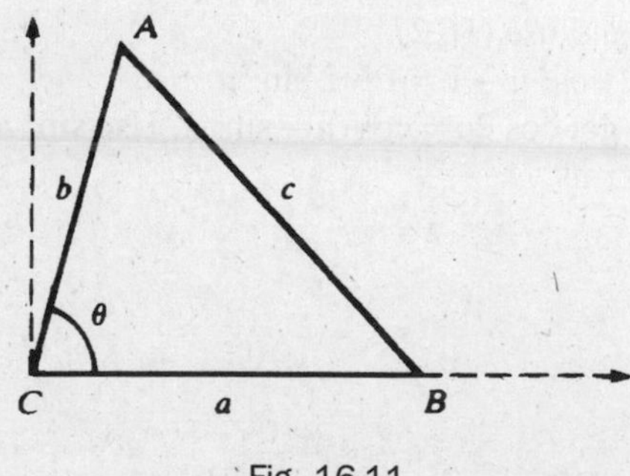

Fig. 16-11

SOLVED PROBLEMS

1. Translate the following degree measures into radian measures: (a) 54°; (b) 120°.

(a) $54° = 54\left(\frac{\pi}{180}\text{ radians}\right) = \frac{3}{10}\pi$ radians

(b) $120° = 120\left(\frac{\pi}{180}\text{ radians}\right) = \frac{2\pi}{3}$ radians

2. Translate the following radian measures into degree measures: (a) $\frac{2\pi}{5}$ radians; (b) $\frac{5\pi}{6}$ radians; (c) 2 radians.

(a) $\frac{2\pi}{5}$ radians $= \frac{2\pi}{5}\left(\frac{180}{\pi}\text{ degrees}\right) = 72°$

(b) $\frac{5\pi}{6}$ radians $= \frac{5\pi}{6}\left(\frac{180}{\pi}\text{ degrees}\right) = 150°$

(c) 2 radians $= 2\left(\frac{180}{\pi}\text{ degrees}\right) = \left(\frac{360}{\pi}\right)^{\circ}$

3. (a) In a circle of radius $r = 3$ centimeters, what arc length s along the circumference corresponds to a central angle θ of $\pi/6$ radians?
(b) In a circle of radius 4 feet, what central angle corresponds to an arc length of 8 feet?

We know that $s = r\theta$, where θ is measured in radians.

(a) $s = 3\left(\frac{\pi}{6}\right) = \frac{\pi}{2}$ centimeters

(b) $\theta = \left(\frac{s}{r}\right) = \frac{8}{4} = 2$ radians

4. What rotations between 0 and 2π radians have the same effect as the rotations with the following measures? (a) $\frac{11\pi}{4}$ radians; (b) 405°; (c) $-\frac{\pi}{3}$ radians; (d) -5π radians.

(a) $\frac{11\pi}{4} = 2\pi + \frac{3\pi}{4}$ So, the equivalent rotation is $\frac{3\pi}{4}$ radians.

(b) $405° = (360 + 45)°$. So, the equivalent rotation is 45°.

(c) $-\frac{\pi}{3} + 2\pi = \frac{5\pi}{3}$. So, the equivalent rotation is $\frac{5\pi}{3}$ radians.

(d) $-5\pi + 6\pi = \pi$. So, the equivalent rotation is π radians.

5. Find $\sin\theta$ if θ is an acute angle such that $\cos\theta = \frac{4}{5}$.
By (16.3), $\frac{4}{5}^2 + \sin^2\theta = 1$. So, $\sin^2\theta = \frac{9}{25}$ and, therefore, $\sin\theta = \pm\frac{3}{5}$. Since θ is acute, $\sin\theta$ is positive. So, $\sin\theta = \frac{3}{5}$.

6. Show that $\sin(\pi - \theta) = \sin\theta$ and $\cos(\pi - \theta) = -\cos\theta$.
By (16.10), $\sin(\pi - \theta) = \sin\pi\cos\theta - \cos\pi\sin\theta = (0)\cos\theta - (-1)\sin\theta = \sin\theta$. By (16.6), $\cos(\pi - \theta) = \cos\pi\cos\theta + \sin\pi\sin\theta = (-1)\cos\theta + (0)\sin\theta = -\cos\theta$.

7. Calculate the following values: (a) $\sin 2\pi/3$; (b) $\sin\frac{7\pi}{3}$; (c) $\cos 9\pi$; (d) $\sin 420°$; (e) $\cos 3\pi/4$; (f) $\cos\pi/12$; (g) $\sin\pi/8$; (h) $\sin 19°$.

(a) By Problem 6, $\sin\frac{2\pi}{3} = \sin\left(\pi - \frac{\pi}{3}\right) = \sin\frac{\pi}{3} = \frac{\sqrt{3}}{2}$

(b) By (16.1), $\sin\frac{7\pi}{3} = \sin\left(2\pi + \frac{\pi}{3}\right) = \sin\frac{\pi}{3} = \frac{\sqrt{3}}{2}$

(c) By (16.1), $\cos 9\pi = \cos(\pi + 8\pi) = \cos\pi = -1$

(d) By (16.1), $\sin 390° = \sin(30 + 360)° = \sin 30° = \frac{1}{2}$

(e) By Problem 6, $\cos\frac{3\pi}{4} = \cos\left(\pi - \frac{\pi}{4}\right) = -\cos\frac{\pi}{4} = -\frac{\sqrt{2}}{2}$

(f) $\cos\frac{\pi}{12} = \cos\left(\frac{\pi}{3} - \frac{\pi}{4}\right) = \cos\frac{\pi}{3}\cos\frac{\pi}{4} + \sin\frac{\pi}{3}\sin\frac{\pi}{4} = \frac{1}{2}\frac{\sqrt{2}}{2} + \frac{\sqrt{3}}{2}\frac{\sqrt{2}}{2} = \frac{\sqrt{2}+\sqrt{6}}{4}$

(g) By (16.14), $\sin^2\left(\frac{\pi}{8}\right) = \frac{1-\cos(\pi/4)}{2} = \frac{1-(\sqrt{2}/2)}{2} = \frac{2-\sqrt{2}}{4}$. Hence, $\sin\frac{\pi}{8} = \pm\frac{\sqrt{2-\sqrt{2}}}{2}$. Since $0 < \frac{\pi}{8} < \frac{\pi}{2}$, $\sin\frac{\pi}{8}$ is positive and, therefore, $\sin\frac{\pi}{8} = \frac{\sqrt{2-\sqrt{2}}}{2}$.

(h) 19° cannot be expressed in terms of more familiar angles (such as 30°, 45°, or 60°) in such a way that any of our formulas are applicable. We must then use the sine table in Appendix A, which gives 0.3256; this is an approximation correct to four decimal places.

8. Prove the result of (16.5): If (r, θ) are polar coordinates of (x, y), then $x = r\cos\theta$ and $y = r\sin\theta$.

Let D be the foot of the perpendicular from $A(x, y)$ to the x axis (see Fig. 16-12). Let F be the point on the ray OA at a unit distance from the origin. Then $F = (\cos\theta, \sin\theta)$. If E is the foot of the perpendicular from F to the x axis, then $\overline{OE} = \cos\theta$ and $\overline{FE} = \sin\theta$ Since ΔADO is similar to ΔFEO (by the AA criterion), we have:

$$\frac{\overline{OD}}{\overline{OE}} = \frac{\overline{OA}}{\overline{OF}} = \frac{\overline{AD}}{\overline{FE}}, \text{ that is, } \frac{x}{\cos\theta} = \frac{r}{1} = \frac{y}{\sin\theta}$$

Hence, $x = r\cos\theta$ and $y = r\sin\theta$. When $A(x, y)$ is in one of the other quadrants, the proof can be reduced to the case where A is in the first quadrant. The case when A is on the x axis or the y axis is very easy.

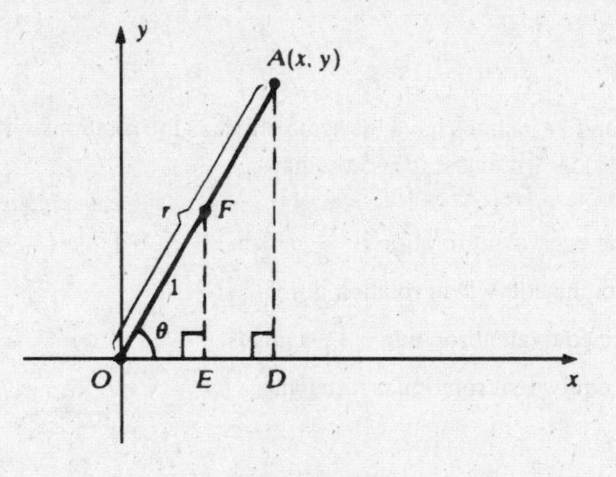

Fig. 16-12

9. Find rectangular coordinates of the point with polar coordinates $r = 3$, $\theta = \pi/6$.

By (16.5), $x = r\cos\theta = 3\cos\frac{\pi}{6} = 3\frac{\sqrt{3}}{2}$, and $y = r\sin\theta = 3\sin\frac{\pi}{2} = 3\left(\frac{1}{2}\right) = \frac{3}{2}$.

10. Find polar coordinates of the point $(1, \sqrt{3})$.

By (16.5), $r^2 = x^2 + y^2 = 1 + 3 = 4$. Then $r = 2$. So, $\cos\theta = \frac{x}{r} = \frac{1}{2}$, and $\sin\theta = \frac{y}{r} = \frac{\sqrt{3}}{2}$. Thus, $\theta = \frac{\pi}{3}$.

11. (a) Prove the law of cosines (16.15(a)). (b) Prove the law of sines (16.15(b)).

(a) See Fig. 16-11. Take a coordinate system with C as origin and B on the positive x axis. Then B has coordinates $(a, 0)$. Let (x, y) be the coordinates of A. By (16.5), $x = b\cos\theta$ and $y = b\sin\theta$. By the distance formula (2.1),

$$c = \sqrt{(x-a)^2 + (y-0)^2} = \sqrt{(x-a)^2 + y^2}$$

Therefore,

$$
\begin{aligned}
c^2 &= (x-a)^2 + y^2 = (b\cos\theta - a)^2 + (b\sin\theta)^2 \\
&= b^2\cos^2\theta - 2ab\cos\theta + a^2 + b^2\sin^2\theta \quad (\text{Algebra}: (u-v)^2 = u^2 - 2uv + v^2) \\
&= a^2 + b^2(\cos^2\theta + \sin^2\theta) - 2ab\cos\theta \\
&= a^2 + b^2 - 2ab\cos\theta \quad (\text{by } (16.3))
\end{aligned}
$$

(b) See Fig. 16-13. Let D be the foot of the perpendicular from A to side BC, and let $h = \overline{AD}$. Then $\sin B = \overline{AD} / \overline{AB} = h/c$. Thus, $h = c \sin B$ and so the area of $\Delta ABC = \frac{1}{2}(\text{base} \times \text{height}) = \frac{1}{2}ah = \frac{1}{2}ac\sin B$. (Verify that this also holds when $\angle B$ is obtuse.) Similarly, $\frac{1}{2}bc\sin A =$ the area of $\Delta ABC = \frac{1}{2}ab\sin C$. Hence, $\frac{1}{2}ac\sin B = \frac{1}{2}bc\sin A = \frac{1}{2}ab\sin C$. Dividing by $\frac{1}{2}abc$, we obtain the law of sines.

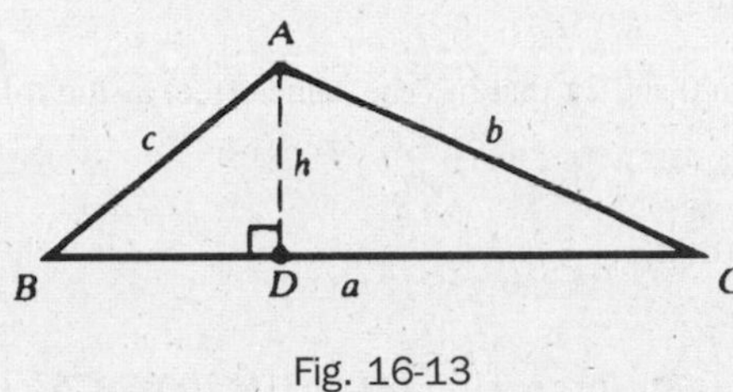

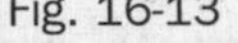
Fig. 16-13

12. Prove the identity (16.6): $\cos(u - v) = \cos u \cos v + \sin u \sin v$.

Consider the case where $0 \le v < u < v + \pi$. (See Fig. 16-14.) By the law of cosines, $BC^2 = 1^2 + 1^2 - 2(1)(1)\cos(\angle BOC)$. Thus,

$$
\begin{aligned}
(\cos u - \cos v)^2 + (\sin u - \sin v)^2 &= 2 - 2\cos(u - v) \\
\cos^2 u - 2\cos u\cos v + \cos^2 v + \sin^2 u - 2\sin u\sin v + \sin^2 v &= 2 - 2\cos(u - v) \\
(\cos^2 u + \sin^2 u) + (\cos^2 v + \sin^2 v) - 2(\cos u\cos v + \sin u\sin v) &= 2 - 2\cos(u - v) \\
1 + 1 - 2(\cos u\cos v + \sin u\sin v) &= 2 - 2\cos(u - v) \\
\cos u\cos v + \sin u\sin v &= \cos(u - v)
\end{aligned}
$$

All the other cases can be derived from the case above.

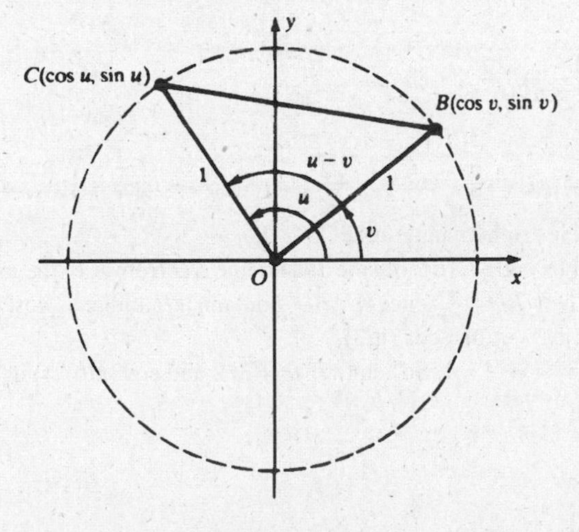

Fig. 16-14

SUPPLEMENTARY PROBLEMS

13. Change the following radian measures into degree measures: (a) 4 radians; (b) $\pi/10$ radians; (c) $11\pi/12$ radians.

Ans. (a) $(720/\pi)°$; (b) 18°; (c) 165°

14. Change the following degree measures into radian measures: (a) 9°; (b) 75°; (c) $(90/\pi)°$.

Ans. (a) $(\pi/20$ radians; (b) $5\pi/12$ radians; (c) $\frac{1}{2}$ radian

15. Refer to the notation of Fig. 16-3. (a) If $r = 7$ and $\theta = \pi/14$, find s; (b) If $\theta = 30°$ and $s = 2$, find r.

Ans. (a) $\pi/2$; (b) $12/\pi$

16. Find the angle of rotation between 0 and 2π that has the same effect as the following rotations: (a) $17\pi/4$; (b) 375°; (c) $-\pi/3$; (d) $-7\pi/2$.

Ans. (a) $\pi/4$; (b) 15°; (c) $5\pi/3$; (d) $\pi/2$

17. Evaluate: (a) cos $(4\pi/3)$; (b) sin$(11\pi/6)$; (c) cos210°; (d) sin315°; (e) cos75°; (f) sin73°.

Ans. (a) $-\frac{1}{2}$; (b) $-\frac{1}{2}$; (c) $-\sqrt{3}/2$; (d) $-\sqrt{2}/2$; (e) $(\sqrt{2-\sqrt{3}})/2$; (f) approximately 0.9563

18. Assume θ is acute and $\sin\theta = \frac{1}{4}$. Evaluate: (a) cos θ; (b) sin 2θ; (c) cos 2θ; (d) cos $(\theta/2)$.

Ans. (a) $\sqrt{15}/4$; (b) $\sqrt{15}/8$; (c) $\frac{7}{8}$; (d) $(\sqrt{8+2\sqrt{15}})/4$

19. Assume θ is in the third quadrant ($\pi < \theta < 3\pi/2$) and $\cos\theta = -\frac{4}{5}$. Find: (a) sin θ; (b) cos 2θ; (c) sin$(\theta/2)$.

Ans. (a) $-\frac{3}{5}$; (b) $\frac{7}{25}$; (c) $(3\sqrt{10})/10$

20. In ΔABC, $\overline{AB} = 5$, $\overline{AC} = 7$, and $\cos(\angle ABC) = \frac{3}{5}$. Find $\overline{BC}$.

Ans. $4\sqrt{2}$

21. Prove the identity $\dfrac{\sin\theta}{\cos\theta} = \dfrac{1-\cos 2\theta}{\sin 2\theta}$.

22. Derive the following values: (a) $\sin\frac{\pi}{4} = \cos\frac{\pi}{4} = \frac{\sqrt{2}}{2}$; (b) $\sin\frac{\pi}{6} = \cos\frac{\pi}{3} = \frac{1}{2}$; (c) $\sin\frac{\pi}{3} = \cos\frac{\pi}{6} = \frac{\sqrt{3}}{2}$

[*Hints*: (*a*) Look at an isosceles right triangle ΔABC.

(b) Consider an equilateral triangle ΔABC of side 1. The line AD from A to the midpoint D of side BC is perpendicular to BC. Then $\overline{BD} = \frac{1}{2}$. Since $\angle ABD$ contains $\pi/3$ radians, $\cos(\pi/3) = \overline{BD}/\overline{AB} = (1/2)/1 = \frac{1}{2}$. By (16.8), sin $(\pi/6)$ = cos $(\pi/2 - \pi/6)$ = cos $(\pi/3)$.

(c) $\sin^2(\pi/3) = 1 - \cos^2(\pi/3) = 1 - \frac{1}{4} = \frac{3}{4}$. So, $\sin(\pi/3) = \sqrt{3}/2$ and cos $(\pi/6)$ = sin $(\pi/3)$ by (16.8).]

Differentiation of Trigonometric Functions

Continuity of $\cos x$ and $\sin x$

It is clear that $\cos x$ and $\sin x$ are continuous functions, that is, for any θ,

$$\lim_{h\to 0} \cos(\theta + h) = \cos\theta \quad \text{and} \quad \lim_{h\to 0} \sin(\theta + h) = \sin\theta$$

To see this, observe that, in Fig. 17-1, as h approaches 0, point C approaches point B. Hence, the x coordinate of C (namely, $\cos(\theta + h)$) approaches the x coordinate of B (namely, $\cos\theta$), and the y coordinate of C (namely, $\sin(\theta + h)$) approaches the y coordinate of B (namely, $\sin\theta$).

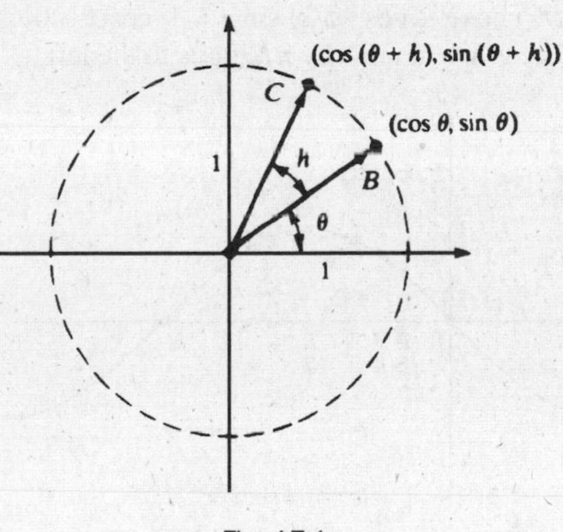

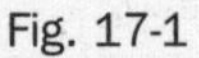
Fig. 17-1

To find the derivative of $\sin x$ and $\cos x$, we shall need the following limits.

(17.1) $\lim_{\theta\to 0} \frac{\sin\theta}{\theta} = 1$

(17.2) $\lim_{\theta\to 0} \frac{1 - \cos\theta}{\theta} = 0$

For a proof of (17.1), see Problem 1. From (17.1), (17.2) is derived as follows:

$$\frac{1-\cos\theta}{\theta} = \frac{1-\cos\theta}{\theta}\cdot\frac{1+\cos\theta}{1+\cos\theta} = \frac{1-\cos^2\theta}{\theta(1+\cos\theta)}$$

$$= \frac{\sin^2\theta}{\theta(1+\cos\theta)} = \frac{\sin\theta}{\theta}\cdot\frac{\sin\theta}{1+\cos\theta}.$$

Hence,

$$\lim_{\theta\to 0}\frac{1-\cos\theta}{\theta}=\lim_{\theta\to 0}\frac{\sin\theta}{\theta}\cdot\lim_{\theta\to 0}\frac{\sin\theta}{1+\cos\theta}=1\cdot\frac{\sin 0}{1+\cos 0}=1\cdot\frac{0}{1+1}=1\cdot 0=0$$

(17.3) $D_x(\sin x)=\cos x$

(17.4) $D_x(\cos x)=-\sin x$

For a proof of (17.3), see Problem 2. From (17.3) we can derive (17.4), with the help of the chain rule and (16.8), as follows:

$$D_x(\cos x)=D_x\left(\sin\left(\frac{\pi}{2}-x\right)\right)=\cos\left(\frac{\pi}{2}-x\right)\cdot(-1)=-\sin x$$

Graph of sin *x*

Since $\sin(x+2\pi)=\sin x$, we need only construct the graph for $0\le x\le 2\pi$. Setting $D_x(\sin x)=\cos x=0$ and noting that $\cos x=0$ in $[0, 2\pi]$ when and only when $x=\pi/2$ or $x=3\pi/2$, we find the critical numbers $\pi/2$ and $3\pi/2$. Since $D_x^2(\sin x)=D_x(\cos x)=-\sin x$, and $-\sin(\pi/2)=-1<0$ and $-\sin(3\pi/2)=1>0$, the second derivative test implies that there is a relative maximum at $(\pi/2, 1)$ and a relative minimum at $(3\pi/2, -1)$. Since $D_x(\sin x)=\cos x$ is positive in the first and fourth quadrants, $\sin x$ is increasing for $0<x<\pi/2$ and for $3\pi/2<x<2\pi$. Since $D_x^2(\sin x)=-\sin x$ is positive in the third and fourth quadrants, the graph is concave upward for $\pi<x<2\pi$. Thus, there will be an inflection point at $(\pi, 0)$, as well as at $(0, 0)$ and $(2\pi, 0)$. Part of the graph is shown in Fig. 17-2.

Graph of cos *x*

Note that $\sin(\pi/2+x)=\sin(\pi/2)\cos x+\cos(\pi/2)\sin x=1\cdot\cos x+0\cdot\sin x=\cos x$. Thus, the graph of $\cos x$ can be drawn by moving the graph of $\sin x$ by $\pi/2$ units to the left, as shown in Fig. 17-3.

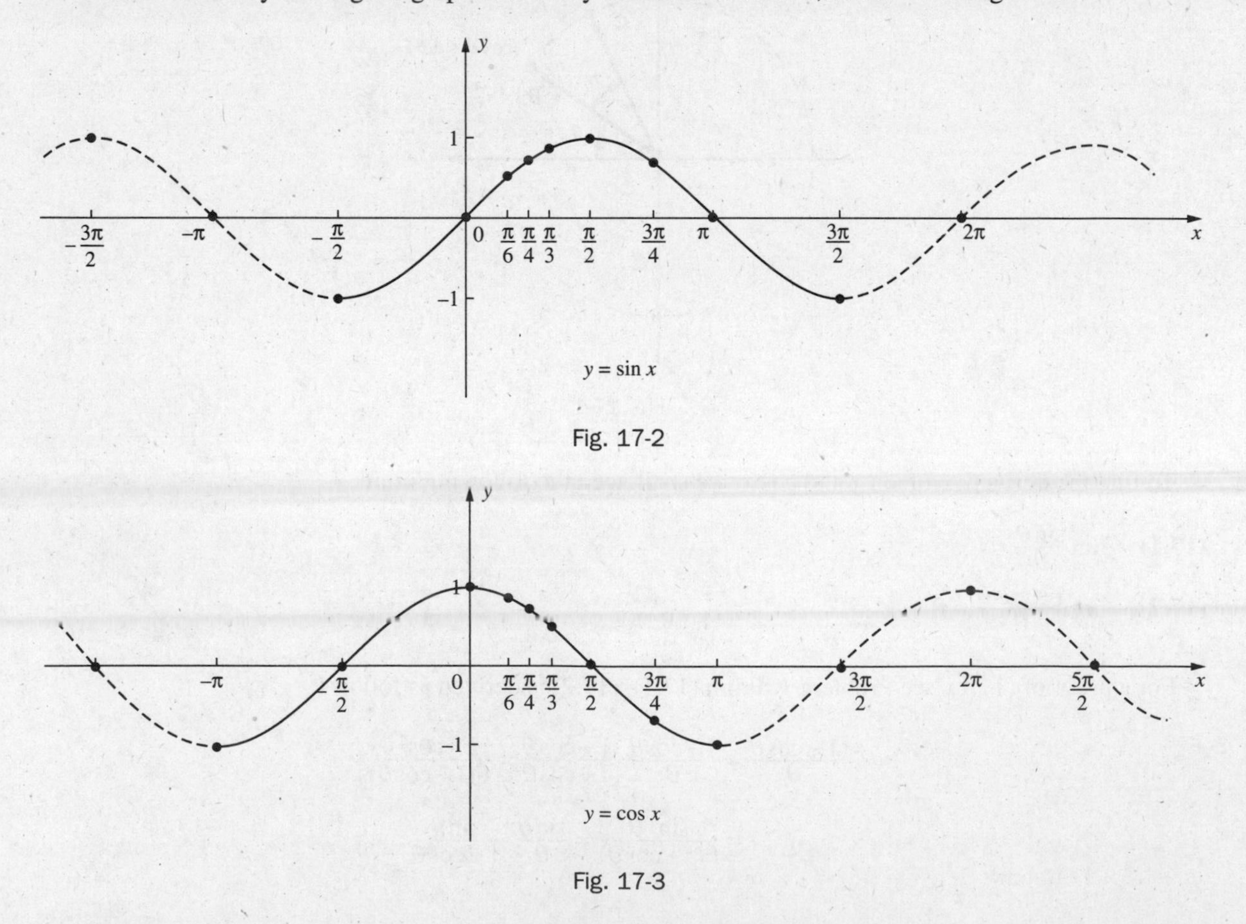

Fig. 17-2

Fig. 17-3

The graphs of $y = \sin x$ and $y = \cos x$ consist of repeated waves, with each wave extending over an interval of length 2π. The length (*period*) and height (*amplitude*) of the waves can be changed by multiplying the argument and the value, respectively, by constants.

EXAMPLE 17.1: Let $y = \cos 3x$. The graph is sketched in Fig. 17-4. Because $\cos 3(x + 2\pi/3) = \cos(3x + 2\pi) = \cos 3x$, the function is of period $p = 2\pi/3$. Hence, the length of each wave is $2\pi/3$. The number of waves over an interval of length 2π (corresponding to one complete rotation of the ray determining the angle x) is 3. This number is called the *frequency* f of $\cos 3x$. In general, pf = (length of each wave) × (number of waves in an interval of 2π) = 2π. Hence, $f = 2\pi/p$.

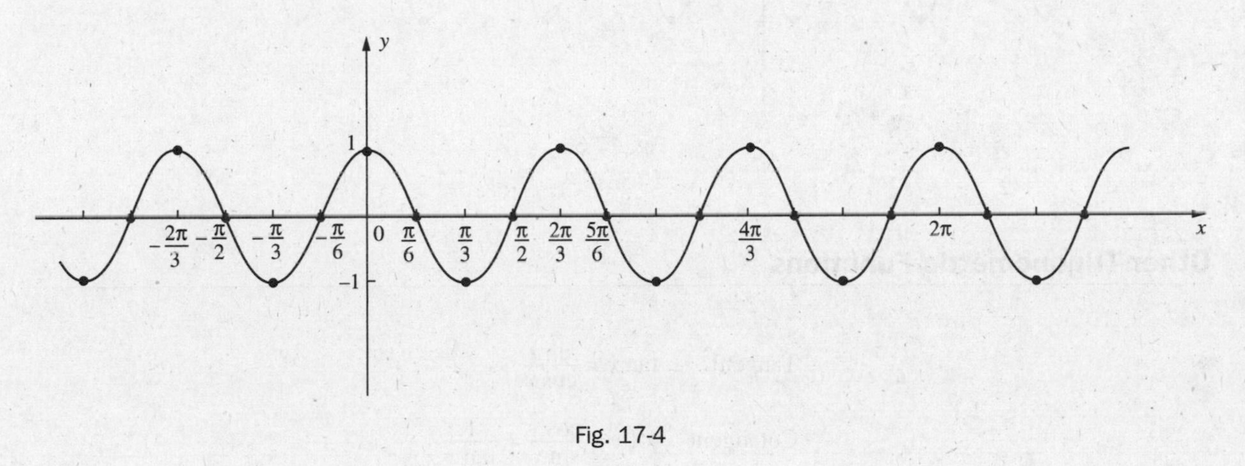

Fig. 17-4

For any $b > 0$, the functions $\sin bx$ and $\cos bx$ have frequency b and period $2\pi/b$.

EXAMPLE 17.2: $y = 2 \sin x$. The graph of this function (see Fig. 17-5) is obtained from that of $y = \sin x$ by doubling the y values. The period and frequency are the same as those of $y = \sin x$, that is, $p = 2\pi$ and $f = 1$. The amplitude, the maximum height of each wave, is 2.

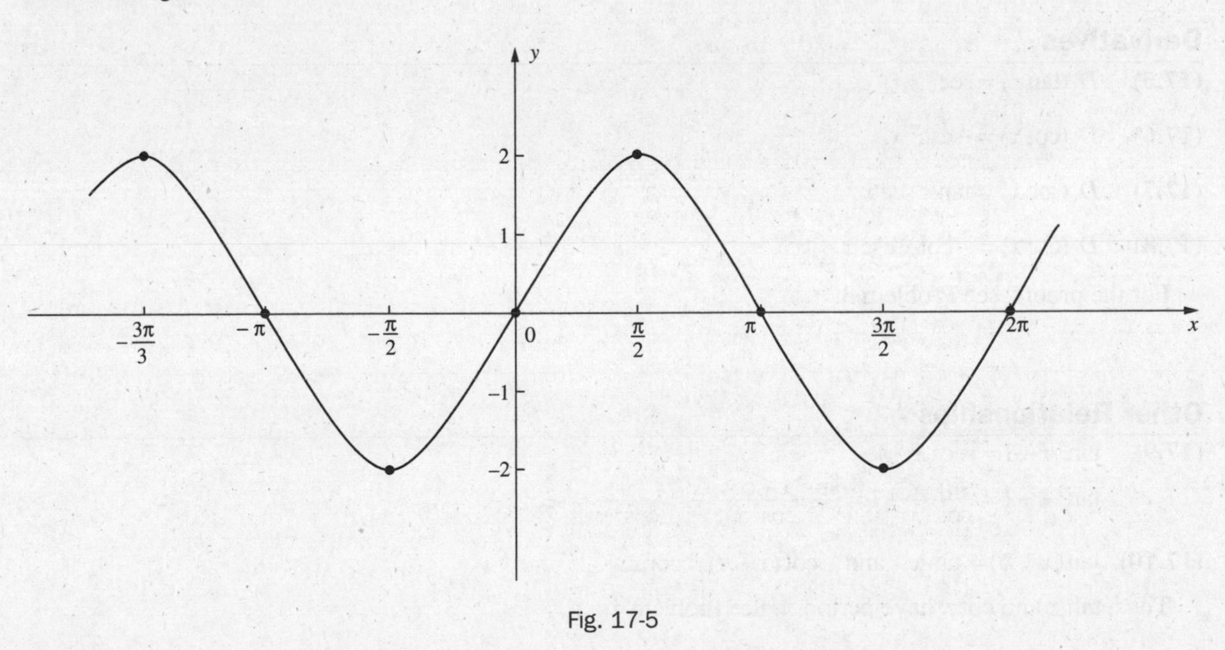

Fig. 17-5

EXAMPLE 17.3: In general, if $b > 0$, then $y = A \sin bx$ and $y = A \cos bx$ have period $2\pi/b$, frequency b, and amplitude $|A|$. Figure 17-6 shows the graph of $y = 1.5 \sin 4x$.

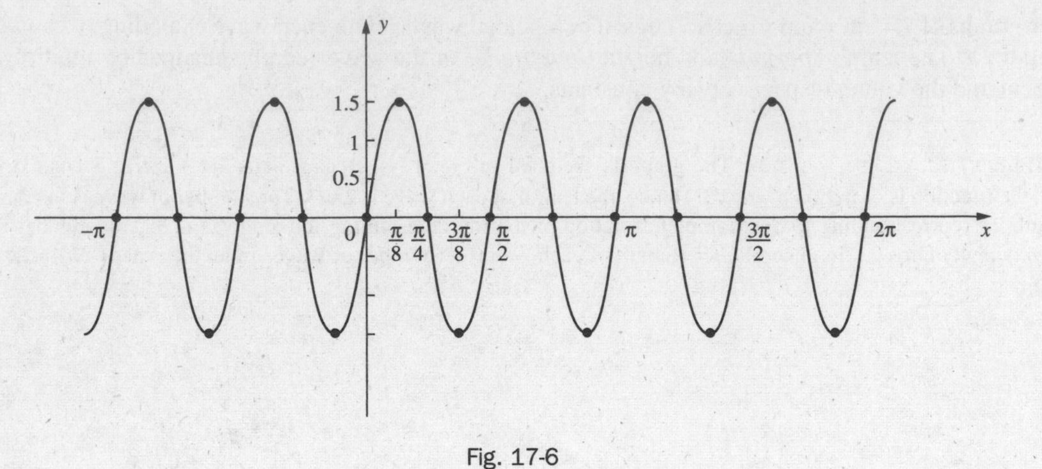

Fig. 17-6

Other Trigonometric Functions

Tangent $\tan x = \dfrac{\sin x}{\cos x}$

Cotangent $\cot x = \dfrac{\cos x}{\sin x} = \dfrac{1}{\tan x}$

Secant $\sec x = \dfrac{1}{\cos x}$

Cosecant $\csc x = \dfrac{1}{\sin x}$

Derivatives

(17.5) $D_x(\tan x) = \sec^2 x$

(17.6) $D_x(\cot x) = -\csc^2 x$

(17.7) $D_x(\sec x) = \tan x \sec x$

(17.8) $D_x(\csc x) = -\cot x \csc x$

For the proofs, see Problem 3.

Other Relationships

(17.9) $\tan^2 x + 1 = \sec^2 x$

$$\tan^2 x + 1 = \frac{\sin^2 x}{\cos^2 x} + 1 = \frac{\sin^2 x + \cos^2 x}{\cos^2 x} = \frac{1}{\cos^2 x} = \sec^2 x$$

(17.10) $\tan(x+\pi) = \tan x$ and $\cot(x+\pi) = \cot x$

Thus, $\tan x$ and $\cot x$ have period π. See Problem 4.

(17.11) $\tan(-x) = -\tan x$ and $\cot(-x) = -\cot x$

$$\tan(-x) = \frac{\sin(-x)}{\cos(-x)} = \frac{-\sin x}{\cos x} = -\frac{\sin x}{\cos x} = -\tan x, \text{ and similarly for } \cot x$$

Graph of $y = \tan x$

Since $\tan x$ has period π, it suffices to determine the graph in $(-\pi/2, \pi/2)$. Since $\tan(-x) = -\tan x$, we need only draw the graph in $(0, \pi/2)$ and then reflect in the origin. Since $\tan x = (\sin x)/(\cos x)$, there will be vertical asymptotes at $x = \pi/2$ and $x = -\pi/2$. By (17.5), $D_x(\tan x) > 0$ and, therefore, $\tan x$ is increasing.

$$D_x^2(\tan x) = D_x(\sec^2 x) = 2\sec x(\tan x \sec x) = 2\tan x \sec^2 x.$$

Thus, the graph is concave upward when $\tan x > 0$, that is, for $0 < x < \pi/2$, and there is an inflection point at (0, 0). Some special values of $\tan x$ are given in Table 17-1, and the graph is shown in Fig. 17-7.

TABLE 17-1

x	$\tan x$
0	0
$\frac{\pi}{6}$	$\frac{\sqrt{3}}{3} \sim 0.58$
$\frac{\pi}{4}$	1
$\frac{\pi}{3}$	$\sqrt{3} \sim 1.73$

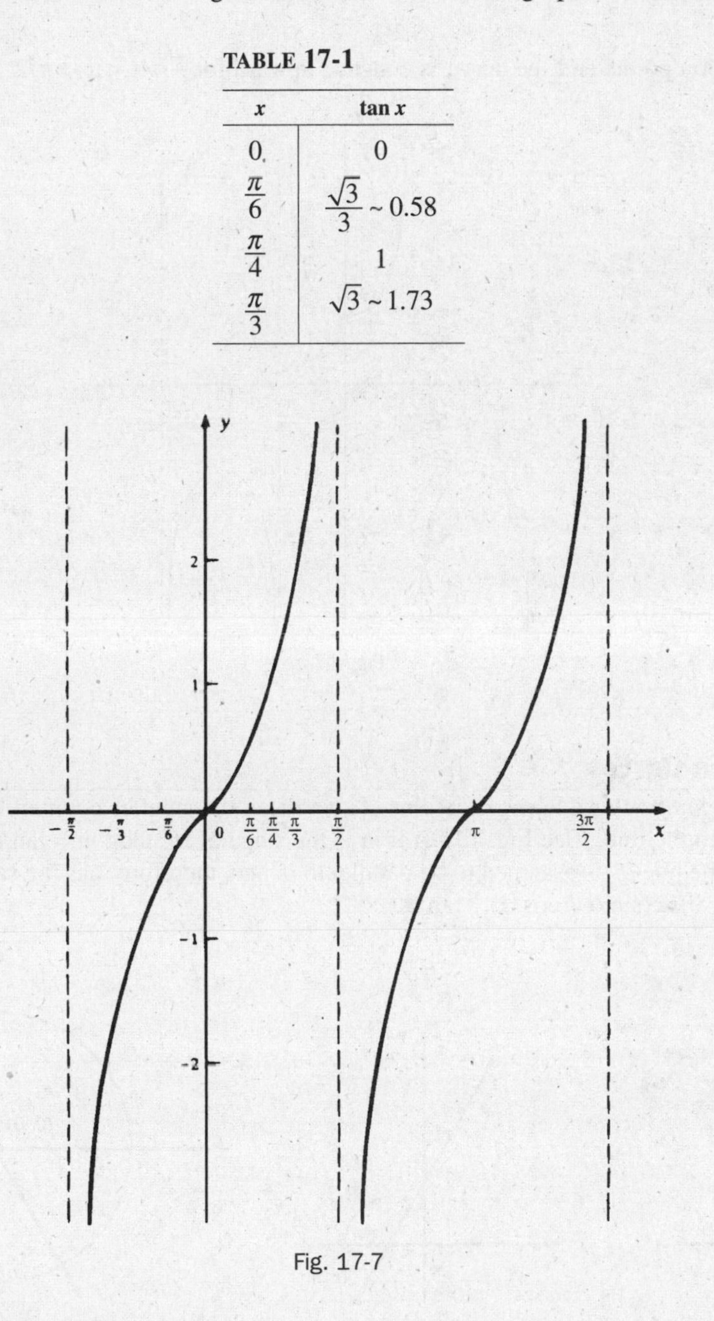

Fig. 17-7

For an acute angle θ of a right triangle,

$$\tan\theta = \frac{\sin\theta}{\cos\theta} = \frac{\text{opposite}}{\text{hypotenuse}} \div \frac{\text{adjacent}}{\text{hypotenuse}} = \frac{\text{opposite}}{\text{adjacent}}$$

Graph of $y = \sec x$

Since $\sec x = 1/(\cos x)$, the graph will have a vertical asymptote $x = x_0$ for all x_0 for which $\cos x_0 = 0$, that is, for $x = (2n + 1)\pi/2$, where n is any integer. Like $\cos x$, $\sec x$ has a period of 2π, and we can confine our attention to $(-\pi, \pi)$. Note that $|\sec x| \geq 1$, since $|\cos x| \leq 1$. Setting $D_x(\sec x) = \tan x \sec x = 0$, we find critical numbers at $x = 0$ and $x = \pi$, and the first derivative test tells us that there is a relative minimum at $x = 0$ and a relative maximum at $x = \pi$.

Since

$$D_x^2(\sec x) = D_x(\tan x \sec x) = \tan x(\tan x \sec x) + \sec x(\sec^2 x) = \sec x(\tan^2 x + \sec^2 x)$$

there are no inflection points and the curve is concave upward for $-\pi/2 < x < \pi/2$. The graph is shown in Fig. 17-8.

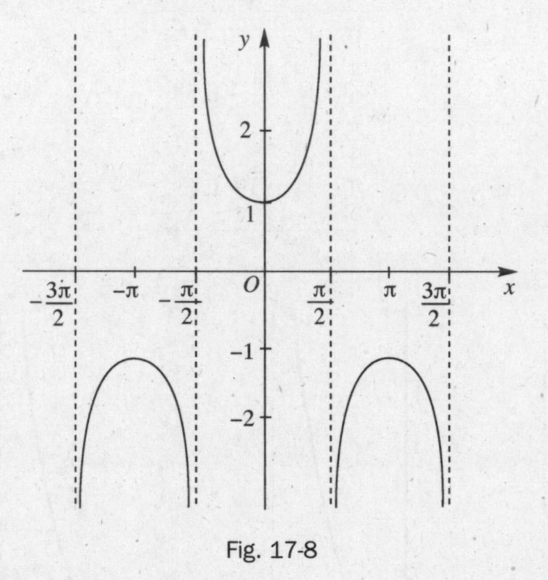

Fig. 17-8

Angles Between Curves

By the *angle of inclination* of a nonvertical line $\mathcal{L}$, we mean the smaller counterclockwise angle α from the positive x axis to the line. (See Fig. 17-9.) If m is the slope of $\mathcal{L}$, then $m = \tan \alpha$. (To see this, look at Fig. 17-10, where the line $\mathcal{L}'$ is assumed to be parallel to $\mathcal{L}$ and, therefore, has the same slope m. Then $m = (\sin \alpha - 0)/(\cos \alpha - 0) = (\sin \alpha)/(\cos \alpha) = \tan \alpha$.)

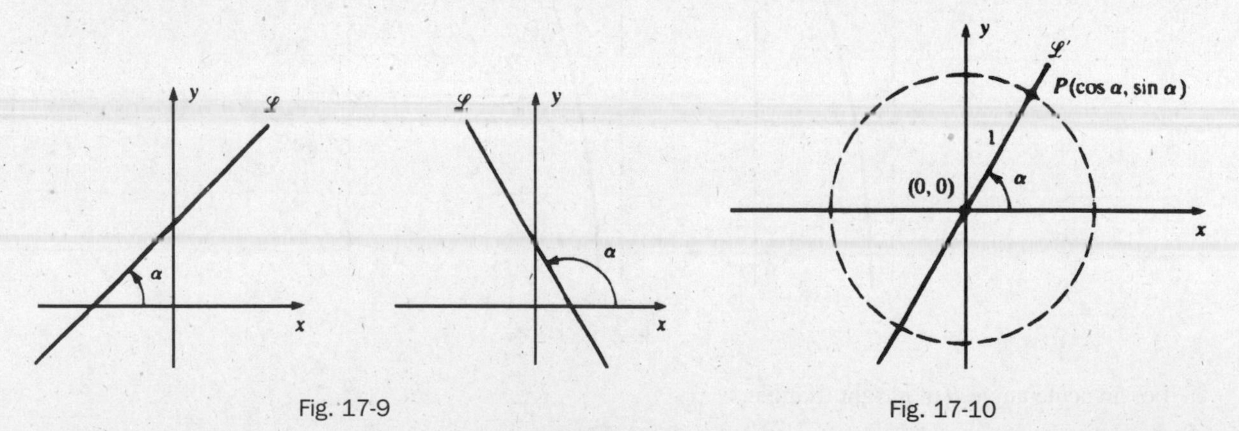

Fig. 17-9

Fig. 17-10

By the *angle between two curves at a point of intersection P*, we mean the smaller of the two angles between the tangent lines to the curves at P. (See Problems 17 and 18.)

SOLVED PROBLEMS

1. Prove (17.1): $\lim_{\theta \to 0} \frac{\sin\theta}{\theta} = 1$.

Since $\frac{\sin(-\theta)}{-\theta} = \frac{\sin\theta}{\theta}$, we need consider only $\theta > 0$. In Fig. 17-11, let $\theta = \angle AOB$ be a small positive central angle of a circle of radius $OA = OB = 1$. Let C be the foot of the perpendicular from B onto OA. Note that $OC = \cos\theta$ and $CB = \sin\theta$. Let D be the intersection of OB and an arc of a circle with center at O and radius OC. So,

$$\text{Area of sector } COD \le \text{area of } \Delta COB \le \text{area of sector } AOB$$

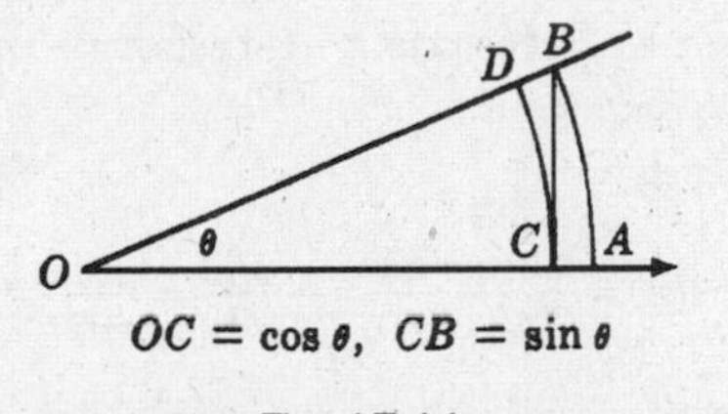

Fig. 17-11

Observe that area of sector $COD = \frac{1}{2}\theta\cos^2\theta$ and that area of sector $AOB = \frac{1}{2}\theta$. (If W is the area of a sector determined by a central angle θ of a circle of radius r, then W/(area of circle) $= \theta/2\pi$. Thus, $W/\pi r^2 = \theta/2\pi$ and, therefore, $W = \frac{1}{2}\theta r^2$.)

Hence,

$$\tfrac{1}{2}\theta\cos^2\theta \le \tfrac{1}{2}\sin\theta\cos\theta \le \tfrac{1}{2}\theta$$

Division by $\frac{1}{2}\theta\cos\theta > 0$ yields

$$\cos\theta \le \frac{\sin\theta}{\theta} \le \frac{1}{\cos\theta}$$

As θ approaches 0^+, $\cos\theta \to 1$, $1/(\cos\theta) \to 1$. Hence,

$$1 \le \lim_{\theta\to 0} \frac{\sin\theta}{\theta} \le 1 \quad \text{Thus} \quad \lim_{\theta\to 0}\frac{\sin\theta}{\theta} = 1$$

2. Prove (17.3): $D_x(\sin x) = \cos x$.

Here we shall use (17.1) and (17.2).
Let $y = \sin x$. Then $y + \Delta y = \sin(x + \Delta x)$ and

$$\Delta y = \sin(x+\Delta x) - \sin x = \cos x \sin\Delta x + \sin x \cos\Delta x - \sin x$$

$$= \cos x \sin\Delta x + \sin x(\cos\Delta x - 1)$$

$$\frac{dy}{dx} = \lim_{\Delta x\to 0}\frac{\Delta y}{\Delta x} = \lim_{\Delta x\to 0}\left(\cos x\frac{\sin\Delta x}{\Delta x} + \sin x\frac{\cos\Delta x - 1}{\Delta x}\right)$$

$$= (\cos x)\lim_{\Delta x\to 0}\frac{\sin\Delta x}{\Delta x} + (\sin x)\lim_{\Delta x\to 0}\frac{\cos\Delta x - 1}{\Delta x}$$

$$= (\cos x)(1) + (\sin x)(0) = \cos x$$

3. Prove: (a) $D_x(\tan x) = \sec^2 x$ (17.5); (b) $D_x(\sec x) = \tan x \sec x$ (17.7).

(a) $$\frac{d}{dx}(\tan x) = \frac{d}{dx}\left(\frac{\sin x}{\cos x}\right) = \frac{\cos x\cos x - \sin x(-\sin x)}{\cos^2 x}$$

$$= \frac{\cos^2 x + \sin^2 x}{\cos^2 x} = \frac{1}{\cos^2 x} = \sec^2 x$$

(b) Differentiating both sides of (17.9), $\tan^2 x + 1 = \sec^2 x$, by means of the chain rule, we get

$$2\tan x \sec^2 x = 2\sec x\, D_x(\sec x).$$

Hence, $D_x(\sec x) = \tan x \sec x$.

4. Prove (17.10): $\tan(x + \pi) = \tan x$.

$$\sin(x+\pi) = \sin x \cos\pi + \cos x \sin\pi = -\sin x$$

$$\cos(x+\pi) = \cos x \cos\pi - \sin x \sin\pi = -\cos x$$

Hence,

$$\tan(x+\pi) = \frac{\sin(x+\pi)}{\cos(x+\pi)} = \frac{-\sin x}{-\cos x} = \frac{\sin x}{\cos x} = \tan x$$

5. Derive $\tan(u - v) = \dfrac{\tan u - \tan v}{1 + \tan u \tan v}$.

$$\tan(u-v) = \frac{\sin(u-v)}{\cos(u-v)} = \frac{\sin u \cos v - \cos u \sin v}{\cos u \cos v + \sin u \sin v}$$

$$= \frac{\dfrac{\sin u}{\cos u} - \dfrac{\sin v}{\cos v}}{1 + \dfrac{\sin u}{\cos u}\dfrac{\sin v}{\cos v}} \quad \text{(divide numerator and denominator by } \cos u \cos v\text{)}$$

$$= \frac{\tan u - \tan v}{1 + \tan u \tan v}$$

6. Calculate the derivatives of the following functions: (a) $2\cos 7x$; (b) $\sin^3(2x)$; (c) $\tan(5x)$; (d) $\sec(1/x)$.

(a) $D_x(2\cos 7x) = 2(-\sin 7x)(7) = -14\sin 7x$
(b) $D_x(\sin^3(2x)) = 3(\sin^2(2x))(\cos(2x))(2) = 6\sin^2(2x)\cos(2x)$
(c) $D_x(\tan(5x)) = (\sec^2(5x))(5) = 5\sec^2(5x)$
(d) $D_x(\sec(1/x)) = \tan(1/x)\sec(1/x)(-1/x^2) = -(1/x^2)\tan(1/x)\sec(1/x)$

7. Find all solutions of the equation $\cos x = \frac{1}{2}$.

Solving $(\frac{1}{2})^2 + y^2 = 1$, we see that the only points on the unit circle with abscissa $\frac{1}{2}$ are $(\frac{1}{2}, \sqrt{3}/2)$ and $(\frac{1}{2}, -\sqrt{3}/2)$. The corresponding central angles are $\pi/3$ and $5\pi/3$. So, these are the solutions in $[0, 2\pi]$. Since $\cos x$ has period 2π, the solutions are $\pi/3 + 2\pi n$ and $5\pi/3 + 2\pi n$, where n is any integer.

8. Calculate the limits (a) $\lim\limits_{x\to 0}\dfrac{\sin 5x}{2x}$; (b) $\lim\limits_{x\to 0}\dfrac{\sin 3x}{\sin 7x}$; (c) $\lim\limits_{x\to 0}\dfrac{\tan x}{x}$

(a) $$\lim_{x\to 0}\frac{\sin 5x}{2x} = \lim_{x\to 0}\frac{5}{2}\frac{\sin 5x}{5x} = \frac{5}{2}\lim_{u\to 0}\frac{\sin u}{u} = \frac{5}{2}(1) = \frac{5}{2}$$

(b) $$\lim_{x\to 0}\frac{\sin 3x}{\sin 7x} = \lim_{x\to 0}\frac{\sin 3x}{3x}\cdot\frac{7x}{\sin 7x}\cdot\frac{3}{7} = \frac{3}{7}\lim_{u\to 0}\frac{\sin u}{u}\lim_{u\to 0}\frac{u}{\sin u}$$

$$= \tfrac{3}{7}(1)(1) = \tfrac{3}{7}$$

(c) $$\lim_{x\to 0}\frac{\tan x}{x} = \lim_{x\to 0}\frac{\sin x}{x}\cdot\frac{1}{\cos x} = \lim_{x\to 0}\frac{\sin x}{x}\cdot\lim_{u\to 0}\frac{1}{\cos x}$$

$$= (1)(\tfrac{1}{1}) = 1$$

9. Let $y = x \sin x$. Find y'''.

$$y' = x\cos x + \sin x$$
$$y'' = x(-\sin x) + \cos x + \cos x = -x\sin x + 2\cos x$$
$$y''' = -x\cos x - \sin x - 2\sin x = -x\cos x - 3\sin x$$

10. Let $y = \tan^2(3x - 2)$. Find y''.

$$y' = 2\tan(3x-2)\sec^2(3x-2)\cdot 3 = 6\tan(3x-2)\sec^2(3x-2)$$
$$y'' = 6[\tan(3x-2)\cdot 2\sec(3x-2)\cdot\sec(3x-2)\tan(3x-2)\cdot 3 + \sec^2(3x-2)\sec^2(3x-2)\cdot 3]$$
$$= 36\tan^2(3x-2)\sec^2(3x-2) + 18\sec^4(3x-2)$$

11. Assume $y = \sin(x + y)$. Find y'.

$$y' = \cos(x+y)\cdot(1+y') = \cos(x+y) + \cos(x+y)\cdot(y')$$

Solving for y',

$$y' = \frac{\cos(x+y)}{1-\cos(x+y)}$$

12. Assume $\sin y + \cos x = 1$. Find y'.

$$\cos y\cdot y' - \sin x = 0. \quad \text{So} \quad y' = \frac{\sin x}{\cos y}$$

$$y'' = \frac{\cos y\cos x - \sin x(-\sin y)\cdot y'}{\cos^2 y} = \frac{\cos x\cos y + \sin x\sin y\cdot y'}{\cos^2 y}$$

$$= \frac{\cos x\cos y + \sin x\sin y(\sin x)/(\cos y)}{\cos^2 y} = \frac{\cos x\cos^2 y + \sin^2 x\sin y}{\cos^3 y}$$

13. A pilot is sighting a location on the ground directly ahead. If the plane is flying 2 miles above the ground at 240 mi/h, how fast must the sighting instrument be turning when the angle between the path of the plane and the line of sight is 30°? See Fig. 17-12.

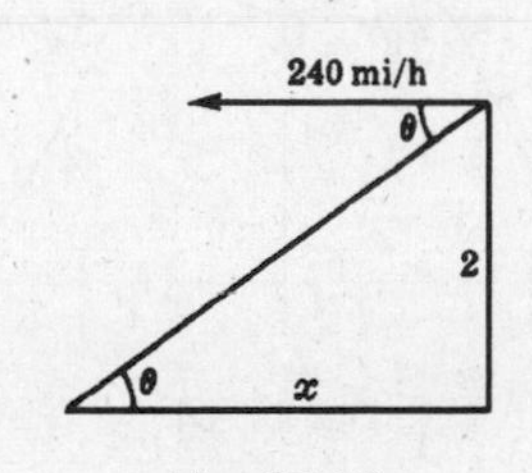

Fig. 17-12

$$\frac{dx}{dt} = -240\,\text{mi/h} \quad \text{and} \quad x = 2\cot\theta$$

From the last equation, $\frac{dx}{dt} = -2\csc^2\theta\frac{d\theta}{dt}$. Thus, $-240 = -2(4)\frac{d\theta}{dt}$ when $\theta = 30°$

$$\frac{d\theta}{dt} = 30\ \text{rad/h} = \frac{3}{2\pi}\,\text{deg/s}$$

14. Sketch the graph of $f(x) = \sin x + \cos x$.

$f(x)$ has a period of 2π. Hence, we need only consider the interval $[0, 2\pi]$. $f'(x) = \cos x - \sin x$, and $f''(x) = -(\sin x + \cos x)$. The critical numbers occur where $\cos x = \sin x$ or $\tan x = 1$, $x = \pi/4$ or $x = 5\pi/4$.

$f''(\pi/4) = -(\sqrt{2}/2 + \sqrt{2}/2) = -\sqrt{2} < 0$. So, there is a relative maximum at $x = \pi/4$, $y = \sqrt{2}$. $f''(5\pi/4) = -(-\sqrt{2}/2 - \sqrt{2}/2) = \sqrt{2} > 0$. Thus, there is a relative minimum at $x = 5\pi/4, y = -\sqrt{2}$. The inflection points occur where $f''(x) = -(\sin x + \cos x) = 0$, $\sin x = -\cos x$, $\tan x = -1$, $x = 3\pi/4$ or $x = 7\pi/4$, $y = 0$. See Fig. 17-13.

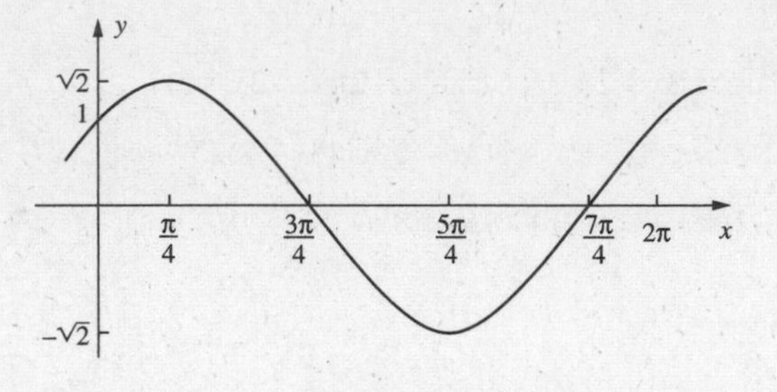

Fig. 17-13

15. Sketch the graph of $f(x) = \cos x - \cos^2 x$.

$$f'(x) = -\sin x - 2(\cos x)(-\sin x) = (\sin x)(2\cos x - 1)$$

and

$$f''(x) = (\sin x)(-2\sin x) + (2\cos x - 1)(\cos x)$$
$$= 2(\cos^2 x - \sin^2 x) - \cos x = 4\cos^2 x - \cos x - 2$$

Since f has period 2π, we need only consider $[-\pi, \pi]$, and since f is even, we have to look at only $[0, \pi]$. The critical numbers are the solutions in $[0, \pi]$ of $\sin x = 0$ or $2\cos x - 1 = 0$. The first equation has solutions 0 and π, and the second is equivalent to $\cos x = \frac{1}{2}$, which has the solution $\pi/3$. $f''(0) = 1 > 0$; so there is a relative minimum at (0, 0). $f''(\pi) = 3 > 0$; so there is a relative minimum at $(\pi, -2)$. $f''(\pi/3) = -\frac{3}{2} < 0$; hence there is a relative maximum at $(\pi/3, \frac{1}{4})$. There are inflection points between 0 and $\pi/3$ and between $\pi/3$ and π; they can be found by using the quadratic formula to solve $4\cos^2 x - \cos x - 2 = 0$ for $\cos x$ and then using a cosine table or a calculator to approximate x. See Fig. 17-14.

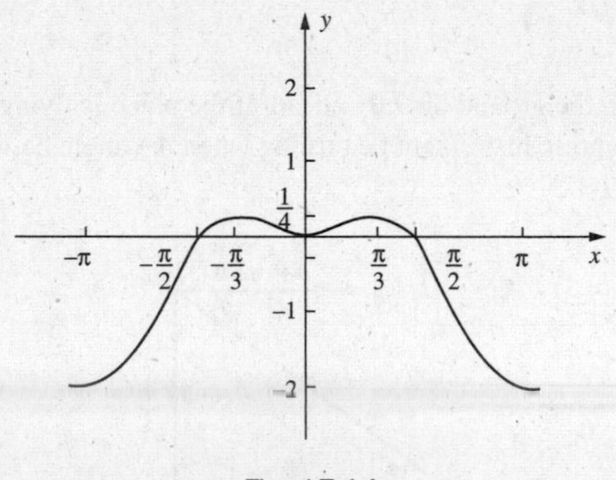

Fig. 17-14

16. Find the absolute extrema of $f(x) = \sin x + x$ on $[0, 2\pi]$.

$f'(x) = \cos x + 1$. Setting $f'(x) = 0$, we get $\cos x = -1$ and, therefore, the only critical number in $[0, 2\pi]$ is $x = \pi$. We list π and the two endpoints 0 and 2π and compute the values of $f(x)$.

x	$f(x)$
π	π
0	0
2π	2π

Hence, the absolute maximum 2π is achieved at $x = 2\pi$, and the absolute minimum 0 at $x = 0$.

17. Find the angle at which the lines $\mathscr{L}_1: y = x + 1$ and $\mathscr{L}_2: y = -3x + 5$ intersect.

Let α_1 and α_2 be the angles of inclination of $\mathscr{L}_1$ and $\mathscr{L}_2$ (see Fig. 17-15), and let m_1 and m_2 be the respective slopes. Then $\tan \alpha_1 = m_1 = 1$ and $\tan \alpha_2 = m_2 = -3$. $\alpha_2 - \alpha_1$ is the angle of intersection. Now, by Problem 5,

$$\tan(\alpha_2 - \alpha_1) = \frac{\tan\alpha_2 - \tan\alpha_1}{1 + \tan\alpha_1 \tan\alpha_2} = \frac{m_2 - m_1}{1 + m_1 m_2} = \frac{-3-1}{1+(-3)(1)}$$

$$= \frac{-4}{-2} = 2$$

From a graphing calculator, $\alpha_2 - \alpha_1 \sim 63.4°$.

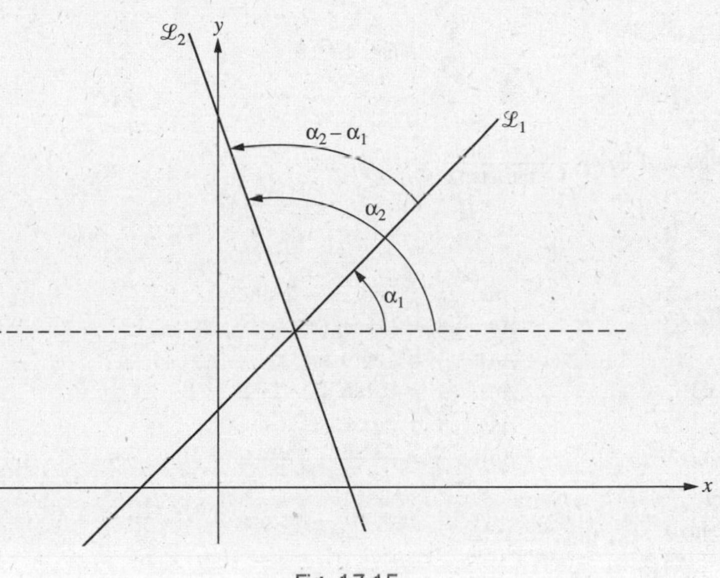

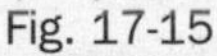
Fig. 17-15

18. Find the angle α between the parabolas $y = x^2$ and $x = y^2$ at (1, 1).

Since $D_x(x^2) = 2x$ and $D_x(\sqrt{x}) = 1/(2\sqrt{x})$, the slopes at (1, 1) are 2 and $\frac{1}{2}$. Hence, $\tan\alpha = \frac{2-(\frac{1}{2})}{1+2(\frac{1}{2})} = \frac{\frac{3}{2}}{2} = \frac{3}{4}$. Thus, using a graphing calculator, we approximate α by 36.9°.

SUPPLEMENTARY PROBLEMS

19. Show that $\cot(x + \pi) = \cot x$, $\sec(x + 2\pi) = \sec x$, and $\csc(x + 2\pi) = \csc x$.

20. Find the period p, frequency f, and amplitude A of $5 \sin(x/3)$ and sketch its graph.

Ans. $p = 6\pi, f = \frac{1}{3}, A = 5$

21. Find all solutions of $\cos x = 0$.

Ans. $x = (2n+1)\frac{\pi}{2}$ for all integers n

22. Find all solutions of $\tan x = 1$.

Ans. $x = (4n+1)\frac{\pi}{4}$ for all integers n

23. Sketch the graph of $f(x) = \frac{\sin x}{2 - \cos x}$.

Ans. See Fig. 17–16.

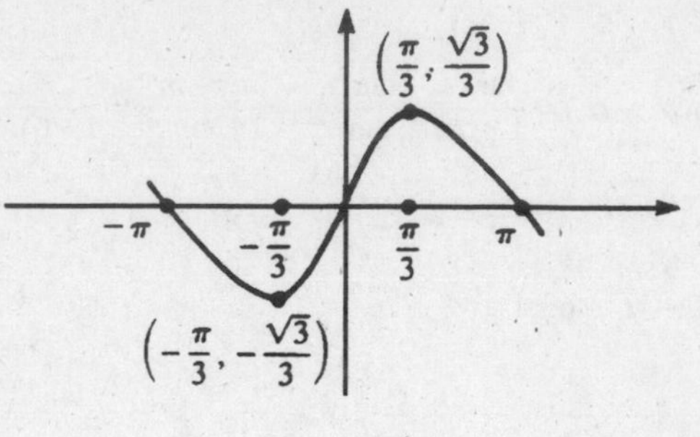

Fig. 17-16

24. Derive the formula $\tan(u+v) = \frac{\tan u + \tan v}{1 - \tan u \tan v}$.

25. Find y'.

(a) $y = \sin 3x + \cos 2x$ *Ans.* $y' = 3 \cos 3x - 2 \sin 2x$

(b) $y = \tan(x^2)$ *Ans.* $y' = 2x \sec^2 (x^2)$

(c) $y = \tan^2 x$ *Ans.* $y' = 2 \tan x \sec^2 x$

(d) $y = \cot(1 - 2x^2)$ *Ans.* $y' = 4x \csc^2 (1 - 2x^2)$

(e) $y = x^2 \sin x$ *Ans.* $y' = x^2 \cos x + 2x \sin x$

(f) $y = \frac{\cos x}{x}$ *Ans.* $y' = \frac{-x \sin x - \cos x}{x^2}$

26. Evaluate: (a) $\lim_{x\to 0} \frac{\sin ax}{\sin bx}$; (b) $\lim_{x\to 0} \frac{\sin^3(2x)}{x \sin^2(3x)}$

Ans. (a) $\frac{a}{b}$; (b) $\frac{8}{9}$

27. If $x = A \sin kt + B \cos kt$, show that $\frac{d^2x}{dt^2} = -k^2 x$.

28. (a) If $y = 3\sin(2x+3)$, show that $y'' + 4y = 0$. (b) If $y = \sin x + 2\cos x$, show that $y''' + y'' + y' + y = 0$.

29. (i) Discuss and sketch the following on the interval $0 \le x < 2\pi$. (ii) (GC) Check your answers to (i) on a graphing calculator.

(a) $y = \frac{1}{2}\sin 2x$

(b) $y = \cos^2 x - \cos x$

(c) $y = x - 2 \sin x$

(d) $y = \sin x(1 + \cos x)$

(e) $y = 4\cos^3 x - 3 \cos x$

Ans. (a) maximum at $x = \pi/4, 5\pi/4$; minimum at $x = 3\pi/4, 7\pi/4$; inflection point at $x = 0, \pi/2, \pi, 3\pi/2$

(b) maximum at $x = 0, \pi$; minimum at $x = \pi/3, 5\pi/3$; inflection point at $x = 32°32', 126°23', 233°37', 327°28'$

(c) maximum at $x = 5\pi/3$; minimum at $x = \pi/3$; inflection point at $x = 0, \pi$

(d) maximum at $x = \pi/3$; minimum at $x = 5\pi/3$; inflection point at $x = 0, \pi, 104°29', 255°31'$

(e) maximum at $x = 0, 2\pi/3, 4\pi/3$; minimum at $x = \pi/3, \pi, 5\pi/3$; inflection point at $x = \pi/2, 3\pi/2, \pi/6, 5\pi/6, 7\pi/6, 11\pi/6$

30. If the angle of elevation of the sun is 45° and is decreasing by $\frac{1}{4}$ radians per hour, how fast is the shadow cast on the ground by a pole 50 ft tall lengthening?

Ans. 25 ft/h

31. Use implicit differentiation to find y': (a) $\tan y = x^2$; (b) $\cos(xy) = 2y$.

Ans. (a) $y'' = 2x \cos^2 y$; (b) $y' = -\dfrac{y \sin(xy)}{2 + x \sin(xy)}$

CHAPTER 18

Inverse Trigonometric Functions

The sine and cosine functions and the other trigonometric functions are not one-to-one and, therefore, do not have inverse functions. However, it is possible to restrict the domain of trigonometric functions so that they become one-to-one.

Looking at the graph of $y = \sin x$ (see Fig. 17-2), we note that on the interval $-\pi/2 \leq x \leq \pi/2$ the restriction of $\sin x$ is one-to-one. We then define $\sin^{-1} x$ to be the corresponding inverse function. The domain of this function is $[-1, 1]$, which is the range of $\sin x$. Thus,

1. $\sin^{-1}(x) = y$ if and only if $\sin y = x$.
2. The domain of $\sin^{-1} x$ is $[-1, 1]$.
3. The range of $\sin^{-1} x$ is $[-\pi/2, \pi/2]$.

The graph of $\sin^{-1} x$ is obtained from the graph of $\sin x$ by reflection in the line $y = x$. See Fig. 18-1.

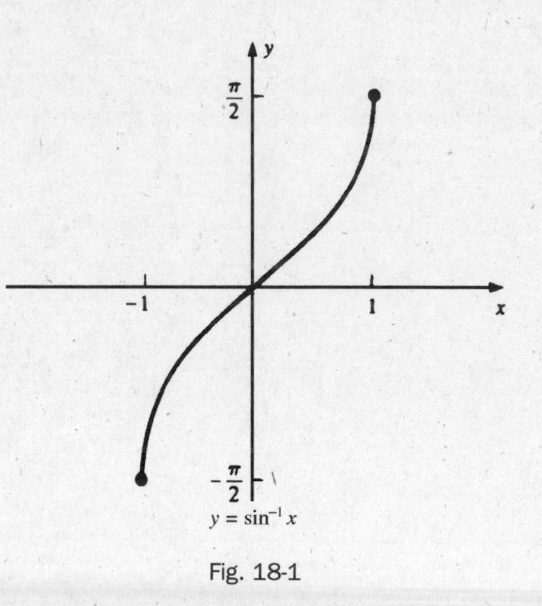

Fig. 18-1

EXAMPLE 18.1: In general, $\sin^{-1} x =$ the number y in $[-\pi/2, \pi/2]$ such that $\sin y = x$. In particular, $\sin^{-1} 0 = 0$, $\sin^{-1} 1 = \pi/2$, $\sin^{-1}(-1) = -\pi/2$, $\sin^{-1}(\frac{1}{2}) = \pi/6$, $\sin^{-1}(\sqrt{2}/2) = \pi/4$, $\sin^{-1}(\sqrt{3}/2) = \pi/3$. Also, $\sin^{-1}(-\frac{1}{2}) = \pi/6$. In general, $\sin^{-1}(-x) = -\sin^{-1} x$, because $\sin(-y) = -\sin y$.

The Derivative of $\sin^{-1} x$

Let $y = \sin^{-1} x$. Since $\sin x$ is differentiable, $\sin^{-1} x$ is differentiable by Theorem 10.2. Now, $\sin y = x$ and, therefore, by implicit differentiation, $(\cos y)\, y' = 1$. Hence, $y' = 1/(\cos y)$. But $\cos^2 y = 1 - \sin^2 y = 1 - x^2$. So, $\cos y = \pm\sqrt{1 - x^2}$. By definition of $\sin^{-1} x$, y is in the interval $[-\pi/2, \pi/2]$ and, therefore, $\cos y \geq 0$.

Hence, $\cos y = \sqrt{1-x^2}$. Thus, $y' = \dfrac{1}{\sqrt{1-x^2}}$. So, we have shown that

(18.1) $D_x(\sin^{-1} x) = \dfrac{1}{\sqrt{1-x^2}}$

The Inverse Cosine Function

If we restrict the domain of $\cos x$ to $[0, \pi]$, we obtain a one-to-one function (with range $[-1, 1]$). So we can define $\cos^{-1} x$ to be the inverse of that restriction.

1. $\cos^{-1}(x) = y$ if and only if $\cos y = x$.
2. The domain of $\cos^{-1} x$ is $[-1, 1]$.
3. The range of $\cos^{-1} x$ is $[0, \pi]$.

The graph of $\cos^{-1} x$ is shown in Fig. 18-2. It is obtained by reflecting the graph of $y = \cos x$ in the line $y = x$.

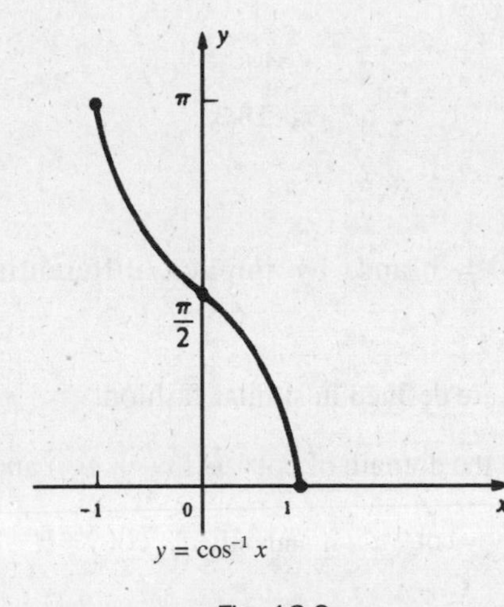

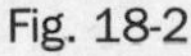

Fig. 18-2

An argument similar to the one above for (18.1) shows that

(18.2) $D_x(\cos^{-1} x) = -\dfrac{1}{\sqrt{1-x^2}}$

The Inverse Tangent Function

Restricting the domain of $\tan x$ to the interval $(-\pi/2, \pi/2)$, we obtain a one-to-one function (with range the set of all real numbers), whose inverse we take to be $\tan^{-1} x$. Then:

1. $\tan^{-1}(x) = y$ if and only if $\tan y = x$.
2. The domain of $\tan^{-1} x$ is $(-\infty, +\infty)$.
3. The range of $\tan^{-1} x$ is $(-\pi/2, \pi/2)$.

EXAMPLE 18.2: In general, $\tan^{-1} x$ = the number y in $(-\pi/2, \pi/2)$ such that $\tan y = x$. In particular, $\tan^{-1} 0 = 0$, $\tan^{-1} 1 = \pi/4$, $\tan^{-1}(\sqrt{3}) = \pi/3$, $\tan^{-1}(\sqrt{3}/3) = \pi/6$. Since $\tan(-x) = -\tan x$, it follows that $\tan^{-1}(-x) = -\tan^{-1} x$. For example, $\tan^{-1}(-1) = -\pi/4$.

The graph of $y = \tan^{-1} x$ is shown in Fig. 18-3. It is obtained from the graph of $y = \tan x$ by reflection in the line $y = x$. Note that $y = \pi/2$ is a horizontal asymptote on the right and $y = -\pi/2$ is a horizontal asymptote on the left.

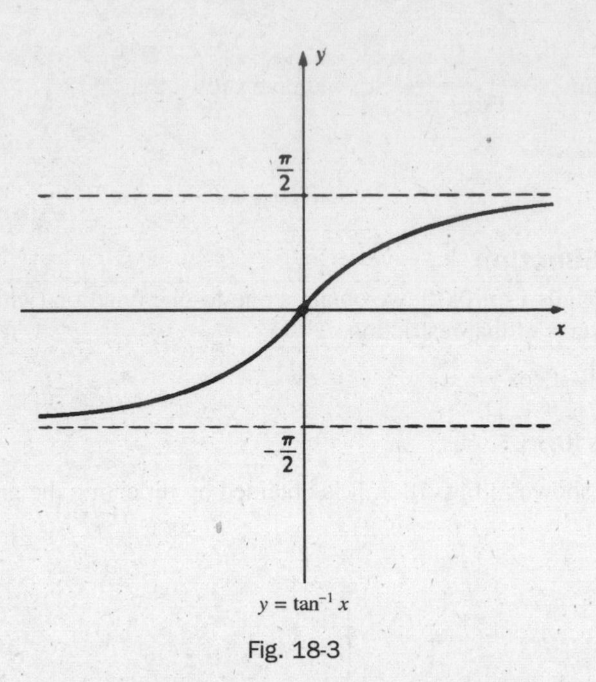

Fig. 18-3

(18.3) $D_x(\tan^{-1} x) = \dfrac{1}{1+x^2}$

In fact, if $y = \tan^{-1} x$, $\tan y = x$ and, by implicit differentiation, $(\sec^2 y)\, y' = 1$. Hence, $y' = \dfrac{1}{\sec^2 y} = \dfrac{1}{1+\tan^2 y} = \dfrac{1}{1+x^2}$.

Inverses of $\cot x$, $\sec x$, and $\csc x$ are defined in similar fashion.

$\cot^{-1} x$. Restrict $\cot x$ to $(0, \pi)$. Then the domain of $\cot^{-1} x$ is $(-\infty, +\infty)$ and

$$y = \cot^{-1} x \quad \text{if and only if} \quad \cot y = x$$

(18.4) $D_x(\cot^{-1} x) = -\dfrac{1}{1+x^2}$

The proof is similar to that of (18.3). The graphs of $\cot x$ and $\cot^{-1} x$ are shown in Fig. 18-4.

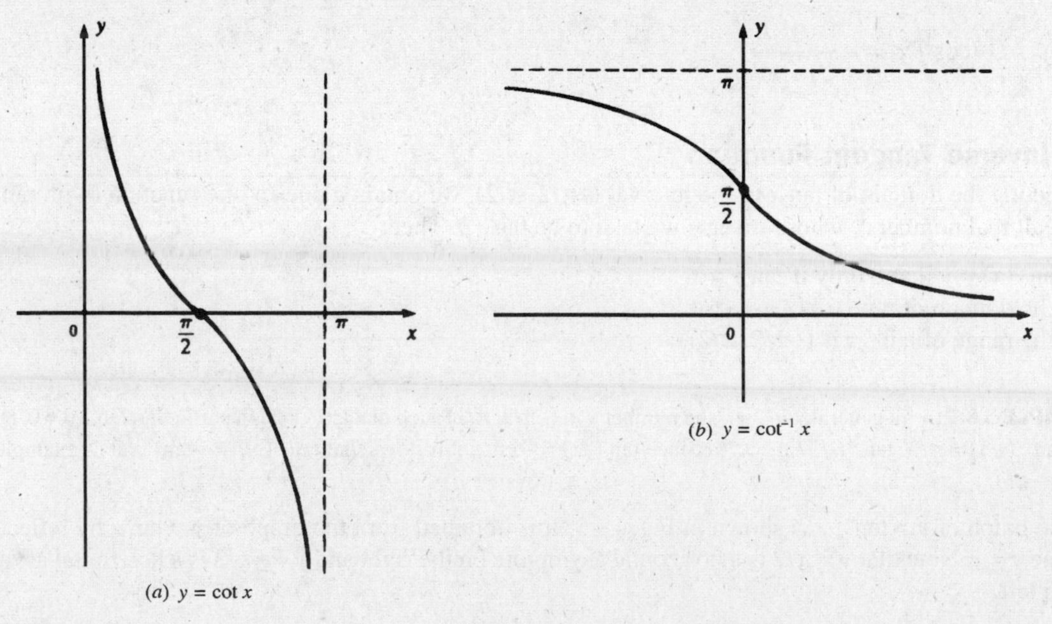

Fig. 18-4

$\sec^{-1} x$. Restrict $\sec x$ to the union of $[0, \pi/2)$ and $[\pi, 3\pi/2)$. Then the domain of $\sec^{-1} x$ consists of all y such that $|y| \geq 1$ and

$$y = \sec^{-1} x \quad \text{if and only if} \quad \sec y = x$$

(18.5) $D_x(\sec^{-1} x) = \dfrac{1}{x\sqrt{x^2 - 1}}$

For the proof, see Problem 1. The graph of $\sec x$ appeared in Fig. 17-8, and that of $\sec^{-1} x$ is shown in Fig. 18-5.

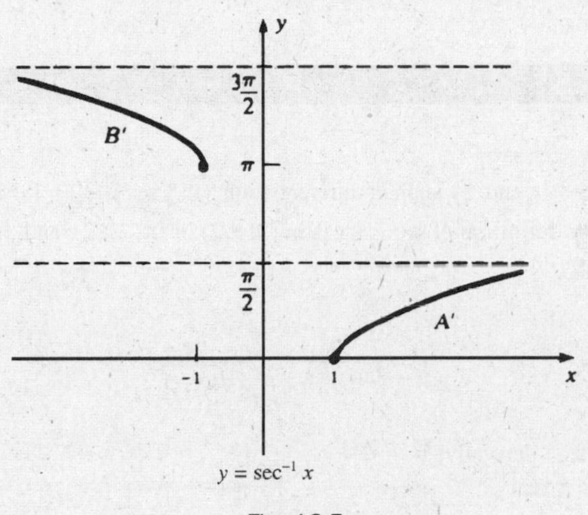

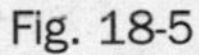

Fig. 18-5

$\csc^{-1} x$. Restrict $\csc x$ to the union of $(0, \pi/2]$ and $(\pi, 3\pi/2]$. Then the domain of $\csc^{-1} x$ consists of all y such that $|y| \geq 1$ and

$$y = \csc^{-1} x \quad \text{if and only if} \quad \csc y = x$$

(18.6) $D_x(\csc^{-1} x) = -\dfrac{1}{x\sqrt{x^2 - 1}}$

The proof is similar to that of (18.5). The graphs of $\csc x$ and $\csc^{-1} x$ are shown in Fig. 18-6.

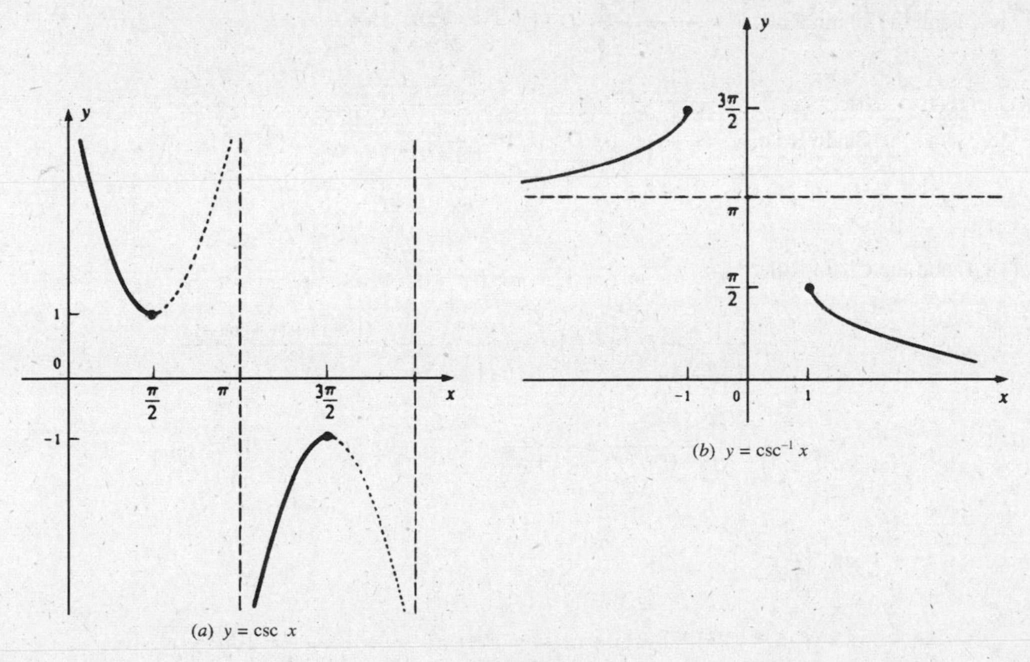

Fig. 18-6

The apparently arbitrary choices of the domains for the inverse trigonometric functions were made in order to obtain simple formulas for the derivatives.

Do not confuse the notation for the inverse trigonometric functions with exponential notation. For example, $\sin^{-1} x$ is not the same as $(\sin x)^{-1}$. To avoid the possibility of such confusion, one can use the following alternative notation for the inverse trigonometric functions:

$$\arcsin x = \sin^{-1} x, \quad \arccos x = \cos^{-1} x, \quad \text{etc.}$$

SOLVED PROBLEMS

1. Prove (18.5): $D_x(\sec^{-1} x) = \dfrac{1}{x\sqrt{x^2-1}}$.

Let $y = \sec^{-1} x$. Then $\sec y = x$ and, by implicit differentiation, $\tan y \sec y\,(y') = 1$. Now $\tan^2 y = \sec^2 y - 1 = x^2 - 1$; hence, $\tan y = \pm\sqrt{x^2-1}$. By definition of $\sec^{-1} x$, y is in $[0, \pi/2)$ or $[\pi, 3\pi/2)$, and, therefore, $\tan y$ is positive. Thus, $\tan y = \sqrt{x^2-1}$ So,

$$y' = \frac{1}{\tan y \sec y} = \frac{1}{x\sqrt{x^2-1}}$$

In Problems 2–8, find the first derivative y'.

2. $y = \sin^{-1}(2x-3)$.

By (18.1) and the Chain Rule,

$$y' = \frac{1}{\sqrt{1-(2x-3)^2}} D_x(2x-3) = \frac{2}{\sqrt{12x-4x^2-8}} = \frac{1}{\sqrt{3x-x^2-2}}$$

3. $y = \cos^{-1}(x^2)$.

By (18.2) and the Chain Rule, $y' = -\dfrac{1}{\sqrt{1-(x^2)^2}} D_x(x)^2 = -\dfrac{2x}{\sqrt{1-x^4}}$.

4. $y = \tan^{-1}(3x^2)$.

By (18.3) and the Chain Rule, $y' = -\dfrac{1}{1+(3x^2)^2} D_x(3x^2) = -\dfrac{6x}{1+9x^4}$.

5. $y = \cot^{-1}\left(\dfrac{1+x}{1-x}\right)$.

By (18.4) and the Chain Rule,

$$y' = -\frac{1}{1+\left(\frac{1+x}{1-x}\right)^2} D_x\left(\frac{1+x}{1-x}\right) = -\frac{1}{1+\left(\frac{1+x}{1-x}\right)^2} \times \frac{(1-x)-(1+x)(-1)}{(1-x)^2}$$

$$= -\frac{2}{(1-x)^2+(1+x)^2} = -\frac{1}{1+x^2}$$

6. $y = x\sqrt{a^2-x^2} + a^2 \sin^{-1}\left(\dfrac{x}{a}\right)$.

$$y' = x[\tfrac{1}{2}(a^2-x^2)^{-1/2}(-2x)] + (a^2-x^2)^{1/2} + a^2 \frac{1}{\sqrt{1-(x/a)^2}}\frac{1}{a} = 2\sqrt{a^2-x^2}$$

7. $y = x\csc^{-1}\left(\frac{1}{x}\right) + \sqrt{1-x^2}$ for $0 < x < 1$.

$$y' = x\left[\frac{1}{\frac{1}{x}\sqrt{\frac{1}{x^2}-1}}\right] + \csc^{-1}\left(\frac{1}{x}\right) + \tfrac{1}{2}(1-x^2)^{1/2}(-2x) = \csc^{-1}\left(\frac{1}{x}\right)$$

8. $y = \frac{1}{ab}\tan^{-1}\left(\frac{b}{a}\tan x\right)$.

$$y' = \frac{1}{ab}\left[\frac{1}{1+\left(\frac{b}{a}\tan x\right)^2}D_x\left(\frac{b}{a}\tan x\right)\right] = \frac{1}{ab}\frac{a^2}{a^2+b^2\tan^2 x}\frac{b}{a}\sec^2 x = \frac{\sec^2 x}{a^2+b^2\tan^2 x}$$

$$= \frac{1}{a^2\cos^2 x + b^2\sin^2 x}$$

9. If $y^2 \sin x + y = \tan^{-1} x$, find y'.
By implicit differentiation, $2yy'\sin x + y^2\cos x + y' = \frac{1}{1+x^2}$ Hence, $y'(2y\sin x + 1) = \frac{1}{1+x^2} - y^2\cos x$ and, therefore,

$$y' = \frac{1-(1+x^2)y^2\cos x}{(1+x^2)(2y\sin x+1)}$$

10. Evaluate: (a) $\sin^{-1}(-\sqrt{2}/2)$ (b) $\cos^{-1}(1)$; (c) $\cos^{-1}(0)$; (d) $\cos^{-1}(\frac{1}{2})$; (e) $\tan^{-1}(-\sqrt{3})$; (f) $\sec^{-1}(2)$; (g) $\sec^{-1}(-2)$

(a) $\sin^{-1}(-\sqrt{2}/2) = -\sin^{-1}(\sqrt{2}/2 = -\pi/4$
(b) $\cos^{-1}(1) = 0$, since $\cos(0) = 1$ and 0 is in $[0, \pi]$
(c) $\cos^{-1}(0) = \pi/2$, since $\cos(\pi/2) = 0$ and $(\pi/2)$ is in $[0, \pi]$
(d) $\cos^{-1}(\frac{1}{2}) = \pi/3$
(e) $\tan^{-1}(-\sqrt{3}) = -\tan^{-1}(\sqrt{3}) = -\pi/3$
(f) $\sec^{-1}(2) = \pi/3$, since

$$\sec\left(\frac{\pi}{3}\right) = \frac{1}{\cos(\pi/3)} = \frac{1}{\frac{1}{2}} = 2$$

(g) $\sec^{-1}(-2) = 4\pi/3$, since $\sec(4\pi/3) = \frac{1}{\cos(4\pi/3)} = \frac{1}{-\frac{1}{2}} = -2$ and $4\pi/3$ is in $[\pi, 3\pi/2)$

11. Show that $\sin^{-1} x + \cos^{-1} x = \frac{\pi}{2}$.

$D_x(\sin^{-1} x + \cos^{-1} x) = \frac{1}{\sqrt{1-x^2}} - \frac{1}{\sqrt{1-x^2}} = 0$. Then, by Problem 15 of Chapter 13, $\sin^{-1} x + \cos^{-1} x$ is a constant. Since $\sin^{-1} 0 + \cos^{-1} 0 = 0 + \frac{\pi}{2} = \frac{\pi}{2}$, that constant is $\frac{\pi}{2}$.

12. (a) Prove: $\sin(\sin^{-1}(y)) = y$; (b) find $\sin^{-1}(\sin \pi)$; (c) prove that $\sin^{-1}(\sin x) = x$ if and only if x is in $[-\pi/2, \pi/2]$.

(a) This follows directly from the definition of $\sin^{-1}(y)$.
(b) $\sin^{-1}(\sin \pi) = \sin^{-1} 0 = 0$.
(c) $\sin^{-1} y$ is equal to that number x in $[-\pi/2, \pi/2]$ such that $\sin x = y$. So, if x is in $[-\pi/2, \pi/2]$, $\sin^{-1}(\sin x) = x$. If x is not in $[-\pi/2, \pi/2]$, then $\sin^{-1}(\sin x) \neq x$, since, by definition, $\sin^{-1}(\sin x)$ must be in $[-\pi/2, \pi/2]$.

13. Evaluate: (a) $\cos(2\sin^{-1}(\frac{2}{5}))$; (b) $\sin(\cos^{-1}(-\frac{3}{4}))$.

(a) By (16.11), $\cos(2\sin^{-1}(\frac{2}{5})) = 1 - 2\sin^2(\sin^{-1}(\frac{2}{5})) = 1 - 2(\frac{2}{5})^2 = 1 - \frac{8}{25} = \frac{17}{25}$.

(b) $\sin^2(\cos^{-1}(-\frac{3}{4})) = 1 - \cos^2(\cos^{-1}(-\frac{3}{4})) = 1 - (-\frac{3}{4})^2 = \frac{7}{16}$.

Hence, $\sin(\cos^{-1}(-\frac{3}{4})) = \pm\sqrt{7}/4$. Since $\cos^{-1}(-\frac{3}{4})$ is in the second quadrant, $\sin(\cos^{-1}(-\frac{3}{4})) > 0$. So, $\sin(\cos^{-1}(-\frac{3}{4})) = \sqrt{7}/4$.

14. The lower edge of a mural, 12 ft high, is 6 ft above an observer's eye. Under the assumption that the most favorable view is obtained when the angle subtended by the mural at the eye is a maximum, at what distance from the wall should the observer stand?

Let θ denote the subtended angle, and let x be the distance from the wall. From Fig. 18-7, $\tan(\theta + \phi) = 18/x$, $\tan\phi = 6/x$, and

$$\tan\theta = \tan[(\theta+\phi) - \phi] = \frac{\tan(\theta+\phi) - \tan\phi}{1 + \tan(\theta+\phi)\tan\phi} = \frac{(18/x) - (6/x)}{1 + (18/x)(6/x)} = \frac{12x}{x^2 + 108}$$

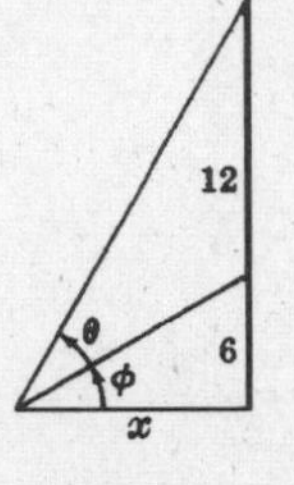

Fig. 18-7

Then

$$\theta = \tan^{-1}\left(\frac{12x}{x^2+108}\right) \quad \text{and} \quad \frac{d\theta}{dx} = \frac{12(-x^2+108)}{x^4 + 360x^2 + 11664}$$

The critical number $x = 6\sqrt{3} \sim 10.4$. By the first derivative test, this yields a relative maximum. The observer should stand about 10.4 ft in front of the wall.

SUPPLEMENTARY PROBLEMS

15. Evaluate: (a) $\sin^{-1}(-\sqrt{3}/2)$; (b) $\cos^{-1}(\sqrt{3}/2)$; (c) $\cos^{-1}(-\sqrt{3}/2)$; (d) $\tan^{-1}(-\sqrt{3}/3)$; (e) $\sec^{-1}(\sqrt{2})$; (f) $\sec^{-1}(-\sqrt{2})$.

Ans. (a) $-\frac{\pi}{3}$; (b) $\frac{\pi}{6}$; (c) $\frac{5\pi}{6}$; (d) $-\frac{\pi}{6}$; (e) $\frac{\pi}{4}$; (f) $\frac{5\pi}{4}$

16. Prove: $\tan^{-1}x + \cot^{-1}x = \frac{\pi}{2}$.

In Problems 17-24, find y'.

17. $y = \sin^{-1}(3x)$ *Ans.* $\frac{3}{\sqrt{1-9x^2}}$

18. $y = \cos^{-1}(\frac{1}{2}x)$ Ans. $-\frac{1}{\sqrt{4-x^2}}$

19. $y = \tan^{-1}\left(\frac{3}{x}\right)$ Ans. $-\frac{3}{x^2+9}$

20. $y = \sin^{-1}(x-1)$ Ans. $\frac{1}{\sqrt{2x-x^2}}$

21. $y = x^2 \cos^{-1}\left(\frac{2}{x}\right)$ Ans. $2x\left(\cos^{-1}\left(\frac{2}{x}\right) + \frac{1}{\sqrt{x^2-4}}\right)$

22. $y = \frac{x}{\sqrt{a^2 - x^2}} - \sin^{-1}(x-a)$ Ans. $\frac{x^2}{(a^2 - x^2)^{3/2}}$

23. $y = (x-a)\sqrt{2ax - x^2} + a^2 \sin^{-1}\left(\frac{x-a}{a}\right)$ Ans. $2\sqrt{2ax - x^2}$

24. $y = \frac{\sqrt{x^2-4}}{x^2} + \frac{1}{2}\sec^{-1}(\frac{1}{2}x)$ Ans. $\frac{8}{x^3\sqrt{x^2-4}}$

25. Prove formulas (18.2), (18.4), and (18.6).

26. Let $\theta = \cos^{-1}(\frac{2}{7})$. Find: (a) sin θ; (b) cos θ; (c) tan θ; (d) cot θ; (e) sec θ; (f) csc θ; (g) cos 2θ; (h) sin 2θ.

Ans. (a) $\frac{3\sqrt{5}}{7}$; (b) $\frac{2}{7}$; (c) $\frac{3\sqrt{5}}{2}$; (d) $\frac{2\sqrt{5}}{15}$; (e) $\frac{7}{2}$; (f) $\frac{7\sqrt{5}}{15}$; (g) $-\frac{41}{49}$; (h) $\frac{12\sqrt{5}}{49}$

27. Let $\theta = \sin^{-1}(-\frac{1}{5})$. Find: (a) sin θ; (b) cos θ; (c) tan θ; (d) cot θ; (e) sec θ; (f) csc θ; (g) cos 2θ; (h) sin 2θ.

Ans. (a) $-\frac{1}{5}$; (b) $\frac{2\sqrt{6}}{5}$; (c) $-\frac{\sqrt{6}}{12}$; (d) $-2\sqrt{6}$; (e) $\frac{5\sqrt{6}}{12}$; (f) -5; (g) $\frac{23}{25}$; (h) $-\frac{4\sqrt{6}}{25}$

28. Prove: $\tan 2\theta = \frac{2\tan\theta}{1-\tan^2\theta}$.

29. Evaluate: (a) $\cos(\sin^{-1}(\frac{3}{11}))$; (b) $\tan(\sec^{-1}(\frac{7}{5}))$; (c) $\sin(\cos^{-1}(\frac{2}{5}) + \sec^{-1} 4)$; (d) $\cos^{-1}\left(\cos\frac{3\pi}{2}\right)$.

Ans. (a) $\frac{4\sqrt{7}}{11}$; (b) $\frac{2\sqrt{6}}{7}$; (c) $\frac{\sqrt{21}}{20} + \frac{\sqrt{15}}{10}$; (d) $\frac{\pi}{2}$

30. Find the domain and range of the function $f(x) = \sin(\sec^{-1} x)$.

Ans. Domain $|x| \geq 1$; range $(-1, 1)$

31. (a) For which values of x is $\tan^{-1}(\tan x) = x$ true?

(b) (GC) Use a graphing calculator to draw the graph of $y = \tan^{-1}(\tan x) - x$ to verify your answer to (*a*).

Ans. (a) $-\frac{\pi}{2} < x < \frac{\pi}{2}$

32. A light is to be placed directly above the center of a circular plot of radius 30 ft, at such a height that the edge of the plot will get maximum illumination. Find the height if the intensity I at any point on the edge is directly proportional to the cosine of the angle of incidence (angle between the ray of light and the vertical) and inversely proportional to the square of the distance from the source.

(*Hint*: Let x be the required height, y the distance from the light to a point on the edge, and θ the angle of incidence. Then $I = k\frac{\cos\theta}{y^2} = \frac{kx}{(x^2+900)^{3/2}}$.)

Ans. $15\sqrt{2}$ ft

33. Show that $\sin^{-1} x = \tan^{-1}\left(\frac{x}{\sqrt{1-x^2}}\right)$ for $|x| < 1$. Examine what happens when $|x| = 1$.

34. (GC) Evalute $\sin^{-1}(\frac{3}{5})$ by using a graphing calculator.

Ans. 0.6435

35. (a) Find $\sec(\tan^{-1}(\frac{5}{7}))$. (b) Find an algebraic formula for $\sec(\tan^{-1}(2x))$. (c) (GC) Verify (a) and (b) on a graphing calculator.

Ans. (a) $\frac{\sqrt{74}}{7}$; (b) $\sqrt{1+4x^2}$

36. Prove: (a) $\sec^{-1}x = \cos^{-1}\left(\frac{1}{x}\right)$ for $x \geq 1$; (b) $\sec^{-1}x = 2\pi - \cos^{-1}\left(\frac{1}{x}\right)$ for $x \leq -1$.

(The formula of part (*a*) would hold in general for $|x| \geq 1$ if we had defined $\sec^{-1}x$ to be the inverse of the restriction of $\sec x$ to $(-\pi/2, \pi/2)$. However, if we had done that, then the formula for $D_x(\sec^{-1}x)$ would have been $1/(|x|\sqrt{x^2-1})$ instead of the simpler formula $1/(x\sqrt{x^2-1})$.)

Rectilinear and Circular Motion

Rectilinear Motion

Rectilinear motion is motion of an object on a straight line. If there is a coordinate system on that line, and s denotes the coordinate of the object at any time t, then the position of the object is given by a function $s = f(t)$. (See Fig. 19-1.)

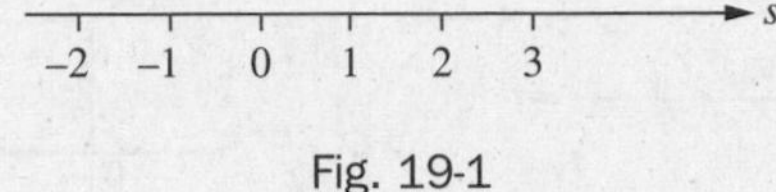

Fig. 19-1

The position at a time $t + \Delta t$, very close to t, is $f(t + \Delta t)$. The "distance" the object travels between time t and time $t + \Delta t$ is $f(t + \Delta t) - f(t)$. The time the object has traveled is Δt. So, the *average velocity* over this period of time is

$$\frac{f(t+\Delta t)-f(t)}{\Delta t}$$

(Note that the "distance" can be negative when the object is moving to the left along the s axis. So the average velocity can be positive or negative or zero.)

As Δt approaches zero, this average velocity approaches what we think of as the *instantaneous velocity* v at time t. So,

$$v = \lim_{\Delta t \to 0} \frac{f(t+\Delta t)-f(t)}{\Delta t} = f'(t)$$

Hence, the instantaneous velocity v is the derivative of the position function s, that is, $v = ds/dt$.

The sign of the instantaneous velocity v tells us in which direction the object is moving along the line. If $v = ds/dt > 0$ on an interval of time, then by Theorem 13.7(*a*), we know that s must be increasing, that is, the object is moving in the direction of increasing s along the line. If $v = ds/dt < 0$, then the object is moving in the direction of decreasing s.

The instantaneous *speed* of the object is defined as the absolute value of the velocity. Thus, the speed indicates how fast the object is moving, but not its direction. In an automobile, the speedometer tells us the instantaneous speed at which the car is moving.

The *acceleration* a of an object moving on a straight line is defined as the rate at which the velocity is changing, that is, the derivative of the velocity:

$$a = \frac{dv}{dt} = \frac{d^2s}{dt^2}$$

EXAMPLE 19.1: Let the position of an automobile on a highway be given by the equation $s = f(t) = t^2 - 5t$, where s is measured in miles and t in hours. Then its velocity $v = 2t - 5$ mi/h and its acceleration $a = 2$ mi/h^2. Thus, its velocity is increasing at the rate of 2 miles per hour per hour.

When an object moving along a straight line changes direction, its velocity $v = 0$. For, a change in direction occurs when the position s reaches a relative extremum, and this occurs only when $ds/dt = 0$. (However, the converse is false; $ds/dt = 0$ does not always indicate a relative extremum. An example is $s = t^3$ at $t = 0$.)

EXAMPLE 19.2: Assume that an object moves along a straight line according to the equation $s = f(t) = (t-2)^2$, where s is measured in feet and t in seconds. (The graph of f is shown in Fig. 19-2.) Then $v = f'(t) = 2(t-2)$ ft/sec and $a = 2$ ft/sec². For $t < 2$, $v < 0$ and the object is moving to the left. (See Fig. 19-3.) For $t > 2$, $v > 0$ and the object is moving to the right. The object changes direction at $t = 2$, where $v = 0$. Note that, although the velocity v is 0 at time $t = 2$, the object is moving at that time; it is not standing still. When we say that an object is *standing still*, we mean that its position is constant over a whole interval of time.

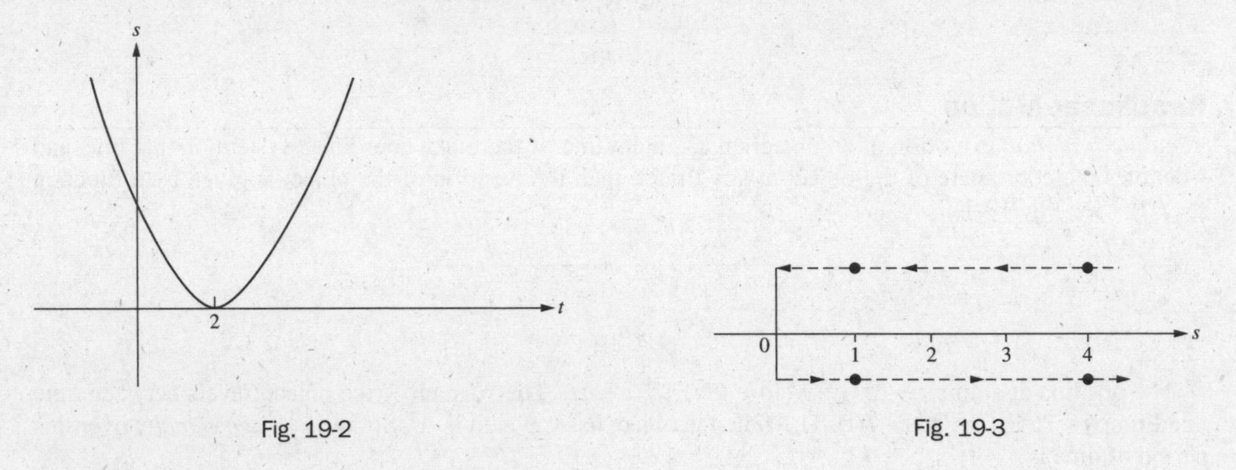

Fig. 19-2

Fig. 19-3

Motion Under the Influence of Gravity

If an object has been thrown straight up or down, or just starts from rest, and the only force acting upon it is the gravitational pull of the earth, then the resulting rectilinear motion is referred to as *free fall*.

Put a coordinate system on the vertical line on which the object is moving. Assume that this s axis is directed upward (see Fig. 19-4), and that ground level (the surface of the earth) corresponds to $s = 0$. It is a fact of physics that the acceleration a is a constant approximately equal to -32 ft/sec². (In the metric system, this constant is -9.8 m/sec². The symbol "m"stands for "meters.") Note that the acceleration is negative because the pull of the earth causes the velocity to decrease.

Since $\frac{dv}{dt} = a = -32$, we have:

Fig. 19-4

(19.1) $v = v_0 - 32t$

where v_0 is the *initial* velocity when $t = 0$.† Now, $v = \frac{ds}{dt}$. Hence,

(19.2) $s = s_0 + v_0 t - 16t^2$

where s_0 is the *initial position*, the value of s when $t = 0$.‡

† In fact, $D_t(v_0 - 32t) = -32 = D_t v$. So, by Chapter 13, Problem 18, v and $v_0 - 32t$ differ by a constant. Since v and $v_0 - 32t$ are equal when $t = 0$, that constant difference is 0.

‡ In fact, $D_t(s_0 + v_0 t - 16t^2) = v_0 - 32t = D_t s$. So, by Chapter 13, Problem 18, s and $s_0 + v_0 t - 16t^2$ differ by a constant. Since s and $s_0 + v_0 t - 16t^2$ are equal when $t = 0$, that constant difference is 0.

Circular Motion

The motion of a particle P along a circle is completely described by an equation $\theta = f(t)$, where θ is the central angle (in radians) swept over in time t by a line joining P to the center of the circle. The x and y coordinates of P are given by $x = r\cos\theta$ and $y = r\sin\theta$.

By the *angular velocity* ω of P at time t, we mean $\dfrac{d\theta}{dt}$.

By the *angular acceleration* α of P at time t, we mean $\dfrac{d\omega}{dt} = \dfrac{d^2\theta}{dt^2}$.

SOLVED PROBLEMS

1. A body moves along a straight line according to the law $s = \frac{1}{2}t^3 - 2t$. Determine its velocity and acceleration at the end of 2 seconds.

$v = \dfrac{ds}{dt} = \frac{3}{2}t^2 - 2$; hence, when $t = 2$, $v = \frac{3}{2}(2)^2 - 2 = 4$ ft/sec.

$a = \dfrac{dv}{dt} = 3t$; hence, when $t = 2$, $a = 3(2) = 6$ ft/sec^2.

2. The path of a particle moving in a straight line is given by $s = t^3 - 6t^2 + 9t + 4$.

(a) Find s and a when $v = 0$.
(b) Find s and v when $a = 0$.
(c) When is s increasing?
(d) When is v increasing?
(e) When does the direction of motion change?

We have

$$v = \frac{ds}{dt} = 3t^2 - 12t + 9 = 3(t-1)(t-3), \qquad a = \frac{dv}{dt} = 6(t-2)$$

(a) When $v = 0$, $t = 1$ and 3. When $t = 1$, $s = 8$ and $a = -6$. When $t = 3$, $s = 4$ and $a = 6$.
(b) When $a = 0$, $t = 2$. At $t = 2$, $s = 6$ and $v = -3$.
(c) s is increasing when $v > 0$, that is, when $t < 1$ and $t > 3$.
(d) v is increasing when $a > 0$, that is, when $t > 2$.
(e) The direction of motion changes when $v = 0$ and $a \neq 0$. From (a), the direction changes when $t = 1$ and $t = 3$.

3. A body moves along a horizontal line according to $s = f(t) = t^3 - 9t^2 + 24t$.
(a) When is s increasing, and when is it decreasing?
(b) When is v increasing, and when is it decreasing?
(c) Find the total distance traveled in the first 5 seconds of motion.

We have

$$v = \frac{ds}{dt} = 3t^2 - 18t + 24 = 3(t-2)(t-4), \qquad a = \frac{dv}{dt} = 6(t-3)$$

(a) s is increasing when $v > 0$, that is, when $t < 2$ and $t > 4$.
s is decreasing when $v < 0$, that is, when $2 < t < 4$.
(b) v is increasing when $a > 0$, that is, when $t > 3$.
v is decreasing when $a < 0$, that is, when $t < 3$.
(c) When $t = 0$, $s = 0$ and the body is at O. The initial motion is to the right ($v > 0$) for the first 2 seconds; when $t = 2$, the body is $s = f(2) = 20$ ft from O.

During the next 2 seconds, it moves to the left, after which it is $s = f(4) = 16$ ft from O.

It then moves to the right, and after 5 seconds of motion in all, it is $s = f(5) = 20$ ft from O. The total distance traveled is $20 + 4 + 4 = 28$ ft (see Fig. 19-5).

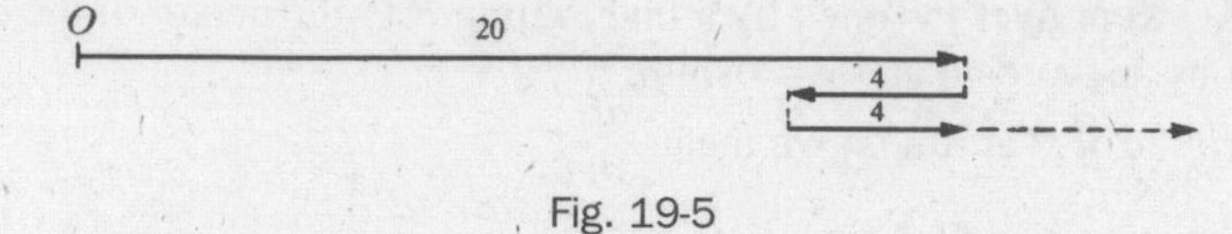

Fig. 19-5

4. A particle moves in a horizontal line according to $s = f(t) = t^4 - 6t^3 + 12t^2 - 10t + 3$.

(a) When is the speed increasing, and when decreasing?
(b) When does the direction of motion change?
(c) Find the total distance traveled in the first 3 seconds of motion.

Here

$$v = \frac{ds}{dt} = 4t^3 - 18t^2 + 24t - 10 = 2(t-1)^2(2t-5), \qquad a = \frac{dv}{dt} = 12(t-1)(t-2)$$

(a) v changes sign at $t = 2.5$, and a changes sign at $t = 1, t = 2$.
For $t < 1$, $v < 0$ and $a > 0$. Since $a > 0$, v is increasing. Since $v < 0$, the speed $|v| = -v$ is decreasing.
For $1 < t < 2$, $v < 0$ and $a < 0$. Since $a < 0$, v is decreasing. Since $v < 0$, the speed $|v| = -v$ is increasing.
For $2 < t < 2.5$, $v < 0$ and $a > 0$. As in the first case, the speed is decreasing.
For $t > 2.5$, $v > 0$ and $a > 0$. v is increasing. Since $v > 0$, the speed $|v| = v$ is increasing.

(b) The direction of motion changes at $t = 2.5$, since, by the second derivative test, s has a relative extremum there.

(c) When $t = 0$, $s = 3$ and the particle is 3 units to the right of O. The motion is to the left until $t = 2.5$, at which time the particle is $\frac{27}{16}$ units to the left of O. When $t = 3$, $s = 0$; the particle has moved $\frac{27}{16}$ units to the right. The total distance traveled is $3 + \frac{27}{16} + \frac{27}{16} = \frac{51}{8}$ units. (See Fig. 19-6.)

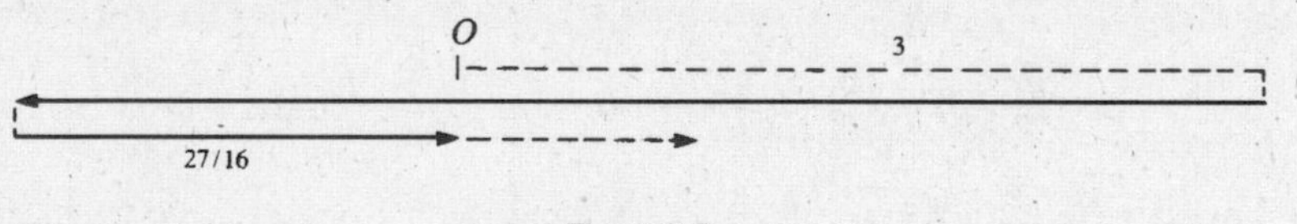

Fig. 19-6

5. A stone, projected vertically upward with initial velocity 112 ft/sec, moves according to $s = 112t - 16t^2$, where s is the distance from the starting point. Compute (a) the velocity and acceleration when $t = 3$ and when $t = 4$, and (b) when the greatest height reached. (c) When will its height be 96 ft?

We have $v = ds/dt = 112 - 32t$ and $a = dv/dt = -32$.

(a) At $t = 3$, $v = 16$ and $a = -32$. The stone is rising at 16 ft/sec.
At $t = 4$, $v = -16$ and $a = -32$. The stone is falling at 16 ft/sec.

(b) At the highest point of the motion, $v = 0$. Solving $v = 0 = 112 - 32t$ yields $t = 3.5$. At this time, $s = 196$ ft.

(c) Letting $96 = 112t - 16t^2$ yields $t^2 - 7t + 6 = 0$, from which $t = 1$ and 6. At the end of 1 second of motion, the stone is at a height of 96 ft and is rising, since $v > 0$. At the end of 6 seconds, it is at the same height but is falling since $v < 0$.

6. A particle rotates counterclockwise from rest according to $\theta = t^3/50 - t$, where θ is in radians and t in seconds. Calculate the angular displacement θ, the angular velocity ω, and the angular acceleration α at the end of 10 seconds.

$$\theta = \frac{t^3}{50} - t = 10 \text{ rad}, \qquad \omega = \frac{d\theta}{dt} = \frac{3t^2}{50} - 1 = 5 \text{ rad/sec}, \qquad \alpha = \frac{d\omega}{dt} = \frac{6t}{50} = \frac{6}{5} \text{ rad/ sec}^2$$

7. At $t = 0$, a stone is dropped from the top of a building 1024 ft high. When does it hit the ground, and with what speed? Find the speed also in miles per hour.

Since $s_0 = 1024$ and $v_0 = 0$, equation (19.2) becomes $s = 1024 - 16t^2$, and the time of hitting the ground is the solution of $1024 - 16t^2 = 0$. This reduces to $t^2 = 64$, yielding $t = \pm 8$. Since the motion occurs when $t \geq 0$, $t = 8$. The equation (19.1) is $v = -32t$, yielding $v = -32(8) = -256$ ft/sec when $t = 8$, that is, when the stone hits the ground. (The velocity is negative because the stone is moving downward.) The speed is 256 ft/sec. To change to miles per hour, note the following:

$$x \text{ feet per second} = 60x \text{ feet per minute} = 60(60x) \text{ feet per hour}$$

$$= \frac{3600x}{5280} \text{ miles per hour} = \frac{15}{22}x \text{ miles per hour.}$$

Thus,

(19.3) x feet per second $= \frac{15}{22}x$ miles per hour.

In particular, when $x = 256$, we get $174\frac{6}{11}$ miles per hour.

8. If a rocket is shot vertically upward from the ground with an initial velocity of 192 ft/sec, when does it reach its maximum height above the ground, and what is that maximum height? Also find how long it takes to reach the ground again and with what speed it hits the ground.

Equations (19.1) and (19.2) are $v = 192 - 32t$ and $s = 192t - 16t^2$. At the maximum height, $v = 0$, and, therefore, $t = 6$. So, it takes 6 seconds to reach the maximum height, which is $192(6) - 16(6)^2 = 576$ ft. The rocket returns to ground level when $0 = 192t - 16t^2$, that is, when $t = 12$. Hence, it took 6 seconds to reach the ground again, exactly the same time it took to reach the maximum height. The velocity when $t = 12$ is $192 - 32(12) = -192$ ft/sec. Thus, its final speed is the same as its initial speed.

SUPPLEMENTARY PROBLEMS

9. Show that, if an object is moving on a straight line, then its speed is increasing when its velocity v and its acceleration a have the same sign, and its speed is decreasing when v and a have opposite sign. (*Hint*: The speed $S = |v|$. When $v > 0$, $S = v$ and $dS/dt = dv/dt = a$. When $v < 0$, $S = -v$ and $dS/dt = -dv/dt = -a$.)

10. An object moves in a straight line according to the equation $s = t^3 - 6t^2 + 9t$, the units being feet and seconds. Find its position, direction, and velocity, and determine whether its speed is increasing or decreasing when (a) $t = \frac{1}{2}$; (b) $t = \frac{3}{2}$; (c) $t = \frac{5}{2}$; (d) $t = 4$.

Ans. (a) $s = \frac{25}{8}$ ft; moving to the right with $v = \frac{15}{4}$ ft/sec; speed decreasing
(b) $s = \frac{27}{8}$ ft; moving to the left with $v = -\frac{9}{4}$ ft/sec; speed increasing
(c) $s = \frac{5}{8}$ ft; moving to the left with $v = -\frac{9}{4}$ ft/sec; speed decreasing
(d) $s = 4$ ft; moving to the right with $v = 9$ ft/sec; speed increasing

11. The distance of a locomotive from a fixed point on a straight track at time t is $3t^4 - 44t^3 + 144t^2$. When is it in reverse?

Ans. $3 < t < 8$

12. Examine, as in Problem 2, each of the following straight line motions: (a) $s = t^3 - 9t^2 + 24t$; (b) $s = t^3 - 3t^2 + 3t + 3$; (c) $s = 2t^3 - 12t^2 + 18t - 5$; (d) $s = 3t^4 - 28t^3 + 90t^2 - 108t$.

Ans. The changes of direction occur at $t = 2$ and $t = 4$ in (a), not at all in (b), at $t = 1$ and $t = 3$ in (c), and at $t = 1$ in (d).

13. An object moves vertically upward from the earth according to the equation $s = 64t - 16t^2$. Show that it has lost one-half its velocity in its first 48 ft of rise.

14. A ball is thrown vertically upward from the edge of a roof in such a manner that it eventually falls to the street 112 ft below. If it moves so that its distance s from the roof at time t is given by $s = 94t - 16t^2$, find (a) the position of the ball, its velocity, and the direction of motion when $t = 2$, and (b) its velocity when it strikes the street (s in feet, and t in seconds).

Ans. (a) 240 ft above the street, 32 ft/sec upward; (b) –128 ft/sec

15. A wheel turns through an angle of θ radians in time t seconds so that $\theta = 128t - 12t^2$. Find the angular velocity and acceleration at the end of 3 seconds.

Ans. $\omega = 56$ rad/sec; $\alpha = -24$ rad/sec^2

16. A stone is dropped down a well that is 144 ft deep. When will it hit the bottom of the well?

Ans. After 3 seconds

17. With what speed in miles per hour does an object dropped from the top of a 10-story building hit the ground? Assume that each story of the building is 10 ft high.

Ans. $54\frac{6}{11}$ m/h

18. An automobile moves along a straight road. If its position is given by $s = 8t^3 - 12t^2 + 6t - 1$, with s in miles and t in hours, what distance does it travel from $t = 0$ to $t = 1$?

Ans. 2 miles

19. Answer the same question as in Problem 18, except that $s = 5t - t^2$ and the car operates from $t = 0$ to $t = 3$.

Ans. 6.5 miles

20. A stone was thrown straight up from the ground. What was its initial velocity in feet per second if it hit the ground after 15 seconds?

Ans. 240 ft/sec

21. (GC) Let the position s of an object moving on a straight line be given by $s = t^4 - 3t^2 + 2t$. Use a graphing calculator to estimate when the object changes direction, when it is moving to the right, and when it is moving to the left. Try to find corresponding exact formulas.

Ans. Change of direction at $t = -1.3660$, 0.3660, and 1. The object moves left for $t < -1.3660$ and for $0.3660 < t < 1$. The exact values of t at which the object changes direction are 1 and $\frac{-1 \pm \sqrt{3}}{2}$.

22. (GC) An object is moving along a straight line according to the equation $s = 3t - t^2$. A second object is moving along the same line according to the equation $s = t^3 - t^2 + 1$. Use a graphing calculator to estimate (a) when they occupy the same position and (b) when they have the same velocity. (c) At the time(s) when they have the same position, are they moving in the same direction?

Ans. (a) 0.3473 and 1.5321; (b) $t = \pm 1$; (c) opposite directions at both intersections.

CHAPTER 20

Related Rates

If a quantity y is a function of time t, the rate of change of y with respect to time is given by dy/dt. When two or more quantities, all functions of the time t, are related by an equation, the relation of their rates of change may be found by differentiating both sides of the equation.

EXAMPLE 20.1: A 25-foot ladder rests against a vertical wall. (See Fig. 20-1.) If the bottom of the ladder is sliding away from the base of the wall at the rate of 3 ft/sec, how fast is the top of the ladder moving down the wall when the bottom of the ladder is 7 feet from the base?

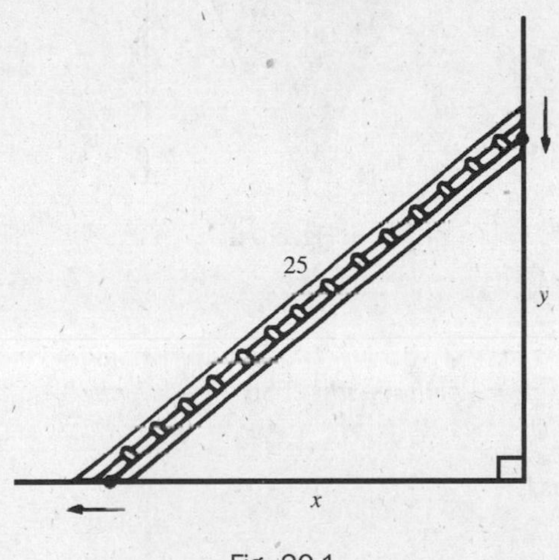

Fig. 20-1

Let x be the distance of the bottom of the ladder from the base of the wall, and let y be the distance of the top of the ladder from the base of the wall. Since the bottom of the ladder is moving away from the base of the wall at the rate of 3 ft/sec, $dx/dt = 3$. We have to find dy/dt when $x = 7$. By the Pythagorean Theorem,

$$x^2 + y^2 = (25)^2 = 625 \tag{20.1}$$

This is the relation between x and y. Differentiating both sides with respect to t, we get

$$2x\frac{dx}{dt} + 2y\frac{dy}{dt} = 0$$

Since $dx/dt = 3$, $6x + 2y\, dy/dt = 0$, whence

$$3x + y\frac{dy}{dt} = 0 \tag{20.2}$$

This is the desired equation for dy/dt. Now, for our particular problem, $x = 7$. Substituting 7 for x in equation (20.1), we get $49 + y^2 = 625$, $y^2 = 576$, $y = 24$. In equation (20.2), we replace x and y by 7 and 24, obtaining $21 + 24\, dy/dt = 0$. Hence, $dy/dt = -\frac{7}{8}$. Since $dy/dt < 0$, we conclude that the top of the ladder is sliding *down* the wall at the rate of $\frac{7}{8}$ ft/sec when the bottom of the ladder is 7 ft from the base of the wall.

SOLVED PROBLEMS

1. Gas is escaping from a spherical balloon at the rate of 2 ft³/min. How fast is the surface area shrinking when the radius is 12 ft?

A sphere of radius r has volume $V = \frac{4}{3}\pi r^3$ and surface area $S = 4\pi r^2$. By hypothesis, $dV/dt = -2$. Now, $dV/dt = 4\pi r^2\, dr/dt$. So, $-2 = 4\pi r^2\, dr/dt$ and, therefore, $dr/dt = -1/(2\pi r^2)$. Also, $dS/dt = 8\pi r\, dr/dt$. Hence, $dS/dt = -8\pi r/2\pi r^2 = -4/r$. So, when $r = 12$, $dS/dt = -\frac{4}{12} = -\frac{1}{3}$. Thus, the surface area is shrinking at the rate of $\frac{1}{3}$ ft²/min.

2. Water is running out of a conical funnel at the rate of 1 in³/sec. If the radius of the base of the funnel is 4 in and the height is 8 in, find the rate at which the water level is dropping when it is 2 in from the top. (The formula for the volume V of a cone is $\frac{1}{3}\pi r^2 h$, where r is the radius of the base and h is the height.)

Let r be the radius and h the height of the surface of the water at time t, and let V be the volume of the water in the cone. (See Fig. 20-2.) By similar triangles, $r/4 = h/8$, whence $r = \frac{1}{2}h$.

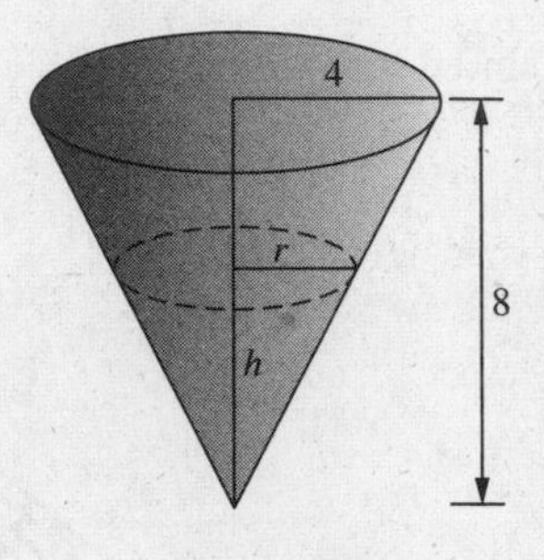

Fig. 20-2

Then

$$V = \tfrac{1}{3}\pi r^2 h = \tfrac{1}{12}\pi h^3. \quad \text{So} \quad \frac{dV}{dt} = \tfrac{1}{4}\pi h^2 \frac{dh}{dt}.$$

By hypothesis, $dV/dt = -1$. Thus,

$$-1 = \tfrac{1}{4}\pi h^2 \frac{dh}{dt}, \quad \text{yielding} \quad \frac{dh}{dt} = \frac{-4}{\pi h^2}.$$

Now, when the water level is 2 in from the top, $h = 8 - 2 = 6$. Hence, at that time, $dh/dt = -1/(9\pi)$, and so the water level is dropping at the rate of $1/(9\pi)$ in/sec.

3. Sand falling from a chute forms a conical pile whose altitude is always equal to $\frac{4}{3}$ the radius of the base. (a) How fast is the volume increasing when the radius of the base is 3 ft and is increasing at the rate of 3 in/min? (b) How fast is the radius increasing when it is 6 ft and the volume is increasing at the rate of 24 ft³/min?

Let r be the radius of the base, and h the height of the pile at time t. Then

$$h = \frac{4}{3}r \quad \text{and} \quad V = \frac{1}{3}\pi r^2 h = \frac{4}{9}\pi r^3. \quad \text{So} \quad \frac{dV}{dt} = \frac{4}{3}\pi r^2 \frac{dr}{dt}.$$

(a) When $r = 3$ and $dr/dt = \frac{1}{4}$, $dV/dt = 3\pi$ ft³/min.

(b) When $r = 6$ and $dV/dt = 24$, $dr/dt = 1/(2\pi)$ ft/min.

4. Ship A is sailing due south at 16 mi/h, and ship B, 32 miles south of A, is sailing due east at 12 mi/h. (a) At what rate are they approaching or separating at the end of 1 hour? (b) At the end of 2 hours? (c) When do they cease to approach each other, and how far apart are they at that time?

Let A_0 and B_0 be the initial positions of the ships, and A_t and B_t their positions t hours later. Let D be the distance between them t hours later. Then (see Fig. 20-3):

$$D^2 = (32-16t)^2 + (12t)^2 \quad \text{and} \quad 2D\frac{dD}{dt} = 2(32-16t)(-16) + 2(12t)(12) = 2(400t-512).$$

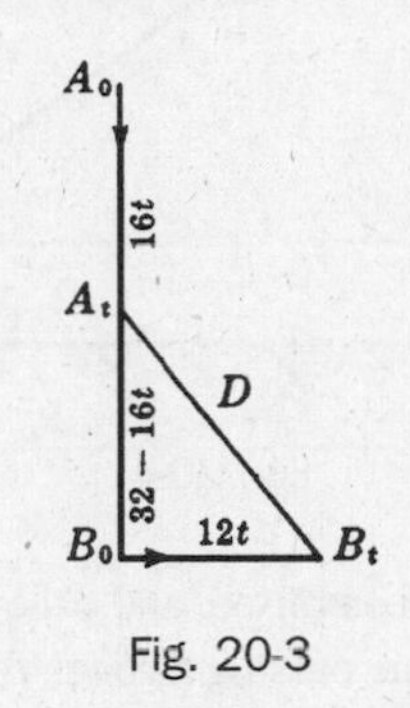

Fig. 20-3

Hence, $\frac{dD}{dt} = \frac{400t-512}{D}$

(a) When $t = 1$, $D = 20$ and $\frac{dD}{dt} = -5.6$. They are approaching at 5.6 mi/h.
(b) When $t = 2$, $D = 24$ and $\frac{dD}{dt} = 12$. They are separating at 12 mi/h.
(c) They cease to approach each other when $\frac{dD}{dt} = 0$, that is, when $t = \frac{512}{400} = 1.28$ h, at which time they are $D = 19.2$ miles apart.

5. Two parallel sides of a rectangle are being lengthened at the rate of 2 in/sec, while the other two sides are shortened in such a way that the figure remains a rectangle with constant area $A = 50$ in². What is the rate of change of the perimeter P when the length of an increasing side is (a) 5 in? (b) 10 in? (c) What are the dimensions when the perimeter ceases to decrease?

Let x be the length of the sides that are being lengthened, and y the length of the other sides, at time t. Then

$$P = 2(x+y), \quad \frac{dp}{dt} = 2\left(\frac{dx}{dt} + \frac{dy}{dt}\right), \quad A = xy = 50, \quad \frac{dA}{dt} = x\frac{dy}{dt} + y\frac{dx}{dt} = 0$$

(a) When $x = 5$, $y = 10$ and $dx/dt = 2$. Then

$$5\frac{dx}{dt} + 10(2) = 0. \text{ So } \frac{dx}{dt} = -4 \quad \text{and} \quad \frac{dp}{dt} = 2(2-4) = -4 \text{ in/sec (decreasing)}$$

(b) When $x = 10$, $y = 5$ and $dx/dt = 2$. Then

$$10\frac{dy}{dt} + 5(2) = 0. \text{ So } \frac{dy}{dt} = -1 \quad \text{and} \quad \frac{dp}{dt} = 2(2-1) = 2 \text{ in/sec (decreasing)}$$

(c) The perimeter will cease to decrease when $dP/dt = 0$, that is, when $dy/dt = -dx/dt = -2$. Then $x(-2) + y(2) = 0$, and the rectangle is a square of side $x = y = 5\sqrt{2}$ in.

6. The radius of a sphere is r when the time is t seconds. Find the radius when the rate of change of the surface area and the rate of change of the radius are equal.

The surface area $S = 4\pi r^2$; hence, $dS/dt = 8\pi r\, dr/dt$. When $dS/dt = dr/dt$, $8\pi r = 1$ and the radius $r = 1/8\pi$.

7. A weight W is attached to a rope 50 ft long that passes over a pulley at a point P, 20 ft above the ground. The other end of the rope is attached to a truck at a point A, 2 ft above the ground, as shown in Fig. 20-4. If the truck moves away at the rate of 9 ft/sec, how fast is the weight rising when it is 6 ft above the ground?

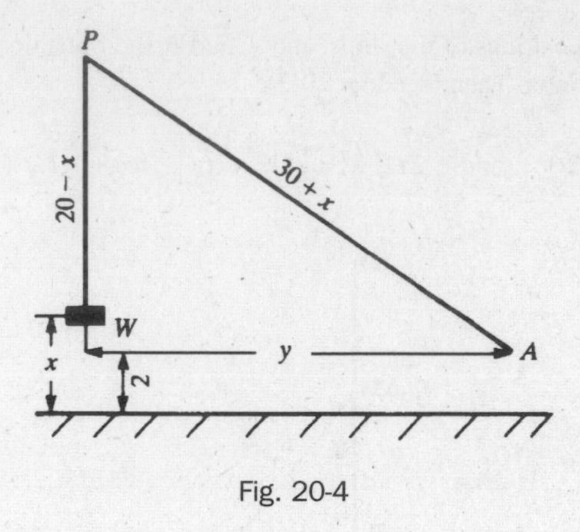

Fig. 20-4

Let x denote the distance the weight has been raised, and y the horizontal distance from point A, where the rope is attached to the truck, to the vertical line passing through the pulley. We must find dx/dt when $dy/dt = 9$ and $x = 6$.

Now

$$y^2 = (30 + x)^2 - (18)^2 \quad \text{and} \quad \frac{dy}{dt} = \frac{30 + x}{y}\frac{dx}{dt}$$

When $x = 6$, $y = 18\sqrt{3}$ and $dy/dt = 9$. Then $9 = \dfrac{30 + 6}{18\sqrt{3}}\dfrac{dx}{dt}$, from which $\dfrac{dx}{dt} = \dfrac{9}{2}\sqrt{3}$ ft/sec.

8. A light L hangs H ft above a street. An object h ft tall at O, directly under the light, moves in a straight line along the street at v ft/sec. Find a formula for the velocity V of the tip of the shadow cast by the object on the street at t seconds. (See Fig. 20-5.)

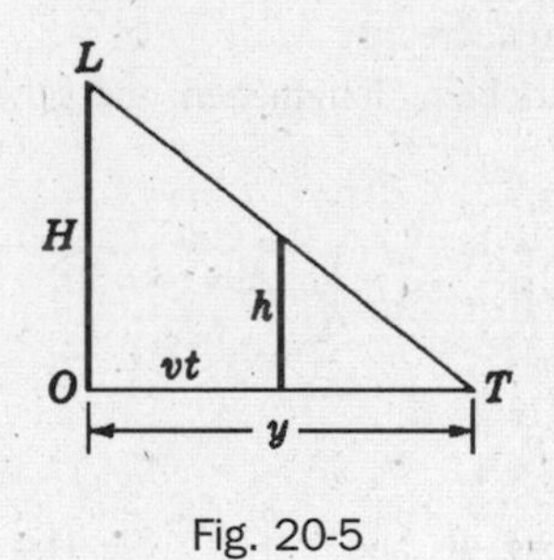

Fig. 20-5

After t seconds, the object has moved a distance vt. Let y be the distance of the tip of the shadow from O. By similar triangles, $(y - vt)/y = h/H$. Hence,

$$y = \frac{Hvt}{H - h} \quad \text{and, therefore,} \quad V = \frac{dy}{dt} = \frac{Hv}{H - h} = \frac{1}{1 - (h/H)}\, v$$

Thus, the velocity of the tip of the shadow is proportional to the velocity of the object, the factor of proportionality depending upon the ratio h/H. As $h \to 0$, $V \to v$, while as $h \to H$, $V \to +\infty$.

SUPPLEMENTARY PROBLEMS

9. A rectangular trough is 8 ft long, 2 ft across the top, and 4 ft deep. If water flows in at a rate of 2 ft³/min, how fast is the surface rising when the water is 1 ft deep?

Ans. $\frac{1}{8}$ ft/min

10. A liquid is flowing into a vertical cylindrical tank of radius 6 ft at the rate of 8 ft^3/min. How fast is the surface rising?

Ans. $2/9\pi$ ft/min

11. A man 5 ft tall walks at a rate of 4 ft/sec directly away from a street light that is 20 ft above the street. (a) At what rate is the tip of his shadow moving? (b) At what rate is the length of his shadow changing?

Ans. (a) $\frac{16}{3}$ ft/sec; (b) $\frac{4}{3}$ ft/sec

12. A balloon is rising vertically over a point *A* on the ground at the rate of 15 ft/sec. A point *B* on the ground is level with and 30 ft from *A*. When the balloon is 40 ft from *A*, at what rate is its distance from *B* changing?

Ans. 12 ft/sec

13. A ladder 20 ft long leans against a house. If the foot of the ladder is moving away from the house at the rate of 2 ft/sec, find how fast (a) the top of the ladder is moving downward, and (b) the slope of the ladder is decreasing, when the foot of the ladder is 12 ft from the house.

Ans. (a) $\frac{3}{2}$ ft/sec; (b) $\frac{25}{72}$ per second

14. Water is being withdrawn from a conical reservoir 3 ft in radius and 10 ft deep at 4 ft^3/min. How fast is the surface falling when the depth of the water is 6 ft? How fast is the radius of this surface diminishing?

Ans. $100/81\pi$ ft/min; $10/27\pi$ ft/min

15. A barge, whose deck is 10 ft below the level of a dock, is being drawn in by means of a cable attached to the deck and passing through a ring on the dock. When the barge is 24 ft away and approaching the dock at $\frac{3}{4}$ ft/sec, how fast is the cable being pulled in? (Neglect any sag in the cable.)

Ans. $\frac{9}{13}$ ft/sec

16. A boy is flying a kite at a height of 150 ft. If the kite moves horizontally away from the boy at 20 ft/sec, how fast is the string being paid out when the kite is 250 ft from him?

Ans. 16 ft/sec

17. One train, starting at 11 A.M., travels east at 45 mi/h while another, starting at noon from the same point, travels south at 60 mi/h. How fast are they separating at 3 P.M.?

Ans. $105\sqrt{2}/2$ mi/h

18. A light is at the top of a pole 80 ft high. A ball is dropped at the same height from a point 20 ft from the light. Assuming that the ball falls according to $s = 16t^2$, how fast is the shadow of the ball moving along the ground 1 second later?

Ans. 200 ft/sec

19. Ship *A* is 15 miles east of *O* and moving west at 20 mi/h; ship *B* is 60 mi south of *O* and moving north at 15 mi/h. (a) Are they approaching or separating after 1 h and at what rate? (b) After 3 h? (c) When are they nearest one another?

Ans. (a) approaching, $115/\sqrt{82}$ mi/h; (b) separating, $9\sqrt{10}/2$ mi/h; (c) 1 h 55 min

20. Water, at a rate of 10 ft^3/min, is pouring into a leaky cistern whose shape is a cone 16 ft deep and 8 ft in diameter at the top. At the time the water is 12 ft deep, the water level is observed to be rising at 4 in/min. How fast is the water leaking away?

Ans. $(10 - 3\pi)$ ft^3/min

21. A solution is passing through a conical filter 24 in deep and 16 in across the top, into a cylindrical vessel of diameter 12 in. At what rate is the level of the solution in the cylinder rising if, when the depth of the solution in the filter is 12 in, its level is falling at the rate 1 in/min?

Ans. $\frac{4}{9}$ in/min

22. Oil from a leaking oil tanker radiates outward in the form of a circular film on the surface of the water. If the radius of the circle increases at the rate of 3 meters per minute, how fast is the area of the circle increasing when the radius is 200 meters?

Ans. 1200π m^2/min

23. A point moves on the hyperbola $x^2 - 4y^2 = 36$ in such a way that the x coordinate increases at a constant rate of 20 units per second. How fast is the y coordinate changing at the point (10, 4)?

Ans. 50 units/sec

24. If a point moves along the curve $y = x^2 - 2x$, at what point is the y coordinate changing twice as fast as the x coordinate?

Ans. (2, 0)

Differentials. Newton's Method

If a function f is differentiable at x, then $f'(x) = \lim_{\Delta x \to 0} \Delta y/\Delta x$, where $\Delta y = f(x + \Delta x) - f(x)$. Hence, for values of Δx close to 0, $\Delta y/\Delta x$ will be close to $f'(x)$. This is often written $\Delta y/\Delta x \sim f'(x)$, whence

$$\Delta y \sim f'(x)\Delta x \tag{21.1}$$

This implies

$$f(x + \Delta x) \sim f(x) + f'(x)\Delta x \tag{21.2}$$

Formula (21.2) can be used to approximate values of a function.

EXAMPLE 21.1: Let us estimate $\sqrt{16.2}$. Let $f(x) = \sqrt{x}$, $x = 16$, and $\Delta x = 0.2$. Then $x + \Delta x = 16.2$, $f(x + \Delta x) = \sqrt{16.2}$, and $f(x) = \sqrt{16} = 4$. Since $f'(x) = D_x(x^{1/2}) = \frac{1}{2}x^{-1/2} = 1/(2\sqrt{x}) = 1/(2\sqrt{16}) = \frac{1}{8}$, formula (21.2) becomes

$$\sqrt{16.2} \sim 4 + \frac{1}{8}(0.2) = 4.025$$

(This approximation turns out to be correct to three decimal places. To four decimal places, the correct value is 4.0249, which can be checked on a graphing calculator.)

EXAMPLE 21.2: Let us estimate sin (0.1). Here, $f(x) = \sin x$, $x = 0$, and $\Delta x = 0.1$. Then $x + \Delta x = 0.1$, $f(x + \Delta x) = \sin(0.1)$, and $f(x) = \sin 0 = 0$. Since $f'(x) = \cos x = \cos 0 = 1$, formula (21.2) yields

$$\sin(0.1) \sim 0 + 1(0.1) = 0.1$$

The actual value turns out to be 0.0998, correct to four decimal places. Note that the method used for this problem shows that $\sin u$ can be approximated by u for values of u close to 0.

A limitation of formula (21.2) is that we have no information about how good the approximation is. For example, if we want the approximation to be correct to four decimal places, we do not know how small Δx should be chosen.

The Differential

The product on the right side of equation (21.1) is called the *differential* of f and is denoted by df.

Definition

The *differential* df of f is defined by

$$df = f'(x)\Delta x$$

Note that df is a function of two variables, x and Δx. If Δx is small, then formula (21.1) becomes

$$f(x+\Delta x) - f(x) \sim df \tag{21.3}$$

This formula is illustrated in Fig. 21-1. Line $\mathcal{L}$ is tangent to the graph of f at P; so its slope is $f'(x)$. Hence, $f'(x) = \overline{RT}/\overline{PR} = \overline{RT}/\Delta x$. Thus, $\overline{RT} = f'(x)\Delta x = df$. For Δx small, Q is close to P on the graph and, therefore $\overline{RT} \sim \overline{RQ}$, that is, $df \sim f(x+\Delta x) - f(x)$, which is formula (21.3).

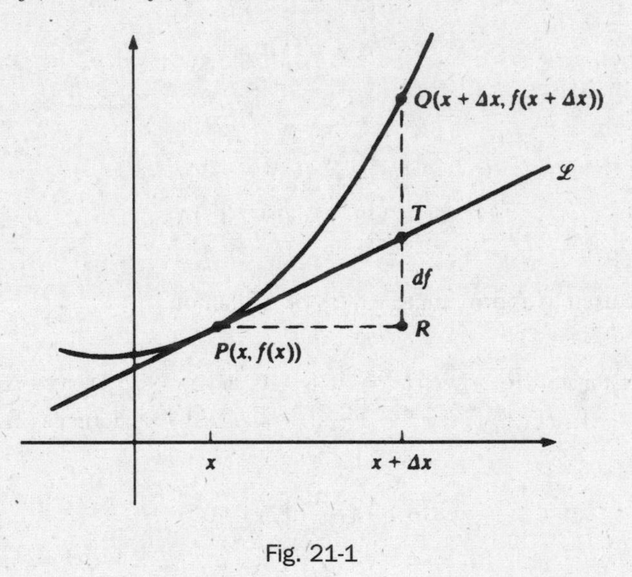

Fig. 21-1

When the function f is given by a formula, say $f(x) = \tan x$, then we often will write df as $d(\tan x)$, Thus,

$$d(\tan x) = df = f'(x)\Delta x = \sec^2 x\,\Delta x$$

Similarly, $d(x^3 - 2x) = (3x^2 - 2)\,\Delta x$. In particular, if $f(x) = x$,

$$dx = df = f'(x)\,\Delta x = (1)\,\Delta x = \Delta x$$

Since $dx = \Delta x$, we obtain $df = f'(x)\,dx$. When $\Delta x \neq 0$, division by Δx yields $df/dx = f'(x)$. When $f(x)$ is written as y, then df is written dy and we get the traditional notation dy/dx for the derivative.

If u and v are functions and c is a constant, then the following formulas are easily derivable:

$$d(c) = 0 \qquad d(cu) = c\,du \qquad d(u+v) = du + dv$$

$$d(uv) = u\,dv + v\,du \qquad d\left(\frac{u}{v}\right) = \frac{v\,du - u\,dv}{v^2}$$

Newton's Method

Assume that we know that x_0 is close to a solution of the equation

$$f(x) = 0 \tag{21.4}$$

where f is a differentiable function. Then the tangent line $\mathcal{T}$ to the graph of f at the point with x coordinate x_0 will ordinarily intersect the x axis at a point whose x coordinate x_1 is closer to the solution of (21.4) than is x_0. (See Fig. 21-2.)

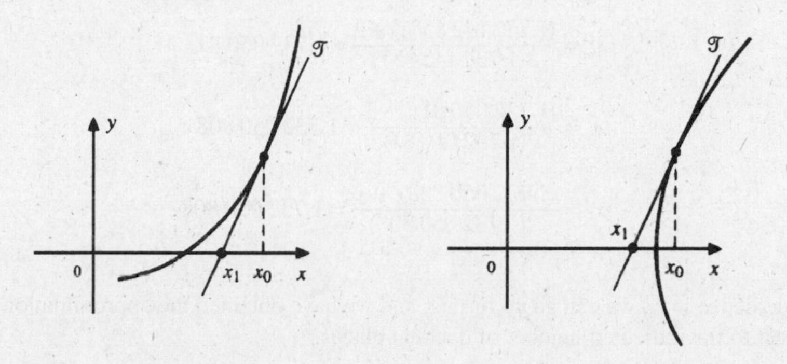

Fig. 21-2

One point–slope equation of the line $\mathcal{T}$ is

$$y - f(x_0) = f'(x_0)(x - x_0)$$

since $f'(x_0)$ is the slope of $\mathcal{T}$. If $\mathcal{T}$ intersects the x axis at $(x_1, 0)$, then

$$0 - f(x_0) = f'(x_0)(x_1 - x_0)$$

If $f'(x_0) \neq 0$,

$$x_1 - x_0 = -\frac{f(x_0)}{f'(x_0)}$$

Hence,

$$x_1 = x_0 - \frac{f(x_0)}{f'(x_0)}$$

Now carry out the same reasoning, but beginning with x_1 instead of x_0. The result is a number x_2 that should be closer to the solution of (21.4) than x_1, where $x_2 = x_1 - f(x_1)/f'(x_1)$. If we keep on repeating this procedure, we would obtain a sequence of numbers $x_0, x_1, x_2, \ldots, x_n, \ldots$ determined by the formula

$$x_{n+1} = x_n - \frac{f(x_n)}{f'(x_n)} \tag{21.5}$$

This is known as *Newton's method* for finding better and better approximations to a solution of the equation $f(x) = 0$. However, the method does not always work. (Some examples of the troubles that can arise are shown in Problems 23 and 24.)

EXAMPLE 21.3: We can approximate $\sqrt{3}$ by applying Newton's method to the function $f(x) = x^2 - 3$. Here, $f'(x) = 2x$ and (21.5) reads

$$x_{n+1} = x_n - \frac{x_n^2 - 3}{2x_n} = \frac{2x_n^2 - (x_n^2 - 3)}{2x_n} = \frac{x_n^2 + 3}{2x_n} \tag{21.6}$$

Let the first approximation x_0 be 1, since we know that $1 < \sqrt{3} < 2$. Successively substituting $n = 0, 1, 2, \dots$ in (21.6),† we get

$$x_1 = \frac{1+3}{2} = 2$$

$$x^2 = \frac{2^2+3}{2(2)} = \frac{7}{4} = 1.75$$

$$x_3 = \frac{(1.75)^2+3}{2(1.75)} = 1.732\,142\,857$$

$$x_4 = \frac{(1.732\,142\,857)^2+3}{2(1.732\,142\,857)} = 1.732\,050\,81$$

$$x_5 = \frac{(1.732\,050\,81)^2+3}{2(1.732\,050\,81)} = 1.732\,050\,808$$

$$x_6 = \frac{(1.732\,050\,808)^2+3}{2(1.732\,050\,808)} = 1.732\,050\,808$$

Since our calculator yielded $x_6 = x_5$, we can go no further, and we have obtained the approximation $\sqrt{3} \sim 1.732\,050\,808$, which is, in fact, correct to the indicated number of decimal places.

SOLVED PROBLEMS

1. Use formula (21.2) to approximate: (a) $\sqrt[3]{124}$; (b) sin 61°.

(a) Let $f(x) = \sqrt[3]{x}$, $x = 125$, and $\Delta x = -1$. Then $x + \Delta x = 124$, $f(x+\Delta x) = \sqrt[3]{124}$, and $f(x) = \sqrt[3]{125} = 5$. Since

$$f'(x) = D_x(x^{1/3}) = \tfrac{1}{3}x^{-2/3} = \frac{1}{3}\frac{1}{(125)^{2/3}} = \frac{1}{3}\frac{1}{5^2} = \frac{1}{75}$$

formula (21.2) yields $\sqrt[3]{124} \sim 5 + (\frac{1}{75})(-1) = 5 - \frac{1}{75} = \frac{374}{75} \sim 4.9867$. (To four decimal places, the correct answer can be shown to be 4.9866.)

(b) Let $f(x) = \sin x$, $x = \pi/3$, and $\Delta x = \pi/180$. Then $x + \Delta x = 61°$, $f(x+\Delta x) = \sin 61°$, and $f(x) = \sqrt{3}/2$. Since $f'(x) = \cos x = \cos(\pi/3) = \frac{1}{2}$, formula (21.2) yields

$$\sin 61° \sim \frac{\sqrt{3}}{2} + \left(\frac{1}{2}\right)\left(\frac{\pi}{180}\right) \sim 0.8660 + 0.0087 = 0.8747$$

(To four decimal places, the correct answer can be shown to be 0.8746.)

2. Approximate the change in the volume V of a cube of side x if the side is increased by 1%.

Here, Δx is $0.01x$, $f(x) = V = x^3$, and $f'(x) = 3x^2$. By formula (21.1), the increase is approximately $(3x^2)(0.01x) = 0.03x^3$. (Thus, the volume increases by roughly 3%.)

3. Find dy for each of the following functions $y = f(x)$:

(a) $y = x^3 + 4x^2 - 5x + 6$.

$dy = d(x^3) + d(4x^2) - d(5x) + d(6) = (3x^2 + 8x - 5)dx$

(b) $y = (2x^3 + 5)^{3/2}$.

$dy = \frac{3}{2}(2x^3+5)^{1/2}d(2x^3+5) = \frac{3}{2}(2x^3+5)^{1/2}(6x^2dx) = 9x^2(2x^3+5)^{1/2}dx$

† The computations are so tedious that a calculator, preferably a programmable calculator, should be used.

(c) $y=\dfrac{x^3+2x+1}{x^2+3}$.

$$dy=\frac{(x^2+3)d(x^3+2x+1)-(x^3+2x+1)d(x^2+3)}{(x^2+3)^2}$$

$$=\frac{(x^2+3)(3x^2+2)dx-(x^3+2x+1)(2x)\,dx}{(x^2+3)^2}=\frac{x^4+7x^2-2x+6}{(x^2+3)^2}dx$$

(d) $y=\cos^2 2x+\sin 3x$.

$$dy=2\cos 2x\,d(\cos 2x)+d(\sin 3x)$$
$$=(2\cos 2x)(-2\sin 2x\,dx)+3\cos 3x\,dx$$
$$=-4\sin 2x\cos 2x\,dx+3\cos 3x\,dx$$
$$=(-2\sin 4x+3\cos 3x)\,dx$$

4. Use differentials to find $\dfrac{dy}{dx}$:

(a) $xy+x-2y=5$.

$$d(xy)+dx-d(2y)=d(5)$$
$$x\,dy+y\,dx+dx-2\,dy=0$$
$$(x-2)\,dy+(y+1)dx=0$$
$$\frac{dy}{dx}=-\frac{y+1}{x-2}$$

(b) $\dfrac{2x}{y}-\dfrac{3y}{x}-8.$

$$2\left(\frac{y\,dx-x\,dy}{y^2}\right)-3\left(\frac{x\,dy-y\,dx}{x^2}\right)=0$$
$$2x^2(y\,dx-x\,dy)-3y^2(x\,dy-y\,dx)=0$$
$$(2x^2y+3y^3)\,dx-(2x^3+3y^2x)dy=0$$
$$\frac{dy}{dx}=\frac{y(2x^2+3y^2)}{x(2x^2+3y^2)}=\frac{y}{x}$$

(c) $x=3\cos\theta-\cos 3\theta,\ y=3\sin\theta-\sin 3\theta$.

$$dx-(-3\sin\theta+3\sin 3\theta)\,d\theta,\ dy=(3\cos\theta-3\cos 3\theta)\,d\theta$$
$$\frac{dy}{dx}=\frac{\cos\theta-\cos 3\theta}{-\sin\theta+\sin 3\theta}$$

5. Approximate the (real) roots of $x^3+2x-5=0$.

Drawing the graphs *of* $y=x^3$ and $y=5-2x$ on the same axes, we see that there must be one root, which lies between 1 and 2. Apply Newton's method, with $x_0=1$. Then $f(x)=x^3+2x-5$ and $f'(x)=3x^2+2$. Equation (21.5) becomes

$$x_{n+1}=x_n-\frac{x_n^3+2x_n-5}{3x_n^2+2}=\frac{2x_n^3+5}{3x_n^2+2}$$

Thus,

$$x_1=\frac{7}{5}=1.4$$
$$x_2\sim 1.330\,964\,467$$
$$x_3\sim 1.328\,272\,82$$
$$x_4\sim 1.328\,268\,856$$
$$x_5\sim 1.328\,268\,856$$

A calculator yields the answer 1.328 2689, which is accurate to the indicated number of places. So, the answer obtained by Newton's method is correct to at least seven decimal places.

6. Approximate the roots of $2 \cos x - x^2 = 0$.

Drawing the graphs of $y = 2 \cos x$ and $y = x^2$, we see that there are two real roots, close to 1 and -1. (Since the function $2 \cos x - x^2$ is even, if r is one root, the other root is $-r$.) Apply Newton's method with $x_0 = 1$. Then $f(x) = 2 \cos x - x^2$ and $f'(x) = -2 \sin x - 2x = -2(x + \sin x)$. Equation (21.5) becomes

$$x_{n+1} = x_n + \frac{1}{2}\frac{2\cos x_n - x_n^2}{x_n + \sin x_n} = \frac{x_n^2 + 2(x_n \sin x_n + \cos x_n)}{2(x_n + \sin x_n)}$$

Then

$$x_1 \sim 1.021\,885\,93$$

$$x_2 \sim 1.021\,689\,97$$

$$x_3 \sim 1.021\,689\,954$$

$$x_4 \sim 1.021\,689\,954$$

A graphing calculator produces 1.021 69, which is correct to the indicated number of places. Thus, the answer obtained by Newton's method is accurate to at least five places.

SUPPLEMENTARY PROBLEMS

7. Use equation (21.2) to approximate: (a) $\sqrt[4]{17}$; (b) $\sqrt[5]{1020}$; (c) cos 59°; (d) tan 44°.

Ans. (a) 2.031 25; (b) 3.996 88; (c) 0.5151; (d) 0.9651

8. Use equation (21.1) to approximate the change in (a) x^3 as x changes from 5 to 5.01; (b) $\frac{1}{x}$ as x changes from 1 to 0.98.

Ans. (a) 0.75; (b) 0.02

9. A circular plate expands under the influence of heat so that its radius increases from 5 to 5.06 inches. Estimate the increase in area.

Ans. 0.6π in^2 ~ 1.88 in^2

10. The radius of a ball of ice shrinks from 10 to 9.8 inches. Estimate the decrease in (a) the volume; (b) the surface area.

Ans. (a) 80π in^3; (b) 16π in^2

11. The velocity attained by an object falling freely a distance h feet from rest is given by $v = \sqrt{64.4h}$ ft/sec. Estimate the error in v due to an error of 0.5 ft when h is measured as 100 ft.

Ans. 0.2 ft/sec

12. If an aviator flies around the world at a distance 2 miles above the equator, estimate how many more miles he will travel than a person who travels along the equator.

Ans. 12.6 miles

13. The radius of a circle is to be measured and its area computed. If the radius can be measured to an accuracy of 0.001 in and the area must be accurate to 0.1 inches^2, estimate the maximum radius for which this process can be used.

Ans. 16 in

14. If $pV = 20$ and p is measured as 5 ± 0.02, estimate V.

Ans. $V = 4 \pm 0.016$

15. If $F = 1/r^2$ and F is measured as 4 ± 0.05, estimate r.

Ans. 0.5 ± 0.003

16. Estimate the change in the total surface of a right circular cone when (a) the radius r remains constant while the height h changes by a small amount Δh; (b) the height remains constant while the radius changes by a small amount Δr.

Ans. (a) $\pi rh\, \Delta h / \sqrt{r^2 + h^2}$; (b) $\pi\left(\dfrac{h^2 + 2r^2}{\sqrt{r^2 + h^2}} + 2r\right)\Delta r$

17. Find dy for each of the following:

(a) $y = (5 - x)^3$ *Ans.* $-3(5 - x)^2\, dx$

(b) $y = \dfrac{\sin x}{x}$ *Ans.* $\dfrac{x \cos x - \sin x}{x^2}\, dx$

(c) $y = \cos^{-1}(2x)$ *Ans.* $\dfrac{-2}{\sqrt{1 - 4x^2}}\, dx$

(d) $y = \cos(bx^2)$ *Ans.* $-2bx \sin(bx^2)\, dx$

18. Find dy/dx in the following examples by using differentials:

(a) $2xy^3 + 3x^2y = 1$ *Ans.* $-\dfrac{2y(y^2 + 3x)}{3x(2y^2 + x)}$

(b) $xy = \sin(x - y)$ *Ans.* $\dfrac{\cos(x - y) - y}{\cos(x - y) + x}$

19. (GC) Use Newton's method to find the solutions of the following equations, to four decimal places:

(a) $x^3 + 3x + 1 = 0$ *Ans.* -0.3222
(b) $x - \cos x = 0$ *Ans.* 0.7391
(c) $x^3 + 2x^2 - 4 = 0$ *Ans.* 1.1304

20. (GC) Use Newton's method to approximate the following to four decimal places:

(a) $\sqrt[4]{3}$ *Ans.* 1.3161
(b) $\sqrt[5]{247}$ *Ans.* 3.0098

21. (a) Verify that Newton's method for calculating $\sqrt{r}$ yields the equation $x_{n+1} = \frac{1}{2}\left(x_n + \frac{r}{x_n}\right)$.
(b) (GC) Apply part (a) to approximate $\sqrt{5}$ to four decimal places.

Ans. (b) 2.2361

22. (GC) Show that $x^3 + x^2 - 3 = 0$ has a unique solution in (1, 2) and use Newton's method to approximate it to four decimal places.

Ans. 1.1746

23. Show that Newton's method does not work if it is applied to the equation $x^{1/3} = 0$, with $x_0 = 1$.

24. Show that Newton's method does not give approximations to the solutions of the following equations, starting with the given initial values, and explain why it does not work in those cases.

(a) $x^3 - 3x^2 + 3x + 2 = 0$, with $x_0 = 1$.
(b) $x^3 - 3x^2 + x - 1 = 0$, with $x_0 = 1$.
(c) $f(x) = \begin{cases} \sqrt{x-2} & \text{for } x \geq 2 \\ -\sqrt{2-x} & \text{for } x < 2 \end{cases}$, with $x_0 = 3$

25. (GC) Approximate π by using Newton's method to find a solution of $\cos x + 1 = 0$.

Ans. 3.141 592 654. (Note how long it takes for the answer to stabilize.)

26. (GC) Use Newton's method to estimate the unique positive solution of $\cos x = \frac{x}{2}$.

Ans. 1.029 866 529

Antiderivatives

If $F'(x) = f(x)$, then F is called an *antiderivative* of f.

EXAMPLE 22.1: x^3 is an antiderivative of $3x^2$, since $D_x(x^3) = 3x^2$. But $x^3 + 5$ is also an antiderivative of $3x^2$, since $D_x(5) = 0$.

(I) In general, if $F(x)$ is an antiderivative of $f(x)$, then $F(x) + C$ is also an antiderivative of $f(x)$, where C is any constant.

(II) On the other hand, if $F(x)$ is an antiderivative of $f(x)$, and if $G(x)$ is any other antiderivative of $f(x)$, then $G(x) = F(x) + C$, for some constant C.

Property (II) follows from Problem 13 of Chapter 18, since $F'(x) = f(x) = G'(x)$.

From Properties (I) and (II) we see that, if $F(x)$ is an antiderivative of $f(x)$, then the antiderivatives of $f(x)$ are precisely those functions of the form $F(x) + C$, for an arbitrary constant C.

Notation: $\int f(x)\,dx$ will denote any antiderivative of $f(x)$. In this notation, $f(x)$ is called the *integrand*.

Terminology: An antiderivative $\int f(x)\,dx$ is also called an *indefinite integral*.

An explanation of the peculiar notation $\int f(x)\,dx$ (including the presence of the differential dx) will be given later.

EXAMPLE 22.2: (a) $\int x\,dx = \frac{1}{2}x^2 + C$; (b) $\int -\sin x\,dx = \cos x + C$.

Laws for Antiderivatives

Law 1. $\int 0\,dx = C$.

Law 2. $\int 1\,dx = x + C$.

Law 3. $\int a\,dx = ax + C$.

Law 4. $\int x^r\,dx = \dfrac{x^{r+1}}{r+1} + C$ for any rational number $r \neq -1$.

(4) follows from the fact that $D_x\left(\dfrac{x^{r+1}}{r+1}\right) = x^r$ for $r \neq -1$.

Law 5. $\int af(x)\,dx = a\int f(x)\,dx$.

Note that $D_x\left(a\int f(x)\,dx\right) = aD_x\left(\int f(x)\,dx\right) = af(x)$.

Law 6. $\int (f(x) + g(x))\,dx = \int f(x) + dx + \int g(x)\,dx$.

Note that $D_x\left(\int f(x)\,dx + \int g(x)\,dx\right) = D_x\left(\int f(x)\,dx\right) + D_x\left(\int g(x)\,dx\right) = f(x) + g(x)$.

Law 7. $\int (f(x) - g(x))\,dx = \int f(x)\,dx - \int g(x)\,dx$.

Note that $D_x\left(\int f(x)\,dx - \int g(x)\,dx\right) = D_x\left(\int f(x)\,dx\right) - D_x\left(\int g(x)\,dx\right) = f(x) - g(x)$.

EXAMPLE 22.3:

(a) $\int \sqrt[3]{x}\,dx = \int x^{1/3} dx = \frac{x^{4/3}}{4/3} + C = \frac{3}{4}x^{4/3} + C$ by Law (4).

(b) $\int \frac{1}{x^2} dx = \int x^{-2} dx = \frac{x^{-1}}{-1} + C = -\frac{1}{x} + C$ by Law (4).

(c) $\int 7x^3 dx = 7\int x^3 dx = 7\left(\frac{x^4}{4}\right) + C = \frac{7}{4}x^4 + C$ by Laws (5), (4).

(d) $\int (x^2 + 4)\,dx = \int x^2 dx + \int 4\,dx = \frac{1}{3}x^3 + 4x + C$ by Laws (6), (4), and (2).

(e) $\int (3x^6 - 4x)\,dx = \int 3x^6 dx - \int 4x\,dx = 3\int x^6\,dx - 4\int x\,dx = 3(\frac{1}{7}x^7) - 4(\frac{1}{2}x^2) + C = \frac{3}{7}x^7 - 2x^2 + C.$

EXAMPLE 22.4: Laws (3)–(7) enable us to compute the antiderivative of any polynomial. For instance,

$$\int (6x^8 - \tfrac{2}{3}x^5 + 7x^4 + \sqrt{3})\,dx = 6(\tfrac{1}{9}x^9) - \tfrac{2}{3}(\tfrac{1}{6}x^6) + 7(\tfrac{1}{5}x^5) + \sqrt{3}x + C$$
$$= \tfrac{2}{3}x^9 - \tfrac{1}{9}x^6 + \tfrac{7}{5}x^5 + \sqrt{3}x + C$$

Law (8). (Quick Formula I)

$$\int (g(x))^r g'(x)\,dx = \frac{1}{r+1}(g(x))^{r+1} + C \quad \text{for any rational number } r \neq -1$$

For verification, $D_x\left(\frac{1}{r+1}(g(x))^{r+1}\right) = \frac{1}{r+1}D_x[(g(x))^{r+1}] = \frac{1}{r+1}(r+1)(g(x))^r g'(x) = (g(x))^r g'(x)$ by the power Chain Rule.

EXAMPLE 22.5: $\int (\frac{1}{3}x^3 + 7)^5 x^2\,dx = \frac{1}{6}(\frac{1}{3}x^3 + 7)^6 + C.$

To see this, let $g(x) = (\frac{1}{3}x^3 + 7)$ and $r = 5$ in Quick Formula I.

EXAMPLE 22.6: $\int (x^2 + 1)^{2/3} x\,dx = \frac{1}{2}\int (x^2+1)^{2/3} 2x\,dx = \frac{1}{2}\left(\frac{1}{5/3}\right)(x^2+1)^{5/3} + C = \frac{3}{10}(x^2+1)^{5/3} + C.$

In this case, we had to insert a factor of 2 in the integrand in order to use Quick Formula I.

Law (9). Substitution Method

$$\int f(g(x))g'(x)\,dx = \int f(u)\,du$$

where u is replaced by $g(x)$ after the right-hand side is evaluated. The "substitution" is carried out on the left-hand side by letting $u = g(x)$ and $du = g'(x)\,dx$. (For justification, see Problem 21.)

EXAMPLE 22.7:

(a) Find $\int x\sin(x^2)\,dx$.

Let $u = x^2$. Then $du = 2x\,dx$. So, $x\,dx = \frac{1}{2}du$. By substitution,

$$\int x\sin(x^2)\,dx = \int \sin u\,(\tfrac{1}{2})du = \tfrac{1}{2}(-\cos u) + C = -\tfrac{1}{2}\cos(x^2) + C$$

(b) Find $\int \sin(x/2)\,dx$.

Let $u = x/2$. Then $du = \frac{1}{2}dx$. So, $dx = 2\,du$. By substitution,

$$\int \sin\left(\frac{x}{2}\right)dx = \int (\sin u)\,2\,du = 2\int \sin u\,du = 2(-\cos u) + C = -2\cos\left(\frac{x}{2}\right) + C$$

Observe that Quick Formula I is just a special case of the Substitution Method, with $u = g(x)$. The advantage of Quick Formula I is that we save the bother of carrying out the substitution.

The known formulas for derivatives of trigonometric and inverse trigonometric functions yield the following formulas for antiderivatives:

$$\int \sin x\,dx = -\cos x + C$$

$$\int \cos x\,dx = \sin x + C$$

$$\int \sec^2 x\,dx = \tan x + C$$

$$\int \tan x \sec x\,dx = \sec x + C$$

$$\int \csc^2 x\,dx = -\cot x + C$$

$$\int \cot x \csc x\,dx = -\csc x + C$$

$$\int \frac{1}{\sqrt{1-x^2}}\,dx = \sin^{-1} x + C$$

$$\int \frac{1}{1+x^2}\,dx = \tan^{-1} x + C$$

$$\int \frac{1}{x\sqrt{x^2-1}}\,dx = \sec^{-1} x + C$$

$$\int \frac{1}{\sqrt{a^2-x^2}}\,dx = \sin^{-1}\left(\frac{x}{a}\right) + C \qquad \text{for } a > 0$$

$$\int \frac{1}{a^2+x^2}\,dx = \frac{1}{a}\tan^{-1}\left(\frac{x}{a}\right) + C \qquad \text{for } a > 0$$

$$\int \frac{1}{x\sqrt{x^2-a^2}}\,dx = \frac{1}{a}\sec^{-1}\left(\frac{x}{a}\right) + C \qquad \text{for } a > 0$$

SOLVED PROBLEMS

In Problems 1–8, evaulate the antiderivative.

1. $\int x^6 dx = \frac{1}{7}x^7 + C$. [Law (4)]

2. $\int \frac{dx}{x^6} = \int x^{-6} dx = \frac{1}{-5}x^{-5} + C = -\frac{1}{5x^5} + C$ [Law (4)]

3. $\int \sqrt[3]{z}\,dz = \int z^{1/3} dz = \frac{1}{4/3} z^{4/3} + C = \frac{3}{4}(\sqrt[3]{z})^4 + C$ [Law (4)]

4. $\int \frac{1}{\sqrt[3]{x^2}}\,dx = \int x^{-2/3} dx = \frac{1}{1/3} x^{1/3} + C = 3\sqrt[3]{x} + C$ [Law (4)]

5. $\int (2x^2 - 5x + 3)\,dx = 2\int x^2 dx - 5\int x\,dx + \int 3\,dx$

$= 2(\frac{1}{3}x^3) - 5(\frac{1}{2}x^2) + 3x + C = \frac{2}{3}x^3 - \frac{5}{2}x^2 + 3x + C$ [Laws (3)–(7)]

6. $\int(1-x)\sqrt{x}\,dx = \int(1-x)x^{1/2}dx = \int(x^{1/2}-x^{3/2})dx$

$$= \int x^{1/2}dx - \int x^{3/2}dx = \frac{1}{3/2}x^{3/2} - \frac{1}{5/2}x^{5/2} + C$$

$$= \tfrac{2}{3}x^{3/2} - \tfrac{2}{5}x^{5/2} + C = 2x^{3/2}(\tfrac{1}{3} - \tfrac{1}{5}x) + C \qquad \text{[Laws (4), (7)]}$$

7. $\int(3s+4)^2 ds = \int(9s^2+24s+16)ds$

$$= 9(\tfrac{1}{3}s^3) + 24(\tfrac{1}{2}s^2) + 16s + C = 3s^3 + 12s^2 + 16s + C \qquad \text{[Laws (3)–(6)]}$$

Note that it would have been easier to use Quick Formula I:

$$\int(3s+4)^2 ds = \tfrac{1}{3}\int(3s+4)^2 3\,ds = \tfrac{1}{3}(\tfrac{1}{3}(3s+4)^3) + C = (\tfrac{1}{9})(3s+4)^3 + C$$

8. $\int \frac{x^3+5x^2-4}{x^2}dx = \int(x+5-4x^{-2})dx = \tfrac{1}{2}x^2 + 5x - 4\left(\frac{1}{-1}x^{-1}\right) + C$

$$= \tfrac{1}{2}x^2 + 5x + \frac{4}{x} + C \qquad \text{[Laws (3)–(7)]}$$

Use Quick Formula I in Problems 9–15.

9. $\int(s^3+2)^2(3s^2)ds = \tfrac{1}{3}(s^3+2)^3 + C$

10. $\int(x^3+2)^{1/2}x^2dx = \tfrac{1}{3}\int(x^3+2)^{1/2}3x^2dx = \tfrac{1}{3}\left(\frac{1}{3/2}(x^3+2)^{3/2}\right) + C = \tfrac{2}{9}(x^3+2)^{3/2} + C$

11. $\int\frac{8x^2}{(x^3+2)^3}dx = \tfrac{8}{3}\int(x^3+2)^{-3}3x^2dx = \tfrac{8}{3}\left(\frac{1}{-2}(x^3+2)^{-2}\right) + C = -\frac{4}{3}\frac{1}{(x^3+2)^2} + C$

12. $\int\frac{x^2dx}{\sqrt[4]{x^3+2}} = \tfrac{1}{3}\int(x^3+2)^{-1/4}3x^2dx = \tfrac{1}{3}\left(\frac{1}{3/4}(x^3+2)^{3/4}\right) + C = \tfrac{4}{9}(x^3+2)^{3/4} + C$

13. $\int 3x\sqrt{1-2x^2}\,dx = -\tfrac{3}{4}\int -4x\sqrt{1-2x^2}\,dx$

$$= -\tfrac{3}{4}\int -4x(1-2x^2)^{1/2}dx = -\tfrac{3}{4}\left(\frac{1}{3/2}(1-2x^2)^{3/2}\right) + C$$

$$= -\tfrac{1}{2}(1-2x^2)^{3/2} + C$$

14. $\int\sqrt[3]{1-x^2}\,x\,dx = -\tfrac{1}{2}\int(1-x^2)^{1/3}(-2x)dx$

$$= -\tfrac{1}{2}\left(\frac{1}{4/3}(1-x^2)^{4/3}\right) + C = -\tfrac{3}{8}(1-x^2)^{4/3} + C$$

15. $\int\sin^2 x\cos x\,dx = \int(\sin x)^2\cos x\,dx = \tfrac{1}{3}(\sin x)^3 + C = \tfrac{1}{3}\sin^3 x + C$

In Problems 16–18, use the Substitution Method.

16. $\int\frac{\cos\sqrt{x}}{\sqrt{x}}dx.$

Let $u = \sqrt{x} = x^{1/2}$. Then $du = \tfrac{1}{2}x^{-1/2}dx$. So, $2\,du = \frac{1}{\sqrt{x}}dx$. Thus,

$$\int\frac{\cos\sqrt{x}}{\sqrt{x}}dx = 2\int\cos u\,du = 2\sin u + C = 2\sin(\sqrt{x}) + C$$

17. $\int x\sec^2(4x^2-5)\,dx$.

Let $u = 4x^2 - 5$. Then $du = 8x\,dx$, $\frac{1}{8}du = x\,dx$. Thus,

$$\int x\sec^2(4x^2-5)\,dx = \tfrac{1}{8}\int \sec^2 u\,du = \tfrac{1}{8}\tan u + C = \tfrac{1}{8}\tan(4x^2-5)+C$$

18. $\int x^2\sqrt{x+1}\,dx$.

Let $u = x+1$. Then $du = dx$ and $x = u-1$. Thus,

$$\int x^2\sqrt{x+1}\,dx = \int (u-1)^2\sqrt{u}\,du = \int (u^2-2u+1)u^{1/2}du$$

$$= \int (u^{5/2} - 2u^{3/2} + u^{1/2})\,du = \tfrac{2}{7}u^{7/2} - 2(\tfrac{2}{5})u^{5/2} + \tfrac{2}{3}u^{3/2} + C$$

$$= 2u^{3/2}(\tfrac{1}{7}u^2 - \tfrac{2}{5}u + \tfrac{1}{3}) + C$$

$$= 2(x+1)^{3/2}[\tfrac{1}{7}(x+1)^2 - \tfrac{2}{5}(x+1) + \tfrac{1}{3}] + C$$

19. A stone is thrown straight up from the ground with an initial velocity of 64 ft/sec. (a) When does it reach its maximum height? (b) What is its maximum height? (c) When does it hit the ground? (d) What is its velocity when it hits the ground?

In free-fall problems, $v = \int a\,dt$ and $s = \int v\,dt$ because $a = \dfrac{dv}{dt}$ and $v = \dfrac{ds}{dt}$. Since $a = -32$ ft/sec^2,

$$v = \int -32\,dt = -32t + C_1$$

Letting $t = 0$, we see that $C_1 = v_0$, the initial velocity at $t = 0$. Thus, $v = -32t + v_0$. Hence,

$$s = \int (-32t + v_0)\,dt = -16t^2 + v_0 t + C_2$$

Letting $t = 0$, we see that $C_2 = s_0$, the initial position at $t = 0$. Hence

$$s = -16t^2 + v_0 t + s_0$$

In this problem, $s_0 = 0$ and $v_0 = 64$. So,

$$v = -32t + 64,\ s = -16t^2 + 64t$$

(a) At the maximum height, $\dfrac{ds}{dt} = v = 0$. So, $-32t + 64 = 0$ and, therefore, $t = 2$ seconds.
(b) When $t = 2$, $s = -16(2)^2 + 64(2) = 64$ ft, the maximum height.
(c) When the stone hits the ground, $0 = s = -16t^2 + 64t$. Dividing by t, $0 = -16t + 64$ and, therefore, $t = 4$.
(d) When $t = 4$, $v = -32(4) + 64 = -64$ ft/sec.

20. Find an equation of the curve passing through the point (3, 2) and having slope $5x^2 - x + 1$ at every point (x, y).

Since the slope is the derivative, $dy/dx = 5x^2 - x + 1$. Hence,

$$y = \int (5x^2 - x + 1)\,dx = \tfrac{5}{3}x^3 - \tfrac{1}{2}x^2 + x + C$$

Since (3, 2) is on the curve, $2 = \frac{5}{3}(3)^3 - \frac{1}{2}(3)^2 + 3 + C = 45 - \frac{9}{2} + 3 + C$. So, $C = -\frac{83}{2}$. Hence, an equation of the curve is

$$y = \tfrac{5}{3}x^3 - \tfrac{1}{2}x^2 + x - \tfrac{83}{2}$$

21. Justify the Substitution Method: $\int f(g(x))g'(x)dx = \int f(u)\,du$.

Here, $u = g(x)$ and $du/dx = g'(x)$. By the Chain Rule,

$$D_x\left(\int f(u)\,du\right) = D_u\left(\int f(u)\,du\right)\cdot\frac{du}{dx} = f(u)\cdot\frac{du}{dx} = f(g(x))\cdot g'(x)$$

SUPPLEMENTARY PROBLEMS

In Problems 22–44, evaluate the given antiderivative.

22. $\int \frac{(1+x^2)}{\sqrt{x}}\,dx.$ *Ans.* $2x^{1/2}(1+\frac{2}{3}x+\frac{1}{5}x^2)+C$

23. $\int \frac{(x^2+2x)}{(x+1)^2}\,dx$ *Ans.* $\frac{x^2}{x+1}+C$

24. $\int \cos 3x\,dx$ *Ans.* $\frac{1}{3}\sin 3x+C$

25. $\int \frac{\sin y\,dy}{\cos^2 y}$ *Ans.* $\sec y + C$

26. $\int \frac{dx}{1+\cos x}$ (*Hint:* Multiply numerator and denominator by $1-\cos x$.)

Ans. $-\cot x + \csc x + C$

27. $\int (\tan 2x + \sec 2x)^2\,dx$ *Ans.* $\tan 2x + \sec 2x - x + C$

28. $\int \frac{dx}{\sqrt{4-x^2}}$ *Ans.* $\sin^{-1}\left(\frac{x}{2}\right)+C$

29. $\int \frac{dx}{9+x^2}$ *Ans.* $\frac{1}{3}\tan^{-1}\left(\frac{x}{3}\right)+C$

30. $\int \frac{dx}{\sqrt{25-16x^2}}$ (*Hint:* Factor 16 out of the radical.)

Ans. $\frac{1}{4}\sin^{-1}\left(\frac{4x}{5}\right)+C$

31. $\int \frac{dx}{4x^2+9}$ (*Hint:* Either factor 4 out of the denominator or make the substitution $u = 2x$.)

Ans. $\frac{1}{6}\tan^{-1}\left(\frac{2x}{3}\right)+C$

32. $\int \frac{dx}{x\sqrt{4x^2-9}}$ (*Hint:* Either factor 4 out of the radical or make the substitution $u = 2x$.)

Ans. $\frac{1}{3}\sec^{-1}\left(\frac{2x}{3}\right)+C$

33. $\int \frac{x^2 dx}{\sqrt{1-x^6}}$ (*Hint.* Substitute $u = x^3$.) *Ans.* $\frac{1}{3}\sin^{-1}(x^3)+C$

34. $\int \frac{x\,dx}{x^4+3}$ (*Hint:* Substitute $u = x^2$.) *Ans.* $\frac{\sqrt{3}}{6}\tan^{-1}\left(\frac{x^2\sqrt{3}}{3}\right)+C$

35. $\int \frac{dx}{x\sqrt{x^4-1}}$ *Ans.* $\frac{1}{2}\cos^{-1}\left(\frac{1}{x^2}\right)+C$

36. $\int \frac{3x^3 - 4x^2 + 3x}{x^2 + 1} dx$ *Ans.* $\frac{3x^2}{2} - 4x + 4\tan^{-1} x + C$

37. $\int \frac{\sec x \tan x \, dx}{9 + 4\sec^2 x}$ *Ans.* $\frac{1}{6}\tan^{-1}\left(\frac{2\sec x}{3}\right) + C$

38. $\int \frac{(x+3)dx}{\sqrt{1-x^2}}$ *Ans.* $-\sqrt{1-x^2} + 3\sin^{-1} x + C$

39. $\int \frac{dx}{x^2 + 10x + 30}$ *Ans.* $\frac{\sqrt{5}}{5}\tan^{-1}\left(\frac{(x+5)\sqrt{5}}{5}\right) + C$

40. $\int \frac{dx}{\sqrt{20 + 8x - x^2}}$ *Ans.* $\sin^{-1}\left(\frac{x-4}{6}\right) + C$

41. $\int \frac{dx}{2x^2 + 2x + 5}$ *Ans.* $\frac{1}{3}\tan^{-1}\left(\frac{2x+1}{3}\right) + C$

42. $\int \frac{dx}{\sqrt{28 - 12x - x^2}}$ *Ans.* $\sin^{-1}\left(\frac{x+6}{8}\right) + C$

43. $\int \frac{x+3}{\sqrt{5 - 4x - x^2}} dx$ *Ans.* $-\sqrt{5 - 4x - x^2} + \sin^{-1}\left(\frac{x+2}{3}\right) + C$

44. $\int \frac{x+2}{\sqrt{4x - x^2}} dx$ *Ans.* $-\sqrt{4x - x^2} + 4\sin^{-1}\left(\frac{x-2}{2}\right) + C$

In Problems 45–52, use Quick Formula I.

45. $\int (x-2)^{3/2} dx$ *Ans.* $\frac{2}{5}(x-2)^{5/2} + C$

46. $\int \frac{dx}{(x-1)^3}$ *Ans.* $-\frac{1}{2(x-1)^2} + C$

47. $\int \frac{dx}{\sqrt{x+3}}$ *Ans.* $2\sqrt{x+3} + C$

48. $\int \sqrt{3x-1} \, dx$ *Ans.* $\frac{2}{9}(3x-1)^{3/2} + C$

49. $\int \sqrt{2-3x} \, dx$ *Ans.* $-\frac{2}{9}(2-3x)^{3/2} + C$

50. $\int (2x^2+3)^{1/3} x \, dx$ *Ans.* $\frac{3}{16}(2x^2+3)^{4/3} + C$

51. $\int \sqrt{1+y^4} \, y^3 \, dy$ *Ans.* $\frac{1}{6}(1+y^4)^{3/2} + C$

52. $\int \frac{x \, dx}{(x^2+4)^3}$ *Ans.* $-\frac{1}{4(x^2+4)^2} + C$

In Problems 53–64, use any method.

53. $\int (x-1)^2 x\,dx$ *Ans.* $\frac{1}{4}x^4 - \frac{2}{3}x^3 + \frac{1}{2}x^2 + C$

54. $\int (x^2-x)^4(2x-1)\,dx$ *Ans.* $\frac{1}{5}(x^2-x)^5 + C$

55. $\int \frac{(x+1)\,dx}{\sqrt{x^2+2x-4}}$ *Ans.* $\sqrt{x^2+2x-4} + C$

56. $\int \frac{(1+\sqrt{x})^2}{\sqrt{x}}\,dx$ *Ans.* $\frac{2}{3}(1+\sqrt{x})^3 + C$

57. $\int \frac{(x+1)(x-2)}{\sqrt{x}}\,dx$ *Ans.* $\frac{2}{5}x^{5/2} - \frac{2}{3}x^{3/2} - 4x^{1/2} + C = 2x^{1/2}(\frac{1}{5}x^2 - \frac{1}{3}x - 2) + C$

58. $\int \sec 3x \tan 3x\,dx$ *Ans.* $\frac{1}{3}\sec 3x + C$

59. $\int \csc^2(2x)\,dx$ *Ans.* $-\frac{1}{2}\cot 2x + C$

60. $\int x \sec^2(x^2)\,dx$ *Ans.* $\frac{1}{2}\tan(x^2) + C$

61. $\int \tan^2 x\,dx$ *Ans.* $\tan x - x + C$

62. $\int \cos^4 x \sin x\,dx$ *Ans.* $-\frac{1}{5}\cos^5 x + C$

63. $\int \frac{dx}{\sqrt{5-x^2}}$ *Ans.* $\sin^{-1}\left(\frac{x\sqrt{5}}{5}\right) + C$

64. $\int \frac{\sec^2 x\,dx}{1-4\tan^2 x}$ *Ans.* $\frac{1}{2}\sin^{-1}(2\tan x) + C$

65. A stone is thrown straight up from a building ledge that is 120 ft above the ground, with an initial velocity of 96 ft/sec. (a) When will it reach its maximum height? (b) What will its maximum height be? (c) When will it hit the ground? (d) With what speed will it hit the ground?

Ans. (a) $t = 3$ sec; (b) 264 ft; (c) $\frac{6+\sqrt{66}}{2} \sim 7.06$ sec; (d) ~129.98 ft/sec

66. An object moves on the x axis with acceleration $a = 3t - 2$ ft/sec². At time $t = 0$, it is at the origin and moving with a speed of 5 ft/sec in the negative direction. (a) Find a formula for its velocity v. (b) Find a formula for its position x. (c) When and where does it change direction? (d) At what times is it moving toward the right?

Ans. (a) $v = \frac{3}{2}t^2 - 2t - 5$; (b) $x = \frac{1}{2}t^3 - t^2 - 5t$; (c) $\frac{2\pm\sqrt{34}}{3}$; (d) $t > \frac{2+\sqrt{34}}{3}$ or $t < \frac{2-\sqrt{34}}{3}$

67. A rocket shot straight up from the ground hits the ground 8 seconds later. (a) What was its initial velocity? (b) What was its maximum height?

Ans. (a) 128 ft/sec; (b) 256 ft

68. A driver applies the brakes on a car going at 55 miles per hour on a straight road. The brakes cause a constant deceleration of 11 ft/sec^2. (a) How soon will the car stop? (b) How far does the car move after the brakes were applied?

Ans. (a) 5 sec; (b) 137.5 ft

69. Find the equation of a curve going through the point (3, 7) and having slope $4x^2 - 3$ at (x, y).

Ans. $y = \frac{4}{3}x^3 - 3x - 20$

CHAPTER 23

The Definite Integral. Area Under a Curve

Sigma Notation

The Greek capital letter Σ denotes repeated addition.

EXAMPLE 23.1:

(a) $\sum_{j=1}^{5} j = 1+2+3+4+5 = 15.$

(b) $\sum_{i=0}^{3} (2i+1) = 1+3+5+7.$

(c) $\sum_{i=2}^{10} i^2 = 2^2 + 3^2 + \cdots + (10)^2$

(d) $\sum_{j=1}^{4} \cos j\pi = \cos \pi + \cos 2\pi + \cos 3\pi + \cos 4\pi$

In general, if f is a function defined on the integers, and if n and k are integers such that $n \geq k$, then:

$$\sum_{j=k}^{n} f(j) = f(k) + f(k+1) + \cdots + f(n)$$

Area Under a Curve

Assume that f is a function such that $f(x) \geq 0$ for all x in a closed interval $[a, b]$. Its graph is a curve that lies on or above the x axis. (See Fig. 23-1.) We have an intuitive idea of the *area* A of the region $\mathcal{R}$ under the graph, above the x axis, and between the vertical lines $x = a$ and $x = b$. We shall specify a method for evaluating A.

Choose points $x_1, x_2, \ldots, x_{n-1}$ between a and b. Let $x_0 = a$ and $x_n = b$. Thus (see Fig. 23-2),

$$a = x_0 < x_1 < x_2 < \cdots < x_{n-1} < x_n = b$$

The interval $[a, b]$ is divided into n subintervals $[x_0, x_1], [x_1, x_2], \ldots, [x_{n-1}, x_n]$. Denote the lengths of these subintervals by $\Delta_1 x, \Delta_2 x, \ldots, \Delta_n x$. Hence, if $1 \leq k \leq n$,

$$\Delta_k x = x_k - x_{k-1}$$

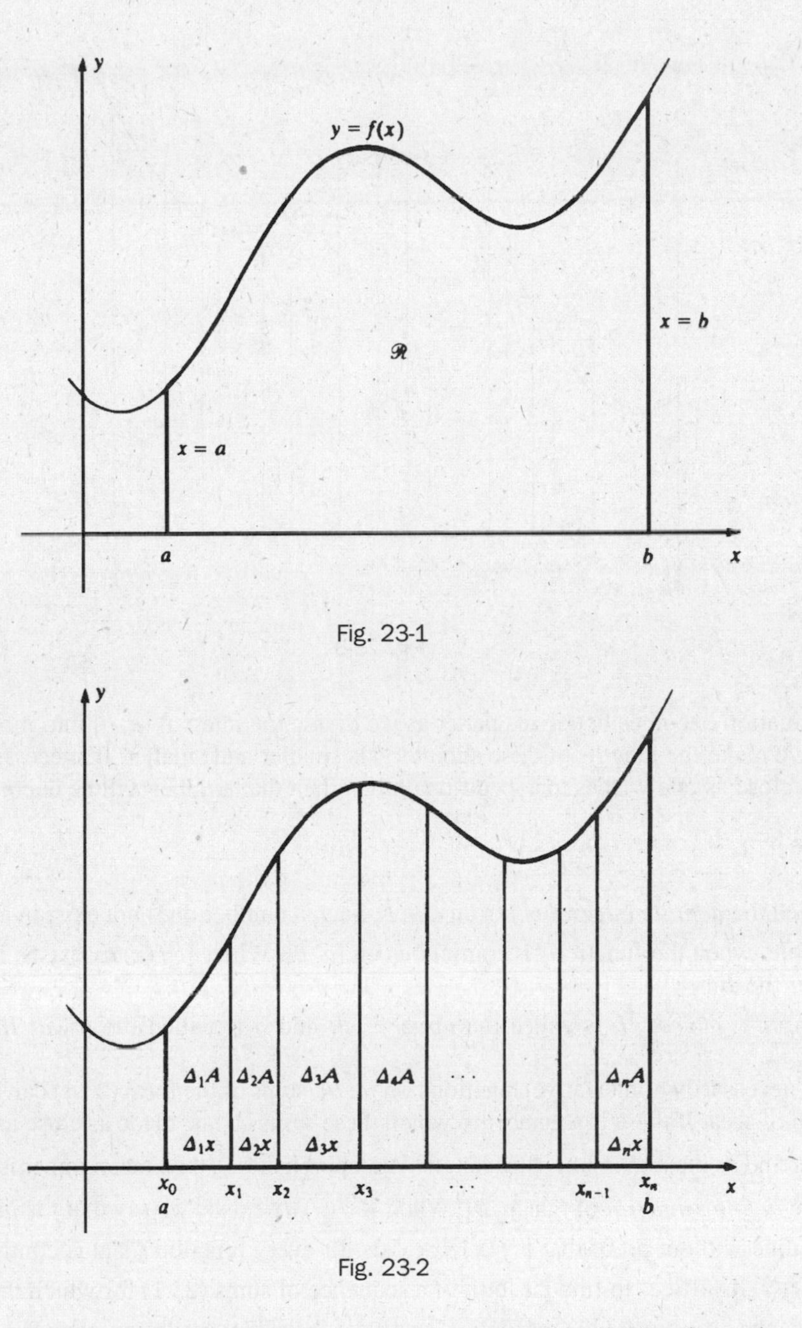

Fig. 23-1

Fig. 23-2

Draw vertical line segments $x = x_k$ from the x axis up to the graph. This divides the region $\mathcal{R}$ into n strips. Letting $\Delta_k A$ denote the area of the kth strip, we obtain

$$A = \sum_{k=1}^{n} \Delta_k A$$

We can approximate the area $\Delta_k A$ in the following manner. Select any point x_k^* in the kth subinterval $[x_{k-1}, x_k]$. Draw the vertical line segment from the point x_k^* on the x axis up to the graph (see the dashed lines in Fig. 23-3); the length of this segment is $f(x_k^*)$. The rectangle with base $\Delta_k x$ and height $f(x_k^*)$ has area $f(x_k^*)\,\Delta_k x$, which is approximately the area $\Delta_k A$ of the kth strip. Hence, the total area A under the curve is approximately the sum

$$\sum_{k=1}^{n} f(x_k^*)\,\Delta_k x = f(x_1^*)\,\Delta_1 x + f(x_2^*)\,\Delta_2 x + \cdots + f(x_n^*)\,\Delta_n x \qquad (23.1)$$

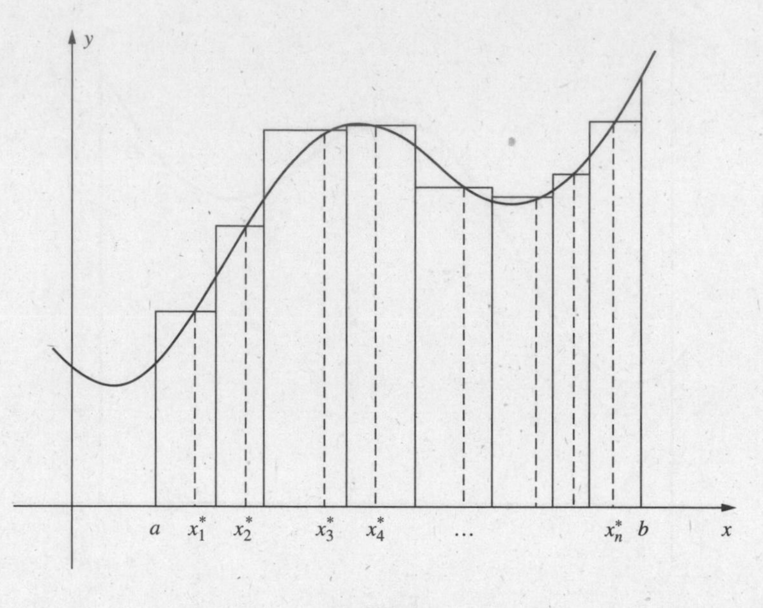

Fig. 23-3

The approximation becomes better and better as we divide the interval $[a, b]$ into more and more subintervals and as we make the lengths of these subintervals smaller and smaller. If successive approximations can be made as close as one wishes to a specific number, then that number will be denoted by

$$\int_a^b f(x)\,dx$$

and will be called the *definite integral* of f from a to b. Such a number does not exist in all cases, but it does exist, for example, when the function f is continuous on $[a, b]$. When $\int_a^b f(x)\,dx$ exists, its value is equal to the area A under the curve.[†]

In the notation $\int_a^b f(x)\,dx$, b is called the *upper limit* and a is called the *lower limit* of the definite integral.

For any (not necessarily nonnegative) function f on $[a, b]$, sums of the form (23.1) can be defined, without using the notion of area. If there is a number to which these sums can be made as close as we wish, as n gets larger and larger and as the maximum of the lengths $\Delta_k x$ approaches 0, then that number is denoted $\int_a^b f(x)\,dx$ and is called the *definite integral* of f on $[a, b]$. When $\int_a^b f(x)\,dx$ exists, we say that f is *integrable* on $[a, b]$.

We shall assume without proof that $\int_a^b f(x)\,dx$ exists for every function f that is continuous on $[a, b]$. To evaluate $\int_a^b f(x)\,dx$, it suffices to find the limit of a sequence of sums (23.1) for which the number n of subintervals approaches infinity and the maximum lengths of the subintervals approach 0.

EXAMPLE 23.2: Let us show that

$$\int_a^b 1\,dx = b - a \tag{23.2}$$

Let $a = x_0 < x_1 < x_2 < \cdots < x_{n-1} < x_n = b$ be a subdivision of $[a, b]$. Then a corresponding sum (23.1) is

$$\sum_{k=1}^{n} f(x_k^*)\Delta_k x = \sum_{k=1}^{n} \Delta_k x \qquad \text{(because } f(x) = 1 \text{ for all } x\text{)}$$

$$= b - a$$

Since every approximating sum is $b - a$, $\int_a^b 1\,dx = b - a$.

[†]The definite integral is also called the *Riemann integral* of f on $[a, b]$, and the sum (23.1) is called a *Riemann sum* for f on $[a, b]$.

An alternative argument would use the fact that the region under the graph of the constant function 1 and above the x axis, between $x = a$ and $x = b$, is a rectangle with base $b - a$ and height 1 (see Fig. 23-4). So, $\int_a^b 1\,dx$, being the area of that rectangle, is $b - a$.

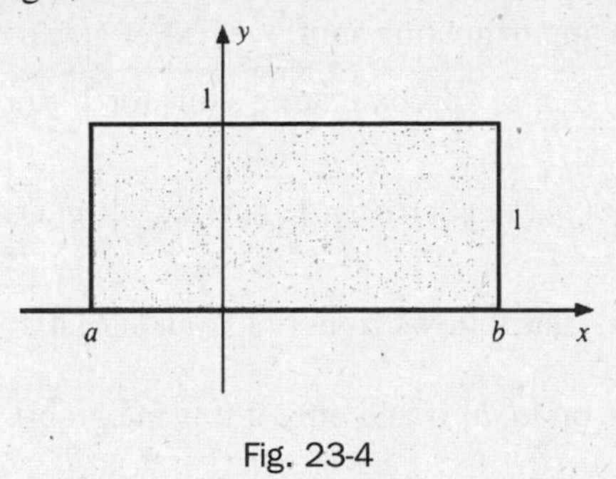

Fig. 23-4

EXAMPLE 23.3: Let us calculate $\int_a^b x\,dx$.

Let $a = x_0 < x_1 < x_2 < \cdots < x_{n-1} < x_n = b$ be a subdivision of $[a, b]$ into n equal subintervals. Thus, each $\Delta_k x = (b - a)/n$. Denote $(b - a)/n$ by Δx. Then $x_1 = a + \Delta x$, $x_2 = a + 2\Delta x$, and, in general, $x_k = a + k\,\Delta x$. In the kth subinterval, $[x_{k-1}, x_k]$, choose x_k^* to be the right-hand endpoint x_k. Then the approximating sum (23.1) has the form

$$
\begin{aligned}
f(x_k^*)\,\Delta_k x &= \sum_{k=1}^{n} x_k^*\,\Delta_k x = \sum_{k=1}^{n} (a + k\,\Delta x)\,\Delta x \\
&= \sum_{k=1}^{n} (a\,\Delta x + k(\Delta x)^2) = \sum_{k=1}^{n} a\,\Delta x + \sum_{k=1}^{n} k(\Delta x)^2 \\
&= n(a\,\Delta x) + (\Delta x)^2 \sum_{k=1}^{n} k = n\left(a\frac{b-a}{n}\right) + \left(\frac{b-a}{n}\right)^2 \left(\frac{n(n+1)}{2}\right) \\
&= a(b-a) + \frac{1}{2}(b-a)^2\,\frac{n+1}{n}
\end{aligned}
$$

Here we have used the fact that $\sum_{k=1}^{n} k = \frac{n(n+1)}{2}$. (See Problem 5.)

Now, as $n \to \infty$, $(n + 1)/n = 1 + 1/n \to 1 + 0 = 1$. Hence, the limit of our approximating sums is

$$
a(b-a) + \tfrac{1}{2}(b-a)^2 = (b-a)\left(a + \frac{b-a}{2}\right) = (b-a)\left(\frac{a+b}{2}\right) = \tfrac{1}{2}(b^2 - a^2)
$$

Thus, $\int_a^b x\,dx = \frac{1}{2}(b^2 - a^2)$.

In the next chapter, we will find a method for calculating $\int_a^b f(x)\,dx$ that will avoid the kind of tedious computation used in this example.

Properties of the Definite Integral

$$
\int_a^b c\,f(x)\,dx = c\int_a^b f(x)\,dx \tag{23.3}
$$

This follows from the fact that an approximating sum $\sum_{k=1}^{n} cf(x_k^*)\,\Delta_k$ for $\int_a^b cf(x)\,dx$ is equal to c times the approximating sum $\sum_{k=1}^{n} f(x_k^*)\,\Delta_k x$ for $\int_a^b f(x)\,dx$, and that the same relation holds for the corresponding limits.

$$
\int_a^b -f(x)\,dx = -\int_a^b f(x)\,dx \tag{23.4}
$$

This is the special case of (23.3) when $c = -1$.

$$\int_a^b (f(x)+g(x))\,dx = \int_a^b f(x)\,dx + \int_a^b g(x)\,dx \tag{23.5}$$

This follows from the fact that an approximating sum $\sum_{k=1}^{n}(f(x_k^*)+g(x_k^*))\,\Delta_k x$ for $\int_a^b (f(x)+g(x))\,dx$ is equal to the sum $\sum_{k=1}^{n} f(x_k^*)\,\Delta_k x + \sum_{k=1}^{n} g(x_k^*)\,\Delta_k x$ of approximating sums for $\int_a^b f(x)\,dx$ and $\int_a^b g(x)\,dx$.

$$\int_a^b (f(x)-g(x))\,dx = \int_a^b f(x)\,dx - \int_a^b g(x)\,dx \tag{23.6}$$

Since $f(x) - g(x) = f(x) + (-g(x))$, this follows from (23.5) and (23.4).

If $a < c < b$, then f is integrable on $[a, b]$ if and only if it is integrable on $[a, c]$ and $[c, b]$. Moreover, if f is integrable on $[a, b]$,

$$\int_a^b f(x)\,dx = \int_c^a f(x)\,dx + \int_c^b f(x)\,dx \tag{23.7}$$

This is obvious when $f(x) \geq 0$ and we interpret the integrals as areas. The general result follows from looking at the corresponding approximating sums, although the case where one of the subintervals of $[a, b]$ contains c requires some extra thought.

We have defined $\int_a^b f(x)\,dx$ only when $a < b$. We can extend the definition to all possible cases as follows:

(i) $\int_a^a f(x)\,dx = 0$

(ii) $\int_b^a f(x)\,dx = -\int_a^b f(x)\,dx$ when $a < b$

In particular, we always have:

$$\int_c^d f(x)\,dx = -\int_d^c f(x)\,dx \quad \text{for any } c \text{ and } d \tag{23.8}$$

It can readily be verified that the laws (23.2)–(23.6), the equation in (23.7), and the result of Example 23.3 all remain valid for arbitrary upper and lower limits in the integrals.

SOLVED PROBLEMS

1. Assume $f(x) \leq 0$ for all x in $[a, b]$. Let A be the area between the graph of f and the x axis, from $x = a$ to $x = b$. (See Fig. 23-5.) Show that $\int_c^b f(x)\,dx = -A$.

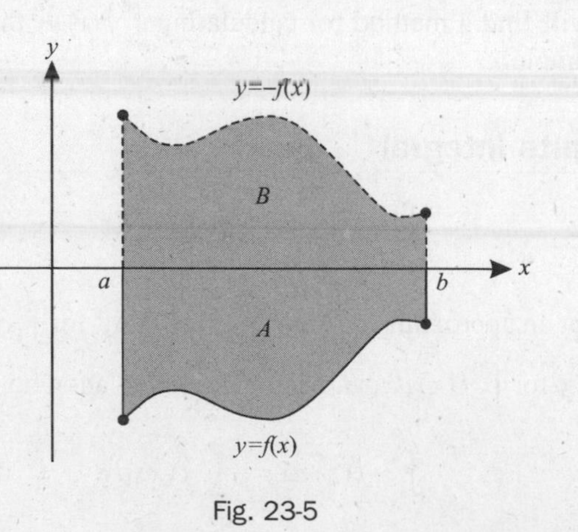

Fig. 23-5

Let B be the area between the graph of $-f$ and the x axis, from $x = a$ to $x = b$. By symmetry, $B = A$. But, $\int_a^b f(x)\,dx = -\int_a^b -f(x)\,dx$ by (23.4).

Since $\quad \int_a^b -f(x)\,dx = B, \quad \int_a^b f(x)\,dx = -B = -A$

2. Consider a function f that, between a and b, assumes both positive and negative values. For example, let its graph be as in Fig. 23-6. Then $\int_a^b f(x)\,dx$ is the difference between the sum of the areas above the x axis and below the graph and the sum of the areas below the x axis and above the graph. In the case of the graph shown in Fig. 23-6,

$$\int_a^b f(x)\,dx = (A_1 + A_3 + A_5) - (A_2 + A_4)$$

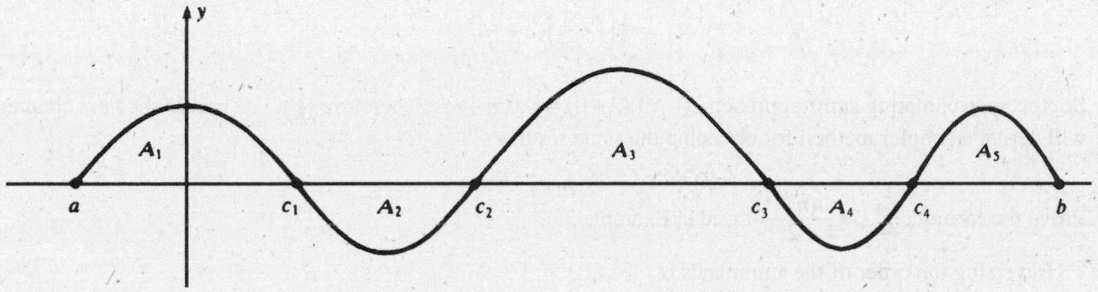

Fig. 23-6

To see this, apply (23.7) and Problem 1:

$$\int_a^b f(x)\,dx = \int_a^{c_1} f(x)\,dx + \int_{c_1}^{c_2} f(x)\,dx + \int_{c_2}^{c_3} f(x)\,dx + \int_{c_3}^{c_4} f(x)\,dx + \int_{c_4}^{b} f(x)\,dx = A_1 - A_2 + A_3 - A_4 + A_5$$

3. Assume that f and g are integrable on $[a, b]$. Prove:

(a) If $f(x) \ge 0$ on $[a, b]$, then $\int_a^b f(x)\,dx \ge 0$.

(b) If $f(x) \le g(x)$ on $[a, b]$, then $\int_a^b f(x)\,dx \le \int_a^b g(x)\,dx$.

(c) If $m \le f(x) \le M$ for all x in $[a, b]$, then $m(b-a) \le \int_a^b f(x)\,dx \le M(b-a)$.

(a) Since every approximating sum $\sum_{k=1}^{n} f(x_k^*)\,\Delta_k x \ge 0$, it follows that

$$\int_a^b f(x)\,dx \ge 0$$

(b) $g(x) - f(x) \ge 0$ on $[a, b]$. So, by (a), $\int_a^b (g(x) - f(x))\,dx \ge 0$. By (23.6), $\int_a^b g(x)\,dx - \int_a^b f(x)\,dx \ge 0$. Hence,

$$\int_a^b f(x)\,dx \le \int_a^b g(x)\,dx$$

(c) By (b), $\int_a^b m\,dx \le \int_a^b f(x) \le \int_a^b M\,dx$. But, by (23.2) and (23.3), $\int_a^b m\,dx = m\int_a^b 1\,dx = m(b-a)$ and $\int_a^b M\,dx = M\int_a^b 1\,dx = M(b-a)$. Hence,

$$m(b-a) \le \int_a^b f(x)\,dx \le M(b-a)$$

4. Evaluate $\int_0^1 x^2\,dx$.

This is the area under the parabola $y = x^2$ from $x = 0$ to $x = 1$. Divide [0, 1] into n equal subintervals. Thus, each $\Delta_k x = 1/n$. In the kth subinterval $\left[\frac{k-1}{n}, \frac{k}{n}\right]$, let x_k^* be the right endpoint k/n. Thus, the approximating sum (23.1) is

$$\sum_{k=1}^{n} f(x_k^*)\,\Delta_k x = \sum_{k=1}^{n}\left(\frac{k}{n}\right)^2\left(\frac{1}{n}\right) = \frac{1}{n^3}\sum_{k=1}^{n} k^2.$$

Now, $\sum_{k=1}^{n} k^2 = \frac{n(n+1)(2n+1)}{6}$ (see Problem 12).

Hence,

$$\sum_{k=1}^{n} f(x_k^*)\,\Delta_k x = \frac{1}{n^3}\frac{n(n+1)(2n+1)}{6} = \frac{1}{6}\left(\frac{n+1}{n}\right)\left(\frac{2n+1}{n}\right)$$

$$= \frac{1}{6}\left(1+\frac{1}{n}\right)\left(2+\frac{1}{n}\right)$$

So, the approximating sums approach $\frac{1}{6}(1+0)(2+0) = \frac{1}{3}$ as $n \to \infty$. Therefore, $\int_0^1 x^2\,dx = \frac{1}{3}$. In the next chapter, we will derive a simpler method for obtaining the same result.

5. Prove the formula $\sum_{k=1}^{n} k = \frac{n(n+1)}{2}$ used in Example 23.3.

Reversing the order of the summands in

$$\sum_{k=1}^{n} k = 1+2+3+\cdots+(n-2)+(n-1)+n$$

we get

$$\sum_{k=1}^{n} k = n+(n-1)+(n-2)+\cdots+3+2+1.$$

Adding the two equations yields

$$2\sum_{k=1}^{n} k = (n+1)+(n+1)+(n+1)+\cdots+(n+1)+(n+1)+(n+1) = n(n+1)$$

since the sum in each column is $n + 1$. Hence, dividing by 2, we get

$$\sum_{k=1}^{n} k = \frac{n(n+1)}{2}.$$

SUPPLEMENTARY PROBLEMS

6. Calculate: (a) $\int_1^4 3\,dx$; (b) $\int_{-2}^5 x\,dx$; (c) $\int_0^1 3x^2\,dx$.

Ans. (a) $3(4-1) = 9$; (b) $\frac{1}{2}(5^2-(-2)^2) = \frac{21}{2}$; (c) $3(\frac{1}{3}) = 1$

7. Find the area under the parabola $y = x^2 - 2x + 2$, above the x axis, and between $x = 0$ and $x = 1$.

Ans. $\frac{1}{3} - 2[\frac{1}{2}(1^2-0^2)] + 2(1-0) = \frac{4}{3}$

8. Evaluate $\int_2^6 (3x+4)\,dx$.

Ans. $3((\frac{1}{2})(6^2-2^2)) + 4(6-2) = 64$

9. For the function f graphed in Fig. 23-7, express $\int_0^3 f(x)\,dx$ in terms of the areas A_1, A_2, and A_3.

Ans. $A_1 - A_2 + A_3$

10. Show that $3 \le \int_1^4 x^3\,dx \le 192$. [*Hint:* Problem 3(*c*).]

11. Evaluate $\int_0^1 \sqrt{1-x^2}\,dx$. (*Hint:* Find the corresponding area by geometric reasoning.)

Ans. $\pi/4$

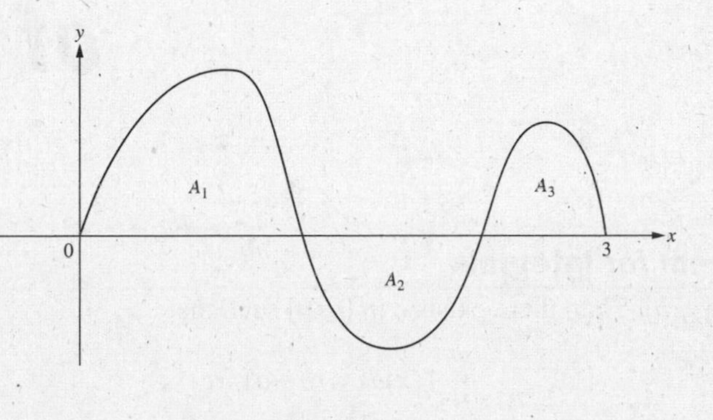

Fig. 23-7

12. Use mathematical induction to prove the formula $\sum_{k=1}^{n} k^2 = \frac{n(n+1)(2n+1)}{6}$ of Problem 4. (Verify it when $n = 1$, and then show that, if it holds for n, then it holds for $n + 1$.)

13. Evaluate (a) $\sum_{j=0}^{2} \cos\frac{j\pi}{6}$; (b) $\sum_{j=0}^{2} (4j+1)$; (c) $\sum_{j=1}^{100} 4j$; (d) $\sum_{j=1}^{18} 2j^2$.

Ans. (a) $\frac{3+\sqrt{3}}{2}$; (b) 15; (c) 20200; (d) 4218

14. Let the graph of f between $x = 1$ and $x = 6$ be as in Fig. 23-8. Evaluate $\int_1^6 f(x)\,dx$.

Ans. $1 - 3 + \frac{1}{2} = -\frac{3}{2}$

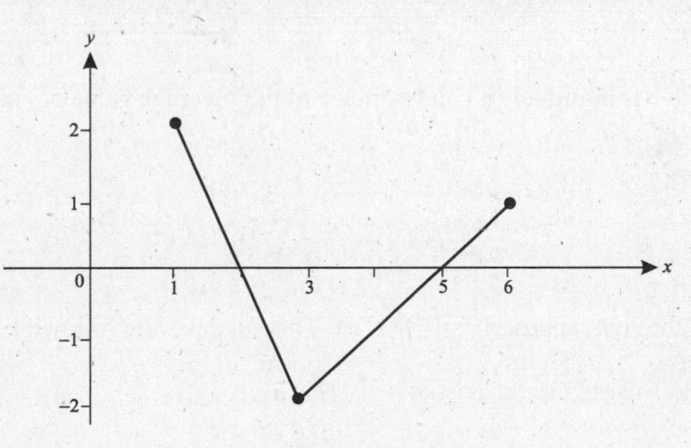

Fig. 23-8

15. If f is continuous on $[a, b]$, $f(x) \ge 0$ on $[a, b]$, and $f(x_0) > 0$ for some x_0 in $[a, b]$, prove that $\int_a^b f(x)\,dx > 0$. [*Hint:* By the continuity of f, $f(x) > \frac{1}{2} f(x_0) > 0$ for all x in some subinterval $[c, d]$. Use (23.7) and Problem 3(*a*, *c*).]

CHAPTER 24

The Fundamental Theorem of Calculus

Mean-Value Theorem for Integrals

Let f be continuous on $[a, b]$. Then there exists c in $[a, b]$ such that

$$\int_b^a f(x)dx = (b-a)f(c) \tag{24.1}$$

To see this, let m and M be the minimum and maximum values of f in $[a, b]$, and apply Problem 3(c) of Chapter 23 to obtain

$$m(b-a) \le \int_a^b f(x)\,dx \le M(b-a) \qquad \text{and, therefore,} \qquad m \le \frac{1}{b-a}\int_a^b f(x)\,dx \le M$$

So, by the intermediate value theorm, $\frac{1}{b-a}\int_a^b f(x)\,dx = f(c)$ for some c in $[a, b]$.

Average Value of a Function on a Closed Interval

Let f be defined on $[a, b]$. Since f may assume infinitely many values on $[a, b]$, we cannot talk about the average of all of the values of f. Instead, divide $[a, b]$ into n equal subintervals, each of $\Delta x = \frac{b-a}{n}$. Select an arbitrary point x_k^* in the kth subinterval. Then the average of the n values $f(x_1^*), f(x_2^*), \ldots, f(x_n^*)$ is

$$\frac{f(x_1^*) + f(x_2^*) + \ldots + f(x_n^*)}{n} = \frac{1}{n}\sum_{k=1}^{n} f(x_k^*)$$

When n is large, this value is intuitively a good estimate of the "average value of f on $[a, b]$." However, since $\frac{1}{n} = \frac{1}{b-a}\Delta x$,

$$\frac{1}{n}\sum_{k=1}^{n} f(x_k^*) = \frac{1}{b-a}\sum_{k=1}^{n} f(x_k^*)\Delta x$$

As $n \to \infty$, the sum on the right approaches $\int_a^b f(x)\,dx$. This suggests the following definition.

Definition: The *average value* of f on $[a, b]$ is $\frac{1}{b-a}\int_a^b f(x)\,dx$.

Let f be continuous on $[a, b]$. If x is in $[a, b]$, then $\int_a^x f(t)\,dt$ is a function of x, and:

$$D_x\left(\int_a^x f(t)\,dt\right) = f(x) \tag{24.2}$$

For a proof, see Problem 4.

Fundamental Theorem of Calculus

Let f be continuous on $[a, b]$, and let $F(x) = \int f(x)\,dx$, that is, F is an antiderivative of f. Then

$$\int_a^b f(x)\,dx = F(b) - F(a) \tag{24.3}$$

To see this, note that, by (24.2), $\int_a^x f(t)\,dt$ and $F(x)$ have the same derivative, $f(x)$. Hence, by Problem 18 of Chapter 13, there is a constant K such that $\int_a^x f(t)\,dt = F(x) + K$. When $x = a$, we get

$$F(a) + K = \int_a^a f(t)\,dt = 0 \quad \text{So,} \quad K = -F(a)$$

Hence, $\int_a^x f(t)\,dt = F(x) - F(a)$. When $x = b$, this yields

$$\int_a^b f(t)\,dt = F(b) - F(a)$$

Equation (24.3) provides a simple way of computing $\int_a^b f(x)\,dx$ when we can find an antiderivative F of f. The expression $F(b) - F(a)$ on the right side of (24.3) is often abbreviated as $F(x)\Big]_a^b$. Then the fundamental theorem of calculus can be written as follows:

$$\int_a^b f(x)\,dx = \int f(x)\,dx\Big]_a^b$$

EXAMPLE 24.1:

(i) The complicated evaluation of $\int_a^b x\,dx$ in Example 23.3 of Chapter 23 can be replaced by the following simple one:

$$\int_a^b x\,dx = \tfrac{1}{2}x^2\Big]_a^b = \tfrac{1}{2}b^2 - \tfrac{1}{2}a^2 = \tfrac{1}{2}(b^2 - a^2)$$

(ii) The very tedious computation of $\int_0^1 x^2\,dx$ in Problem 4 of Chapter 23 can be replaced by

$$\int_0^1 x^2\,dx = \tfrac{1}{3}x^3\Big]_0^1 = \tfrac{1}{3}1^3 - \tfrac{1}{3}0^3 = \tfrac{1}{3}$$

(iii) In general, $\int_a^b x^r\,dx = \frac{1}{r+1}x^{r+1}\Big]_a^b = \frac{1}{r+1}(b^{r+1} - a^{r+1}) \quad \text{for } r \neq -1$

Change of Variable in a Definite Integral

In the computation of a definite integral by the fundamental theorem, an antiderivative $\int f(x)\,dx$ is required. In Chapter 22, we saw that substitution of a new variable u is sometimes useful in finding $\int f(x)\,dx$. When the substitution also is made in the definite integral, the limits of integration must be replaced by the corresponding values of u.

EXAMPLE 24.2: Evaluate $\int_1^9 \sqrt{5x+4}\, dx$.

Let $u = 5x + 4$. Then $du = 5\, dx$. When $x = 1$, $u = 9$, and when $x = 9$, $u = 49$. Hence,

$$\begin{aligned}\int_1^9 \sqrt{5x+4}\, dx &= \int_9^{49} \sqrt{u}\, \tfrac{1}{5}\, du = \tfrac{1}{5}\int_9^{49} u^{1/2}\, du \\ &= \tfrac{1}{5}\left(\tfrac{2}{3}u^{3/2}\right)\Big]_9^{49} \quad \text{(by the fundamental theorem)} \\ &= \tfrac{2}{15}(49^{3/2} - 9^{3/2}) = \tfrac{2}{15}[(\sqrt{49})^3 - (\sqrt{9})^3] \\ &= \tfrac{2}{15}(7^3 - 3^3) = \tfrac{2}{15}(316) = \tfrac{632}{15}\end{aligned}$$

For justification of this method, see Problem 5.

SOLVED PROBLEMS

1. Evaluate $\int_0^{\pi/2} \sin^2 x \cos x\, dx$.

$\int \sin^2 x \cos x\, dx = \frac{1}{3}\sin^3 x$ by Quick Formula I. Hence, by the fundamental theorem,

$$\int_0^{\pi/2} \sin^2 x \cos x\, dx = \tfrac{1}{3}\sin^3 x\Big]_0^{\pi/2} = \tfrac{1}{3}\left[\left(\sin\frac{\pi}{2}\right)^3 - (\sin 0)^3\right] = \tfrac{1}{3}(1^3 - 0^3) = \tfrac{1}{3}$$

2. Find the area under the graph of $f(x) = \frac{1}{\sqrt{4-x^2}}$, above the x axis, and between 0 and 1.

The area is $\int_0^1 \frac{1}{\sqrt{4-x^2}}\, dx = \sin^{-1}\left(\frac{x}{2}\right)\Big]_0^1 = \sin^{-1}\left(\frac{1}{2}\right) - \sin^{-1}(0) = \frac{\pi}{6} - 0 = \frac{\pi}{6}$.

3. Find the average value of $f(x) = 4 - x^2$ on $[0, 2]$.

The average value is

$$\frac{1}{b-a}\int_a^b f(x)\, dx = \tfrac{1}{2}\int_0^2 (4 - x^2)\, dx = \tfrac{1}{2}\left(4x - \frac{x^3}{3}\right)\Big]_0^2 = \tfrac{1}{2}[(8 - \tfrac{8}{3}) - (0 - 0)] = \tfrac{8}{3}$$

4. Prove formula (24.2): $D_x\left(\int_a^x f(t)\, dt\right) = f(x)$

Let $h(x) = \int_a^x f(t)\, dt$. Then:

$$\begin{aligned}h(x+\Delta x) - h(x) &= \int_a^{x+\Delta x} f(t)\, dt - \int_a^x f(t)\, dt \\ &= \int_a^x f(t)\, dt + \int_x^{x+\Delta x} f(t)\, dt - \int_a^x f(t)\, dt \qquad \text{(by 23.7)} \\ &= \int_x^{x+\Delta x} f(t)\, dt \\ &= \Delta x \cdot f(x^*) \qquad \text{for some } x^* \text{ between } x \text{ and } x+\Delta x \text{ (by the mean value theorem for integrals)}\end{aligned}$$

Thus, $\frac{h(x+\Delta x) - h(x)}{\Delta x} = f(x^*)$ and therefore,

$$D_x\left(\int_a^x f(t)\, dt\right) = D_x(h(x)) = \lim_{\Delta x \to 0} \frac{h(x+\Delta x) - h(x)}{\Delta x} = \lim_{\Delta x \to 0} f(x^*)$$

But, as $\Delta x \to 0$, $x + \Delta x \to x$ and so, $x^* \to x$ (since x^* is between x and $x + \Delta x$). Since f is continuous, $\lim_{\Delta x \to 0} f(x^*) = f(x)$.

5. Justify a change of variable in a definite integral in the following precise sense. Given $\int_a^b f(x)\,dx$, let $x = g(u)$ where, as x varies from a to b, u increases or decreases from c to d. (See Fig. 24-1 for the case where u is increasing.) Show that

$$\int_a^b f(x)\,dx = \int_c^d f(g(u))g'(u)\,du$$

(The right side is obtained by substituting $g(u)$ for x, $g'(u)\,du$ for dx, and changing the limits of integration from a and b to c and d.)

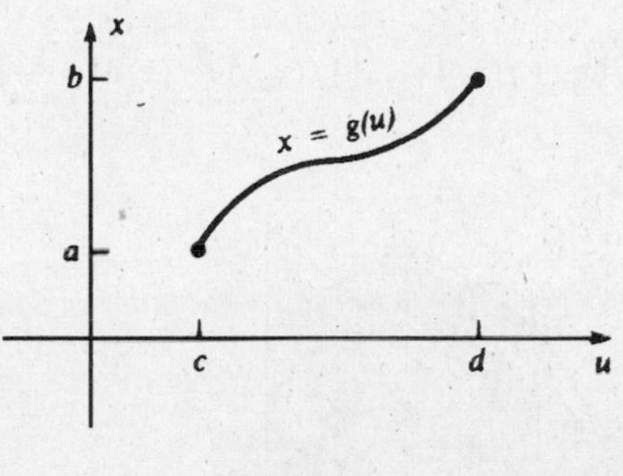

Fig. 24-1

Let $F(x) = \int f(x)\,dx$, that is, $F'(x) = f(x)$. By the Chain Rule,

$$D_u(F(g(u)) = F'(g(u)) \cdot g'(u) = f(g(u))g'(u) \qquad \text{Thus,} \qquad \int f(g(u))g'(u)\,du = F(g(u))$$

So, by the fundamental theorem,

$$\int_c^d f(g(u))g'(u)\,du = F(g(u))\Big]_c^d = F(g(d)) - F(g(c))$$

$$= F(b) - F(a) = \int_a^b f(x)\,dx$$

6. (a) If f is an even function, show that, for $a > 0$, $\int_{-a}^a f(x)\,dx = 2\int_0^a f(x)\,dx$.

(b) If f is an odd function, show that, for $a > 0$, $\int_{-a}^a f(x)\,dx = 0$.

Let $u = -x$. Then $du = -dx$, and

$$\int_{-a}^0 f(x)\,dx = \int_a^0 f(-u)(-1)\,du = -\int_a^0 f(-u)\,du = \int_0^a f(-u)\,du$$

Rewriting u as x in the last integral, we have:

$$\int_{-a}^0 f(x)\,dx = \int_0^a f(-x)\,dx \qquad (*)$$

Thus,

$$\begin{aligned}\int_{-a}^a f(x)\,dx &= \int_{-a}^0 f(x)\,dx + \int_0^a f(x)\,dx \quad \text{(by (23.7))}\\ &= \int_0^a f(-x)\,dx + \int_0^a f(x)\,dx \quad \text{(by (*))}\\ &= \int_0^a f(-x) + f(x)\,dx \quad \text{(by 23.5))}\end{aligned}$$

(a) If f is even, $f(-x) + f(x) = 2f(x)$, whence $\int_{-a}^a f(x)\,dx = \int_0^a 2f(x)\,dx = 2\int_0^a f(x)\,dx$.

(b) If f is odd, $f(-x) + f(x) = 0$, whence $\int_{-a}^a f(x)\,dx = \int_0^a 0\,dx = 0\int_0^a 1\,dx = 0$.

7. Trapezoidal Rule

(a) Let $f(x) \geq 0$ on $[a, b]$. Divide [a, b] into n equal parts, each of length $\Delta x = \frac{b-a}{n}$, by means of points $x_1, x_2, \ldots, x_{n-1}$. (See Fig. 24-2(*a*).) Prove the following *trapezoidal rule*: $\int_a^b f(x)\,dx \sim \frac{\Delta x}{2}\left(f(a) + 2\sum_{k=1}^{n-1} f(x_k) + f(b)\right)$

(b) Use the trapezoidal rule with $n = 10$ to approximate $\int_0^1 x^2\,dx$.

(a) The area of the strip, over $[x_{k-1}, x_k]$, is approximately the area of trapezoid *ABCD* (in Fig. 24-2(*b*)):, $\frac{1}{2}\Delta x(f(x_{k-1}) + f(x_k))$† (Remember that $x_0 = a$ and $x_n = b$.) So, the area under the curve is approximated by the sum of the trapezoidal areas,

$$\frac{\Delta x}{2}\{[f(x_0) + f(x_1)] + [f(x_1) + f(x_2)] + \ldots + [f(x_{n-1}) + f(x_n)]\} = \frac{\Delta x}{2}[f(a) + 2\sum_{k=1}^{n-1} f(x_k) + f(b)]$$

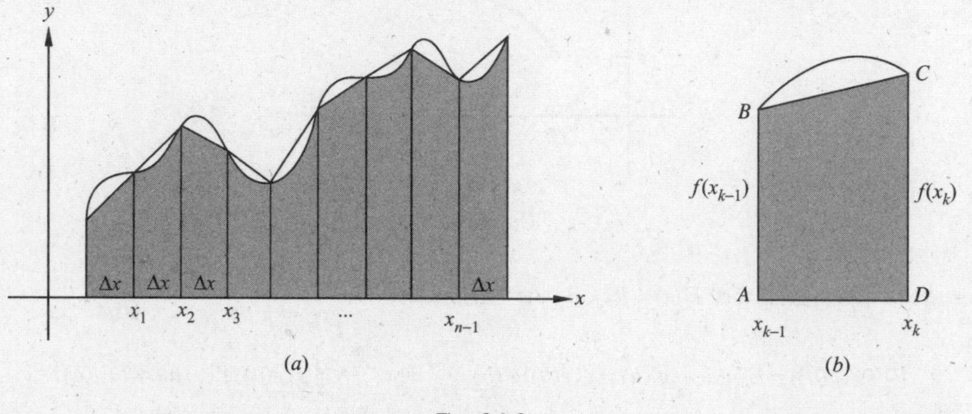

Fig. 24-2

(b) With $n = 10$, $a = 0$, $b = 1$, $\Delta x = \frac{1}{10}$ and $x_k = k/10$, we get

$$\int_0^1 x^2\,dx \sim \frac{1}{20}\left(0^2 + 2\sum_{k=1}^{9}\frac{k^2}{100} + 1^2\right) = \frac{1}{20}\left(\frac{2}{100}\sum_{k=1}^{9} k^2 + 1\right)$$

$$= \frac{1}{20}\left[\frac{2}{100}(285) + 1\right] \qquad \text{(by Problem 12 of Chapter 23)}$$

$$= 0.335$$

The exact value is $\frac{1}{3}$ (by Example 24.1 (ii)).

SUPPLEMENTARY PROBLEMS

In Problems 8–22, use the fundamental theorem of calculus to evaluate the definite integral.

8. $\int_{-1}^{1}(2x^2 - x^3)\,dx$ *Ans.* $\frac{4}{3}$

9. $\int_{-3}^{-1}\left(\frac{1}{x^2} - \frac{1}{x^3}\right)dx$ *Ans.* $\frac{10}{9}$

10. $\int_1^4 \frac{dx}{\sqrt{x}}$ *Ans.* 2

† Recall that the area of a trapezoid of height h and bases b_1 and b_2 is $\frac{1}{2}h(b_1 + b_2)$.

11. $\int_{\pi/2}^{3\pi/4} \sin x\, dx$ *Ans.* $\frac{\sqrt{2}}{2}$

12. $\int_0^2 (2+x)\, dx$ *Ans.* 6

13. $\int_0^2 (2-x)^2\, dx$ *Ans.* $\frac{8}{3}$

14. $\int_0^3 (3-2x+x^2)\, dx$ *Ans.* 9

15. $\int_{-1}^2 (1-t^2)t\, dt$ *Ans.* $-\frac{9}{4}$

16. $\int_1^4 (1-u)\sqrt{u}\, du$ *Ans.* $-\frac{116}{15}$

17. $\int_1^8 \sqrt{1+3x}\, dx$ *Ans.* 26

18. $\int_0^2 x^2(x^3+1)\, dx$ *Ans.* $\frac{40}{3}$

19. $\int_0^3 \frac{1}{\sqrt{1+x}}\, dx$ *Ans.* 2

20. $\int_0^1 x(1-\sqrt{x})^2\, dx$ *Ans.* $\frac{1}{30}$

21. $\int_4^8 \frac{x}{\sqrt{x^2-15}}\, dx$ *Ans.* 6

22. $\int_0^{2\pi} \sin\frac{t}{2}\, dt$ *Ans.* 4

In Problems 23–26, use Problem 6(*a, b*).

23. $\int_{-2}^2 \frac{dx}{x^2+4}\, dx$ *Ans.* $\frac{\pi}{4}$

24. $\int_{-2}^2 (x^3-x^5)\, dx$ *Ans.* 0

25. $\int_{-3}^3 \sin\frac{x}{5}\, dx$ *Ans.* 0

26. $\int_{-\pi/2}^{\pi/2} \cos x\, dx$ *Ans.* 2

27. Prove: $D_x\left(\int_x^b f(t)\, dt\right) = -f(x)$.

28. Prove $D_x\left(\int_{h(x)}^{g(x)} f(t)\, dt\right) = f(g(x))g'(x) - f(h(x))h'(x)$.

In Problems 29–32, use Problems 27–28 and (24.2) to find the given derivative.

29. $D_x\left(\int_1^x \sin t\, dt\right)$ *Ans.* $\sin x$

30. $D_x\left(\int_x^0 t^2\, dt\right)$ *Ans.* $-x^2$

31. $D_x\left(\int_0^{\sin x} t^3\,dt\right)$ *Ans.* $\sin^3 x \cos x$

32. $D_x\left(\int_{x^2}^{4x} \cos t\,dt\right)$ *Ans.* $4 \cos 4x - 2x \cos x^2$

33. Compute the average value of the following functions on the indicated intervals.

(a) $f(x) = \sqrt[5]{x}$ on $[0, 1]$ *Ans.* $\frac{5}{6}$

(b) $f(x) = \sec^2 x$ on $\left[0, \frac{\pi}{3}\right]$ *Ans.* $\frac{3\sqrt{3}}{\pi}$

(c) $f(x) = 3x^2 - 1$ on $[-1, 4]$ *Ans.* 12

(d) $f(x) = \sin x - \cos x$ on $[0, \pi]$ *Ans.* $\frac{2}{\pi}$

34. Use the change-of-variables method to find $\int_{1/2}^{3} \sqrt{2x+3}\, x\,dx$.

Ans. $\frac{58}{5}$

35. An object moves along the x axis for a period of time T. If its initial position is x_1 and its final position is x_2, show that its average velocity was $\frac{x_2 - x_1}{T}$.

36. Let $f(x) = \begin{cases} \cos x & \text{for } x < 0 \\ 1 - x & \text{for } x \geq 0 \end{cases}$. Evaluate $\int_{-\pi/2}^{1} f(x)\,dx$.

Ans. $\frac{3}{2}$

37. Evaluate $\lim_{h \to 0} \frac{1}{h} \int_3^{3+h} \frac{5}{x^3 + 7}\,dx$.

Ans. $\frac{5}{34}$

38. **(Midpoint Rule)** In an approximating sum (23.1) $\sum_{k=1}^{n} f(x_k^*)\Delta_k x$, if we select x_k^* to be the midpoint of the kth subinterval, then the sum is said to be obtained by the *midpoint rule*. Apply the midpoint rule to approximate $\int_0^1 x^2\,dx$, using a division into five equal subintervals, and compare with the exact result of $\frac{1}{3}$.

Ans. 0.33

39. **(Simpson's Rule)** If we divide $[a, b]$ into n equal subintervals, where n is even, the following approximating sum for $\int_a^b f(x)\,dx$,

$$\frac{b-a}{3n}[f(x_0) + 4f(x_1) + 2f(x_2) + 4f(x_3) + 2f(x_4) + \cdots + 4f(x_{n-1}) + f(x_n)]$$

is said to be obtained by *Simpson's rule*. Except for the first and last terms, the coefficients consist of alternating 4s and 2s. (The basic idea is to use parabolas as approximating arcs instead of line segments as in the trapezoidal rule. Simpson's rule is usually much more accurate than the midpoint or trapezoidal rule.)

Apply Simpson's rule to approximate (a) $\int_0^1 x^2\,dx$ and (b) $\int_0^\pi \sin x\,dx$ with $n = 4$, and compare the results with the answers obtained by the fundamental theorem.

Ans. (a) $\frac{1}{3}$, which is the exact answer; (b) $\frac{\pi}{6}(2\sqrt{2}+1) \sim 2.0046$ as compared to 2

40. Consider $\int_0^1 x^3\,dx$. (a) Show that the fundamental theorem yields the answer $\frac{1}{4}$. (b) (GC) With $n = 10$, approximate (to four decimal places) the integral by the trapezoidal, midpoint, and Simpson's rules.

Ans. Trapezoidal 0.2525; midpoint 0.2488; Simpson's 0.2500

41. Evaluate:

(a) $\lim_{n\to+\infty} \frac{1}{n}\left(\cos\frac{\pi}{n} + \cos\frac{2\pi}{n} + \cdots + \cos\frac{n\pi}{n}\right)$

(b) $\lim_{n\to+\infty} \frac{\pi}{6n}\left[\sec^2\left(\frac{\pi}{6n}\right) + \sec^2\left(2\frac{\pi}{6n}\right) + \cdots + \sec^2\left((n-1)\frac{\pi}{6n}\right) + \frac{4}{3}\right]$

Ans. (a) $\frac{1}{\pi}\int_0^{\pi} \cos x\,dx = 0$; (b) $\int_0^{\pi/6} \sec^2 x\,dx = \frac{\sqrt{3}}{3}$

42. (a) Use a substitution to evaluate $\int_1^2 \frac{x}{\sqrt{x+1}}\,dx$ (to eight decimal places).

(b) (GC) Use a graphing calculator to estimate the integral of (a).

Ans. (a) $\frac{2}{3}\sqrt{2}$; (b) 0.39052429

43. (GC) Estimate $\int_0^{\pi/4} x\sin^3(\tan x)\,dx$ (to four decimal places).

Ans. 0.0710

44. (GC) Consider $\int_1^2 x\sqrt[3]{x^5 + 2x^2 - 1}\,dx$. Estimate (to six decimal places) its value using the trapezoidal and Simpson's rule (both with $n = 4$), and compare with the value given by a graphing calculator.

Ans. trapezoidal 3.599492; Simpson's 3.571557; graphing calculator 3.571639

CHAPTER 25

The Natural Logarithm

The traditional way of defining a logarithm, $\log_a b$, is to define it as that number u such that $a^u = b$. For example, $\log_{10} 100 = 2$ because $10^2 = 100$. However, this definition has a theoretical gap. The flaw is that we have not yet defined a^u when u is an irrational number, for example, $\sqrt{2}$ or π. This gap can be filled in, but that would require an extensive and sophisticated detour.† Instead, we take a different approach that will eventually provide logically unassailable definitions of the logarithmic and exponential functions. A temporary disadvantage is that the motivation for our initial definition will not be obvious.

The Natural Logarithm

We are already familiar with the formula

$$\int x^r dx = \frac{x^{r+1}}{r+1} + C \quad (r \neq -1)$$

The problem remains of finding out what happens when $r = -1$, that is, of finding the antiderivative of x^{-1}.

The graph of $y = 1/t$, for $t > 0$, is shown in Fig. 25-1. It is one branch of a hyperbola. For $x > 1$, the definite integral

$$\int_1^x \frac{1}{t}\,dt$$

is the value of the area under the curve $y = 1/t$ and above the t axis, between $t = 1$ and $t = x$.

Definition

$\ln x = \int_1^x \frac{1}{t}\,dt \quad$ for $x > 0$

The function $\ln x$ is called the *natural logarithm*. The reasons for referring to it as a logarithm will be made clear later. By (24.2),

(25.1) $\quad D_x(\ln x) = \frac{1}{x} \quad$ for $x > 0$

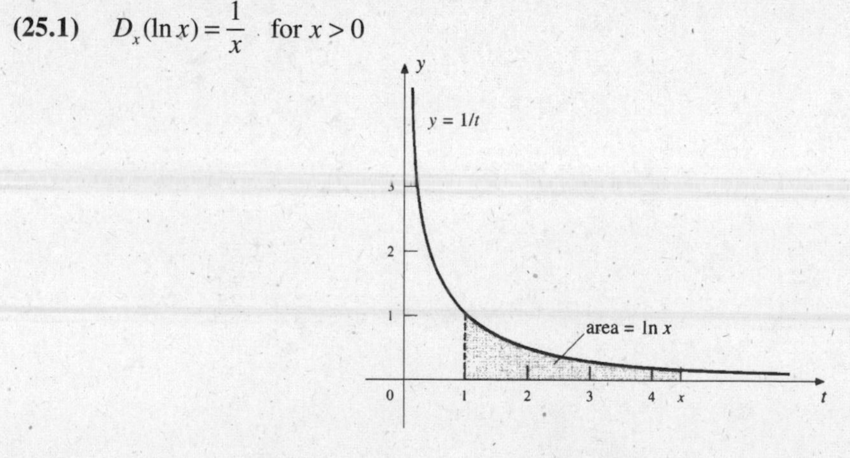

Fig. 25-1

† Some calculus textbooks just ignore the difficulty. They assume that a^u is defined when $a > 0$ and u is *any* real number and that the usual laws for exponents are valid.

Hence, the natural logarithm is the antiderivative of x^{-1}, but only on the interval $(0, +\infty)$. An antiderivative for all $x \neq 0$ will be constructed below in (25.5).

Properties of the Natural Logarithm

(25.2) $\ln 1 = 0$, since $\ln 1 = \int_1^1 \frac{1}{t}\,dt = 0$.
(25.3) If $x > 1$, then $\ln x > 0$.

This is true by virtue of the fact that $\int_1^x \frac{1}{t}\,dt$ represents an area, or by Problem 15 of Chapter 23.

(25.4) If $0 < x < 1$, then $\ln x < 0$.

$\ln x = \int_1^x \frac{1}{t}\,dt = -\int_x^1 \frac{1}{t}\,dt$ by (23.8). Now, for $0 < x < 1$, if $x \le t \le 1$, then $1/t > 0$ and, therefore, by Problem 15 of Chapter 23, $\int_x^1 \frac{1}{t}\,dt > 0$.

(25.5) (a) $D_x(\ln|x|) = \frac{1}{x}$ for $x \neq 0$

(b) $\int \frac{1}{x}\,dx = \ln\,|x| + C$ for $x \neq 0$

The argument is simple. For $x > 0$, $|x| = x$, and so $D_x(\ln|x|) = D_x(\ln x) = 1/x$ by (25.1). For $x < 0$, $|x| = -x$, and so

$$D_x(\ln|x|) = D_x(\ln(-x)) = D_u(\ln u)D_x(u) \qquad \text{(Chain Rule, with } u = -x > 0)$$
$$= \left(\frac{1}{u}\right)(-1) = \frac{1}{-u} = \frac{1}{x}$$

EXAMPLE 25.1: $D_x(\ln|3x+2|) = \frac{1}{3x+2}D_x(3x+2)$ (Chain Rule)

$$= \frac{3}{3x+2}$$

(25.6) $\ln uv = \ln u + \ln v$

Note that

$$D_x(\ln(ax)) = \frac{1}{ax}D_x(ax) \qquad \text{(by the Chain Rule and (25.1))}$$
$$= \frac{1}{ax}(a) = \frac{1}{x} = D_x(\ln x)$$

Hence, $\ln(ax) = \ln x + K$ for some constant K (by Problem 18 of Chapter 13). When $x = 1$, $\ln a = \ln 1 + K = 0 + K = K$. Thus, $\ln(ax) = \ln x + \ln a$. Replacing a and x by u and v yields (25.6).

(25.7) $\ln\left(\frac{u}{v}\right) = \ln u - \ln v$

In (25.6), replace u by $\frac{u}{v}$.

(25.8) $\ln\frac{1}{v} = -\ln v$

In (25.7), replace u by 1 and use (25.2).

(25.9) $\ln(x^r) = r \ln x$ for any rational number r and $x > 0$.

By the Chain Rule, $D_x(\ln(x^r)) = \frac{1}{x^r}(rx^{r-1}) = \frac{r}{x} = D_x(r\ln x)$. So, by Problem 18 of Chapter 13, $\ln(x^r) = r\ln x + K$ for some constant K. When $x = 1$, $\ln 1 = r \ln 1 + K$. Since $\ln 1 = 0$, $K = 0$, yielding (25.9).

EXAMPLE 25.2: $\ln\sqrt[3]{2x-5} = \ln(2x-5)^{1/3} = \frac{1}{3}\ln(2x-5)$.

(25.10) $\ln x$ is an increasing function.

$D_x(\ln x) = \frac{1}{x} > 0$ since $x > 0$. Now use Theorem 13.7.

(25.11) $\ln u = \ln v$ implies $u = v$.

This is a direct consequence of (25.10). For, if $u \neq v$, then either $u < v$ or $v < u$ and, therefore, either $\ln u < \ln v$ or $\ln v < \ln u$.

(25.12) $\frac{1}{2} < \ln 2 < 1$

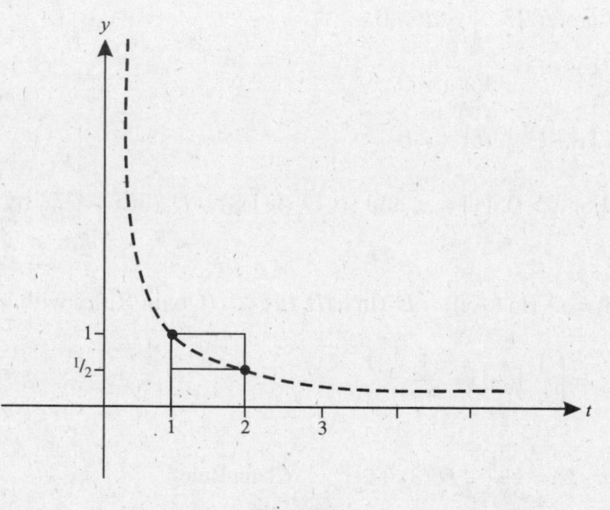

Fig. 25-2

The area under the graph of $y = 1/t$, between $t = 1$ and $t = 2$, and above the t axis, is greater than the area $\frac{1}{2}$ of the rectangle with base [1, 2] and height $\frac{1}{2}$. (See Fig. 25-2.) It is also less than the area 1 of the rectangle with base [1, 2] and height 1. (A more rigorous argument would use Problems 3(c) and 15 of Chapter 23.)

(25.13) $\lim_{x\to+\infty} \ln x = +\infty$

Let k be any positive integer. Then, for $x > 2^{2k}$,
$\ln x > \ln(2^{2k}) = 2k \ln 2 > 2k(\frac{1}{2}) = k$

by (25.10) and (25.9). Thus, as $x \to +\infty$, $\ln x$ eventually exceeds every positive integer

(25.14) $\lim_{x\to 0^+} \ln x = -\infty$

Let $u = 1/x$. As $x \to 0^+$, $u \to +\infty$. Hence,

$$\lim_{x\to 0^+} \ln x = \lim_{u\to+\infty} \ln\left(\frac{1}{u}\right) = \lim_{u\to+\infty} -\ln u \quad \text{(by (25.8))}$$

$$= -\lim_{u\to+\infty} \ln u = -\infty \quad \text{(by (25.13))}$$

(25.15) Quick Formula II: $\int \frac{g'(x)}{g(x)}\,dx = \ln|g(x)| + C$

By the Chain Rule and (25.5) (a), $D_x(\ln|g(x)|) = \frac{1}{g(x)}g'(x)$.

EXAMPLE 25.3:

(a) $\int \frac{2x}{x^2+1} dx = \ln|x^2+1| + C = \ln(x^2+1) + C$

The absolute value sign was dropped because $x^2 + 1 \geq 0$. In the future, we shall do this without explicit mention.

(b) $\int \frac{x^2}{x^3+5} dx = \frac{1}{3}\int \frac{3x^2}{x^3+5} dx = \frac{1}{3}\ln|x^3+5| + C$

SOLVED PROBLEMS

1. Evaluate: (a) $\int \tan x\, dx$; (b) $\int \cot x\, dx$; (c) $\int \sec x\, dx$.

(a) $\int \tan x\, dx = \int \frac{\sin x}{\cos x} dx = -\int \frac{-\sin x}{\cos x} dx$

$= -\ln|\cos x| + C$ by Quick Formula II.

$= -\ln\left|\frac{1}{\sec x}\right| + C = -(-\ln|\sec x|) + C = \ln|\sec x| + C$

(25.16) $\int \tan x\, dx = \ln|\sec x| + C$

(b) $\int \cot x\, dx = \int \frac{\cos x}{\sin x} dx = \ln|\sin x| + C$ by Quick Formula II.

(25.17) $\int \cot x\, dx = \ln|\sin x| + C$

(c) $\int \sec x\, dx = \int \sec x \frac{\sec x + \tan x}{\sec x + \tan x} dx$

$= \int \frac{\sec^2 x + \sec x \tan x}{\sec x + \tan x} dx = \ln|\sec x + \tan x| + C$ by Quick Formula II.

(25.18) $\int \sec x\, dx = \ln|\sec x + \tan x| + C$

2. (GC) Estimate the value of ln 2.

A graphing calculator yields the value $\ln 2 \sim 0.6931471806$. Later we shall find another method for calculating ln 2.

3. (GC) Sketch the graph of $y = \ln x$.

A graphing calculator yields the graph shown in Fig. 25-3. Note by (25.10) that $\ln x$ is increasing. By (25.13), the graph increases without bound on the right, and, by (25.14), the negative y axis is a vertical asymptote. Since

$$D_x^2(\ln x) = D_x(x^{-1}) = -x^{-2} = -\frac{1}{x^2} < 0$$

the graph is concave downward. By (25.13) and (25.14), and the intermediate value theorem, the range of $\ln x$ is the set of all real numbers.

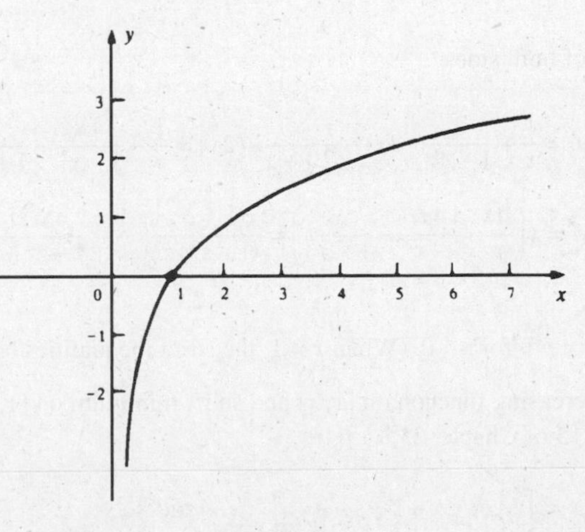

Fig. 25-3

4. Find: (a) $D_x(\ln(x^4+7x))$; (b) $D_x(\ln(\cos 2x))$; (c) $D_x(\cos(\ln 2x))$.

(a) $D_x(\ln(x^4+7x)) = \dfrac{1}{x^4+7x}(4x^3+7) = \dfrac{4x^3+7}{x^4+7x}$

(b) $D_x(\ln(\cos 2x)) = \dfrac{1}{\cos 2x}(-\sin 2x)(2) = -\dfrac{2\sin 2x}{\cos 2x}$

$= -2\tan 2x$

(c) $D_x(\cos(\ln 2x)) = (-\sin(\ln 2x))\left(\dfrac{1}{2x}\right)(2) = -\dfrac{\sin(\ln 2x)}{x}$

5. Find the following antiderivatives. Use Quick Formula II when possible.

(a) $\int \frac{1}{8x-3}dx$; (b) $\int \frac{4x^7}{3x^8-2}dx$; (c) $\int \frac{x-4}{x^2+5}dx$; (d) $\int \frac{x}{x^2-4x+5}dx$

(a) $\int \frac{1}{8x-3}dx = \frac{1}{8}\int \frac{8}{8x-3}dx = \frac{1}{8}\ln|8x-3| + C$

(b) $\int \frac{4x^7}{3x^8-2}dx = \frac{1}{6}\int \frac{24x^7}{3x^8-2}dx = \frac{1}{6}\ln|3x^8-2| + C$

(c) $\int \frac{x-4}{x^2+5}dx = \int \frac{x}{x^2+5}dx - \int \frac{4}{x^2+5}dx$

$$= \frac{1}{2}\int \frac{2x}{x^2+5}dx - 4\frac{1}{\sqrt{5}}\tan^{-1}\left(\frac{x}{\sqrt{5}}\right)$$

$$= \frac{1}{2}\ln(x^2+5) - \frac{4\sqrt{5}}{5}\tan^{-1}\left(\frac{x}{\sqrt{5}}\right) + C$$

(d) Complete the square in the denominator: $\int \frac{x}{x^2-4x+5}dx = \int \frac{x}{(x-2)^2+1}dx$.
Let $u = x-2$, $du = dx$.

$$\int \frac{x}{(x-2)^2+1}dx = \int \frac{u+2}{u^2+1}du = \int \frac{u}{u^2+1}du + \int \frac{2}{u^2+1}du$$

$$= \tfrac{1}{2}\ln(u^2+1) + 2\tan^{-1}u + C = \tfrac{1}{2}\ln(x^2-4x+5) + 2\tan^{-1}(x-2) + C$$

6. Logarithmic Differentiation. Find the derivative of $y = \dfrac{x(1-x^2)^2}{(1+x^2)^{1/2}}$.

First take the natural logarithms of the absolute values of both sides:

$$\ln|y| = \ln\left|\frac{x(1-x^2)^2}{(1+x^2)^{1/2}}\right| = \ln|x(1-x^2)^2| - \ln|(1+x^2)^{1/2}|$$

$$= \ln|x| + \ln|(1-x^2)^2| - \tfrac{1}{2}\ln(1+x^2)$$

$$= \ln|x| + 2\ln|1-x^2| - \tfrac{1}{2}\ln(1+x^2)$$

Now take the derivatives of both sides:

$$\frac{1}{y}y' = \frac{1}{x} + \frac{2}{1-x^2}(-2x) - \frac{1}{2}\frac{1}{1+x^2}(2x) = \frac{1}{x} - \frac{4x}{1-x^2} - \frac{x}{1+x^2}$$

$$y' = y\left(\frac{1}{x} - \frac{4x}{1-x^2} - \frac{x}{1+x^2}\right) = \frac{x(1-x^2)^2}{(1+x^2)^{1/2}}\left(\frac{1}{x} - \frac{4x}{1-x^2} - \frac{x}{1+x^2}\right)$$

7. Show that $1 - \dfrac{1}{x} \le \ln x \le x - 1$ for $x > 0$. (When $x \ne 1$, the strict inequalities hold.)

When $x > 1$, $1/t$ is a decreasing function on $[1, x]$ and so its minimum on $[1, x]$ is $1/x$ and its maximum is 1. So, by Problems 3(c) and 15 of Chapter 23,

$$\frac{1}{x}(x-1) < \ln x = \int_1^x \frac{1}{t}dt < x-1 \qquad \text{and so} \qquad 1 - \frac{1}{x} < \ln x < x - 1.$$

For $0 < x < 1$, $-\frac{1}{t}$ is increasing on $[x, 1]$. Then, by Problems 3(*c*) and 15 of Chapter 23,

$$-\frac{1}{x}(1-x) < \ln x = \int_1^x \frac{1}{t}\,dt = \int_x^1 \left(-\frac{1}{t}\right) dt < -1(1-x)$$

Hence, $1 - \frac{1}{x} < \ln x < x - 1$. When $x = 1$, the three terms are all equal to 0.

SUPPLEMENTARY PROBLEMS

8. Find the derivatives of the following functions.

(a) $y = \ln (x+3)^2 = 2\ln (x+3)$.

Ans. $y' = \frac{2}{x+3}$

(b) $y = (\ln (x+3))^2$

Ans. $y' = 2\ln (x+3)\frac{1}{x+3} = \frac{2\ln (x+3)}{x+3}$

(c) $y = \ln [(x^3+2)(x^2+3)] = \ln (x^3+2) + \ln (x^2+3)$

Ans. $y' = \frac{1}{x^3+2}(3x^2) + \frac{1}{x^2+3}(2x) = \frac{3x^2}{x^3+2} + \frac{2x}{x^2+3}$

(d) $y = \ln \frac{x^4}{(3x-4)^2} = \ln x^4 - \ln (3x-4)^2 = 4\ln x - 2\ln (3x-4)$

Ans. $y' = \frac{4}{x} - \frac{2}{3x-4}(3) = \frac{4}{x} - \frac{6}{3x-4}$

(e) $y = \ln \sin 5x$

Ans. $y' = \frac{1}{\sin 5x}\cos(5x)(5) = 5\cot 5x$

(f) $y = \ln (x + \sqrt{1+x^2})$

Ans. $y' = \frac{1+\frac{1}{2}(1+x^2)^{-1/2}(2x)}{x+(1+x^2)^{1/2}} = \frac{1+x(1+x^2)^{-1/2}}{x+(1+x^2)^{1/2}}\frac{(1+x^2)^{1/2}}{(1+x^2)^{1/2}} = \frac{1}{\sqrt{1+x^2}}$

(g) $y = \ln \sqrt{3-x^2} = \ln (3-x^2)^{1/2} = \frac{1}{2}\ln (3-x^2)$

Ans. $y' = \frac{1}{2}\frac{1}{3-x^2}(-2x) = -\frac{x}{3-x^2}$

(h) $y = x\ln x - x$

Ans. $y' = \ln x$

(i) $y = \ln (\ln (\tan x))$

Ans. $y' = \frac{\tan x + \cot x}{\ln (\tan x)}$

9. Find the following antiderivatives. Use Quick Formula II when possible.

(a) $\int \frac{1}{7x} dx$

Ans. $\frac{1}{7} \ln |x| + C$

(b) $\int \frac{x^8}{x^9 - 1} dx$

Ans. $\frac{1}{9} \ln |x^9 - 1| + C$

(c) $\int \frac{\sqrt{\ln x + 3}}{x} dx$

Ans. Use Quick Formula I. $\frac{2}{3}(\ln x + 3)^{3/2} + C$

(d) $\int \frac{dx}{x \ln x}$

Ans. $\ln |\ln x| + C$

(e) $\int \frac{\sin 3x}{1 - \cos 3x} dx$

Ans. $\frac{1}{3} \ln |1 - \cos 3x| + C$

(f) $\int \frac{2x^4 - x^2}{x^3} dx$

Ans. $x^2 - \ln |x| + C$

(g) $\int \frac{\ln x}{x} dx$

Ans. $\frac{1}{2}(\ln x)^2 + C$

(h) $\int \frac{dx}{\sqrt{x}(1 - \sqrt{x})}$

Ans. $-2 \ln |1 - \sqrt{x}| + C$

10. Use logarithmic differentiation to calculate y'.

(a) $y = x^4 \sqrt{2 - x^2}$

Ans. $y' = x^4 \sqrt{2 - x^2} \left(\frac{4}{x} - \frac{x}{2 - x^2} \right) = 4x^3 \sqrt{2 - x^2} - \frac{x^5}{\sqrt{2 - x^2}}$

(b) $y = \frac{(x-1)^5 \sqrt[4]{x+2}}{\sqrt{x^2 + 7}}$

Ans. $y' = y \left(\frac{5}{x-1} + \frac{1}{4} \frac{1}{x+2} - \frac{x}{x^2 + 1} \right)$

(c) $y = \dfrac{\sqrt{x^2+3}\cos x}{(3x-5)^3}$

Ans. $y' = y\left(\dfrac{x}{x^2+3} - \tan x - \dfrac{1}{3x-5}\right)$

(d) $y = \sqrt[4]{\dfrac{2x+3}{2x-3}}$

Ans. $y' = -\dfrac{3y}{4x^2-9}$

11. Express in terms of ln 2 and ln 3: (a) $\ln(3^7)$; (b) $\ln\dfrac{2}{27}$.

Ans. (a) 7 ln 3; (b) ln 2–3 ln 3

12. Express in terms of ln 2 and ln 5: (a) ln 50; (b) $\ln\frac{1}{4}$; (c) $\ln\sqrt{5}$; (d) $\ln\frac{1}{40}$.

Ans. (a) ln 2 + 2 ln 5; (b) – 2 ln 2; (c) $\frac{1}{2}$ ln 5; (d) – (3 ln 2 + ln 5)

13. Find the area under the curve $y = \dfrac{1}{x}$ and above the x axis, between $x = 2$ and $x = 4$.

Ans. ln 2

14. Find the average value of $\dfrac{1}{x}$ on [3, 5].

Ans. $\frac{1}{2}\ln\frac{5}{3}$

15. Use implicit differentiation to find y': (a) $y^3 = \ln(x^3 + y^3)$; (b) $3y - 2x = 1 + \ln xy$.

Ans. (a) $y' = \dfrac{x^2}{y^2(x^3+y^3-1)}$; (b) $y' = \dfrac{y^2x+1}{x^3y-1}$

16. Evaluate $\lim\limits_{h\to 0}\dfrac{1}{h}\ln\dfrac{2+h}{2}$.

Ans. $\frac{1}{2}$

17. Check the formula $\int \csc x\,dx = \ln|\csc x - \cot x| + C$.

18. (GC) Approximate $\ln 2 = \int_1^2 \frac{1}{t}\,dt$ to six decimal places by (a) the trapezoidal rule; (b) the midpoint rule; (c) Simpson's rule, in each case with $n = 10$.

Ans. (a) 0.693771; (b) 0.692835; (c) 0.693147

19. (GC) Use Newton's method to approximate the root of $x^2 + \ln x = 2$ to four decimal places.

Ans. 1.3141

CHAPTER 26

Exponential and Logarithmic Functions

From Chapter 25, we know that the natural logarithm $\ln x$ is an increasing differentiable function with domain the set of all positive real numbers and range the set of all real numbers. Since it is increasing, it is a one-to-one function and, therefore, has an inverse function, which we shall denote by e^x.

Definition

e^x is the inverse of $\ln x$.

It follows that the domain of e^x is the set of all real numbers and its range is the set of all positive real numbers. Since e^x is the inverse of $\ln x$, the graph of e^x can be obtained from that of $\ln x$ by reflection in the line $y = x$. See Fig. 26-1.

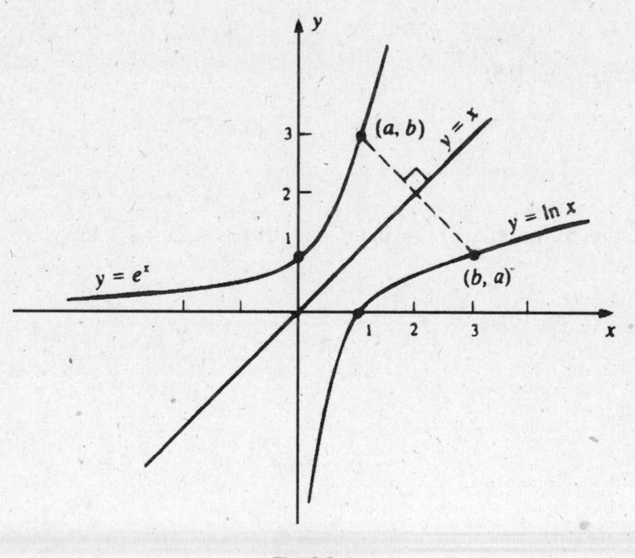

Fig 26-1

Our notation may be confusing. It should not be assumed from the notation that e^x is an ordinary power of base e with exponent x. Later in this chapter, we will find out that this is indeed true, but we do not know it yet.

Properties of e^x

(26.1) $e^x > 0$ for all x

The range of e^x is the set of positive real numbers.

(26.2) $\ln (e^x) = x$

(26.3) $e^{\ln x} = x$

Properties (26.2) and (26.3) follow from the fact that e^x and $\ln x$ are inverses of each other.

(26.4) e^x is an increasing function.
Assume $u < v$. Since $u = \ln(e^u)$ and $v = \ln(e^v)$, $\ln(e^u) < \ln(e^v)$. But, since $\ln x$ is increasing, $e^u < e^v$. [For, if $e^v \le e^u$, then $\ln(e^v) \le \ln(e^u)$.]

(26.5) $D_x(e^x) = e^x$
Let $y = e^x$. Then $\ln y = x$. By implicit differentiation, $\frac{1}{y}y' = 1$ and, therefore, $y' = y = e^x$. For a more rigorous argument, let $f(x) = \ln x$ and $f^{-1}(y) = e^y$. Note that $f'(x) = \frac{1}{x}$. By Theorem 10.2(b),

$$(f^{-1})'(y) = \frac{1}{f'(f^{-1}(y))}, \quad \text{that is,} \quad D_y(e^y) = \frac{1}{1/e^y} = e^y$$

EXAMPLE 26.1: $D_x(e^{\sin x}) = D_u(e^u)D_x(u)$ (Chain Rule, with $u = \sin x$)

$$= e^u(\cos x) = e^{\sin x}(\cos x)$$

(26.6) $\int e^x\,dx = e^x + C$

EXAMPLE 26.2: To find $\int xe^{x^2}\,dx$, let $u = x^2$, $du = 2x\,dx$. Then

$$\int xe^{x^2}\,dx = \tfrac{1}{2}\int e^u\,du = \tfrac{1}{2}e^u + C = \tfrac{1}{2}e^{x^2} + C$$

(26.7) $\int e^{-x}dx = -e^{-x} + C$
Let $u = -x$, $du = -dx$. Then $\int e^{-x}dx = -\int e^u du = -e^u + C = -e^{-x} + C$.

(26.8) $e^0 = 1$
By (26.3), $1 = e^{\ln 1} = e^0$.

(26.9) $e^{u+v} = e^u e^v$
$\ln(e^{u+v}) = u + v = \ln(e^u) + \ln(e^v) = \ln(e^u e^v)$ by (25.6). Hence, $e^{u+v} = e^u e^v$ because $\ln x$ is a one-to-one function.

(26.10) $e^{u-v} = \dfrac{e^u}{e^v}$
By (26.9), $e^{u-v}e^v = e^{(u-v)+v} = e^u$. Now divide by e^v.

(26.11) $e^{-v} = \dfrac{1}{e^v}$
Replace u by 0 in (26.10) and use (26.8).

(26.12) $x < e^x$ for all x
By Problem 7 of Chapter 25, $\ln x \le x - 1 < x$. By (26.3) and (26.4), $x = e^{\ln x} < e^x$.

(26.13) $\lim_{x\to+\infty} e^x = +\infty$
This follows from (26.4) and (26.12).

(26.14) $\lim_{x\to-\infty} e^x = 0$
Let $u = -x$. As $x \to \infty$, $u \to +\infty$ and, by (26.13), $e^u \to +\infty$. Then, by (26.11), $e^x = e^{-u} = \dfrac{1}{e^u} \to 0$.

The mystery of the letter e in the expression e^x can now be cleared up.

Definition

Let e be the number such that $\ln e = 1$.

Since $\ln x$ is a one-to-one function from the set of positive real numbers onto the set of all real numbers, there must be exactly one number x such that $\ln x = 1$. That number is designated e.

Since, by (25.12), $\ln 2 < 1 < 2\ln 2 = \ln 4$, we know that $2 < e < 4$.

(26.15) (GC) $e \sim 2.718281828$
This estimate can be obtained from a graphing calculator. Later we will find out how to approximate e to any degree of accuracy.

Now we can show that the notation e^x is not misleading, that is, that e^x actually is a power of e. First of all, this can be proved for positive integers x by mathematical induction. [In fact, by (26.3), $e = e^{\ln e} = e^1$. So, by (26.9), $e^{n+1} = e^n e^1 = e^n e$ for any positive integer n and therefore, if we assume by inductive hypothesis that e^n represents the produce of e by itself n times, then e^{n+1} is the product of e by itself $n + 1$ times.] By (26.8) $e^0 = 1$, which corresponds to the standard definition of e^0. If n is a positive integer, e^{-n} would ordinarily be defined by $1/e^n$ and this is identical to the function value given by (26.11). If k and n are positive integers, then the power $e^{k/n}$ is ordinarily defined as $\sqrt[n]{e^k}$. Now, in fact, by (26.9), the product $e^{k/n}e^{k/n}\ldots e^{k/n}$, where there are n factors, is equal to $e^{k/n+k/n+\cdots+k/n} = e^k$. Thus, the function value $e^{k/n}$ is identical to the nth root of e^k. For negative fractions, we again apply (26.11) to see that the function value is identical to the value specified by the usual definition. Hence, the function value e^x is the usual power of e when x is any rational number. Since our function e^x is continuous, the value of e^x when x is irrational is the desired limit of e^r for rational numbers r approaching x.

The graph of $y = e^x$ is shown in Fig. 26-2. By (26.13), the graph rises without bound on the right and, by (26.14), the negative x axis is a horizontal asymptote on the left. Since $D_x^2(e^x) = D_x(e^x) = e^x > 0$, the graph is concave upward everywhere. The graph of $y = e^{-x}$ is also shown in Fig. 26-2. It is obtained from the graph of $y = e^x$ by reflection in the y axis.

(26.16) $e^x = \lim_{n\to+\infty}\left(1+\frac{x}{n}\right)^n$

For a proof, see Problem 5.

(26.17) $e = \lim_{n\to+\infty}\left(1+\frac{1}{n}\right)^n$

This is a special case of (26.16) when $x = 1$. We can use this formula to approximate e, although the convergence to e is rather slow. For example, when $n = 100$, we get 2.7169 and, when $n = 10\,000$, we get 2.7181, which is correct only to three decimal places.

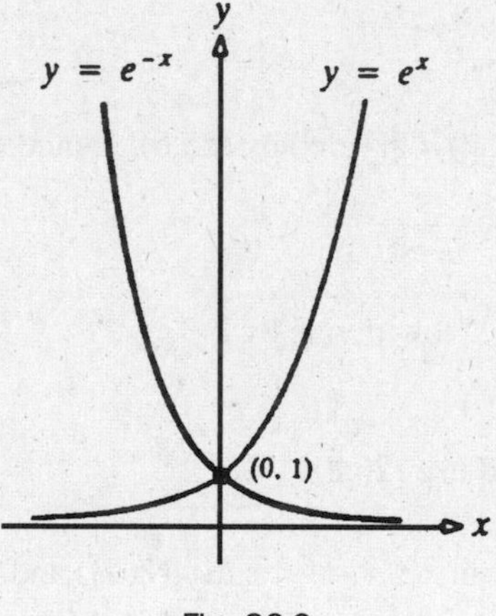

Fig. 26-2

The General Exponential Function

Let $a > 0$. Then we can define a^x as follows:

Definition

$a^x = e^{x\ln a}$

Note that this is consistent with the definition of e^u since, when $u = e$, $\ln u = 1$.

(26.18) $D_x(a^x) = (\ln a)a^x$

In fact,

$$D_x(e^{x\ln a}) = D_u(e^u)D_x u \qquad \text{(chain rule with } u = x\ln a)$$

$$= e^u(\ln a) = e^{x\ln a}(\ln a) = a^x(\ln a)$$

EXAMPLE 26.3: $D_x(2^x) = (\ln 2)2^x$.

(26.19) $\int a^x dx = \frac{1}{\ln a} a^x + C$

This is a direct consequence of (26.18).

EXAMPLE 26.4: $\int 10^x = \frac{1}{\ln 10} 10^x + C$

We can derive the usual properties of powers.

(26.20) $a^0 = 1$

$a^0 = e^{0 \ln a} = e^0 = 1$

(26.21) $a^{u+v} = a^u a^v$

$a^{u+v} = e^{(u+v)\ln a} = e^{u \ln a + v \ln a} = e^{u \ln a} e^{v \ln a} = a^u a^v$

(26.22) $a^{u-v} = \frac{a^u}{a^v}$

By (26.21), $a^{u-v}a^v = a^{(u-v)+v} = a^u$. Now divide by a^v.

(26.23) $a^{-v} = \frac{1}{a^v}$

Replace u by 0 in (26.22) and use (26.20).

(26.24) $a^{uv} = (a^u)^v$

$(a^u)^v = e^{v \ln(a^u)} = e^{v(u(\ln a))} = e^{(uv)\ln a} = a^{uv}$

(26.25) $(ab)^u = a^u b^u$

$a^u b^u = e^{u \ln a} e^{u \ln b} = e^{u \ln a + u \ln b} = e^{u(\ln a + \ln b)} = e^{u \ln(ab)} = (ab)^u$

Recall that we know that $D_x(x^r) = rx^{r-1}$ for rational numbers r. Now we are able to prove that formula for any real number r.

(26.26) $D_x(x^r) = rx^{r-1}$

Since $x^r = e^{r \ln x}$,

$$D_x(x^r) = D_x(e^{r \ln x}) = D_u(e^u)D_x(u) \quad \text{(Chain Rule with } u = r \ln x)$$

$$= e^u\left(r\left(\frac{1}{x}\right)\right) = r(x^r)\left(\frac{1}{x}\right) = r\frac{x^r}{x^1} = rx^{r-1}$$

General Logarithmic Functions

Let $a > 0$. We want to define a function $\log_a x$ that plays the role of the traditional logarithm to the base a. If $y = \log_a x$, then $a^y = x$ and, therefore, $\ln(a^y) = \ln x$, $y \ln a = \ln x$, $y = \frac{\ln x}{\ln a}$.

Definition

$$\log_a x = \frac{\ln x}{\ln a}.$$

(26.27) $y = \log_a x$ is equivalent to $a^y = x$

$$y = \log_a x \Leftrightarrow y = \frac{\ln x}{\ln a} \Leftrightarrow y \ln a = \ln x$$

$\Leftrightarrow \ln(a^y) = \ln x \Leftrightarrow a^y = x$ (The symbol $\Leftrightarrow$ is the symbol for equivalence, that is, *if and only if*.)

Thus, the general logarithmic function with base a is the inverse of the general exponential function with base a.

(26.28) $a^{\log_a x} = x$

(26.29) $\log_a(a^x) = x$

These follow from (26.27). See Problem 6.

The usual properties of logarithm can easily be derived. See Problem 7.

Notice that $\log_e x = \frac{\ln x}{\ln e} = \frac{\ln x}{1} = \ln x$. Thus, the natural logarithm turns out to be a logarithm in the usual sense, with base e.

SOLVED PROBLEMS

1. Evaluate: *(a)* $\ln(e^3)$; *(b)* $e^{7\ln 2}$; *(c)* $e^{(\ln 3)-2}$; *(e)* 1^u.

(a) $\ln(e^3) = 3$ by (26.2)

(b) $e^{7\ln 2} = (e^{\ln 2})^7 = 2^7 = 128$ by (26.24) and (26.3)

(c) $e^{(\ln 3)-2} = \frac{e^{\ln 3}}{e^2} = \frac{3}{e^2}$ by (26.10)

(d) $1^u = e^{u\ln 1} = e^{u(0)} = e^0 = 1$ by (26.8)

2. Find the derivatives of: (a) e^{3x+1}; (b)5^{3x}; (c)$3x^{\pi}$; (d) x^2e^x.

(a) $D_x(e^{3x+1}) = e^{3x+1}(3) = 3e^{3x+1}$ by the Chain Rule

(b) $D_x(5^{3x}) = D_u(5^u)D(u)$ (chain rule with $u = 3x$)

$= (\ln 5)5^u(3)$ by (26.18)

$= 3(\ln 5)5^{3x}$

(c) $D_x(3x^{\pi}) = 3(\pi x^{\pi-1}) = 3\pi x^{\pi-1}$ by (26.26)

(d) $D_x(x^2e^x) = x^2D_x(e^x) + e^xD_x(x^2)$ by the product rule

$= x^2e^x + e^x(2x) = xe^x(x+2)$

3. Find the following antiderivative: (a) $\int 3(2^x)\,dx$; (b) $\int x^2e^{x^3}\,dx$.

(a) $\int 3(2^x)\,dx = 3\int 2^x\,dx = 3\frac{1}{\ln 2}2^x + C = \frac{3}{\ln 2}2^x + C$

(b) Let $u = x^3$, $du = 3x^2\,dx$. Then $\int x^2e^{x^3}dx = \frac{1}{3}\int e^u du = \frac{1}{3}e^u + C = \frac{1}{3}e^{x^3} + C$

4. Solve the following equations for x: (a) $\ln x^3 = 2$; (b) $\ln(\ln x) = 0$; (c) $e^{2x-1} = 3$; (d) $e^x - 3e^{-x} = 2$.

In general, $\ln A = B$ is equivalent to $A = e^B$, and $e^C = D$ is equivalent to $C = \ln D$.

(a) $\ln x^3 = 3\ln x$. Hence, $\ln x^3 = 2$ yields $3\ln x = 2$, $\ln x = \frac{2}{3}$, $x = e^{2/3}$.

(b) $\ln(\ln x) = 0$ is equivalent to $\ln x = e^0 = 1$, which, in turn, is equivalent to $x = e^1 = e$.

(c) $e^{2x-1} = 3$ is equivalent to $2x - 1 = \ln 3$, and then to $x = \frac{\ln 3 + 1}{2}$.

(d) Multiply both sides by e^x: $e^{2x} - 3 = 2e^x$, $e^{2x} - 2e^x - 3 = 0$. Letting $u = e^x$ yields the quadratic equation $u^2 - 2u - 3 = 0$; $(u - 3)(u + 1) = 0$, with solutions $u = 3$ and $u = -1$. Hence, $e^x = 3$ or $e^x = -1$. The latter is impossible since e^x is always positive. Hence, $e^x = 3$ and, therefore, $x = \ln 3$.

5. Prove (26.16): $e^u = \lim_{n\to+\infty}\left(1+\frac{u}{n}\right)^n$.

Let $a_n = \left(1+\frac{u}{n}\right)^n$. Then

$$\ln a_n = n\ln\left(1+\frac{u}{n}\right) = u\left(\frac{\ln(1+u/n) - \ln 1}{u/n}\right)$$

The expression $\left(\frac{\ln(1+u/n) - \ln 1}{u/n}\right)$ is a difference quotient for $D_x(\ln x)$ at $x = 1$, with $\Delta x = u/n$. As $n \to +\infty$, $u/n \to 0$. So, that difference quotient approaches $D_x(\ln x)\big|_{x=1} = (1/x)\big|_{x=1} = 1$. Hence, $\lim_{n\to+\infty}\ln a_n = u(1) = u$. So, $\lim_{n\to+\infty}a_n = \lim_{n\to+\infty}e^{\ln a_n} = e^u$.

6. Prove (26.28) $a^{\log_a x} = x$ and (26.29) $\log_a (a^x) = x$.
Substituting $\log_a x$ for y in (26.27), we get $a^{\log_a x} = x$.
Substituting a^y for x in (26.27), we get $y = \log_a (a^y)$.

7. Derive the following properties of $\log_a x$:

(a) $\log_a 1 = 0$.

$$\log_a 1 = \frac{\ln 1}{\ln a} = \frac{0}{\ln a} = 0$$

(b) $\log_a a = 1$.

$$\log_a a = \frac{\ln a}{\ln a} = 1$$

(c) $\log_a uv = \log_a u + \log_a v$.

$$\log_a uv = \frac{\ln uv}{\ln a} = \frac{\ln u + \ln v}{\ln a} = \frac{\ln u}{\ln a} + \frac{\ln v}{\ln a} = \log_a u + \log_a v$$

(d) $\log_a \frac{u}{v} = \log_a u - \log_a v$.

Replace u in (c) by $\frac{u}{v}$.

(e) $\log_a \frac{1}{v} = -\log_a v$.

Replace u by 1 in (d).

(f) $\log_a (u^r) = r \log_a u$.

$$\log_a (u^r) = \frac{\ln(u^r)}{\ln a} = \frac{r \ln u}{\ln a} = r \log_a u$$

(g) $D_x(\log_a x) = \frac{1}{\ln a}\frac{1}{x}$.

$$D_x(\log_a x) = D_x\left(\frac{\ln x}{\ln a}\right) = \frac{1}{\ln a} D_x(\ln x) = \frac{1}{\ln a}\frac{1}{x}$$

SUPPLEMENTARY PROBLEMS

8. Calculate the derivatives of the following functions:

(a) $y = e^{5x}$ *Ans.* $y' = 5e^{5x}$

(b) $y = e^{\tan 3x}$ *Ans.* $y' = 3\sec^2(3x)e^{\tan 3x}$

(c) $y = e^{-x\cos x}$ *Ans.* $y' = -e^{-x}(\cos x + \sin x)$

(d) $y = 3^{-x^2}$ *Ans.* $y' = -2x(\ln 3)3^{-x^2}$

(e) $y = \sin^{-1}(e^x)$ *Ans.* $y' = \dfrac{e^x}{\sqrt{1-e^{2x}}}$

(f) $y = e^{e^x}$ *Ans.* $y' = e^{x+e^x}$

(g) $y = x^x$ *Ans.* $y' = x^x(1 + \ln x)$

(h) $y = \log_{10}(3x^2 - 5)$ *Ans.* $y' = \dfrac{1}{\ln 10}\dfrac{6x}{3x^2 - 5}$

9. Find the following antiderivatives:

(a) $\int 3^{2x}\,dx$ *Ans.* $\dfrac{1}{2\ln 3}3^{2x} + C$

(b) $\int \dfrac{e^{1/x}}{x^2}\,dx$ *Ans.* $-e^{1/x} + C$

(c) $\int (e^x + 1)^3 e^x\,dx$ *Ans.* $\dfrac{(e^x + 1)^4}{4} + C$

(d) $\int \dfrac{dx}{e^x + 1}$ *Ans.* $x - \ln(e^x + 1) +$

(e) $\int \dfrac{e^{1/x^2}}{x^3}\,dx$ *Ans.* $-\frac{1}{2}e^{1/x^2} + C$

(f) $\int e^{-x^2+2}\,x\,dx$ *Ans.* $-\dfrac{1}{2}e^{-x^2+2} + C$

(g) $\int (e^x+1)^2\,dx$ *Ans.* $\frac{1}{2}e^{2x}+2e^x+x+C$

(h) $\int (e^x-x^e)\,dx$ *Ans.* $e^x-\frac{x^{e+1}}{e+1}+C$

(i) $\int \frac{e^{2x}}{e^{2x}+3}\,dx$ *Ans.* $\frac{1}{2}\ln(e^{2x}+3)+C$

(j) $\int \frac{e^x\,dx}{\sqrt{1-e^{2x}}}$ *Ans.* $\sin^{-1}(e^x)+C$

(k) $\int x^3(5^{x^4+1})\,dx$ *Ans.* $\frac{1}{4\ln 5}5^{x^4+1}+C$

(l) $\int \frac{\log_{10} x}{x}\,dx$ *Ans.* $\frac{1}{2\ln 10}(\ln x)^2+C=\frac{\ln 10}{2}(\log_{10} x)^2+C$

10. (**Hyperbolic Functions**) Define

$$\sinh x=\frac{e^x-e^{-x}}{2},\quad \cosh x=\frac{e^x+e^{-x}}{2},\quad \tanh x=\frac{\sinh x}{\cosh x},\quad \operatorname{sec}h=\frac{1}{\cosh x}$$

Derive the following results:

(a) $D_x(\sinh x)=\cosh x$ and $D_x(\cosh x)=\sinh x$.
(b) $D_x(\tanh x)=\operatorname{sech}^2 x$ and $D_x(\operatorname{sech} x)=-\operatorname{sech} x\tanh x$.
(c) $\cosh^2 x-\sinh^2 x=1$.
(d) $\sinh(x+y)=\sinh x\cosh y+\cosh x\sinh y$.
(e) $\cosh(x+y)=\cosh x\cosh y+\sinh x\sinh y$.
(f) $\sinh 2x=2\sinh x\cosh x$.
(g) $\cosh 2x=\cosh^2 x+\sinh^2 x=2\cosh^2 x-1=2\sinh^2 x+1$.
(h) (GC) Sketch the graph of $y=2\cosh(x/2)$ (called a "catenary"), and find its minimum point.

Ans. (0, 2)

11. Solve the following equations for x.

(a) $e^{3x}=2$ *Ans.* $\frac{1}{3}\ln 2$
(b) $\ln(x^4)=-1$ *Ans.* $e^{-1/4}$
(c) $\ln(\ln x)=2$ *Ans.* e^{e^2}
(d) $e^x-4e^{-x}=3$ *Ans.* $2\ln 2$
(e) $e^x+12e^{-x}=7$ *Ans.* $2\ln 2$ and $\ln 3$
(f) $5^x=7$ *Ans.* $\frac{\ln 7}{\ln 5}=\log_5 7$
(g) $\log_2(x+3)=5$ *Ans.* 29
(h) $\log_2 x^2+\log_2 x=4$ *Ans.* $\sqrt[3]{16}$
(i) $\log_2(2^{4x})=20$ *Ans.* 5
(j) $e^{-2x}-7e^{-x}=8$ *Ans.* $-3\ln 2$
(k) $x^x=x^3$ *Ans.* 1 and 3

12. Evaluate (a) $\lim_{h\to 0}\frac{e^h-1}{h}$; (b) $\lim_{h\to 0}\frac{e^{h^2}-1}{h}$.

Ans. (a) 1; (b) 0

13. Evaluate: (a) $\int_0^{\ln 2}\frac{e^x}{e^x+2}\,dx$; (b) $\int_1^e \frac{2+\ln x}{x}\,dx$

Ans. (a) $\ln\frac{4}{3}$; (b) $\frac{5}{2}$

14. (GC) Use Newton's method to approximate (to four decimal places) a solution of $e^x=\frac{1}{x}$.

Ans. 0.5671

15. (GC) Use Simpson's rule with $n = 4$ to approximate $\int_0^1 e^{-x^2/2}\,dx$ to four decimal places.

Ans. 0.8556

16. If interest is paid at r percent per year and is compounded n times per year, then P dollars become $P\left(1+\frac{r}{100n}\right)^n$ dollars after 1 year. If $n \to +\infty$, then the interest is said to be *compounded continuously.*

(a) If compounded continuously at r percent per year, show that P dollars becomes $Pe^{r/100}$ dollars after 1 year, and $Pe^{rt/100}$ dollars after t years.
(b) At r percent compounded continuously, how many years does it take for a given amount of money to double?
(c) (GC) Estimate to two decimal places how many years it would take to double a given amount of money compounded continuously at 6% per year?
(d) (GC) Compare the result of compounding continuously at 5% with that obtained by compounding once a year.

Ans. (b) $\frac{100(\ln 2)}{r} \sim \frac{69.31}{r}$; (c) about 11.55 years;
(d) After 1 year, \$1 becomes \$1.05 when compounded once a year, and about \$1.0512 when compounded continuously.

17. Find $(\log_{10} e) \cdot \ln 10$.

Ans. 1

18. Write as a single logarithm with base a: $3 \log_a 2 + \log_a 40 - \log_a 16$

Ans. $\log_a 20$

19. (GC) Estimate $\log_2 7$ to eight decimal places.

Ans. 2.80735492

20. Show that $\log_b x = (\log_a x)(\log_b a)$.

21. (GC) Graph $y = e^{-x^2/2}$. Indicate absolute extrema, inflection points, asymptotes, and any symmetry.

Ans. Absolute maximum at (0, 1), inflection points at $x = \pm 1$, x axis is a horizontal asymptote on the left and right, symmetric with respect to the y axis.

22. Given $e^{xy} - x + y^2 = 1$, find $\frac{dy}{dx}$ by implicit differentiation.

Ans. $\frac{1 - ye^{xy}}{2y + xe^{xy}}$

23. (GC) Graph $y = \sinh x = \frac{e^x - e^{-x}}{2}$.

24. Evaluate $\int \frac{e^x - e^{-x}}{e^x + e^{-x}}\,dx$.

Ans. $\ln(e^x + e^{-x}) + C$

25. Use logarithmic differentiation to find the derivative of $y = x^{3/x}$.

Ans. $\frac{3y(1 - \ln x)}{x^2}$

CHAPTER 27

L'Hôpital's Rule

Limits of the form $\lim \frac{f(x)}{g(x)}$ can be evaluated by the following theorem in the *indeterminate cases* where $f(x)$ and $g(x)$ both approach 0 or both approach $\pm\infty$.

L'Hôpital's Rule

If $f(x)$ and $g(x)$ either both approach 0 or both approach $\pm\infty$, then

$$\lim \frac{f(x)}{g(x)} = \lim \frac{f'(x)}{g'(x)}$$

Here, "lim" stands for any of

$$\lim_{x\to+\infty}, \quad \lim_{x\to-\infty}, \quad \lim_{x\to a}, \quad \lim_{x\to a^+}, \quad \lim_{x\to a^-}$$

For a sketch of the proof, see Problems 1, 11, and 12. It is assumed, in the case of the last three types of limits, that $g'(x) \neq 0$ for x sufficiently close to a, and in the case of the first two limits, that $g'(x) \neq 0$ for sufficiently large or sufficiently small values of x. (The corresponding statements about $g(x) \neq 0$ follow by Rolle's Theorem.)

EXAMPLE 27.1: Since $\ln x$ approaches $+\infty$ as x approaches $+\infty$, L'Hôpital's Rule implies that

$$\lim_{x\to+\infty} \frac{\ln x}{x} = \lim_{x\to+\infty} \frac{1/x}{1} = \lim_{x\to+\infty} \frac{1}{x} = 0$$

EXAMPLE 27.2: Since e^x approaches $+\infty$ as x approaches $+\infty$, L'Hôpital's Rule implies that

$$\lim_{x\to+\infty} \frac{x}{e^x} = \lim_{x\to+\infty} \frac{1}{e^x} = 0$$

EXAMPLE 27.3: We already know from Problem 13(*a*) of Chapter 7 that

$$\lim_{x\to+\infty} \frac{3x^2+5x-8}{7x^2-2x+1} = \frac{3}{7}$$

Since both $3x^2 + 5x - 8$ and $7x^2 - 2x + 1$ approach $+\infty$ as x approaches $+\infty$, L'Hôpital's Rule tells us that

$$\lim_{x\to+\infty} \frac{3x^2+5x-8}{7x^2-2x+1} = \lim_{x\to+\infty} \frac{6x+5}{14x-2}$$

and another application of the rule tells us that

$$\lim_{x\to+\infty} \frac{6x+5}{14x-2} = \lim_{x\to+\infty} \tfrac{6}{14} = \tfrac{6}{14} = \tfrac{3}{7}$$

EXAMPLE 27.4: Since $\tan x$ approaches 0 as x approaches 0, L'Hôpital's Rule implies that

$$\lim_{x\to 0} \frac{\tan x}{x} = \lim_{x\to 0} \frac{\sec^2 x}{1} = \lim_{x\to 0} \frac{1}{\cos^2 x} = \frac{1}{1^2} = 1$$

Indeterminate Type $0 \cdot \infty$

If $f(x)$ approaches 0 and $g(x)$ approaches $\pm\infty$, we do not know how to find $\lim f(x)g(x)$. Sometimes such a problem can be transformed into a problem to which L'Hôpital's Rule is applicable.

EXAMPLE 27.5: As x approaches 0 from the right, $\ln x$ approaches $-\infty$. So, we do not know how to find $\lim_{x\to 0^+} x \ln x$. But as x approaches 0 from the right, $1/x$ approaches $+\infty$. So, by L'Hôpital's Rule,

$$\lim_{x\to 0^+} x \ln x = \lim_{x\to 0^+} \frac{\ln x}{1/x} = \lim_{x\to 0^+} \frac{1/x}{-1/x^2} = \lim_{x\to 0^+} -x = 0$$

Indeterminate Type $\infty - \infty$

If $f(x)$ and $g(x)$ both approach ∞, we do not know what happens to $\lim(f(x) - g(x))$. Sometimes we can transform the problem into a L'Hôpital's-type problem.

EXAMPLE 27.6: $\lim_{x\to 0}\left(\csc x - \frac{1}{x}\right)$ is a problem of this kind. But,

$$\lim_{x\to 0}\left(\csc x - \frac{1}{x}\right) = \lim_{x\to 0}\left(\frac{1}{\sin x} - \frac{1}{x}\right) = \lim_{x\to 0} \frac{x - \sin x}{x \sin x}$$

Since $x - \sin x$ and $x \sin x$ both approach 0, L'Hôpital's Rule applies and we get $\lim_{x\to 0} \frac{1-\cos x}{x\cos x + \sin x}$. Here both numerator and denominator approach 0 and L'Hôpital's Rule yeilds

$$\lim_{x\to 0} \frac{\sin x}{-x\sin x + \cos x + \cos x} = \frac{0}{0+1+1} = \frac{0}{2} = 0$$

Indeterminate Types 0^0, ∞^0, and 1^∞

If lim y is of one of these types, then lim (ln y) will be of type $0 \cdot \infty$.

EXAMPLE 27.7: In $\lim_{x\to 0^+} x^{\sin x}$, $y = x^{\sin x}$ is of type 0^0 and we do not know what happens in the limit. But $\ln y = \sin x \ln x = \frac{\ln x}{\csc x}$ and $\ln x$ and $\csc x$ approach $\pm\infty$. So, by L'Hôpital's Rule,

$$\begin{aligned}\lim_{x\to 0^+} \ln y &= \lim_{x\to 0^+} \frac{1/x}{-\csc x \cot x} = \lim_{x\to 0^+} -\frac{\sin^2 x}{x\cos x} = -\lim_{x\to 0^+} \frac{\sin x}{x}\,\frac{\sin x}{\cos x} \\ &= -\lim_{x\to 0^+} \frac{\sin x}{x} \lim_{x\to 0^+} \tan x = -(1)(0) = 0\end{aligned}$$

Here, we used the fact that $\lim_{x\to 0}((\sin x)/x) = 1$ (Problem 1 of Chapter 17). Now, since $\lim_{x\to 0^+} \ln y = 0$,

$$\lim_{x\to 0^+} y = \lim_{x\to 0^+} e^{\ln y} = e^0 = 1$$

EXAMPLE 27.8: In $\lim_{x\to 0^+} |\ln x|^x$, $y = |\ln x|^x$ is of type ∞^0, and it is not clear what happens in the limit. But $\ln y = x \ln |\ln x| = \frac{\ln |\ln x|}{1/x}$ and both $\ln |\ln x|$ and $1/x$ approach $+\infty$. So L'Hôpital's Rule yields

$$\lim_{x\to 0^+} \ln y = \lim_{x\to 0^+} \left(\frac{1}{x \ln x}\right) \bigg/ \left(-\frac{1}{x^2}\right) = \lim_{x\to 0^+} -\frac{x}{\ln x} = 0,$$

since

$$\lim_{x\to 0^+} \frac{1}{\ln x} = 0. \quad \text{Hence,} \quad \lim_{x\to 0^+} y = \lim_{x\to 0^+} e^{\ln y} = e^0 = 1$$

EXAMPLE 27.9: In $\lim_{x\to 1} x^{1/(x-1)}$, $y = x^{1/(x-1)}$ is of type 1^∞ and we cannot see what happens in the limit. But $\ln y = \frac{\ln x}{x-1}$ and both the numerator and the denominator approach 0. So by L'Hôpital's Rule, we get

$$\lim_{x\to 1} \ln y = \lim_{x\to 1} \frac{1/x}{1} = 1. \quad \text{Hence,} \quad \lim_{x\to 1} y = \lim_{x\to 1} e^{\ln y} = e^1 = e$$

SOLVED PROBLEMS

1. Prove the following $\frac{0}{0}$ form of L'Hôpital's Rule. Assume $f(x)$ and $g(x)$ are differentiable and $g'(x) \neq 0$ in some open interval (a, b) and $\lim_{x\to a^+} f(x) = 0 = \lim_{x\to a^+} g(x)$. Then, if $\lim_{x\to a^+} \frac{f'(x)}{g'(x)}$ exists,

$$\lim_{x\to a^+} \frac{f(x)}{g(x)} = \lim_{x\to a^+} \frac{f'(x)}{g'(x)}$$

Since $\lim_{x\to a^+} f(x) = 0 = \lim_{x\to a^+} g(x)$, we may assume that $f(a)$ and $g(a)$ are defined and that $f(a) = g(a) = 0$. Replacing b by x in the Extended Law of the Mean (Theorem 13.5), and using the fact that $f(a) = g(a) = 0$, we obtain

$$\frac{f(x)}{g(x)} = \frac{f(x) - f(a)}{g(x) - g(a)} = \frac{f'(x_0)}{g'(x_0)}$$

for some x_0 with $a < x_0 < x$. So, $x_0 \to a^+$ as $x \to a^+$. Hence,

$$\lim_{x\to a^+} \frac{f(x)}{g(x)} = \lim_{x\to a^+} \frac{f'(x)}{g'(x)}$$

We also can obtain the $\frac{0}{0}$ form of L'Hôpital's Rule for $\lim_{x\to a^-}$ (simply let $u = -x$), and then the results for $\lim_{x\to a^-}$ and $\lim_{x\to a^+}$ yield the $\frac{0}{0}$ form of L'Hôpital's Rule $\lim_{x\to a}$.

2. We already know by Examples 1 and 2 that $\lim_{x\to +\infty} \frac{\ln x}{x} = 0$ and $\lim_{x\to +\infty} \frac{x}{e^x} = 0$. Show further that $\lim_{x\to +\infty} \frac{(\ln x)^n}{x} = 0$ and $\lim_{x\to +\infty} \frac{x^n}{e^x} = 0$ for all positive integers n.

Use mathematical induction. Assume these results for a given $n \geq 1$. By L'Hôpital's Rule,

$$\lim_{x\to +\infty} \frac{(\ln x)^{n+1}}{x} = \lim_{x\to +\infty} \frac{(n+1)(\ln x)^n (1/x)}{1} = (n+1) \lim_{x\to +\infty} \frac{(\ln x)^n}{x} = (n+1)(0) = 0$$

Likewise,

$$\lim_{x\to +\infty} \frac{x^{n+1}}{e^x} = \lim_{x\to +\infty} \frac{(n+1)x^n}{e^x} = (n+1) \lim_{x\to +\infty} \frac{x^n}{e^x} = (n+1)(0) = 0$$

3. Use L'Hôpital's Rule one or more times to evaluate the following limits. Always check that the appropriate assumptions hold.

(a) $\lim_{x\to 0} \dfrac{x+\sin 2x}{x-\sin 2x}$.

We get $\lim_{x\to 0} \dfrac{1+2\cos 2x}{1-2\cos 2x} = \dfrac{1+2(1)}{1-2(1)} = -3$.

(b) $\lim_{x\to 0^+} \dfrac{e^x-1}{x^2}$.

We get $\lim_{x\to 0^+} \dfrac{e^x}{2x} = \frac{1}{2}\lim_{x\to 0^+} \dfrac{e^x}{x} = +\infty$ by Example 2.

(c) $\lim_{x\to 0} \dfrac{e^x+e^{-x}-x^2-2}{\sin^2 x - x^2}$.

We obtain $\lim_{x\to 0} \dfrac{e^x-e^{-x}-2x}{2\sin x\cos x - 2x} = \lim_{x\to 0} \dfrac{e^x-e^{-x}-2x}{\sin 2x - 2x}$.

By repeated uses of L'Hôpital's Rule, we get

$$\lim_{x\to 0} \frac{e^x+e^{-x}-2}{2\cos 2x - 2} = \lim_{x\to 0} \frac{e^x-e^{-x}}{-4\sin 2x} =$$

$$\lim_{x\to 0} \frac{e^x+e^{-x}}{-8\cos 2x} = \frac{1+1}{-8(1)} = -\frac{2}{8} = -\frac{1}{4}$$

(d) $\lim_{x\to \pi^+} \dfrac{\sin x}{\sqrt{x-\pi}}$.

We get $\lim_{x\to \pi^+} \dfrac{\cos x}{1/[2(x-\pi)^{1/2}]} = \lim_{x\to \pi^+} 2(x-\pi)^{1/2}\cos x = 0$.

(e) $\lim_{x\to 0^+} \dfrac{\ln \sin x}{\ln \tan x}$.

One obtains $\lim_{x\to 0^+} \dfrac{(\cos x)/(\sin x)}{(\sec^2 x)/(\tan x)} = \lim_{x\to 0^+} \cos^4 x = 1$

(f) $\lim_{x\to 0} \dfrac{\cot x}{\cot 2x}$.

The direct use of L'Hôpital's Rule

$$\lim_{x\to 0} \frac{-\csc^2 x}{-2\csc^2(2x)} = \frac{1}{4}\lim_{x\to 0} \frac{2\csc^2 x(\cot x)}{(\csc^2(2x))(\cot 2x)}$$

leads us to ever more complicated limits. Instead, if we change from cot to tan, we get

$$\lim_{x\to 0} \frac{\cot x}{\cot 2x} = \lim_{x\to 0} \frac{\tan 2x}{\tan x} = \lim_{x\to 0} \frac{2\sec^2(2x)}{\sec^2 x} = 2\lim_{x\to 0} \frac{\cos^2 x}{\cos^2(2x)} = 2\frac{1}{1} = 2$$

(g) $\lim_{x\to 0^+} x^2 \ln x$.

This is of type $0\cdot\infty$. Then L'Hôspiutal's Rule can be brought in as follows:

$$\lim_{x\to 0^+} \frac{\ln x}{1/x^2} = \lim_{x\to 0^+} \frac{1/x}{-2/x^3} = \lim_{x\to 0^+} -\tfrac{1}{2}x^2 = 0$$

(h) $\lim_{x\to \pi/4} (1-\tan x)\sec 2x$.

This is of type $0\cdot\infty$. However, it is equal to

$$\lim_{x\to \pi/4} \frac{1-\tan x}{\cos 2x} = \lim_{x\to \pi/4} \frac{-\sec^2 x}{-2\sin 2x} = \frac{-2}{-2} = 1$$

$$\left(\text{Here we used the value } \cos\frac{\pi}{4} = \frac{1}{\sqrt{2}}.\right)$$

(i) $\lim_{x\to 0}\left(\dfrac{1}{x} - \dfrac{1}{e^x-1}\right)$.

This is type $\infty-\infty$. But it is equal to

$$\lim_{x\to 0} \frac{e^x-1-x}{x(e^x-1)} = \lim_{x\to 0} \frac{e^x-1}{xe^x+e^x-1} = \lim_{x\to 0} \frac{e^x}{xe^x+2e^x} = \frac{1}{0+2} = \frac{1}{2}$$

(j) $\lim_{x\to 0}(\csc x - \cot x)$.

This is of type $\infty-\infty$. But it is equal to

$$\lim_{x\to 0}\left(\frac{1}{\sin x} - \frac{\cos x}{\sin x}\right) = \lim_{x\to 0} \frac{1-\cos x}{\sin x} = \lim_{x\to 0} \frac{\sin x}{\cos x} = 0$$

(k) $\lim_{x\to (\pi/2)^-} (\tan x)^{\cos x}$.

This if of type ∞^0. Let $y = (\tan x)^{\cos x}$. Then $\ln y = (\cos x)(\ln \tan x) = \dfrac{\ln \tan x}{\sec x}$.

So

$$\lim_{x\to(\pi/2)^-} \ln y = \lim_{x\to(\pi/2)^-} \frac{\ln \tan x}{\sec x} = \lim_{x\to(\pi/2)^-} (\sec^2 x/\tan x)/(\sec x \tan x) = \lim_{x\to(\pi/2)^-} \frac{\cos x}{\sin^2 x} = \frac{0}{1} = 1$$

(l) $\lim_{x\to+\infty} \frac{\sqrt{2+x^2}}{x}$.

We get $\lim_{x\to+\infty} \frac{x}{\sqrt{2+x^2}} = \lim_{x\to+\infty} \frac{\sqrt{2+x^2}}{x}$ and we are going around in a circle. So, L'Hôpital's Rule is of no use. But,

$$\lim_{x\to+\infty} \frac{\sqrt{2+x^2}}{x} = \lim_{x\to+\infty} \sqrt{\frac{2+x^2}{x^2}} = \lim_{x\to+\infty} \sqrt{\frac{2}{x^2}+1}$$

$$= \sqrt{0+1} = 1$$

4. Criticize the following use of L'Hôpital's Rule:

$$\lim_{x\to2} \frac{x^3-x^2-x-2}{x^3-3x^2+3x-2} = \lim_{x\to2} \frac{3x^2-2x-1}{3x^2-6x+3} = \lim_{x\to2} \frac{6x-2}{6x-6} = \lim_{x\to2} \frac{6}{6} = 1$$

The second equation is an incorrect use of L'Hôpital's Rule, since $\lim_{x\to2} (3x^2-2x-1) = 7$ and $\lim_{x\to2} (3x^2-6x+3) = 3$. So, the correct limit should be $\frac{7}{3}$.

5. (GC) Sketch the graph of $y = xe^{-x} = \frac{x}{e^x}$.

See Fig 27-1. By Example 2, $\lim_{x\to+\infty} y = 0$. So, the positive x axis is a horizontal asyomptote. Since $\lim_{x\to-\infty} e^{-x} = +\infty$, $\lim_{x\to-\infty} y = -\infty$. $y' = e^{-x}(1-x)$ and $y'' = e^{-x}(x-2)$. Then $x = 1$ is a critical number. By the second derivative test, there is a relative maximum at $(1, 1/e)$ since $y'' < 0$ at $x = 0$. The graph is concave downward for $x < 2$ (where $y'' < 0$) and concave upward for $x > 2$ (where $y'' > 0$). $(2, 2/e^2)$ is an inflection point. The graphing calculator gives us the estimates $1/e \sim 0.37$ and $2/e^2 \sim 0.27$.

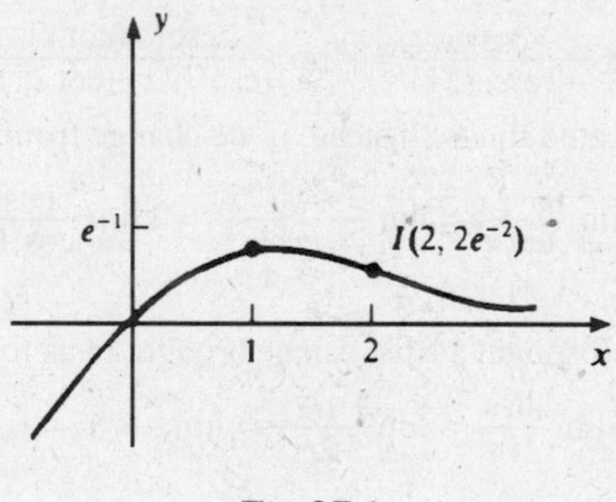

Fig. 27-1

6. (GC) Sketch the graph $y = x \ln x$.

See Fig. 27-2. The graph is defined only for $x > 0$. Clearly, $\lim_{x\to+\infty} y = +\infty$. By Example 5, $\lim_{x\to0^+} y = 0$. Since $y' = 1 + \ln x$ and $y'' = 1/x > 0$, the critical number at $x = 1/e$ (where $y' = 0$) yields, by the second derivative test, a relative minimum at $(1/e, -1/e)$. The graph is concave upward everywhere.

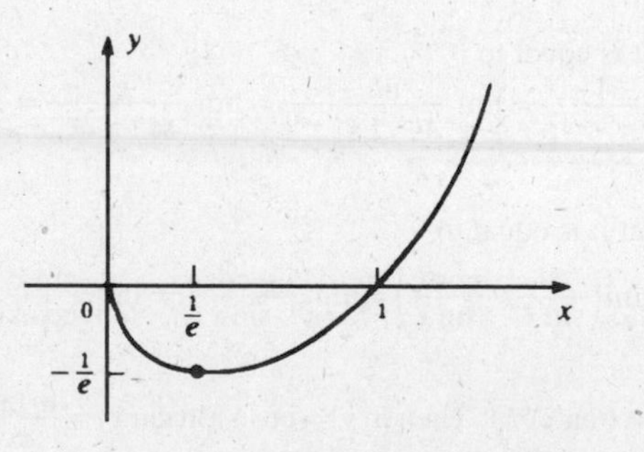

Fig. 27-2

SUPPLEMENTARY PROBLEMS

7. Show that $\lim_{x\to-\infty} x^n e^x = 0$ for all positive integers x.

8. Find $\lim_{x\to+\infty} x\sin\dfrac{\pi}{x}$.

Ans. π

9. Sketch the graphs of the following functions: (a) $y = x - \ln x$; (b) $y = \dfrac{\ln x}{x}$; (c) $y = x^2 e^x$

Ans. See Fig. 27-3.

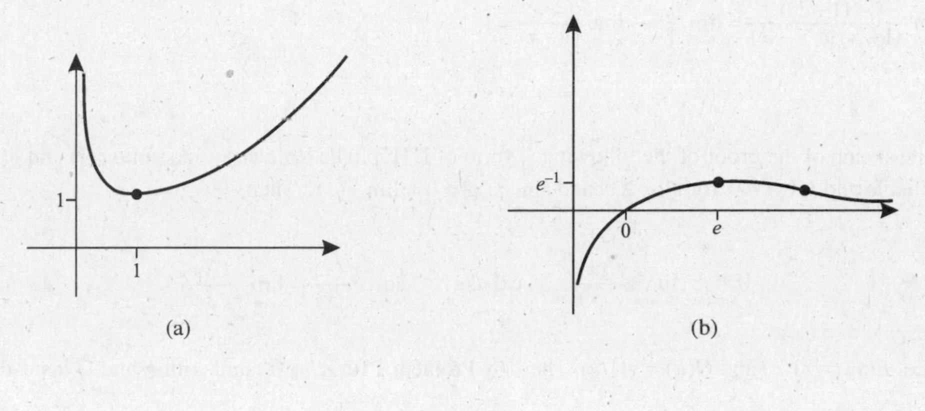

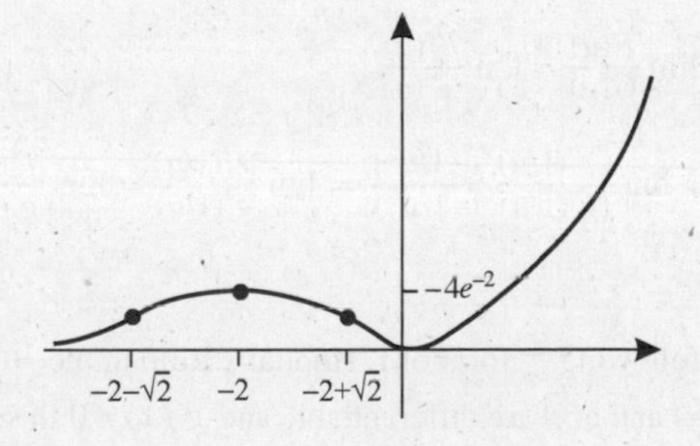

(c)

Fig. 27-3

10. Evaluate the following limits:

(a) $\lim_{x\to4} \dfrac{x^4-256}{x-4} = 256$

(b) $\lim_{x\to4} \dfrac{x^4-256}{x^2-16} = 32$

(c) $\lim_{x\to3} \dfrac{x^2-3x}{x^2-9} = \dfrac{1}{2}$

(d) $\lim_{x\to2} \dfrac{e^x-e^2}{x-2} = e^2$

(e) $\lim_{x\to0} \dfrac{xe^x}{1-e^x} = -1$

(f) $\lim_{x\to0} \dfrac{e^x-1}{\tan 2x} = \dfrac{1}{2}$

(g) $\lim_{x\to-1} \dfrac{\ln(2+x)}{x+1} = 1$

(h) $\lim_{x\to0} \dfrac{\cos x - 1}{\cos 2x - 1} = \dfrac{1}{4}$

(i) $\lim_{x\to0} \dfrac{e^{2x}-e^{-2x}}{\sin x} = 4$

(j) $\lim_{x\to0} \dfrac{8^x-2^x}{4x} = \dfrac{1}{2}\ln 2$

(k) $\lim_{x\to0} \dfrac{2\tan^{-1}x - x}{2x - \sin^{-1}x} = 1$

(l) $\lim_{x\to0} \dfrac{\ln\sec 2x}{\ln\sec x} = 4$

(m) $\lim_{x\to0} \dfrac{\ln\cos x}{x^2} = -\dfrac{1}{2}$

(n) $\lim_{x\to0} \dfrac{\cos 2x - \cos x}{\sin^2 x} = -\dfrac{3}{2}$

(o) $\lim_{x\to+\infty} \dfrac{\ln x}{\sqrt{x}} = 0$

(p) $\lim_{x\to\frac{1}{2}\pi} \dfrac{\csc 6x}{\csc 2x} = \dfrac{1}{3}$

(q) $\lim_{x\to+\infty} \dfrac{5x + 2\ln x}{x + 3\ln x} = 5$

(r) $\lim_{x\to+\infty} \dfrac{x^4+x^2}{e^x+1} = 0$

(s) $\lim_{x\to0^+} \dfrac{\ln\cot x}{e^{\csc^2 x}} = 0$

(t) $\lim_{x\to0^+} \dfrac{e^x+3x^3}{4e^x+2x^2} = \dfrac{1}{4}$

(u) $\lim_{x\to0} (e^x-1)\cos x = 1$

(v) $\lim_{x\to-\infty} x^2 e^x = 0$

(w) $\lim_{x\to0} x\csc x = 1$

(x) $\lim_{x\to1} \csc\pi x \ln x = -1/\pi$

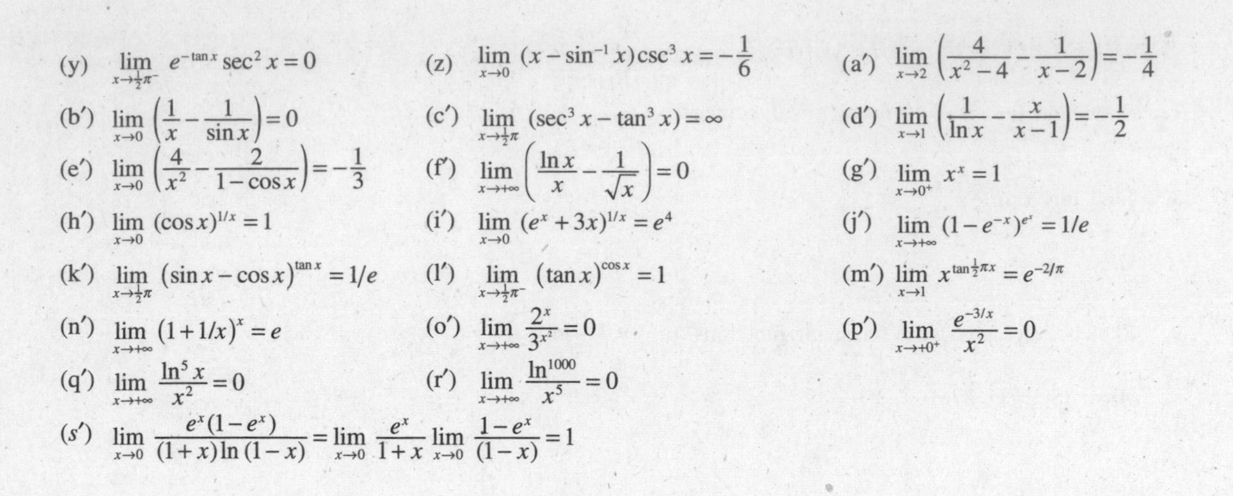

(y) $\lim\limits_{x\to\frac{1}{2}\pi^-} e^{-\tan x}\sec^2 x = 0$ (z) $\lim\limits_{x\to 0}(x-\sin^{-1}x)\csc^3 x = -\frac{1}{6}$ (a′) $\lim\limits_{x\to 2}\left(\frac{4}{x^2-4}-\frac{1}{x-2}\right) = -\frac{1}{4}$

(b′) $\lim\limits_{x\to 0}\left(\frac{1}{x}-\frac{1}{\sin x}\right) = 0$ (c′) $\lim\limits_{x\to\frac{1}{2}\pi}(\sec^3 x - \tan^3 x) = \infty$ (d′) $\lim\limits_{x\to 1}\left(\frac{1}{\ln x}-\frac{x}{x-1}\right) = -\frac{1}{2}$

(e′) $\lim\limits_{x\to 0}\left(\frac{4}{x^2}-\frac{2}{1-\cos x}\right) = -\frac{1}{3}$ (f′) $\lim\limits_{x\to+\infty}\left(\frac{\ln x}{x}-\frac{1}{\sqrt{x}}\right) = 0$ (g′) $\lim\limits_{x\to 0^+} x^x = 1$

(h′) $\lim\limits_{x\to 0}(\cos x)^{1/x} = 1$ (i′) $\lim\limits_{x\to 0}(e^x+3x)^{1/x} = e^4$ (j′) $\lim\limits_{x\to+\infty}(1-e^{-x})^{e^x} = 1/e$

(k′) $\lim\limits_{x\to\frac{1}{2}\pi}(\sin x - \cos x)^{\tan x} = 1/e$ (l′) $\lim\limits_{x\to\frac{1}{2}\pi^-}(\tan x)^{\cos x} = 1$ (m′) $\lim\limits_{x\to 1} x^{\tan\frac{1}{2}\pi x} = e^{-2/\pi}$

(n′) $\lim\limits_{x\to+\infty}(1+1/x)^x = e$ (o′) $\lim\limits_{x\to+\infty}\frac{2^x}{3^{x^2}} = 0$ (p′) $\lim\limits_{x\to+0^+}\frac{e^{-3/x}}{x^2} = 0$

(q′) $\lim\limits_{x\to+\infty}\frac{\ln^5 x}{x^2} = 0$ (r′) $\lim\limits_{x\to+\infty}\frac{\ln^{1000}}{x^5} = 0$

(s′) $\lim\limits_{x\to 0}\frac{e^x(1-e^x)}{(1+x)\ln(1-x)} = \lim\limits_{x\to 0}\frac{e^x}{1+x}\lim\limits_{x\to 0}\frac{1-e^x}{(1-x)} = 1$

11. Verify the sketch of the proof of the following $\frac{0}{0}$ form of L'Hôpital's Rule at $+\infty$. Assume $f(x)$ and $g(x)$ are differentiable and $g'(x)\neq 0$ for all $x \geq c$, and $\lim\limits_{x\to+\infty} f(x) = 0 = \lim\limits_{x\to+\infty} g(x)$. Then,

$$\text{if} \quad \lim_{x\to+\infty}\frac{f'(x)}{g'(x)} \quad \text{exists,} \quad \lim_{x\to+\infty}\frac{f(x)}{g(x)} = \lim_{x\to+\infty}\frac{f'(x)}{g'(x)}$$

Proof: Let $F(u) = f(1/u)$ and $G(u) = g(1/u)$. Then, by Problem 1 for $a \to 0^+$, and with F and G instead of f and g,

$$\lim_{x\to+\infty}\frac{f(x)}{g(x)} = \lim_{u\to 0^+}\frac{F(u)}{G(u)} = \lim_{u\to 0^+}\frac{F'(u)}{G'(u)}$$

$$= \lim_{u\to 0^+}\frac{(f'(1/u)\cdot(-1/u^2))}{(g'(1/u)\cdot(-1/u^2))} = \lim_{u\to 0^+}\frac{f'(1/u)}{g'(1/u)} = \lim_{x\to+\infty}\frac{f'(x)}{g'(x)}$$

12. Fill in the gaps in the proof of the following $\frac{\infty}{\infty}$ form of L'Hôpital's Rule in the $\lim\limits_{x\to a^+}$ case. (The other cases follow easy as in the $\frac{0}{0}$ form.) Assume $f(x)$ and $g(x)$ are differentiable and $g'(x)\neq 0$ in some open interval (a, b) and $\lim\limits_{x\to a^+} f(x) = \pm\infty = \lim\limits_{x\to a^+} g(x)$. Then,

$$\text{if} \quad K = \lim_{x\to a^+}\frac{f'(x)}{g'(x)} \quad \text{exists,} \quad \lim_{x\to a^+}\frac{f(x)}{g(x)} = \lim_{x\to a^+}\frac{f'(x)}{g'(x)}$$

Proof: Assume $\epsilon > 0$ and choose c so that $|K - (f'(x)/g'(x))| < \epsilon/2$ for $a < x < c$. Fix d in (a, c). Let $a < y < d$. By the extended mean value theorem, there exists x^* such that

$$y < x^* < d \quad \text{and} \quad \frac{f(d)-f(y)}{g(d)-g(y)} = \frac{f'(x^*)}{g'(x^*)}$$

Then

$$\left|K - \frac{f(d)-f(y)}{g(d)-g(y)}\right| < \frac{\epsilon}{2} \quad \text{and so} \quad \left|K - \left[\left(\frac{f(y)}{g(y)} - \frac{f(d)}{g(y)}\right)\Big/\left(1-\frac{g(d)}{g(y)}\right)\right]\right| < \frac{\epsilon}{2}$$

Now we let $y \to a^+$. Since $g(y) \to \pm\infty$ and $f(d)$ and $g(d)$ are constant, $f(d)/g(y) \to 0$ and $1 - g(d)/g(y) \to 1$. So, for y close to a,

$$\left|K - \frac{f(y)}{g(y)}\right| < \epsilon. \quad \text{Hence,} \quad \lim_{y\to a^+}\frac{f(y)}{g(y)} = K$$

13. (GC) In the following cases, try to find the limit by analytic methods, and then check by estimating the limit on a graphing calculator: (a) $\lim_{x\to 0^+} x^{1/x}$; (b) $\lim_{x\to +\infty} x^{1/x}$; (c) $\lim_{x\to 0} (1-\cos x)^x$; (d) $\lim_{x\to +\infty}\left(\sqrt{x^2+3x}-x\right)$.

Ans. (a) 0; (b) 1; (c) 1; (d) $\frac{3}{2}$

14. The current in a coil containing a resistance R, an inductance, L, and a constant electromotive force, E, at time t is given by $i = \frac{E}{R}(1-e^{-Rt/L})$. Obtain a formula for estimating i when R is very close to 0.

Ans. $\frac{Et}{L}$

CHAPTER 28

Exponential Growth and Decay

Assume that a quantity y varies with time and that

$$\frac{dy}{dx} = ky \tag{28.1}$$

for some nonzero constant k. Let $F(t) = y/e^{kt}$. Then, by the Quotient Rule,

$$\frac{dF}{dt} = \frac{e^{kt}D_t y - y D_t e^{kt}}{e^{2kt}} = \frac{e^{kt}ky - ye^{kt}k}{e^{2kt}} = \frac{0}{e^{2kt}} = 0$$

Hence, $F(t)$ must be a constant C. (Why?) Thus, $y/e^{kt} = C$ and, therefore, $y = Ce^{kt}$. To evaluate C, let $t = 0$. Then $y(0) = Ce^0 = C(1) = C$. If we designate $y(0)$ by y_0, then $C = y_0$ and we have obtained the general form of the solution of equation (28.1):

$$y = y_0 e^{kt} \tag{28.2}$$

If $k > 0$, we say that y *grows exponentially* and k is called the *growth constant*. If $k < 0$, we say that y *decays exponentially*, and k is called the *decay constant*. The constant y_0 is called the *initial value*.

From Problem 2 of Chapter 27, we know that $\lim_{u\to+\infty} \frac{u^n}{e^u} = 0$. So, when $k > 0$, $\lim_{t\to+\infty} \frac{t^n}{e^{kt}} = 0$. Thus, a quantity that grows exponentially grows much more rapidly than any power of t. There are many natural processes, such as bacterial growth or radioactive decay, in which quantities increase or decrease at an exponential rate.

Half-Life

Assume that a quantity y of a certain substance decays exponentially, with decay constant k. Let y_0 be the quantity at time $t = 0$. At what time T will only half of the original quantity remain?

By (28.2), we get the equation $y = y_0 e^{kt}$. Hence, at time T,

$$\tfrac{1}{2}y_0 = y_0 e^{kT}$$

$$\tfrac{1}{2} = e^{kT}$$

$$\ln(\tfrac{1}{2}) = \ln(e^{kT}) = kT$$

$$-\ln 2 = kT \tag{28.3}$$

$$T = -\frac{\ln 2}{k} \tag{28.4}$$

Note that the same value T is obtained for *any* original amount y_0. T is called the *half-life* of the substance. It is related to the decay constant k by the equation (28.3). So, if we know the value of either k or T, we can compute the value of the other. Also observe that, in (28.4), $k < 0$, so that $T > 0$.

The value of k can be obtained by experiment. For a given initial value y_0 and a specific positive time t_0, we observe the value of y, substitute in the equation (28.2), and solve for k.

SOLVED PROBLEMS

1. Given that the half-life T of radium is 1690 years, how much will remain of one gram of radium after 1000 years?

From (28.3), $k = -\frac{\ln 2}{T} = -\frac{\ln 2}{1690}$ and the quantity of radium is given by $y = y_0 e^{-(\ln 2)t/1690}$. Noting that $y_0 = 1$ and substituting 1000 for t, we get the quantity

$$y = e^{-(\ln 2)1000/1690} \sim e^{-693.1/1690} \sim e^{-0.4101} \sim 0.6636 \text{ grams}$$

Thus, about 663.6 milligrams are left after 1000 years.

2. If 20% of a radioactive substance disappears in one year, find its half-life T. Assume exponential decay.

By (28.2), $0.8y_0 = y_0 e^{k(1)} = y_0 e^k$. So, $0.8 = e^k$ whence, $k = \ln(0.8) = \ln(\frac{4}{5}) = \ln 4 - \ln 5$. From (28.4), $T = -\frac{\ln 2}{k} = \frac{\ln 2}{\ln 5 - \ln 4} \sim 3.1063$ years.

3. Assume that the number of bacteria in a culture grows exponentially with a growth constant of 0.02, time being measured in hours. (Although the number of bacteria must be a nonnegative integer, the assumption that the number is a continuous quantity always seems to lead to results that are experimentally verified.)

(a) How many bacteria will be present after 1 hour if there are initially 1000?
(b) Given the same initial 1000 bacteria, in how many hours will there be 100 000 bacteria?

(a) From (28.2), $y = 1000e^{0.02} \sim 1000(1.0202) = 1020.2 \sim 1020$
(b) From (28.2),

$$100\,000 = 1000e^{0.02t}$$

$$100 = e^{0.02t}$$

$$\ln 100 = 0.02t$$

$$2\ln 10 = 0.02t \quad (\text{since } \ln 100 = \ln(10)^2 = 2\ln 10)$$

$$t = 100 \ln 10 \sim 100(2.0326) = 203.26 \text{ hours}$$

Note: Sometimes, instead of giving the growth constant, say $k = 0.02$, one gives a corresponding rate of increase per unit time (in our case, 2% per hour.) This is not quite accurate. A rate of increase of r% per unit time is approximately the same as a value of $k = 0.0r$ when r is relatively small (say, $r \le 3$). In fact, with an r% rate of growth, $y = y_0(1 + 0.0r)$ after one unit of time. Since $y = y_0 e^k$ when $t = 1$, we get $1 + 0.0r = e^k$ and, therefore, $k = \ln(1 + 0.0r)$. This is close to $0.0r$, since $\ln(1 + x) \sim x$ for small positive x. (For example, $\ln 1.02 \sim 0.0198$ and $\ln 1.03 \sim 0.02956$.) For that reason, many textbooks often interpret a rate of increase of r% to mean that $k = 0.0r$.

4. If a quantity y increases or decreases exponentially, find a formula for the average value of y over a time interval $[0, b]$.

By definition, the average value $y_{av} = \frac{1}{b-0}\int_0^b y\,dt = \frac{1}{bk}\int_0^b ky\,dt$ (where k is the growth or decay constant). By (28.1), $ky = \frac{dy}{dt}$ and, therefore, $y_{av} = \frac{1}{bk}\int_0^b \frac{dy}{dt}\,dt$. By the Fundamental Theorem of Calculus,

$$\int_0^b \frac{dy}{dt}\,dt = y(b) - y(0) = y(b) - y_0. \quad \text{Thus,} \quad y_{av} = \frac{1}{bk}(y(b) - y_0)$$

5. If the population of a country is 100 million people and the population is increasing exponentially with a growth constant $k = \ln 2$, calculate precisely the population after 5 years.

By (28.2), the population $y = y_0 e^{kt} = 10^8 e^{(\ln 2)5} = 10^8 (e^{\ln 2})^5 = 10^8 (2^5) = 32(10^8)$. Thus, the population will reach 3.2 billion people in 5 years.

6. **Carbon-Dating.** A certain isotope ^{14}C of carbon occurs in living organisms in a fixed proportion to ordinary carbon. When that organism dies, its ^{14}C decays exponentially, and its half-life is 5730 years. Assume that a piece of charcoal from a wood fire was found in cave and contains only 9% of the ^{14}C expected in a corresponding piece of wood in a live tree. (This figure is obtained by measuring the amount of ordinary carbon in the piece of charcoal.) How long ago was the wood burned to form that charcoal?

If y is the amount of ^{14}C present in the piece of charcoal, we have $y = y_0 e^{kt}$. The present quantity $0.09y_0 = y_0 e^{k\tau}$, where τ is the elapsed time. Thus, $0.09 = e^{k\tau}$, $\ln(0.09) = k\tau$, $\tau = (\ln(0.09))/k$. Since the half-life $T = 5730$ and $k = -(\ln 2)/T = -(\ln 2)/5730$, we obtain

$$\tau = -\frac{5730 \ln(0.09)}{\ln 2} = \frac{5730(\ln 100 - \ln 9)}{\ln 2} \sim 19906 \text{ years}$$

7. **Newton's Law of Cooling:** The rate of change of the temperature of an object is proportional to the difference between the object's temperature and the temperature of the surrounding medium.

Assume that a refrigerator is maintained at a constant temperature of 45°F and that an object having a temperature of 80°F is placed inside the refrigerator. If the temperature of the object drops from 80°F to 70°F in 15 minutes, how long will it take for the object's temperature to decrease to 60°F?

Let u be the temperature of the object. Then, by Newton's Law of Cooling, $du/dt = k(u - 45)$, for some (negative) constant k. Let $y = u - 45$. Then $dy/dt = du/dt = ky$. Thus, by (28.2), $y = y_0 e^{kt}$. Since u is initially 80°F, $y_0 = 80 - 45 = 35$. So, $y = 35e^{kt}$. When $t = 15$, $u = 70$ and $y = 25$. Hence, $25 = 35e^{15k}$, $5 = 7e^{15k}$ and, therefore, $15k = \ln(\frac{5}{7}) = \ln 5 - \ln 7$. Thus, $k = \frac{1}{15}(\ln 5 - \ln 7)$. When the object's temperature is 60°F, $y = 15$. So, $15 = 35e^{kt}$, $3 = 7e^{kt}$ and therefore, $kt = \ln(\frac{3}{7}) = \ln 3 - \ln 7$. Thus,

$$t = \frac{\ln 3 - \ln 7}{k} = 15\frac{\ln 3 - \ln 7}{\ln 5 - \ln 7} \sim 37.7727 \text{ minutes}$$

Hence, it would take about 22.7727 minutes for the object's temperature to drop from 70° to 60°.

8. **Compound Interest.** Assume that a savings account earns interest at a rate of r% per year. So, after one year, an amount of P dollars would become $P\left(1+\frac{r}{100}\right)$ dollars and, after t years, it would become $P\left(1+\frac{r}{100}\right)^t$ dollars. However, if the interest is calculated n times a year instead of once a year, then in each period the interest rate would be (r/n)%; after t years, there would have been nt such periods and the final amount would be $P\left(1+\frac{r}{100n}\right)^{nt}$. If we let $n \to +\infty$, then we say that the interest is *compounded continuously*. In such a case, the final amount would be

$$\lim_{n\to+\infty} P\left(1+\frac{r}{100n}\right)^{nt} = P\left[\lim_{n\to+\infty}\left(1+\frac{r}{100n}\right)^n\right]^t = Pe^{0.01rt} \quad \text{by (26.16)}$$

Let \$100 be deposited in a savings account paying an interest rate of 4% per year. After 5 years, how much would be in the account if:

(a) The interest is calculated once a year?
(b) The interest is calculated quarterly (that is, four times per year)?
(c) The interest is compounded continuously?

(a) $100(1.04)^5 \sim 121.6653$ dollar.
(b) $100(1.01)^{20} \sim 122.0190$ dollar.
(c) $100e^{0.04(5)} = 100e^{0.2} \sim 122.1403$ dollar.

SUPPLEMENTARY PROBLEMS

9. Assume that, in a chemical reaction, a certain substance decomposes at a rate proportional to the amount present. Assume that an initial quantity of 10,000 grams is reduced to 1000 grams in 5 hours. How much would be left of an initial quantity of 20,000 grams after 15 hours?

Ans. 20 grams

10. A container with a maximum capacity of 25,000 fruit flies initially contains 1000 fruit flies. If the population grows exponentially with a growth constant of (ln 5)/10 fruit flies per day, in how many days will the container be full?

Ans. 20 days

11. The half-life of radium is 1690 years. How much will be left of 32 grams of radium after 6760 years?

Ans. 2 grams

12. If a population grows exponentially and increases at the rate of 2.5% per year, find the growth constant k.

Ans. $\ln 1.025 \sim 0.0247$

13. A saltwater solution initially contains 5 lb of salt in 10 gal of fluid. If water flows in at the rate of $\frac{1}{2}$ gal/min and the mixture flows out at the same rate, how much salt is present after 20 min?

Ans. $\frac{dS}{dt} = -\frac{1}{2}\left(\frac{S}{10}\right)$. At $t = 20$, $S = 5e^{-1} \sim 1.8395$ lb.

14. Fruit flies in an enclosure increase exponentially in such a way that their population doubles in 4 hours. How many times the initial number will there be after 12 hours?

Ans. 8

15. (GC) If the world population in 1990 was 4.5 billion and it is growing exponentially with growth constant $k = (\ln 3)/8$, estimate the world population in the years (a) 2014; (b) 2020.

Ans. (a) 111.5 billion; (b) 277.0 billion

16. (GC) If a thermometer with a reading of 65°F is taken into the outside air where the temperature is a constant 25°F, the thermometer reading decreases to 50°F in 2.0 minutes.

(a) Find the thermometer reading after one more minute.
(b) How much longer (after 3.0 minutes) will it take for the thermometer reading to reach 32°F?

Use Newton's Law of Cooling.

Ans. (a) 45°F; (b) about 4.4 minutes more

17. (GC) Under continuous compounding at a rate of r% per year:

(a) How long does it take for a given amount of money P to double?
(b) If a given amount P doubles in 9 years, what is r?
(c) If $r = 8$, how much must be deposited now to yield $100,000 in 17 years?

Ans. (a) $\frac{100 \ln 2}{r} \sim \frac{69.31}{r}$; (b) about 7.7; (c) about $25,666

18. An object cools from 120°F to 95°F in half an hour when surrounded by air whose temperature is 70°F. Use Newton's Law of Cooling to find its temperature at the end of another half an hour.

Ans. 82.5°F

19. If an amount of money receiving interest of 8% per year is compounded continuously, what is the equivalent yearly rate of return?

Ans. about 8.33%

20. How long does it take for 90% of a given quantity of the radioactive element cobalt-60 to decay, given that its half-life is 5.3 years?

Ans. about 17.6 years

21. A radioactive substance decays exponentially. If we start with an initial quantity of y_0, what is the average quantity present over the first half-life?

Ans. $\dfrac{y_0}{2\ln 2}$

CHAPTER 29

Applications of Integration I: Area and Arc Length

Area Between a Curve and the y Axis

We already know how to find the area of a region like that shown in Fig. 29-1, bounded below by the x axis, above by a curve $y = f(x)$, and lying between $x = a$ and $x = b$. The area is the definite integral $\int_a^b f(x)\,dx$.

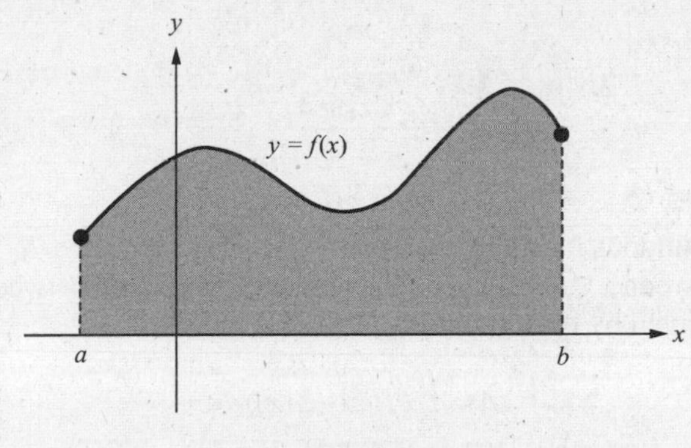

Fig. 29-1

Now consider a region like that shown in Fig. 29-2, bounded on the left by the y axis, on the right by a curve $x = g(y)$, and lying between $y = c$ and $y = d$. Then, by an argument similar to that for the case shown in Fig. 29-1, the area of the region is the definite integral $\int_c^d g(y)\,dy$.

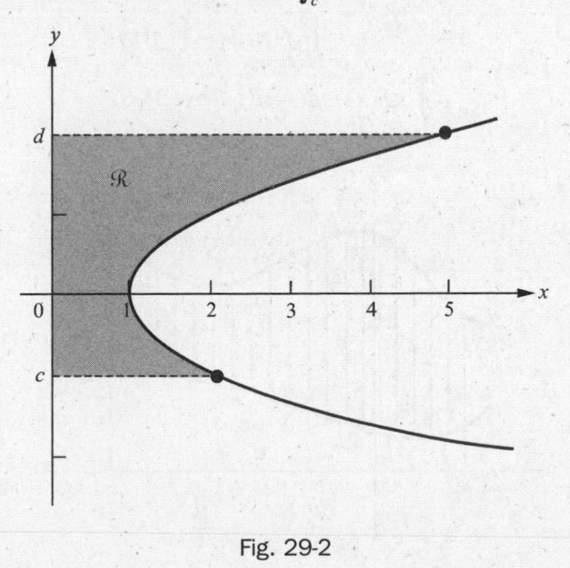

Fig. 29-2

EXAMPLE 29.1: Consider the region bounded on the right by the parabola $x = 4 - y^2$, on the left by the y axis, and above and below by $y = 2$ and $y = -1$. See Fig. 29-3. Then the area of this region is $\int_{-1}^{2}(4 - y^2)\,dy$. By the Fundamental Theorem of Calculus, this is

$$(4y - \tfrac{1}{3}y^3)\big]_{-1}^{2} = (8 - \tfrac{8}{3}) - (-4 - (-\tfrac{1}{3})) = 12 - \tfrac{9}{3} = 12 - 3 = 9$$

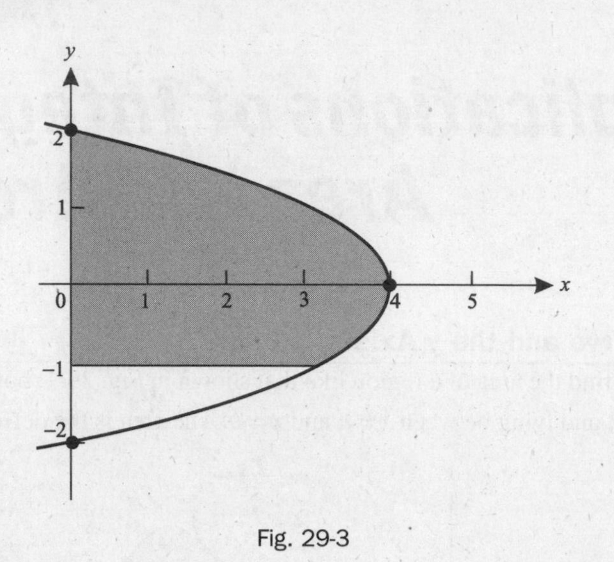

Fig. 29-3

Areas Between Curves

Assume that f and g are continuous functions such that $g(x) \le f(x)$ for $a \le x \le b$. Then the curve $y = f(x)$ lies above the curve $y = g(x)$ between $x = a$ and $x = b$. The area A of the region between the two curves and lying between $x = a$ and $x = b$ is given by the formula

$$A = \int_a^b (f(x) - g(x))\,dx \tag{29.1}$$

To see why this formula holds, first look at the special case where $0 \le g(x) \le f(x)$ for $a \le x \le b$. (See Fig. 29-4.) Clearly, the area is the difference between two areas, the area A_f of the region under the curve $y = f(x)$ and above the x axis, and the area A_g of the region under the curve $y = g(x)$ and above the x axis. Since $A_f = \int_a^b f(x)\,dx$ and $A_g = \int_a^b g(x)\,dx$,

$$A = A_f - A_g = \int_a^b f(x)\,dx - \int_a^b g(x)\,dx$$
$$= \int_a^b (f(x) - g(x))\,dx \quad \text{by (23.6)}$$

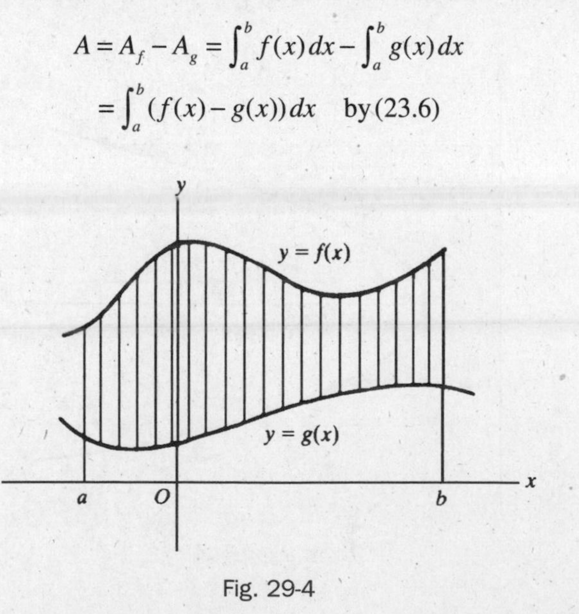

Fig. 29-4

Now look at the general case (see Fig. 29-5), when one or both of the curves $y = f(x)$ and $y = g(x)$ may lie below the x axis. Let $m < 0$ be the absolute minimum of g on $[a, b]$. Raise both curves by $|m|$ units. The new graphs, shown in Fig. 29-6, are on or above the x axis and enclose the same area A as the original graphs. The upper curve is the graph of $y = f(x) + |m|$ and the lower curve is the graph of $y = g(x) + |m|$. Hence, by the special case above,

$$A = \int_a^b ((f(x) + |m| - (g(x) + |m|))\,dx = \int_a^b (f(x) - g(x))\,dx$$

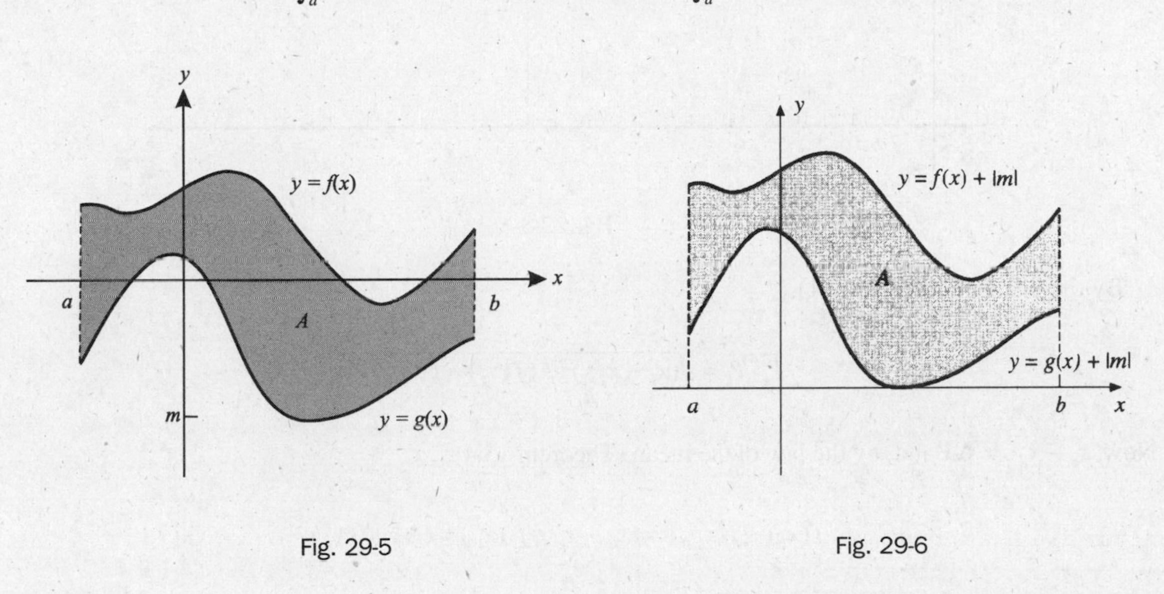

Fig. 29-5 Fig. 29-6

EXAMPLE 29.2: Find the area A of the region $\mathcal{R}$ under the line $y = \frac{1}{2}x + 2$, above the parabola $y = x^2$, and between the y axis and $x = 1$. (See the shaded region in Fig. 29-7.) By (29.1),

$$A = \int_0^1 \left(\left(\frac{1}{2}x + 2\right) - x^2\right)dx = \left(\frac{1}{4}x^2 + 2x - \frac{1}{3}x^3\right)\Big]_0^1 = \left(\frac{1}{4} + 2 - \frac{1}{3}\right) - (0 + 0 - 0) = \frac{3}{12} + \frac{24}{12} - \frac{4}{12} = \frac{23}{12}$$

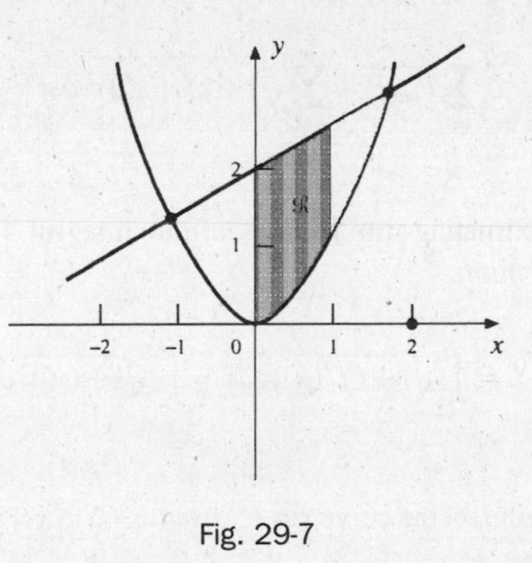

Fig. 29-7

Arc Length

Let f be differentiable on $[a, b]$. Consider the part of the graph of f from $(a, f(a))$ to $(b, f(b))$. Let us find a formula for the length L of this curve. Divide [a, b] into n equal subintervals, each of length Δx. To each point x_k in this subdivision there corresponds a point $P_k(x_k, f(x_k))$ on the curve. (See Fig. 29-8.) For large n, the sum $\overline{P_0P_1} + \overline{P_1P_2} + \ldots + \overline{P_{n-1}P_n} = \sum_{k=1}^{n} \overline{P_{k-1}P_k}$ of the lengths of the line segments $P_{k-1}P_k$ is an approximation to the length of the curve.

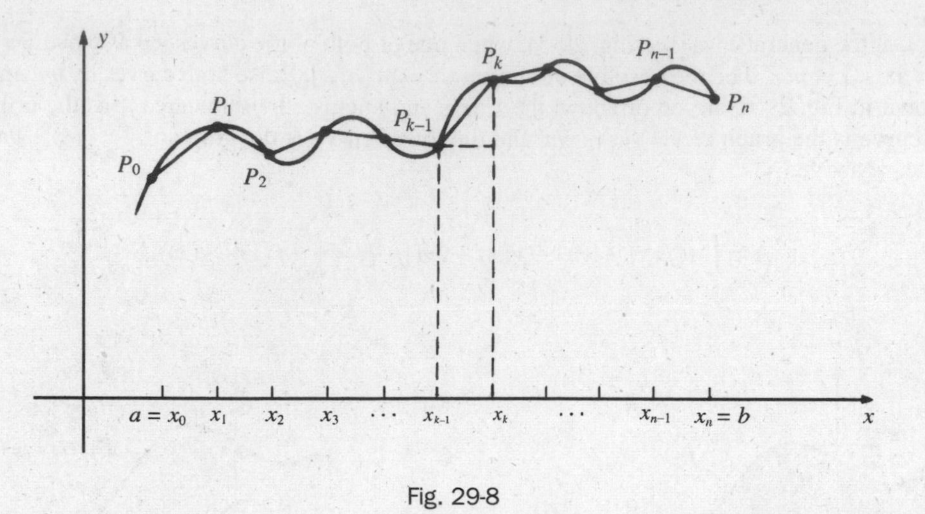

Fig. 29-8

By the distance formula (2.1),

$$\overline{P_{k-1}P_k} = \sqrt{(x_k - x_{k-1})^2 + (f(x_k) - f(x_{k-1}))^2}$$

Now, $x_k - x_{k-1} = \Delta x$ and, by the law of the mean (Theorem 13.4),

$$f(x_k) - f(x_{k-1}) = (x_k - x_{k-1})f'(x_k^*) = (\Delta x)f'(x_k^*)$$

for some x_k^* in (x_{k-1}, x_k). Thus,

$$\overline{P_{k-1}P_k} = \sqrt{(\Delta x)^2 + (\Delta x)^2(f'(x_k^*))^2} = \sqrt{(1+(f'(x_k^*))^2)(\Delta x)^2}$$
$$= \sqrt{1+(f'(x_k^*))^2}\sqrt{(\Delta x)^2} = \sqrt{1+(f'(x_k^*))^2}\,\Delta x$$

So,
$$\sum_{k=1}^{n} \overline{P_{k-1}P_k} = \sum_{k=1}^{n} \sqrt{1+(f'(x_k^*))^2}\,\Delta x$$

The right-hand sum is an approximating sum for the definite integral $\int_a^b \sqrt{1+(f'(x))^2}\,dx$. Therefore, letting $n \to +\infty$, we get the *arc length formula:*

$$L = \int_a^b \sqrt{1+(f'(x))^2}\,dx = \int_a^b \sqrt{1+(y')^2}\,dx \tag{29.2}$$

EXAMPLE 29.3: Find the arc length L of the curve $y = x^{3/2}$ from $x = 0$ to $x = 5$.

By (29.2), since $y' = \frac{3}{2}x^{1/2} = \frac{3}{2}\sqrt{x}$,

$$L = \int_0^5 \sqrt{1+(y')^2}\,dx = \int_0^5 \sqrt{1+\tfrac{9}{4}x}\,dx$$
$$= \tfrac{4}{9}\int_0^5 (1+\tfrac{9}{4}x)^{1/2}(\tfrac{9}{4})dx = \tfrac{4}{9}\tfrac{2}{3}(1+\tfrac{9}{4}x)^{3/2}\Big]_0^5 \quad \text{(by Quick Formula I and the Fundamental Theorem of Calculus)}$$
$$= \tfrac{8}{27}((\tfrac{49}{4})^{3/2} - 1^{3/2}) = \tfrac{8}{27}(\tfrac{343}{8} - 1) = \tfrac{335}{27}$$

SOLVED PROBLEMS

1. Find the area bounded by the parabola $x = 8 + 2y - y^2$, the y axis, and the lines $y = -1$ and $y = 3$.

Note, by completing the square, that $x = -(y^2 - 2y - 8) = -((y-1)^2 - 9) = 9 - (y-1)^2 = (4-y)(2+y)$. Hence, the vertex of the parabola is (9, 1) and the parabola cuts the y axis at $y = 4$ and $y = -2$. We want the area of the shaded region in Fig. 29-9, which is given by

$$\int_{-1}^{3}(8 + 2y - y^2)\,dy = \left(8y + y^2 - \tfrac{1}{3}y^3\right)\Big]_{-1}^{3} = (24 + 9 - 9) - (-8 + 1 - \tfrac{1}{3}) = \tfrac{92}{3}$$

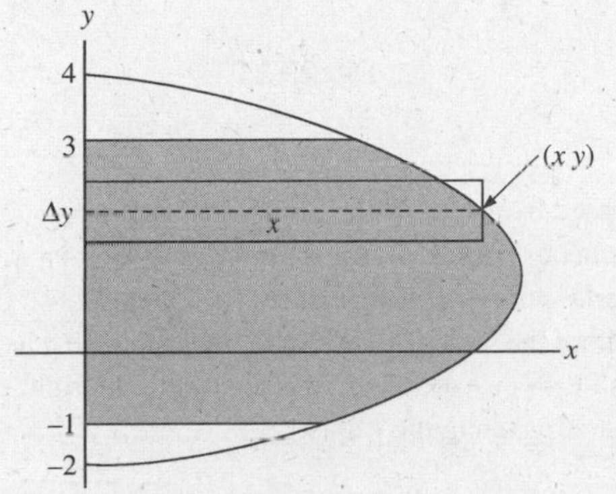

Fig. 29-9

2. Find the area of the region between the curves $y = \sin x$ and $y = \cos x$ from $x = 0$ to $x = \pi/4$.

The curves intersect at $(\pi/4, \sqrt{2}/2)$, and $0 \le \sin x < \cos x$ for $0 \le x < \pi/4$. (See Fig. 29-10.) Hence, the area is

$$\int_{0}^{\pi/4}(\cos x - \sin x)\,dx = (\sin x + \cos x)\Big]_{0}^{\pi/4} = \left(\frac{\sqrt{2}}{2} + \frac{\sqrt{2}}{2}\right) - (0 + 1) = \sqrt{2} - 1$$

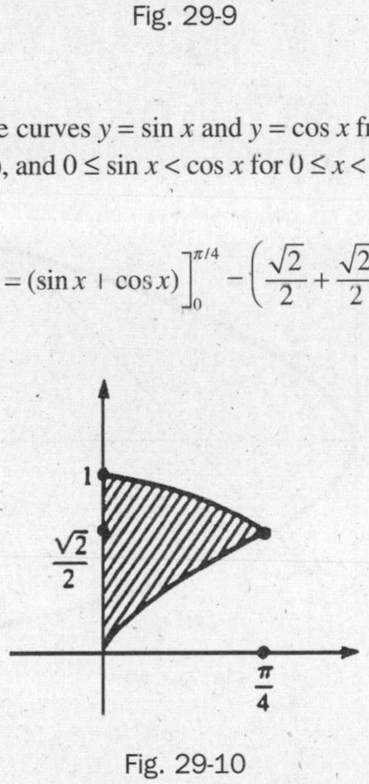

Fig. 29-10

3. Find the area of the region bounded by the parabolas $y = 6x - x^2$ and $y = x^2 - 2x$.

By solving $6x - x^2 = x^2 - 2x$, we see that the parabolas intersect when $x = 0$ and $x = 4$, that is, at (0, 0) and (4, 8). (See Fig. 29-11.) By completing the square, the first parabola has the equation $y = 9 - (x-3)^2$; therefore, it has its vertex at (3, 9) and opens downward. Likewise, the second parabola has the equation $y = (x-1)^2 - 1$; therefore, its vertex is at (1, −1) and it opens upward. Note that the first parabola lies above the second parabola in the given region. By (29.1), the required area is

$$\int_{0}^{4}((6x - x^2) - (x^2 - 2x))\,dx = \int_{0}^{4}(8x - 2x^2)\,dx = \left(4x^2 - \tfrac{2}{3}x^3\right)\Big]_{0}^{4} = \left(64 - \tfrac{128}{3}\right) = \tfrac{64}{3}$$

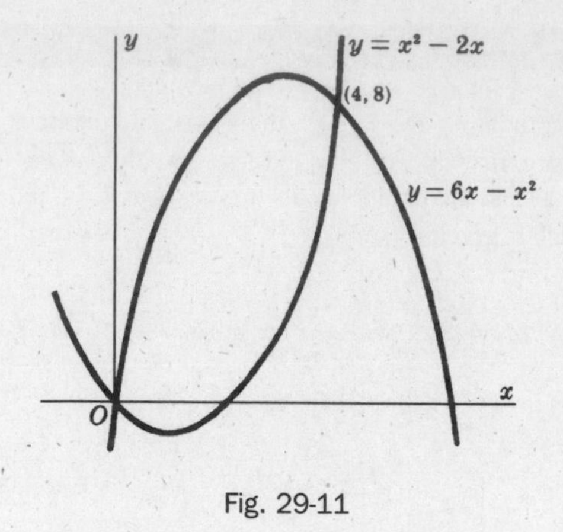

Fig. 29-11

4. Find the area of the region bounded by the parabola $y^2 = 4x$ and the line $y = 2x - 4$.

Solving the equations simultaneously, we get $(2x-4)^2 = 4x$, $x^2 - 4x + 4 = x$, $x^2 - 5x + 4 = 0$, $(x-1)(x-4) = 0$. Hence, the curves intersect when $x = 1$ or $x = 4$, that is, at $(1, -2)$ and $(4, 4)$. (See Fig. 29-12.) Note that neither curve is above the other throughout the region. Hence, it is better to take y as the independent variable and rewrite the curves as $x = \frac{1}{4}y^2$ and $x = \frac{1}{2}(y+4)$. The line is always to the right of the parabola.

The area is obtained by integrating along the y axis:

$$\int_{-2}^{4}\left(\tfrac{1}{2}(y+4) - \tfrac{1}{4}y^2\right)dy = \tfrac{1}{4}\int_{-2}^{4}(2y + 8 - y^2)\,dy$$

$$= \tfrac{1}{4}\left(y^2 + 8y - \tfrac{1}{3}y^3\right)\Big]_{-2}^{4} = \tfrac{1}{4}\left((16 + 32 - \tfrac{64}{3}) - (4 - 16 + \tfrac{8}{3})\right) = 9$$

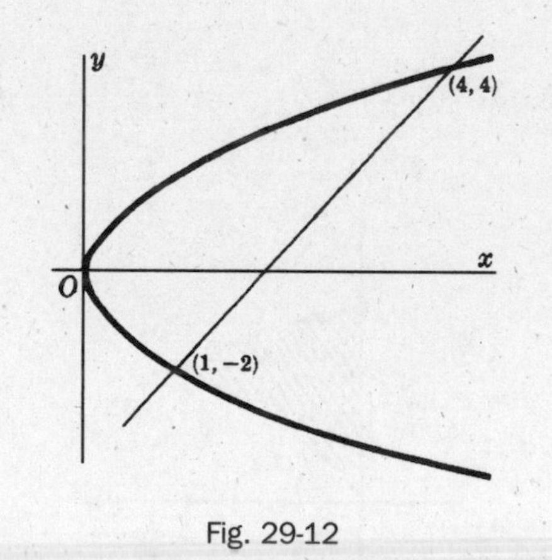

Fig. 29-12

5. Find the area of the region between the curve $y = x^3 - 6x^2 + 8x$ and the x axis.

Since $x^3 - 6x^2 + 8x = x(x^2 - 6x + 8) = x(x-2)(x-4)$, the curve crosses the x axis at $x = 0$, $x = 2$, and $x = 4$. The graph looks like the curve shown in Fig. 29-13. (By applying the quadratic formula to y', we find that the maximum and minimum values occur at $x = 2 \pm \frac{2}{3}\sqrt{3}$.) Since the part of the region with $2 \le x \le 4$ lies below the x axis, we must calculate two separate integrals, one with respect to y between $x = 0$ and $x = 2$, and the other with respect to $-y$ between $x = 2$ and $x = 4$. Thus, the required area is

$$\int_0^2 (x^3 - 6x^2 + 8x)\,dx - \int_2^4 (x^3 - 6x^2 + 8x)\,dx = \left(\tfrac{1}{4}x^4 - 2x^3 + 4x^2\right)\Big]_0^2 - \left(\tfrac{1}{4}x^4 - 2x^3 + 4x^2\right]_2^4 = 4 + 4 = 8$$

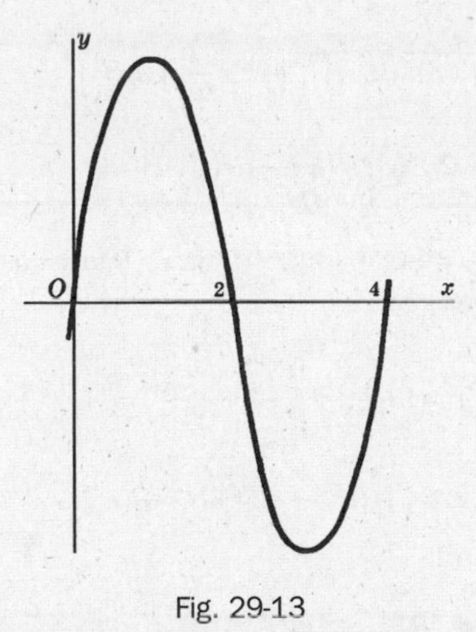

Fig. 29-13

Note that, if we had made the mistake of simply calculating the integral $\int_0^4 (x^3 - 6x^2 + 8x)\,dx$, we would have got the incorrect answer 0.

6. Find the area enclosed by the curve $y^2 = x^2 - x^4$.

The curve is symmetric with respect to the coordinate axes. Hence the required area is four times the portion lying in the first quadrant. (See Fig. 29-14.) In the first quadrant, $y = \sqrt{x^2 - x^4} = x\sqrt{1 - x^2}$ and the curve intersects the x axis at $x = 0$ and $x = 1$. So, the required area is

$$\begin{aligned} 4\int_0^1 x\sqrt{1-x^2}\,dx &= -2\int_0^1 (1-x^2)^{1/2}(-2x)\,dx \\ &= -2\left(\tfrac{2}{3}\right)(1-x^2)^{3/2}\Big]_0^1 \quad \text{(by Quick Formula I)} \\ &= -\tfrac{4}{3}(0 - 1^{3/2}) = -\tfrac{4}{3}(-1) = \tfrac{4}{3} \end{aligned}$$

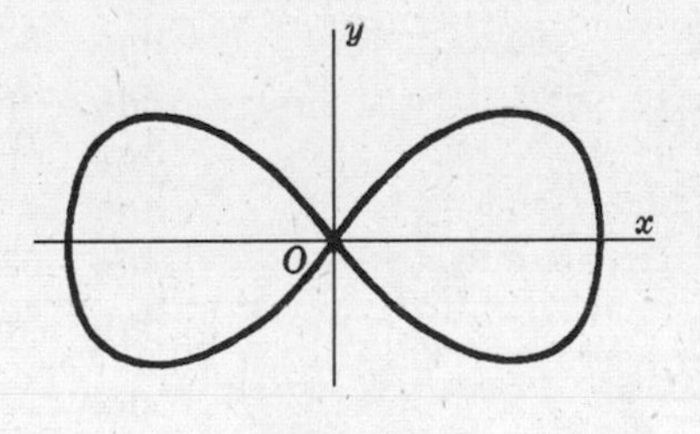

Fig. 29-14

7. Find the arc length of the curve $x = 3y^{3/2} - 1$ from $y = 0$ to $y = 4$.

We can reverse the roles of x and y in the arc length formula (29.2): $L = \int_c^d \sqrt{1 + \left(\frac{dx}{dy}\right)^2}\,dy$. Since $\frac{dx}{dy} = \frac{9}{2}y^{1/2}$,

$$L = \int_0^4 \sqrt{1 + \tfrac{81}{4}y}\,dy = \tfrac{4}{81}\int_0^4 (1 + \tfrac{81}{4}y)^{1/2}(\tfrac{81}{4})\,dy = \tfrac{4}{81}(\tfrac{2}{3})(1 + \tfrac{81}{4}y)^{3/2}\Big]_0^4 = \tfrac{8}{243}((82)^{3/2} - 1^{3/2}) = \tfrac{8}{243}(82\sqrt{82} - 1)$$

8. Find the arc length of the curve $24xy = x^4 + 48$ from $x = 2$ to $x = 4$.

$y = \frac{1}{24}x^3 + 2x^{-1}$. Hence, $y' = \frac{1}{8}x^2 - 2/x^2$. Thus,

$$\begin{aligned} (y')^2 &= \tfrac{1}{64}x^4 - \tfrac{1}{2} + \frac{4}{x^4} \\ 1 + (y')^2 &= \tfrac{1}{64}x^4 + \tfrac{1}{2} + \frac{4}{x^4} = \left(\tfrac{1}{8}x^2 + \frac{2}{x^2}\right)^2 \end{aligned}$$

So,
$$L = \int_2^4 \sqrt{1+(y')^2}\,dx = \int_2^4 \left(\tfrac{1}{8}x^2 + \frac{2}{x^2}\right)dx = \int_2^4 \left(\tfrac{1}{8}x^2 + 2x^{-2}\right)dx$$
$$= \left(\tfrac{1}{24}x^3 - 2x^{-1}\right)\Big]_2^4 = \left(\tfrac{8}{3} - \tfrac{1}{2}\right) - \left(\tfrac{1}{3} - 1\right) = \tfrac{17}{6}$$

9. Find the arc length of the catenary $y = \frac{a}{2}(e^{x/a} + e^{-x/a})$ from $x = 0$ to $x = a$.

$y' = \frac{1}{2}(e^{x/a} + e^{-x/a})$ and, therefore,

$$1 + (y')^2 = 1 + \tfrac{1}{4}(e^{2x/a} - 2 + e^{-2x/a}) = \tfrac{1}{4}(e^{x/a} + e^{-x/a})^2$$

So,
$$L = \tfrac{1}{2}\int_0^a (e^{x/a} + e^{-x/a})\,dx = \frac{a}{2}(e^{x/a} - e^{-x/a})\Big]_0^a = \frac{a}{2}(e - e^{-1})$$

SUPPLEMENTARY PROBLEMS

10. Find the area of the region lying above the x axis and under the parabola $y = 4x - x^2$.

Ans. $\frac{32}{3}$

11. Find the area of the region bounded by the parabola $y = x^2 - 7x + 6$, the x axis, and the lines $x = 2$ and $x = 6$.

Ans. $\frac{56}{3}$

12. Find the area of the region bounded by the given curves.

(a) $y = x^2, y = 0, x = 2, x = 5$ — *Ans.* 39
(b) $y = x^3, y = 0, x = 1, x = 3$ — *Ans.* 20
(c) $y = 4x - x^2, y = 0, x = 1, x = 3$ — *Ans.* $\frac{22}{3}$
(d) $x = 1 + y^2, x = 10$ — *Ans.* 36
(e) $x = 3y^2 - 9, x = 0, y = 0, y = 1$ — *Ans.* 8
(f) $x = y^2 + 4y, x = 0$ — *Ans.* $\frac{32}{3}$
(g) $y = 9 - x^2, y = x + 3$ — *Ans.* $\frac{125}{6}$
(h) $y = 2 - x^2 y = -x$ — *Ans.* $\frac{9}{2}$
(i) $y = x^2 - 4, y = 8 - 2x^2$ — *Ans.* 32
(j) $y = x^4 - 4x^2, y = 4x^2$ — *Ans.* $\frac{512}{15}\sqrt{2}$
(k) $y = e^x, y = e^{-x}, x = 0, x = 2$ — *Ans.* $\frac{e^2+1}{e^2-2}$
(l) $y = e^{x/a} + e^{-x/a}, y = 0, x = \pm a$ — *Ans.* $2a\left(\frac{e-1}{e}\right)$
(m) $xy = 12, y = 0, x = 1, x = e^2$ — *Ans.* 24
(n) $y = \frac{1}{1+x^2}, y = 0, x = \pm 1$ — *Ans.* $\frac{\pi}{2}$
(o) $y = \tan x, x = 0, x = \frac{\pi}{4}$ — *Ans.* $\frac{1}{2}\ln 2$
(p) $y = 25 - x^2, 256x = 3y^2, 16y = 9x^2$ — *Ans.* $\frac{98}{3}$

13. Find the length of the indicated arc of the given curve.

(a) $y^3 = 8x^2$ from $x = 1$ to $x = 8$ — *Ans.* $\left(104\sqrt{13} - 125\right)/27$
(b) $6xy = x^4 + 3$ from $x = 1$ to $x = 2$ — *Ans.* $\frac{17}{12}$
(c) $27y^2 = 4(x - 2)^3$ from (2, 0) to (11, $6\sqrt{3}$ — *Ans.* 14

(d) $y = \frac{1}{2}x^2 - \frac{1}{4}\ln x$ from $x = 1$ to $x = e$ *Ans.* $\frac{1}{2}e^2 - \frac{1}{4}$

(e) $y = \ln \cos x$ from $x = \frac{\pi}{6}$ to $x = \frac{\pi}{4}$ *Ans.* $\ln\left(\frac{1+\sqrt{2}}{\sqrt{3}}\right)$

(f) $x^{2/3} + y^{2/3} = 4$ from $x = 1$ to $x = 8$ *Ans.* 9

14. (GC) Estimate the arc length of $y = \sin x$ from $x = 0$ to $x = \pi$ to an accuracy of four decimal places. (Use Simpson's Rule with $n = 10$.)

Ans. 3.8202

CHAPTER 30

Applications of Integration II: Volume

A *solid of revolution* is obtained by revolving a region in a plane about a line that does not intersect the region. The line about which the rotation takes place is called the *axis of revolution*.

Let f be a continuous function such that $f(x) \geq 0$ for $a \leq x \leq b$. Consider the region $\mathcal{R}$ under the graph of f, above the x axis, and between $x = a$ and $x = b$. (See Fig. 30-1.) If $\mathcal{R}$ is revolved about the x axis, the resulting solid is a solid of revolution. The generating regions $\mathcal{R}$ for some familiar solids are shown in Fig. 30-2.

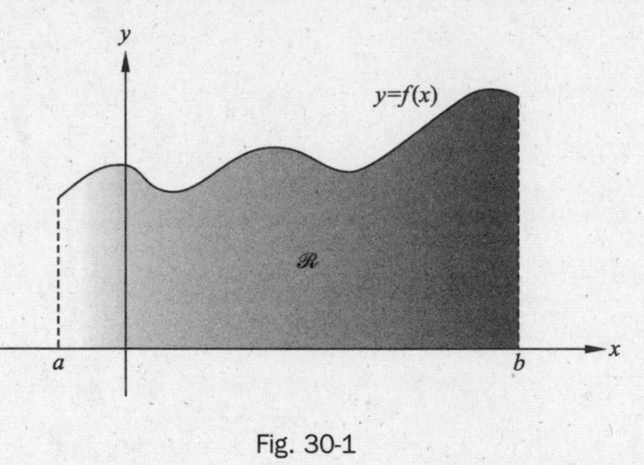

Fig. 30-1

Disk Formula

The volume V of the solid of revolution obtained by revolving the region $\mathcal{R}$ of Fig. 30-1 about the x axis is given by

$$V = \pi \int_a^b (f(x))^2 dx = \pi \int_a^b y^2 dx \qquad \text{(disk formula)}$$

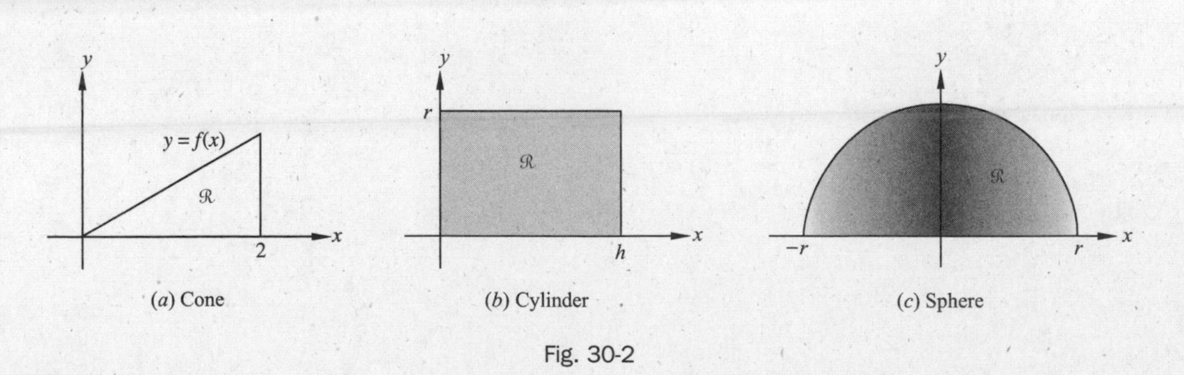

Fig. 30-2

See Problem 9 for a sketch of the proof of this formula.

Similarly, when the axis of rotation is the y axis and the region that is revolved lies between the y axis and a curve $x = g(y)$ and between $y = c$ and $y = d$ (see Fig. 30-3), then the volume V of the resulting solid of revolution is given by the formula

$$V = \pi \int_c^d (g(y))^2\, dy = \pi \int_a^b x^2\, dy \qquad \text{(disk formula)}$$

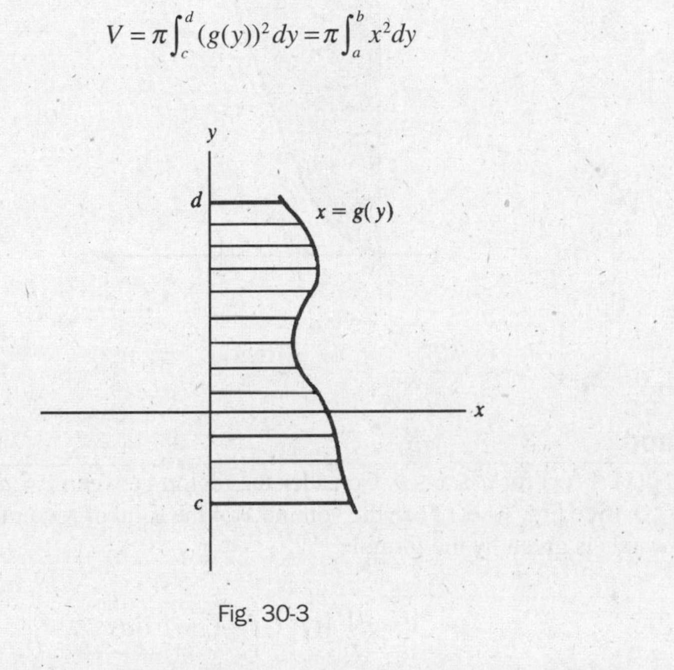

Fig. 30-3

EXAMPLE 30.1: Consider the solid of revolution obtained by revolving about the x axis the region in the first quadrant bounded by the parabola $y^2 = 8x$ and the line $x = 2$. (See Fig. 30-4.) By the disk formula, the volume is

$$V = \pi \int_0^2 y^2 dx = \pi \int_0^2 8x\, dx = \pi(4x^2)\Big]_0^2 = \pi(16 - 0) = 16\pi$$

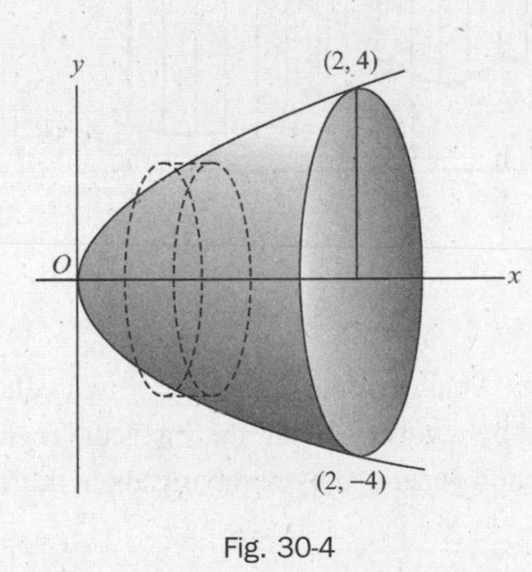

Fig. 30-4

EXAMPLE 30.2: Consider the solid of revolution obtained by revolving about the y axis the region bounded by the parabola $y = 4x^2$ and the lines $x = 0$ and $y = 16$. (See Fig. 30-5.) To find its volume, we use the version of the disk formula in which we integrate along the y axis. Thus,

$$V = \pi \int_0^{16} x^2 dy = \pi \int_0^{16} \frac{y}{4} dy = \frac{\pi}{8} y^2\Big]_0^{16} = \frac{\pi}{8}(256 - 0) = 32\pi$$

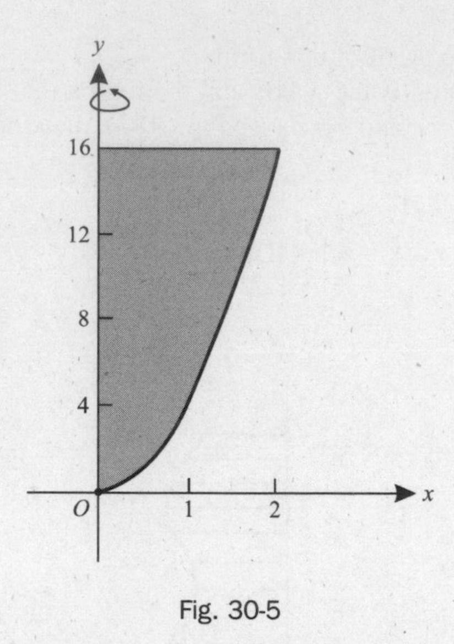

Fig. 30-5

Washer Method

Assume that $0 \le g(x) \le f(x)$ for $a \le x \le b$. Consider the region between $x = a$ and $x = b$ and lying between $y = g(x)$ and $y = f(x)$. (See Fig. 30-6.) Then the volume V of the solid of revolution obtained by revolving this region about the x axis is given by the formula

$$V = \pi \int_a^b [(f(x))^2 - (g(x))^2]\,dx \qquad \text{(washer formula)}^\dagger$$

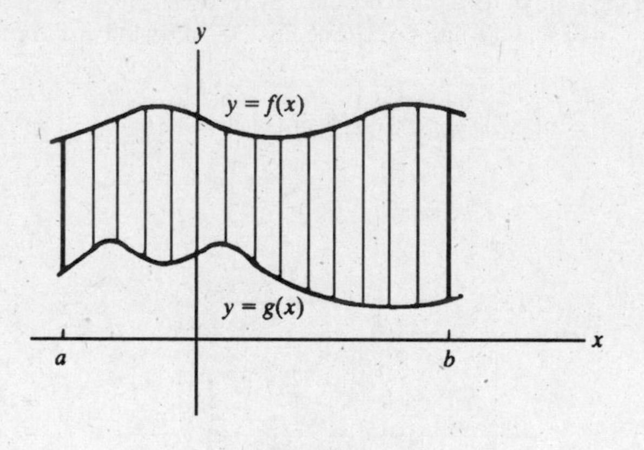

Fig. 30-6

The justification is clear. The desired volume is the difference of two volumes, the volumes $\pi \int_a^b (f(x))^2 dx$ of the solid of revolution generated by revolving about the x axis the region under $y = f(x)$ and the volume $\pi \int_a^b (g(x))^2 dx$ of the solid of revolution generated by revolving about the x axis the region under $y = g(x)$.

A similar formula

$$V = \pi \int_c^d [(f(y))^2 - (g(y))^2]\,dy \qquad \text{(washer formula)}$$

holds when the region lies between the two curves $x = f(y)$ and $x = g(y)$ and between $y = c$ and $y = d$, and it is revolved about the y axis. (It is assumed that $0 \le g(y) \le f(y)$ for $c \le y \le d$.)

† The word "washer" is used because each thin vertical strip of the region being revolved produces a solid that resembles a plumbing part called a *washer* (a small cylindrical disk with a hole in the middle).

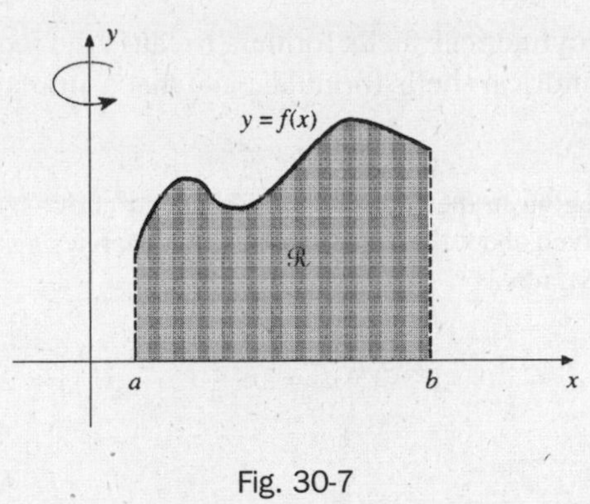

Fig. 30-7

EXAMPLE 30.3: Consider the solid of revolution obtained by revolving about the x axis the region bounded by the curves, $y = 4x^2$, $x = 0$, and $y = 16$. (The same region as in Fig. 30-5.) Here the upper curve is $y = 16$ and the lower curve is $y = 4x^2$. Hence, by the washer formula,

$$V = \pi\int_0^2 [16^2 - (4x^2)^2]\,dx = \pi\int_0^2 [256 - 16x^4]\,dx = \pi\left(256x - \tfrac{16}{5}x^5\right)\Big]_0^2 = \pi\left(512 - \frac{512}{5}\right) = \frac{2048\pi}{5}$$

Cylindrical Shell Method

Consider the solid of revolution obtained by revolving about the y axis the region $\mathcal{R}$ in the first quadrant between the x axis and the curve $y = f(x)$, and lying between $x = a$ and $x = b$. (See Fig. 30-7.) Then the volume of the solid is given by

$$V = 2\pi\int_a^b xf(x)\,dx = 2\pi\int_a^b xy\,dx \qquad \text{(cylindrical shell formula)}$$

See Problem 10 for the justification of this formula.

A similar formula holds when the roles of x and y are reversed, that is, the region $\mathcal{R}$ in the first quadrant between the y axis and the curve $x = f(y)$, and lying between $y = c$ and $y = d$, is revolved about the x axis

$$V = 2\pi\int_c^d yf(y)\,dy = 2\pi\int_c^d yx\,dy$$

EXAMPLE 30.4: Revolve about the y axis the region above the x axis and below $y = 2x^2$, and between $x = 0$ and $x = 5$. By the cylindrical shell formula, the resulting solid has volume

$$2\pi\int_0^5 xy\,dx = 2\pi\int_0^5 x(2x^2)\,dx = 4\pi\int_0^5 x^3 dx = \pi(x^4)\Big]_0^5 = 625\pi$$

Note that the volume could also have been computed by the washer formula, but the calculation would have been somewhat more complicated.

Difference of Shells Formula

Assume that $0 \le g(x) \le f(x)$ on an interval $[a, b]$ with $a \ge 0$. Let $\mathcal{R}$ be the region in the first quadrant between the curves $y = f(x)$ and $y = g(x)$ and between $x = a$ and $x = b$. Then the volume of the solid of revolution obtained by revolving $\mathcal{R}$ about the y axis is given by

$$V = 2\pi\int_a^b x(f(x) - g(x))\,dx \qquad \text{(difference of shells formula)}$$

This obviously follows from the cylindrical shells formula because the required volume is the difference of two volumes obtained by the cylindrical shells formula. Note that a similar formula holds when the roles of x and y are reversed.

EXAMPLE 30.5: Consider the region in the first quadrant bounded above by $y = x^2$, below by $y = x^3$, and lying between $x = 0$ and $x = 1$. When revolved about the y axis, this region generates a solid of revolution whose volume, according to the difference of shells formula, is

$$2\pi \int_0^1 x(x^2 - x^3)\, dx = 2\pi \int_0^1 (x^3 - x^4)\, dx = 2\pi \left(\frac{1}{4}x^4 - \frac{1}{5}x^5\right)\Big]_0^1 = 2\pi\left(\frac{1}{4} - \frac{1}{5}\right) = \frac{\pi}{10}$$

Cross-Section Formula (Slicing Formula)

Assume that a solid lies entirely between the plane perpendicular to the x axis at $x = a$ and the plane perpendicular to the x axis at $x = b$. For each x such that $a \le x \le b$, assume that the plane perpendicular to the x axis at that value of x intersects the solid in a region of area $A(x)$. (See Fig. 30-8.) Then the volume V of the solid is given by

$$V = \int_a^b A(x)\, dx \qquad \text{(cross-section formula)}^\dagger$$

For justification, see Problem 11.

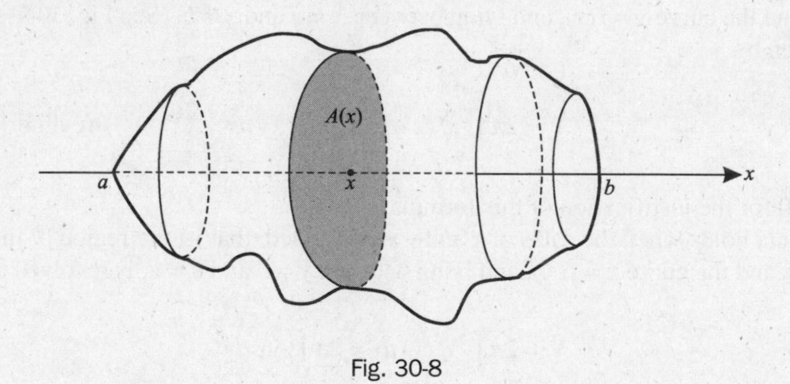

Fig. 30-8

EXAMPLE 30.6: Assume that half of a salami of length h is such that a cross-section perpendicular to the axis of the salami at a distance x from the end O is a circle of radius $\sqrt{x}$. (See Fig. 30-9.) Hence, the area $A(x)$ of the cross-section is $\pi(\sqrt{x})^2 = \pi x$. So, the cross-section formula yields

$$V = \int_0^h A(x)\, dx = \int_0^h \pi x\, dx = \frac{\pi}{2}x^2\Big]_0^h = \frac{\pi h^2}{2}$$

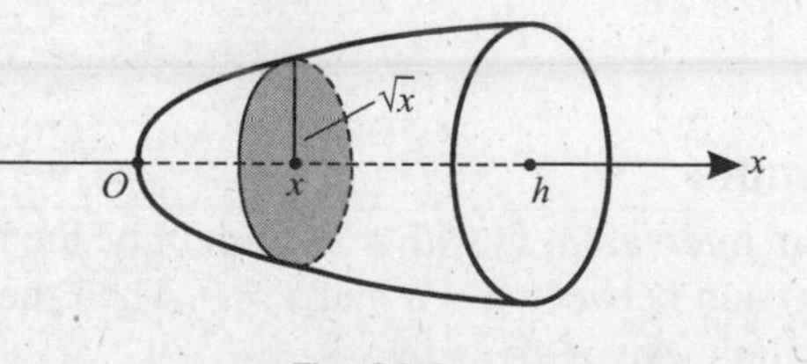

Fig. 30-9

†This formula is also called the *slicing formula* because each cross-sectional area $A(x)$ is obtained by slicing through the solid.

SOLVED PROBLEMS

1. Find the volume of a cone that has height h and whose base has radius r.

The cone is generated by revolving about the x axis the region between the line $y = \frac{r}{h}x$ and the x axis, between $x = 0$ and $x = h$. (See Fig. 30-2(*a*).) By the disk formula, the volume of the cone is

$$\pi \int_0^h y^2\,dx = \pi \int_0^h \frac{r^2}{h^2}x^2\,dx = \frac{\pi r^2}{h^2}\left(\tfrac{1}{3}x^3\right)\Big]_0^h = \frac{\pi r^2}{h^2}\left(\tfrac{1}{3}h^3\right) = \tfrac{1}{3}\pi r^2 h$$

2. Find the volume of the cylinder of height h and radius r.

The cylinder is generated by revolving about the x axis the region between the line $y = r$ and the x axis, between $x = 0$ and $x = h$. (See Fig. 30-2(*b*).) By the disk formula, the volume of the cylinder is

$$V = \pi \int_0^h y^2\,dx = \pi \int_0^h r^2\,dx = \pi r^2 x\Big]_0^h = \pi r^2 h.$$

3. Find the volume of a sphere of radius r.

The sphere is generated by revolving about the x axis the region between the semicircle $y = \sqrt{r^2 - x^2}$ and the x axis, between $x = -r$ and $x = r$. (See Fig. 30-2(*c*).) By the symmetry with respect to the y axis, we can use the part of the given region between $x = 0$ and $x = r$ and then double the result. Hence, by the disk formula, the volume of the sphere is

$$V = 2\pi \int_0^r y^2\,dx = 2\pi \int_0^r (r^2 - x^2)\,dx = 2\pi\left(r^2x - \tfrac{1}{3}x^3\right)\Big]_0^r = 2\pi\left(r^3 - \frac{r^3}{3}\right) = 2\pi\left(\tfrac{2}{3}r^3\right) = \tfrac{4}{3}\pi r^3$$

4. Let $\mathcal{R}$ be the region between the x axis, the curve $y = x^3$, and the line $x = 2$. (See Fig. 30-10.)

(a) Find the volume of the solid obtained by revolving $\mathcal{R}$ about the x axis.
(b) Find the volume of the solid obtained by revolving $\mathcal{R}$ about the y axis.

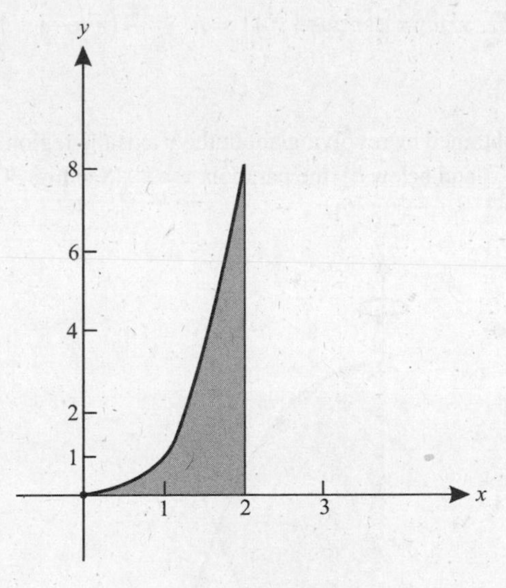

Fig. 30-10

(a) The disk formula yields the volume

$$V = \pi \int_0^2 y^2\,dx = \pi \int_0^2 (x^3)^2 dx = \pi \int_0^2 x^6\,dx = \frac{\pi}{7}x^7\Big]_0^2 = \frac{128\pi}{7}$$

(b) (*First solution*) The cylindrical shells formula yields the volume

$$V = 2\pi\int_0^2 xy\,dx = 2\pi\int_0^2 x(x^3)\,dx = 2\pi\int_0^2 x^4\,dx = 2\pi\left(\tfrac{1}{5}x^5\right)\Big]_0^2 = \frac{64\pi}{5}$$

(*Second solution*) Integrating along the y axis and using the washer formula yields the volume

$$V = \pi\int_0^8\left[2^2 - \left(\sqrt[3]{y}\right)^2\right]dy = \pi\int_0^8\left[4 - y^{2/3}\right]dy = \pi\left(4y - \frac{3}{5}y^{5/3}\right)\Big]_0^8 = \pi\left(32 - \left(\frac{3}{5}\right)32\right) = \frac{64\pi}{5}$$

5. Find the volume of the solid obtained by revolving about the y axis the region in the first quadrant inside the circle $x^2 + y^2 = r^2$, and between $y = a$ and $y = r$ (where $0 < a < r$). See Fig. 30-11. (The solid is a "polar cap" of a sphere of radius r.)

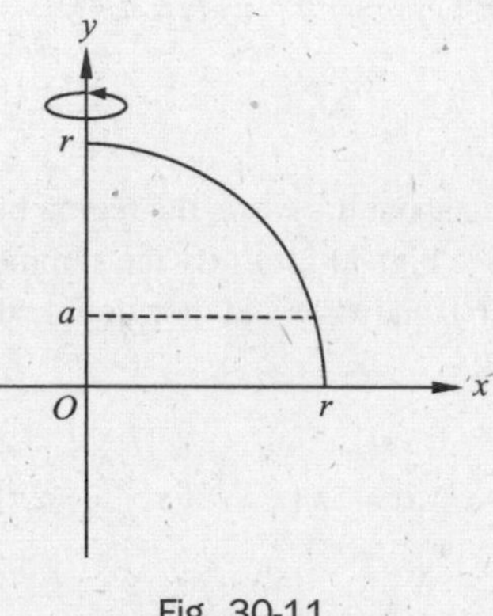

Fig. 30-11

Integrating along the y axis, the disk formula yields the volume

$$V = \pi\int_a^r x^2\,dy = \pi\int_a^r (r^2 - y^2)\,dy = \pi\left(r^2y - \tfrac{1}{3}y^3\right)\Big]_a^r = \pi\left(\tfrac{2}{3}r^3 - \left(r^2a - \tfrac{1}{3}a^3\right)\right) = \frac{\pi}{3}(2r^3 - 3r^2a + a^3)$$

6. Find the volume of the solid obtained by revolving about the y axis the region in the first quadrant bounded above by the parabola $y = 2 - x^2$ and below by the parabola $y = x^2$. (See Fig. 30-12.)

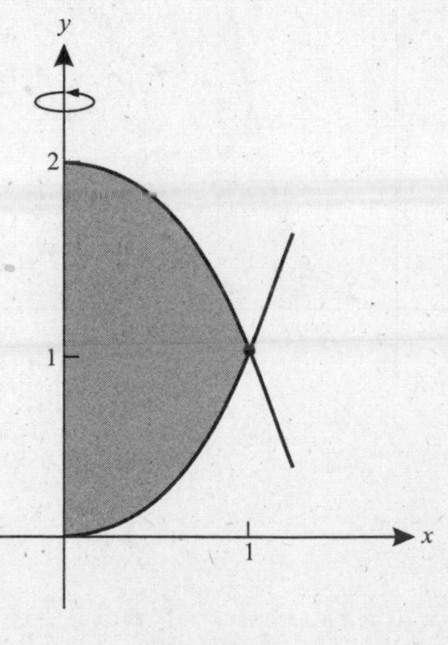

Fig. 30-12

The curves intersect at (1,1). By the difference of cylindrical shells formula, the volume is

$$V = 2\pi \int_0^1 x((2-x^2)-x^2)\,dx = 4\pi \int_0^1 (x-x^3)\,dx = 4\pi\left(\tfrac{1}{2}x^2 - \tfrac{1}{4}x^4\right)\Big]_0^1 = 4\pi\left(\tfrac{1}{2}-\tfrac{1}{4}\right) = \pi$$

7. Consider the region $\mathcal{R}$ bounded by the parabola $y = 4x^2$ and the lines $x = 0$ and $y = 16$. (See Fig. 30-5.) Find the volume of the solid obtained by revolving $\mathcal{R}$ about the line $y = -2$.

To solve this problem, we reduce it to the case of a revolution about the x axis. Raise the region $\mathcal{R}$ vertically upward through a distance of 2 units. This changes $\mathcal{R}$ into a region $\mathcal{R}^*$ that is bounded below by the parabola $y = 4x^2 + 2$, on the left by the y axis, and above by the line $y = 18$. (See Fig. 30-13.) Then the original solid of revolution has the same volume as the solid of revolution obtained by revolving R^* around the x axis. The latter volume is obtained by the washer formula:

$$V = \pi \int_0^2 (18^2 - (4x^2+2)^2)\,dx = \pi \int_0^2 (256 - 16x^4 - 16x^2 - 4)\,dx$$

$$= \pi\left(252x - \tfrac{16}{5}x^5 - \tfrac{16}{3}x^3\right)\Big]_0^2 = \pi\left(504 - \tfrac{512}{5} - \tfrac{128}{3}\right) = \frac{5384\pi}{15}$$

8. As in Problem 7, consider the region $\mathcal{R}$ bounded by the parabola $y = 4x^2$ and the lines $x = 0$ and $y = 16$. (See Fig. 30-5.) Find the volume of the solid obtained by revolving $\mathcal{R}$ about the line $x = -1$.

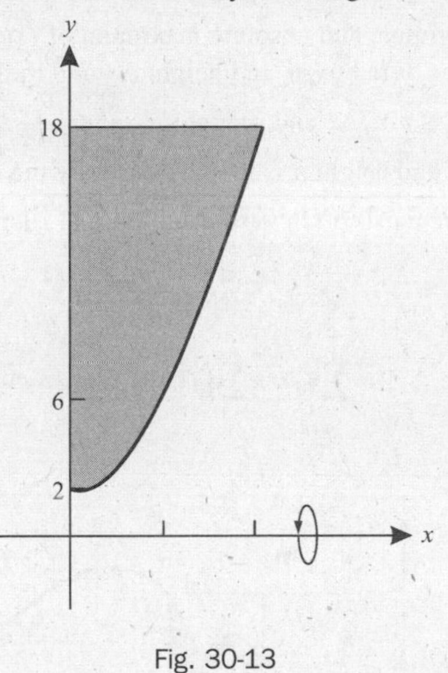

Fig. 30-13

To solve this problem, we reduce it to the case of a revolution about the y axis. Move the region $\mathcal{R}$ to the right through a distance of 1 unit. This changes $\mathcal{R}$ into a region $\mathcal{R}^*$ that is bounded on the right by the parabola $y = 4(x-1)^2$, above by $y = 16$, and on the left by $x = 1$. (See Fig. 30-14.) The desired volume is the same as that obtained when we revolve $\mathcal{R}^*$ about the y axis. The latter volume is got by the difference of cylindrical shells formula:

$$V = 2\pi \int_1^3 x(16 - 4(x-1)^2)\,dx = 2\pi \int_1^3 x(16 - 4x^2 + 8x - 4)\,dx$$

$$= 2\pi \int_1^3 (16x - 4x^3 + 8x^2 - 4x)\,dx = 2\pi\left(8x^2 - x^4 + \tfrac{8}{3}x^3 - 2x^2\right)\Big]_1^3$$

$$= 2\pi\left[(72 - 81 + 72 - 18) - \left(8 - 1 + \tfrac{8}{3} - 2\right)\right] = \tfrac{112}{3}$$

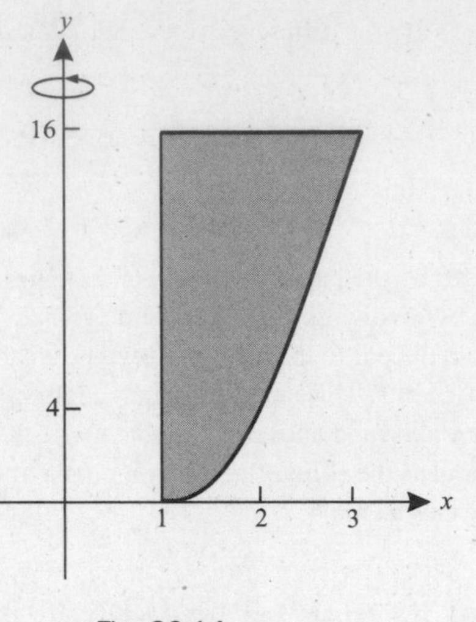

Fig. 30-14

9. Justify the disk formula: $V = \pi\int_a^b (f(x))^2\,dx$.

Divide the interval $[a, b]$ into n equal subintervals, each of length $\Delta x = \dfrac{b-a}{n}$. (See Fig. 30-15.) Consider the volume V_i obtained by revolving the region $\mathscr{R}_i$ above the ith subinterval about the x axis. If m_i and M_i are the absolute minimum and absolute maximum of f on the ith subinterval, then V_i lies between the volume of a cylinder of radius m_i and height Δx and the volume of a cylinder of radius M_i and height Δx. Thus, $\pi m_i^2 \Delta x \le V_i \le \pi M_i^2 \Delta x$ and, therefore, $m_i^2 \le \dfrac{V_i}{\pi\Delta x} \le M_i^2$. (We have assumed that the volume of a cylinder of radius r and height h is $\pi r^2 h$.) Hence, by the intermediate value theorem for the continuous function $(f(x))^2$, there exists x_i^* in the ith subinterval such that $\dfrac{V_i}{\pi\Delta x} = \left(f(x_i^*)\right)^2$ and, therefore, $V_i = \pi\left(f(x_i^*)\right)^2 \Delta x$. Thus,

$$V = \sum_{i=1}^{n} V_i = \pi\sum_{i=1}^{n}\left(f(x_i^*)\right)^2 \Delta x \qquad \text{Letting } n \to +\infty \text{, we obtain the disk formula.}$$

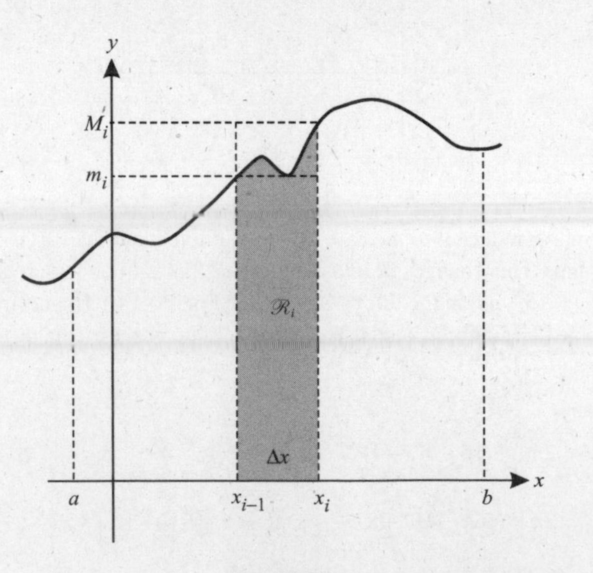

Fig. 30-15

10. Justify the cylindrical shells formula: $V = 2\pi\int_a^b xf(x)\,dx$.

Divide $[a, b]$ into n equal subintervals, each of length Δx. (See Fig. 30-16.) Let $\mathcal{R}_i$ be the region above the ith subinterval. Let x_i^* be the midpoint $\frac{x_{i-1}+x_i}{2}$ of the ith interval. The solid obtained by revolving the region $\mathcal{R}_i$ about the y axis is approximately the solid obtained by revolving the rectangle with base Δx and height $y_i^* = f(x_i^*)$. The latter solid is a cylindrical shell, that is, it lies between the cylinders obtained by revolving the rectangles with the same height $f(x_i^*)$ and with bases $[0, x_{i-1}]$ and $[0, x_i]$. Hence, it has volume

$$\begin{aligned}\pi x_i^2 f(x_i^*) - \pi x_{i-1}^2 f(x_i^*) &= \pi f(x_i^*)(x_i^2 - x_{i-1}^2)\\ &= \pi f(x_i^*)(x_i - x_{i-1})(x_i + x_{i-1}) = \pi f(x_i^*)(2x_i^*)(\Delta x) = 2\pi x_i^* f(x_i^*)(\Delta x)\end{aligned}$$

Thus, the total V is approximated by $2\pi\sum_{i=1}^{n} x_i^* f(x_1^*)\Delta x$ which approaches $2\pi\int_a^b xf(x)dx$ as $n \to +\infty$.

11. Justify the cross-section formula: $V = \int_a^b A(x)\,dx$.

Divide $[a, b]$ into n equal subintervals $[x_{i-1}, x_i]$, and choose a point x_i^* in $[x_{i-1}, x_i]$. If n is large, Δx is small and the piece of the solid between x_{i-1} and x_i, will be close to a (noncircular) disk of thickness Δx and base area $A(x_i^*)$. (See Fig. 30-17.) This disk has volume $A(x_i^*)\Delta x$. So V is approximated by $\sum_{i=1}^{n} A(x_i^*)\Delta x$, which approaches $\int_a^b A(x)\,dx$ as $n \to +\infty$.

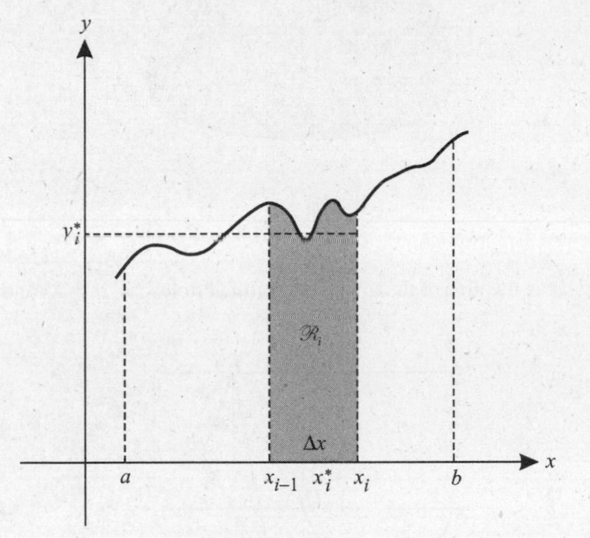

Fig. 30-16

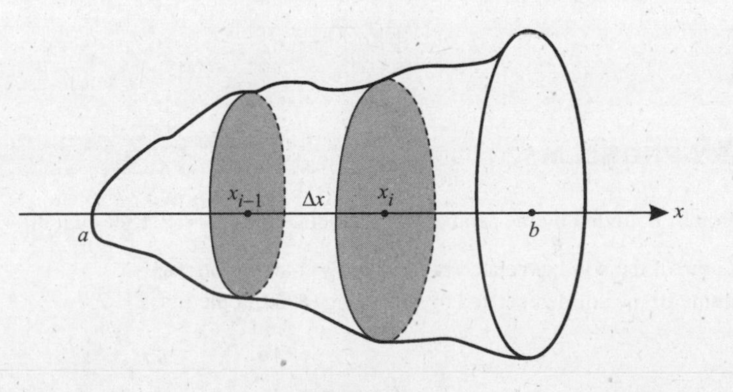

Fig. 30-17

12. A solid has a circular base of radius 4 units. Find the volume of the solid if every plane section perpendicular to a particular fixed diameter is an equilateral triangle.

Take the circle as in Fig. 30-18, with the fixed diameter on the x axis. The equation of the circle is $x^2 + y^2 = 16$. The cross-section ABC of the solid is an equilateral triangle of side $2y$ and area $A(x) = \sqrt{3}y^2 = \sqrt{3}(16 - x^2)$. Then, by the cross-section formula,

$$V = \sqrt{3}\int_{-4}^{4}(16 - x^2)\,dx = \sqrt{3}\left(16x - \tfrac{1}{3}x^3\right)\Big]_{-4}^{4} = \tfrac{256}{3}\sqrt{3}$$

13. A solid has a base in the form of an ellipse with major axis 10 and minor axis 8. Find its volume if every section perpendicular to the major axis is an isosceles triangle with altitude 6.

Take the ellipse as in Fig. 30-19, with equation $\frac{x^2}{25} + \frac{y^2}{16} = 1$. The section ABC is an isosceles triangle of base $2y$, altitude 6, and area $A(x) = 6y = 6\left(\frac{4}{5}\sqrt{25 - x^2}\right)$. Hence,

$$V = \tfrac{24}{5}\int_{-5}^{5}\sqrt{25 - x^2}\,dx = 60\pi$$

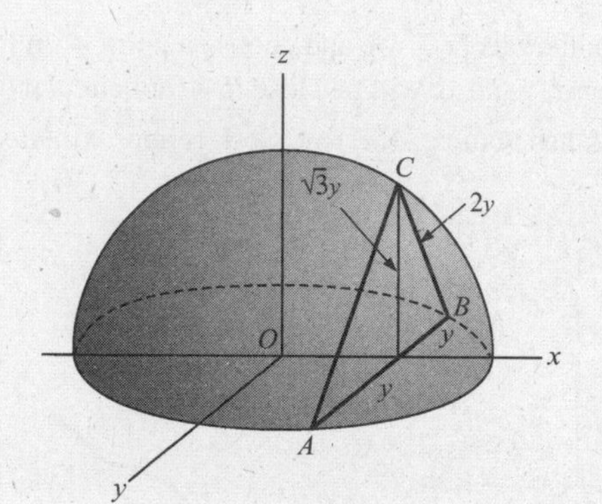

Fig. 30-18

(Note that $\int_{-5}^{5}\sqrt{25 - x^2}\,dx$ is the area of the upper half of the circle $x^2 + y^2 = 25$ and, therefore, is equal to $25\pi/2$.)

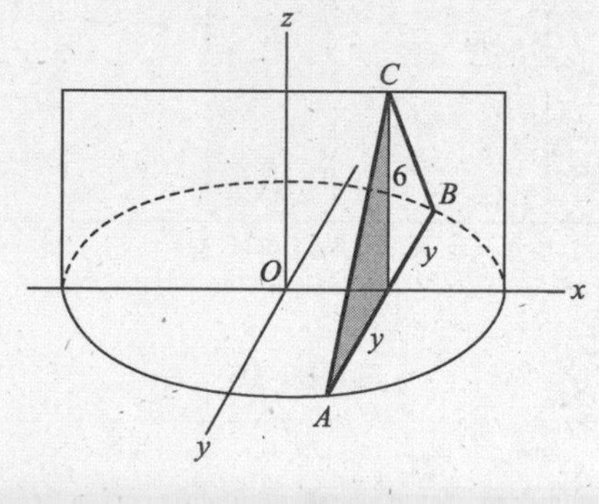

Fig. 30-19

SUPPLEMENTARY PROBLEMS

14. Consider the region $\mathcal{R}$ bounded by the parabola $y^2 = 8x$ and the line $x = 2$. (See Fig. 30-4.)

(a) Find the volume of the solid generated by revolving $\mathcal{R}$ about the y axis.
(b) Find the volume of the solid generated by revolving $\mathcal{R}$ about the line $x = 2$.

Ans. (a) $\frac{128\pi}{5}$; (b) $\frac{256\pi}{15}$

15. Find the volume of the solid generated by revolving the region between the x axis and the parabola $y = 4x - x^2$ about the line $y = 6$.

Ans. $\dfrac{1408\pi}{15}$

16. Find the volume of the torus (doughnut) generated by revolving the circle $(x - a)^2 + y^2 = b^2$ about the y axis, where $0 < b < a$.

Ans. $2\pi^2 ab^2$

17. Consider the region $\mathcal{R}$ bounded by $y = -x^2 - 3x + 6$ and $x + y = 3$. Find the volume of the solid generated by revolving $\mathcal{R}$ about:

(a) the x axis; (b) the line $x = 3$.

Ans. (a) $\dfrac{1792\pi}{15}$; (b) $\dfrac{256\pi}{3}$

In Problems 18–26, find the volume generated when the given region is revolved about the given line. Use the disk formula.

18. The region bounded by $y = 2x^2$, $y = 0$, $x = 0$, $x = 5$, about the x axis.

Ans. 2500π

19. The region bounded by $x^2 - y^2 = 16$, $y = 0$, $x = 8$, about the x axis.

Ans. $\dfrac{256\pi}{3}$

20. The region bounded by $y = 4x^2$, $x = 0$, $y = 16$, about $y = 16$. (See Fig. 30-5.)

Ans. $\dfrac{4096\pi}{15}$

21. The region bounded by $y^2 = x^3$, $y = 0$, $x = 2$, about the x axis.

Ans. 4π

22. The region bounded by $y = x^3$, $y = 0$, $x = 2$, about $x = 2$.

Ans. $\dfrac{16\pi}{5}$

23. The region within the curve $y^2 = x^4(1 - x^2)$, about the x axis.

Ans. $\dfrac{4\pi}{35}$

24. The region within the ellipse $4x^2 + 9y^2 = 36$, about the x axis.

Ans. 16π

25. The region within the ellipse $4x^2 + 9y^2 = 36$, about the y axis.

Ans. 24π

26. The region within the parabola $x = 9 - y^2$ and between $y = x - 7$ and the y axis, about the y axis.

Ans. $\frac{963\pi}{5}$

In Problems 27–32, find the volume of the solid generated by revolving the given region about the given line. Use the washer formula.

27. The region bounded by $y = 2x^2$, $y = 0$, $x = 0$, $x = 5$, about the y axis.

Ans. 625π

28. The region bounded by $x^2 - y^2 = 16$, $y = 0$, $x = 8$, about the y axis.

Ans. $128\sqrt{3}\pi$

29. The region bounded by $y = x^3$, $x = 0$, $y = 8$, about $x = 2$.

Ans. $\frac{144\pi}{5}$

30. The region bounded by $y = x^2$, $y = 4x - x^2$, about the x axis.

Ans. $\frac{32\pi}{3}$

31. The region bounded by $y = x^2$, $y = 4x - x^2$, about $y = 6$.

Ans. $\frac{64\pi}{3}$

32. The region bounded by $x = 9 - y^2$, $y = x - 7$, about $x = 4$.

Ans. $\frac{153\pi}{5}$

In Problems 33–37, find the volume of the solid generated by revolving the given region about the given line. Use the cylindrical shells formula.

33. The region bounded by $y = 2x^2$, $y = 0$, $x = 0$, $x = 5$, about $x = 6$.

Ans. 375π

34. The region bounded by $y = x^3$, $y = 0$, $x = 2$, about $y = 8$.

Ans. $\frac{320\pi}{7}$

35. The region bounded by $y = x^2$, $y = 4x - x^2$, about $x = 5$.

Ans. $\frac{64\pi}{3}$

36. The region bounded by $y = x^2 - 5x + 6$ and $y = 0$, about the y axis.

Ans. $\frac{5\pi}{6}$

37. The region bounded by $x = 9 - y^2$, $y = x - 7$, $x = 0$, about $y = 3$.

Ans. $\dfrac{369\pi}{2}$

In Problems 38–42, find the volume generated by revolving the given region about the given line. Use any appropriate method.

38. The region bounded by $y = e^{-x^2}$, $y = 0$, $x = 0$, $x = 1$, about the y axis.

Ans. $\pi(1 - e^{-1})$

39. The region bounded by $y = 2x^2$, $y = 2x + 4$, about $x = 2$.

Ans. 27π

40. The region bounded by $y = 2x$, $y = 0$, $x = 0$, $x = 1$, about the y axis.

Ans. $\dfrac{4\pi}{3}$

41. The region bounded by $y = x^2$, $x = y^2$, about the x axis.

Ans. $\dfrac{3\pi}{10}$

42. The region bounded by $xy = 4$, $y = (x - 3)^2$, about the x axis.

Ans. $\dfrac{27\pi}{5}$

43. Find the volume of the frustum of a cone whose lower base is of radius R, upper base is of radius r, and altitude is h.

Ans. $\frac{1}{3}\pi h(r^2 + rR + R^2)$

44. A solid has a circular base of radius 4 units. Find the volume of the solid if every plane perpendicular to a fixed diameter (the x axis of Fig. 30-18) is: (a) a semicircle; (b) a square; (c) an isosceles right triangle with the hypotenuse in the plane of the base.

Ans. (a) $\dfrac{128\pi}{3}$; (b) $\dfrac{1024}{3}$; (c) $\dfrac{256}{3}$

45. A solid has a base in the form of an ellipse with major axis 10 and minor axis 8. Find its volume if every section perpendicular to the major axis is an isosceles right triangle with one leg in the plane of the base.

Ans. $\dfrac{640}{3}$

46. The base of a solid is the first-quadrant region bounded by the line $4x + 5y = 20$ and the coordinate axes. Find its volume if every plane section perpendicular to the x axis is a semicircle.

Ans. $\dfrac{10\pi}{3}$

47. The base of a solid is the circle $x^2 + y^2 = 16x$, and every plane section perpendicular to the x axis is a rectangle whose height is twice the distance of the plane of the section from the origin. Find its volume.

Ans. 1024π

48. The section of a certain solid cut by any plane perpendicular to the x axis is a circle with the ends of a diameter lying on the parabolas $y^2 = 4x$ and $x^2 = 4y$. Find its volume.

Ans. $\dfrac{6561\pi}{280}$

49. The section of a certain solid cut by any plane perpendicular to the x axis is a square with the ends of a diagonal lying on the parabolas $y^2 = 4x$ and $x^2 = 4y$. Find its volume.

Ans. $\dfrac{144}{35}$

50. A hole of radius 1 unit is bored through a sphere of radius 3 units, the axis of the hole being a diameter of the sphere. Find the volume of the remaining part of the sphere.

Ans. $\dfrac{64\pi\sqrt{2}}{3}$

CHAPTER 31

Techniques of Integration I: Integration by Parts

If u and v are functions, the product rule yields

$$D_x(uv) = uv' + vu'$$

which can be rewritten in terms of antiderivatives as follows:

$$uv = \int uv'\,dx + \int vu'\,dx$$

Now, $\int uv'\,dx$ can be written as $\int u\,dv$, and $\int vu'\,dx$ can be written as $\int v\,du$.[†] Thus, $uv = \int u\,dv + \int v\,du$ and, therefore,

$$\int u\,dv = uv - \int v\,du \qquad \text{(integration by parts)}$$

The purpose of integration by parts is to replace a "difficult" integration $\int u\,dv$ by an "easy" integration $\int v\,du$.

EXAMPLE 31.1: Find $\int x \ln x\,dx$.

In order to use the integration by parts formula, we must divide the integrand $x \ln x\,dx$ into two "parts" u and dv so that we can easily find v by an integration and also easily find $\int v\,du$. In this example, let $u = \ln x$ and $dv = x\,dx$. Then we can set $v = \frac{1}{2}x^2$ and note that $du = \frac{1}{x}dx$. So, the integration by parts formula yields:

$$\int x \ln x\,dx = \int u\,dv = uv - \int v\,du = (\ln x)(\tfrac{1}{2}x^2) - \int \tfrac{1}{2}x^2\left(\frac{1}{x}dx\right)$$

$$= \tfrac{1}{2}x^2 \ln x - \tfrac{1}{2}\int x\,dx = \tfrac{1}{2}x^2 \ln x - \tfrac{1}{4}x^2 + C$$

$$= \tfrac{1}{4}x^2(2\ln x - 1) + C$$

Integration by parts can be made easier to apply by setting up a rectangle such as the following one for Example 1.

$u = \ln x$	$dv = x\,dx$
$du = \frac{1}{x}dx$	$v = \frac{1}{2}x^2$

[†] $\int uv'\,dx = \int u\,dv$, where, after the integration on the right, the variable v is replaced by the corresponding function of x. In fact, by the Chain Rule, $D_x\left(\int u\,dv\right) = D_v\left(\int u\,dv\right)\cdot D_x v = u \cdot v'$. Hence, $\int u\,dv = \int uv'\,dx$. Similarly, $\int v\,du = \int vu'\,dx$.

In the first row, we place u and dv. In the second row, we place the results of computing du and v. The desired result of the integration parts formula $uv - \int v\,du$ can be obtained by first multiplying the upper-left corner u by the lower-right corner v, and then subtracting the integral of the product $v\,du$ of the two entries v and du in the second row.

EXAMPLE 31.2: Find $\int xe^x\,dx$.

Let $u = x$ and $dv = e^x\,dx$. We can picture this in the box below.

$$\begin{array}{|cc|} u = x & dv = e^x dx \\ du = dx & v = e^x \end{array}$$

Then,

$$\int xe^x dx = uv - \int v\,du = xe^x - \int e^x\,dx = xe^x - e^x + C$$

$$= e^x(x-1) + C$$

EXAMPLE 31.3: Find $\int e^x \cos x\,dx$.

Let $u = e^x$ and $dv = \cos x\,dx$. Then we get the box

$$\begin{array}{|cc|} u = e^x & dv = \cos x\,dx \\ du = e^x\,dx & v = \sin x \end{array}$$

So,

$$\int e^x \cos x\,dx = uv - \int v\,du = e^x \sin x - \int e^x \sin x\,dx \qquad (1)$$

Now we have the problem of finding $\int e^x \sin x\,dx$, which seems to be just as hard as the original integral $\int e^x \cos x\,dx$. However, let us try to find $\int e^x \sin x\,dx$ by another integration by parts. This time, let $u = e^x$ and $dv = \sin x\,dx$.

$$\begin{array}{|cc|} u = e^x & dv = \sin x\,dx \\ du = e^x\,dx & v = -\cos x \end{array}$$

Then,

$$\int e^x \sin x\,dx = -e^x \cos x - \int -e^x \cos x\,dx$$

$$= -e^x \cos x + \int e^x \cos x\,dx$$

Substituting in formula (1) above, we get:

$$\int e^x \cos x\,dx = e^x \sin x - \left(-e^x \cos x + \int e^x \cos x\,dx\right)$$

$$= e^x \sin x + e^x \cos x - \int e^x \cos x\,dx$$

Adding $\int e^x \cos x\,dx$ to both sides yields $2\int e^x \cos x\,dx = e^x \sin x + e^x \cos x$. So,

$$\int e^x \cos x\,dx = \tfrac{1}{2}(e^x \sin x + e^x \cos x)$$

We must add an arbitrary constant:

$$\int e^x \cos x\,dx = \tfrac{1}{2}(e^x \sin x + e^x \cos x) + C$$

Notice that this example required an iterated application of integration by parts.

SOLVED PROBLEMS

1. Find $\int x^3 e^{x^2} dx$.

Let $u = x^2$ and $dv = xe^{x^2} dx$. Note that v can be evaluated by using the substitution $w = x^2$. (We get $v = \frac{1}{2}\int e^w dw = \frac{1}{2}e^w = \frac{1}{2}e^{x^2}$.)

$$\begin{array}{|ll|} u = x^2 & dv = xe^{x^2} dx \\ du = 2x\,dx & v = \frac{1}{2}e^{x^2} \end{array}$$

Hence,

$$\int x^3 e^{x^2} dx = \tfrac{1}{2}x^2 e^{x^2} - \int xe^{x^2} dx$$
$$= \tfrac{1}{2}x^2 e^{x^2} - \tfrac{1}{2}e^{x^2} + C$$
$$= \tfrac{1}{2}e^{x^2}(x^2 - 1) + C$$

2. Find $\int \ln(x^2 + 2)\,dx$.

Let $u = \ln(x^2 + 2)$ and $dv = dx$.

$$\begin{array}{|ll|} u = \ln(x^2+2) & dv = dx \\ du = \dfrac{2x}{x^2+2}dx & v = x \end{array}$$

So,

$$\int \ln(x^2+2)\,dx = x\ln(x^2+2) - 2\int \frac{x^2}{x^2+2}dx$$
$$= x\ln(x^2+2) - 2\int \left(1 - \frac{2}{x^2+2}\right)dx$$
$$= x\ln(x^2+2) - 2x + \frac{4}{\sqrt{2}}\tan^{-1}\left(\frac{x}{\sqrt{2}}\right) + C$$
$$= x(\ln(x^2+2) - 2) + 2\sqrt{2}\tan^{-1}\left(\frac{x}{\sqrt{2}}\right) + C$$

3. Find $\int \ln x\,dx$.

Let $u = \ln x$ and $dv = dx$.

$$\begin{array}{|ll|} u = \ln x & dv = dx \\ du = \dfrac{1}{x}dx & v = x \end{array}$$

So,

$$\int \ln x\,dx = x\ln x - \int 1\,dx = x\ln x - x + C$$
$$= x(\ln x - 1) + C$$

4. Find $\int x\sin x\,dx$.

We have three choices: (a) $u = x\sin x$, $dv = dx$; (b) $u = \sin x$, $dv = x\,dx$; (c) $u = x$, $dv = \sin x\,dx$.

(a) Let $u = x\sin x$, $dv = dx$. Then $du = (\sin x + x\cos x)\,dx$, $v = x$, and

$$\int x\sin x\,dx = x \cdot x\sin x - \int x(\sin x + x\cos x)\,dx$$

The resulting integral is not as simple as the original, and this choice is discarded.

(b) Let $u = \sin x$, $dv = x\,dx$. Then $du = \cos x\,dx$, $v = \frac{1}{2}x^2$, and

$$\int x \sin x\,dx = \tfrac{1}{2}x^2 \sin x - \int \tfrac{1}{2}x^2 \cos x\,dx$$

The resulting integral is not as simple as the original, and this choice too is discarded.

(c) Let $u = x$, $dv = \sin x\,dx$. Then $du = dx$, $v = -\cos x$, and

$$\int x \sin x\,dx = -x\cos x - \int -\cos x\,dx = -x\cos x + \sin x + C$$

5. Find $\int x^2 \ln x\,dx$.
Let $u = \ln x$, $dv = x^2\,dx$. Then $du = \frac{dx}{x}$, $v = \frac{x^3}{3}$, and

$$\int x^2 \ln x\,dx = \frac{x^3}{3}\ln x - \int \frac{x^3}{3}\frac{dx}{x} = \frac{x^3}{3}\ln x - \tfrac{1}{3}\int x^2 dx = \frac{x^3}{3}\ln x - \tfrac{1}{9}x^3 + C$$

6. Find $\int \sin^{-1} x\,dx$.
Let $u = \sin^{-1} x$, $dv = dx$.

$u = \sin^{-1} x$	$dv = dx$
$du = \frac{1}{\sqrt{1-x^2}}dx$	$v = x$

So,

$$\int \sin^{-1} x\,dx = x\sin^{-1} x - \int \frac{x}{\sqrt{1-x^2}}dx$$

$$= x\sin^{-1} x + \tfrac{1}{2}\int (1-x^2)^{-1/2}(-2x)dx$$

$$= x\sin^{-1} x + \tfrac{1}{2}(2(1-x^2)^{1/2}) + C \qquad \text{(by Quick Formula I)}$$

$$= x\sin^{-1} x + (1-x^2)^{1/2} + C = x\sin^{-1} x + \sqrt{1-x^2} + C$$

7. Find $\int \tan^{-1} x\,dx$.
Let $u = \tan^{-1} x$, $dv = dx$.

$u = \tan^{-1} x$	$dv = dx$
$du = \frac{1}{1+x^2}dx$	$v = x$

So,

$$\int \tan^{-1} x\,dx = x\tan^{-1} x - \int \frac{x}{1+x^2}dx = x\tan^{-1} x - \tfrac{1}{2}\int \frac{2x}{1+x^2}dx$$

$$= x\tan^{-1} x - \tfrac{1}{2}\ln(1+x^2) + C \qquad \text{(by Quick Formula II)}$$

8. Find $\int \sec^3 x\,dx$.
Let $u = \sec x$, $dv = \sec^2 x\,dx$.

$u = \sec x$	$dv = \sec^2 x\,dx$
$du = \sec x \tan x\,dx$	$v = \tan x$

Thus,
$$\int \sec^3 x\,dx = \sec x \tan x - \int \sec x \tan^2 x\,dx$$
$$= \sec x \tan x - \int \sec x(\sec^2 x - 1)\,dx$$
$$= \sec x \tan x - \int \sec^3 x\,dx + \int \sec x\,dx$$
$$= \sec x \tan x - \int \sec^3 x\,dx + \ln|\sec x + \tan x|$$

Then,
$$2\int \sec^3 x\,dx = \sec x \tan x + \ln|\sec x + \tan x|$$

Hence,
$$\int \sec^3 x\,dx = \tfrac{1}{2}(\sec x \tan x + \ln|\sec x + \tan x|) + C$$

9. Find $\int x^2 \sin x\,dx$.
Let $u = x^2$, $dv = \sin x\,dx$. Thus, $du = 2x\,dx$ and $v = -\cos x$. Then

$$\int x^2 \sin x\,dx = -x^2 \cos x - \int -2x \cos x\,dx$$
$$= -x^2 \cos x + 2\int x \cos x\,dx$$

Now apply integration by parts to $\int x \cos x\,dx$, with $u = x$ and $dv = \cos x\,dx$, getting

$$\int x \cos x = x \sin x - \int \sin x\,dx = x \sin x + \cos x$$

Hence,
$$\int x^2 \sin x\,dx = -x^2 \cos x + 2(x \sin x + \cos x) + C$$

10. Find $\int x^3 e^{2x} dx$.
Let $u = x^3$, $dv = e^{2x}\,dx$. Then $du = 3x^2\,dx$, $v = \frac{1}{2}e^{2x}$, and

$$\int x^3 e^{2x} dx = \tfrac{1}{2}x^3 e^{2x} - \tfrac{3}{2}\int x^2 e^{2x} dx$$

For the resulting integral, let $u = x^2$ and $dv = e^{2x}\,dx$. Then $du = 2x\,dx$, $v = \frac{1}{2}e^{2x}$, and

$$\int x^3 e^{2x} dx = \tfrac{1}{2}x^3 e^{2x} - \tfrac{3}{2}\left(\tfrac{1}{2}x^2 e^{2x} - \int x e^{2x} dx\right) = \tfrac{1}{2}x^3 e^{2x} - \tfrac{3}{4}x^2 e^{2x} + \tfrac{3}{2}\int x e^{2x} dx$$

For the resulting integral, let $u = x$ and $dv = e^{2x}\,dx$. Then $du = dx$, $v = \frac{1}{2}e^{2x}$, and

$$\int x^3 e^{2x} dx = \tfrac{1}{2}x^3 e^{2x} - \tfrac{3}{4}x^2 e^{2x} + \tfrac{3}{2}\left(\tfrac{1}{2}x e^{2x} - \tfrac{1}{2}\int e^{2x} dx\right) = \tfrac{1}{2}x^3 e^{2x} - \tfrac{3}{4}x^2 e^{2x} + \tfrac{3}{4}x e^{2x} - \tfrac{3}{8}e^{2x} + C$$

11. Derive the following *reduction formula* for $\int \sin^m x\,dx$.

$$\int \sin^m x\,dx = -\frac{\sin^{m-1} x \cos x}{m} + \frac{m-1}{m}\int \sin^{m-2} x\,dx$$

Let $u = \sin^{m-1} x$ and $dv = \sin x\,dx$.

$u = \sin^{m-1} x$	$dv = \sin x\,dx$
$du = (m-1)\sin^{m-2} x\,dx$	$v = -\cos x$

Then
$$\int \sin^m x\,dx = -\cos x \sin^{m-1} x + (m-1)\int \sin^{m-2} x \cos^2 x\,dx$$
$$= -\cos x \sin^{m-1} x + (m-1)\int \sin^{m-2} x(1-\sin^2 x)\,dx$$
$$= -\cos x \sin^{m-1} x + (m-1)\int \sin^{m-2} x\,dx - (m-1)\int \sin^m x\,dx$$

Hence,
$$m\int \sin^m x\,dx = -\cos x \sin^{m-1} x + (m-1)\int \sin^{m-2} x\,dx$$
and division by m yields the required formula.

12. Apply the reduction formula of Problem 11 to find $\int \sin^2 x\,dx$.
When $m = 2$, we get

$$\int \sin^2 x\,dx = -\frac{\sin x \cos x}{2} + \tfrac{1}{2}\int \sin^0 x\,dx$$
$$= -\frac{\sin x \cos x}{2} + \tfrac{1}{2}\int 1\,dx$$
$$= -\frac{\sin x \cos x}{2} + \frac{x}{2} + C = \frac{x - \sin x \cos x}{2} + C$$

13. Apply the reduction formula of Problem 11 to find $\int \sin^3 x\,dx$.
When $m = 3$, we get

$$\int \sin^3 x\,dx = -\frac{\sin^2 x \cos x}{3} + \tfrac{2}{3}\int \sin x\,dx$$
$$= -\frac{\sin^2 x \cos x}{3} - \tfrac{2}{3}\cos x + C$$
$$= -\frac{\cos x}{3}(2 + \sin^2 x) + C$$

SUPPLEMENTARY PROBLEMS

In Problems 14–21, use integration by parts to verify the specified formulas.

14. $\int x \cos x\,dx = x \sin x + \cos x + C$

15. $\int x \sec^2 3x\,dx = \tfrac{1}{3}x \tan 3x - \tfrac{1}{9}\ln|\sec x| + C$

16. $\int \cos^{-1} 2x\,dx = x\cos^{-1} 2x - \tfrac{1}{2}\sqrt{1-4x^2} + C$

17. $\int x \tan^{-1} x\,dx = \tfrac{1}{2}(x^2+1)\tan^{-1} x - \tfrac{1}{2}x + C$

18. $\int x^2 e^{-3x}\,dx = -\tfrac{1}{3}e^{-3x}(x^2 + \tfrac{2}{3}x + \tfrac{2}{9}) + C$

19. $\int x^3 \sin x\,dx = -x^3 \cos x + 3x^2 \sin x + 6x \cos x - 6 \sin x + C$

20. $\int x\sin^{-1}(x^2)\,dx = \frac{1}{2}x^2\sin^{-1}(x^2) + \frac{1}{2}\sqrt{1-x^4} + C$

21. $\int \frac{\ln x}{x^2}\,dx = -\frac{\ln x + 1}{x} + C$

22. Show that $\int_0^{2\pi} x\sin nx\,dx = -\frac{2\pi}{n}$ for any positive integer n.

23. Prove the following reduction formula: $\int \sec^n x\,dx = \frac{\tan x\sec^{n-2} x}{n-1} + \frac{n-2}{n-1}\int \sec^{n-2} x\,dx$.

24. Apply Problem 23 to find $\int \sec^4 x\,dx$.

Ans. $\frac{1}{3}\tan x(\sec^2 x + 2) + C$

25. Prove the reduction formula:

$$\int \frac{x^2}{(a^2+x^2)^n}\,dx = \frac{1}{2n-2}\left(-\frac{x}{(a^2+x^2)^{n-1}} + \int \frac{dx}{(a^2+x^2)^{n-1}}\right)$$

26. Apply Problem 25 to find $\int \frac{x^2}{(a^2+x^2)^2}\,dx$.

Ans. $\frac{1}{2}\left(-\frac{x}{a^2+x^2} + \frac{1}{a}\tan^{-1}\frac{x}{a}\right) + C$

27. Prove $\int x^n \ln x\,dx = \frac{x^{n+1}}{(n+1)^2}[(n+1)\ln x - 1)] + C$ for $n \neq -1$.

28. Prove the reduction formula: $\int x^n e^{ax} dx = \frac{1}{a}x^n e^{ax} - \frac{n}{a}\int x^{n-1}e^{ax} dx$.

29. Use Problem 28 and Example 2 to show that: $\int x^2 e^x dx = e^x(x^2 - 2x + 2) + C$.

CHAPTER 32

Techniques of Integration II: Trigonometric Integrands and Trigonometric Substitutions

Trigonometric Integrands

1. Let us consider integrals of the form $\int \sin^k x \cos^n x\, dx$, where k and n are nonnegative integers.

   Type 1. At least one of sin x and cos x occurs to an odd power: Then a substitution for the other function works.

EXAMPLE 32.1: $\int \sin^3 x \cos^2 x\, dx$.

Let $u = \cos x$. Then $du = -\sin x\, dx$. Hence,

$$\begin{aligned}\int \sin^3 x \cos^2 x\, dx &= \int \sin^2 x \cos^2 x \sin x\, dx \\ &= \int (1 - \cos^2 x)\cos^2 x \sin x\, dx \\ &= -\int (1 - u^2)u^2\, du = \int (u^4 - u^2)\, du \\ &= \tfrac{1}{5}u^5 - \tfrac{1}{3}u^3 + C = \tfrac{1}{5}\cos^5 x - \tfrac{1}{3}\cos^3 x + C\end{aligned}$$

EXAMPLE 32.2: $\int \sin^4 x \cos^7 x\, dx$.

Let $u = \sin x$. Then $du = \cos x\, dx$, and

$$\begin{aligned}\int \sin^4 x \cos^7 x\, dx &= \int \sin^4 x \cos^6 x \cos x\, dx \\ &= \int u^4(1 - u^2)^3\, du = \int u^4(1 - 3u^2 + 3u^4 - u^6)\, du \\ &= \int (u^5 - 3u^6 + 3u^8 - u^{10})\, du \\ &= \tfrac{1}{6}u^6 - \tfrac{3}{7}u^7 + \tfrac{1}{3}u^9 - \tfrac{1}{11}u^{11} + C \\ &= \tfrac{1}{6}\sin^6 x - \tfrac{3}{7}\sin^7 x + \tfrac{1}{3}\sin^9 x - \tfrac{1}{11}\sin^{11} x + C\end{aligned}$$

EXAMPLE 32.3: $\int \sin^5 x\, dx$.

Let $u = \cos x$. Then $du = -\sin x\, dx$ and

$$\int \sin^5 x\,dx = \int \sin^4 x \sin x\,dx = \int (1-\cos^2 x)^2 \sin x\,dx$$
$$= -\int (1-u^2)^2\,du = -\int (1-2u^2+u^4)\,du$$
$$= -(u - \tfrac{2}{3}u^3 + \tfrac{1}{5}u^5) + C$$
$$= -\tfrac{1}{5}\cos^5 x + \tfrac{2}{3}\cos^3 x - \cos x + C$$

Type 2. Both powers of $\sin x$ *and* $\cos x$ *are even*: This always involves a more tedious computation, using the identities

$$\cos^2 x = \frac{1+\cos 2x}{2} \quad \text{and} \quad \sin^2 x = \frac{1-\cos 2x}{2}$$

EXAMPLE 32.4:

$$\int \cos^2 x \sin^4 x\,dx = \int (\cos^2 x)(\sin^2 x)^2\,dx$$
$$= \int \left(\frac{1+\cos 2x}{2}\right)\left(\frac{1-\cos 2x}{2}\right)^2 dx$$
$$= \int \left(\frac{1+\cos 2x}{2}\right)\left(\frac{1-2\cos 2x+\cos^2 2x}{4}\right) dx$$
$$= \frac{1}{8}\int (1(1-2\cos 2x+\cos^2 2x) + (\cos 2x)(1-2\cos 2x+\cos^2 2x))\,dx$$
$$= \frac{1}{8}\int (1-2\cos 2x+\cos^2 2x+\cos 2x-2\cos^2 2x+\cos^3 2x)\,dx$$
$$= \frac{1}{8}\int (1-\cos 2x-\cos^2 2x+\cos^3 2x)\,dx$$
$$= \frac{1}{8}\left(\int 1\,dx - \int \cos 2x\,dx - \int \cos^2 2x\,dx + \int \cos^3 2x\,dx\right)$$
$$= \frac{1}{8}\left(x - \frac{\sin 2x}{2} - \int \frac{1+\cos 4x}{2}\,dx + \int (\cos 2x)(1-\sin^2 2x)\,dx\right)$$
$$= \frac{1}{8}\left(x - \frac{\sin 2x}{2} - \frac{1}{2}\left(x + \frac{\sin 4x}{4}\right) + \int \cos 2x\,dx - \frac{1}{2}\int u^2\,du\right) \quad [\text{letting } u = \sin 2x]$$
$$= \frac{1}{8}\left(x - \frac{\sin 2x}{2} - \frac{x}{2} - \frac{\sin 4x}{8} + \frac{\sin 2x}{2} - \frac{1}{2}\frac{\sin^3 2x}{3}\right) + C$$
$$= \frac{1}{8}\left(\frac{x}{2} - \frac{\sin 4x}{8} - \frac{\sin^3 2x}{6}\right) + C$$
$$= \frac{x}{16} - \frac{\sin 4x}{64} - \frac{\sin^3 2x}{48} + C$$

2. Let us consider integrals of the form $\int \tan^k x \sec^n x\,dx$, where k and n are nonnegative integers. Recall that $\sec^2 x = 1 + \tan^2 x$.

Type 1. n is even: Substitute $u = \tan x$.

EXAMPLE 32.5: $\int \tan^2 x \sec^4 x\,dx$

Let $u = \tan x$, $du = \sec^2 x\,dx$. So,

$$\int \tan^2 x \sec^4 x\,dx = \int \tan^2 x(1+\tan^2 x)\sec^2 x\,dx = \int u^2(1+u^2)\,du$$
$$= \int (u^4+u^2)\,du = \tfrac{1}{5}u^5 + \tfrac{1}{3}u^3 + C = \tfrac{1}{5}\tan^5 x + \tfrac{1}{3}\tan^3 x + C$$

Type 2. *n is odd and k is odd*: Substitute $u = \sec x$.

EXAMPLE 32.6: $\int \tan^3 x \sec x\, dx$

Let $u = \sec x$, $du = \sec x \tan x\, dx$. So,

$$\int \tan^3 x \sec x\, dx = \int \tan^2 x \sec x \tan x\, dx = \int (\sec^2 x - 1) \sec x \tan x\, dx$$

$$= \int (u^2 - 1)\, du = \tfrac{1}{3}u^3 - u + C = \tfrac{1}{3}\sec^3 x - \sec x + C$$

Type 3. *n is odd and k is even*: This case usually requires a tedious calculation.

EXAMPLE 32.7:

$$\int \tan^2 x \sec x\, dx = \int (\sec^2 x - 1)\sec x\, dx = \int (\sec^3 x - \sec x)\, dx$$

$$= \frac{1}{2}(\sec x \tan x + \ln|\sec x + \tan x|) - \ln|\sec x + \tan x| + C \quad \text{(by Problem 8 of Chapter 31)}$$

$$= \tfrac{1}{2}(\sec x \tan x - \ln|\sec x + \tan x|) + C$$

3. Let us consider integrals of the form $\int \sin Ax \cos Bx\, dx$, $\int \sin Ax \sin Bx\, dx$, and $\int \cos Ax \cos Bx\, dx$. We shall need the identities

$$\sin Ax \cos Bx = \tfrac{1}{2}(\sin(A+B)x + \sin(A-B)x)$$

$$\sin Ax \sin Bx = \tfrac{1}{2}(\cos(A-B)x - \cos(A+B)x)$$

$$\cos Ax \cos Bx = \tfrac{1}{2}(\cos(A-B)x + \cos(A+B)x)$$

EXAMPLE 32.8:

$$\int \sin 7x \cos 3x\, dx = \int \tfrac{1}{2}(\sin(7+3)x + \sin(7-3)x)\, dx = \int \tfrac{1}{2}(\sin 10x + \sin 4x)\, dx$$

$$= \tfrac{1}{2}(-\tfrac{1}{10}\cos 10x - \tfrac{1}{4}\cos 4x) + C = -\tfrac{1}{40}(2\cos 10x + 5\cos 4x) + C$$

EXAMPLE 32.9:

$$\int \sin 7x \sin 3x\, dx = \int \tfrac{1}{2}(\cos(7-3)x - \cos(7+3)x)\, dx = \int \tfrac{1}{2}(\cos 4x - \cos 10x)\, dx$$

$$= \tfrac{1}{2}(\tfrac{1}{4}\sin 4x - \tfrac{1}{10}\sin 10x) + C = \tfrac{1}{40}(5\sin 4x - 2\sin 10x) + C$$

EXAMPLE 32.10:

$$\int \cos 7x \cos 3x\, dx = \int \tfrac{1}{2}(\cos(7-3)x + \cos(7+3)x)\, dx = \int \tfrac{1}{2}(\cos 4x + \cos 10x)\, dx$$

$$= \tfrac{1}{2}(\tfrac{1}{4}\sin 4x + \tfrac{1}{10}\sin 10x) + C = \tfrac{1}{40}(5\sin 4x + 2\sin 10x) + C$$

Trigonometric Substitutions

There are three principal kinds of trigonometric substitutions. We shall introduce each one by means of a typical example.

EXAMPLE 32.11: Find $\int \frac{dx}{x^2\sqrt{4+x^2}}$.

Let $x = 2\tan\theta$, that is, $\theta = \tan^{-1}(x/2)$. Then

$$dx = 2\sec^2\theta\, d\theta \quad \text{and} \quad \sqrt{4+x^2} = \sqrt{4+4\tan^2\theta} = 2\sqrt{1+\tan^2\theta} = 2\sqrt{\sec^2\theta} = 2\,|\sec\theta|$$

By definition of the inverse tangent, $-\pi/2 < \theta < \pi/2$. So, $\cos\theta > 0$ and, therefore, $\sec\theta > 0$. Thus, $\sec\theta = |\sec\theta| = \sqrt{4+x^2}/2$. Hence,

$$\int \frac{dx}{z^2\sqrt{4+x^2}} = \int \frac{2\sec^2\theta\, d\theta}{4\tan^2\theta(2\sec\theta)}$$

$$= \frac{1}{4}\int \frac{\sec\theta\, d\theta}{\tan^2\theta} = \frac{1}{4}\int \frac{\cos\theta\, d\theta}{\sin^2\theta} = \frac{1}{4}\int (\sin\theta)^{-2}\cos\theta\, d\theta$$

$$= \tfrac{1}{4}(-(\sin\theta)^{-1}) + C = -\frac{1}{4\sin\theta} + C$$

Now we must evaluate $\sin\theta$.

Analytic method: $\sin\theta = \dfrac{\tan\theta}{\sec\theta} = \dfrac{x/2}{\sqrt{4+x^2}/2} = \dfrac{x}{\sqrt{4+x^2}}$.

Geometric method: Draw the right triangle shown in Fig. 32-1. From this triangle we see that $\sin\theta = x/\sqrt{4+x^2}$. (Note that it follows also for $\theta < 0$.)

Hence, $$\int \frac{dx}{x^2\sqrt{4+x^2}} = -\frac{\sqrt{4+x^2}}{4x} + C$$

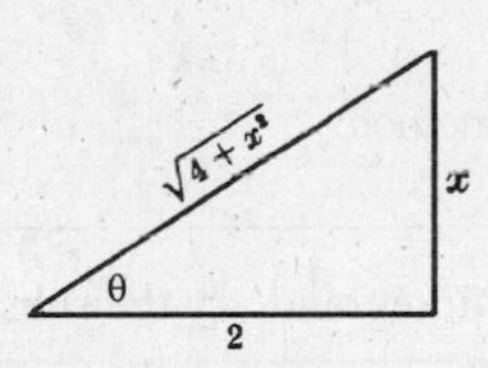

Fig. 32-1

This example illustrates the following general rule:

Strategy I. If $\sqrt{a^2+x^2}$ occurs in an integrand, try the substitution $x = a\tan\theta$.

EXAMPLE 32.12: Find $\displaystyle\int \frac{dx}{x^2\sqrt{9-x^2}}$.

Let $x = 3\sin\theta$, that is, $\theta = \sin^{-1}(x/3)$. Then $dx = 3\cos\theta\, d\theta$ and

$$\sqrt{9-x^2} = \sqrt{9-9\sin^2\theta} = 3\sqrt{\sin^2\theta} = 3\sqrt{\cos^2\theta} = 3|\cos\theta|$$

By definition of the inverse sine, $-\pi/2 < \theta < \pi/2$ and, therefore, $\cos\theta > 0$. Thus, $\cos\theta = |\cos\theta| = \sqrt{9-x^2}/3$. Now,

$$\int \frac{dx}{x^2\sqrt{9-x^2}} = \int \frac{3\cos\theta\, d\theta}{9\sin^2\theta\,(3\cos\theta)} = \frac{1}{9}\int \csc^2\theta\, d\theta$$

$$= -\frac{1}{9}\cot\theta + C = -\frac{1}{9}\frac{\cos\theta}{\sin\theta} + C = -\frac{1}{9}\frac{\sqrt{9-x^2}/3}{x/3} + C = -\frac{1}{9}\frac{\sqrt{9-x^2}}{x} + C$$

This example illustrates the following general method:

Strategy II. If $\sqrt{a^2-x^2}$ occurs in an integrand, try the substitution $x = a\sin\theta$.

EXAMPLE 32.13: Find $\displaystyle\int \frac{x^2}{\sqrt{x^2-4}}\, dx$.

Let $x = 2\sec\theta$, that is, $\theta = \sec^{-1}(x/2)$. Then $dx = 2\sec\theta\, d\theta$ and

$$\sqrt{x^2-4} = \sqrt{4\sec^2\theta - 4} = 2\sqrt{\sec^2\theta - 1} = 2\sqrt{\tan^2\theta} = 2|\tan\theta|$$

By definition of the inverse secant, θ is in the first or third quadrant and, therefore, $\tan\theta > 0$. So, $\tan\theta = |\tan\theta| = \sqrt{x^2-4}/2$. Now,

$$\int \frac{x^2}{\sqrt{x^2-4}}\,dx = \int \frac{4\sec^2\theta(2\sec\theta\tan\theta)}{2\tan\theta}\,d\theta$$

$$= 4\int \sec^3\theta\, d\theta = 2(\sec\theta\tan\theta + \ln|\sec\theta + \tan\theta| + C) \text{ (by Problem 8 of Chapter 31)}$$

$$= 2\left(\frac{x}{2}\frac{\sqrt{x^2-4}}{2} + \ln\left|\frac{x}{2} + \frac{\sqrt{x^2-4}}{2}\right|\right) + C$$

$$= \frac{x\sqrt{x^2-4}}{2} + 2\ln\left|\frac{x+\sqrt{x^2-4}}{2}\right| + C$$

$$= \frac{x\sqrt{x^2-4}}{2} + 2\ln\left|x+\sqrt{x^2-4}\right| + K \quad \text{where} \quad K = C - 2\ln 2$$

This example illustrates the following general method:

Strategy III. If $\sqrt{x^2-a^2}$ occurs in an integrand, try the substitution $x = a\sec\theta$.

SOLVED PROBLEMS

In Problems 1–23, verify the given solutions. Recall the identities

$$\sin^2 u = \tfrac{1}{2}(1-\cos 2u) \quad \cos^2 u \tfrac{1}{2}(1+\cos 2u) \quad \sin 2x = 2\sin x\cos x$$

1. $\int \sin^2 x\, dx = \int \frac{1}{2}(1-\cos 2x)\, dx = \frac{1}{2}(x - \frac{1}{2}\sin 2x) + C = \frac{1}{2}(x - \sin x\cos x) + C.$

2. $\int \cos^2(3x)\, dx = \int \frac{1}{2}(1+\cos 6x)\, dx = \frac{1}{2}(x + \frac{1}{6}\sin 6x) + C.$

3. $\int \sin^3 x\, dx = \int \sin^2 x \sin x\, dx = \int (1-\cos^2 x)\sin x\, dx$

$$= \int \sin x\, dx + \int \cos^2 x(-\sin x)\, dx$$

$$= -\cos x + \tfrac{1}{3}\cos^3 x + C \quad \text{(by Quick Formula I)}$$

4. $\int \sin^2 x\cos^3 x\, dx = \int \sin^2 x\cos^2 x\cos x\, dx$

$$= \int \sin^2 x(1-\sin^2 x)\cos x\, dx$$

$$= \int \sin^2 x\cos x\, dx - \int \sin^4 x\cos x\, dx$$

$$= \tfrac{1}{3}\sin^3 x - \tfrac{1}{5}\sin^5 x + C \quad \text{(by Quick Formula I)}$$

5. $\int \sin^3(3x)\cos^5(3x)\,dx = \int (1-\cos^2(3x))\cos^5(3x)\sin(3x)\,dx$

$$= \int \cos^5(3x)\sin(3x)\,dx - \int \cos^7(3x)\sin 3x\,dx$$

$$= -\frac{1}{3}\int \cos^5(3x)(-3\sin(3x))\,dx + \frac{1}{3}\int \cos^7(3x)(-3\sin(3x))\,dx$$

$$= -\tfrac{1}{3}\tfrac{1}{6}\cos^6(3x) + \tfrac{1}{3}\tfrac{1}{8}\cos^8(3x) + C \quad \text{(by Quick Formula I)}$$

$$= \tfrac{1}{72}(3\cos^8(3x) - 4\cos^6(3x)) + C$$

6. $\int \cos^3\left(\frac{x}{3}\right) dx = \int \left(1-\sin^2\left(\frac{x}{3}\right)\right)\cos\frac{x}{3}\,dx$

$$= \int \left(1-\sin^2\left(\frac{x}{3}\right)\right)\cos\frac{x}{3}\,dx = \int \cos\frac{x}{3}\,dx - \int \sin^2\left(\frac{x}{3}\right)\cos\frac{x}{3}\,dx$$

$$= 3\sin\frac{x}{3} - 3\int \sin^2\left(\frac{x}{3}\right)\left(\frac{1}{3}\cos\frac{x}{3}\right) dx$$

$$= 3\sin\frac{x}{3} - 3\tfrac{1}{3}\sin^3\left(\frac{x}{3}\right) + C \quad \text{(by Quick Formula I)}$$

$$= 3\sin\frac{x}{3} - \sin^3\left(\frac{x}{3}\right) + C$$

7. $\int \sin^4 x\,dx = \int (\sin^2 x)^2\,dx = \frac{1}{4}\int (1-\cos(2x))^2\,dx$

$$= \frac{1}{4}\int 1\,dx - \frac{1}{2}\int \cos(2x)\,dx + \frac{1}{4}\int \cos^2(2x)\,dx$$

$$= \frac{1}{4}x - \frac{1}{4}\sin(2x) + \frac{1}{8}\int (1+\cos 4x))\,dx$$

$$= \tfrac{1}{4}x - \tfrac{1}{4}\sin(2x) + \tfrac{1}{8}(x + \tfrac{1}{4}\sin(4x)) + C$$

$$= \tfrac{3}{8}x - \tfrac{1}{4}\sin(2x) + \tfrac{1}{32}\sin(4x) + C$$

8. $\int \sin^2 x\cos^2 x\,dx = \frac{1}{4}\int \sin^2(2x)\,dx = \frac{1}{8}\int (1-\cos(4x))\,dx$

$$= \tfrac{1}{8}(x - \tfrac{1}{4}\sin(4x)) + C = \tfrac{1}{8}x - \tfrac{1}{32}\sin(4x) + C$$

9. $\int \sin^4(3x)\cos^2(3x)\,dx = \int (\sin^2(3x)\cos^2(3x))\sin^2(3x)\,dx$

$$= \frac{1}{8}\int \sin^2(6x)(1-\cos(6x))\,dx$$

$$= \frac{1}{8}\int \sin^2(6x)\,dx - \frac{1}{8}\int \sin^2(6x)\cos(6x)\,dx$$

$$= \frac{1}{16}\int (1-\cos(12x))\,dx - \frac{1}{48}\int \sin^2(6x)(6\cos(6x))\,dx$$

$$= \tfrac{1}{16}(x - \tfrac{1}{12}\sin(12x)) - \tfrac{1}{144}\sin^3(6x) + C \quad \text{(by Quick Formula I)}$$

$$= \tfrac{1}{16}x - \tfrac{1}{192}\sin(12x)) - \tfrac{1}{144}\sin^3(6x) + C$$

10. $\int \sin 3x\sin 2x\,dx = \int \tfrac{1}{2}(\cos(3x-2x) - \cos(3x+2x))\,dx$

$$= \tfrac{1}{2}\int (\cos x - \cos 5x))\,dx = \tfrac{1}{2}(\sin x - \tfrac{1}{5}\sin 5x) + C$$

$$= \tfrac{1}{2}\sin x - \tfrac{1}{10}\sin 5x + C$$

11. $\int \sin 3x \cos 5x\, dx = \int \frac{1}{2}(\sin(3x-5x)+\sin(3x+5x))\, dx$

$$= \frac{1}{2}\int(\sin(-2x)+\sin(8x))\, dx = \frac{1}{2}\int(-\sin(2x)+\sin(8x))\, dx$$

$$= \tfrac{1}{2}(\tfrac{1}{2}\cos(2x) - \tfrac{1}{8}\cos(8x)) + C = \tfrac{1}{4}\cos(2x) - \tfrac{1}{16}\cos(8x) + C$$

12. $\int \cos 4x \cos 2x\, dx = \frac{1}{2}\int(\cos(2x)+\cos(6x))\, dx$

$$= \tfrac{1}{2}(\tfrac{1}{2}\sin(2x) + \tfrac{1}{6}\sin(6x)) + C = \tfrac{1}{4}\sin(2x) + \tfrac{1}{12}\sin(6x) + C$$

13. $\int \sqrt{1-\cos x}\, dx = \sqrt{2}\int \sin\left(\frac{x}{2}\right) dx \quad \left(\text{by } \sin^2\left(\frac{x}{2}\right) = \frac{1-\cos x}{2}\right)$

$$= \sqrt{2}\left(-2\cos\left(\frac{x}{2}\right)\right) + C = -2\sqrt{2}\cos\left(\frac{x}{2}\right) + C$$

14. $\int (1+\cos 3x)^{3/2}\, dx = 2\sqrt{2}\int \cos^3\left(\frac{3x}{2}\right) dx \quad \text{since } \cos^2\left(\frac{3x}{2}\right) = \frac{1+\cos(3x)}{2}$

$$= 2\sqrt{2}\int\left(1-\sin^2\left(\frac{3x}{2}\right)\right)\cos\left(\frac{3x}{2}\right) dx$$

$$= 2\sqrt{2}\left[\int \cos\left(\frac{3x}{2}\right) dx - \int \sin^2\left(\frac{3x}{2}\right)\cos\left(\frac{3x}{2}\right) dx\right]$$

$$= 2\sqrt{2}\left[\frac{2}{3}\sin\left(\frac{3x}{2}\right) - \frac{2}{3}\int \sin^2\left(\frac{3x}{2}\right)\left(\frac{3}{2}\cos\left(\frac{3x}{2}\right)\right) dx\right]$$

$$= 2\sqrt{2}\left[\frac{2}{3}\sin\left(\frac{3x}{2}\right) - \frac{2}{3}\frac{1}{3}\sin^3\left(\frac{3x}{2}\right)\right] + C$$

$$= \frac{4\sqrt{2}}{9}\left[3\sin\left(\frac{3x}{2}\right) - \sin^3\left(\frac{3x}{2}\right)\right] + C$$

15. $\int \frac{dx}{\sqrt{1-\sin 2x}} = \int \frac{dx}{\sqrt{1-\cos\left(\frac{\pi}{2}-2x\right)}}$

$$= \frac{\sqrt{2}}{2}\int \frac{dx}{\sin\left(\frac{\pi}{4}-x\right)} \quad \left(\text{since } \sin^2\left(\frac{\pi}{4}-x\right) = \frac{1-\cos\left(\frac{\pi}{2}-2x\right)}{2}\right)$$

$$= \frac{\sqrt{2}}{2}\int \csc\left(\frac{\pi}{4}-x\right) dx = -\frac{\sqrt{2}}{2}\ln\left|\csc\left(\frac{\pi}{4}-x\right) - \cot\left(\frac{\pi}{4}-x\right)\right| + C$$

16. $\int \tan^4 x\, dx = \int \tan^2 x \tan^2 x\, dx = \int \tan^2 x(\sec^2 x - 1)\, dx$

$$= \int \tan^2 x \sec^2 x\, dx - \int \tan^2 x\, dx$$

$$= \tfrac{1}{3}\tan^3 x - \int(\sec^2 x - 1)dx \quad \text{(by Quick Formula I)}$$

$$= \tfrac{1}{3}\tan^3 x - (\tan x - x) + C$$

$$= \tfrac{1}{3}\tan^3 x - \tan x + x + C$$

17. $\int \tan^5 x\,dx = \int \tan^3 x \tan^2 x\,dx = \int \tan^3 x(\sec^2 x - 1)\,dx$

$= \int \tan^3 x \sec^2 x\,dx - \int \tan^3 dx$

$= \frac{1}{4}\tan^4 x - \int \tan x(\sec^2 x - 1)\,dx$ (by Quick Formula I)

$= \frac{1}{4}\tan^4 x - \int \tan x \sec^2 x\,dx + \int \tan x\,dx$

$= \frac{1}{4}\tan^4 x - \frac{1}{2}\tan^2 x + \ln|\sec x| + C$ (by Quick Formula I)

18. $\int \sec^4(2x)\,dx = \int \sec^2(2x)\sec^2(2x)\,dx$

$= \int \sec^2(2x)(1 + \tan^2(2x))\,dx$

$= \int \sec^2(2x)\,dx + \int \sec^2(2x)\tan^2(2x))\,dx$

$= \frac{1}{2}\tan(2x) + \frac{1}{2}\int \tan^2(2x)(2\sec^2(2x))\,dx$

$= \frac{1}{2}\tan(2x) + \frac{1}{2}\frac{1}{3}\tan^3(2x) + C$ (by Quick Formula I)

$= \frac{1}{2}\tan(2x) + \frac{1}{6}\tan^3(2x) + C$

19. $\int \tan^3(3x)\sec^4(3x)\,dx = \int \tan^3(3x)(1 + \tan^2(3x))\sec^2(3x)\,dx$

$= \int \tan^3(3x)\sec^2(3x)\,dx + \int \tan^5(3x)\sec^2(3x)\,dx$

$= \frac{1}{3}\frac{1}{4}\tan^4(3x) + \frac{1}{3}\frac{1}{6}\tan^6(3x) + C$

$= \frac{1}{12}\tan^4(3x) + \frac{1}{18}\tan^6(3x) + C$

20. $\int \cot^3(2x)\,dx = \int \cot(2x)(\csc^2(2x) - 1)\,dx$

$= -\frac{1}{4}\cot^2(2x) + \frac{1}{2}\ln|\csc(2x)| + C$

21. $\int \cot^4(3x)\,dx = \int \cot^3(3x)(\csc^2(3x) - 1)\,dx$

$= \int \cot^2(3x)\csc^2(3x)\,dx - \int \cot^2(3x)\,dx$

$= -\frac{1}{9}\cot^3(3x) - \int (\csc^2(3x) - 1)\,dx$

$= -\frac{1}{9}\cot^3(3x) + \frac{1}{3}\cot(3x) + x + C$

22. $\int \csc^6 x\,dx = \int \csc^2 x(1 + \cot^2 x)^2\,dx$

$= \int \csc^2 x\,dx + 2\int \cot^2 x \csc^2 x\,dx + \int \cot^4 x \csc^2 x\,dx$

$= -\cot x - \frac{2}{3}\cot^3 x - \frac{1}{5}\cot^5 x + C$

23. $\int \cot^3 x \csc^5 x\,dx = \int \cot^2 x \csc^4 \csc x \cot x\,dx$

$= \int (\csc^2 x - 1)\csc^4 x \csc x \cot x\,dx$

$= \int \csc^6 x \csc x \cot x\,dx - \int \csc^4 x \csc x \cot x\,dx$

$= -\frac{1}{7}\csc^7 x + \frac{1}{5}\csc^5 x + C$

24. Find $\int \frac{\sqrt{9-4x^2}}{x}\,dx$.

$\sqrt{9-4x^2} = 2\sqrt{\frac{9}{4}-x^2}$. So, let $x = \frac{3}{2}\sin\theta$. Then

$$dx = \tfrac{3}{2}\cos\theta\, d\theta \quad \text{and} \quad \sqrt{9-4x^2} = \sqrt{9-9\sin^2\theta} = 3\sqrt{\cos^2\theta} = 3|\cos\theta| = 3\cos\theta$$

Hence,

$$\int \frac{\sqrt{9-4x^2}}{x}\,dx = \int \frac{3\cos\theta(\frac{3}{2}\cos\theta)\,d\theta}{\frac{3}{2}\sin\theta} = 3\int \frac{\cos^2\theta}{\sin\theta}\,d\theta = 3\int \frac{1-\sin^2\theta}{\sin\theta}\,d\theta$$

$$= 3\int (\csc\theta - \sin\theta)\,d\theta = 3\ln|\csc\theta - \cot\theta| + 3\cos\theta + C$$

But

$$\csc\theta = \frac{1}{\sin\theta} = \frac{3}{2x} \quad \text{and} \quad \cot\theta = \frac{\cos\theta}{\sin\theta} = \frac{\sqrt{9-4x^2}/3}{2x/3} = \frac{\sqrt{9-4x^2}}{2x}$$

So,

$$\int \frac{\sqrt{9-4x^2}}{x}\,dx = 3\ln\left|\frac{3-\sqrt{9-4x^2}}{x}\right| + \sqrt{9-4x^2} + K \quad \text{where} \quad K = C - 3\ln 2$$

25. Find $\int \frac{dx}{x\sqrt{9-4x^2}}$.

Let $x = \frac{3}{2}\tan\theta$. (See Fig. 32-2.) Then $dx = \frac{3}{2}\sec^2\theta$ and $\sqrt{9-4x^2} = 3\sec\theta$. Hence,

$$\int \frac{dx}{x\sqrt{9+4x^2}} = \int \frac{\frac{3}{2}\sec^2\theta\, d\theta}{(\frac{3}{2}\tan\theta)(3\sec\theta)}$$

$$= \frac{1}{3}\int \csc\theta\, d\theta = \tfrac{1}{3}\ln|\csc\theta - \cot\theta| + C$$

$$= \tfrac{1}{3}\ln\left|\frac{\sqrt{9+4x^2}-3}{x}\right| + K$$

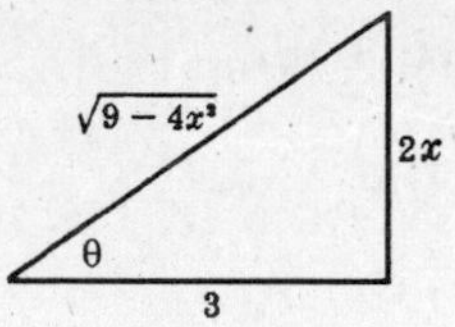

Fig. 32-2

26. Find $\int \frac{(16-9x^2)^{3/2}}{x^6}\,dx$.

Let $x = \frac{4}{3}\sin\theta$. (See Fig. 32-3.) Then $dx = \frac{4}{3}\cos\theta\, d\theta$ and $\sqrt{16-9x^2} = 4\cos\theta$. Hence,

$$\int \frac{(16-9x^2)^{3/2}}{x^6}\,dx = \int \frac{(64\cos^3\theta)(\frac{4}{3}\cos\theta\, d\theta)}{\frac{4096}{729}\sin^6\theta}$$

$$= \frac{243}{16}\int \cot^4\theta\csc^2\theta\, d\theta = -\frac{243}{80}\cot^5\theta + C$$

$$= -\frac{243}{80}\frac{(16-9x^2)^{5/2}}{243x^5} + C = -\frac{1}{80}\frac{(16-9x^2)^{5/2}}{x^5} + C$$

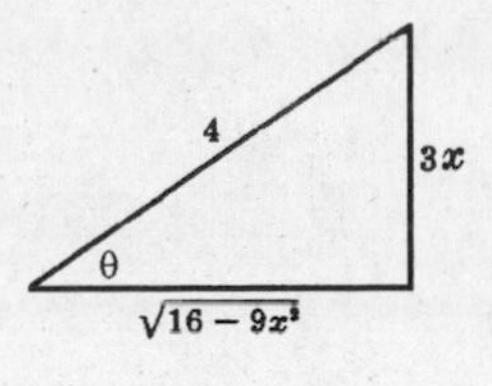

Fig. 32-3

27. Find $\int \frac{x^2\,dx}{\sqrt{2x-x^2}} = \int \frac{x^2\,dx}{\sqrt{1-(x-1)^2}}$.

Let $x-1 = \sin\theta$. (See Fig. 32-4.) Then $dx = \cos\theta\,d\theta$ and $\sqrt{2x-x^2} = \cos\theta$. Hence,

$$\begin{aligned}\int \frac{x^2\,dx}{\sqrt{2x-x^2}} &= \int \frac{(1+\sin\theta)^2}{\cos\theta}\cos\theta\,d\theta \\ &= \int (1+\sin\theta)^2\,d\theta = \int \left(\tfrac{3}{2} + 2\sin\theta - \tfrac{1}{2}\cos 2\theta\right)d\theta \\ &= \tfrac{3}{2}\theta - 2\cos\theta - \tfrac{1}{4}\sin 2\theta + C \\ &= \tfrac{3}{2}\sin^{-1}(x-1) - 2\sqrt{2x-x^2} - \tfrac{1}{2}(x-1)\sqrt{2x-x^2} + C \\ &= \tfrac{3}{2}\sin^{-1}(x-1) - \tfrac{1}{2}(x+3)\sqrt{2x-x^2} + C\end{aligned}$$

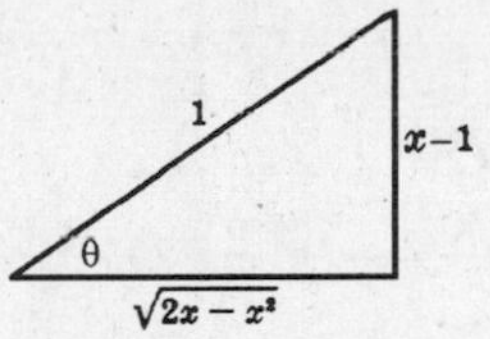

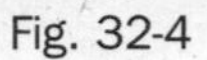

Fig. 32-4

28. Find $\int \frac{dx}{(4x^2-24x+27)^{3/2}} = \int \frac{dx}{(4(x-3)^2-9)^{3/2}}$.

Let $x-3 = \frac{3}{2}\sec\theta$. (See Fig. 32-5.) Then $dx = \frac{3}{2}\sec\theta\tan\theta\,d\theta$ and $\sqrt{4x^2-24x+27} = 3\tan\theta$. So,

$$\begin{aligned}\int \frac{dx}{(4x^2-24x+27)^{3/2}} &= \int \frac{\frac{3}{2}\sec\theta\tan\theta\,d\theta}{27\tan^3\theta} = \frac{1}{18}\int \csc\theta\cot\theta\,d\theta \\ &= -\frac{1}{18}\csc\theta + C = -\frac{1}{9}\frac{x-3}{\sqrt{4x^2-24x+27}} + C \quad \text{(from Fig. 32-5)}\end{aligned}$$

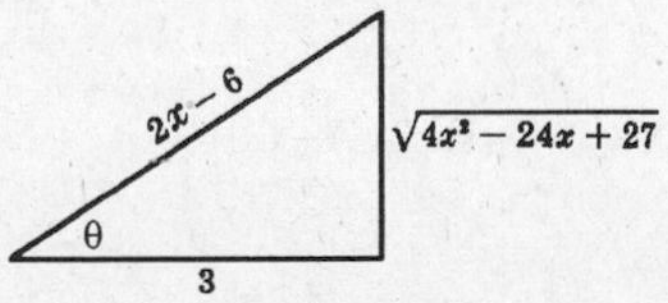

Fig. 32-5

SUPPLEMENTARY PROBLEMS

29. $\int \cos^2 x\,dx = \frac{1}{2}x + \frac{1}{4}\sin 2x + C$

30. $\int \sin^3 2x\,dx = \frac{1}{6}\cos^3 2x - \frac{1}{2}\cos 2x + C$

31. $\int \sin^4 2x\,dx = \frac{3}{8}x - \frac{1}{8}\sin 4x + \frac{1}{64}\sin 8x + C$

32. $\int \cos^4 \frac{1}{2}x\,dx = \frac{3}{8}x + \frac{1}{2}\sin x + \frac{1}{16}\sin 2x + C$

33. $\int \sin^7 x\,dx = \frac{1}{7}\cos^7 x - \frac{3}{5}\cos^5 x + \cos^3 x - \cos x + C$

34. $\int \cos^6 \frac{1}{2}x\,dx = \frac{5}{16}x + \frac{1}{2}\sin x + \frac{3}{32}\sin 2x - \frac{1}{24}\sin^3 x + C$

35. $\int \sin^2 x\cos^5 x\,dx = \frac{1}{3}\sin^3 x - \frac{2}{5}\sin^5 x + \frac{1}{7}\sin^7 x + C$

36. $\int \sin^3 x\cos^2 x\,dx = \frac{1}{5}\cos^5 x - \frac{1}{3}\cos^3 x + C$

37. $\int \sin^3 x\cos^3 x\,dx = \frac{1}{48}\cos^3 2x - \frac{1}{16}\cos 2x + C$

38. $\int \sin^4 x\cos^4 x\,dx = \frac{1}{128}(3x - \sin 4x + \frac{1}{8}\sin 8x + C$

39. $\int \sin 2x\cos 4x\,dx = \frac{1}{4}\cos 2x - \frac{1}{12}\cos 6x + C$

40. $\int \cos 3x\cos 2x\,dx = \frac{1}{2}\sin x + \frac{1}{10}\sin 5x + C$

41. $\int \sin 5x\sin x\,dx = \frac{1}{8}\sin 4x - \frac{1}{12}\sin 6x + C$

42. $\int \frac{\cos^3 x\,dx}{1 - \sin x} = \sin x + \frac{1}{2}\sin^2 x + C$

43. $\int \frac{\cos^{2/3} x}{\sin^{8/3} x}\,dx = -\frac{3}{5}\cot^{5/3} x + C$

44. $\int \frac{\cos^3 x}{\sin^4 x}\,dx = \csc x - \frac{1}{3}\csc^3 x + C$

45. $\int x(\cos^3 x^2 - \sin^3 x^2)\,dx = \frac{1}{12}(\sin x^2 + \cos x^2)(4 + \sin 2x^2) + C$

46. $\int \tan^3 x\,dx = \frac{1}{2}\tan^2 x + \ln|\cos x| + C$

47. $\int \tan^3 3x \sec 3x\,dx = \frac{1}{9}\sec^3 3x - \frac{1}{3}\sec 3x + C$

48. $\int \tan^{3/2} x\sec^4 x\,dx = \frac{2}{5}\tan^{5/2} x + \frac{2}{9}\tan^{9/2} x + C$

49. $\int \tan^4 x\sec^4 x\,dx = \frac{1}{7}\tan^7 x + \frac{1}{5}\tan^5 x + C$

50. $\int \cot^3 x\, dx = -\frac{1}{2}\cot^2 x - \ln|\sin x| + C$

51. $\int \cot^3 x \csc^4 x\, dx = -\frac{1}{4}\cot^4 x - \frac{1}{6}\cot^6 x + C$

52. $\int \cot^3 x \csc^3 x\, dx = -\frac{1}{5}\csc^5 x + \frac{1}{3}\csc^3 x + C$

53. $\int \csc^4 2x\, dx = -\frac{1}{2}\cot 2x - \frac{1}{6}\cot^3 2x + C$

54. $\int \left(\frac{\sec x}{\tan x}\right)^4 dx = -\frac{1}{3\tan^3 x} - \frac{1}{\tan x} + C$

55. $\int \frac{\cot^3 x}{\csc x} dx = -\sin x - \csc x + C$

56. $\int \tan x \sqrt{\sec x}\, dx = 2\sqrt{\sec x} + C$

57. $\int \frac{dx}{(4 - x^2)^{3/2}} = \frac{x}{4\sqrt{4 - x^2}} + C$

58. $\int \frac{\sqrt{25 - x^2}}{x} dx = 5\ln\left|\frac{5 - \sqrt{25 - x^2}}{x}\right| + \sqrt{25 - x^2} + C$

59. $\int \frac{dx}{x^2\sqrt{a^2 - x^2}} = -\frac{\sqrt{a^2 - x^2}}{a^2 x} + C$

60. $\int \sqrt{x^2 + 4}\, dx = \frac{1}{2}x\sqrt{x^2 + 4} + 2\ln(x + \sqrt{x^2 + 4}) + C$

61. $\int \frac{x^2\, dx}{(a^2 - x^2)^{3/2}} = \frac{x}{\sqrt{a^2 - x^2}} - \sin^{-1}\left(\frac{x}{a}\right) + C$

62. $\int \sqrt{x^2 - 4}\, dx = \frac{1}{2}x\sqrt{x^2 - 4} - 2\ln\left|x + \sqrt{x^2 - 4}\right| + C$

63. $\int \frac{\sqrt{x^2 + a^2}}{x} dx = \sqrt{x^2 + a^2} + \frac{a}{2}\ln\frac{\sqrt{a^2 + x^2} - a}{\sqrt{a^2 + x^2} + a} + C$

64. $\int \frac{x^2 dx}{(4 - x^2)^{5/2}} = \frac{x^3}{12(4 - x^2)^{3/2}} + C$

65. $\int \frac{dx}{(a^2 + x^2)^{3/2}} = \frac{x}{a^2\sqrt{a^2 + x^2}} + C$

66. $\int \frac{dx}{x^2\sqrt{9 - x^2}} = -\frac{\sqrt{9 - x^2}}{9x} + C$

67. $\int \frac{x^2 dx}{\sqrt{x^2 - 16}} = \frac{1}{2}x\sqrt{x^2 - 16} + 8\ln\left|x + \sqrt{x^2 - 16}\right| + C$

68. $\int x^3\sqrt{a^2-x^2}\,dx = \frac{1}{5}(a^2-x^2)^{5/2} - \frac{a^2}{3}(a^2-x^2)^{3/2} + C$

69. $\int \frac{dx}{\sqrt{x^2-4x+13}} = \ln(x-2+\sqrt{x^2-4x+13}) + C$

70. $\int \frac{dx}{(4x-x^2)^{3/2}} = \frac{x-2}{4\sqrt{4x-x^2}} + C$

71. $\int \frac{dx}{(9+x^2)^2} = \frac{1}{54}\tan^{-1}\left(\frac{x}{3}\right) + \frac{x}{18(9+x^2)} + C$

In Problems 72 and 73, first apply integration by parts.

72. $\int x\sin^{-1}x\,dx = \frac{1}{4}(2x^2-1)\sin^{-1}x + \frac{1}{4}x\sqrt{1-x^2} + C$

73. $\int x\cos^{-1}x\,dx = \frac{1}{4}(2x^2-1)\cos^{-1}x - \frac{1}{4}x\sqrt{1-x^2} + C$

CHAPTER 33

Techniques of Integration III: Integration by Partial Fractions

We shall give a general method for finding antiderivatives of the form $\int \frac{N(x)}{D(x)}dx$, where $N(x)$ and $D(x)$ are polynomials. A function of the form $\frac{N(x)}{D(x)}$ is called a *rational function*. ($N(x)$ is the numerator and $D(x)$ is the denominator.) As examples, consider

$$\int \frac{x-1}{x^3+8}dx \quad \text{and} \quad \int \frac{x^3-x}{x+2}dx$$

Two restrictions will be assumed, neither of which limits the applicability of our method: (i) the leading coefficient (the coefficient of the highest power of x) in $D(x)$ is +1; (ii) $N(x)$ is of lower degree than $D(x)$. A quotient $N(x)/D(x)$ that satisfies (ii) is called a *proper* rational function. Let us see that the restrictions (i)–(ii) are not essential.

EXAMPLE 33.1: Consider the case where $\frac{N(x)}{D(x)}$ is $\frac{2x^3}{5x^8+3x-4}$. Here, our first restriction is not satisfied. However, note that

$$\int \frac{2x^3}{5x^8+3x-4}dx = \frac{1}{5}\int \frac{2x^3}{x^8+\frac{3}{5}x-\frac{4}{5}}dx$$

The integral on the right side satisfies restrictions (i) and (ii).

EXAMPLE 33.2: Consider the case where $\frac{N(x)}{D(x)}$ is $\frac{2x^5+7}{x^2+3}$. Here, our second restriction is not satisfied. But we can divide $N(x)$ by $D(x)$:

$$\frac{2x^5+7}{x^2+3} = 2x^3 - 6x + \frac{18x+7}{x^2+3}$$

Hence,

$$\int \frac{2x^5+7}{x^2+3}dx = \frac{1}{2}x^4 - 3x^2 + \int \frac{18x+7}{x^2+3}dx$$

and the problem is reduced to evaluating $\int \frac{18x+7}{x^2+3}dx$, which satisfies our restrictions.

A polynomial is said to be *irreducible* if it is not the product of two polynomials of lower degree.

Any linear polynomial $f(x) = ax + b$ is automatically irreducible, since polynomials of lower degree than $f(x)$ are constants and $f(x)$ is not the product of two constants.

Now consider any quadratic polynomial $g(x) = ax^2 + bx + c$. Then

$$g(x) \text{ is irreducible if and only if } b^2 - 4ac < 0$$

To see why this is so, assume that $g(x)$ is reducible. Then $g(x) = (Ax + B)(Cx + D)$. Hence, $x = -B/A$ and $x = -D/C$ are roots of $g(x)$. The quadratic formula

$$x = \frac{-b \pm \sqrt{b^2 - 4ac}}{2a}$$

should yield these roots. Therefore, $b^2 - 4ac$ cannot be negative. Conversely, assume $b^2 - 4ac \geq 0$. Then the quadratic formula yields two roots of $g(x)$. But, if r is a root of $g(x)$, then $g(x)$ is divisible by $x - r$.† Hence, $g(x)$ is reducible.

EXAMPLE 33.3:
(a) $x^2 + 4$ is irreducible, since $b^2 - 4ac = 0 - 4(1)(4) = -16 < 0$.
(b) $x^2 + x - 4$ is reducible, since $b^2 - 4ac = 1 - 4(1)(-4) = 17 \geq 0$.

We will assume without proof the following fairly deep property of polynomials with real coefficients.

THEOREM 33.1: Any polynomial $D(x)$ with leading coefficient 1 can be expressed as a product of linear factors of the form $x - a$ and of irreducible quadratic factors of the form $x^2 + bx + c$. (Repetition of factors is permitted.)

EXAMPLE 33.4:
(a) $x^3 - 4x = x(x^2 - 4) = x(x - 2)(x + 2)$
(b) $x^3 + 4x = x(x^2 + 4)$ ($x^2 + 4$ is irreducible.)
(c) $x^4 - 9 = (x^2 - 3)(x^2 + 3) = (x - \sqrt{3})(x + \sqrt{3})(x^2 + 3)$ ($x^2 + 3$ is irreducible.)
(d) $x^3 - 3x^2 - x + 3 = (x + 1)(x - 2)^2$

Method of Partial Fractions

Assume that we wish to evaluate $\int \frac{N(x)}{D(x)}dx$, where $\frac{N(x)}{D(x)}$ is a proper rational function and $D(x)$ has leading coefficient 1. First, write $D(x)$ as a product of linear and irreducible quadratic factors.

Our method will depend on this factorization. We will consider various cases and, in each case, we will first explain the method by means of an example and then state the general procedure.

Case I

$D(x)$ is a product of distinct linear factors.

EXAMPLE 33.5: Find $\int \frac{dx}{x^2 - 4}$.

In this case, $D(x) = x^2 - 4 = (x - 2)(x + 2)$. Write

$$\frac{1}{(x-2)(x+2)} = \frac{A}{x-2} + \frac{B}{x+2}$$

It is assumed that A and B are certain constants, that we must now evaluate. Clear the denominators by multiplying both sides by $(x - 2)(x + 2)$:

$$1 = A(x + 2) + B(x - 2) \qquad (1)$$

First, substitute -2 for x in (1): $1 = A(0) + B(-4) = -4B$. Thus, $B = -\frac{1}{4}$.
Second, substitute 2 for x in (1): $1 = A(4) + B(0) = 4A$. Thus, $A = \frac{1}{4}$. Hence,

$$\frac{1}{(x-2)(x+2)} = \frac{1}{4}\frac{1}{x-2} - \frac{1}{4}\frac{1}{x+2}$$

† In general, if a polynomial $h(x)$ has r as a root, then $h(x)$ must be divisible by $x - r$.

So, $$\int \frac{dx}{x^2-4} = \int \left(\frac{1}{4}\frac{1}{x-2} - \frac{1}{4}\frac{1}{x+2} \right) dx = \tfrac{1}{4}\ln|x-2| - \tfrac{1}{4}\ln|x+2| + C$$

$$= \tfrac{1}{4}(\ln|x-2| - \ln|x+2|) + C$$

$$= \tfrac{1}{4}\ln\left|\frac{x-2}{x+2}\right| + C$$

EXAMPLE 33.6: Find $\int \frac{(x+1)\,dx}{x^3+x^2-6x}$.

Factoring the denominator yields $x(x^2+x-6) = x(x-2)(x+3)$. The integrand is $\frac{x+1}{x(x-2)(x+3)}$. Represent it in the following form:

$$\frac{x+1}{x(x-2)(x+3)} = \frac{A}{x} + \frac{B}{x-2} + \frac{C}{x-3}$$

Clear the denominators by multiplying by $x(x-2)(x+3)$:

$$x+1 = A(x-2)(x+3) + Bx(x+3) + Cx(x-2) \tag{2}$$

Let x be 0 in (2): $1 = A(-2)(3) + B(0)(3) + C(0)(-2) = -6A$. So, $A = -\frac{1}{6}$.

Let x be 2 in (2): $3 = A(0)(5) + B(2)(5) + C(2)(0) = 10B$. So, $B = \frac{3}{10}$.

Let x be -3 in (2): $-2 = A(-5)(0) + B(-3)(0) + C(-3)(-5) = 15C$. So, $C = -\frac{2}{15}$.

Hence, $$\int \frac{(x+1)\,dx}{x^3+x^2-6x} = \int \left(-\frac{1}{6}\frac{1}{x} + \frac{3}{10}\frac{1}{x+2} - \frac{2}{15}\frac{1}{x+3} \right) dx$$

$$= -\tfrac{1}{6}\ln|x| + \tfrac{3}{10}\ln|x+2| - \tfrac{2}{15}\ln|x+3| + C$$

General Rule for Case I

Represent the integrand as a sum of terms of the form $\frac{A}{x-a}$ for each linear factor $x-a$ of the denominator, where A is an unknown constant. Solve for the constants. Integrating yields a sum of terms of the form $A\ln|x-a|$.

Remark: We assume without proof that the integrand always has a representation of the required kind. For every particular problem, this can be verified at the end of the calculation.

Case II

$D(x)$ is a product of linear factors, some of which occur more than once.

EXAMPLE 33.7: Find $\int \frac{(3x+5)dx}{x^3-x^2-x+1}$.

First factor the denominator:†

$$x^3 - x^2 - x + 1 = (x+1)(x-1)^2$$

Then represent the integrand $\frac{3x+5}{x^3-x^2-x+1}$ as a sum of the following form:

$$\frac{3x+5}{x^3-x^2-x+1} = \frac{A}{x+1} + \frac{B}{x-1} + \frac{C}{(x-1)^2}$$

† In trying to find linear factors of a denominator that is a polynomial with integral coefficients, test each of the divisors r of the constant term to see whether it is a root of the polynomial. If it is, then $x-r$ is a factor of the polynomial. In the given example, the constant term is 1. Both of its divisors, 1 and -1, turn out to be roots.

Note that, for the factor $(x-1)$ that occurs twice, there are terms with both $(x-1)$ and $(x-1)^2$ in the denominator. Now clear the denominators by multiplying both sides by $(x+1)(x-1)^2$:

$$3x+5 = A(x-1)^2 + B(x+1)(x-1) + C(x+1) \tag{1}$$

Let $x=1$. Then $8=(0)A+(2)(0)B+(2)C=2C$. Thus, $C=4$.
Let $x=-1$. Then $2=(4)A+(0)(-2)B+(0)C=4A$. Thus, $A=\frac{1}{2}$.

To find B, compare the coefficients of x^2 on both sides of (1). On the left it is 0, and on the right it is $A+B$. Hence, $A+B=0$. Since $A=\frac{1}{2}$, $B=-\frac{1}{2}$. Thus,

$$\frac{3x+5}{x^3-x^2-x+1}=\frac{1}{2}\frac{1}{x+1}-\frac{1}{2}\frac{1}{x-1}+4\frac{1}{(x-2)^2}$$

Therefore,

$$\int\frac{(3x+5)\,dx}{x^3-x^2-x+1}=\tfrac{1}{2}\ln|x+1|-\tfrac{1}{2}\ln|x-1|+4\int\frac{dx}{(x-1)^2}$$

By Quick Formula I,

$$\int\frac{dx}{(x-1)^2}=\int(x-1)^{-2}\,dx=-(x-1)^{-1}=-\frac{1}{x-1}$$

So,

$$\int\frac{(3x+5)\,dx}{x^3-x^2-x+1}=\tfrac{1}{2}\ln|x+1|-\tfrac{1}{2}\ln|x-1|-4\frac{1}{x-1}+C$$

$$=\tfrac{1}{2}\ln\frac{|x+1|}{|x-1|}-\frac{4}{x-1}+C$$

EXAMPLE 33.8: Find $\int\frac{(x+1)\,dx}{x^3(x-2)^2}$.

Represent the integrand $\frac{(x+1)}{x^3(x-2)^2}$ in the following form:

$$\frac{(x+1)}{x^3(x-2)^2}=\frac{A}{x}+\frac{B}{x^2}+\frac{C}{x^3}+\frac{D}{x-2}+\frac{E}{(x-2)^2}$$

Clear denominators by multiplying by $x^3(x-2)^2$:

$$x+1=Ax^2(x-2)^2+Bx(x-2)^2+C(x-2)^2+Dx^3(x-2)+Ex^3$$

Let $x=0$. Then $1=4C$. So, $C=\frac{1}{4}$.
Let $x=2$. Then $3=8E$. So, $E=\frac{3}{8}$.
Compare coefficients of x. Then $1=4B-4C$. Since $C=\frac{1}{4}$, $B=\frac{1}{2}$.
Compare coefficients of x^2. Then $0=4A-4B+C$. Since $B=\frac{1}{2}$ and $C=\frac{1}{4}$, $A=\frac{7}{16}$.
Compare coefficients of x^4. Then $0=A+D$. So, $D=-\frac{7}{16}$.

Hence,

$$\frac{(x+1)}{x^3(x-2)^2}=\frac{7}{16}\frac{1}{x}+\frac{1}{2}\frac{1}{x^2}+\frac{1}{4}\frac{1}{x^3}-\frac{7}{16}\frac{1}{x-2}+\frac{3}{8}\frac{1}{(x-2)^2}.$$

Thus,

$$\int\frac{(x+1)}{x^3(x-2)^2}\,dx=\frac{7}{16}\ln|x|-\frac{1}{2}\frac{1}{x}-\frac{1}{8}\frac{1}{x^2}-\frac{7}{16}\ln|x-2|-\frac{3}{8}\frac{1}{x-2}+C^*$$

$$=\frac{7}{16}\ln\left|\frac{x}{x-2}\right|-\frac{4x+1}{8x^2}-\frac{3}{8}\frac{1}{x-2}+C^*$$

General Rule for Case II

For each repeated linear factor $(x-r)$ that occurs k times in the denominator, use $\frac{A_1}{x-r}+\frac{A_2}{(x-r)^2}+\cdots+\frac{A_k}{(x-r)^k}$ as part of the representation of the integrand. Every linear factor that occurs only once is handled as in Case I.

Case III

$D(x)$ is a product of one or more distinct irreducible quadratic factors and possibly also some linear factors (that may occur more than once).

General Rule for Case III

Linear factors are handled as in Cases I–II. For each irreducible quadratic factor $x^2 + bx + c$, place a term $\frac{Ax+B}{x^2+bx+c}$ in the representation of the integrand.

EXAMPLE 33.9: Find $\int \frac{(x-1)\,dx}{x(x^2+1)(x^2+2)}$.

Represent the integrand as follows:

$$\frac{(x-1)}{x(x^2+1)(x^2+2)}=\frac{A}{x}+\frac{Bx+C}{x^2+1}+\frac{Dx+E}{x^2+2}$$

Clear the denominators by multiplying by $x(x^2+1)(x^2+2)$.

$$x-1=A(x^2+1)(x^2+2)+(Bx+C)x(x^2+2)+(Dx+E)x(x^2+1)$$

Multiply out on the right:

$$x-1=(A+B+D)x^4+(B+E)x^3+(3A+C+D)x^2+(2C+E)x+2A$$

Comparing coefficients, we get:

$$2A=-1, \qquad 2C+E=1, \qquad 3A+2B+D=0, \qquad C+E=0, \qquad A+B+D=0$$

So, $A=-\frac{1}{2}$. From $2C+E=1$ and $C+E=0$, it follows that $C=1$ and $E=-1$.

From $3A+2B+D=0$ and $A+B+D=0$, we get $2A+B=0$. Since $A=-\frac{1}{2}$, $B=1$.

From $A+B+D=0$, $-\frac{1}{2}+1+D=0$. So, $D=-\frac{1}{2}$.

Thus,
$$\frac{(x-1)}{x(x^2+1)(x^2+2)}=-\frac{1}{2}\frac{1}{x}+\frac{x+1}{x^2+1}-\frac{\frac{1}{2}x+1}{x^2+2}.$$

Then the antiderivative of the left side is equal to

$$-\frac{1}{2}\ln|x|+\int\frac{x\,dx}{x^2+1}+\int\frac{dx}{x^2+1}-\frac{1}{2}\int\frac{x\,dx}{x^2+2}-\int\frac{dx}{x^2+2}$$

$$=-\frac{1}{2}\ln|x|+\frac{1}{2}\ln(x^2+1)+\tan^{-1}x-\frac{1}{4}\ln(x^2+2)-\frac{1}{2}\sqrt{2}\,\tan^{-1}(x/\sqrt{2})+(\text{constant}).$$

Case IV

$D(x)$ is a product of zero or more linear factors and one or more irreducible quadratic factors.

General Rule for Case IV

Linear factors are handled as in Cases I–II. For each irreducible quadratic factor $x^2 + bx + c$ that occurs to the kth power, insert as part of the representation of the integrand.

$$\frac{A_1x+B_1}{x^2+bx+c}+\frac{A_2x+B_2}{(x^2+bx+c)^2}+\cdots+\frac{A_kx+B_k}{(x^2+bx+c)^k}$$

EXAMPLE 33.10: Find $\int \frac{2x^2+3}{(x^2+1)^2}\,dx$.

Let $\frac{2x^2+3}{(x^2+1)^2}=\frac{Ax+B}{x^2+1}+\frac{Cx+D}{(x^2+1)^2}$. Then

$$2x^2+3=(Ax+B)(x^2+1)+Cx+D=Ax^3+Bx^2+(A+C)x+(B+D)$$

Compare coefficients: $A = 0$, $B = 2$, $A + C = 0$, $B + D = 3$. Hence, $C = 0$, $D = 1$. Thus,

$$\int \frac{2x^2+3}{(x^2+1)^2}\,dx=\int\frac{2}{x^2+1}\,dx+\int\frac{1}{(x^2+1)^2}\,dx$$

$$=2\tan^{-1}x+\int\frac{1}{(x^2+1)^2}\,dx$$

In the second integral, let $x = \tan\theta$. Then

$$\int\frac{1}{(x^2+1)^2}\,dx=\int\frac{\sec^2\theta\,d\theta}{\sec^4\theta}=\int\cos^2\theta\,d\theta=\tfrac{1}{2}(\theta+\sin\theta\cos\theta)$$

$$=\frac{1}{2}\left(\theta+\frac{\tan\theta}{\tan^2\theta+1}\right)=\frac{1}{2}\left(\tan^{-1}x+\frac{x}{x^2+1}\right)$$

Thus,

$$\int\frac{2x^2+3}{(x^2+1)^2}\,dx=\tfrac{5}{2}\tan^{-1}x+\frac{1}{2}\frac{x}{x^2+1}+C$$

SOLVED PROBLEMS

1. Find $\int\frac{x^4-x^3-x-1}{x^3-x^2}\,dx$.

   The integrand is an improper fraction. By division,

$$\frac{x^4-x^3-x-1}{x^3-x^2}=x-\frac{x+1}{x^3-x^2}=x-\frac{x+1}{x^2(x-1)}$$

   We write $\frac{x+1}{x^2(x-1)}=\frac{A}{x}+\frac{B}{x^2}+\frac{C}{x-1}$ and obtain

$$x+1=Ax(x-1)+B(x-1)+Cx^2$$

For $x = 0$, $1 = -B$ and $B = -1$. For $x = 1$, $2 = C$. For $x = 2$, $3 = 2A + B + 4C$ and $A = -2$. Thus,

$$\int \frac{x^4 - x^3 - x - 1}{x^3 - x^2}\,dx = \int x\,dx + 2\int \frac{dx}{x} + \int \frac{dx}{x^2} - 2\int \frac{dx}{x-1}$$

$$= \tfrac{1}{2}x^2 + 2\ln|x| - \frac{1}{x} - 2\ln|x-1| + C = \tfrac{1}{2}x^2 - \frac{1}{x} + 2\ln\left|\frac{x}{x-1}\right| + C$$

2. Find $\int \frac{x\,dx}{(x+2)(x+3)}$.

Let $\frac{x}{(x+2)(x+3)} = \frac{A}{x+2} + \frac{B}{x+3}$. Clear the denominators:

$$x = A(x+3) + B(x+2)$$

Let $x = -2$. Then $-2 = A$. Let $x = -3$. Then $-3 = -B$. So, $B = 3$.

$$\int \frac{x\,dx}{(x+2)(x+3)} = -2\int \frac{1}{x+2}\,dx + 3\int \frac{1}{x+3}\,dx$$

$$= -2\ln|x+2| + 3\ln|x+3| + C = -\ln((x+2)^2) + \ln(|x+3|)^3 + C$$

$$= \ln\left|\frac{(x+3)^3}{(x+2)^2}\right| + C$$

3. Find $\int \frac{x^2+2}{x(x+2)(x-1)}\,dx$.

Let $\frac{x^2+2}{x(x+2)(x-1)} = \frac{A}{x} + \frac{B}{x+2} + \frac{C}{x-1}$. Clear the denominators:

$$x^2 + 2 = A(x+2)(x-1) + Bx(x-1) + Cx(x+2)$$

Let $x = 0$. Then $2 = -2A$. So, $A = -1$. Let $x = -2$. Then $6 = 6B$. So, $B = 1$. Let $x = 1$. Then $3 = 3C$. So, $C = 1$. Hence,

$$\int \frac{x^2+2}{x(x+2)(x-1)}\,dx = -\int \frac{1}{x}\,dx + \int \frac{1}{x+2}\,dx + \int \frac{1}{x-1}\,dx$$

$$= -\ln|x| + \ln|x+2| + \ln|x-1| + C = \ln\left|\frac{(x+2)(x-1)}{x}\right| + C$$

4. Find $\int \frac{x^3+1}{(x+2)(x-1)^3}\,dx$.

Let $\frac{x^3+1}{(x+2)(x-1)^3} = \frac{A}{x+2} + \frac{B}{x-1} + \frac{C}{(x-1)^2} + \frac{D}{(x-1)^3}$. Clear the denominators:

$$x^3 + 1 = A(x-1)^3 + B(x+2)(x-1)^2 + C(x+2)(x-1) + D(x+2)$$

Let $x = -2$. Then $-7 = -27A$. So, $A = \frac{7}{27}$. Let $x = 1$. Then $2 = 3D$. So, $D = \frac{2}{3}$. Compare coefficients of x^3. Then $1 = A + B$. Since $A = \frac{7}{27}$, $B = \frac{20}{27}$. Compare coefficients of x^2. $0 = -3A + C$. Since $A = \frac{7}{27}$, $C = \frac{7}{9}$.

Thus,

$$\int \frac{x^3+1}{(x+2)(x-1)^3}\,dx = \frac{7}{27}\int \frac{1}{x+2}\,dx + \frac{20}{27}\int \frac{1}{x-1}\,dx + \frac{7}{9}\int \frac{1}{(x-1)^2}\,dx + \frac{2}{3}\int \frac{1}{(x-1)^3}\,dx$$

$$= \frac{7}{27}\ln|x+2| + \frac{20}{27}\ln|x-1| - \frac{7}{9}\frac{1}{x-1} - \frac{1}{3}\frac{1}{(x-1)^2} + C$$

5. Find $\int \frac{x^3+x^2+x+2}{x^4+3x^2+2}\,dx$.

$x^4+3x^2+2=(x^2+1)(x^2+2)$. We write $\frac{x^3+x^2+x+2}{x^4+3x^2+2}=\frac{Ax+B}{x^2+1}+\frac{Cx+D}{x^2+2}$ and obtain

$$\begin{aligned} x^3+x^2+x+2 &= (Ax+B)(x^2+2)+(Cx+D)(x^2+1) \\ &= (A+C)x^3+(B+D)x^2+(2A+C)x+(2B+D) \end{aligned}$$

Hence $A+C=1$, $B+D=1$, $2A+C=1$, and $2B+D=2$. Solving simultaneously yields $A=0$, $B=1$, $C=1$, $D=0$. Thus,

$$\begin{aligned} \int \frac{x^3+x^2+x+2}{x^4+3x^2+2}\,dx &= \int \frac{1}{x^2+1}\,dx+\int \frac{x}{x^2+2}\,dx \\ &= \tan^{-1}x+\tfrac{1}{2}\ln(x^2+2)+C \end{aligned}$$

6. Find $\int \frac{x^5-x^4+4x^3-4x^2+8x-4}{(x^2+2)^3}\,dx$.

We write $\frac{x^5-x^4+4x^3-4x^2+8x-4}{(x^2+2)^3}=\frac{Ax+B}{x^2+2}+\frac{Cx+D}{(x^2+2)^2}+\frac{Ex+F}{(x^2+2)^3}$. Then

$$\begin{aligned} x^5-x^4+4x^3-4x^2+8x-4 &= (Ax+B)(x^2+2)^2+(Cx+D)(x^2+2)+Ex+F \\ &= Ax^5+Bx^4+(4A+C)x^3+(4B+D)x^2+(4A+2C+E)x \\ &\quad +(4B+2D+F) \end{aligned}$$

from which $A=1$, $B=-1$, $C=0$, $D=0$, $E=4$, $F=0$. Thus the given integral is equal to

$$\int \frac{(x-1)\,dx}{x^2+2}+4\int \frac{x\,dx}{(x^2+2)^3}=\int \frac{x\,dx}{x^2+2}-\int \frac{dx}{x^2+2}+4\int \frac{x\,dx}{(x^2+2)^3}$$

By Quick Formula II,

$$\int \frac{x\,dx}{x^2+2}=\frac{1}{2}\int \frac{2x\,dx}{x^2+2}=\tfrac{1}{2}\ln(x^2+2)$$

and by Quick Formula I,

$$\int \frac{x\,dx}{(x^2+2)^3}=\frac{1}{2}\int (x^2+2)^{-3}(2x)\,dx=\tfrac{1}{2}(-\tfrac{1}{2})(x^2+2)^{-2}=-\frac{1}{4}\frac{1}{(x^2+2)^2}$$

So, $$\int \frac{x^5-x^4+4x^3-4x^2+8x-4}{(x^2+2)^3}\,dx=\tfrac{1}{2}\ln(x^2+2)-\frac{\sqrt{2}}{2}\tan^{-1}\left(\frac{x}{\sqrt{2}}\right)-\frac{1}{(x^2+2)^2}+C$$

SUPPLEMENTARY PROBLEMS

In Problems 7–25, evaluate the given integrals.

7. $\int \frac{dx}{x^2-9}=\frac{1}{6}\ln\left|\frac{x-3}{x+3}\right|+C$

8. $\int \frac{x\,dx}{x^2-3x-4}=\tfrac{1}{5}\ln|(x+1)(x-4)^4|+C$

9. $\int \frac{x^2-3x-1}{x^3+x^2-2x}\,dx=\ln\left|\frac{x^{1/2}(x+2)^{3/2}}{x-1}\right|+C$

10. $\int \frac{dx}{x^2+7x+6} = \frac{1}{5}\ln\left|\frac{x+1}{x+6}\right| + C$

11. $\int \frac{x^2+3x-4}{x^2-2x-8}\,dx = x + \ln|(x+2)(x-4)^4| + C$

12. $\int \frac{x\,dx}{(x-2)^2} = \ln|x-2| - \frac{2}{x-2} + C$

13. $\int \frac{x^4}{(1-x)^3}\,dx = -\frac{1}{2}x^2 - 3x - \ln(1-x)^6 - \frac{4}{1-x} + \frac{1}{2(1-x)^2} + C$

14. $\int \frac{dx}{x^3+x} = \ln\left|\frac{x}{\sqrt{x^2+1}}\right| + C$

15. $\int \frac{x^3+x^2+x+3}{(x^2+1)(x^2+3)}\,dx = \ln\sqrt{x^2+3} + \tan^{-1}x + C$

16. $\int \frac{x^4-2x^3+3x^2-x+3}{x^3-2x^2+3x}\,dx = \frac{1}{2}x^2 + \ln\left|\frac{x}{\sqrt{x^2-2x+3}}\right| + C$

17. $\int \frac{2x^3\,dx}{(x^2+1)^2} = \ln(x^2+1) + \frac{1}{x^2+1} + C$

18. $\int \frac{2x^3+x^2+4}{(x^2+4)^2}\,dx = \ln(x^2+4) + \frac{1}{2}\tan^{-1}\left(\frac{x}{2}\right) + \frac{4}{x^2+4} + C$

19. $\int \frac{x^3+x-1}{(x^2+1)^2}\,dx = \ln\sqrt{x^2+1} - \frac{1}{2}\tan^{-1}x - \frac{1}{2}\frac{x}{x^2+1} + C$

20. $\int \frac{x^4+8x^3-x^2+2x+1}{(x^2+3)(x^3+1)}\,dx = \frac{1}{168}\left[-126\ln|x+1| + 51\ln(x^2+3) + 96\ln(x^2-x+1) + 422\sqrt{3}\tan^{-1}\left(\frac{x}{\sqrt{3}}\right) - 96\sqrt{3}\tan^{-1}\left(\frac{2x-1}{\sqrt{3}}\right)\right]$

21. $\int \frac{x^3+x^2-5x+15}{(x^2+5)(x^2+2x+3)}\,dx = \ln\sqrt{x^2+2x+3} + \frac{5}{\sqrt{2}}\tan^{-1}\left(\frac{x+1}{\sqrt{2}}\right) - \sqrt{5}\tan^{-1}\left(\frac{x}{\sqrt{5}}\right) + C$

22. $\int \frac{x^6+7x^5+15x^4+23x^2+25x-3}{(x^2+x+2)^2(x^2+1)^2}\,dx = \frac{1}{x^2+x+2} - \frac{3}{x^2+1} + \ln\frac{x^2+1}{x^2+x+2} + C$

23. $\int \frac{dx}{e^{2x}-3e^x} = \frac{1}{3e^x} + \frac{1}{9}\ln\left|\frac{e^x-3}{e^x}\right| + C$ (*Hint:* Let $e^x = u$.)

24. $\int \frac{\sin x\,dx}{\cos x(1+\cos^2 x)} = \ln\left|\frac{\sqrt{1+\cos^2 x}}{\cos x}\right| + C$ (*Hint:* Let $\cos x = u$.)

25. $\int \frac{(2+\tan^2\theta)\sec^2\theta}{1+\tan^3\theta}\,d\theta = \ln|1+\tan\theta| + \frac{2}{\sqrt{3}}\tan^{-1}\left(\frac{2\tan\theta-1}{\sqrt{3}}\right) + C$

CHAPTER 34

Techniques of Integration IV: Miscellaneous Substitutions

I. Assume that, in a rational function, a variable is replaced by one of the following radicals.

1. $\sqrt[n]{ax+b}$. Then the substitution $ax + b = z^n$ will produce a rational function. (See Problems 1–3.)
2. $\sqrt{q+px+x^2}$. Then the substitution $q + px + x^2 = (z - x)^2$ will yield a rational function. (See Problem 4.)
3. $\sqrt{q+px-x^2} = \sqrt{(\alpha+x)(\beta-x)}$. Then the substitution $q + px - x^2 = (\alpha + x)^2 z^2$ will produce a rational function. (See Problem 5.)

II. Assume that, in a rational function, some variables are replaced by $\sin x$ and/or $\cos x$. Then the substitution $x = 2\tan^{-1} z$ will produce an integral of a rational function of z.

The reason that this will happen is that

$$\sin x = \frac{2z}{1+z^2}, \qquad \cos x = \frac{1-z^2}{1+z^2}, \qquad dx = \frac{2\,dz}{1+z^2} \tag{34.1}$$

(See Problem 6 for a derivation of the first two equations.)

In the final result, replace z by $\tan(x/2)$. (See Problems 7–10.)

SOLVED PROBLEMS

1. Find $\int \frac{dx}{x\sqrt{1-x}}$.

Let $1 - x = z^2$. Then $x = 1 - z^2$, $dx = -2z\,dz$, and

$$\int \frac{dx}{x\sqrt{1-x}} = \int \frac{-2z\,dz}{(1-z^2)z} = -2\int \frac{dz}{1-z^2}$$

By integration by partial fractions, one obtains

$$-2\int \frac{dz}{1-z^2} = -\ln\left|\frac{1+z}{1-z}\right| + C. \quad \text{Hence,} \quad \int \frac{dx}{x\sqrt{1-x}} = \ln\left|\frac{1-\sqrt{1-x}}{1+\sqrt{1-x}}\right| + C$$

2. Find $\int \frac{dx}{(x-2)\sqrt{x+2}}$.

Let $x + 2 = z^2$. Then $x = z^2 - 2$, $dx = 2z\,dz$, and

$$\int \frac{dx}{(x-2)\sqrt{x+2}} = \int \frac{2z\,dz}{z(z^2-4)} = 2\int \frac{dz}{z^2-4}$$

By integration by partial fractions, we get

$$2\int \frac{dz}{z^2-4} = \frac{1}{2}\ln\left|\frac{z-2}{z+2}\right| + C = \frac{1}{2}\ln\left|\frac{\sqrt{x+2}-2}{\sqrt{x+2}+2}\right| + C$$

3. Find $\int \frac{dx}{x^{1/2}-x^{1/4}}$.

Let $x = z^4$. Then $dx = 4z^3\,dz$ and

$$\int \frac{dx}{x^{1/2}-x^{1/4}} = \int \frac{4z^3\,dz}{z^2-z} = 4\int \frac{z^2\,dz}{z-1}$$

$$= 4\int \frac{(z^2-1)+1}{z-1}dz = 4\int \frac{(z-1)(z+1)+1}{z-1}dz = 4\int \left(z+1+\frac{1}{z-1}\right)dz$$

$$= 4(\tfrac{1}{2}z^2 + z + \ln|z-1|) + C = 2\sqrt{x} + 4\sqrt[4]{x} + 4\ln(\sqrt[4]{x}-1) + C$$

4. Find $\int \frac{dx}{x\sqrt{x^2+x+2}}$.

Let $x^2 + x + 2 = (z-x)^2$. Then

$$x = \frac{z^2-2}{1+2z}, \quad dx = \frac{2(z^2+z+2)dz}{(1+2z)^2}, \quad \sqrt{x^2+x+2} = \frac{z^2+z+2}{1+2z}$$

and

$$\int \frac{dx}{x\sqrt{x^2+x+2}} = \int \frac{\frac{2(z^2+z+2)}{(1+2z)^2}}{\frac{z^2-2}{1+2z}\,\frac{z^2+z+2}{1+2z}}dz = 2\int \frac{dz}{z^2-2} = \frac{1}{\sqrt{2}}\ln\left|\frac{z-\sqrt{2}}{z+\sqrt{2}}\right| + C$$

$$= \frac{1}{\sqrt{2}}\ln\left|\frac{\sqrt{x^2+x+2}+x-\sqrt{2}}{\sqrt{x^2+x+2}+x+\sqrt{2}}\right| + C$$

The equation $2\int \frac{dz}{z^2-2} = \frac{1}{\sqrt{2}}\ln\left|\frac{z-\sqrt{2}}{z+\sqrt{2}}\right| + C$ was obtained by integration by partial fractions.

5. Find $\int \frac{x\,dx}{(5-4x-x^2)^{3/2}}$.

Let $5 - 4x - x^2 = (5+x)(1-x) = (1-x)^2z^2$. Then

$$x = \frac{z^2-5}{1+z^2}, \quad dx = \frac{12z\,dz}{(1+z^2)^2}, \quad \sqrt{5-4x-x^2} = (1-x)z = \frac{6z}{1+z^2}$$

and

$$\int \frac{x\,dx}{(5-4x-x^2)^{3/2}} = \int \frac{\frac{z^2-5}{1+z^2}\,\frac{12z}{(1+z^2)^2}}{\frac{216z^3}{(1+z^2)^3}}dz = \frac{1}{18}\int\left(1-\frac{5}{z^2}\right)dz$$

$$= \frac{1}{18}\left(z+\frac{5}{z}\right) + C = \frac{5-2x}{9\sqrt{5-4x-x^2}} + C$$

6. Given $z = \tan\left(\frac{x}{2}\right)$, that is, $x = 2\tan^{-1} z$, show that

$$\sin x = \frac{2z}{1+z^2} \quad \text{and} \quad \cos x = \frac{1-z^2}{1+z^2}$$

Since

$$\frac{1+\cos x}{2} = \cos^2\left(\frac{x}{2}\right) = \frac{1}{\sec^2(x/2)} = \frac{1}{1+\tan^2(x/2)} = \frac{1}{1+z^2}$$

solving for cos x yields $\cos x = \dfrac{2}{1+z^2} - 1 = \dfrac{1-z^2}{1+z^2}$. Also,

$$\sin x = 2\sin\left(\frac{x}{2}\right)\cos(x/2) = 2\frac{\tan(x/2)}{\sec^2(x/2)} = 2\frac{\tan(x/2)}{1+\tan^2(x/2)} = \frac{2z}{1+z^2}$$

7. Find $\int \dfrac{dx}{1+\sin x - \cos x}$.

Let $x = 2\tan^{-1} z$. Using equations (34.1), we get

$$\int \frac{dx}{1+\sin x - \cos x} = \int \frac{\frac{2}{1+z^2}}{1+\frac{2z}{1+z^2} - \frac{1-z^2}{1+z^2}} dz$$

$$= \int \frac{dz}{z(1+z)} = \int \left(\frac{1}{z} - \frac{1}{1+z}\right) dz = \ln|z| - \ln|1+z| + C = \ln\left|\frac{z}{1+z}\right| + C$$

$$= \ln\left|\frac{\tan(x/2)}{1+\tan(x/2)}\right| + C$$

8. Find $\int \dfrac{dx}{3-2\cos x}$.

Let $x = 2\tan^{-1} z$. Using equations (34.1), we get

$$\int \frac{\frac{2}{1+z^2}}{3-2\frac{1-z^2}{1+z^2}} dz = \int \frac{2\,dz}{1+5z^2} = \frac{2}{\sqrt{5}}\tan^{-1}(z\sqrt{5}) + C$$

$$= \frac{2\sqrt{5}}{5}\tan^{-1}\left(\sqrt{5}\tan\left(\frac{x}{2}\right)\right) + C$$

9. Find $\int \dfrac{dx}{2+\cos x}$.

Let $x = 2\tan^{-1} z$. Using equations (34.1), we obtain

$$\int \frac{dx}{2+\cos x} = \int \frac{\frac{2}{1+z^2}}{2+\frac{1-z^2}{1+z^2}} dz = \int \frac{2\,dz}{3+z^2} = \frac{2}{\sqrt{3}}\tan^{-1}\left(\frac{z}{\sqrt{3}}\right) + C = \frac{2\sqrt{3}}{3}\tan^{-1}\left(\frac{\sqrt{3}}{3}\tan\left(\frac{x}{2}\right)\right) + C$$

10. Find $\int \dfrac{dx}{5+4\sin x}$.

Let $x = 2\tan^{-1} z$. Using equations (34.1), we obtain

$$\int \frac{dx}{5+4\sin x} = \int \frac{\frac{2}{1+z^2}}{5+4\frac{2z}{1+z^2}} dz = \int \frac{2\,dz}{5+8z+5z^2}$$

$$= \frac{2}{5}\int \frac{dz}{(z+\frac{4}{5})^2 + \frac{9}{25}} = \tfrac{2}{3}\tan^{-1}\left(\frac{z+(\frac{4}{5})}{\frac{3}{5}}\right) + C = \tfrac{2}{3}\tan^{-1}\left(\frac{5\tan(x/2)+4}{3}\right) + C$$

11. Use the substitution $1 - x^3 = z^2$ to find $\int x^5\sqrt{1-x^3}\,dx$.

The substitution yields $x^3 = 1 - z^2$, $3x^2\,dx = -2z\,dz$, and

$$\int x^5\sqrt{1-x^3}\,dx = \int x^3\sqrt{1-x^3}\,(x^2\,dx) = \int (1-z^2)z(-\tfrac{2}{3}z\,dz) = -\frac{2}{3}\int (1-z^2)z^2\,dz$$

$$= -\frac{2}{3}\left(\frac{z^3}{3} - \frac{z^5}{5}\right) + C = -\frac{2}{45}(1-x^3)^{3/2}(2+3x^3) + C$$

12. Use $x = \frac{1}{z}$ to find $\int \frac{\sqrt{x - x^2}}{x^4}\, dx$.

The substitution yields $dx = -dz/z^2$, $\sqrt{x - x^2} = \sqrt{z - 1}/z$, and

$$\int \frac{\sqrt{x - x^2}}{x^4}\, dx = \int \frac{\frac{\sqrt{z-1}}{z}\left(-\frac{dz}{z^2}\right)}{1/z^4} = -\int z\sqrt{z-1}\, dz$$

Let $z - 1 = s^2$. Then

$$-\int z\sqrt{z-1}\, dz = -\int (s^2+1)(s)(2s\, ds) = -2\left(\frac{s^5}{5} + \frac{s^3}{3}\right) + C$$

$$= -2\left[\frac{(z-1)^{5/2}}{5} + \frac{(z-1)^{3/2}}{3}\right] + C = -2\left[\frac{(1-x)^{5/2}}{5x^{5/2}} + \frac{(1-x)^{3/2}}{3x^{3/2}}\right] + C$$

13. Find $\int \frac{dx}{x^{1/2} + x^{1/3}}$.

Let $u = x^{1/6}$ so that $x = u^6$, $dx = 6u^5\, du$, $x^{1/2} = u^3$, and $x^{1/3} = u^2$. Then we obtain

$$\int \frac{6u^5 du}{u^3 + u^2} = 6\int \frac{u^3}{u+1}\, du = 6\int \left(u^2 - u + 1 - \frac{1}{u+1}\right) du = 6\left(\frac{1}{3}u^3 - \frac{1}{2}u^2 + u - \ln|u+1|\right) + C$$

$$= 2x^{1/2} - 3x^{1/3} + x^{1/6} - \ln|x^{1/6} + 1| + C$$

SUPPLEMENTARY PROBLEMS

In Problems 14–39, evaluate the given integral.

14. $\int \frac{\sqrt{x}}{1+x}\, dx - 2\sqrt{x} - 2\tan^{-1}\sqrt{x} + C$

15. $\int \frac{dx}{\sqrt{x}(1+\sqrt{x})} = 2\ln(1+\sqrt{x}) + C$

16. $\int \frac{dx}{3+\sqrt{x+2}} = 2\sqrt{x+2} - 6\ln(3+\sqrt{x+2}) + C$

17. $\int \frac{1-\sqrt{3x+2}}{1+\sqrt{3x+2}}\, dx = -x + \frac{4}{3}[\sqrt{3x+2} - \ln(1+\sqrt{3x+2})] + C$

18. $\int \frac{dx}{\sqrt{x^2 - x + 1}} = \ln\left|2\sqrt{x^2-x+1} + 2x - 1\right| + C$

19. $\int \frac{dx}{x\sqrt{x^2+x-1}} = 2\tan^{-1}(\sqrt{x^2+x-1} + x) + C$

20. $\int \frac{dx}{\sqrt{6+x-x^2}} = \sin^{-1}\left(\frac{2x-1}{5}\right) + C$

21. $\int \frac{\sqrt{4x-x^2}}{x^3}\, dx = -\frac{(4x-x)^{3/2}}{6x^3} + C$

22. $\int \frac{dx}{(x+1)^{1/2} + (x+1^{1/4})} = 2(x+1)^{1/2} - 4(x+1)^{1/4} + 4\ln(1+(x+1)^{1/4}) + C$

23. $\int \frac{dx}{2+\sin x} = \frac{2}{\sqrt{3}} \tan^{-1} \frac{2\tan(x/2)+1}{\sqrt{3}} + C$

24. $\int \frac{dx}{1-2\sin x} = \frac{\sqrt{3}}{3} \ln \left| \frac{\tan\frac{1}{2}x - 2 - \sqrt{3}}{\tan\frac{1}{2}x - 2 + \sqrt{3}} \right| + C$

25. $\int \frac{dx}{3+5\sin x} = \frac{1}{4} \ln \left| \frac{3\tan\frac{1}{2}x + 1}{\tan\frac{1}{2}x + 3} \right| + C$

26. $\int \frac{dx}{\sin x - \cos x - 1} = \ln\left|\tan\frac{1}{2}x - 1\right| + C$

27. $\int \frac{dx}{5+3\sin x} = \frac{1}{2} \tan^{-1} \frac{5\tan(x/2)+3}{4} + C$

28. $\int \frac{\sin x\, dx}{1+\sin^2 x} = \frac{\sqrt{2}}{4} \ln \left| \frac{\tan^2\frac{1}{2}x + 3 - 2\sqrt{2}}{\tan^2\frac{1}{2}x + 3 + 2\sqrt{2}} \right| + C$

29. $\int \frac{dx}{1+\sin x + \cos x} = \ln\left|1 + \tan\frac{1}{2}x\right| + C$

30. $\int \frac{dx}{2-\cos x} = \frac{2}{\sqrt{3}} \tan^{-1}\left(\sqrt{3}\tan\left(\frac{x}{2}\right)\right) + C$

31. $\int \sin\sqrt{x}\, dx = -2\sqrt{x}\cos\sqrt{x} + 2\sin\sqrt{x} + C$

32. $\int \frac{dx}{x\sqrt{3x^2+2x-1}} = -\sin^{-1}\left(\frac{1-x}{2x}\right) + C.$ (*Hint:* Let $x = 1/z$.)

33. $\int \frac{(e^x-2)e^x}{e^x+1} dx = e^x - 3\ln(e^x+1) + C.$ (*Hint:* Let $e^x + 1 = z$.)

34. $\int \frac{\sin x \cos x}{1-\cos x} dx = \cos x + \ln(1-\cos x) + C.$ (*Hint:* Let $\cos x = z$.)

35. $\int \frac{dx}{x^2\sqrt{4-x^2}} = -\frac{\sqrt{4-x^2}}{4x} + C.$ (*Hint:* Let $x = 2/z$.)

36. $\int \frac{dx}{x^2(4+x^2)} = -\frac{1}{4x} + \frac{1}{8}\tan^{-1}\left(\frac{2}{x}\right) + C$

37. $\int \sqrt{1+\sqrt{x}}\, dx = \frac{4}{5}(1+\sqrt{x})^{5/2} - \frac{4}{3}(1+\sqrt{x})^{3/2} + C$

38. $\int \frac{dx}{3(1-x^2) - (5+4x)\sqrt{1-x^2}} = \frac{2\sqrt{1+x}}{3\sqrt{1+x} - \sqrt{1-x}} + C$

39. $\int \frac{x^{1/2}}{x^{1/5}+1} dx = 10[\frac{1}{13}x^{13/10} - \frac{1}{11}x^{11/10} + \frac{1}{9}x^{9/10} - \frac{1}{7}x^{7/10} + \frac{1}{5}x^{1/2} - \frac{1}{3}x^{3/10} + x^{1/10} - \tan^{-1}(x^{1/10})] + C$
(*Hint:* Let $u = x^{1/10}$.)

40. (GC) Use a graphing calculator to approximate (to eight decimal places) $\int_0^{\pi/3} \frac{\sin x\, dx}{3-2\cos x}$ and compare your result with the value obtained by the methods of this chapter.

41. (GC) Use a graphing calculator to approximate (to eight decimal places) $\int_2^4 \frac{dx}{x\sqrt{x-1}}$ and compare your result with the value obtained by the methods of this chapter.

Improper Integrals

For a definite integral $\int_c^b f(x)\,dx$ to be defined, it suffices that a and b are real numbers and that $f(x)$ is continuous on $[a, b]$. We shall now study two different kinds of integrals that we shall call *improper integrals*.

Infinite Limits of Integration

(a) $\int_a^{+\infty} f(x)\,dx = \lim_{c \to +\infty} \int_a^c f(x)\,dx$

See Problems 1–3, 5, and 6.

(b) $\int_{-\infty}^b f(x)\,dx = \lim_{c \to -\infty} \int_c^b f(x)\,dx$

See Problem 4.

(c) $\int_{-\infty}^{+\infty} f(x)\,dx = \int_a^{+\infty} f(x)\,dx + \int_{-\infty}^a f(x)\,dx$

provided that *both* limits on the right exist. See Problem 7.

Discontinuities of the Integrand

(a) If f is continuous on $[a, b]$ except that it is not continuous from the right at a, then

$$\int_a^b f(x)\,dx = \lim_{u \to a^+} \int_u^b f(x)\,dx$$

See Problem 16.

(b) If f is continuous on $[a, b]$ except that it is not continuous from the left at b, then

$$\int_a^b f(x)\,dx = \lim_{u \to b^-} \int_a^u f(x)\,dx$$

See Problems 9, 10, 12, 14, and 15.

(c) If f is continuous on $[a, b]$ except at a point c in (a, b), then

$$\int_a^b f(x)\,dx = \lim_{u \to c^-} \int_a^u f(x)\,dx + \lim_{u \to c^+} \int_u^b f(x)\,dx$$

provided that *both* integrals on the right exist. See Problems 11 and 13.

When the limit defining an improper integral exists, we say that the integral is *convergent*. In the opposite case, we say that the integral is *divergent*. If the integral is divergent, we say that it is equal to $+\infty$ (respectively $-\infty$) if the limit defining the improper integral approaches $+\infty$ (respectively $-\infty$).

SOLVED PROBLEMS

1. Evaluate $\int_1^{+\infty} \frac{1}{x^2}\,dx$.

$$\int_1^{+\infty} \frac{1}{x^2}\,dx = \lim_{c\to+\infty} \int_1^c \frac{1}{x^2}\,dx = \lim_{c\to+\infty} -\frac{1}{x}\bigg]_1^c$$

$$= \lim_{c\to+\infty} -\left(\frac{1}{c}-1\right) = -(0-1) = 1$$

Note: The integral $\int_1^{+\infty} \frac{1}{x^2}\,dx$ can be interpreted as the area of the region under the curve $y = 1/x^2$ and above the x axis, for $x > 1$. Thus, a region that is infinite (in the sense of being unbounded) can have a finite area.

2. Evaluate $\int_1^{+\infty} \frac{1}{x}\,dx$.

$$\int_1^{+\infty} \frac{1}{x}\,dx = \lim_{c\to+\infty} \int_1^c \frac{1}{x}\,dx = \lim_{c\to+\infty} \ln x\bigg]_1^c$$

$$= \lim_{c\to+\infty} -(\ln c - 0) = +\infty$$

Thus, the integral diverges to $+\infty$.

3. Show that $\int_1^{+\infty} \frac{1}{x^p}\,dx$ converges for $p > 1$ and diverges to $+\infty$ for $p \le 1$.

$$\int_1^{+\infty} \frac{1}{x^p}\,dx = \lim_{c\to+\infty} \int_1^c \frac{1}{x^p}\,dx = \lim_{c\to+\infty} \frac{1}{1-p}\frac{1}{x^{p-1}}\bigg]_1^c$$

Assume $p > 1$. Then we have $\lim_{c\to+\infty} \frac{1}{1-p}\left(\frac{1}{c^{p-1}}-1\right) = \frac{1}{1-p}(0-1) = \frac{1}{p-1}$.

By Problem 2, we already know that $\int_1^{+\infty} \frac{1}{x}\,dx$ diverges to $+\infty$. So, assume $p < 1$. Then we have

$$\lim_{c\to+\infty} \frac{1}{1-p}\left(\frac{1}{c^{p-1}}-1\right) = \lim_{c\to+\infty} \frac{1}{1-p}(c^{1-p}-1) = +\infty \quad \text{since } 1-p>0$$

4. Evaluate $\int_{-\infty}^0 e^{rx}\,dx$ for $r > 0$.

$$\int_{-\infty}^0 e^{rx}\,dx = \lim_{c\to+\infty} \int_c^0 e^{rx}\,dx = \lim_{c\to-\infty} \frac{1}{r}e^{rx}\bigg]_c^0$$

$$= \frac{1}{r}\lim_{c\to-\infty} (1-e^{rc}) = \frac{1}{r}(1-0) = \frac{1}{r}$$

5. Evaluate $\int_0^{+\infty} \frac{1}{x^2+4}\,dx$.

$$\int_0^{+\infty} \frac{1}{x^2+4}\,dx = \lim_{c\to+\infty} \int_0^c \frac{1}{x^2+4}\,dx = \lim_{c\to+\infty} \frac{1}{2}\tan^{-1}\left(\frac{x}{2}\right)\bigg]_0^c$$

$$= \lim_{c\to+\infty} \frac{1}{2}\left(\tan^{-1}\left(\frac{c}{2}\right)-0\right) = \frac{1}{2}\left(\frac{\pi}{2}\right) = \frac{\pi}{4}$$

6. Evaluate $\int_0^{+\infty} e^{-x}\sin x\,dx$.

$$\int_0^{+\infty} e^{-x}\sin x\,dx = \lim_{c\to+\infty} \int_0^c e^{-x}\sin x\,dx$$

$$= \lim_{c\to+\infty} (-\tfrac{1}{2}e^{-x}(\sin x + \cos x))\bigg]_0^c \quad \text{(by integration by parts)}$$

$$= \lim_{c\to+\infty} [(-\tfrac{1}{2}e^{-c}(\sin c + \cos c)) + \tfrac{1}{2}]$$

As $c \to +\infty$, $e^{-c} \to 0$, while $\sin c$ and $\cos c$ oscillate between -1 and 1. Hence, $\lim_{c\to+\infty} e^{-c}(\sin c + \cos c) = 0$ and, therefore,

$$\int_0^{+\infty} e^{-x} \sin x \, dx = \tfrac{1}{2}$$

7. Evaluate $\int_{-\infty}^{+\infty} \frac{dx}{e^x + e^{-x}} = \int_{-\infty}^{+\infty} \frac{e^x\,dx}{e^{2x}+1}$.

$$\int_0^{+\infty} \frac{e^x\,dx}{e^{2x}+1} = \lim_{c\to+\infty} \int_0^c \frac{e^x\,dx}{e^{2x}+1}$$

$$= \lim_{c\to+\infty} \int_1^{e^c} \frac{du}{u^2+1} \quad \text{(by the substitution } u = e^x\text{)}$$

$$= \lim_{c\to+\infty} \tan^{-1} u \Big]_1^{e^c} = \lim_{c\to+\infty} (\tan^{-1}(e^c) - \tan^{-1}(1))$$

$$= \lim_{c\to+\infty} \left(\tan^{-1}(e^c) - \frac{\pi}{4}\right) = \frac{\pi}{2} - \frac{\pi}{4} = \frac{\pi}{4}$$

Similarly,

$$\int_{-\infty}^0 \frac{e^x\,dx}{e^{2x}+1} = \lim_{c\to-\infty} \int_c^0 \frac{e^x\,dx}{e^{2x}+1}$$

$$= \lim_{c\to-\infty} \int_{e^c}^1 \frac{du}{u^2+1} = \lim_{c\to-\infty} \tan^{-1} u \Big]_{e^c}^1$$

$$= \lim_{c\to+\infty} \left(\frac{\pi}{4} - \tan^{-1}(e^c)\right) = \frac{\pi}{4} - \lim_{c\to-\infty} \tan^{-1}(e^c) = \frac{\pi}{4} - 0 = \frac{\pi}{4}$$

Thus,

$$\int_{-\infty}^{+\infty} \frac{dx}{e^x + e^{-x}} = \int_0^{+\infty} \frac{e^x\,dx}{e^{2x}+1} + \int_{-\infty}^0 \frac{e^x\,dx}{e^{2x}+1}$$

$$= \frac{\pi}{4} + \frac{\pi}{4} = \frac{\pi}{2}$$

8. Find the area of the region lying to the right of $x = 3$ and between the curve $y = \frac{1}{x^2-1}$ and the x axis.

The area

$$\int_3^{+\infty} \frac{dx}{x^2-1} = \lim_{c\to+\infty} \int_3^c \frac{dx}{x^2-1}$$

$$= \frac{1}{2} \lim_{c\to+\infty} \ln \frac{x-1}{x+1} \Big]_3^c \quad \text{(by the integration by partial fractions)}$$

$$= \frac{1}{2} \lim_{c\to+\infty} \left(\ln \frac{c-1}{c+1} - \ln \tfrac{1}{2}\right) = \frac{1}{2} \lim_{c\to+\infty} \left(\ln \frac{1-(1/c)}{1+(1/c)} - \ln \tfrac{1}{2}\right)$$

$$= \tfrac{1}{2}(\ln 1 + \ln 2) = \frac{\ln 2}{2}$$

9. Evaluate $\int_0^3 \frac{dx}{\sqrt{9-x^2}}$.

The integrand is discontinuous at $x = 3$. So,

$$\int_0^3 \frac{dx}{\sqrt{9-x^2}} = \lim_{u\to3^-} \int_0^u \frac{dx}{\sqrt{9-x^2}} = \lim_{u\to3^-} \sin^{-1}\left(\frac{x}{3}\right)\Big]_0^u$$

$$= \lim_{u\to3^-} \left(\sin^{-1}\left(\frac{u}{3}\right) - \sin^{-1} 0\right) = \lim_{u\to3^-} \left(\sin^{-1}\left(\frac{u}{3}\right) - 0\right)$$

$$= \sin^{-1} 1 = \frac{\pi}{2}$$

10. Evaluate $\int_0^2 \frac{dx}{2-x}$.

The integrand is discontinuous at $x = 2$.

$$\int_0^2 \frac{dx}{2-x} = \lim_{u\to 2^-} \int_0^u \frac{dx}{2-x} = \lim_{u\to 2^-} -\ln(2-x)\Big]_0^u$$

$$= \lim_{u\to 2^-} -(\ln(2-u) - \ln 2)) = +\infty$$

Hence, the integral diverges to $+\infty$.

11. Evaluate $\int_0^4 \frac{dx}{(x-1)^2}$.

The integrand is discontinuous at $x = 1$, which is inside (0, 4). (See Fig. 35-1.)

$$\lim_{u\to 1^-} \int_0^u \frac{dx}{(x-1)^2} = \lim_{u\to 1^-} -\frac{1}{x-1}\Big]_0^u$$

$$= \lim_{u\to 1^-} -\left(\frac{1}{u-1} - (-1)\right) = \lim_{u\to 1^-} -\left(\frac{1}{u-1} + 1\right) = +\infty$$

Hence, $\int_0^4 \frac{dx}{(x-1)^2}$ is divergent. (We do not have to consider $\lim_{u\to 1^+} \int_0^4 \frac{dx}{(x-1)^2}$ at all. For $\int_0^4 \frac{dx}{(x-1)^2}$ to be convergent, both $\lim_{u\to 1^-} \int_0^u \frac{dx}{(x-1)^2}$ and $\lim_{u\to 1^+} \int_u^4 \frac{dx}{(x-1)^2}$ must exist.)

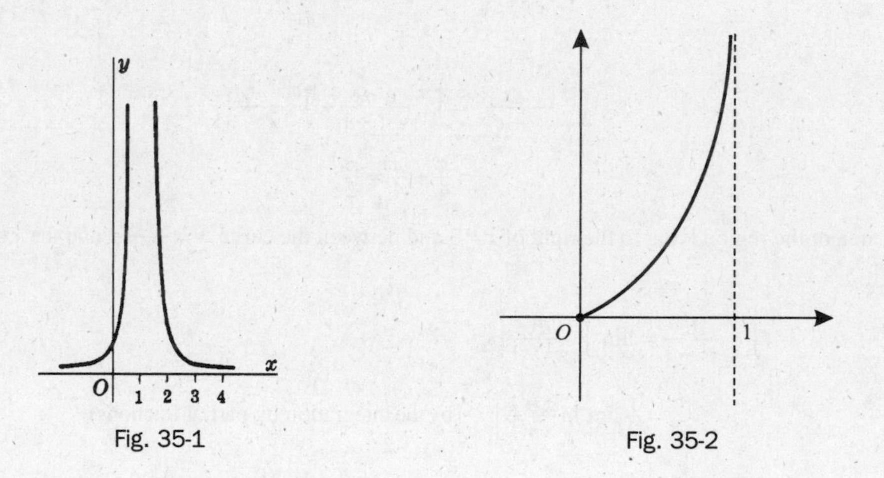

Fig. 35-1

Fig. 35-2

12. Find the area of the region between the curve $y = \frac{x}{\sqrt{1-x^2}}$, the x axis, and $x = 0$ and $x = 1$. (See Fig. 35-2.)

The area is

$$\int_0^1 \frac{x}{\sqrt{1-x^2}}\,dx = \lim_{u\to 1^-} \int_0^u \frac{x}{\sqrt{1-x^2}}\,dx$$

$$= \lim_{u\to 1^-} -\frac{1}{2}\int_0^u (1-x^2)^{1/2}(-2x)\,dx$$

$$= \lim_{u\to 1^-} -(1-x^2)^{1/2}\Big]_0^u \quad \text{(by Quick Formula I)}$$

$$= \lim_{u\to 1^-} -[\sqrt{1-u^2} - 1] = 1$$

13. Evaluate $\int_0^4 \frac{dx}{\sqrt[3]{x-1}}$.

The integrand is discontinuous at $x = 1$, which lies inside (0, 4).

$$\lim_{u\to 1^-}\int_0^u \frac{dx}{\sqrt[3]{x-1}} = \lim_{u\to 1^-}\int_0^u (x-1)^{1/3}\,dx$$

$$= \lim_{u\to 1^-} \tfrac{3}{2}(x-1)^{2/3}\Big]_0^u = \lim_{u\to 1^-}\tfrac{3}{2}[(u-1)^{2/3}-1] = -\tfrac{3}{2}$$

On the other hand,

$$\lim_{u\to 1+}\int_0^4 \frac{dx}{\sqrt[3]{x-1}} = \lim_{u\to 1^+}\int_0^4 (x-1)^{1/3}\,dx$$

$$= \lim_{u\to 1^+} \tfrac{3}{2}(x-1)^{2/3}\Big]_0^4 = \lim_{u\to 1^+}\tfrac{3}{2}[\sqrt[3]{9}-(u-1)^{2/3}-1] = \tfrac{3}{2}\sqrt[3]{9}$$

Hence,

$$\int_0^4 \frac{dx}{\sqrt[3]{x-1}} = \lim_{u\to 1^-}\int_0^u \frac{dx}{\sqrt[3]{x-1}} + \lim_{u\to 1^+}\int_u^4 \frac{dx}{\sqrt[3]{x-1}} = -\tfrac{3}{2}+\tfrac{3}{2}\sqrt[3]{9}$$

$$= \tfrac{3}{2}(\sqrt[3]{9}-1)$$

14. Evaluate $\int_0^{\pi/2} \sec x\,dx$.

The integrand is discontinuous at $x = \frac{\pi}{2}$.

$$\int_0^{\pi/2} \sec x\,dx = \lim_{u\to\pi/2}\int_0^u \sec x\,dx$$

$$= \lim_{u\to\pi/2^-} \ln(\sec x + \tan x)\Big]_0^u$$

$$= \lim_{u\to\pi/2^-} [\ln(\sec u + \tan u) - \ln(1+0)]$$

$$= \lim_{u\to\pi/2^-} \ln(\sec u + \tan u) = +\infty$$

since
$$\lim_{u\to\pi/2^-} \sec u = +\infty \text{ and } \lim_{u\to\pi/2^-} \tan u = +\infty$$

Thus, $\int_0^{\pi/2} \sec x\,dx$ diverges to $+\infty$.

15. Evaluate $\int_0^{\pi/2} \frac{\cos x}{\sqrt{1-\sin x}}\,dx$.

The integrand is discontinuous at $x = \frac{\pi}{2}$.

$$\int_0^{\pi/2} \frac{\cos x}{\sqrt{1-\sin x}}\,dx = \lim_{u\to\pi/2^-}\int_0^u \frac{\cos x}{\sqrt{1-\sin x}}\,dx$$

$$= \lim_{u\to\pi/2^-} -\int_0^u (1-\sin x)^{-1/2}(-\cos x)\,dx$$

$$= \lim_{u\to\pi/2^-} -2(1-\sin x)^{1/2}\Big]_0^u = \lim_{u\to\pi/2^-} -2[(1-\sin u)^{1/2}-1] = 2$$

16. Evaluate $\int_0^1 \frac{1}{x^2}\,dx$.

The integrand is discontinuous at $x = 0$.

$$\int_0^1 \frac{1}{x^2}\,dx = \lim_{u\to 0^+}\int_u^1 \frac{1}{x^2}\,dx = \lim_{u\to 0^+} -\frac{1}{x}\Big]_u^1$$

$$= \lim_{u\to 0^+} -\left(1-\frac{1}{u}\right) = +\infty$$

SUPPLEMENTARY PROBLEMS

17. Evaluate the given integrals:

(a) $\int_0^1 \frac{dx}{\sqrt{x}} = 2$ (b) $\int_0^4 \frac{1}{4-x}\,dx = +\infty$ (c) $\int_0^4 \frac{1}{\sqrt{4-x}}\,dx = 4$

(d) $\int_0^4 \frac{1}{(4-x)^{3/2}}\,dx = +\infty$ (e) $\int_{-2}^2 \frac{1}{\sqrt{4-x}}\,dx = \pi$ (f) $\int_{-1}^8 \frac{1}{x^3}\,dx = \frac{9}{2}$

(g) $\int_0^4 \frac{dx}{(x-2)^{2/3}} = 6\sqrt[3]{2}$ (h) $\int_{-1}^1 \frac{dx}{x^4} = +\infty$ (i) $\int_0^1 \ln x\,dx = -1$

(j) $\int_0^1 x\ln x\,dx = -\frac{1}{4}$

18. Find the area of the region between the given curve and its asymptotes:

(a) $y^2 = \frac{x^4}{4-x^2}$; (b) $y^2 = \frac{4-x}{x}$; (c) $y^2 = \frac{1}{x(1-x)}$

Ans. (a) 4π; (b) 4π; (c) 2π

19. Evaluate the given integrals:

(a) $\int_1^{+\infty} \frac{dx}{x^2} = 1$ (b) $\int_{-\infty}^0 \frac{dx}{(4-x)^2} = \frac{1}{4}$ (c) $\int_0^{+\infty} e^{-x}\,dx = 1$

(d) $\int_{-\infty}^6 \frac{dx}{(4-x)^2} = +\infty$ (e) $\int_2^{+\infty} \frac{dx}{x\ln^2 x} = \frac{1}{\ln 2}$ (f) $\int_1^{+\infty} \frac{e^{-\sqrt{x}}}{\sqrt{x}}\,dx = \frac{2}{e}$

(g) $\int_{-\infty}^{+\infty} xe^{-x^2}\,dx = 0$ (h) $\int_{-\infty}^{+\infty} \frac{dx}{1+4x^2} = \frac{\pi}{2}$ (i) $\int_{-\infty}^0 xe^x\,dx = -1$

(j) $\int_0^{+\infty} x^3e^{-x}\,dx = 6$

20. Find the area of the region between the given curve and its asymptote:

(a) $y = \frac{8}{x^2+4}$; (b) $y = \frac{x}{(4+x^2)^2}$; (c) $y = xe^{-x^2/2}$

Ans. (a) 4π; (b) $\frac{1}{4}$; (c) 2

21. Find the area of the following regions:

(a) Above the x axis, under $y = \frac{1}{x^2-4}$ and to the right of $x = 3$.

(b) Above the x axis, under $y = \frac{1}{x(x-1)^2}$ and to the right of $x = 2$.

Ans. (a) $\frac{1}{4}\ln 5$; (b) $1 - \ln 2$

22. Show that the areas of the following regions are infinite:

(a) Above the x axis, under $y = \frac{1}{4 - x^2}$ from $x = -2$ to $x = 2$.

(b) Above the x axis, under $xy = 9$ and to the right of $x = 1$.

23. Show that the area of the region in the first quadrant under $y = e^{-2x}$ is $\frac{1}{2}$, and that the volume generated by revolving that region about the x axis is $\frac{\pi}{4}$.

24. Find the length of the indicated arc: (a) $9y^2 = x(3 - x)^2$, a loop; (b) $x^{2/3} + y^{2/3} = a^{2/3}$, entire length; (c) $9y^2 = x^2(2x + 3)$, a loop

Ans. (a) $4\sqrt{3}$ units; (b) $6a$ units; (c) $2\sqrt{3}$ units

25. Show that $\int_a^b \frac{dx}{(x-b)^p}$ converges for $p < 1$ and diverges to $+\infty$ for $p \geq 1$.

26. Let $0 \leq f(x) \leq g(x)$ for $a \leq x < b$. Assume that $\lim_{x \to b^-} f(x) = +\infty$ and $\lim_{x \to b^-} g(x) = +\infty$. (See Fig. 35-3.) It is not hard to show that, if $\int_a^b g(x)\,dx$ converges, then so does $\int_a^b f(x)\,dx$ and, equivalently, if $\int_a^b f(x)\,dx$ does not converge, then neither does $\int_a^b g(x)\,dx$. A similar result also holds for $a < x \leq b$, with $\lim_{x \to a^+}$ replacing $\lim_{x \to b^-}$.

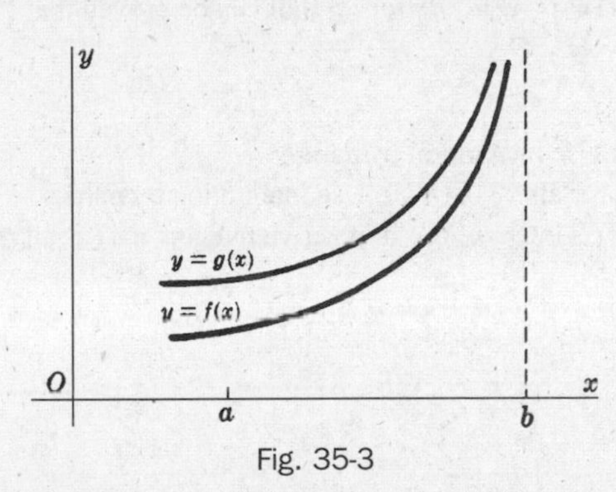

Fig. 35-3

As an example, consider $\int_0^1 \frac{dx}{1-x^4}$. For $0 \leq x < 1$,

$$1 - x^4 = (1-x)(1+x)(1+x^2) < 4(1-x) \quad \text{and} \quad \frac{1}{4}\frac{1}{1-x} < \frac{1}{1-x^4}$$

Since $\frac{1}{4}\int_0^1 \frac{dx}{1-x}$ does not converge, neither does $\int_0^1 \frac{dx}{1-x^4}$.

Now consider $\int_0^1 \frac{dx}{x^2+\sqrt{x}}$. For $0 < x \leq 1$, $\frac{1}{x^2+\sqrt{x}} < \frac{1}{\sqrt{x}}$. Since $\int_0^1 \frac{1}{\sqrt{x}}dx$ converges, so does $\int_0^1 \frac{dx}{x^2+\sqrt{x}}$.

Determine whether each of the following converges:

(a) $\int_0^1 \frac{e^x dx}{x^{1/3}}$; (b) $\int_0^{\pi/4} \frac{\cos x}{x}dx$; (c) $\int_0^{\pi/4} \frac{\cos x}{\sqrt{x}}dx$

Ans. (a) and (c) converge

27. Assume that $0 \leq f(x) \leq g(x)$ for $x \geq a$. Assume also that $\lim_{x \to +\infty} f(x) = \lim_{x \to +\infty} g(x) = 0$. (See Fig. 35-4.) It is not hard to show that, if $\int_a^{+\infty} g(x)\,dx$ converges, so does $\int_a^{+\infty} f(x)\,dx$ (and, equivalently, that, if $\int_a^{+\infty} f(x)\,dx$ does not converge, then neither does $\int_a^{+\infty} g(x)\,dx$).

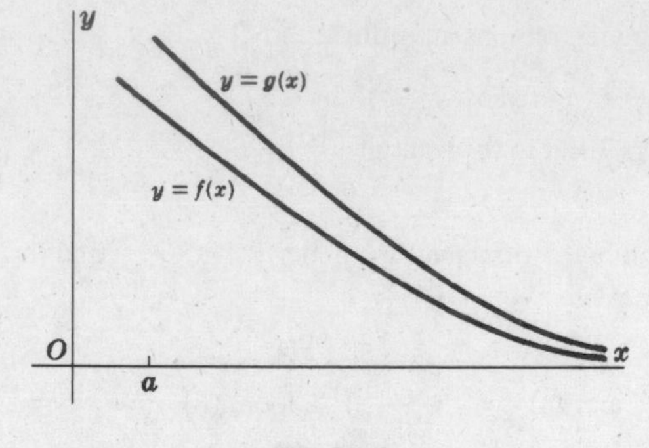

Fig. 35-4

As an example, consider $\int_1^{+\infty} \frac{dx}{\sqrt{x^4+2x+6}}$. For $x \geq 1$, $\frac{1}{\sqrt{x^4+2x+6}} < \frac{1}{x^2}$. Since $\int_1^{+\infty} \frac{dx}{x^2}$ converges, so does $\int_1^{+\infty} \frac{dx}{\sqrt{x^4+2x+6}}$.

Determine whether or not each of the following converges:

(a) $\int_2^{+\infty} \frac{dx}{\sqrt{x^3+2x}}$; (b) $\int_1^{+\infty} e^{-x^2} dx$; (c) $\int_0^{+\infty} \frac{dx}{\sqrt{x+x^4}}$

Ans. all converge

28. Define the gamma function $\Gamma(t) = \int_0^{+\infty} x^{t-1} e^{-x} dx$ for $t > 0$. It can be proved that $\Gamma(t)$ is convergent. (This is left as a project for the student.)

(a) Show that $\Gamma(1) = 1$.
(b) Show that $\Gamma(2) = 1$. (*Hint*: Use integration by parts.)
(c) Prove that $\Gamma(t+1) = t\Gamma(t)$ for all $t > 0$. (*Hint*: Use integration by parts.)
(d) Use part (c) to show that $\Gamma(n+1) = n!$ for all positive integers n. (Recall that $n! = 1 \cdot 2 \cdot 3 \cdot 4 \cdot \cdots \cdot n$.)

CHAPTER 36

Applications of Integration III: Area of a Surface of Revolution

If an arc of a curve is revolved about a line that does not intersect the arc, then the resulting surface is called a *surface of revolution*. By the *surface area* of that surface, we mean the area of its outer surface.

Let f be a continuous function on $[a, b]$ that is differentiable in (a, b) and such that $f(x) \geq 0$ for $a \leq x \leq b$. Then the surface area S of the surface of revolution generated by revolving the graph of f on $[a, b]$ about the x axis is given by the formula

$$S = 2\pi \int_a^b y\sqrt{1+\left(\frac{dy}{dx}\right)^2}\,dx = 2\pi \int_a^b f(x)\sqrt{1+(f'(x))^2}\,dx \tag{36.1}$$

For a justification of this formula, see Problem 11.

There is another formula like (36.1) that is obtained when we exchange the roles of x and y. Let g be a continuous function on $[c, d]$ that is differentiable on (c, d) and such that $g(y) \geq 0$ for $c \leq y \leq d$. Then the surface area S of the surface of revolution generated by revolving the graph of g on $[c,d]$ about the y axis is given by the formula:

$$S = 2\pi \int_c^d x\sqrt{1+\left(\frac{dx}{dy}\right)^2}\,dy = 2\pi \int_c^d g(y)\sqrt{1+(g'(y))^2}\,dy \tag{36.2}$$

Similarly, if a curve is given by parametric equations $x = f(u)$, $y = g(u)$ (see Chapter 37), and, if the arc from $u = u_1$ to $u = u_2$ is revolved about the x axis, then the surface area of the resulting surface of revolution is given by the formula

$$S = 2\pi \int_{u_1}^{u_2} y\sqrt{\left(\frac{dx}{du}\right)^2+\left(\frac{dy}{du}\right)^2}\,du \tag{36.3}$$

Here, we have assumed that f and g are continuous on $[u_1, u_2]$ and differentiable on (u_1, u_2), and that $y = g(u) \geq 0$ on $[u_1, u_2]$. Another such formula holds in the case of a revolution around the y axis.

SOLVED PROBLEMS

1. Find the area S of the surface of revolution generated by revolving about the x axis the arc of the parabola $y^2 = 12x$ from $x = 0$ to $x = 3$.

By implicit differentiation,

$$\frac{dy}{dx} = \frac{6}{y} \quad \text{and} \quad 1+\left(\frac{dy}{dx}\right)^2 = \frac{y^2+36}{y^2}$$

By (36.1),

$$S = 2\pi\int_0^3 y\frac{\sqrt{y^2+36}}{y}\,dx = 2\pi\int_0^3 \sqrt{12x+36}\,dx$$

$$= 2\pi(8(12x+36)^{3/2})]_0^3 = 24(2\sqrt{2}-1)\pi$$

2. Find the area S of the surface of revolution generated by revolving about the y axis the arc of $x = y^3$ from $y = 0$ to $y = 1$.

$$\frac{dx}{dy} = 3y^2 \text{ and } 1+\left(\frac{dx}{dy}\right)^2 = 1+9y^4. \text{ So, by (36.2),}$$

$$S = 2\pi\int_0^1 x\sqrt{1+9y^4}\,dy = 2\pi\int_0^1 y^3\sqrt{1+9y^4}\,dy$$

$$= \frac{\pi}{18}\int_0^1 (1+9y^4)^{1/2}(36y^3)\,dy$$

$$= \frac{\pi}{18}\frac{2}{3}(1+9y^4)^{3/2}]_0^1$$

$$= \frac{\pi}{27}(10\sqrt{10}-1)$$

3. Find the area of the surface of revolution generated by revolving about the y-axis the arc of $y^2 + 4x = 2\ln y$ from $y = 1$ to $y = 3$.

$$S = 2\pi\int_e^d y\sqrt{1+\left(\frac{dx}{dy}\right)^2}\,dy = 2\pi\int_1^3 y\frac{1+y^2}{2y}\,dy = \pi\int_1^3 (1+y^2)\,dy = \frac{32}{3}\pi$$

4. Find the area of the surface of revolution generated by revolving a loop of the curve $8a^2y^2 = a^2x^2 - x^4$ about the x axis. (See Fig. 36-1.)

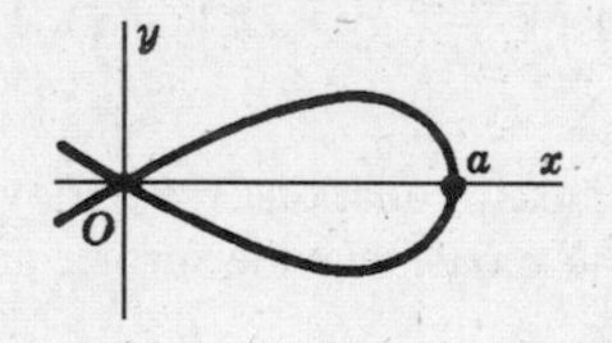

Fig. 36-1

Here $\dfrac{dy}{dx} = \dfrac{a^2x-2x^3}{8a^2y}$ and $1+\left(\dfrac{dy}{dx}\right)^2 = 1+\dfrac{(a^2-2x^2)^2}{8a^2(a^2-x^2)} = \dfrac{(3a^2-2x^2)^2}{8a^2(a^2-x^2)}$

Hence $$S = 2\pi\int_0^a y\sqrt{1+\left(\frac{dy}{dx}\right)^2}\,dx = 2\pi\int_0^a \frac{x\sqrt{a^2-x^2}}{2a\sqrt{2}}\,\frac{3a^2-2x^2}{2a\sqrt{2}\sqrt{a^2-x^2}}\,dx$$

$$= \frac{\pi}{4a^2}\int_0^a (3a^2-2x^2)x\,dx = \frac{1}{4}\pi a^2$$

5. Find the area of the surface of revolution generated by revolving about the x axis the ellipse $\dfrac{x^2}{16}+\dfrac{y^2}{4}=1$.

$$S = 2\pi\int_{-4}^4 y\frac{\sqrt{16y^2+x^2}}{4y}\,dx = \frac{\pi}{2}\int_{-4}^4 \sqrt{64-3x^2}\,dx$$

$$= \frac{\pi}{2\sqrt{3}}\left(\frac{x\sqrt{3}}{2}\sqrt{64-3x^2}+32\sin^{-1}\left(\frac{x\sqrt{3}}{8}\right)\right)\Bigg]_{-4}^4 = 8\pi\left(1+\frac{4\sqrt{3}}{9}\pi\right)$$

6. Find the area of the surface of revolution generated by revolving about the x axis the hypocycloid $x = a\cos^3\theta$, $y = a\sin^3\theta$.

The required surface is generated by revolving the arc from $\theta = 0$ to $\theta = \pi$. We have $\frac{dx}{d\theta} = -3a\cos^2\theta\sin\theta$, $\frac{dy}{d\theta} = 3a\sin^2\theta\cos\theta$, and $\left(\frac{dx}{d\theta}\right)^2 + \left(\frac{dy}{d\theta}\right)^2 = 9a^2\cos^2\theta\sin^2\theta$. Then

$$S = 2(2\pi)\int_0^{\pi/2} y\sqrt{\left(\frac{dx}{d\theta}\right)^2 + \left(\frac{dy}{d\theta}\right)^2}\,d\theta = 2(2\pi)\int_0^{\pi/2}(a\sin^3\theta)3a\cos\theta\sin\theta\,d\theta$$

$$= \frac{12a^2\pi}{5} \quad \text{(square units)}$$

7. Find the area of the surface of revolution generated by revolving about the x axis the cardioid $x = 2\cos\theta - \cos 2\theta$, $y = 2\sin\theta - \sin 2\theta$.

The required surface is generated by revolving the arc from $\theta = 0$ to $\theta = \pi$. (See Fig. 36-2.) We have

$$\frac{dx}{d\theta} = -2\sin\theta + 2\sin 2\theta, \qquad \frac{dy}{d\theta} = 2\cos\theta - 2\cos 2\theta,$$

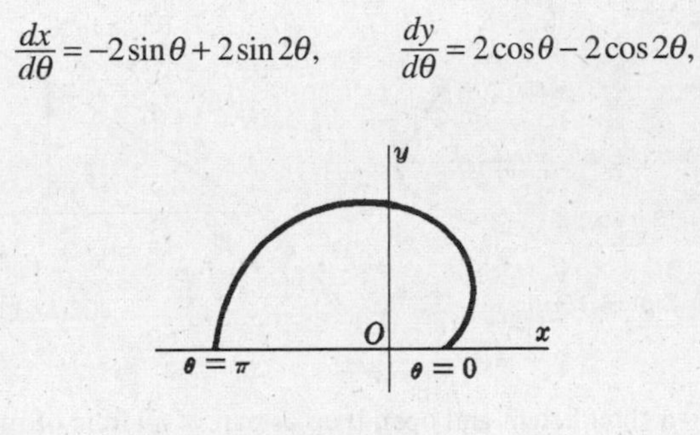

Fig. 36-2

and

$$\left(\frac{dx}{d\theta}\right)^2 + \left(\frac{dy}{d\theta}\right)^2 = 8(1 - \sin\theta\sin 2\theta - \cos\theta\cos 2\theta) = 8(1 - \cos\theta)$$

Then

$$S = 2\pi\int_0^{\pi}(2\sin\theta - \sin 2\theta)(2\sqrt{2}\sqrt{1-\cos\theta})\,d\theta$$

$$= 8\sqrt{2}\pi\int_0^{\pi}\sin\theta(1-\cos\theta)^{3/2}\,d\theta = \left(\frac{16\sqrt{2}}{5}\pi(1-\cos\theta)^{5/2}\right)\Bigg]_0^{\pi}$$

$$= \frac{128\pi}{5} \quad \text{(square units)}$$

8. Show that the surface area of a cylinder of radius r and height h is $2\pi rh$.

The surface is generated by revolving about the x axis the curve $y = r$ from $x = 0$ to $x = h$. Since $\frac{dy}{dx} = 0$, $1 + \left(\frac{dy}{dx}\right)^2 = 1$. Then, by (36.1),

$$S = 2\pi\int_0^h r\,dx = 2\pi(rx)\Big]_0^h = 2\pi rh$$

9. Show that the surface area of a sphere of radius r is $4\pi r^2$.

The surface area is generated by revolving about the x axis the semicircle $y = \sqrt{r^2 - x^2}$ from $x = -r$ to $x = r$. By symmetry, this is double the surface area from $x = 0$ to $x = r$. Since $y^2 = r^2 - x^2$,

$$2y\frac{dy}{dx} = -2x \quad \text{and therefore} \quad \frac{dy}{dx} = -\frac{x}{y} \quad \text{and} \quad 1 + \left(\frac{dy}{dx}\right)^2 = 1 + \frac{x^2}{y^2} = \frac{x^2 + y^2}{y^2} = \frac{r^2}{y^2}$$

Hence, by (36.1),

$$S = 2 \cdot 2\pi \int_0^r y\sqrt{\frac{r^2}{y^2}}\,dx = 4\pi r \int_0^r 1\,dx = 4\pi r x\Big]_0^r = 4\pi r^2$$

10. (a) Show that the surface area of a cone with base of radius r and with slant height s (see Fig. 36-3) is πrs.

(b) Show that the surface area of a frustum of a cone having bases of radius r_1 and r_2 and slant height u (see Fig. 36-4) is $\pi(r_1 + r_2)u$. (Note that the *frustum* is obtained by revolving the right-hand segment of the slant height around the base of the triangle.)

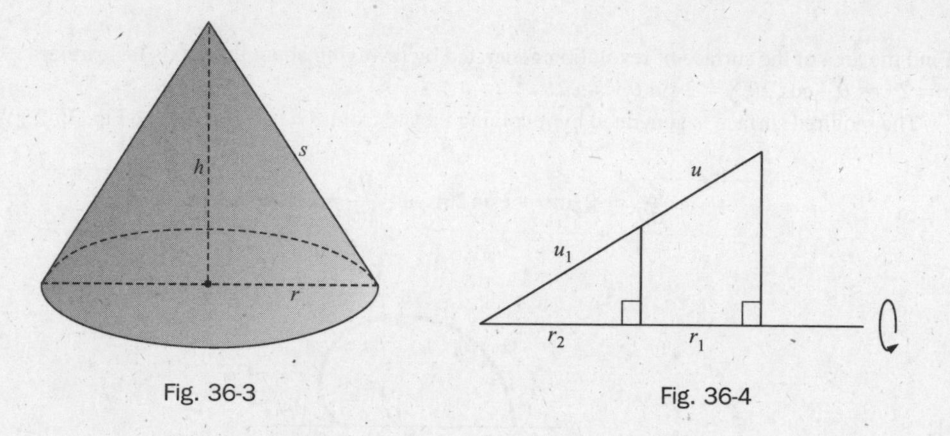

Fig. 36-3 Fig. 36-4

(a) Cut open the cone along a slant height and open it up as part of a circle of radius s (as shown in Fig. 36-5). Note that the portion of the circumference cut off by this region is $2\pi r$ (the circumference of the base of the cone.) Now the desired area S is the difference between πs^2 (the area of the circle in Fig. 36-5) and the area A_1 of the circular sector with central angle θ. This area A_1 is $\frac{\theta}{2\pi}(\pi s^2) = \frac{1}{2}\theta s^2$. Since the arc cut off by θ is $2\pi s - 2\pi r$, we get $\theta = \frac{2\pi s - 2\pi r}{s}$. Thus, $A_1 = \pi(s - r)s$. Hence, $S = \pi s^2 - \pi(s - r)s = \pi rs$ square units.

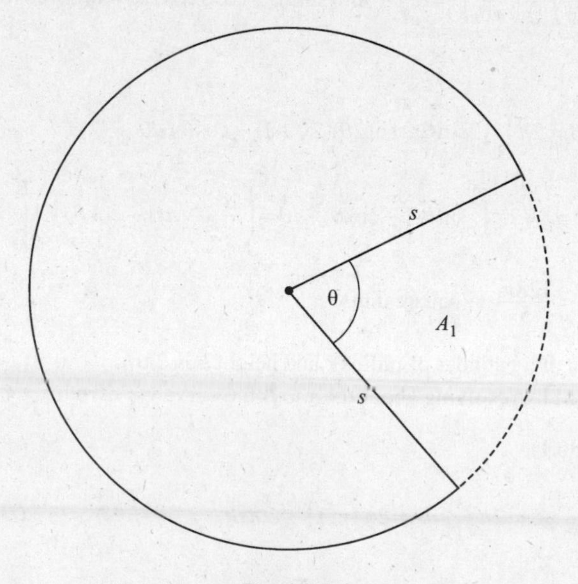

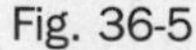
Fig. 36-5

(b) From the similar triangles in Fig. 36-4, we get $\frac{u_1}{r_1} = \frac{u_1 + u}{r_2}$. Then $r_2 u_1 = r_1 u_1 + r_1 u$. So, $u_1 = \frac{r_1 u}{r_2 - r_1}$. Now, by part (a), the surface area of the frustum is $\pi r_2(u_1 + u) - \pi r_1 u_1 = \pi(r_2 - r_1)u_1 + \pi r_2 u = \pi r_1 u + \pi r_2 u = \pi(r_1 + r_2)u$ square units.

11. Sketch a derivation of formula (36.1).

Assume that $[a, b]$ is divided into n equal subintervals, $[x_{k-1}, x_k]$, each of length $\Delta x = \frac{b-a}{n}$. The total surface area S is the sum of the surface areas S_k generated by the arcs between the points $(x_{k-1}, f(x_{k-1}))$ and $(x_k, f(x_k))$, each of which is approximated by the surface area generated by the line segment between $(x_{k-1}, f(x_{k-1}))$ and $(x_k, f(x_k))$. The latter is the area of a frustum of a cone. In the notation of Fig. 36-6, this is, by virtue of Problem 10(*b*):

$$\pi\left(f(x_{k-1})+f(x_k)\right)\sqrt{(\Delta x)^2+(\Delta y)^2} = 2\pi\left(\frac{f(x_{k-1})+f(x_k)}{2}\right)\sqrt{(\Delta x)^2+(\Delta y)^2}$$

Now, $\frac{f(x_{k-1})+f(x_k)}{2}$, being the average of $f(x_{k-1})$ and $f(x_k)$, is between those two values and, by the intermediate value theorem, is equal to $f(x_k^*)$ for some x_k^* in (x_{k-1}, x_k). Also, $\sqrt{(\Delta x)^2+(\Delta y)^2} = \sqrt{1+\left(\frac{\Delta y}{\Delta x}\right)^2}\,\Delta x$. By the mean value, theorem $\frac{\Delta y}{\Delta x} = f'(x_k^{\#})$ for some $x_k^{\#}$ in (x_{k-1}, x_k). Thus, S is approximated by the sum

$$\sum_{k=1}^{n} 2\pi f(x_k^*)\sqrt{1+\left(f'(x_k^{\#})\right)^2}\,\Delta x$$

and it can be shown that this sum can be made arbitrarily close to $2\pi\int_a^b f(x)\sqrt{1+\left(f'(x)\right)^2}\,dx$.† Hence, the latter is equal to S.

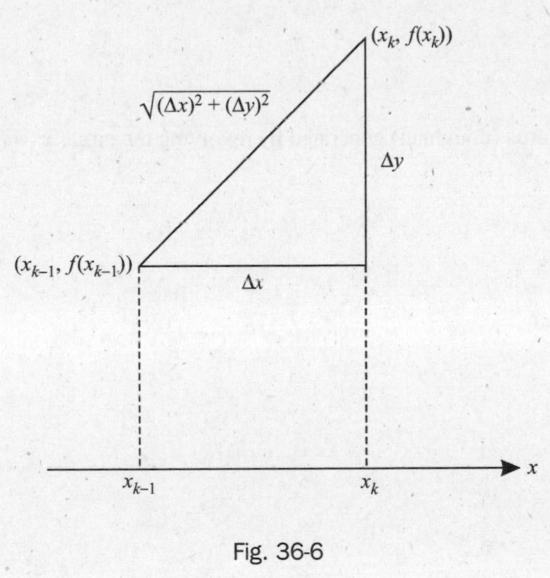

Fig. 36-6

SUPPLEMENTARY PROBLEMS

In Problems 12–20, find the area of the surface of revolution generated by revolving the given arc about the given axis:

12. $y = mx$ from $x = 0$ to $x = 2$; x axis *Ans.* $4m\pi\sqrt{1+m^2}$

13. $y = \frac{1}{3}x^3$ from $x = 0$ to $x = 3$; x axis *Ans.* $\pi(82\sqrt{82} - 1)/9$

---

†In general, the following result can be proved:

Bliss's Theorem: Assume f and g are continuous on $[a, b]$. Divide $[a, b]$ into subintervals $[x_{k-1}, x_k]$ with $a = x_0 < x_1 < \cdots < x_n < b$, and let $\Delta_k x = x_k - x_{k-1}$. In each $[x_{k-1}, x_k]$, choose x_k^* and $x_k^{\#}$. Then the approximating sum $\sum_{k=1}^{n} f(x_k^*)g(x_k^{\#})\Delta_k$ can be made arbitrarily close to $\int_a^b f(x)g(x)\,dx$ by letting $n \to +\infty$ and making the maximum lengths of the subintervals approach 0.

14. $y = \frac{1}{3}x^3$ from $x = 0$ to $x = 3$; y axis *Ans.* $\frac{1}{2}\pi\left[9\sqrt{82} + \ln\left(9 + \sqrt{82}\right)\right]$

15. One loop of $8y^2 = x^2(1 - x^2)$; x axis *Ans.* $\frac{1}{4}\pi$

16. $y = x^3/6 + 1/2x$ from $x = 1$ to $x = 2$; y axis *Ans.* $\left(\frac{15}{4} + \ln 2\right)\pi$

17. $y = \ln x$ from $x = 1$ to $x = 7$; y axis *Ans.* $\left[34\sqrt{2} + \ln\left(3 + 2\sqrt{2}\right)\right]\pi$

18. One loop of $9y^2 = x(3 - x)^2$; y axis *Ans.* $28\pi\sqrt{3}/5$

19. An arch of $x = a(\theta - \sin\theta)$, $y = a(1 - \cos\theta)$; x axis *Ans.* $64\pi a^2/3$

20. $x = e^t \cos t$, $y = e^t \sin t$ from $t = 0$ to $t = \frac{1}{2}\pi$; x axis Ans. $2\pi\sqrt{2}(2e^{\pi} + 1)/5$

21. Find the surface area of a zone cut from a sphere of radius r by two parallel planes, each at a distance $\frac{1}{2}a$ from the center.

Ans. $2\pi ar$

22. Find the surface area of a torus (doughnut) generated by revolving the circle $x^2 + (y - b)^2 = a^2$ about the x axis. Assume $0 < a < b$.

Ans. $4\pi^2 ab$

CHAPTER 37

Parametric Representation of Curves

Parametric Equations

If the coordinates (x, y) of a point P on a curve are given as functions $x = f(u)$, $y = g(u)$ of a third variable or *parameter*, u, the equations $x = f(u)$ and $y = g(u)$ are called *parametric equations* of the curve.

EXAMPLE 37.1:

(a) $x = \cos\theta$, $y = 4\sin^2\theta$ are parametric equations, with parameter θ, of the parabola $4x^2 + y = 4$, since $4x^2 + y = 4\cos^2\theta + 4\sin^2\theta = 4$.

(b) $x = \frac{1}{2}t$, $y = 4 - t^2$ is another parametric representation, with parameter t, of the same curve.

It should be noted that the first set of parametric equations represents only a portion of the parabola (Fig. 37-1(*a*)), whereas the second represents the entire curve (Fig. 37-1(*b*)).

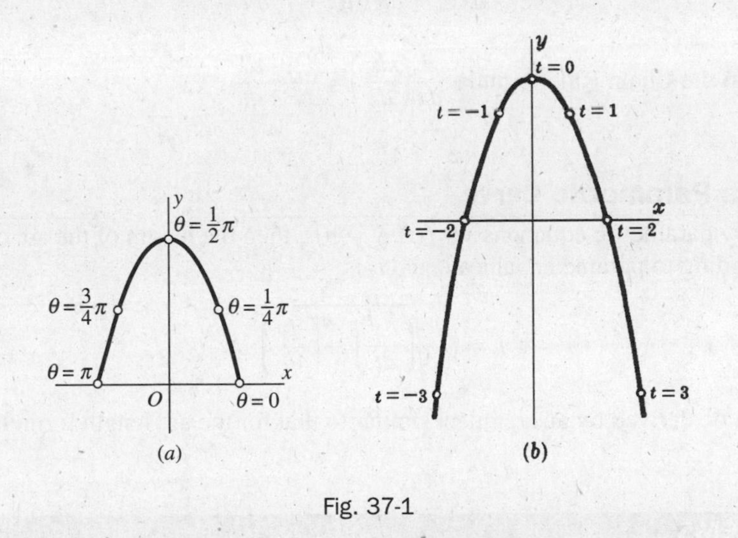

Fig. 37-1

EXAMPLE 37.2:

(a) The equations $x = r\cos\theta$, $y = r\sin\theta$ represent the circle of radius r with center at the origin, since $x^2 + y^2 = r^2\cos^2\theta + r^2\sin^2\theta = r^2(\cos^2\theta + \sin^2\theta) = r^2$. The parameter θ can be thought of as the angle from the positive x axis to the segment from the origin to the point P on the circle (Fig. 37-2).

(b) The equations $x = a + r\cos\theta$, $y = b + r\sin\theta$ represents the circle of radius r with center at (a, b), since $(x - a)^2 + (y - b)^2 = r^2\cos^2\theta + r^2\sin^2\theta = r^2(\cos^2\theta + \sin^2\theta) = r^2$.

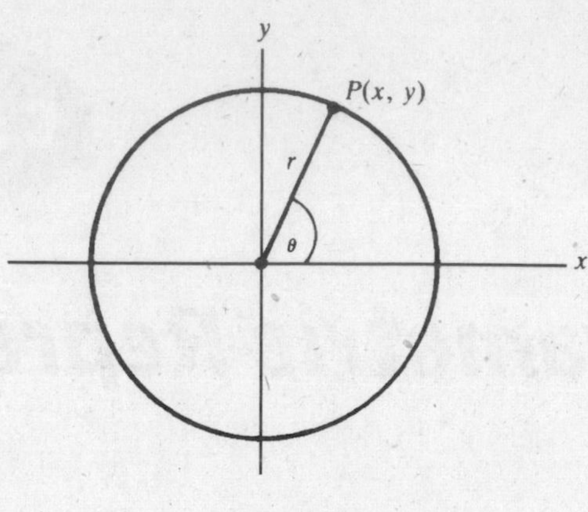

Fig. 37-2

Assume that a curve is specified by means of a pair of parametric equations $x = f(u)$ and $y = g(u)$. Then the first and second derivatives $\frac{dy}{dx}$ and $\frac{d^2y}{dx^2}$ are given by the following formulas.

(37.1) First Derivative

$$\frac{dy}{dx} = \left(\frac{dy}{du}\right) \Big/ \left(\frac{dx}{du}\right)$$

This follows from the Chain Rule formula $\frac{dy}{du} = \frac{dy}{dx} \cdot \frac{dx}{du}$.

(37.2) Second Derivative

$$\frac{d^2y}{dx^2} = \left(\frac{d}{du}\left(\frac{dy}{dx}\right)\right) \Big/ \frac{dx}{du}$$

This follows from the Chain Rule formula $\frac{d}{du}\left(\frac{dy}{dx}\right) = \frac{d^2y}{dx^2} \cdot \frac{dx}{du}$.

Arc Length for a Parametric Curve

If a curve is given by parametric equations $x = f(t)$, $y = g(t)$, then the length of the arc of the curve between the points corresponding to parameter values t_1 and t_2 is

$$L = \int_{t_1}^{t_2} \sqrt{\left(\frac{dx}{dt}\right)^2 + \left(\frac{dy}{dt}\right)^2}\, dt$$

This formula can be derived by an argument similar to that for the arc length formula (29.2).

SOLVED PROBLEMS

1. Find $\frac{dy}{dx}$ and $\frac{d^2y}{dx^2}$ if $x = t - \sin t$, $y = 1 - \cos t$.

$\frac{dx}{dt} = 1 - \cos t$ and $\frac{dy}{dt} = \sin t$. By (37.1), $\frac{dy}{dx} = \frac{\sin t}{1 - \cos t}$. Then

$$\frac{d}{dt}\left(\frac{dy}{dx}\right) = \frac{(1 - \cos t)(\cos t) - (\sin t)(\sin t)}{(1 - \cos t)^2}$$

$$= \frac{\cos t - (\cos^2 t + \sin^2 t)}{(1 - \cos t)^2} = \frac{\cos t - 1}{(1 - \cos t)^2} = \frac{1}{\cos t - 1}$$

Hence, by (37.2),

$$\frac{d^2y}{dx^2} = \frac{1}{\cos t - 1} \Big/ (1 - \cos t) = -\frac{1}{(1-\cos t)^2}$$

2. Find $\frac{dy}{dx}$ and $\frac{d^2y}{dx^2}$ if $x = e^t \cos t,\ y = e^t \sin t$.

$\frac{dx}{dt} = e^t(\cos t - \sin t)$ and $\frac{dy}{dt} = e^t(\cos t + \sin t)$. By (37.1), $\frac{dy}{dx} = \frac{\cos t + \sin t}{\cos t - \sin t}$. Then,

$$\frac{d}{dt}\left(\frac{dy}{dx}\right) = \frac{(\cos t - \sin t)^2 - (\cos t + \sin t)(-\sin t - \cos t)}{(\cos t - \sin t)^2}$$

$$= \frac{(\cos t - \sin t)^2 + (\cos t + \sin t)^2}{(\cos t - \sin t)^2} = \frac{2(\cos^2 t + \sin^2 t)}{(\cos t - \sin t)^2}$$

$$= \frac{2}{(\cos t - \sin t)^2}$$

So, by (37.2),

$$\frac{d^2y}{dx^2} = \frac{2}{(\cos t - \sin t)^2} \Big/ e^t(\cos t - \sin t) = \frac{2}{e^t(\cos t - \sin t)}$$

3. Find an equation of the tangent line to the curve $x = \sqrt{t},\ y = t - \frac{1}{\sqrt{t}}$ at the point where $t = 4$.

$\frac{dx}{dt} = \frac{1}{2\sqrt{t}}$ and $\frac{dy}{dt} = 1 + \frac{1}{2t^{3/2}}$. By (37.1), $\frac{dy}{dx} = 2\sqrt{t} + \frac{1}{t}$. So, the slope of the tangent line when $t = 4$ is $2\sqrt{4} + \frac{1}{4} = \frac{17}{4}$. When $t = 4$, $x = 2$ and $y = \frac{7}{2}$. An equation of the tangent line is $y - \frac{7}{2} = \frac{17}{4}(x - 2)$.

4. The position of a particle that is moving along a curve is given at time t by the parametric equations $x = 2 - 3\cos t,\ y = 3 + 2\sin t$, where x and y are measured in feet and t in seconds. (See Fig. 37-3.) Note that $\frac{1}{9}(x-2)^2 + \frac{1}{4}(y-3)^2 = 1$, so that the curve is an ellipse. Find: (a) the time rate of change of x when $t = \pi/3$; (b) the time rate of change of y when $t = 5\pi/3$; (c) the time rate of change of the angle of inclination θ of the tangent line when $t = 2\pi/3$.

$\frac{dy}{dt} = 3\sin t$ and $\frac{dy}{dt} = 2\cos t$. Then $\tan\theta = \frac{dy}{dx} = \frac{2}{3}\cot t$.

(a) When $t = \frac{\pi}{3}$, $\frac{dx}{dt} = \frac{3\sqrt{3}}{2}$ ft/sec

(b) When $t = \frac{5\pi}{3}$, $\frac{dy}{dt} = 2(\frac{1}{2}) = 1$ ft/sec

(c) $\theta = \tan^{-1}(\frac{2}{3}\cot t)$. So, $\frac{d\theta}{dt} = \frac{-\frac{2}{3}\csc^2 t}{1 + \frac{4}{9}\cot t^2 t} = \frac{-6\csc^2 t}{9 + 4\cot^2 t}$.

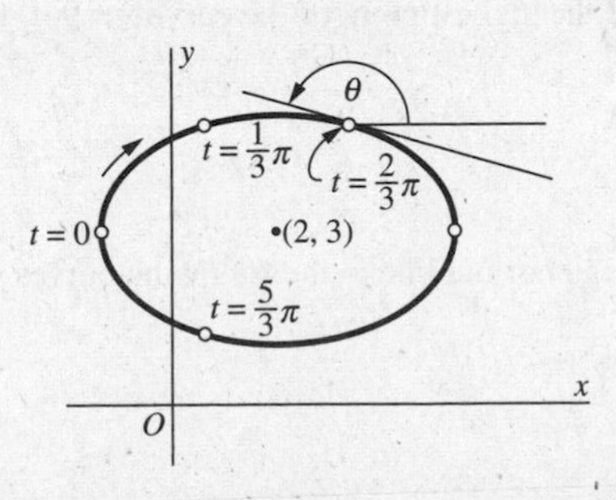

Fig. 37-3

When $t = \frac{2\pi}{3}$, $\frac{d\theta}{dt} = \frac{-6(2/\sqrt{3})^2}{9+4(-1/\sqrt{3})^2} = -\frac{24}{31}$. Thus, the angle of inclination of the tangent line is decreasing at the rate of $\frac{24}{31}$ radians per second.

5. Find the arc length of the curve $x = t^2$, $y = t^3$ from $t = 0$ to $t = 4$.

$\frac{dx}{dt} = 2t$, $\frac{dy}{dt} = 3t^2$ and $\left(\frac{dx}{dt}\right)^2 + \left(\frac{dy}{dt}\right)^2 = 4t^2 + 9t^4 = 4t^2(1+\frac{9}{4}t^2)$.

Then

$$L = \int_0^4 2t\sqrt{1+\tfrac{9}{4}t^2}\,dt = \tfrac{4}{9}\int_0^4 (1+\tfrac{9}{4}t^2)^{1/2}(\tfrac{9}{2}t)\,dt$$
$$= \tfrac{4}{9}\tfrac{2}{3}(1+\tfrac{9}{4}t^2)^{3/2}\Big]_0^4 = \tfrac{8}{27}(37\sqrt{37}-1)$$

6. Find the length of an arch of the cycloid $x = \theta - \sin\theta$, $y = 1 - \cos\theta$ between $\theta = 0$ and $\theta = 2\pi$.

$\frac{dx}{d\theta} = 1-\cos\theta$, $\frac{dy}{d\theta} = \sin\theta$ and $\left(\frac{dx}{d\theta}\right)^2 + \left(\frac{dy}{d\theta}\right)^2 = (1-\cos\theta)^2 + \sin^2\theta = 2(1-\cos\theta) = 4\sin^2\left(\frac{\theta}{2}\right)$. Then

$$L = 2\int_0^4 \sin\left(\frac{\theta}{2}\right)d\theta = -4\cos\left(\frac{\theta}{2}\right)\Big]_0^{2\pi} = -4(\cos\pi - \cos 0) = 8$$

SUPPLEMENTARY PROBLEMS

In Problems 7–11, find: *(a)* $\frac{dy}{dx}$; *(b)* $\frac{d^2y}{dx^2}$.

7. $x = 2 + t$, $y = 1 + t^2$ — *Ans.* (a) $2t$; (b) 2

8. $x = t + 1/t$, $y = t + 1$ — *Ans.* (a) $t^2/(t^2 - 1)$; (b) $-2t^3/(t^2 - 1)^3$

9. $x = 2 \sin t$, $y = \cos 2t$ — *Ans.* (a) $-2 \sin t$; (b) -1

10. $x = \cos^3\theta$, $y = \sin^3\theta$ — *Ans.* (a) $-\tan\theta$; (b) $1/(3\cos^4\theta\sin\theta)$

11. $x = a(\cos\phi + \phi\sin\phi)$, $y = a(\sin\phi - \phi\cos\phi)$ — *Ans.* (a) $\tan\phi$; (b) $1/(a\phi\cos^3\phi)$

12. Find the slope of the curve $x = e^{-t}\cos 2t$, $y = e^{-2t}\sin 2t$ at the point $t = 0$.

Ans. -2

13. Find the rectangular coordinates of the highest point of the curve $x = 96t$, $y = 96t - 16t^2$. (*Hint:* Find t for maximum y.)

Ans. (288, 144)

14. Find equations of the tangent line and normal line to the following curves at the points determined by the given value of the parameter:

(a) $x = 3e^t$, $y = 5e^{-t}$ at $t = 0$
(b) $x = a\cos^4\theta$, $y = a\sin^4\theta$ at $\theta = \frac{\pi}{4}$

Ans. (a) $3y + 5x = 30$, $5y - 3x = 16$; (b) $2x + 2y = a$, $y = x$

15. Find an equation of the tangent line at any point $P(x, y)$ of the curve $x = a\cos^3 t$, $y = a\sin^3 t$. Show that the length of the segment of the tangent line intercepted by the coordinate axes is a.

Ans. $x\sin t + y\cos t = \frac{a}{2}\sin 2t$

16. For the curve $x = t^2 - 1$, $y = t^3 - t$, locate the points where the tangent line is (a) horizontal, and (b) vertical. Show that, at the point where the curve crosses itself, the two tangent lines are mutually perpendicular.

Ans. (a) $t = \pm\frac{\sqrt{3}}{3}$; (b) $t = 0$

In Problems 17–20, find the length of the specified arc of the given curve.

17. The circle $x = a\cos\theta$, $y = a\sin\theta$ from $\theta = 0$ to $\theta = 2\pi$.

Ans. $2\pi a$

18. $x = e^t\cos t$, $y = e^t\sin t$ from $t = 0$ to $t = 4$.

Ans. $\sqrt{2}(e^4 - 1)$

19. $x = \ln\sqrt{1+t^2}$, $y = \tan^{-1} t$ from $t = 0$ to $t = 1$.

Ans. $\ln(1+\sqrt{2})$

20. $x = 2\cos\theta + \cos 2\theta + 1$, $y = 2\sin\theta + \sin 2\theta$.

Ans. 16

21. The position of a point at time t is given as $x = \frac{1}{2}t^2$, $y = \frac{1}{9}(6t+9)^{3/2}$. Find the distance the point travels from $t = 0$ to $t = 4$.

Ans. 20

22. Identify the curves given by the following parametric equations and write equations for the curves in terms of x and y:

(a) $x = 3t + 5$, $y = 4t - 1$ *Ans.* Straight line: $4x - 3y = 23$
(b) $x = t + 2$, $y = t^2$ *Ans.* Parabola: $y = (x - 2)^2$
(c) $x = t - 2$, $y = \frac{t}{t-2}$ *Ans.* Hyperbola: $y = \frac{2}{x} + 1$
(d) $x = 5\cos t$, $y = 5\sin t$ *Ans.* Circle: $x^2 + y^2 = 25$

23. (GC) Use a graphing calculator to find the graphs of the following parametric curves:

(a) $x = \theta + \sin\theta$, $y = 1 - \cos\theta$ (cycloid)
(b) $x = 3\cos^3\theta$, $y = 3\sin^3\theta$ (hypocycloid)
(c) $x = 2\cot\theta$, $y = 2\sin^2\theta$ (witch of Agnesi)
(d) $x = \frac{3\theta}{(1+\theta^3)}$, $y = \frac{3\theta^2}{(1+\theta^3)}$ (folium of Descartes)

CHAPTER 38

Curvature

Derivative of Arc Length

Let $y=f(x)$ have a continuous first derivative. Let $A(x_0, y_0)$ be a fixed point on its graph (see Fig. 38-1) and denote by s the arc length measured from A to any other point $P(x, y)$ on the curve. We know that, by formula (29.2),

$$s=\int_{x_0}^{x}\sqrt{1+\left(\frac{dy}{dx}\right)^2}\,dx$$

if s is chosen so as to increase with x. Let $Q(x+\Delta x, y+\Delta y)$ be a point on the curve near P. Let Δs denote the arc length from P to Q. Then

$$\frac{ds}{dx}=\lim_{\Delta x\to 0}\frac{\Delta s}{\Delta x}=\pm\sqrt{1+\left(\frac{dy}{dx}\right)^2}$$

and, similarly,

$$\frac{ds}{dy}=\lim_{\Delta y\to 0}\frac{\Delta s}{\Delta y}=\pm\sqrt{1+\left(\frac{dx}{dy}\right)^2}$$

The plus or minus sign is to be taken in the first formula according as s increases or decreases as x increases, and in the second formula according as s increases or decreases as y increases.

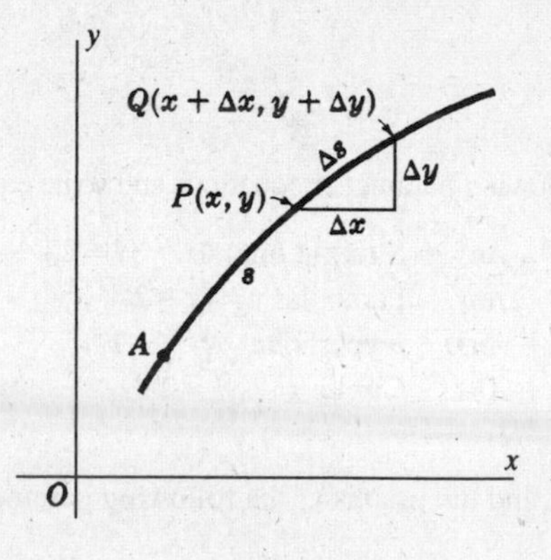

Fig. 38-1

When a curve is given by parametric equations $x=f(u)$, $y=g(u)$,

$$\frac{ds}{du}=\lim_{\Delta u\to 0}\frac{\Delta s}{\Delta u}=\pm\sqrt{\left(\frac{dx}{du}\right)^2+\left(\frac{dy}{du}\right)^2}$$

Here the plus or minus sign is to be taken according as s increases or decreases as u increases.

To avoid the repetition of ambiguous signs, we shall assume hereafter that the direction on each arc has been established so that the derivative of arc length will be positive.

Curvature

The curvature K of a curve $y = f(x)$ at any point P on it is defined to be the rate of change of the direction of the curve at P, that is, of the angle of inclination of the tangent line at P, with respect to the arc length s. (See Fig. 38-2.) Intuitively, the curvature tells us how fast the tangent line is turning. Thus, the curvature is large when the curve bends sharply.

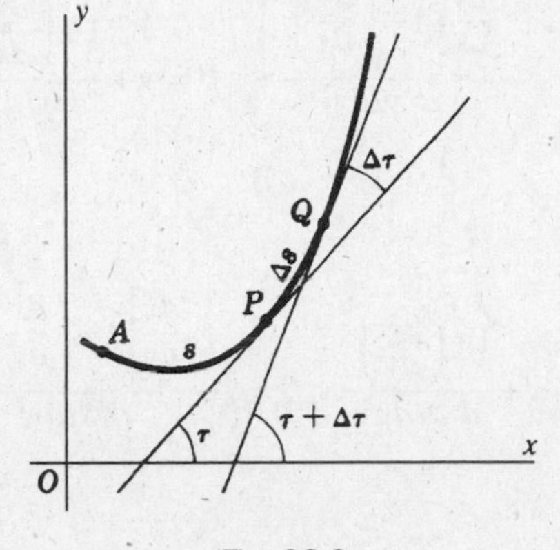

Fig. 38-2

As formulas for the curvature, we get:

$$K = \frac{d\tau}{ds} = \lim_{\Delta s \to 0} \frac{\Delta \tau}{\Delta s} = \frac{\dfrac{d^2y}{dx^2}}{\left(1 + \left(\dfrac{dy}{dx}\right)^2\right)^{3/2}} \tag{38.1}$$

or, in terms of y,

$$K = \frac{-\dfrac{d^2x}{dy^2}}{\left(1 + \left(\dfrac{dx}{dy}\right)^2\right)^{3/2}} \tag{38.2}$$

For a derivation, see Problem 13.

K is sometimes defined so as to be positive. If this is assumed, then the sign of K should be ignored in what follows.

The Radius of Curvature

The radius of curvature R at a point P on a curve is defined by $R = \left|\frac{1}{K}\right|$, provided that $K \neq 0$.

The Circle of Curvature

The circle of curvature, or *osculating circle* of a curve at a point P on it, is the circle of radius R lying on the concave side of the curve and tangent to it at P. (See Fig. 38-3.)

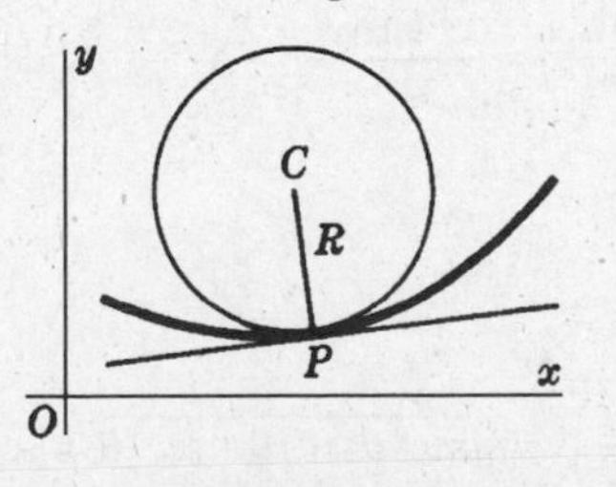

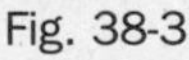

Fig. 38-3

To construct the circle of curvature, on: the concave side of the curve, construct the normal line at P and on it lay off a segment PC of length R. The point C is the center of the required circle.

The Center of Curvature

The center of curvature for a point $P(x, y)$ of a curve is the center C of the circle of curvature at P. The co-ordinates (,) of the center of curvature are given by

$$\alpha = x - \frac{\frac{dy}{dx}\left[1+\left(\frac{dy}{dx}\right)^2\right]}{d^2y/dx^2} \qquad \beta = y + \frac{1+\left(\frac{dy}{dx}\right)^2}{d^2y/dx^2}$$

or by

$$\alpha = x + \frac{1+\left(\frac{dx}{dy}\right)^2}{d^2x/dy^2} \qquad \beta = y - \frac{\frac{dx}{dy}\left[1+\left(\frac{dx}{dy}\right)^2\right]}{d^2x/dy^2}$$

See Problem 9 for details.

The Evolute

The evolute of a curve is the locus of the centers of curvature of the given curve. (See Problems 11–12.)

SOLVED PROBLEMS

1. Find $\frac{ds}{dx}$ at $P(x, y)$ on the parabola $y = 3x^2$.

$$\frac{ds}{dx} = \sqrt{1+\left(\frac{dy}{dx}\right)^2} = \sqrt{1+(6x)^2} = \sqrt{1+36x^2}$$

2. Find $\frac{ds}{dx}$ and $\frac{ds}{dy}$ at $P(x, y)$ on the ellipse $x^2 + 4y^2 = 8$.

Since $2x + 8y\frac{dy}{dx} = 0$, $\frac{dy}{dx} = -\frac{x}{4y}$ and $\frac{dx}{dy} = -\frac{4y}{x}$. Then

$$1+\left(\frac{dy}{dx}\right)^2 = 1+\frac{x^2}{16y^2} = \frac{x^2+16y^2}{16y^2} = \frac{32-3x^2}{32-4x^2} \quad \text{and} \quad \frac{ds}{dx} = \sqrt{\frac{32-3x^2}{32-4x^2}}$$

$$1+\left(\frac{dx}{dy}\right)^2 = 1+\frac{16y^2}{x^2} = \frac{x^2+16y^2}{x^2} = \frac{2+3y^2}{2-y^2} \quad \text{and} \quad \frac{ds}{dy} = \sqrt{\frac{2+3y^2}{2-y^2}}$$

3. Find $\frac{ds}{d\theta}$ at P( ) on the curve $x = \sec$, $y = \tan$.

$$\frac{ds}{d\theta} = \sqrt{\left(\frac{dx}{d\theta}\right)^2 + \left(\frac{dy}{d\theta}\right)^2} = \sqrt{\sec^2\theta\tan^2\theta + \sec^4\theta} = |\sec\theta|\sqrt{\tan^2\theta + \sec^2\theta}$$

4. The coordinates (x, y) in feet of a moving particle P are given by $x = \cos t - 1$, $y = 2\sin t + 1$, where t is the time in seconds. At what rate is P moving along the curve when (a) $t = 5\pi/6$, (b) $t = 5\pi/3$, and (c) P is moving at its fastest and slowest?

$$\frac{ds}{dt} = \sqrt{\left(\frac{dx}{dt}\right)^2 + \left(\frac{dy}{dt}\right)^2} = \sqrt{\sin^2 t + 4\cos^2 t} = \sqrt{1 + 3\cos^2 t}$$

(a) When $t = 5\pi/6$, $ds/dt = \sqrt{1+3(\frac{3}{4})} = \sqrt{13}/2$ ft/sec.

(b) When $t = 5\pi/3$, $ds/dt = \sqrt{1+3(\frac{1}{4})} = \sqrt{7}/2$ ft/sec.

(c) Let $S = \frac{ds}{dt} = \sqrt{1+3\cos^2 t}$. Then $\frac{dS}{dt} = \frac{-3\cos t \sin t}{S}$. Solving $dS/dt = 0$ gives the critical numbers $t = 0$, $\pi/2$, π, $3\pi/2$.

When $t = 0$ and π, the rate $ds/dt = \sqrt{1+3(1)} = 2$ ft/sec is fastest. When $t = \pi/2$ and $3\pi/2$, the rate $ds/dt = \sqrt{1+3(0)} = 1$ ft/sec is slowest. The curve is shown in Fig. 38-4.

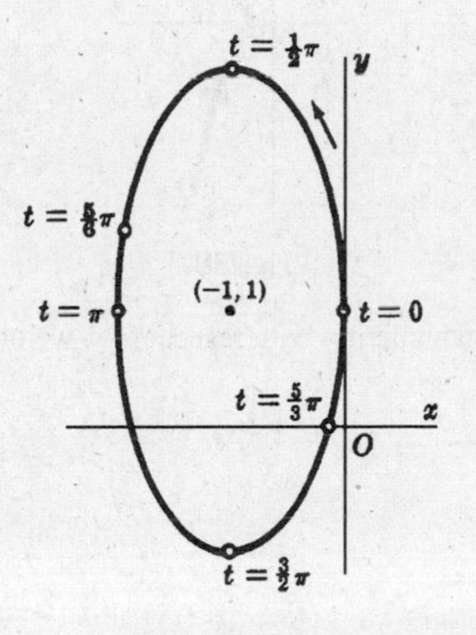

Fig. 38-4

5. Find the curvature of the parabola $y^2 = 12x$ at the points: (a) $(3, 6)$; (b) $(\frac{3}{4}, -3)$; (c) $(0, 0)$.

$$\frac{dy}{dx} = \frac{6}{y}; \text{ so } 1+\left(\frac{dy}{dx}\right)^2 = 1 + \frac{36}{y^2} \text{ and } \frac{d^2y}{dx^2} = -\frac{6}{y^2}\frac{dy}{dx} = -\frac{36}{y^3}$$

(a) At $(3, 6)$: $1+\left(\frac{dy}{dx}\right)^2 = 2$ and $\frac{d^2y}{dx^2} = -\frac{1}{6}$, so $K = \frac{-1/6}{2^{3/2}} = -\frac{\sqrt{2}}{24}$.

(b) At $(\frac{3}{4}, -3)$: $1+\left(\frac{dy}{dx}\right)^2 = 5$ and $\frac{d^2y}{dx^2} = \frac{4}{3}$, so $K = \frac{4/3}{5^{3/2}} = \frac{4\sqrt{5}}{75}$.

(c) At $(0, 0)$, $\frac{dy}{dx}$ is undefined. But $\frac{dx}{dy} = \frac{y}{6} = 0$, $1+\left(\frac{dx}{dy}\right)^2 = 1$, $\frac{d^2x}{dy^2} = \frac{1}{6}$, and $K = -\frac{1}{6}$.

6. Find the curvature of the cycloid $x = \quad - \sin \quad , y = 1 - \cos \quad$ at the highest point of an arch. (See Fig. 38-5.)

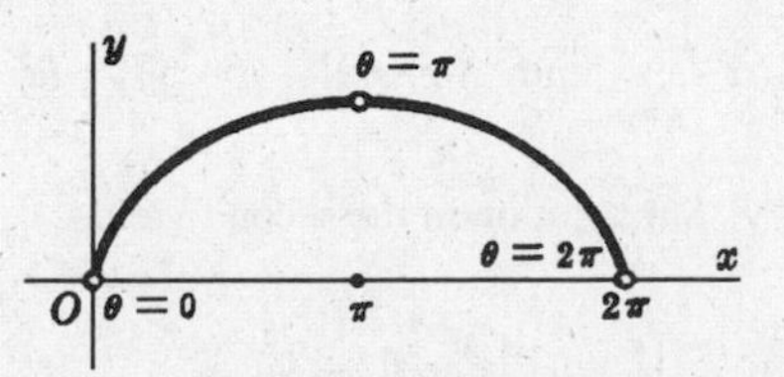

Fig. 38-5

To find the highest point on the interval $0 < x < 2\pi$; dy/d = sin , so that the critical number on the interval is $x = \pi$. Since d^2y/d ^{2} = cos < 0 when $= \pi$, the point $= \pi$ is a relative maximum point and is the highest point of the curve on the interval.

To find the curvature,

$$\frac{dx}{d\theta} = 1 - \cos\theta, \qquad \frac{dy}{d\theta} = \sin\theta, \qquad \frac{dy}{dx} = \frac{\sin\theta}{1-\cos\theta}, \qquad \frac{d^2y}{dx^2} = \frac{d}{d\theta}\left(\frac{\sin\theta}{1-\cos\theta}\right)\frac{d\theta}{dx} = -\frac{1}{(1-\cos\theta)^2}$$

At $= \pi$, $dy/dx = 0$, $d^2y/dx^2 = -\frac{1}{4}$, and $K = -\frac{1}{4}$.

7. Find the curvature of the cissoid $y^2(2 - x) = x^3$ at the point (1, 1). (See Fig. 38-6.)

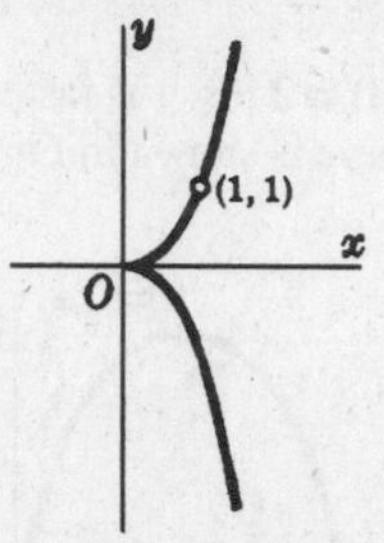

Fig. 38-6

Differentiating the given equation implicitly with respect to x, we obtain

$$-y^2 + (2-x)2yy' = 3x^2 \tag{1}$$

and

$$-2yy' + (2-x)2yy'' + (2-x)2(y')^2 - 2yy' = 6x \tag{2}$$

From (1), for $x = y = 1$, $-1 + 2y' = 3$ and $y' = 2$. Similarly, from (2), for $x = y = 1$ and $y' = 2$, we find $y'' = 3$. Then $K = 3/(1+4)^{3/2} = 3\sqrt{5}/25$.

8. Find the point of greatest curvature on the curve $y = \ln x$.

$$\frac{dy}{dx} = \frac{1}{x} \quad \text{and} \quad \frac{d^2y}{dx^2} = -\frac{1}{x^2}. \quad \text{So,} \quad K = \frac{-x}{(1+x^2)^{3/2}} \quad \text{and} \quad \frac{dK}{dx} = \frac{2x^2-1}{(1+x^2)^{5/2}}$$

The critical number is, therefore, $x = \frac{1}{\sqrt{2}}$. The required point is $\left(\frac{1}{\sqrt{2}}, -\frac{\ln 2}{2}\right)$.

9. Find the coordinates of the center of curvature C of the curve $y = f(x)$ at a point $P(x, y)$ at which $y' \neq 0$. (See Fig. 38-3.)

The center of curvature C(,) lies: (1) on the normal line at P and (2) at a distance R from P measured toward the concave side of the curve. These conditions give, respectively,

$$\beta - y = -\frac{1}{y'}(\alpha - x) \quad \text{and} \quad (\alpha - x)^2 + (\beta - y)^2 = R^2 = \frac{[1+(y')^2]^3}{(y'')^2}$$

From the first, $- x = - y'($ $- y)$. Substitution in the second yields

$$(\beta - y)^2[1 + (y')^2] = \frac{[1+(y')^2]^3}{(y'')^2} \quad \text{and, therefore,} \quad \beta - y = \pm\frac{1+(y')^2}{y''}$$

To determine the correct sign, note that, when the curve is concave upward, $y'' > 0$ and, since C then lies above P, $\quad - y > 0$. Thus, the proper sign in this case is +. (You should show that the sign is also + when $y'' < 0$.) Thus,

$$\beta = y + \frac{1+(y')^2}{y''} \quad \text{and} \quad \alpha = x - \frac{y'[1+(y')^2]}{y''}$$

10. Find the equation of the circle of curvature of $2xy + x + y = 4$ at the point (1, 1).

Differentiating yields $2y + 2xy' + 1 + y' = 0$. At (1, 1), $y' = -1$ and $1 + (y')^2 = 2$. Differentiating again yields $4y' + 2xy'' + y'' = 0$. At (1, 1), $y'' = \frac{4}{3}$. Then

$$K = \frac{4/3}{2\sqrt{2}}, \qquad R = \frac{3\sqrt{2}}{2}, \qquad \alpha = 1 - \frac{-1(2)}{4/3} = \frac{5}{2}, \qquad \beta = 1 + \frac{2}{4/3} = \frac{5}{2}$$

The required equation is $(x - \quad)^2 + (y - \quad)^2 = R^2$ or $(x - \frac{5}{2})^2 + (y - \frac{5}{2})^2 = \frac{9}{2}$.

11. Find the equation of the evolute of the parabola $y^2 = 12x$.

At $P(x, y)$:

$$\frac{dy}{dx} = \frac{6}{y} = \frac{\sqrt{3}}{\sqrt{x}}, \qquad 1 + \left(\frac{dy}{dx}\right)^2 = 1 + \frac{36}{y^2} = 1 + \frac{3}{x}, \qquad \frac{d^2y}{dx^2} = -\frac{36}{y^3} = -\frac{\sqrt{3}}{2x^{3/2}}$$

Then

$$\alpha = x - \frac{\sqrt{3/x}\,(1+3/x)}{-\sqrt{3}/2x^{3/2}} = x + \frac{2\sqrt{3}(x+3)}{\sqrt{3}} - 3x + 6$$

and

$$\beta = y + \frac{1+36/y^2}{-36/y^3} = y - \frac{y^3 + 36y}{36} = -\frac{y^3}{36}$$

The equations $\quad = 3x + 6$, $\quad = -y^3/36$ may be regarded as parametric equations of the evolute with x and y, connected by the equation of the parabola, as parameters. However, it is relatively simple in this problem to eliminate the parameters. Thus, $x = (\quad - 6)/3$, $y = -\sqrt[3]{36\beta}$, and substituting in the equation of the parabola, we have

$$(36\beta)^{2/3} = 4(\alpha - 6) \text{ or } 81\beta^2 = 4(\alpha - 6)^3$$

The parabola and its evolute are shown in Fig. 38-7.

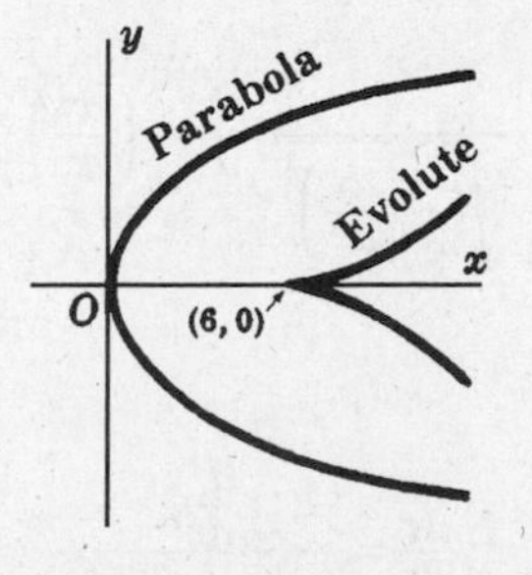

Fig. 38-7

12. Find the equation of the evolute of the curve $x = \cos \quad + \quad \sin \quad , y = \sin \quad - \quad \cos \quad .$

At $P(x, y)$:

$$\frac{dx}{d\theta} = \theta\cos\theta, \qquad \frac{dy}{d\theta} = \theta\sin\theta, \qquad \frac{dy}{dx} = \tan\theta, \qquad \frac{d^2y}{dx^2} = \frac{\sec^2\theta}{\theta\cos\theta} = \frac{\sec^3\theta}{\theta}$$

Then

$$\alpha = x - \frac{\tan\theta\sec^2\theta}{(\sec^3\theta)/\theta} = x - \theta\sin\theta = \cos\theta$$

and

$$\beta = y + \frac{\sec^2\theta}{(\sec^3\theta)/\theta} = y + \theta\cos\theta = \sin\theta$$

and $= \cos \quad , \quad = \sin$ are parametric equations of the evolute (see Fig. 38-8).

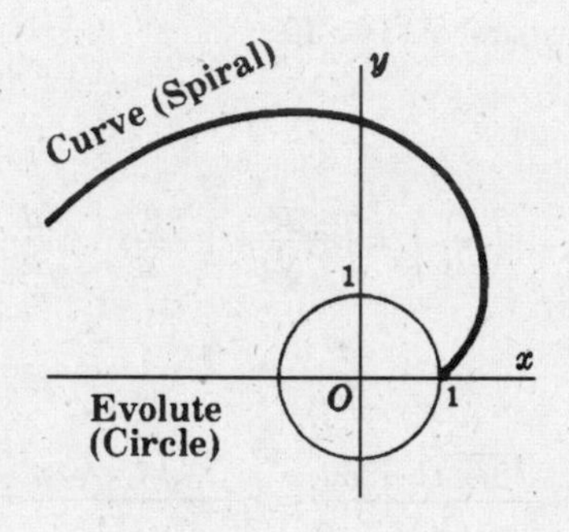

Fig. 38-8

13. Derive formula (38.1).

tan is the slope of the tangent line and, therefore,

$$\frac{dy}{dx} = \tan\tau. \quad \text{So,} \quad \frac{d}{ds}\left(\frac{dy}{dx}\right) = \frac{d}{d\tau}\left(\frac{dy}{dx}\right)\cdot\frac{d\tau}{ds}$$

Hence

$$\frac{d}{dx}\left(\frac{dy}{dx}\right)\cdot\frac{dx}{ds} = \sec^2\tau\cdot\frac{d\tau}{ds}$$

This yields

$$\frac{d^2y}{dx^2}\cdot\frac{1}{\sqrt{1+\left(\frac{dy}{dx}\right)^2}} = \left(1+\left(\frac{dy}{dx}\right)^2\right)\cdot\frac{d\tau}{ds}$$

from which

$$\frac{d\tau}{ds} = \frac{\frac{d^2y}{dx^2}}{\left(1+\left(\frac{dy}{dx}\right)^2\right)^{3/2}}$$

SUPPLEMENTARY PROBLEMS

In Problems 14–16, find $\frac{ds}{dx}$ and $\frac{ds}{dy}$.

14. $x^2 + y^2 = 25$ *Ans.* $\frac{ds}{dx} = \frac{5}{\sqrt{25 - x^2}}, \frac{ds}{dy} - \frac{5}{\sqrt{25 - y^2}}$

15. $y^2 = x^3$ *Ans.* $\frac{ds}{dx} = \frac{1}{2}\sqrt{4 + 9x}, \frac{ds}{dy} = \frac{\sqrt{4 + 9y^{2/3}}}{3y^{1/3}}$

16. $x^{2/3} + y^{2/3} = a^{2/3}$ *Ans.* $\frac{ds}{dx} = (a/x)^{1/3}, \frac{ds}{dy} = \left(\frac{a}{y}\right)^{1/3}$

In Problems 17 and 18, find $\frac{ds}{dx}$.

17. $6xy = x^4 + 3$ *Ans.* $\frac{ds}{dx} = \frac{x^4 + 1}{2x^2}$

18. $27ay^2 = 4(x - a)^3$ *Ans.* $\frac{ds}{dx} = \sqrt{(x + 2a)/3a}$

In Problems 19–22, find $\frac{ds}{dt}$.

19. $x = t^2, y = t^3$ *Ans.* $t\sqrt{4 + 9t^2}$

20. $x = 2 \cos t, y = 3 \sin t$ *Ans.* $\sqrt{4 + 5\cos^2 t}$

21. $x = \cos t, y = \sin t$ *Ans.* 1

22. $x = \cos^3 t, y = \sin^3 t$ *Ans.* $\frac{3}{2}\sin 2t$

23. Find the curvature of each curve at the given points:

(a) $y = x^3/3$ at $x = 0, x = 1, x = -2$ (b) $x^2 = 4ay$ at $x = 0, x = 2a$
(c) $y = \sin x$ at $x = 0, x = \frac{1}{2}\pi$ (d) $y = e^{-x^2}$ at $x = 0$

Ans. (a) $0, \sqrt{2}/2, -4\sqrt{17}/289$; (b) $1/2a, \sqrt{2}/8a$; (c) $0, -1$; (d) -2

24. Show (a) the curvature of a straight line is 0; (b) the curvature of a circle is numerically the reciprocal of its radius.

25. Find the points of maximum curvature of (a) $y = e^x$; (b) $y = \frac{1}{3}x^3$.

Ans. (a) $x = -\frac{1}{2}\ln 2$; (b) $x = \frac{1}{5^{1/4}}$

26. Find the radius of curvature of

(a) $x^3 + xy^2 - 6y^2 = 0$ at (3, 3).
(b) $x = 2a \tan\ , y = a \tan^2\ $ at (x, y).
(c) $x = a \cos^4\ , y = a \sin^4\ $ at (x, y).

Ans. (a) $5\sqrt{5}$; (b) $2a|\sec^3\theta|$; (c) $2a(\sin^4\ + \cos^4\ )^{3/2}$

27. Find the center of curvature of (a) Problem 26(a); (b) $y = \sin x$ at a maximum point.

Ans. (a) $C(-7, 8)$; (b) $C\left(\frac{\pi}{2}, 0\right)$

28. Find the equation of the circle of curvature of the parabola $y^2 = 12x$ at the points (0, 0) and (3, 6).

Ans. $(x-6)^2 + y^2 = 36$; $(x-15)^2 + (y+6)^2 = 288$

29. Find the equation of the evolute of (a) $b^2x^2 + a^2y^2 = a^2b^2$; (b) $x^{2/3} + y^{2/3} + a^{2/3}$; (c) $x = 2\cos t + \cos 2t$, $y = 2\sin t + \sin 2t$.

Ans. (a) $(a\quad)^{2/3} + (b\quad)^{2/3} = (a^2 - b^2)^{2/3}$; (b) $(\quad+\quad)^{2/3} + (\quad-\quad)^{2/3} = 2a^{2/3}$; (c) $\alpha = \frac{1}{3}(2\cos t - \cos 2t)$, $\beta = \frac{1}{3}(2\sin t - \sin 2t)$

CHAPTER 39

Plane Vectors

Scalars and Vectors

Quantities such as time, temperature, and speed, which have magnitude only, are called *scalars*. Quantities such as force, velocity, and acceleration, which have both magnitude and direction, are called *vectors*. Vectors are represented geometrically by directed line segments (arrows). The direction of the arrow (the angle that it makes with some fixed directed line of the plane) is the direction of the vector, and the length of the arrow represents the magnitude of the vector.

Scalars will be denoted by letters $a, b, c, \ldots$ in ordinary type; vectors will be denoted in bold type by letters **a**, **b**, **c**, . . ., or by an expression of the form **OP** (where it is assumed that the vector goes from O to P. (See Fig. 39-1(a).) The magnitude (length) of a vector **a** or **OP** will be denoted $|\mathbf{a}|$ or $|\mathbf{OP}|$.

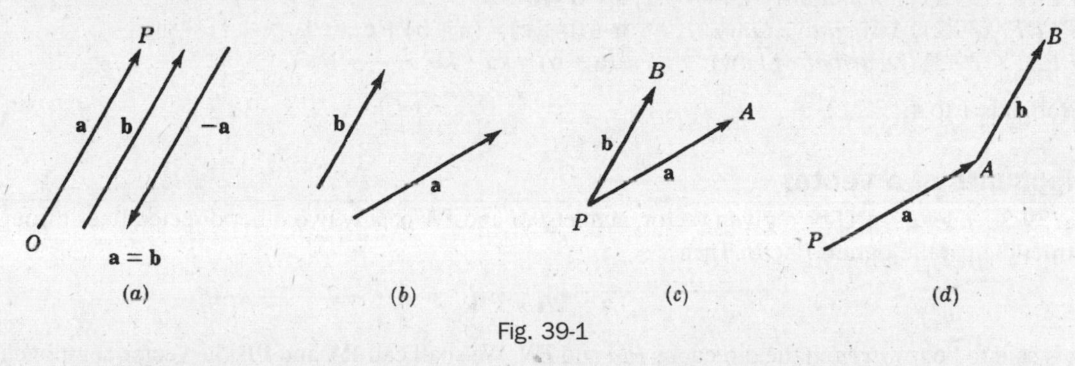

Fig. 39-1

Two vectors **a** and **b** are said to be equal (and we write $\mathbf{a} = \mathbf{b}$) if they have the same direction and magnitude. A vector whose magnitude is that of **a**, but whose direction is opposite that of **a**, is called the *negative* of **a** and is denoted $-\mathbf{a}$. (See Fig. 39-1(a).)

If **a** is a vector and k is a positive scalar, then $k\mathbf{a}$ is defined to be a vector whose direction is that of **a** and whose magnitude is k times that of **a**. If k is a negative scalar, then $k\mathbf{a}$ has direction opposite that of **a** and has magnitude $|k|$ times that of **a**.

We also assume a *zero vector* **0** with magnitude 0 and no direction. We define $-\mathbf{0} = \mathbf{0}$, $0\mathbf{a} = \mathbf{0}$, and $k\mathbf{0} = \mathbf{0}$.

Unless indicated otherwise, a given vector has no fixed position in the plane and so may be moved under parallel displacement at will. In particular, if **a** and **b** are two vectors (Fig. 39-1(b)), they may be placed so as to have a common initial or beginning point P (Fig. 39-1(c)) or so that the initial point of **b** coincides with the terminal or endpoint of **a** (Fig. 39-1 (d)).

Sum and Difference of Two Vectors

If **a** and **b** are the vectors of Fig. 39-1(b), their *sum* $\mathbf{a} + \mathbf{b}$ is to be found in either of two equivalent ways:

1. By placing the vectors as in Fig. 39-1(c) and completing the parallelogram $PAQB$ of Fig. 39-2(a). The vector **PQ** is the required sum.
2. By placing the vectors as in Fig. 39-1(d) and completing the triangle PAB of Fig. 39-2(b). Here, the vector **PB** is the required sum.

It follows from Fig. 39-2(b) that three vectors may be displaced to form a triangle, provided that one of them is either the sum or the negative of the sum of the other two.

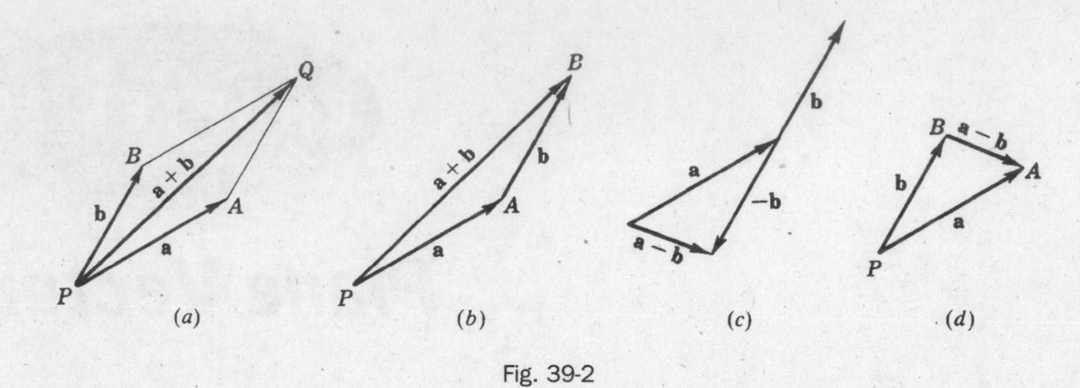

Fig. 39-2

If **a** and **b** are the vectors of Fig. 39-1(*b*), their difference **a** − **b** is to be found in either of two equivalent ways:

1. From the relation **a** − **b** = **a** + (− **b**) as in Fig. 39-2(*c*).
2. By placing the vectors as in Fig. 39-1(*c*) and completing the triangle. In Fig. 39-2(*d*), the vector

   BA = **a** − **b**.

If **a**, **b**, and **c** are vectors, the following laws are valid.

PROPERTY (39.1) (*Commutative Law*) **a** + **b** = **b** + **a**
PROPERTY (39.2) (*Associative Law*) **a** + (**b** + **c**) = (**a** + **b**) + **c**
PROPERTY (39.3) (*Distributive Law*) $k(\mathbf{a} + \mathbf{b}) = k\mathbf{a} + k\mathbf{b}$

See Problems 1 to 4.

Components of a Vector

In Fig. 39-3(*a*), let **a** = **PQ** be a given vector, and let *PM* and *PN* be any two other directed lines through *P*. Construct the parallelogram *PAQB*. Then

$$\mathbf{a} = \mathbf{PA} + \mathbf{PB}$$

and **a** is said to be *resolved* in the directions *PM* and *PN*. We shall call **PA** and **PB** the vector components of **a** in the pair of directions *PM* and *PN*.

Consider next the vector **a** in a rectangular coordinate system (Fig. 39-3(*b*)), having equal units of measure on the two axes. Denote by **i** the vector from (0, 0) to (1, 0), and by **j** the vector from (0, 0) to (0, 1). The direction of **i** is that of the positive *x* axis, the direction of **j** is that of the positive *y* axis, and both are *unit vectors*, that is, vectors of magnitude 1.

From the initial point *P* and the terminal point *Q* of **a**, drop perpendiculars to the *x* axis, meeting it at *M* and *N*, respectively, and to the *y* axis, meeting it at *S* and *T*, respectively. Now, $MN = a_1\mathbf{i}$, with a_1 positive, and $ST = a_2\mathbf{j}$, with a_2 negative. Then: $\mathbf{MN} = \mathbf{RQ} = a_1\mathbf{i}$, $\mathbf{ST} = \mathbf{PR} = a_2\mathbf{j}$, and

$$\mathbf{a} = a_1\mathbf{i} + a_2\mathbf{j} \tag{39.1}$$

(*a*)

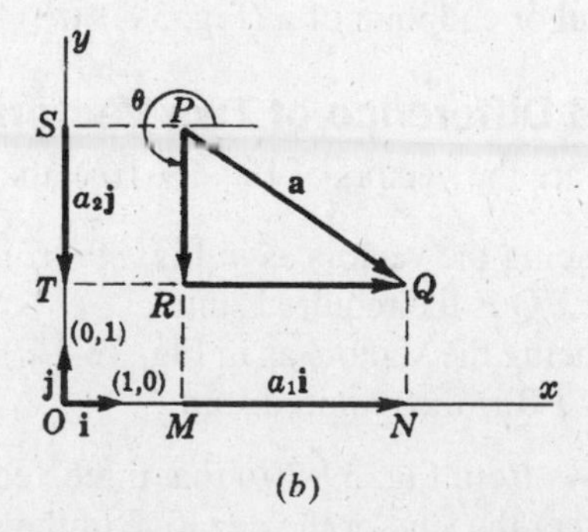

(*b*)

Fig. 39-3

Let us call $a_1\mathbf{i}$ and $a_2\mathbf{j}$ the *vector components* of $\mathbf{a}$.[†] The scalars a_1 and a_2 will be called the *scalar components* (or the *x component* and *y component*, or simply the *components*) of $\mathbf{a}$. Note that $\mathbf{0} = 0\mathbf{i} + 0\mathbf{j}$.

Let the direction of $\mathbf{a}$ be given by the angle θ, with $0 \le \theta < 2\pi$, measured counterclockwise from the positive x axis to the vector. Then

$$|\mathbf{a}| = \sqrt{a_1^2 + a_2^2} \tag{39.2}$$

and

$$\tan\theta = \frac{a_2}{a_1} \tag{39.3}$$

with the quadrant of θ being determined by

$$a_1 = |\mathbf{a}|\cos\theta, \quad a_2 = |\mathbf{a}|\sin\theta$$

If $\mathbf{a} = a_1\mathbf{i} + a_2\mathbf{j}$ and $\mathbf{b} = b_1\mathbf{j} + b_2\mathbf{j}$, then the following hold.

PROPERTY (39.4) $\mathbf{a} = \mathbf{b}$ if and only if $a_1 = b_1$ and $a_2 = b_2$
PROPERTY (39.5) $k\mathbf{a} = ka_1\mathbf{i} + ka_2\mathbf{j}$
PROPERTY (39.6) $\mathbf{a} + \mathbf{b} = (a_1 + b_1)\mathbf{i} + (a_2 + b_2)\mathbf{j}$
PROPERTY (39.7) $\mathbf{a} - \mathbf{b} = (a_1 - b_1)\mathbf{i} + (a_2 - b_2)\mathbf{j}$

Scalar Product (or Dot Product)

The *scalar product* (or *dot product*) of vectors $\mathbf{a}$ and $\mathbf{b}$ is defined by

$$\mathbf{a}\cdot\mathbf{b} = |\mathbf{a}||\mathbf{b}|\cos\theta \tag{39.4}$$

where θ is the smaller angle between the two vectors when they are drawn with a common initial point (see Fig. 39-4). We also define: $\mathbf{a}\cdot\mathbf{0} = \mathbf{0}\cdot\mathbf{a} = 0$.

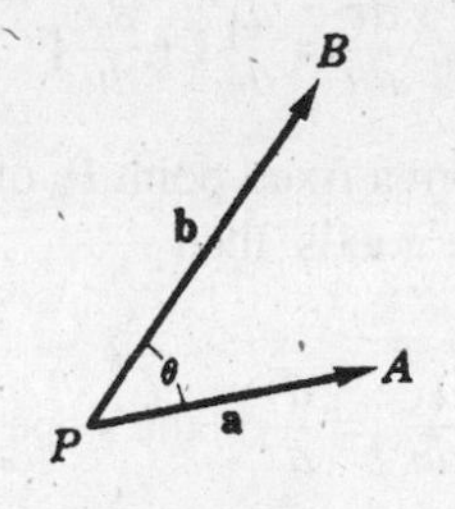

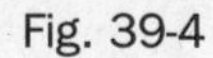
Fig. 39-4

From the definitions, we can derive the following properties of the scalar product.

PROPERTY (39.8) (*Commutative Law*) $\mathbf{a}\cdot\mathbf{b} = \mathbf{b}\cdot\mathbf{a}$
PROPERTY (39.9) $\mathbf{a}\cdot\mathbf{a} = |\mathbf{a}|^2$ and $|\mathbf{a}| = \sqrt{\mathbf{a}\cdot\mathbf{a}}$
PROPERTY (39.10) $\mathbf{a}\cdot\mathbf{b} = 0$ if and only if ($\mathbf{a} = 0$ or $\mathbf{b} = 0$ or $\mathbf{a}$ is perpendicular to $\mathbf{b}$)
PROPERTY (39.11) $\mathbf{i}\cdot\mathbf{i} = \mathbf{j}\cdot\mathbf{j} = 1$ and $\mathbf{i}\cdot\mathbf{j} = 0$
PROPERTY (39.12) $\mathbf{a}\cdot\mathbf{b} = (a_1\mathbf{i} + a_2\mathbf{j})\cdot(b_1\mathbf{i} + b_2\mathbf{j}) = a_1b_1 + a_2b_2$
PROPERTY (39.13) (*Distributive Law*) $\mathbf{a}\cdot(\mathbf{b}+\mathbf{c}) = \mathbf{a}\cdot\mathbf{b} + \mathbf{a}\cdot\mathbf{c}$
PROPERTY (39.14) $(\mathbf{a}+\mathbf{b})\cdot(\mathbf{c}+\mathbf{d}) = \mathbf{a}\cdot\mathbf{c} + \mathbf{a}\cdot\mathbf{d} + \mathbf{b}\cdot\mathbf{c} + \mathbf{b}\cdot\mathbf{d}$

[†]A pair of directions (such as *OM* and *OT*) need not be mentioned, since they are determined by the coordinate system.

Scalar and Vector Projections

In equation (39.1), the scalar a_1 is called the *scalar projection* of **a** on any vector whose direction is that of the positive x axis, while the vector $a_1\mathbf{i}$ is called the *vector projection* of **a** on any vector whose direction is that of the positive x axis. In general, for any nonzero vector **b** and any vector **a**, we define we define $\mathbf{a}\cdot\frac{\mathbf{b}}{|\mathbf{b}|}$ to be the scalar projection of **a** on **b**, and $\left(\mathbf{a}\cdot\frac{\mathbf{b}}{|\mathbf{b}|}\right)\frac{\mathbf{b}}{|\mathbf{b}|}$ to be the vector projection of **a** on **b**. (See Problem 7.) Note that, when **b** has the direction of the positive x axis, $\frac{\mathbf{b}}{|\mathbf{b}|}=\mathbf{i}$.

PROPERTY (39.15) $\mathbf{a}\cdot\mathbf{b}$ is the product of the length of **a** and the scalar projection of **b** on **a**. Likewise, $\mathbf{a}\cdot\mathbf{b}$ is the product of the length of **b** and the scalar projection of **a** on **b**. (See Fig. 39-5.)

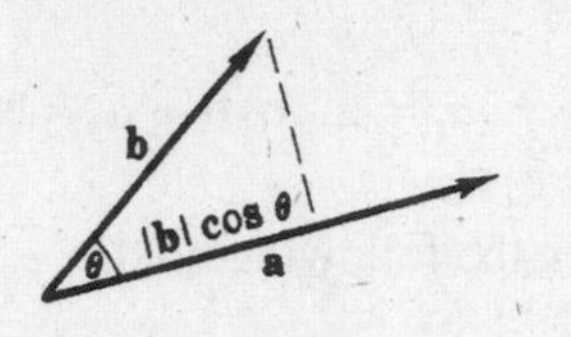

Fig. 39-5

Differentiation of Vector Functions

Let the curve of Fig. 39-6 be given by the parametric equations $x=f(u)$ and $y=g(u)$. The vector

$$\mathbf{r}=x\mathbf{i}+y\mathbf{j}=f(u)\mathbf{i}+g(u)\mathbf{j}$$

joining the origin to the point $P(x, y)$ of the curve is called the *position vector* or the *radius vector* of P. It is a function of u. (From now on, the letter **r** will be used exclusively to denote position vectors. Thus, $\mathbf{a}=3\mathbf{i}+4\mathbf{j}$ is meant to be a "free" vector, whereas $\mathbf{r}=3\mathbf{i}+4\mathbf{j}$ is meant to be the vector joining the origin to $P(3, 4)$.)

The derivative $\frac{d\mathbf{r}}{du}$ of the function **r** with respect to u is defined to be $\lim_{\Delta u\to 0}\frac{\mathbf{r}(u+\Delta u)-\mathbf{r}(u)}{\Delta u}$.

Straightforward computation yields:

$$\frac{d\mathbf{r}}{du}=\frac{dx}{du}\mathbf{i}+\frac{dy}{du}\mathbf{j} \tag{39.5}$$

Let s denote the arc length measured from a fixed point P_0 of the curve so that s increases with u. If τ is the angle that $d\mathbf{r}/du$ makes with the positive x axis, then

$$\tan\tau=\left(\frac{dy}{du}\right)\Big/\left(\frac{dx}{du}\right)=\frac{dy}{dx}=\text{the slope of the curve at } P$$

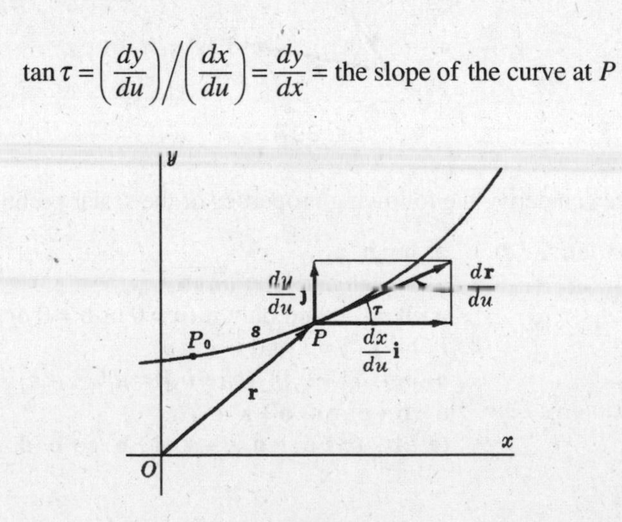

Fig. 39-6

Moreover, $\frac{d\mathbf{r}}{du}$ is a vector of magnitude

$$\left|\frac{d\mathbf{r}}{du}\right| = \sqrt{\left(\frac{dx}{du}\right)^2 + \left(\frac{dy}{du}\right)^2} = \frac{ds}{du} \tag{39.6}$$

whose direction is that of the tangent line to the curve at P. It is customary to show this vector with P as its initial point.

If now the scalar variable u is taken to be the arc length s, then equation (39.5) becomes

$$\mathbf{t} = \frac{d\mathbf{r}}{ds} = \frac{dx}{ds}\mathbf{i} + \frac{dy}{ds}\mathbf{j} \tag{39.7}$$

The direction of $\mathbf{t}$ is τ, while its magnitude is $\sqrt{\left(\frac{dx}{ds}\right)^2 + \left(\frac{dy}{ds}\right)^2}$, which is equal to 1. Thus, $\mathbf{t} = d\mathbf{r}/ds$ is the *unit tangent vector* to the curve at P.

Since $\mathbf{t}$ is a unit vector, $\mathbf{t}$ and $d\mathbf{t}/ds$ are perpendicular. (See Problem 10.) Denote by $\mathbf{n}$ a unit vector at P having the direction of $d\mathbf{t}/ds$. As P moves along the curve shown in Fig. 39-7, the magnitude of $\mathbf{t}$ remains constant; hence $d\mathbf{t}/ds$ measures the rate of change of the direction of $\mathbf{t}$. Thus, the magnitude of $d\mathbf{t}/ds$ at P is the absolute value of the curvature at P, that is, $|d\mathbf{t}/ds| = |K|$, and

$$\frac{d\mathbf{t}}{ds} = |K|\,\mathbf{n} \tag{39.8}$$

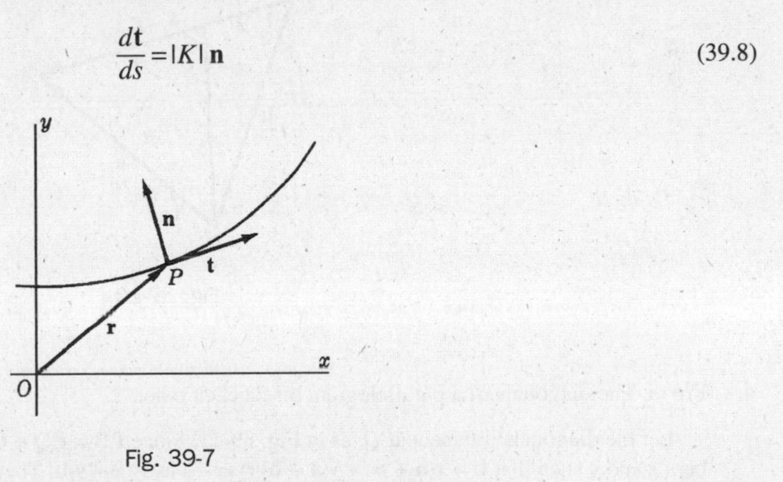

Fig. 39-7

SOLVED PROBLEMS

1. Prove $\mathbf{a} + \mathbf{b} = \mathbf{b} + \mathbf{a}$.

From Fig. 39-8, $\mathbf{a} + \mathbf{b} = \mathbf{PQ} = \mathbf{b} + \mathbf{a}$.

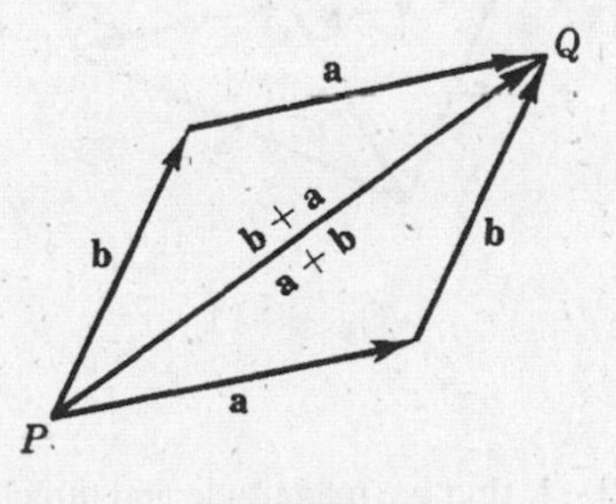

Fig. 39-8

2. Prove $(\mathbf{a} + \mathbf{b}) + \mathbf{c} = \mathbf{a} + (\mathbf{b} + \mathbf{c})$.

From Fig. 39-9, $\mathbf{PC} = \mathbf{PB} + \mathbf{BC} = (\mathbf{a} + \mathbf{b}) + \mathbf{c}$. Also, $\mathbf{PC} = \mathbf{PA} + \mathbf{AC} = \mathbf{a} + (\mathbf{b} + \mathbf{c})$.

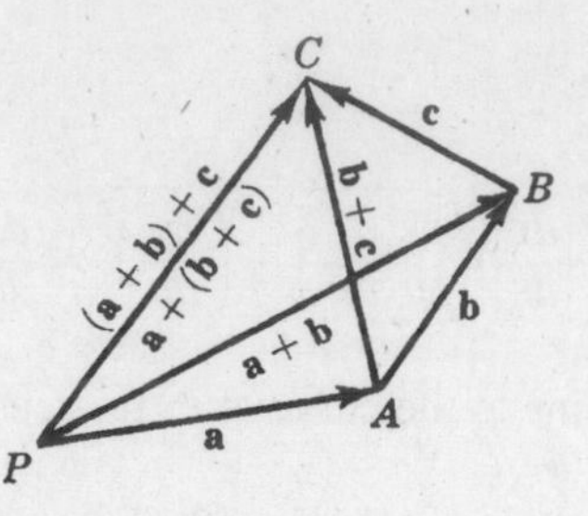

Fig. 39-9

3. Let **a**, **b**, and **c** be three vectors issuing from P such that their endpoints A, B, and C lie on a line, as shown in Fig. 39-10. If C divides BA in the ratio $x:y$, where $x + y = 1$, show that $\mathbf{c} = x\mathbf{a} + y\mathbf{b}$.

Just note that

$$\mathbf{c} = \mathbf{PB} + \mathbf{BC} = \mathbf{b} + x(\mathbf{a} - \mathbf{b}) = x\mathbf{a} + (1 - x)\mathbf{b} = x\mathbf{a} + y\mathbf{b}$$

As an example, if C bisects BA, then $\mathbf{c} = \frac{1}{2}(\mathbf{a} + \mathbf{b})$ and $\mathbf{BC} = \frac{1}{2}(\mathbf{a} - \mathbf{b})$.

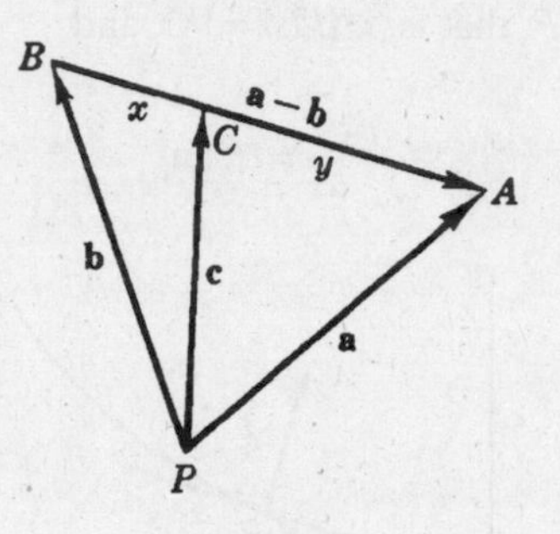

Fig. 39-10

4. Prove: The diagonals of a parallelogram bisect each other.

Let the diagonals intersect at Q, as in Fig. 39-11. Since $\mathbf{PB} = \mathbf{PQ} + \mathbf{QB} = \mathbf{PQ} - \mathbf{BQ}$, there are positive numbers x and y such that $\mathbf{b} = x(\mathbf{a} + \mathbf{b}) - y(\mathbf{a} - \mathbf{b}) = (x - y)\mathbf{a} + (x + y)\mathbf{b}$. Then $x + y = 1$ and $x - y = 0$. Hence, $x = y = \frac{1}{2}$, and Q is the midpoint of each diagonal.

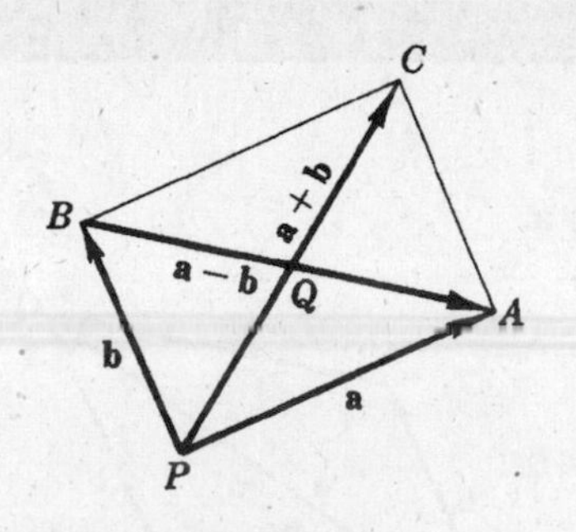

Fig. 39-11

5. For the vectors $\mathbf{a} = 3\mathbf{i} + 4\mathbf{j}$ and $\mathbf{b} = 2\mathbf{i} - \mathbf{j}$, find the magnitude and direction of (a) **a** and **b**; (b) $\mathbf{a} + \mathbf{b}$; (c) $\mathbf{b} - \mathbf{a}$.

(a) For $\mathbf{a} = 3\mathbf{i} + 4\mathbf{j}$: $|\mathbf{a}| = \sqrt{a_1^2 + a_2^2} = \sqrt{3^2 + 4^2} = 5$; $\tan\theta = a_2/a_1 = \frac{4}{3}$ and $\cos\theta = a_1/|\mathbf{a}| = \frac{3}{5}$; then θ is a first quadrant angle and is $53°8'$.

For $\mathbf{b} = 2\mathbf{i} - \mathbf{j}$: $|\mathbf{b}| = \sqrt{4 + 1} = \sqrt{5}$; $\tan\theta = -\frac{1}{2}$ and $\cos\theta = 2/\sqrt{5}$; $\theta = 360° - 26°34' = 333°26'$.

(b) $\mathbf{a}+\mathbf{b}=(3\mathbf{i}+4\mathbf{j})+(2\mathbf{i}-\mathbf{j})=5\mathbf{i}+3\mathbf{j}$. Then $|\mathbf{a}+\mathbf{b}|=\sqrt{5^2+3^2}=\sqrt{34}$. Since $\tan\theta=\frac{3}{5}$ and $\cos\theta=5/\sqrt{34}$, $\theta=30°58'$.

(c) $\mathbf{b}-\mathbf{a}=(2\mathbf{i}-\mathbf{j})-(3\mathbf{i}+4\mathbf{j})=-\mathbf{i}-5\mathbf{j}$. Then $|\mathbf{b}-\mathbf{a}|=\sqrt{26}$. Since $\tan\theta=5$ and $\cos\theta=-1/\sqrt{26}$, $\theta=258°41'$.

6. Prove: The median to the base of an isosceles triangle is perpendicular to the base. (See Fig. 39-12, where $|\mathbf{a}|=|\mathbf{b}|$.)

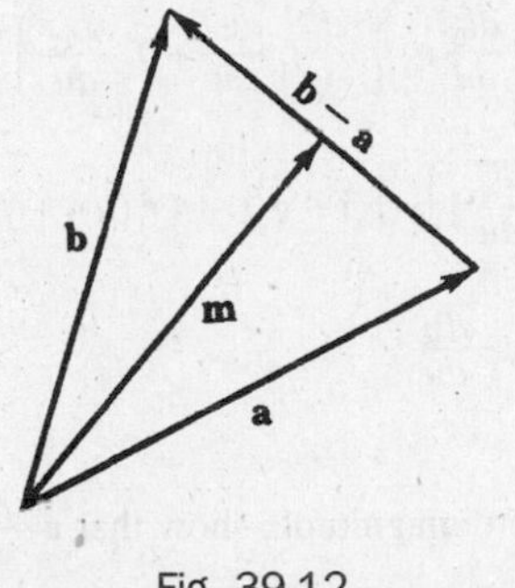

Fig. 39-12

From Problem 3, since $\mathbf{m}$ bisects the base, $\mathbf{m}=\frac{1}{2}(\mathbf{a}+\mathbf{b})$. Then

$$\mathbf{m}\cdot(\mathbf{b}-\mathbf{a})=\tfrac{1}{2}(\mathbf{a}+\mathbf{b})\cdot(\mathbf{b}-\mathbf{a})$$
$$=\tfrac{1}{2}(\mathbf{a}\cdot\mathbf{b}-\mathbf{a}\cdot\mathbf{a}+\mathbf{b}\cdot\mathbf{b}-\mathbf{b}\cdot\mathbf{a})=\tfrac{1}{2}(\mathbf{b}\cdot\mathbf{b}-\mathbf{a}\cdot\mathbf{a})=0$$

Thus, the median is perpendicular to the base.

7. If $\mathbf{b}$ is a nonzero vector, resolve a vector $\mathbf{a}$ into components $\mathbf{a}_1$ and $\mathbf{a}_2$, respectively parallel and perpendicular to $\mathbf{b}$.

In Fig. 39-13, we have $\mathbf{a}=\mathbf{a}_1+\mathbf{a}_2$, $\mathbf{a}_1=c\mathbf{b}$, and $\mathbf{a}_2\cdot\mathbf{b}=0$. Hence, $\mathbf{a}_2=\mathbf{a}-\mathbf{a}_1=\mathbf{a}-c\mathbf{b}$. Moreover, $\mathbf{a}_2\cdot\mathbf{b}=(\mathbf{a}-c\mathbf{b})\cdot\mathbf{b}=\mathbf{a}\cdot\mathbf{b}-c|\mathbf{b}|^2=0$, whence $c=\dfrac{\mathbf{a}\cdot\mathbf{b}}{|\mathbf{b}|^2}$. Thus,

$$\mathbf{a}_1=c\mathbf{b}=\frac{\mathbf{a}\cdot\mathbf{b}}{|\mathbf{b}|^2}\mathbf{b} \quad\text{and}\quad \mathbf{a}_2=\mathbf{a}-c\mathbf{b}=\mathbf{a}-\frac{\mathbf{a}\cdot\mathbf{b}}{|\mathbf{b}|^2}\mathbf{b}$$

The scalar $\mathbf{a}\cdot\dfrac{\mathbf{b}}{|\mathbf{b}|}$ is the scalar projection of $\mathbf{a}$ on $\mathbf{b}$. The vector $\left(\mathbf{a}\cdot\dfrac{\mathbf{b}}{|\mathbf{b}|}\right)\dfrac{\mathbf{b}}{|\mathbf{b}|}$ is the vector projection of $\mathbf{a}$ on $\mathbf{b}$.

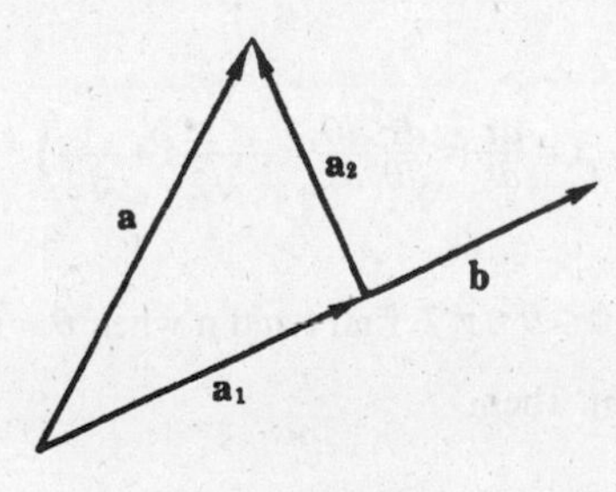

Fig. 39-13

8. Resolve $\mathbf{a}=4\mathbf{i}+3\mathbf{j}$ into components $\mathbf{a}_1$ and $\mathbf{a}_2$, parallel and perpendicular, respectively, to $\mathbf{b}=3\mathbf{i}+\mathbf{j}$.

From Problem 7, $c=\dfrac{\mathbf{a}\cdot\mathbf{b}}{|\mathbf{b}|^2}=\dfrac{12+3}{10}=\dfrac{3}{2}$. Then

$$\mathbf{a}_1=c\mathbf{b}=\tfrac{9}{2}\mathbf{i}+\tfrac{3}{2}\mathbf{j} \text{ and } \mathbf{a}_2=\mathbf{a}-\mathbf{a}_1=-\tfrac{1}{2}\mathbf{i}+\tfrac{3}{2}\mathbf{j}$$

9. If $\mathbf{a} = f_1(u)\mathbf{i} + f_2(u)\mathbf{j}$ and $\mathbf{b} = g_1(u)\mathbf{i} + g_2(u)\mathbf{j}$, show that $\frac{d}{du}(\mathbf{a}\cdot\mathbf{b}) = \frac{d\mathbf{a}}{du}\cdot\mathbf{b} + \mathbf{a}\cdot\frac{d\mathbf{b}}{du}$.

By Property 39.12, $\mathbf{a}\cdot\mathbf{b} = (f_1(u)\mathbf{i} + f_2(u)\mathbf{j})\cdot(g_1(u)\mathbf{i} + g_2(u)\mathbf{j}) = f_1g_1 + f_2g_2$. Then

$$\begin{aligned}\frac{d}{du}(\mathbf{a}\cdot\mathbf{b}) &= \frac{df_1}{du}g_1 + f_1\frac{dg_1}{du} + \frac{df_2}{du}g_2 + f_2\frac{dg_2}{du}\\ &= \left(\frac{df_1}{du}g_1 + \frac{df_2}{du}g_2\right) + \left(f_1\frac{dg_1}{du} + f_2\frac{dg_2}{du}\right)\\ &= \left(\frac{df_1}{du}\mathbf{i} + \frac{df_2}{du}\mathbf{j}\right)\cdot(g_1\mathbf{i} + g_2\mathbf{j}) + (f_1(u)\mathbf{i} + f_2(u)\mathbf{j})\cdot\left(\frac{dg_1}{du}\mathbf{i} + \frac{dg_2}{du}\mathbf{j}\right)\\ &= \frac{d\mathbf{a}}{du}\cdot\mathbf{b} + \mathbf{a}\cdot\frac{d\mathbf{b}}{du}\end{aligned}$$

10. If $\mathbf{a} = f_1(u)\mathbf{i} + f_2(u)\mathbf{j}$ is of constant nonzero magnitude, show that $\mathbf{a}\cdot\frac{d\mathbf{a}}{du} = 0$ and, therefore, when $\frac{d\mathbf{a}}{du}$ is not zero, $\mathbf{a}$ and $\frac{d\mathbf{a}}{du}$ are perpendicular.

Let $|\mathbf{a}| = c$. Thus, $\mathbf{a}\cdot\mathbf{a} = c^2$. By Problem 9,

$$\frac{d}{du}(\mathbf{a}\cdot\mathbf{a}) = \frac{d\mathbf{a}}{du}\cdot\mathbf{a} + \mathbf{a}\cdot\frac{d\mathbf{a}}{du} = 2\mathbf{a}\cdot\frac{d\mathbf{a}}{du} = 0$$

Then $\mathbf{a}\cdot\frac{d\mathbf{a}}{du} = 0$.

11. Given $\mathbf{r} = (\cos^2\theta)\mathbf{i} + (\sin^2\theta)\mathbf{j}$, for $0 \le \theta \le \pi/2$, find $\mathbf{t}$.

Since $\frac{d}{d\theta}\cos^2\theta = -2\cos\theta\sin\theta = -\sin 2\theta$ and $\frac{d}{d\theta}\sin^2\theta = 2\sin\theta\cos\theta = \sin 2\theta$, equation (39.5) yields

$$\frac{d\mathbf{r}}{d\theta} = -(\sin 2\theta)\mathbf{i} + (\sin 2\theta)\mathbf{j}$$

Therefore, by equation (39.6),

$$\frac{ds}{d\theta} = \left|\frac{d\mathbf{r}}{d\theta}\right| = \sqrt{\frac{d\mathbf{r}}{d\theta}\cdot\frac{d\mathbf{r}}{d\theta}} = \sqrt{2}\sin 2\theta$$

by Property 39.12. So,

$$\mathbf{t} = \frac{d\mathbf{r}}{ds} = \frac{d\mathbf{r}}{d\theta}\frac{d\theta}{ds} = -\frac{1}{\sqrt{2}}\mathbf{i} + \frac{1}{\sqrt{2}}\mathbf{j}$$

12. Given $x = a\cos^3\theta$, $y = a\sin^3\theta$, with $0 \le \theta \le \pi/2$, find $\mathbf{t}$ and $\mathbf{n}$ when $\theta = \pi/4$.

We have $\mathbf{r} = a(\cos^3\theta)\mathbf{i} + a(\sin^3\theta)\mathbf{j}$. Then

$$\frac{d\mathbf{r}}{d\theta} = -3a(\cos^2\theta)(\sin\theta)\mathbf{i} + 3a(\sin^2\theta)(\cos\theta)\mathbf{j} \quad\text{and}\quad \frac{ds}{d\theta} = \left|\frac{d\mathbf{r}}{d\theta}\right| = 3a\sin\theta\cos\theta$$

Hence,

$$\mathbf{t} = \frac{d\mathbf{r}}{ds} = \frac{d\mathbf{r}}{d\theta}\frac{d\theta}{ds} = -(\cos\theta)\mathbf{i} + (\sin\theta)\mathbf{j} \quad\text{and}\quad \frac{d\mathbf{t}}{ds} = ((\sin\theta)\mathbf{i} + (\cos\theta)\mathbf{j})\frac{d\theta}{ds}$$

$$= \frac{1}{3a\cos\theta}\mathbf{i} + \frac{1}{3a\sin\theta}\mathbf{j}$$

At $\theta = \pi/4$,

$$\mathbf{t} = -\frac{1}{\sqrt{2}}\mathbf{i} + \frac{1}{\sqrt{2}}\mathbf{j},\ \frac{d\mathbf{t}}{ds} = \frac{\sqrt{2}}{3a}\mathbf{i} + \frac{\sqrt{2}}{3a}\mathbf{j},\ |K| = \left|\frac{d\mathbf{t}}{ds}\right| = \frac{2}{3a} \quad \text{and} \quad \mathbf{n} = \frac{1}{|K|}\frac{d\mathbf{t}}{ds} = \frac{1}{\sqrt{2}}\mathbf{i} + \frac{1}{\sqrt{2}}\mathbf{j}$$

13. Show that the vector $\mathbf{a} = a\mathbf{i} + b\mathbf{j}$ is perpendicular to the line $ax + by + c = 0$.

Let $P_1(x_1, y_1)$ and $P_2(x_2, y_2)$ be two distinct points on the line. Then $ax_1 + by_1 + c = 0$ and $ax_2 + by_2 + c = 0$. Subtracting the first from the second yields

$$a(x_2 - x_1) + b(y_2 - y_1) = 0 \tag{1}$$

Now

$$a(x_2 - x_1) + b(y_2 - y_1) = (a\mathbf{i} + b\mathbf{j}) \cdot [(x_2 - x_1)\mathbf{i} + (y_2 - y_1)\mathbf{j}]$$
$$= \mathbf{a} \cdot \mathbf{P_1P_2}$$

By (1), the left side is zero. Thus, **a** is perpendicular (normal) to the line.

14. Use vector methods to find:

(a) The equation of the line through $P_1(2, 3)$ and perpendicular to the line $x + 2y + 5 = 0$.
(b) The equation of the line through, $P_1(2, 3)$ and $P_2(5, -1)$.

Take $P(x, y)$ to be any other point on the required line.

(a) By Problem 13, the vector $\mathbf{a} = \mathbf{i} + 2\mathbf{j}$ is normal to the line $x + 2y + 5 = 0$. Then $\mathbf{P_1P} = (x - 2)\mathbf{i} + (y - 3)\mathbf{j}$ is parallel to **a** if $(x - 2)\mathbf{i} + (y - 3)\mathbf{j} = k(\mathbf{i} + 2\mathbf{j})$ for some scalar k. Equating components, we get $x - 2 = k$ and $y - 3 = 2k$. Eliminating k, we obtain the required equation $y - 3 = 2(x - 2)$, or, equivalently, $2x - y - 1 = 0$.
(b) We have $\mathbf{P_1P} = (x - 2)\mathbf{i} + (y - 3)\mathbf{j}$ and $\mathbf{P_1P_2} = 3\mathbf{i} - 4\mathbf{j}$. Now $\mathbf{a} = 4\mathbf{i} + 3\mathbf{j}$ is perpendicular to $\mathbf{P_1P_2}$ and, hence, to $\mathbf{P_1P}$. Thus, $0 = \mathbf{a} \cdot \mathbf{P_1P} = (4\mathbf{i} + 3\mathbf{j}) \cdot [(x - 2)\mathbf{i} + (y - 3)\mathbf{j}]$ and, equivalently, $4x + 3y - 17 = 0$.

15. Use vector methods to find the distance of the point $P_1(2, 3)$ from the line $3x + 4y - 12 = 0$.

At any convenient point on the line, say $A(4, 0)$, construct the vector $\mathbf{a} = 3\mathbf{i} + 4\mathbf{j}$ perpendicular to the line. The required distance is $d = |\mathbf{AP_1}| \cos\theta$ in Fig. 39-14. Now, $\mathbf{a} \cdot \mathbf{AP_1} = |\mathbf{a}|\,|\mathbf{AP_1}| \cos\theta = |\mathbf{a}|\, d$. Hence,

$$d = \frac{\mathbf{a} \cdot \mathbf{AP_1}}{|\mathbf{a}|} = \frac{(3\mathbf{i} + 4\mathbf{j}) \cdot (-2\mathbf{i} + 3\mathbf{j})}{5} = \frac{-6 + 12}{5} = \frac{6}{5}$$

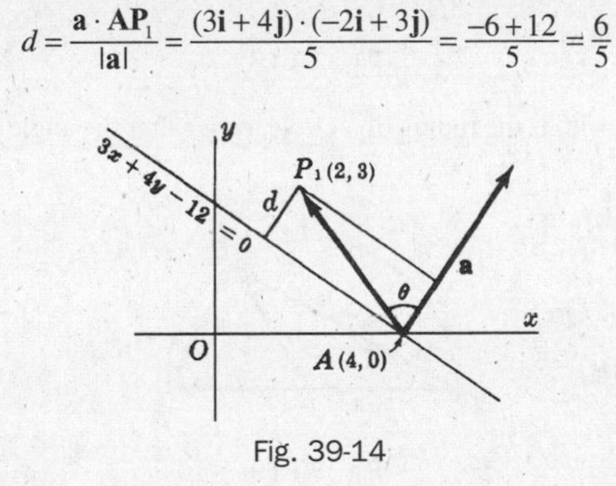

Fig. 39-14

16. The *work* done by a force expressed as a vector **b** in moving an object along a vector **a** is defined as the product of the magnitude of **b** in the direction of **a** and the distance moved. Find the work done in moving an object along the vector $\mathbf{a} = 3\mathbf{i} + 4\mathbf{j}$ if the force applied is $\mathbf{b} = 2\mathbf{i} + \mathbf{j}$.

The work done is

(magnitude of **b** in the direction of **a**) $\cdot$ (distance moved) $= (|\mathbf{b}| \cos\theta)\,|\mathbf{a}| = \mathbf{b} \cdot \mathbf{a} = (2\mathbf{i} + \mathbf{j}) \cdot (3\mathbf{i} + 4\mathbf{j}) = 10$

SUPPLEMENTARY PROBLEMS

17. Given the vectors **a**, **b**, **c** in Fig. 39-15, construct (a) 2**a**; (b) −3**b**; (c) **a** + 2**b**; (d) **a** + **b** − **c**; (e) **a** − 2**b** + 3**c**.

18. Prove: The line joining the midpoints of two sides of a triangle is parallel to and one-half the length of the third side. (See Fig. 39-16.)

19. If **a**, **b**, **c**, **d** are consecutive sides of a quadrilateral (see Fig. 39-17), show that **a** + **b** + **c** + **d** = 0. (*Hint:* Let P and Q be two nonconsecutive vertices.) Express **PQ** in two ways.

Fig. 39-15

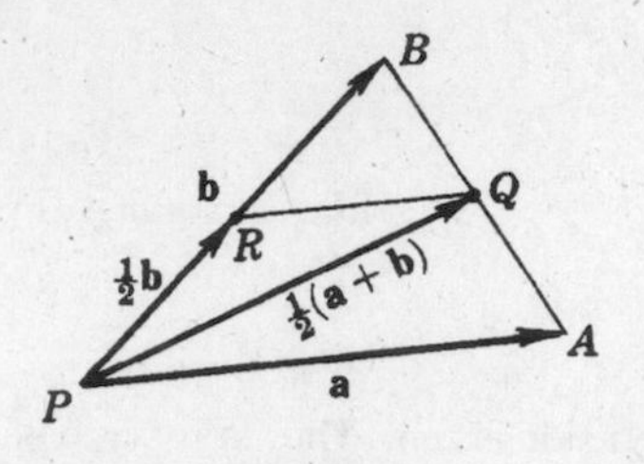

Fig. 39-16

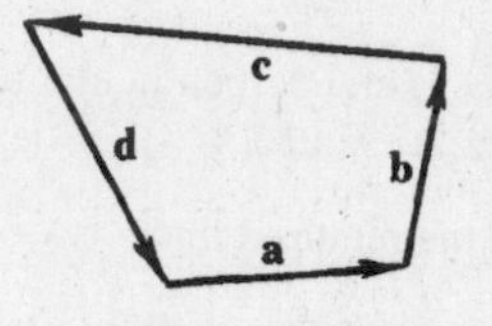

Fig. 39-17

20. Prove: If the midpoints of the consecutive sides of any quadrilateral are joined, the resulting quadrilateral is a parellelogram. (See Fig. 39-18.)

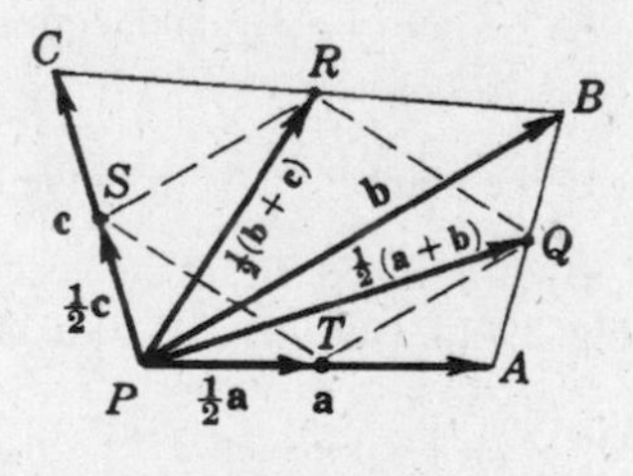

Fig. 39-18

21. Using Fig. 39-19, in which |**a**| = |**b**| is the radius of a circle, prove that the angle inscribed in a semicircle is a right angle.

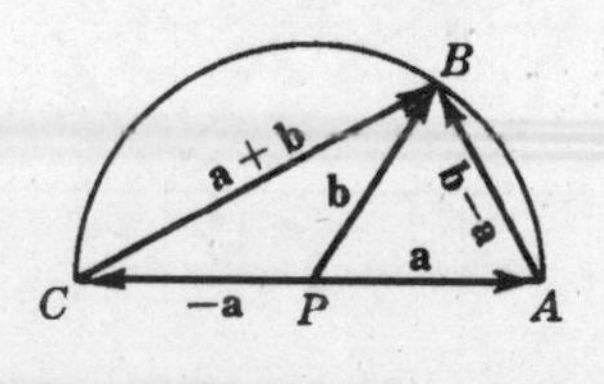

Fig. 39-19

22. Find the length of each of the following vectors and the angle it makes with the positive x axis: (a) **i** + **j**; (b) −**i** + **j**; (c) $\mathbf{i} + \sqrt{3}\mathbf{j}$; (d) $\mathbf{i} - \sqrt{3}\mathbf{j}$.

Ans. (a) $\sqrt{2}$, $\theta = \frac{1}{4}\pi$; (b) $\sqrt{2}$, $\theta = 3\pi/4$; (c) 2, $\theta = \pi/3$; (d) 2, $\theta = 5\pi/3$

23. Prove: If **u** is obtained by rotating the unit vector **i** counterclockwise about the origin through the angle θ, then $\mathbf{u} = \mathbf{i}\cos\theta + \mathbf{j}\sin\theta$.

24. Use the law of cosines for triangles to obtain $\mathbf{a} \cdot \mathbf{b} = |\mathbf{a}||\mathbf{b}| \cos\theta = \frac{1}{2}(|\mathbf{a}|^2 + |\mathbf{b}|^2 - |\mathbf{c}|^2)$.

25. Write each of the following vectors in the form $a\mathbf{i} + b\mathbf{j}$.

(a) The vector joining the origin to $P(2, -3)$; (b) The vector joining $P_1(2, 3)$ to $P_2(4, 2)$;
(c) The vector joining $P_2(4, 2)$ to P_1 $(2, 3)$; (d) The unit vector in the direction of $3\mathbf{i} + 4\mathbf{j}$;
(e) The vector having magnitude 6 and direction 120°

Ans. (a) $2\mathbf{i} - 3\mathbf{j}$; (b) $2\mathbf{i} - \mathbf{j}$; (c) $-2\mathbf{i} + \mathbf{j}$; (d) $\frac{3}{5}\mathbf{i} + \frac{4}{5}\mathbf{j}$; (e) $-3\mathbf{i} + 3\sqrt{3}\mathbf{j}$

26. Using vector methods, derive the formula for the distance between $P_1(x_1, y_1)$ and $P_2(x_2, y_2)$.

27. Given $O(0, 0)$, $A(3, 1)$, and $B(1, 5)$ as vertices of the parallelogram $OAPB$, find the coordinates of P.

Ans. (4, 6)

28. (a) Find k so that $\mathbf{a} = 3\mathbf{i} + 2\mathbf{j}$ and $\mathbf{b} = \mathbf{i} + k\mathbf{j}$ are perpendicular. (b) Write a vector perpendicular to $\mathbf{a} = 2\mathbf{i} + 5\mathbf{j}$.

29. Prove Properties (39.8) to (39.15).

30. Find the vector projection and scalar projection of $\mathbf{b}$ on $\mathbf{a}$, given: (a) $\mathbf{a} = \mathbf{i} - 2\mathbf{j}$ and $\mathbf{b} = -3\mathbf{i} + \mathbf{j}$; (b) $\mathbf{a} = 2\mathbf{i} + 3\mathbf{j}$ and $\mathbf{b} = 10\mathbf{i} + 2\mathbf{j}$.

Ans. (a) $-\mathbf{i} + 2\mathbf{j}$, $-\sqrt{5}$; (b) $4\mathbf{i} + 6\mathbf{j}$, $2\sqrt{13}$

31. Prove: Three vectors $\mathbf{a}$, $\mathbf{b}$, $\mathbf{c}$ will, after parallel displacement, form a triangle provided (a) one of them is the sum of the other two or (b) $\mathbf{a} + \mathbf{b} + \mathbf{c} = \mathbf{0}$.

32. Show that $\mathbf{a} = 3\mathbf{i} - 6\mathbf{j}$, $\mathbf{b} = 4\mathbf{i} + 2\mathbf{j}$, and $\mathbf{c} = -7\mathbf{i} + 4\mathbf{j}$ are the sides of a right triangle. Verify that the midpoint of the hypotenuse is equidistant from the vertices.

33. Find the unit tangent vector $\mathbf{t} = d\mathbf{r}/ds$, given: (a) $\mathbf{r} = 4\mathbf{i} \cos\theta + 4\mathbf{j} \sin\theta$; (b) $\mathbf{r} = e^{\theta}\mathbf{i} + e^{-\theta}\mathbf{j}$; (c) $\mathbf{r} = \theta\mathbf{i} + \theta^2\mathbf{j}$.

Ans. (a) $-\mathbf{i} \sin\theta + \mathbf{j} \cos\theta$; (b) $\dfrac{e^{\theta}\mathbf{i} - e^{-\theta}\mathbf{j}}{\sqrt{e^{2\theta} + e^{-2\theta}}}$; (c) $\dfrac{\mathbf{i} + 2\theta\mathbf{j}}{\sqrt{1 + 4\theta^2}}$

34. (a) Find $\mathbf{n}$ for the curve of Problem 33(a); (b) Find $\mathbf{n}$ for the curve of Problem 33(c); (c) Find $\mathbf{t}$ and $\mathbf{n}$ given $x = \cos\theta + \theta \sin\theta$, $y = \sin\theta - \theta\cos\theta$.

Ans. (a) $\mathbf{i} \cos\theta - \mathbf{j} \sin\theta$; (b) $\dfrac{-2\theta}{\sqrt{1 + 4\theta^2}}\mathbf{i} + \dfrac{1}{\sqrt{1 + 4\theta^2}}\mathbf{j}$; (c) $\mathbf{t} = \mathbf{i} \cos\theta + \mathbf{j} \sin\theta$, $\mathbf{n} = -\mathbf{i} \sin\theta + \mathbf{j} \cos\theta$

CHAPTER 40

Curvilinear Motion

Velocity in Curvilinear Motion

Consider a point $P(x, y)$ moving along a curve with the equations $x = f(t)$, $y = g(t)$, where t is time. By differentiating the position vector

$$\mathbf{r} = x\mathbf{i} + y\mathbf{j} \tag{40.1}$$

with respect to t, we obtain the *velocity vector*

$$\mathbf{v} = \frac{d\mathbf{r}}{dt} = \frac{dx}{dt}\mathbf{i} + \frac{dy}{dt}\mathbf{j} = v_x\mathbf{i} + v_y\mathbf{j} \tag{40.2}$$

where $v_x = \frac{dx}{dt}$ and $v_y = \frac{dy}{dt}$.

The magnitude of $\mathbf{v}$ is called the *speed* and is given by

$$|\mathbf{v}| = \sqrt{\mathbf{v} \cdot \mathbf{v}} = \sqrt{v_x^2 + v_y^2} = \frac{ds}{dt}$$

The direction of $\mathbf{v}$ at P is along the tangent line to the curve at P, as shown in Fig. 40-1. If τ denotes the direction of $\mathbf{v}$ (the angle between $\mathbf{v}$ and the positive x axis), then $\tan \tau = v_y/v_x$, with the quadrant being determined by $v_x = |\mathbf{v}| \cos \tau$ and $v_y = |\mathbf{v}| \sin \tau$.

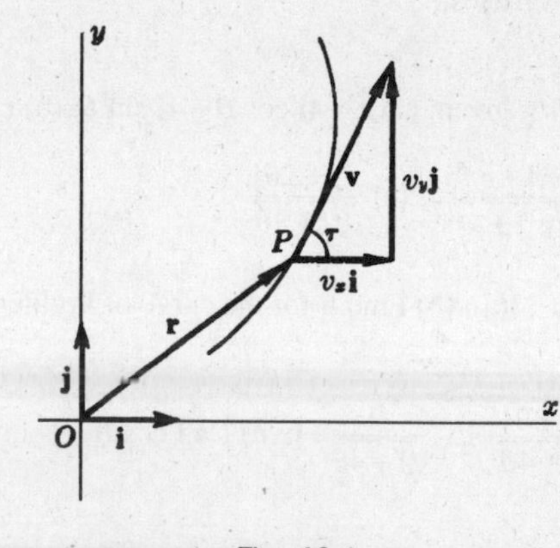

Fig. 40-1

Acceleration in Curvilinear Motion

Differentiating (40.2) with respect to t, we obtain the *acceleration vector*

$$\mathbf{a} = \frac{d\mathbf{v}}{dt} = \frac{d^2\mathbf{r}}{dt^2} = \frac{d^2x}{dt^2}\mathbf{i} + \frac{d^2y}{dt^2}\mathbf{j} = a_x\mathbf{i} + a_y\mathbf{j} \tag{40.3}$$

where $a_x = \dfrac{d^2x}{dt^2}$ and $a_y = \dfrac{d^2y}{dt^2}$. The magnitude of **a** is given by

$$|\mathbf{a}| = \sqrt{\mathbf{a}\cdot\mathbf{a}} = \sqrt{a_x^2 + a_y^2}$$

The direction ϕ of **a** is given by $\tan\phi = a_y/a_x$, with the quadrant being determined by $a_x = |\mathbf{a}|\cos\phi$ and $a_y = |\mathbf{a}|\sin\phi$. (See Fig. 40-2.)

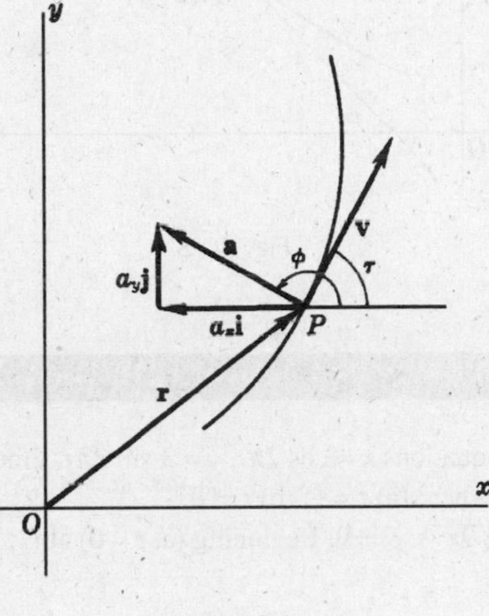

Fig. 40-2

Tangential and Normal Components of Acceleration

By equation (39.7),

$$\mathbf{v} = \frac{d\mathbf{r}}{dt} = \frac{d\mathbf{r}}{ds}\frac{ds}{dt} = \mathbf{t}\frac{ds}{dt} \tag{40.4}$$

Then

$$\mathbf{a} = \frac{d\mathbf{v}}{dt} = \mathbf{t}\frac{d^2s}{dt^2} + \frac{d\mathbf{t}}{dt}\frac{ds}{dt} = \mathbf{t}\frac{d^2s}{dt^2} + \frac{d\mathbf{t}}{ds}\left(\frac{ds}{dt}\right)^2$$

$$= \mathbf{t}\frac{d^2s}{dt^2} + |K|\,\mathbf{n}\left(\frac{ds}{dt}\right)^2 \tag{40.5}$$

by (39.8).

Equation (40.5) resolves the acceleration vector at P along the tangent and normal vectors there. Denoting the components by a_t and a_n, respectively, we have, for their magnitudes,

$$|a_t| = \left|\frac{d^2s}{dt^2}\right| \quad \text{and} \quad |a_n| = \frac{1}{R}\left(\frac{ds}{dt}\right)^2 = \frac{|\mathbf{v}|^2}{R}$$

where R is the radius of curvature of the curve at P. (See Fig. 40-3.)

Since $|\mathbf{a}|^2 = a_x^2 + a_y^2 = a_t^2 + a_n^2$, we obtain

$$a_n^2 = |\mathbf{a}|^2 - a_t^2$$

as a second way of determining $|a_n|$.

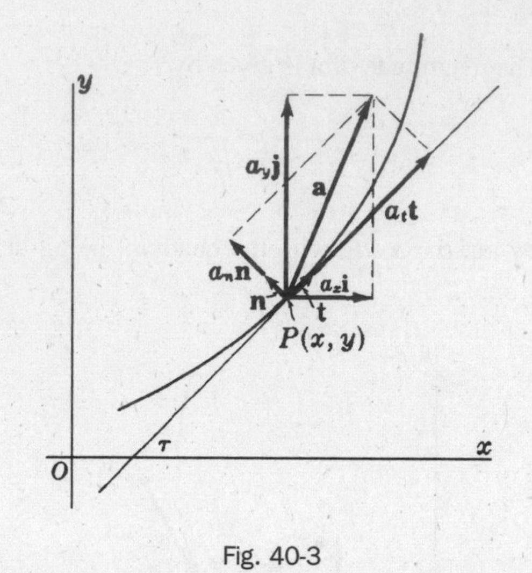

Fig. 40-3

SOLVED PROBLEMS

1. Discuss the motion given by the equations $x = \cos 2\pi t$, $y = 3 \sin 2\pi t$. Find the magnitude and direction of the velocity and acceleration vectors when: (a) $t = \frac{1}{6}$; (b) $t = \frac{2}{3}$.

The motion is along the ellipse $9x^2 + y^2 = 9$. Beginning (at $t = 0$) at (1, 0), the moving point traverses the curve counterclockwise.

$$\mathbf{r} = x\mathbf{i} + y\mathbf{j} = (\cos 2\pi t)\mathbf{i} + (3\sin 2\pi t)\mathbf{j}$$

$$\mathbf{v} = \frac{d\mathbf{r}}{dt} = v_x\mathbf{i} + v_y\mathbf{j} = -(2\pi \sin 2\pi t)\mathbf{i} + (6\pi \cos 2\pi t)\mathbf{j}$$

$$\mathbf{a} = \frac{d\mathbf{v}}{dt} = a_x\mathbf{i} + a_y\mathbf{j} = -(4\pi^2 \cos 2\pi t)\mathbf{i} - (12\pi^2 \sin 2\pi t)\mathbf{j}$$

(a) At $t = \frac{1}{6}$:

$$\mathbf{v} = -\sqrt{3}\pi\mathbf{i} + 3\pi\mathbf{j} \quad \text{and} \quad \mathbf{a} = -2\pi^2\mathbf{i} - 6\sqrt{3}\pi^2\mathbf{j}$$

$$|\mathbf{v}| = \sqrt{\mathbf{v}\cdot\mathbf{v}} = \sqrt{(-\sqrt{3}\pi)^2 + (3\pi)^2} = 2\sqrt{3}\pi$$

$$\tan\tau = \frac{v_y}{v_x} = -\sqrt{3}, \quad \cos\tau = \frac{v_x}{|\mathbf{v}|} = -\frac{1}{2}$$

So, $\tau = 120°$.

$$|\mathbf{a}| = \sqrt{\mathbf{a}\cdot\mathbf{a}} = \sqrt{(-2\pi^2)^2 + (-6\sqrt{3}\pi^2)^2} = 4\sqrt{7}\pi^2$$

$$\tan\phi = \frac{a_y}{a_x} = 3\sqrt{3}, \quad \cos\phi = \frac{a_x}{|\mathbf{a}|} = -\frac{1}{2\sqrt{7}}$$

So, $\phi = 259° 6'$.

(b) At $t = \frac{2}{3}$: $\quad \mathbf{v} = \sqrt{3}\pi\mathbf{i} - 3\pi\mathbf{j}$ and $\mathbf{a} = 2\pi^2\mathbf{i} + 6\sqrt{3}\pi^2\mathbf{j}$

$$|\mathbf{v}| = 2\sqrt{3}\pi, \quad \tan\tau = -\sqrt{3}\cos\tau = \tfrac{1}{2}$$

So, $\tau = \frac{5\pi}{3}$.

$$|\mathbf{a}| = 4\sqrt{7}\pi^2, \quad \tan\phi = 3\sqrt{3}, \quad \cos\phi = \frac{1}{2\sqrt{7}}$$

So, $\phi = 79° \ 6'$.

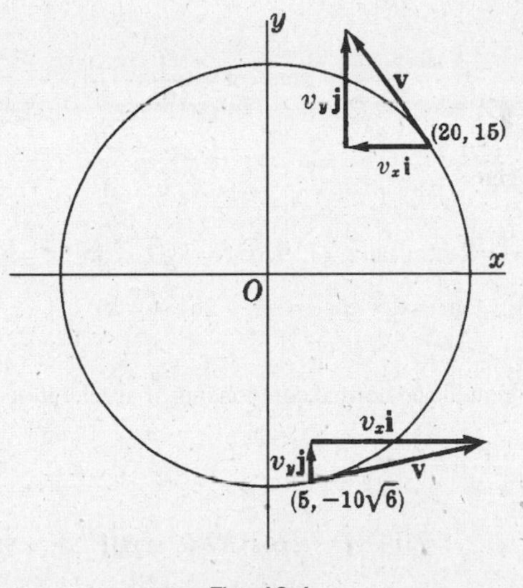

Fig. 40-4

2. A point travels counterclockwise about the circle $x^2 + y^2 = 625$ at a rate $|\mathbf{v}| = 15$. Find τ, $|\mathbf{a}|$, and ϕ at (a) the point (20, 15) and (b) the point $(5, -10\sqrt{6})$. Refer to Fig. 40-4.

Using the parametric equations $x = 25 \cos \theta$, $y = 25 \sin \theta$, we have at $P(x, y)$:

$$\mathbf{r} = (25 \cos \theta)\mathbf{i} + (25 \sin \theta)\mathbf{j}$$

$$\mathbf{v} = \frac{d\mathbf{r}}{dt} = [(-25 \sin \theta)\mathbf{i} + (25 \cos \theta)\mathbf{j}]\frac{d\theta}{dt}$$

$$= (-15 \sin \theta)\mathbf{i} + 15 \cos \theta)\mathbf{j}$$

$$\mathbf{a} = \frac{d\mathbf{v}}{dt} = [(-15 \cos \theta)\mathbf{i} - (15 \sin \theta)\mathbf{j}]\frac{d\theta}{dt}$$

$$= (-9 \cos \theta)\mathbf{i} - (9 \sin \theta)\mathbf{j}$$

since $|\mathbf{v}| = 15$ is equivalent to a constant angular speed of $\frac{d\theta}{dt} = \frac{3}{5}$.

(a) At the point (20, 15), $\sin\theta = \frac{3}{5}$ and $\cos\theta = \frac{4}{5}$. Thus,

$$\mathbf{v} = -9\mathbf{i} + 12\mathbf{j}, \quad \tan\tau = -\tfrac{4}{3}, \quad \cos\tau = -\tfrac{3}{5}. \quad \text{So} \quad \tau = 126° \ 52'$$

$$\mathbf{a} = -\tfrac{36}{5}\mathbf{i} - \tfrac{27}{3}\mathbf{j}, \quad |\mathbf{a}| = 9, \quad \tan\phi = \tfrac{3}{4}, \quad \cos\phi = -\tfrac{4}{5}. \quad \text{So} \quad \phi = 216° \ 52'$$

(b) At the point $(5, -10\sqrt{6})$, $\sin\theta = -\frac{2}{5}\sqrt{6}$ and $\cos\theta = \frac{1}{5}$. Thus,

$$\mathbf{v} = 6\sqrt{6}\mathbf{i} + 3\mathbf{j}, \quad \tan\tau = \sqrt{6}/12, \quad \cos\tau = \tfrac{2}{5}\sqrt{6}. \quad \text{So} \quad \tau = 11° \ 32'$$

$$\mathbf{a} = -\tfrac{9}{5}\mathbf{i} + \tfrac{18}{5}\sqrt{6}\mathbf{j}, \quad |\mathbf{a}| = 9, \quad \tan\phi = -2\sqrt{6}, \quad \cos\phi = -\tfrac{1}{5}. \quad \text{So} \quad \phi = 101° \ 32'$$

3. A particle moves on the first-quadrant arc of $x^2 = 8y$ so that $v_y = 2$. Find $|\mathbf{v}|$, τ, $|\mathbf{a}|$, and ϕ at the point (4, 2).

Using the parametric equations $x = 4\theta$, $y = 2\theta^2$, we have

$$\mathbf{r} = 4\theta\mathbf{i} + 2\theta^2\mathbf{j} \quad \text{and} \quad \mathbf{v} = 4\frac{d\theta}{dt}\mathbf{i} + 4\theta\frac{d\theta}{dt}\mathbf{j}$$

Since, $v_y = 4\theta\frac{d\theta}{dt} = 2$ and $\frac{d\theta}{dt} = \frac{1}{2\theta}$, we have

$$\mathbf{v} = \frac{2}{\theta}\mathbf{i} + 2\mathbf{j} \quad \text{and} \quad \mathbf{a} = -\frac{1}{\theta^3}\mathbf{i}$$

At the point (4, 2), $\theta = 1$. Then

$$\mathbf{v} = 2\mathbf{i} + 2\mathbf{j}, \quad |\mathbf{v}| = 2\sqrt{2}, \quad \tan\tau = 1, \quad \cos\tau = \tfrac{1}{2}\sqrt{2}. \quad \text{So} \quad \tau = \tfrac{1}{4}\pi$$

$$\mathbf{a} = -\mathbf{i}, \quad |\mathbf{a}| = 1, \quad \tan\phi = 0, \quad \cos\phi = -1. \quad \text{So} \quad \phi = \pi$$

4. Find the magnitudes of the tangential and normal components of acceleration for the motion $x = e^t \cos t$, $y = e^t \sin t$ at any time t.

We have:

$$\mathbf{r} = x\mathbf{i} + y\mathbf{j} = (e^t \cos t)\mathbf{i} + (e^t \sin t)\mathbf{j}$$

$$\mathbf{v} = e^t(\cos t - \sin t)\mathbf{i} + e^t(\sin t + \cos t)\mathbf{j}$$

$$\mathbf{a} = -2e^t(\sin t)\mathbf{i} + 2e^t(\cos t)\mathbf{j}$$

Then $|\mathbf{a}| = 2e^t$. Also, $\frac{ds}{dt} = |\mathbf{v}| = \sqrt{2}e^t$ and $|a_t| = \left|\frac{d^2s}{dt^2}\right| = \sqrt{2}e^t$. Finally,

$$|a_n| = \sqrt{|\mathbf{a}|^2 - a_t^2} = \sqrt{2}e^t$$

5. A particle moves from left to right along the parabola $y = x^2$ with constant speed 5. Find the magnitude of the tangential and normal components of the acceleration at (1, 1).

Since the speed is constant, $|a_t| = \left|\frac{d^2s}{dt^2}\right| = 0$. At (1, 1), $y' = 2x = 2$ and $y'' = 2$. The radius of curvature at (1, 1) is then $R = \frac{(1+(y')^2)^{3/2}}{|y''|} = \frac{5\sqrt{5}}{2}$. Hence, $|a_n| = \frac{|\mathbf{v}|^2}{R} = 2\sqrt{5}$.

6. The centrifugal force F (in pounds) exerted by a moving particle of weight W (in pounds) at a point in its path is given by the equation $F = \frac{W}{g}|a_n|$. Find the centrifugal force exerted by a particle, weighing 5 pounds, at the ends of the major and minor axes as it traverses the elliptical path $x = 20 \cos t$, $y = 15 \sin t$, the measurements being in feet and seconds. Use $g = 32$ ft/sec^2.

We have:

$$\mathbf{r} = (20\cos t)\mathbf{i} + (15\sin t)\mathbf{j}$$

$$\mathbf{v} = (-20\sin t)\mathbf{i} + (15\cos t)\mathbf{j}$$

$$\mathbf{a} = -20(\cos t)\mathbf{i} - 15(\sin t)\mathbf{j}$$

Then

$$\frac{ds}{dt} = |\mathbf{v}| = \sqrt{400\sin^2 t + 225\cos^2 t} \quad \text{and} \quad \frac{d^2s}{dt^2} = \frac{175\sin t\cos t}{\sqrt{400\sin^2 t + 225\cos^2 t}}$$

At the ends of the major axis ($t = 0$ or $t = \pi$):

$$|\mathbf{a}| = 20, \quad |a_t| = \left|\frac{d^2s}{dt^2}\right| = 0, \quad |a_n| = \sqrt{20^2 - 0^2} = 20 \quad \text{and} \quad F = \tfrac{5}{32}(20) = \tfrac{25}{8}\text{ pounds}$$

At the ends of the minor axis $\left(t = \dfrac{\pi}{2} \text{ or } t = \dfrac{3\pi}{2}\right)$:

$$|\mathbf{a}| = 15, \quad |a_t| = 0, \quad |a_n| = 15 \qquad \text{and} \qquad F = \tfrac{5}{32}(15) = \tfrac{75}{32} \text{ pounds}$$

7. Assuming the equations of motion of a projectile to be $x = v_0 t \cos\psi$, $y = v_0 t \sin\psi - \frac{1}{2}gt^2$, where v_0 is the initial velocity, ψ is the angle of projection, $g = 32$ ft/sec², and x and y are measured in feet and t in seconds, find: (a) the equation of motion in rectangular coordinates; (b) the range; (c) the angle of projection for maximum range; and (d) the speed and direction of the projectile after 5 sec of flight if $v_0 = 500$ ft/sec and $\psi = 45°$. (See Fig. 40-5.)

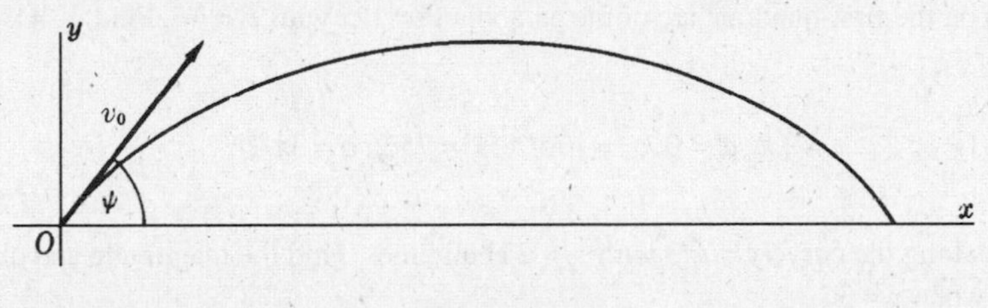

Fig. 40-5

(a) We solve the first of the equations for $t = \dfrac{x}{v_0 \cos\psi}$ and substitute in the second:

$$y = v_0 \frac{x}{v_0 \cos\psi} \sin\psi - \tfrac{1}{2} g \left(\frac{x}{v_0 \cos\psi}\right)^2 = x \tan\psi - \frac{gx^2}{2v_0^2 \cos^2\psi}$$

(b) Solving $y = v_0 t \sin\psi - \frac{1}{2}gt^2 = 0$ for t, we get $t = 0$ and $t = (2v_0 \sin\psi)/g$. For the latter, we have

$$\text{Range} = x = v_0 \cos\psi \frac{2v_0 \sin\psi}{g} = \frac{v_0^2 \sin 2\psi}{g}$$

(c) For x a maximum, $\dfrac{dx}{d\psi} = \dfrac{2v_0^2 \cos 2\psi}{g} = 0$; hence $\cos 2\psi = 0$ and $\psi = \frac{1}{4}\pi$.

(d) For $v_0 = 500$ and $\psi = \frac{1}{4}\pi$, $x = 250\sqrt{2}t$ and $y = 250\sqrt{2}t - 16t^2$. Then

$$v_x = 250\sqrt{2} \quad \text{and} \quad v_y = 250\sqrt{2} - 32t$$

When $t = 5$, $v_x = 250\sqrt{2}$ and $v_y = 250\sqrt{2} - 160$. Then

$$\tan\tau = \frac{v_y}{v_x} = 0.5475. \quad \text{So} \quad \tau = 28°\ 42', \quad \text{and} \quad |\mathbf{v}| = \sqrt{v_x^2 + v_y^2} = 403 \text{ ft/sec}$$

8. A point P moves on a circle $x = r\cos\beta$, $y = r\sin\beta$ with constant speed v. Show that, if the radius vector to P moves with angular velocity ω and angular acceleration α, (a) $v = r\omega$ and (b) $a = r\sqrt{\omega^4 + \alpha^2}$.

(a) $v_x = -r\sin\beta \dfrac{d\beta}{dt} = -r\omega \sin\beta \qquad \text{and} \qquad v_y = r\cos\beta \dfrac{d\beta}{dt} = r\omega\cos\beta$

Then $$v = \sqrt{v_x^2 + v_y^2} = \sqrt{(r^2\sin^2\beta + r^2\cos^2\beta)\omega^2} = r\omega$$

(b) $a_x = \dfrac{dv_x}{dt} = -r\omega\cos\beta \dfrac{d\beta}{dt} - r\sin\beta \dfrac{d\omega}{dt} = -r\omega^2\cos\beta - r\alpha\sin\beta$

$a_y = \dfrac{dv_y}{dt} = -r\omega\sin\beta \dfrac{d\beta}{dt} + r\cos\beta \dfrac{d\omega}{dt} = -r\omega^2\sin\beta + r\alpha\cos\beta$

Then $$a = \sqrt{a_x^2 + a_y^2} = \sqrt{r^2(\omega^4 + a^2)} = r\sqrt{\omega^4 + \alpha^2}$$

SUPPLEMENTARY PROBLEMS

9. Find the magnitude and direction of velocity and acceleration at time t, given

(a) $x = e^t, y = e^{2t} - 4e^t + 3$; at $t = 0$ *Ans.* (a) $|\mathbf{v}| = \sqrt{5}, \tau = 296°34'$; $|\mathbf{a}| = 1, \phi = 0$

(b) $x = 2 - t, y = 2t^3 - t$; at $t = 1$ *Ans.* (b) $|\mathbf{v}| = \sqrt{26}, \tau = 101°19'$; $|\mathbf{a}| = 12, \phi = \frac{1}{2}\pi$

(c) $x = \cos 3t, y = \sin t$; at $t = \frac{1}{4}\pi$ *Ans.* (c) $|\mathbf{v}| = \sqrt{5}, \tau = 161°34'$; $|\mathbf{a}| = \sqrt{41}, \phi = 353°\ 40'$

(d) $x = e^t \cos t, y = e^t \sin t$; at $t = 0$ *Ans.* (d) $|\mathbf{v}| = \sqrt{2}, \tau = \frac{1}{4}\pi$; $|\mathbf{a}| = 2, \phi = \frac{1}{2}\pi$

10. A particle moves on the first-quadrant arc of the parabola $y^2 = 12x$ with $v_x = 15$. Find v_y, $|\mathbf{v}|$, and τ; and a_x, a_y, $|\mathbf{a}|$, and ϕ at (3, 6).

Ans. $v_y = 15, |\mathbf{v}| = 15\sqrt{2}, \tau = \frac{1}{4}\pi$; $a_x = 0, a_y = -75/2, |\mathbf{a}| = 75/2, \phi = 3\pi/2$

11. A particle moves along the curve $y = x^3/3$ with $v_x = 2$ at all times. Find the magnitude and direction of the velocity and acceleration when $x = 3$.

Ans. $|\mathbf{v}| = 2\sqrt{82}, \tau = 83°40'$; $|\mathbf{a}| = 24, \phi = \frac{1}{2}\pi$

12. A particle moves around a circle of radius 6 ft at the constant speed of 4 ft/sec. Determine the magnitude of its acceleration at any position.

Ans. $|a_t| = 0$, $|\mathbf{a}| = |a_n| = 8/3$ ft/sec^2

13. Find the magnitude and direction of the velocity and acceleration, and the magnitudes of the tangential and normal components of acceleration at time t, for the motion:

(a) $x = 3t, y = 9t - 3t^2$; at $t = 2$
(b) $x = \cos t + t \sin t, y = \sin t - t \cos t$; at $t = 1$

Ans. (a) $|\mathbf{v}| = 3\sqrt{2}, \tau = 7\pi/4$; $|\mathbf{a}| = 6, \phi = 3\pi/2$; $|a_t| = |a_n| = 3\sqrt{2}$
(b) $|\mathbf{v}| = 1, \tau = 1$; $|\mathbf{a}| = \sqrt{2}, \phi = 102°\ 18'$; $|a_t| = |a_n| = 1$

14. A particle moves along the curve $y = \frac{1}{2}x^2 - \frac{1}{4}\ln x$ so that $x = \frac{1}{2}t^2$, for $t > 0$. Find v_x, v_y, $|\mathbf{v}|$, and τ; a_x, a_y, $|\mathbf{a}|$, and ϕ; $|a_t|$ and $|a_n|$ when $t = 1$.

Ans. $v_x = 1, v_y = 0, |\mathbf{v}| = 1, \tau = 0$; $a_x = 1, a_y = 2, |\mathbf{a}| = \sqrt{5}, \phi = 63°\ 26'$; $|a_t| = 1, |a_n| = 2$

15. A particle moves along the path $y = 2x - x^2$ with $v_x = 4$ at all times. Find the magnitudes of the tangential and normal components of acceleration at the position (a) (1, 1) and (b) (2, 0).

Ans. (a) $|a_t| = 0, |a_n| = 32$; (b) $|a_t| = 64/\sqrt{5}, |a_n| = 32\sqrt{5}$

16. If a particle moves on a circle according to the equations $x = r\cos\omega t, y = r\sin\omega t$, show that its speed is ωr.

17. Prove that if a particle moves with constant speed, then its velocity and acceleration vectors are perpendicular; and, conversely, prove that if its velocity and acceleration vectors are perpendicular, then its speed is constant.

Polar Coordinates

The position of a point P in a plane may be described by its coordinates (x, y) with respect to a given rectangular coordinate system. Its position may also be described by choosing a fixed point O and specifying the directed distance $\rho = OP$ and the angle θ that OP makes with a fixed half-line OX. (See Fig. 41-1.) This is the *polar coordinate system*. The point O is called the *pole*, and OX is called the *polar axis*.

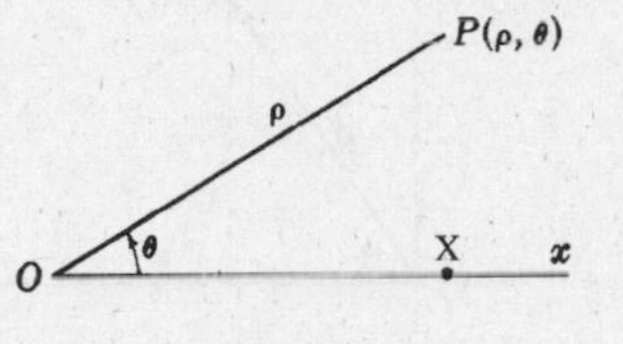

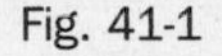
Fig. 41-1

To each number pair (ρ, θ) there corresponds one and only one point. The converse is not true. For example, $(1, 0)$ and $(1, 2\pi)$ describe the same point, on the polar axis and at a distance 1 from the pole. That same point also corresponds to $(-1, \pi)$. (When ρ is negative, the point corresponding to (ρ, θ) is obtained as follows: Rotate the polar axis OX through θ radians (counterclockwise if θ is positive and clockwise if θ is negative) to a new position OX' and then move $|\rho|$ units on the half-line opposite to OX'.)

In general, a point P with polar coordinates (ρ, θ) also can be described by $(\rho, \theta \pm 2n\pi)$ and $(-\rho, \theta \pm (2n+1)\pi)$, where n is any nonnegative integer. In addition, the pole itself corresponds to $(0, \theta)$, with arbitrary θ.

EXAMPLE 41.1: In Fig. 41-2, several points and their polar coordinates are shown. Note that point C has polar coordinates $\left(1, \frac{3\pi}{2}\right)$.

A *polar equation* of the form $\rho = f(\theta)$ or $F(\rho, \theta) = 0$ determines a curve, consisting of those points corresponding to pairs (ρ, θ) that satisfy the equation. For example, the equation $\rho = 2$ determines the circle with center at the pole and radius 2. The equation $\rho = -2$ determines the same set of points. In general, an equation $\rho = c$, where c is a constant, determines the circle with center at the pole and radius $|c|$. An equation $\theta = c$ determines the line through the pole obtained by rotating the polar axis through c radians. For example, $\theta = \pi/2$ is the line through the pole and perpendicular to the polar axis.

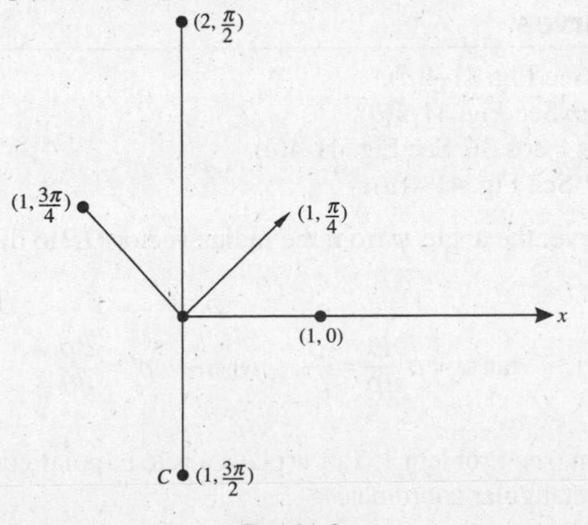

Fig. 41-2

Polar and Rectangular Coordinates

Given a pole and polar axis, set up a rectangular coordinate system by letting the polar axis be the positive x axis and letting the y axis be perpendicular to the x axis at the pole. (See Fig. 41-3.) Then the pole is the origin of the rectangular system. If a point P has rectangular coordinates (x, y) and polar coordinates (ρ, θ), then

$$x = \rho\cos\theta \quad \text{and} \quad y = \rho\sin\theta \tag{41.1}$$

These equations entail

$$\rho^2 = x^2 + y^2 \quad \text{and} \quad \tan\theta = \frac{y}{x} \tag{41.2}$$

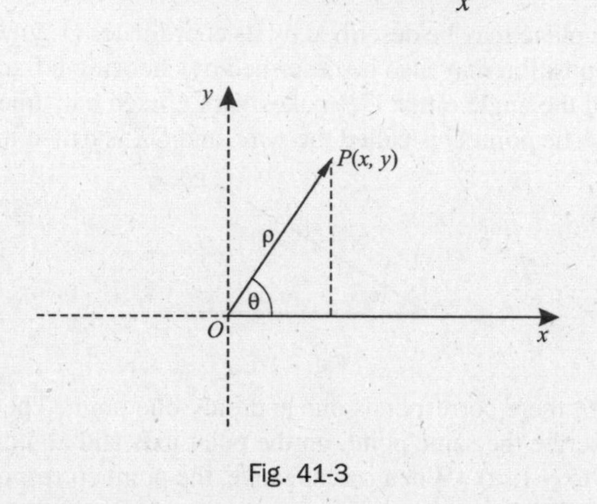

Fig. 41-3

EXAMPLE 41.2: Consider the polar curve $\rho = \cos\theta$.

Multiplying by ρ, we get $\rho^2 = \rho\cos\theta$. Hence, $x^2 + y^2 = x$ holds for the rectangular coordinates of points on the curve. That is equivalent to $x^2 - x + y^2 = 0$ and completion of the square with respect to x yields $(x - \frac{1}{2})^2 + y^2 = \frac{1}{4}$. Hence, the curve is the circle with center at $(\frac{1}{2}, 0)$ and radius $\frac{1}{2}$. Note that, as θ varies from 0 to $\pi/2$, the upper semicircle is traced out from (1, 0) to (0, 0), and then, as θ varies from $\frac{\pi}{2}$ to π, the lower semicircle is traced out from (0, 0) back to (1, 0). This whole path is retraced once more as θ varies from π to 2π. Since $\cos\theta$ has a period of 2π, we have completely described the curve.

EXAMPLE 41.3: Consider the parabola $y = x^2$. In polar coordinates, we get $\rho\sin\theta = \rho^2\cos^2\theta$, and, therefore, $\rho = \tan\theta\sec\theta$, which is a polar equation of the parabola.

Some Typical Polar Curves

(a) Cardioid: $\rho = 1 + \sin\theta$. See Fig. 41-4(*a*).
(b) Limaçon: $\rho = 1 + 2\cos\theta$. See Fig. 41-4(*b*).
(c) Rose with three petals: $\rho = \cos 3\theta$. See Fig. 41-4(*c*).
(d) Lemniscate: $\rho^2 = \cos 2\theta$. See Fig. 41-4(*d*).

At a point P on a polar curve, the angle ψ from the radius vector OP to the tangent PT to the curve (see Fig. 41-5) is given by

$$\tan\psi = \rho\frac{d\theta}{d\rho} = \frac{\rho}{\rho'}, \quad \text{where} \quad \rho' = \frac{d\rho}{d\theta} \tag{41.3}$$

For a proof of this equation, see Problem 1. Tan ψ plays a role in polar coordinates similar to that of the slope of the tangent line in rectangular coordinates.

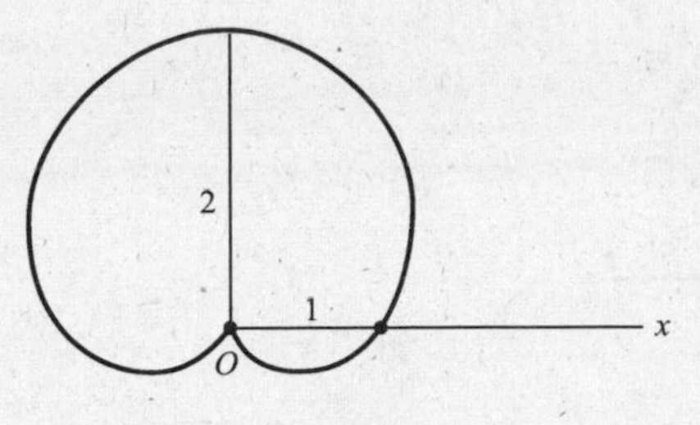

(a)

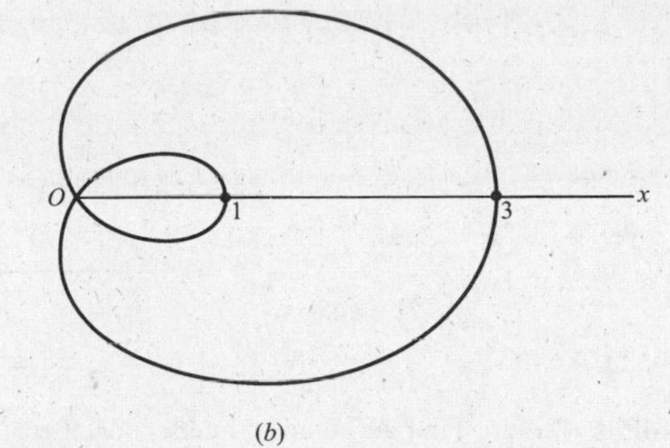

(b)

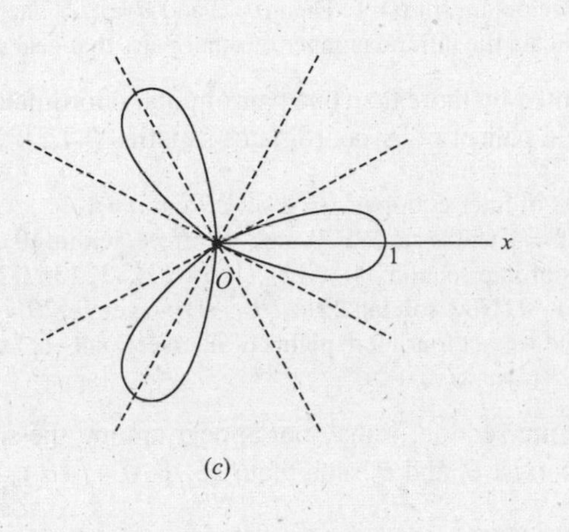

(c)

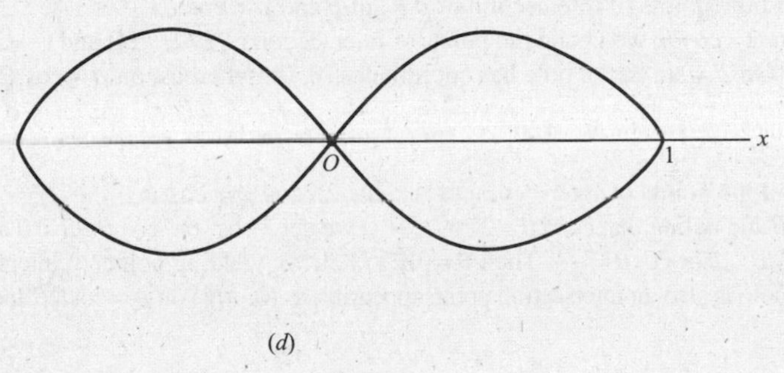

(d)

Fig. 41-4

Angle of Inclination

The angle of inclination τ of the tangent line to a curve at a point $P(\rho, \theta)$ on it (see Fig. 41-5) is given by

$$\tan \tau = \frac{\rho \cos \theta + \rho' \sin \theta}{-\rho \sin \theta + \rho' \cos \theta} \tag{41.4}$$

For a proof of this equation, see Problem 4.

Points of Intersection

Some or all of the points of intersection of two polar curves $\rho = f_1(\theta)$ and $\rho = f_2(\theta)$ (or equivalent equations) may be found by solving

$$f_1(\theta) = f_2(\theta) \tag{41.5}$$

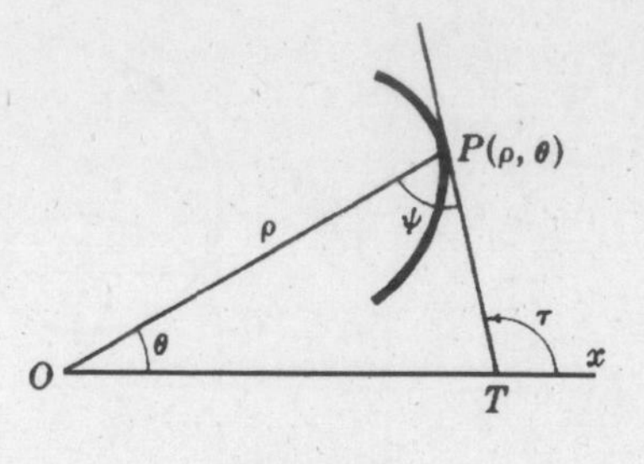

Fig. 41-5

EXAMPLE 41.4: Find the points of intersection of $\rho = 1 + \sin\theta$ and $\rho = 5 - 3\sin\theta$.
Setting $1 + \sin\theta = 5 - 3\sin\theta$, we obtain $\sin\theta = 1$. Then $\rho = 2$ and $\theta = \pi/2$. The only point of intersection is $(2, \pi/2)$. Note that we need not indicate the infinite number of other pairs that designate the same point.

Since a point may be represented by more than one pair of polar coordinates, the intersection of two curves may contain points that no single pair of polar coordinates satisfies (41.5).

EXAMPLE 41.5: Find the points of intersection of $\rho = 2\sin 2\theta$ and $\rho = 1$.
Solution of the equation $2\sin 2\theta = 1$ yields $\sin 2\theta = \frac{1}{2}$, and, therefore, within $[0, 2\pi)$, $\theta = \pi/12, 5\pi/12, 13\pi/12, 17\pi/12$. We have found four points of intersection: $(1, \pi/12)$, $(1, 5\pi/12)$, $(1, 13\pi/12)$, and $(1, 17\pi/12)$. But the circle $\rho = 1$ also can be represented as $\rho = -1$. Now solving $2\sin 2\theta = -1$, we get $\sin 2\theta = -\frac{1}{2}$ and, therefore, $\theta = 7\pi/12, 11\pi/12, 19\pi/12$, and $23\pi/12$. Hence we get four more points of intersection $(-1, 7\pi/12)$, $(-1, 11\pi/12)$, $(-1, 19\pi/12)$, and $(-1, 23\pi/12)$.

When the pole is a point of intersection, it may not appear among the solutions of (41.5). The pole is a point of intersection when there exist θ_1 and θ_2 such that $f_1(\theta_1) = 0 = f_2(\theta_2)$.

EXAMPLE 41.6: Find the points of intersection of $\rho = \sin\theta$ and $\rho = \cos\theta$.
From the equation $\sin\theta = \cos\theta$, we obtain the points of intersection $(\sqrt{2}/2, \pi/4)$ and $(-\sqrt{2}/2, 5\pi/4)$. However, both curves contain the pole. On $\rho = \sin\theta$, the pole has coordinates $(0, 0)$, whereas, on $\rho = \cos\theta$, the pole has coordinates $(0, \pi/2)$.

EXAMPLE 41.7: Find the points of intersection of $\rho = \cos 2\theta$ and $\rho = \cos\theta$.
Setting $\cos 2\theta = \cos\theta$ and noting that $\cos 2\theta = 2\cos^2\theta - 1$, we get $2\cos^2\theta - \cos\theta - 1 = 0$ and, therefore, $(\cos\theta - 1)(2\cos\theta + 1) = 0$. So, $\cos\theta = 1$ or $\cos\theta = -\frac{1}{2}$. Then $\theta = 0, 2\pi/3, 4\pi/3$, yielding points of intersection $(1, 0)$, $(-\frac{1}{2}, 2\pi/3)$, and $(-\frac{1}{2}, 4\pi/3)$. But the pole is also an intersection point, appearing as $(0, \pi/4)$ on $\rho = \cos 2\theta$ and as $(0, \pi/2)$ on $\rho = \cos\theta$.

Angle of Intersection

The angle of intersection, ϕ, of two curves at a common point $P(\rho, \theta)$, not the pole, is given by

$$\tan\phi = \frac{\tan\psi_1 - \tan\psi_2}{1 + \tan\psi_1 \tan\psi_2} \tag{41.6}$$

where ψ_1 and ψ_2 are the angles from the radius vector OP to the respective tangent lines to the curves at P. (See Fig. 41-6.) This formula follows from the trigonometric identity for $\tan(\psi_1 - \psi_2)$, since $\phi = \psi_1 - \psi_2$.

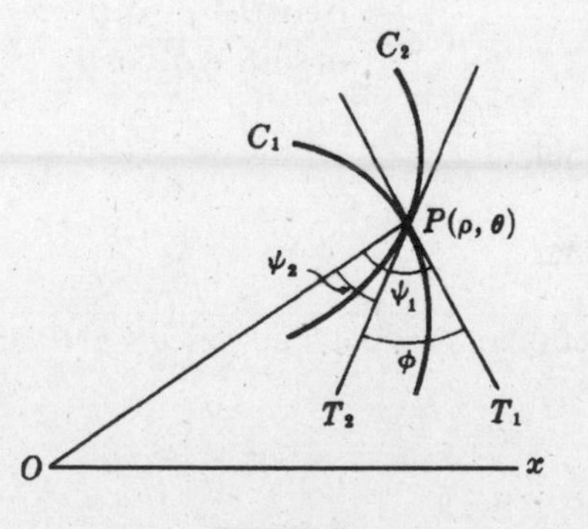

Fig. 41-6

EXAMPLE 41.8: Find the (acute) angles of intersection of $\rho = \cos 2\theta$ and $\rho = \cos\theta$.

The points of intersection were found in Example 7. We also need $\tan\psi_1$ and $\tan\psi_2$. For $\rho = \cos\theta$, formula (41.3) yields $\tan\psi_1 = -\cot\theta$. For $\rho = \cos 2\theta$, formula (41.3) yields $\tan\psi_2 = -\frac{1}{2}\cot 2\theta$.

At the point (1, 0), $\tan\psi_1 = -\cot 0 = \infty$ and, likewise, $\tan\psi_2 = \infty$. Then $\psi_1 = \psi_2 = \pi/2$ and, therefore, $\phi = 0$.

At the point $\left(-\frac{1}{2}, \frac{2\pi}{3}\right)$, $\tan\psi_1 = \sqrt{3}/3$ and $\tan\psi_2 = -\sqrt{3}/6$. So, by (41.6),

$$\tan\phi = \frac{(\sqrt{3}/3) + (\sqrt{3}/6)}{1 - (1/6)} = \frac{3\sqrt{3}}{5}$$

and, therefore, the acute angle of intersection $\phi \approx 46°\ 6'$. By symmetry, this is also the acute angle of intersection at the point $(-\frac{1}{2}, 4\pi/3)$.

At the pole, on $\rho = \cos\theta$, the pole is given by $\theta = \pi/2$. On $\rho = \cos 2\theta$, the pole is given by $\theta = \pi/4$ and $\theta = 3\pi/4$. Thus, at the pole there are two intersections, the acute angle being $\pi/4$ for each.

The Derivative of the Arc Length

The derivative of the arc length is given by

$$\frac{ds}{d\theta} = \sqrt{\rho^2 + (\rho')^2} \tag{41.7}$$

where $\rho' = \dfrac{d\rho}{d\theta}$ and it is understood that s increases with θ.

For a proof, see Problem 20.

Curvature

The curvature of a polar curve is given by

$$K = \frac{\rho^2 + 2(\rho')^2 - \rho\rho''}{[\rho^2 + (\rho')^2]^{3/2}} \tag{41.8}$$

For a proof, see Problem 17.

SOLVED PROBLEMS

1. Derive formula (41.3): $\tan\psi = \rho\dfrac{d\theta}{d\rho} = \dfrac{\rho}{\rho'}$, where $\rho' \dfrac{d\rho}{d\theta}$.

In Fig. 41-7, $Q(\rho + \Delta\rho, \theta + \Delta\theta)$ is a point on the curve near P. From the right triangle PSQ,

$$\tan\lambda = \frac{SP}{SQ} = \frac{SP}{OQ - OS} = \frac{\rho\sin\Delta\theta}{\rho + \Delta\rho - \rho\cos\Delta\theta} = \frac{\rho\sin\Delta\theta}{\rho(1 - \cos\Delta\theta) + \Delta\rho} = \frac{\rho\dfrac{\sin\Delta\theta}{\Delta\theta}}{\rho\dfrac{1 - \cos\Delta\theta}{\Delta\theta} + \dfrac{\Delta\rho}{\Delta\theta}}$$

Now as $Q \to P$ along the curve, $\Delta\theta \to 0$, $OQ \to OP$, $PQ \to PT$, and $\angle\lambda \to \angle\psi$.

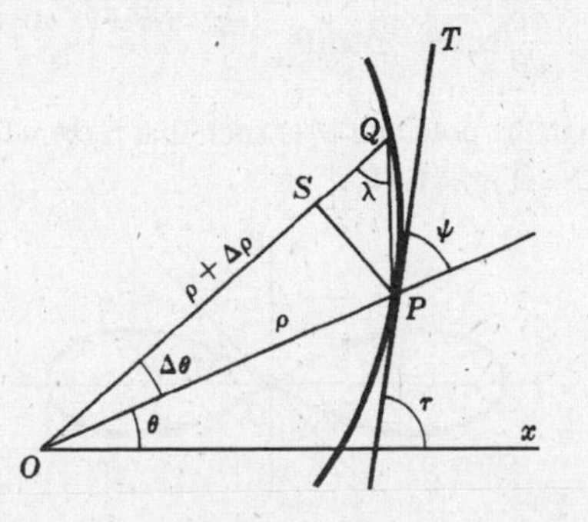

Fig. 41-7

As $\Delta\theta \to 0$, $\dfrac{\sin\Delta\theta}{\Delta\theta} \to 1$ and $\dfrac{1-\cos\Delta\theta}{\Delta\theta} \to 0$. Thus,

$$\tan\psi = \lim_{\Delta\theta\to 0} \tan\lambda = \frac{\rho}{d\rho/d\theta} = \rho\frac{d\theta}{d\rho}$$

In Problems 2 and 3, use formula (41.3) to find tan ψ for the given curve at the given point.

2. $\rho = 2+\cos\theta$ at $\theta = \frac{\pi}{3}$. (See Fig. 41-8.)

At $\theta = \frac{\pi}{3}$, $\rho = 2+\frac{1}{2} = \frac{5}{2}$, $\rho' = -\sin\theta = -\frac{\sqrt{3}}{2}$, and $\tan\psi = \frac{\rho}{\rho'} = -\frac{5}{\sqrt{3}}$.

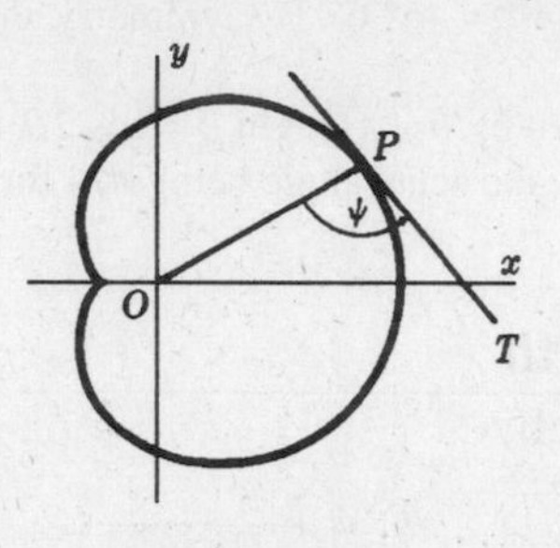

Fig. 41-8

3. $\rho = 2\sin 3\theta$ at $\theta = \frac{\pi}{4}$. (See Fig. 41-9.)

At $\theta = \frac{\pi}{4}$, $\rho = 2\frac{1}{\sqrt{2}} = \sqrt{2}$, $\rho' = 6\cos 3\theta = 6\left(-\frac{1}{\sqrt{2}}\right) = -3\sqrt{2}$ and $\tan\psi = \frac{\rho}{\rho'} = -\frac{1}{3}$.

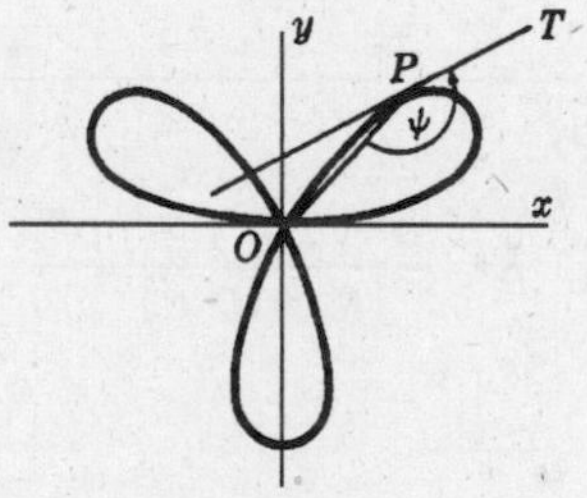

Fig. 41-9

4. Derive formula (41.4): $\tan\tau = \dfrac{\rho\cos\theta + \rho'\sin\theta}{-\rho\sin\theta + \rho\cos\theta}$.

From Fig. 41-7, $\tau = \psi + \theta$ and

$$\tan\tau = \tan(\psi+\theta) = \frac{\tan\psi + \tan\theta}{1-\tan\psi\tan\theta} = \frac{\rho\dfrac{d\theta}{d\rho} + \dfrac{\sin\theta}{\cos\theta}}{1-\rho\dfrac{d\theta}{d\rho}\dfrac{\sin\theta}{\cos\theta}}$$

$$= \frac{\rho\cos\theta + \dfrac{d\rho}{d\theta}\sin\theta}{\dfrac{d\rho}{d\theta}\cos\theta - \rho\sin\theta} = \frac{\rho\cos\theta + \rho'\sin\theta}{-\rho\sin\theta + \rho'\cos\theta}$$

5. Show that, if $\rho = f(\theta)$ passes through the pole and θ_1 is such that $f(\theta_1) = 0$, then the direction of the tangent line to the curve at the pole $(0, \theta_1)$ is θ_1. (See Fig. 41-10.)

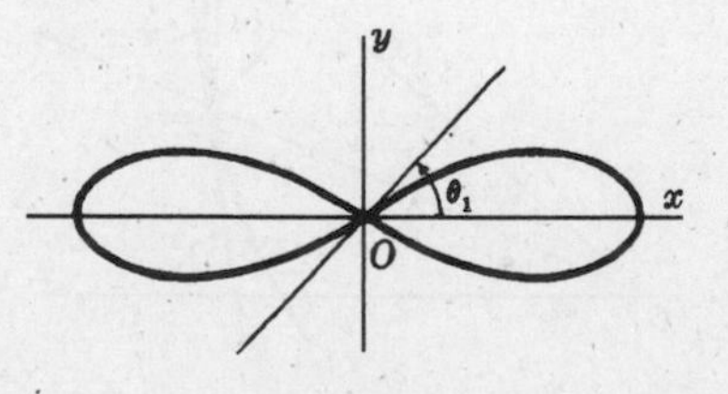

Fig. 41-10

At $(0, \theta_1)$, $\rho = 0$, and $\rho' = f'(\theta_1)$. If $\rho' \neq 0$

$$\tan\tau = \frac{\rho\cos\theta + \rho'\sin\theta}{-\rho\sin\theta + \rho'\cos\theta} = \frac{0 + f'(\theta_1)\sin\theta_1}{0 + f'(\theta_1)\cos\theta_1} = \tan\theta_1$$

If $\rho' = 0$,

$$\tan\tau = \lim_{\theta\to\theta_1}\frac{f'(\theta)\sin\theta}{f'(\theta)\cos\theta} = \tan\theta_1$$

In Problems 6–8, find the slope of the given curve at the given point.

6. $\rho = 1 - \cos\theta$ at $\theta = \frac{\pi}{2}$. (See Fig. 41-11.)

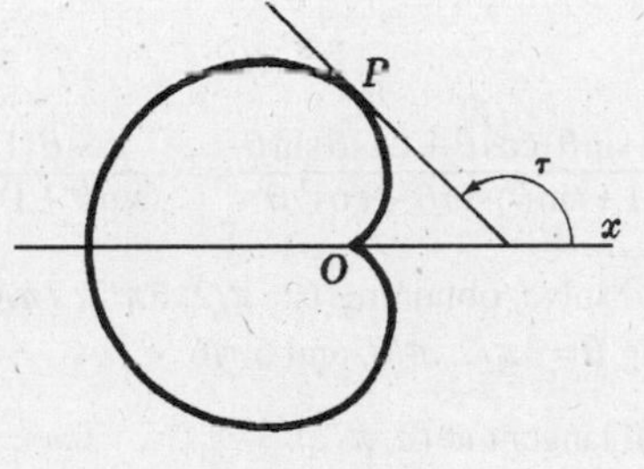

Fig. 41-11

At $\theta = \frac{\pi}{2}$,

$$\sin\theta = 1, \quad \cos\theta = 0, \quad \rho = 1, \quad \rho' = \sin\theta = 1$$

and

$$\tan\tau = \frac{\rho\cos\theta + \rho'\sin\theta}{-\rho\sin\theta + \rho'\cos\theta} = \frac{1\cdot 0 + 1\cdot 1}{-1\cdot 1 + 1\cdot 0} = -1$$

7. $\rho = \cos 3\theta$ at the pole. (See Fig. 41-12.)

When $\rho = 0$, $\cos 3\theta = 0$. Then $3\theta = \pi/2, 3\pi/2, 5\pi/2$, and $\theta = \pi/6, \pi/2, 5\pi/6$. By Problem 5, $\tan\tau = 1/\sqrt{3}, \infty$, and $-1\sqrt{3}$.

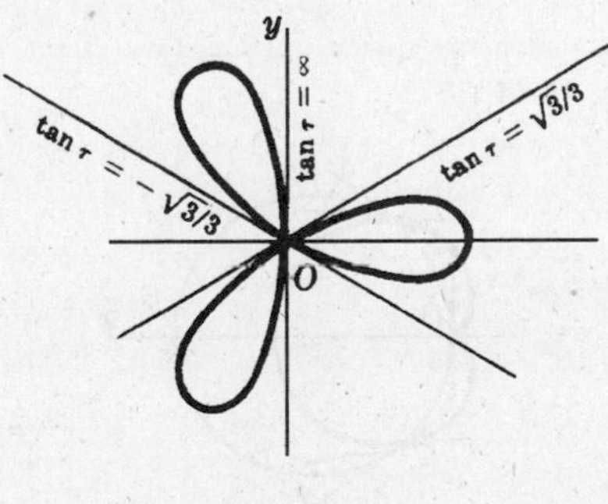

Fig. 41-12

8. $\rho\theta = a$ at $\theta = \frac{\pi}{3}$.

At $\theta = \pi/3$: $\sin\theta = \sqrt{3}/2$, $\cos\theta = \frac{1}{2}$, $\rho = 3a/\pi$, and $\rho' = -a/\theta^2 = -9a/\pi^2$. Then

$$\tan\tau = \frac{\rho\cos\theta + \rho'\sin\theta}{-\rho\sin\theta + \rho'\cos\theta} = -\frac{\pi - 3\sqrt{3}}{\sqrt{3}\pi + 3}$$

9. Investigate $\rho = 1 + \sin\theta$ for horizontal and vertical tangents. (See Fig. 41-13.)

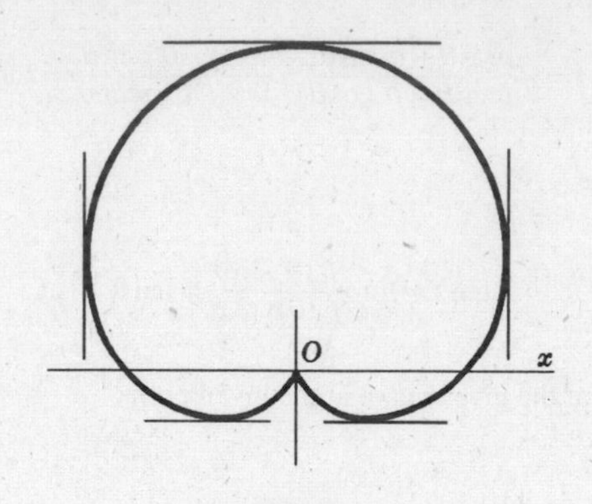

Fig. 41-13

At $P(\rho, \theta)$:

$$\tan\tau = \frac{(1+\sin\theta)\cos\theta + \cos\theta\sin\theta}{-(1+\sin\theta)\sin\theta + \cos^2\theta} = -\frac{\cos\theta(1+2\sin\theta)}{(\sin\theta+1)(2\sin\theta-1)}$$

We set $\cos\theta(1 + 2\sin\theta) = 0$ and solve, obtaining $\theta = \pi/2, 3\pi/2, 7\pi/6$, and $11\pi/6$. We also set $(\sin\theta + 1)(2\sin\theta - 1) = 0$ and solve, obtaining $\theta = 3\pi/2, \pi/6$, and $5\pi/6$.

For $\theta = \pi/2$: There is a horizontal tangent at $(2, \pi/2)$.
For $\theta = 7\pi/6$ and $11\pi/6$: There are horizontal tangents at $(\frac{1}{2}, 7\pi/6)$ and $(\frac{1}{2}, 11\pi/6)$.
For $\theta = \pi/6$ and $5\pi/6$: There are vertical tangents at $(\frac{3}{2}, \pi/6)$ and $(\frac{3}{2}, 5\pi/6)$.
For $\theta = 3\pi/2$: By Problem 5, there is a vertical tangent at the pole.

10. Show that the angle that the radius vector to any point of the cardioid $\rho = a(1 - \cos\theta)$ makes with the curve is one-half that which the radius vector makes with the polar axis.

At any point $P(\rho, \theta)$ on the cardioid,

$$\rho' = a\sin\theta \quad \text{and} \quad \tan\psi = \frac{\rho}{\rho'} = \frac{1-\cos\theta}{\sin\theta} = \tan\frac{\theta}{2}.$$

So $\psi = \frac{1}{2}\theta$.

In Problems 11–13, find the angles of intersection of the given pair of curves.

11. $\rho = 3\cos\theta$, $\rho = 1 + \cos\theta$. (See Fig. 41-14.)

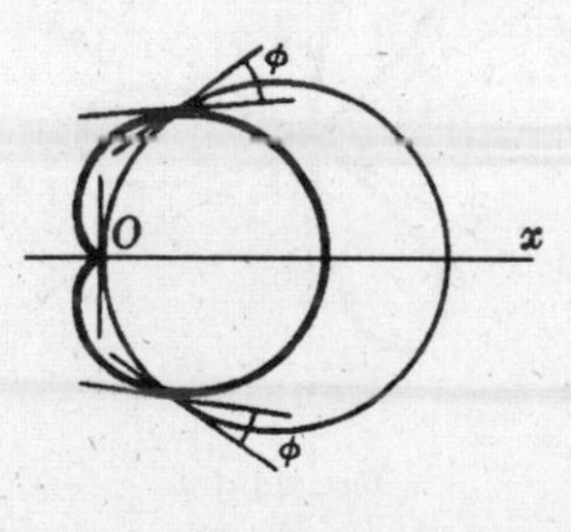

Fig. 41-14

Solve $3\cos\theta = 1 + \cos\theta$ for the points of intersection, obtaining $(3/2, \pi/3)$ and $(3/2, 5\pi/3)$. The curves also intersect at the pole.

For $\rho = 3\cos\theta$: $\quad \rho' = -3\sin\theta \quad$ and $\quad \tan\psi_1 = -\cot\theta$
For $\rho = 1 + \cos\theta$: $\quad \rho' = -\sin\theta \quad$ and $\quad \tan\psi_2 = -\dfrac{1+\cos\theta}{\sin\theta}$

At $\theta = \pi/3$, $\tan\psi_1 = -1\sqrt{3}$, $\tan\psi_2 = -\sqrt{3}$, and $\tan\phi = 1/\sqrt{3}$. The acute angle of intersection at $(\frac{3}{2}, \pi/3)$ and, by symmetry, at $(\frac{3}{2}, 5\pi/3)$ is $\pi/6$.

At the pole, either a diagram or the result of Problem 5 shows that the curves are orthogonal.

12. $\rho = \sec^2 \frac{1}{2}\theta$, $\rho = 3\csc^2 \frac{1}{2}\theta$.

Solve $\sec^2 \frac{1}{2}\theta = 3\csc^2 \frac{1}{2}\theta$ for the points of intersection, obtaining $(4, 2\pi/3)$ and $(4, 4\pi/3)$.

For $\rho = \sec^2 \frac{1}{2}\theta$: $\quad \rho' = \sec^2 \frac{1}{2}\theta \tan \frac{1}{2}\theta \quad$ and $\quad \tan\psi_1 = \cot \frac{1}{2}\theta$

For $\rho = 3\csc^2 \frac{1}{2}\theta$: $\quad \rho' = -3\csc^2 \frac{1}{2}\theta \cot \frac{1}{2}\theta \quad$ and $\quad \tan\psi_2 = -\tan \frac{1}{2}\theta$

At $\theta = 2\pi/3$, $\tan\psi_1 = 1/\sqrt{3}$, and $\tan\pi_2 = -\sqrt{3}$, and $\phi = \frac{1}{2}\pi$; the curves are orthogonal. Likewise, the curves are orthogonal at $\theta = 4\pi/3$.

13. $\rho = \sin 2\theta$, $\rho = \cos\theta$. (See Fig. 41-15.)

The curves intersect at the points $(\sqrt{3}/2, \pi/6)$ and $(-\sqrt{3}/2, 5\pi/6)$ and the pole.

For $\rho = \sin 2\theta$: $\quad \rho' = 2\cos 2\theta \quad$ and $\quad \tan\psi_1 = \frac{1}{2}\tan 2\theta$

For $\rho = \cos\theta$: $\quad \rho' = -\sin\theta \quad$ and $\quad \tan\psi_2 = -\cot\theta$

At $\theta = \pi/6$, $\tan\psi_1 = \sqrt{3}/2$, $\tan\psi_2 = -\sqrt{3}$, and $\tan\phi_1 = -3\sqrt{3}$. The acute angle of intersection at the point $(\sqrt{3}/2, \pi/6)$ is $\phi = \tan^{-1} 3\sqrt{3} = 79° 6'$. Similarly, at $\theta = 5\pi/6$, $\tan\psi_1 = -\sqrt{3}/2$, $\tan\psi_2 = \sqrt{3}$, and the angle of intersection is $\tan^{-1} 3\sqrt{3}$.

At the pole, the angles of intersection are 0 and $\pi/2$.

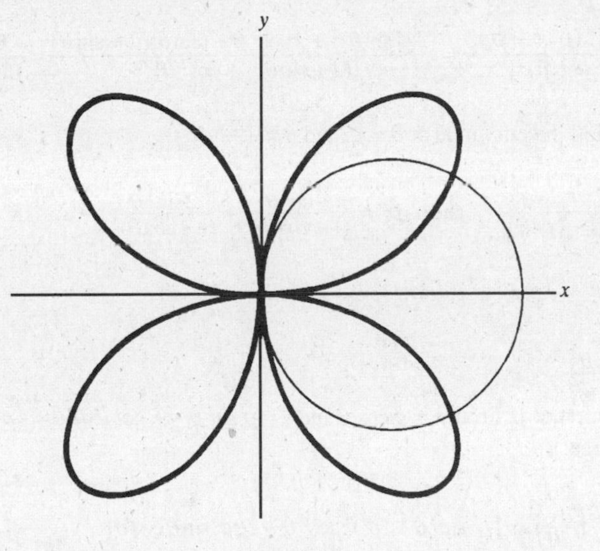

Fig. 41-15

In Problems 14–16, find $\frac{ds}{d\theta}$ at the point $P(\rho, \theta)$.

14. $\rho = \cos 2\theta$.

$$\rho' = -2\sin 2\theta \text{ and } \frac{ds}{d\theta} = \sqrt{\rho^2 + (\rho')^2} = \sqrt{\cos^2 2\theta + 4\sin^2 2\theta} = \sqrt{1 + 3\sin^2 2\theta}$$

15. $\rho(1 + \cos\theta) = 4$.

Differentiation yields $-\rho\sin\theta + \rho'(1 + \cos\theta) = 0$. Then

$$\rho' = \frac{\rho\sin\theta}{1+\cos\theta} = \frac{4\sin\theta}{(1+\cos\theta)^2} \quad \text{and} \quad \frac{ds}{d\theta} = \sqrt{\rho^2 + (\rho')^2} = \frac{4\sqrt{2}}{(1+\cos\theta)^{3/2}}$$

16. $\rho = \sin^3\left(\frac{\theta}{3}\right)$. (Also evaluate $\frac{ds}{d\theta}$ at $\theta = \frac{\pi}{2}$.)

$$\rho' = \sin^2 \tfrac{1}{3}\theta \cos \tfrac{1}{3}\theta \quad \text{and} \quad \frac{ds}{d\theta} = \sqrt{\sin^6 \tfrac{1}{3}\theta + \sin^4 \tfrac{1}{3}\theta \cos^2 \tfrac{1}{3}\theta} = \sin^2 \tfrac{1}{3}\theta$$

At $\theta = \frac{1}{2}\pi$, $ds/d\theta = \sin^2 \frac{1}{6}\pi = \frac{1}{4}$.

17. Derive formula (41.8): $K = \frac{\rho^2 + 2(\rho')^2 - \rho\rho''}{[\rho^2 + (\rho')^2]^{3/2}}$.

By definition, $K = \frac{d\tau}{ds}$. Now, $\tau = \theta + \psi$ and, therefore,

$$\frac{d\tau}{ds} = \frac{d\theta}{ds} + \frac{d\psi}{ds} = \frac{d\theta}{ds} + \frac{d\psi}{d\theta}\frac{d\theta}{ds} = \frac{d\theta}{ds}\left(1 + \frac{d\psi}{d\theta}\right)$$

where $\psi = \tan^{-1}\left(\frac{\rho}{\rho'}\right)$. Also,

$$\frac{d\psi}{d\theta} = \frac{[(\rho')^2 - \rho\rho'']/(\rho')^2}{1 + (\rho/\rho')^2} = \frac{(\rho')^2 - \rho\rho''}{\rho^2 + (\rho')^2}; \quad \text{so} \quad 1 + \frac{d\psi}{d\theta} = 1 + \frac{(\rho')^2 - \rho\rho''}{\rho^2 + (\rho')^2} = \frac{\rho^2 + 2(\rho')^2 - \rho\rho''}{\rho^2 + (\rho')^2}$$

Thus,

$$K = \frac{d\theta}{ds}\left(1 + \frac{d\psi}{d\theta}\right) = \frac{1 + d\psi/d\theta}{ds/d\theta} = \frac{1 + d\psi/d\theta}{\sqrt{\rho^2 + (\rho')^2}} = \frac{\rho^2 + 2(\rho')^2 - \rho\rho''}{[\rho^2 + (\rho')^2]^{3/2}}$$

18. Let $\rho = 2 + \sin\theta$. Find the curvature at the point $P(\rho, \theta)$

$$K = \frac{\rho^2 + 2(\rho')^2 - \rho\rho''}{[\rho^2 + (\rho')^2]^{3/2}} = \frac{(2 + \sin\theta)^2 + 2\cos^2\theta + (\sin\theta)(2 + \sin\theta)}{[(2 + \sin\theta)^2 + \cos^2\theta]^{3/2}} = \frac{6(1 + \sin\theta)}{(5 + 4\sin\theta)^{3/2}}$$

19. Let $\rho(1 - \cos\theta) = 1$. Find the curvature at $\theta = \frac{\pi}{2}$ and $\theta = \frac{4\pi}{3}$.

$$\rho' = \frac{-\sin\theta}{(1 - \cos\theta)^2} \quad \text{and} \quad \rho'' = \frac{-\cos\theta}{(1 - \cos\theta)^2} + \frac{2\sin^2\theta}{(1 - \cos\theta)^3}; \quad \text{so} \quad K = \sin^3\frac{\theta}{2}$$

At $\theta = \pi/2$, $K = (1/\sqrt{2})^3 = \sqrt{2}/4$; at $\theta = 4\pi/3$, $K = (\sqrt{3}/2)^3 = 3\sqrt{3}/8$.

20. Derive formula (41.7): $\frac{ds}{d\theta} = \sqrt{\rho^2 + (\rho')^2}$.

Consider ρ as a function of θ. From $x = \rho\cos\theta$ and $y = \rho\sin\theta$, we get $dx/d\theta = -\rho\sin\theta + (\cos\theta)\rho'$ and $dy/d\theta = \rho\cos\theta + (\sin\theta)\rho'$. Hence,

$$\left(\frac{dx}{d\theta}\right)^2 = [\rho^2\sin^2\theta + (\rho')^2\cos^2\theta - 2\rho\rho'\sin\theta\cos\theta]$$

and

$$\left(\frac{dy}{d\theta}\right)^2 = [\rho^2\cos^2\theta + (\rho')^2\sin^2\theta + 2\rho\rho'\sin\theta\cos\theta]$$

Thus,

$$\left(\frac{ds}{d\theta}\right)^2 = \left(\frac{dx}{d\theta}\right)^2 + \left(\frac{dy}{d\theta}\right)^2 = \rho^2 + (\rho')^2$$

Since s increases with θ, $\frac{ds}{d\theta} > 0$ and we obtain formula (41.7).

21. For $\rho = \cos 2\theta$, find $\frac{ds}{d\theta}$ at $\theta = \frac{\pi}{4}$. (Assume as usual that s increases with θ.)

$\rho' = \frac{d\rho}{d\theta} = -2\sin 2\theta$. By Formula (41.7),

$$\frac{ds}{d\theta} = \sqrt{\cos^2(2\theta) + 4\sin^2(2\theta)} = \sqrt{1 + 3\sin^2(2\theta)}$$

$$= \sqrt{1 + 3\sin^2(\pi/2)} = 2$$

SUPPLEMENTARY PROBLEMS

In Problems 22–25, find tan ψ for the given curve at the given points.

22. $\rho = 3 - \sin\theta$ at $\theta = 0$, $\theta = 3\pi/4$ *Ans.* -3; $3\sqrt{2} - 1$

23. $\rho = a(1 - \cos\theta)$ at $\theta = \pi/4$, $\theta = 3\pi/2$ *Ans.* $\sqrt{2} - 1$; -1

24. $\rho(1 - \cos\theta) = a$ at $\theta = \pi/3$, $\theta = 5\pi/4$ *Ans.* $-\sqrt{3}/3$; $1 + \sqrt{2}$

25. $\rho^2 = 4\sin 2\theta$ at $\theta = 5\pi/12$, $\theta = 2\pi/3$ *Ans.* $-1\sqrt{3}$; $\sqrt{3}$

In Problems 26–29, find tan τ for the given curve at the given point.

26. $\rho = 2 + \sin\theta$ at $\theta = \pi/6$ *Ans.* $-3\sqrt{3}$

27. $\rho^2 = 9\cos 2\theta$; at $\theta = \pi/6$ *Ans.* 0

28. $\rho = \sin^3(\theta/3)$ at $\theta = \pi/2$ *Ans.* $-\sqrt{3}$

29. $2\rho(1 - \sin\theta) = 3$ at $\theta = \pi/4$ *Ans.* $1 + \sqrt{2}$

30. Investigate $\rho = \sin 2\theta$ for horizontal and vertical tangents.

Ans. horizontal tangents at $\theta = 0, \pi$, 54°44′, 125°16′, 234°44′, 305°16′; vertical tangents at $\theta = \pi/2, 3\pi/2$, 35°16′, 144°44′, 215°16′, 324°44′

In Problems 31–33, find the acute angles of intersection of each pair of curves.

31. $\rho = \sin\theta$, $\rho = \sin 2\theta$ *Ans.* $\phi = 79°6'$ at $\theta = \pi/3$ and $5\pi/3$; $\phi = 0$ at the pole

32. $\rho = \sqrt{2}\sin\theta$, $\rho^2 = \cos 2\theta$ *Ans.* $\phi = \pi/3$ at $\theta = \pi/6, 5\pi/6$; $\phi = \pi/4$ at the pole

33. $\rho^2 = 16\sin 2\theta$, $\rho^2 = 4\csc 2\theta$ *Ans.* $\phi = \pi/3$ at each intersection

34. Show that each pair of curves intersects at right angles at all points of intersection.

(a) $\rho = 4\cos\theta$, $p = 4\sin\theta$ (b) $\rho = e^{\theta}$, $\rho = e^{-\theta}$
(c) $\rho^2\cos 2\theta = 4$, $\rho^2\sin 2\theta = 9$ (d) $\rho = 1 + \cos\theta$, $\rho = 1 - \cos\theta$

35. Find the angle of intersection of the tangents to $\rho = 2 - 4\sin\theta$ at the pole.

Ans. $2\pi/3$

36. Find the curvature of each of these curves at $P(\rho, \theta)$: (a) $\rho = e^{\theta}$; (b) $\rho = \sin\theta$; (c) $\rho^2 = 4\cos 2\theta$; (d) $\rho = 3\sin\theta + 4\cos\theta$.

Ans. (a) $1/(\sqrt{2}e^{\theta})$; (b) 2; (c) $\frac{3}{2}\sqrt{\cos 2\theta}$; (d) $\frac{2}{5}$

37. Find $\frac{ds}{d\theta}$ for the curve $\rho = a\cos\theta$.

Ans. a

38. Find $\frac{ds}{d\theta}$ for the curve $\rho = a(1 + \cos\theta)$.

Ans. $a\sqrt{2+2\cos\theta}$

39. Suppose a particle moves along a curve $\rho = f(\theta)$ with its position at any time t given by $\rho = g(t)$, $\theta = h(t)$.

(a) Multiply the equation $\left(\frac{ds}{d\theta}\right)^2 = \rho^2 + (\rho')^2$ obtained in Problem 20 by $\left(\frac{d\theta}{dt}\right)^2$ to obtain $v^2 = \left(\frac{ds}{dt}\right)^2 = \rho^2\left(\frac{d\theta}{dt}\right)^2 + \left(\frac{d\rho}{dt}\right)^2$.

(b) From $\tan\psi = \rho\frac{d\theta}{d\rho} = \rho\frac{d\theta/dt}{d\rho/dt}$, obtain $\sin\psi = \frac{\rho}{v}\frac{d\theta}{dt}$ and $\cos\psi = \frac{1}{v}\frac{d\rho}{dt}$.

In Problems 40–43, find all points of intersection of the given equations.

40. $\rho = 3\cos\theta$, $\rho = 3\sin\theta$ *Ans.* $(0, 0)$, $(3\sqrt{2}/2, \pi/4)$

41. $\rho = \cos\theta$, $\rho = 1 - \cos\theta$ *Ans.* $(0, 0)$, $(\frac{1}{2}, \pi/3)$, $(\frac{1}{2}, -\pi/3)$

42. $\rho = \theta$, $\rho = \pi$ *Ans.* (π, π), $(-\pi, -\pi)$

43. $\rho = \sin 2\theta$, $\rho = \cos 2\theta$ *Ans.* $(0, 0)$, $\left(\frac{\sqrt{2}}{2}, \frac{(2n+1)\pi}{6}\right)$ for $n = 0, 1, 2, 3, 4, 5$

44. (GC) Sketch the curves in Problems 40–43, find their graphs on a graphing calculator, and check your answers to Problems 40–43.

45. (GC) Sketch the graphs of the following equations and then check your answers on a graphing calculator:

(a) $\rho = 2\cos 4\theta$ (b) $\rho = 2\sin 5\theta$ (c) $\rho^2 = 4\sin 2\theta$
(d) $\rho = 2(1 - \cos\theta)$ (e) $\rho = \frac{2}{1+\cos\theta}$ (f) $\rho^2 = \frac{1}{\theta}$
(g) $\rho = 2 - \sec\theta$ (h) $\rho = \frac{2}{\theta}$
(In parts (*g*) and (*h*), look for asymptotes.)

46. Change the following rectangular equations to polar equations and sketch the graphs:

(a) $x^2 - 4x + y^2 = 0$ (b) $4x = y^2$ (c) $xy = 1$
(d) $x = a$ (e) $y = b$ (f) $y = mx + b$

Ans. (a) $\rho = 4\cos\theta$; (b) $\rho = 4\cot\theta\csc\theta$; (c) $\rho^2 = \sec\theta\csc\theta$; (d) $\rho = a\sec\theta$; (e) $\rho = b\csc\theta$; (f) $\rho = \frac{b}{\sin\theta - m\cos\theta}$

47. (GC) Change the following polar equations to rectangular coordinates and then sketch the graph. (Verify on a graphing calculator.) (a) $\rho = 2c \sin \theta$; (b) $\rho = \theta$; (c) $\rho = 7 \sec \theta$

Ans. (a) $x^2 + (y - c)^2 = c^2$; (b) $y = x \tan(\sqrt{x^2 + y^2})$; (c) $x = 7$

48. (a) Show that the distance between two points with polar coordinates (ρ_1, θ_1) and (ρ_2, θ_2) is

$$\sqrt{\rho_1^2 + \rho_2^2 - 2\rho_1\rho_2 \cos(\theta_1 - \theta_2)}$$

(b) When $\theta_1 = \theta_2$, what does the distance simplify to? Explain why this is so.

Ans. $|\rho_1 - \rho_2|$

(c) When $\theta_1 - \theta_2 = \frac{\pi}{2}$, what does the formula yield? Explain the signficance of the result.

Ans. $\sqrt{\rho_1^2 + \rho_2^2}$

(d) Find the distance between the points with the polar coordinates (1, 0) and $\left(1, \frac{\pi}{4}\right)$.

Ans. $\sqrt{2 - \sqrt{2}}$

49. (a) Let f be a continuous function such that $f(\theta) \geq 0$ for $\alpha < \theta < \beta$. Let A be the area of the region bounded by the lines $\theta = \alpha$ and $\theta = \beta$, and the polar curve $\rho = f(\theta)$. Derive the formula $A = \frac{1}{2}\int_\alpha^\beta (f(\theta))^2 d\theta = \frac{1}{2}\int_\alpha^\beta \rho^2 d\theta$. (*Hint:* Divide $[\alpha, \beta]$ into n equal parts, each equal to $\Delta\theta$. Each resulting subregion has area approximately equal to $\frac{1}{2}\Delta\theta(f(\theta_i^*))^2$, where θ_i^* is in the ith subinterval.)

(b) Find the area inside the cardioid $\rho = 1 + \sin \theta$.

(c) Find the area of one petal of the rose with three petals, $\rho = \cos 3\theta$. (*Hint*: Integrate from $-\frac{\pi}{6}$ to $\frac{\pi}{6}$.)

Infinite Sequences

Infinite Sequences

An infinite sequence $\langle s_n \rangle$ is a function whose domain is the set of positive integers; s_n is the value of this function for a given positive integer n. Sometimes we indicate $\langle s_n \rangle$ just by writing the first few terms of the sequence $s_1, s_2, s_3, \ldots, s_n, \ldots$. We shall consider only sequences where the values s_n are real numbers.

EXAMPLE 42.1:

(a) $\left\langle \frac{1}{n} \right\rangle$ is the sequence $1, \frac{1}{2}, \frac{1}{3}, \ldots, \frac{1}{n}, \ldots$.

(b) $\left\langle \left(\frac{1}{2}\right)^n \right\rangle$ is the sequence $\frac{1}{2}, \frac{1}{4}, \frac{1}{8}, \ldots, \frac{1}{2^n}, \ldots$.

(c) $\langle n^2 \rangle$ is the sequence of squares $1, 4, 9, 16, \ldots, n^2, \ldots$.

(d) $\langle 2n \rangle$ is the sequence of positive even integers $2, 4, 6, 8, \ldots, 2n, \ldots$.

(e) $\langle 2n-1 \rangle$ is the sequence of positive odd integers $1, 3, 5, 7, \ldots$.

Limit of a Sequence

If $\langle s_n \rangle$ is an infinite sequence and L is a number, then we say that $\lim_{n\to+\infty} s_n = L$ if s_n gets arbitrarily close to L as n increases without bound.

From a more precise standpoint, $\lim_{n\to+\infty} s_n = L$ means that, for any positive real number $\epsilon > 0$, there exists a positive integer n_0 such that, whenever $n \geq n_0$, we have $|s_n - L| < \epsilon$. To illustrate what this means, place the points L, $L-\epsilon$, and $L+\epsilon$ on a coordinate line (see Fig. 42-1), where ϵ is some positive real number. Now, if we place the points $s_1, s_2, s_3, \ldots$ on the coordinate line, there will eventually be an index n_0 such that $s_{n_0}, s_{n_0+1}, s_{n_0+2}, s_{n_0+3}, \ldots$ and all subsequent terms of the sequence will lie inside the interval $(L-\epsilon, L+\epsilon)$.

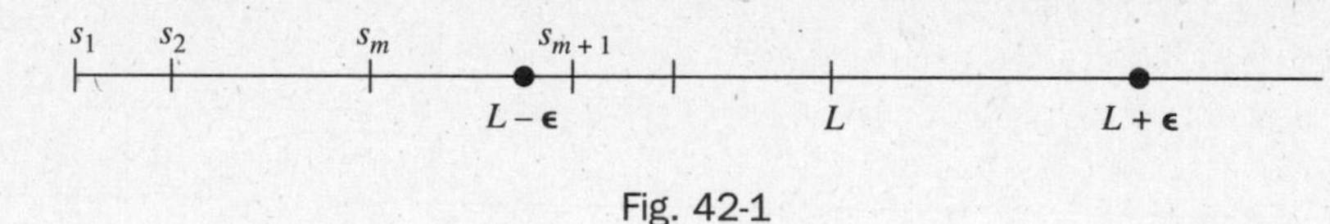

Fig. 42-1

If $\lim_{n\to+\infty} s_n = L$, then we say that the sequence $\langle s_n \rangle$ *converges* to L. If there is a number L such that $\langle s_n \rangle$ converges to L, then we say that $\langle s_n \rangle$ is *convergent*. When $\langle s_n \rangle$ is not convergent, then we say that $\langle s_n \rangle$ is *divergent*.

EXAMPLE 42.2: $\left\langle \frac{1}{n} \right\rangle$ is convergent, since $\lim_{n\to+\infty} \frac{1}{n} = 0$. To see this, observe that $1/n$ can be made arbitrarily close to 0 by making n large enough. To get an idea of why this is so, note that $1/10 = 0.1$, $1/100 = 0.01$, $1/1000 = 0.001$, and so on. To check that the precise definition is satisfied, let ϵ be any positive number. Take n_0 to be the smallest positive integer greater than $1/\epsilon$. So, $1/\epsilon < n_0$. Hence, if $n \geq n_0$, then $n > 1/\epsilon$ and, therefore, $1/n < \epsilon$. Thus, if $n \geq n_0$, $|1/n - 0| < \epsilon$. This proves $\lim_{n\to+\infty} \frac{1}{n} = 0$.

EXAMPLE 42.3: $\langle 2n \rangle$ is a divergent sequence, since $\lim_{n\to+\infty} 2n \neq L$ for each real number L. In fact, $2n$ gets arbitrarily large as n increases.

We write $\lim_{n\to+\infty} s_n = +\infty$ if s_n gets arbitrarily large as n increases. In such a case, we say that $\langle s_n \rangle$ diverges to $+\infty$. More precisely, $\lim_{n\to+\infty} s_n = +\infty$ if and only if, for any number c, no matter how large, there exists a positive integer n_0 such that, whenever $n \geq n_0$, we have $s_n > c$.

Likewise, we write $\lim_{n\to+\infty} s_n = -\infty$ if s_n gets arbitrarily small as n increases. In such a case, we say that $\langle s_n \rangle$ diverges to $-\infty$. More precisely, $\lim_{n\to+\infty} s_n = -\infty$ if and only if, for any number c, no matter how small, there exists a positive integer n_0 such that, whenever $n \geq n_0$, we have $s_n < c$.

We shall write $\lim_{n\to+\infty} s_n = \infty$ if $\lim_{n\to+\infty} |s_n| = +\infty$, that is, the magnitude of s_n gets arbitrarily large as n increases.

EXAMPLE 42.4: (a) $\lim_{n\to+\infty} 2n = +\infty$; (b) $\lim_{n\to+\infty} (1-n)^3 = -\infty$; (c) $\lim_{n\to+\infty} (-1)^n(n^2) = \infty$. Note that, in case (c), the sequence converges neither to $+\infty$ nor to $-\infty$.

EXAMPLE 42.5: The sequence $\langle(-1)^n\rangle$ is divergent, but it diverges neither to $+\infty$, nor to $-\infty$, nor to ∞. Its values oscillate between 1 and -1.

A sequence $\langle s_n \rangle$ is said to be *bounded above* if there is a number c such that $s_n \leq c$ for all n, and $\langle s_n \rangle$ is said to be *bounded below* if there is a number b such that $b \leq s_n$ for all n. A sequence $\langle s_n \rangle$ is said to be *bounded* if it is bounded both above and below. It is clear that a sequence $\langle s_n \rangle$ is bounded if and only if there is a number d such that $|s_n| \leq d$ for all n.

EXAMPLE 42.6: (a) The sequence $\langle 2n \rangle$ is bounded below (for example, by 0) but is not bounded above. (b) The sequence $\langle(-1)^n\rangle$ is bounded. Note that $\langle(-1)^n\rangle$ is $-1, 1, -1, \ldots$ So, $|(-1)^n| \leq 1$ for all n.

Theorem 42.1: Every convergent sequence is bounded.

For a proof, see Problem 5.

The converse of Theorem 42.1 is false. For example, the sequence $\langle(-1)^n\rangle$ is bounded but not convergent.

Standard arithmetic operations on convergent sequences yield convergent sequences, as the following intuitively obvious results show.

Theorem 42.2: Assume $\lim s_n = c$ and $\lim t_n = d$. Then:

(a) $\lim_{n\to+\infty} k = k$, where k is a constant.
(b) $\lim_{n\to+\infty} ks_n = kc$, where k is a constant.
(c) $\lim_{n\to+\infty} (s_n + t_n) = c + d$.
(d) $\lim_{n\to+\infty} (s_n - t_n) = c - d$.
(e) $\lim_{n\to+\infty} (s_n t_n) = cd$.
(f) $\lim_{n\to+\infty} (s_n / t_n) = c/d$ provided that $d \neq 0$ and $t_n \neq 0$ for all n.

For proofs of parts (c) and (e), see Problem 10.

The following facts about sequences are intuitively clear.

Theorem 42.3: If $\lim_{n\to+\infty} s_n = \infty$ and $s_n \neq 0$ for all n, then $\lim_{n\to+\infty} \frac{1}{s_n} = 0$.

For a proof, see Problem 7.

Theorem 42.4:

(a) If $|a| > 1$, then $\lim_{n\to+\infty} a^n = \infty$.

In particular, if $a > 1$, then $\lim_{n\to+\infty} a^n = +\infty$.

(b) If $|r| < 1$, then $\lim_{n\to+\infty} r^n = 0$.

For proofs, see Problem 8.

Theorem 42.5 (Squeeze Theorem): If $\lim_{n\to+\infty} s_n = L = \lim_{n\to+\infty} u_n$, and there is an integer m such that $s_n \leq t_n \leq u_n$ for all $n \geq m$, then $\lim_{n\to+\infty} t_n = L$.

For a proof, see Problem 11.

Corollary 42.6: If $\lim_{n\to+\infty} u_n = 0$ and there is an integer m such that $|t_n| \le |u_n|$ for all $n \ge m$, then $\lim_{n\to+\infty} t_n = 0$.

This is a consequence of Theorem 42.5 and the fact that $\lim_{n\to+\infty} a_n = 0$ is equivalent to $\lim_{n\to+\infty} |a_n| = 0$.

EXAMPLE 42.7: $\lim_{n\to+\infty} (-1)^n \frac{1}{n^2} = 0$. To see this, use Corollary 42.6, noting that $\left|(-1)^n \frac{1}{n^2}\right| \le \frac{1}{n}$ and $\lim_{n\to+\infty} \frac{1}{n} = 0$.

Theorem 42.7: Assume that f is a function that is continuous at c, and assume that $\lim_{n\to+\infty} s_n = c$, where all the terms s_n are in the domain of f. Then $\lim_{n\to+\infty} f(s_n) = f(c)$.

See Problem 33.

It is clear that whether or not a sequence converges would not be affected by deleting, adding, or altering a finite number of terms at the beginning of the sequence. Convergence depends on what happens "in the long run."

We shall extend the notion of infinite sequence to the case where the domain of a sequence is allowed to be the set of nonnegative integers or any set consisting of all integers greater than or equal to a fixed integer. For example, if we take the domain to be the set of nonnegative integers, then $\langle 2n+1 \rangle$ would denote the sequence of positive odd integers, and $\langle 1/2^n \rangle$ would denote the sequence $1, \frac{1}{2}, \frac{1}{4}, \frac{1}{8}, \ldots$.

Monotonic Sequences

(a) A sequence $\langle s_n \rangle$ is said to be *nondecreasing* if $s_n \le s_{n+1}$ for all n.
(b) A sequence $\langle s_n \rangle$ is said to be *increasing* if $s_n < s_{n+1}$ for all n.
(c) A sequence $\langle s_n \rangle$ is said to be *nonincreasing* if $s_n \ge s_{n+1}$ for all n.
(d) A sequence $\langle s_n \rangle$ is said to be *decreasing* if $s_n > s_{n+1}$ for all n.
(e) A sequence is said to be *monotonic* if it is either nondecreasing or nonincreasing.

Clearly, every increasing sequence is nondecreasing (but not conversely), and every decreasing sequence is nonincreasing (but not conversely).

EXAMPLE 42.8: (a) The sequence 1, 1, 2, 2, 3, 3, 4, 4, . . . is nondecreasing, but not increasing. (b) −1, −1, −2, −2, −3, −3, −4, −4, . . . is nonincreasing, but not decreasing.

An important basic property of the real number system is given by the following result. Its proof is beyond the scope of this book.

Theorem 42.8: Every bounded monotonic sequence is convergent.

There are several methods for showing that a given sequence $\langle s_n \rangle$ is nondecreasing, increasing, nonincreasing, or decreasing. Let us concentrate on the property that $\langle s_n \rangle$ is increasing.

Method 1: *Show that* $s_{n+1} - s_n > 0$.

EXAMPLE 42.9: Consider $s_n = \frac{3n}{4n+1}$. Then $s_{n+1} = \frac{3(n+1)}{4(n+1)+1} = \frac{3n+3}{4n+5}$. So,

$$s_{n+1} - s_n = \frac{3n+3}{4n+5} - \frac{3n}{4n+1} = \frac{(12n^2+15n+3)-(12n^2+15n)}{(4n+5)(4n+1)}$$

$$= \frac{3}{(4n+5)(4n+1)} > 0$$

since $4n+5>0$ and $4n+1>0$.

Method 2: *When all* $s_n > 0$, *show that* $s_{n+1}/s_n > 1$.

EXAMPLE 42.10: Using the same example $s_n = \frac{3n}{4n+1}$ as above,

$$\frac{s_{n+1}}{s_n} = \left(\frac{3n+3}{4n+5}\right)\bigg/\left(\frac{3n}{4n+1}\right) = \frac{3n+3}{3n}\,\frac{4n+1}{4n+5} = \frac{12n^2+15n+3}{12n^2+15n} > 1,$$

since $12n^2 + 15n + 3 > 12n^2 + 15n > 0$.

Method 3: Find a differentiable function $f(x)$ such that $f(n) = s_n$ for all n, and show that $f'(x) > 0$ for all $x \geq 1$ (and, hence, that f is an increasing function for $x \geq 1$).

EXAMPLE 42.11: Consider $s_n = \frac{3n}{4n+1}$ again. Let $f(x) = \frac{3x}{4x+1}$. Then $f'(x) = \frac{3}{(4x+1)^2} > 0$ for all x.

SOLVED PROBLEMS

1. For each of the following sequences, write a formula for the nth term and determine the limit (if it exists). It is assumed that $n = 1, 2, 3, \ldots$.

(a) $\frac{1}{2}, \frac{1}{4}, \frac{1}{6}, \frac{1}{8}, \ldots$ (b) $\frac{1}{2}, \frac{2}{3}, \frac{3}{4}, \frac{4}{5}, \ldots$

(c) $1, -\frac{1}{2}, \frac{1}{3}, -\frac{1}{4}, \frac{1}{5}, -\frac{1}{6}, \ldots$ (d) $0.9, 0.99, 0.999, 0.9999, \ldots$

(e) $\sin\frac{\pi}{2}, \sin\pi, \sin\frac{3\pi}{2}, \sin 2\pi, \sin\frac{5\pi}{2}, \ldots$ (f) $\frac{2}{1}, \left(\frac{3}{2}\right)^2, \left(\frac{4}{3}\right)^3, \left(\frac{5}{4}\right)^4, \ldots$

(a) $s_n = \frac{1}{2_n}$; $\lim_{n\to+\infty} \frac{1}{2n} = 0$.

(b) $s_n = \frac{n}{n+1}$; $\lim_{n\to+\infty} \frac{n}{n+1} = \lim_{n\to+\infty}\left(1 - \frac{1}{n+1}\right) = 1 - \lim_{n\to+\infty} \frac{1}{n+1} = 1 - 0 = 1$.

(c) $s_n = \frac{(-1)^{n+1}}{n}$; $\lim_{n\to+\infty} \frac{(-1)^{n+1}}{n} = 0$. This is intuitively clear, but one can also apply Theorem 42.3 to the sequence $\langle(-1)^{n+1}n\rangle$, since $\lim_{n\to+\infty} (-1)^{n+1}n = \infty$.

(d) $s_n = 1 - \frac{1}{10^n}$; $\lim_{n\to+\infty}\left(1 - \frac{1}{10^n}\right) = 1 - \lim_{n\to+\infty} \frac{1}{10^n} = 1 - 0 = 1$.

Note that $\lim_{n\to+\infty} \frac{1}{10^n} = 0$ by virtue of Theorem 42.4(*b*).

(e) $s_n = \sin\frac{n\pi}{2}$. Note that the sequence consists of repetitions of the cycle 1, 0, −1, 0 and has no limit.

(f) $s_n = \left(\frac{n+1}{n}\right)^n$; $\lim_{n\to+\infty}\left(\frac{n+1}{n}\right)^n = \lim_{n\to+\infty}\left(1 + \frac{1}{n}\right)^n = e$ by (26.17).

2. Evaluate $\lim_{n\to+\infty} s_n$ in the following cases:

(a) $s_n = \frac{5n^2 - 4n + 13}{3n^2 - 95n - 7}$ (b) $s_n = \frac{8n^2 - 3}{2n + 5}$ (c) $\frac{3n+7}{n^3 - 2n - 9}$

(a) Recall that $\lim_{x\to+\infty} \frac{5x^2 - 4x + 13}{3x^2 - 95x - 7} = \frac{5}{3}$ by Chapter 7, Problem 13. Therefore, $\lim_{n\to+\infty} \frac{5n^2 - 4n + 13}{3n^2 - 95n - 7} = \frac{5}{3}$. A similar result holds whenever s_n is a quotient of polynomials of the same degree.

(b) Recall that $\lim_{x\to+\infty} \frac{8x^2 - 3}{2x + 5} = +\infty$ by Chapter 7, Problem 13. Therefore, $\lim_{n\to+\infty} \frac{8n^2 - 3}{2n + 5} = +\infty$. A similar result holds whenever s_n is a rational function whose numerator has greater degree than the denominator (and whose leading coefficients have the same sign).

(c) Recall that $\lim_{x\to+\infty} \frac{3x+7}{x^3 - 2x - 9} = 0$ by Chapter 7, Problem 13. Therefore, $\lim_{n\to+\infty} \frac{3n+7}{n^3 - 2n - 9} = 0$. The same result holds whenever s_n is a rational function whose denominator has greater degree than the numerator.

3. For each of the following sequencies, determine whether it is nondecreasing, increasing, nonincreasing, decreasing, or none of these. Then determine its limit, if it exists.

(a) $s_n = \frac{5n-2}{7n+3}$ (b) $s_n = \frac{n}{2^n}$ (c) $\frac{1}{3^n}$

(a) Let $f(x) = \frac{5x-2}{7x+3}$. Then $f'(x) = \frac{(7x+3)(5) - (5x-2)(7)}{(7x+3)^2} = \frac{29}{(7x+3)^2} > 0$.

Hence, $f(x)$ is an increasing function and, therefore, $\langle s_n\rangle$ is an increasing sequence.

(b) Let $f(x) = \frac{x}{2^x}$. Then $f'(x) = \frac{2^x - x(\ln x)2^x}{2^{2x}} = \frac{1 - x(\ln 2)}{2^x}$.

Since $\ln 2 > \frac{1}{2}$ (by (*25.12*)), $x(\ln 2) > x/2 \geq 1$, when $x \geq 2$. Thus, $1 - x(\ln 2) < 0$ when $x \geq 2$ and, therefore, $f'(x) < 0$ when $x \geq 2$. So, $f(x)$ is decreasing for $x \geq 2$ and this implies that s_n is decreasing for $n \geq 2$. Note that $s_1 = \frac{1}{2} = s_2$. Hence, $\langle s_n \rangle$ is nonincreasing. Now let us find the limit. By L'Hôpital's Rule,

$$\lim_{x\to+\infty} \frac{x}{2^x} = \lim_{x\to+\infty} \frac{1}{(\ln 2)2^x} = 0 \text{ and, therefore, } \lim_{n\to+\infty} \frac{n}{2^n} = 0$$

(c) $\dfrac{s_{n+1}}{s_n} = \left(\dfrac{1}{3^{n+1}}\right) \Big/ \left(\dfrac{1}{3^n}\right) = \dfrac{1}{3} < 1$. Hence, $\langle s_n \rangle$ is decreasing.

Theorem 42.4(*b*) tells us that $\lim_{n\to+\infty} \dfrac{1}{3^n} = \lim_{n\to+\infty} \left(\dfrac{1}{3}\right)^n = 0$.

4. Show that the sequence $s_n = \dfrac{1\cdot 3\cdot 5\cdot 7\cdots(2n-1)}{2\cdot 4\cdot 6\cdot 8\cdots(2n)}$ is convergent.

Let us use Theorem 42.8. $\langle s_n \rangle$ is bounded, since $0 < s_n < 1$. Let us show that $\langle s_n \rangle$ is decreasing. Note that

$$s_{n+1} = \frac{1\cdot 3\cdot 5\cdot 7\cdots(2n+1)}{2\cdot 4\cdot 6\cdot 8\cdots(2n+2)} = s_n \frac{2n+1}{2n+2} < s_n$$

5. Prove Theorem 42.1: Every convergent sequence $\langle s_n \rangle$ is bounded.

Let $\lim_{n\to+\infty} s_n = L$. Take $\epsilon = 1$. Then there exists a positive integer n_0 such that, whenever $n \geq n_0$, we have $|s_n - L| < 1$. Hence, for $n \geq n_0$, using the triangle inequality, we get

$$|s_n| = |(s_n - L) + L| \leq |s_n - L| + |L| < 1 + |L|$$

So, if we take M to be the maximum of $1 + |L|$ and $|s_1|, |s_2|, |s_3|, \ldots, |s_{n_0}|$, then $|s_n| \leq M$ for all n. Thus, $\langle s_n \rangle$ is bounded.

6. Show that the sequence $\left\langle \dfrac{n!}{2^n} \right\rangle$ is divergent.

Since $\dfrac{n!}{2^n} = \dfrac{1\cdot 2\cdot 3\cdots n}{2\cdot 2\cdot 2\cdots 2} = \dfrac{1}{2}\,\dfrac{3}{2}\,\dfrac{4}{2}\cdots\dfrac{n}{2} > \dfrac{n}{2}$ for $n > 4$, the sequence is not bounded. So, by Theorem 42.1, the sequence cannot be convergent.

7. Prove Theorem 42.3: If $\lim_{n\to+\infty} s_n = \infty$ and $s_n \neq 0$ for all n, then $\lim_{n\to+\infty} \dfrac{1}{s^n} = 0$.

Consider any $\epsilon > 0$. Since $\lim_{n\to+\infty} s_n = \infty$, there exists some positive integer m such that, whenever $n \geq m$,

$$|s_n| > \frac{1}{\epsilon} \quad \text{and, therefore,} \quad \left|\frac{1}{s_n} - 0\right| = \left|\frac{1}{s_n}\right| < \epsilon. \quad \text{So,} \quad \lim_{n\to+\infty} \frac{1}{s_n} = 0.$$

8. Prove Theorem 42.4: (a) if $|a| > 1$, then $\lim_{n\to+\infty} a^n = \infty$; (b) If $|r| < 1$, then $\lim_{n\to+\infty} r^n = 0$.

(a) Let $M > 0$, and let $|a| = 1 + b$. So, $b > 0$. Now, $|a|^n = (1+b)^n = 1 + nb + \cdots > 1 + nb > M$ when $n \geq \dfrac{M}{b}$.

(b) Let $a = 1/r$. Since $|r| < 1$, $|a| > 1$. By part (*a*), $\lim_{n\to+\infty} a^n = \infty$. Hence, $\lim_{n\to+\infty} (1/r^n) = \infty$. So, by Theorem 42.3, $\lim_{n\to+\infty} r^n = 0$.

9. Prove: $\lim_{n\to+\infty} \dfrac{1}{2^n} = 0$.

$\lim_{n\to+\infty} 2^n = \infty$ by Theorem 42.4(*a*). Hence, $\lim_{n\to+\infty} \dfrac{1}{2^n} = 0$ by Theorem 42.3.

10. Prove Theorem 42.2(*c*) and (*e*).

Assume $\lim_{n\to+\infty} s_n = c$ and $\lim_{n\to+\infty} t_n = d$.

(c) $\lim_{n\to+\infty}(s_n+t_n)=c+d$. Let $\epsilon>0$. Then there exist integers m_1 and m_2 such that $|s_n-c|<\epsilon/2$ for $n\ge m_1$ and $|t_n-d|<\epsilon/2$ for $n\ge m_2$. Let m be the maximum of m_1 and m_2. So, for $n\ge m$, $|s_n-c|<\epsilon/2$ and $|t_n-d|<\epsilon/2$. Hence, for $n\ge m$,

$$|(s_n+t_n)-(c+d)| = |(s_n-c)+(t_n-d)| \le |s_n-c|+|t_n-d| < \frac{\epsilon}{2}+\frac{\epsilon}{2}=\epsilon$$

(e) $\lim_{n\to+\infty}(s_nt_n)=cd$. Since $\langle s_n\rangle$ is convergent, it is bounded, by Theorem 42.1 and, therefore, there is a positive number M such that $|s_n|\le M$ for all n. Let $\epsilon>0$. If $d\ne0$, there exists an integer m_1 such that $|s_n-c|<\epsilon/2|d|$ for $n\ge m_1$ and, therefore, $|d||s_n-c|<\epsilon/2$ for $n\ge m_1$. If $d=0$, then we can choose $m_1=1$ and we would still have $|d||s_n-c|<\epsilon/2$ for $n\ge m_1$. There also exists m_2 such that $|t_n-d|<\epsilon/2M$ for $n\ge m_2$. Let m be the maximum of m_1 and m_2. If $n\ge m$,

$$|s_nt_n-cd| = |s_n(t_n-d)+d(s_n-c)| \le |s_n(t_n-d)|+|d(s_n-c)|$$

$$= |s_n||t_n-d|+|d||s_n-c| \le M\left(\frac{\epsilon}{2M}\right)+\frac{\epsilon}{2}=\epsilon$$

11. Prove the Squeeze Theorem: If $\lim_{n\to+\infty}s_n=L=\lim_{n\to+\infty}u_n$, and there is an integer m such that $s_n\le t_n\le u_n$ for all $n\ge m$, then $\lim_{n\to+\infty}t_n=L$.

Let $\epsilon>0$. There is an integer $m_1\ge m$ such that $|s_n-L|<\epsilon/4$ and $|u_n-L|<\epsilon/4$ for $n\ge m_1$. Now assume $n\ge m_1$. Since $s_n\le t_n\le u_n$, $|t_n-s_n|\le|u_n-s_n|$. But

$$|u_n-s_n| = |(u_n-L)+(L-s_n)| \le |u_n-L|+|L-s_n| < \frac{\epsilon}{4}+\frac{\epsilon}{4}=\frac{\epsilon}{2}$$

Thus, $|t_n-s_n|<\epsilon/2$. Hence,

$$|t_n-L| = |t_n-s_n|+(s_n-L)| \le |t_n-s_n|+|s_n-L| < \frac{\epsilon}{2}+\frac{\epsilon}{4}<\epsilon$$

SUPPLEMENTARY PROBLEMS

In each of Problems 12–29, determine for each given sequence $\langle s_n\rangle$ whether it is bounded and whether it is nondecreasing, increasing, nonincreasing, or decreasing. Also determine whether it is convergent and, if possible, find its limit. (*Note*: If the sequence has a finite limit, it must be bounded. If it has an infinite limit, it must be unbounded.)

12. $\left\langle n+\frac{2}{n}\right\rangle$ *Ans.* nondecreasing; increasing for $n\ge2$; limit $+\infty$

13. $\left\langle \sin\frac{n\pi}{4}\right\rangle$ *Ans.* bounded; no limit

14. $\langle\sqrt[3]{n^2}\rangle$ *Ans.* increasing; limit $+\infty$

15. $\left\langle\frac{n!}{10^n}\right\rangle$ *Ans.* increasing for $n\ge10$; limit $+\infty$

16. $\left\langle\frac{\ln n}{n}\right\rangle$ *Ans.* decreasing for $n\ge3$; limit 0

17. $\langle\frac{1}{2}(1+(-1)^{n+1})\rangle$ *Ans.* bounded; no limit

18. $\left\langle\ln\frac{n+1}{n}\right\rangle$ *Ans.* decreasing; limit 0

19. $\left\langle \frac{2^n}{n!} \right\rangle$ *Ans.* nonincreasing; decreasing for $n \geq 2$; limit 0

20. $\langle \sqrt[n]{n} \rangle$ *Ans.* decreasing for $n \geq 3$; limit 1

21. $\left\langle \frac{3n}{n+2} \right\rangle$ *Ans.* increasing; limit 3

22. $\left\langle \cos\frac{\pi}{n} \right\rangle$ *Ans.* increasing, limit 1

23. $\left\langle \frac{4n+5}{n^3-2n+3} \right\rangle$ *Ans.* decreasing; limit 0

24. $\left\langle \frac{\sin n}{n} \right\rangle$ *Ans.* limit 0

25. $\langle \sqrt{n+1}-\sqrt{n} \rangle$ *Ans.* decreasing; limit 0

26. $\left\langle \frac{2^n}{3^n-4} \right\rangle$ *Ans.* decreasing; limit 0

27. $\left\langle n\sin\frac{\pi}{n} \right\rangle$ *Ans.* increasing, limit π

28. $\left\langle \frac{1}{\sqrt{n^2+1}-n} \right\rangle$ *Ans.* increasing; limit $+\infty$

29. $\left\langle \frac{n^n}{n!} \right\rangle$ *Ans.* increasing; limit $+\infty$

In each of Problems 30–32, find a plausible formula for a sequence whose first few terms are given. Find the limit (if it exists) of your sequence.

30. $1, \frac{3}{2}, \frac{9}{4}, \frac{27}{6}, \frac{81}{8}, \ldots$ *Ans.* $s_n = \frac{3^{n-1}}{2(n-1)}$; limit is $+\infty$

31. $-1, 1, -1, 1, -1, 1, \ldots$ *Ans.* $s_n = (-1)^n$; no limit

32. $\frac{3}{1}, \frac{7}{4}, \frac{11}{7}, \frac{3}{2}, \frac{19}{11}, \ldots$ *Ans.* $s_n = \frac{4n-1}{3n-2}$; decreasing, limit is $\frac{4}{3}$

33. Prove Theorem 42.7. (*Hint*: Let $\epsilon > 0$. Choose $\delta > 0$ such that, for x in the domain of f for which $|x-c| < \delta$, we have $|f(x) - f(c)| < \epsilon$. Choose m so that $n \geq m$ implies $|s_n - c| < \delta$.)

34. Show that $\lim_{n\to+\infty} \sqrt[n]{1/n^p} = 1$ for $p > 0$. (*Hint*: $n^{p/n} = e^{(p \ln n)/n}$.)

35. (GC) Use a graphing calculator to investigate $s_n = \frac{n^2+5}{\sqrt{4n^4+n}}$ for $n = 1$ to $n = 5$. Then determine analytically the behavior of the sequence.

Ans. decreasing; limit is $\frac{1}{2}$

36. (GC) Use a graphing calculator to investigate $s_n = \frac{n^5}{2^n}$ for $n = 1$ to $n = 10$. Then determine analytically the behavior of the sequence.

Ans. decreasing for $n \geq 7$; limit is 0

37. Prove that $\lim_{n\to+\infty} a_n = 0$ is equivalent to $\lim_{n\to+\infty} |a_n| = 0$.

38. If $s_n > 0$ for all n and $\lim_{n\to+\infty} s_n^2 = c$, prove that $\lim_{n\to+\infty} s_n = \sqrt{c}$.

39. (GC) Define s_n by recursion as follows: $s_1 = 2$ and $s_{n+1} = \frac{1}{2}\left(s_n + \frac{2}{s_n}\right)$ for $n \geq 1$.

(a) Use a graphing calculator to estimate s_n for $n = 2, \ldots, 5$.
(b) Show that, if $\lim_{n\to+\infty} s_n$ exists, then $\lim_{n\to+\infty} s_n = \sqrt{2}$.
(c) Prove that $\lim_{n\to+\infty} s_n$ exists.

40. Define s_n by recursion as follows: $s_1 = 3$, and $s_{n+1} = \frac{1}{2}(s_n + 6)$ for $n \geq 1$.

(a) Prove $s_n < 6$ for all n.
(b) Show that $< s_n >$ is increasing.
(c) Prove that $\lim_{n\to+\infty} s_n$ exists.
(d) Evaluate $\lim_{n\to+\infty} s_n$.

Ans. (d) 6

41. Prove Theorem 42.2, parts (a), (b), (d), (f).

CHAPTER 43

Infinite Series

Let $\langle s_n \rangle$ be an infinite sequence. We can form the infinite sequence of *partial sums* $\langle S_n \rangle$ as follows:

$$S_1 = s_1$$
$$S_2 = s_1 + s_2$$
$$S_3 = s_1 + s_2 + s_3$$
$$\vdots$$
$$S_n = s_1 + s_2 + \cdots + s_n$$
$$\vdots$$

We usually will designate the sequence $\langle S_n \rangle$ by the notation

$$\sum s_n = s_1 + s_2 + \cdots + s_n + \cdots$$

The numbers $s_1, s_2, \ldots, s_n, \ldots$ will be called the *terms* of the series.

If S is a number such that $\lim_{n\to+\infty} S_n = S$, then the series $\sum s_n$ is said to *converge* and S is called the *sum* of the series. We usually designate S by $\sum_{n=1}^{+\infty} s_n$.

If there is no number S such that $\lim_{n\to+\infty} S_n = S$, then the series $\sum s_n$ is said to *diverge*. If $\lim_{n\to+\infty} S_n = +\infty$, then the series is said to diverge to $+\infty$ and we write $\sum_{n=1}^{+\infty} s_n = +\infty$. Similarly, if $\lim_{n\to+\infty} S_n = -\infty$, then the series is said to diverge to $-\infty$ and we write $\sum_{n=1}^{+\infty} s_n = -\infty$.

EXAMPLE 43.1: Consider the sequence $\langle(-1)^{n+1}\rangle$. The terms are $s_1 = 1$, $s_2 = -1$, $s_3 = 1$, $s_4 = -1$, and so on. Hence, the partial sums begin with $S_1 = 1$, $S_2 = 1 + (-1) = 0$, $S_3 = 1 + (-1) + 1 = 1$, $S_4 = 1 + (-1) + (1) + (-1) = 0$, and continue with alternating 1s and 0s. So, $\lim_{n\to+\infty} S_n$ does not exist and the series diverges (but not to $+\infty$ or $-\infty$).

Geometric Series

Consider the sequence $\langle ar^{n-1}\rangle$, which consists of the terms $a, ar, ar^2, ar^3, \ldots$.

The series $\sum ar^{n-1}$ is called a *geometric series* with *ratio r* and *first term a*. Its nth partial sum S_n is given by

$$S_n = a + ar + ar^2 + \cdots + ar^{n-1}$$

Multiply by r: $$rS_n = ar + ar^2 + \cdots + ar^{n-1} + ar^n$$

Subtract: $$S_n - rS_n = a - ar^n$$

Hence, $$(1-r)S_n = a(1-r^n)$$

$$S_n = \frac{a(1-r^n)}{1-r}$$

Everything now depends on the ratio r. If $|r|<1$, then $\lim_{n\to+\infty} r^n = 0$ (by Theorem 42.4(b)) and, therefore, $\lim_{n\to+\infty} S_n = a/(1-r)$. If $|r|>1$, then $\lim_{n\to+\infty} r^n = \infty$ (by Theorem 42.4(a)) and, therefore, $\lim_{n\to+\infty} S_n = \infty$. (A trivial exception occurs when $a=0$. In that case, all terms are 0, the series converges, and its sum is 0.) These results are summarized as follows:

Theorem 43.1: Given a geometric series $\sum ar^{n-1}$:

(a) If $|r| < 1$, the series converges and has sum $\frac{a}{1-r}$.
(b) If $|r| > 1$ and $a \neq 0$, the series diverges to ∞.

EXAMPLE 43.2: Take the geometric series $\sum (\frac{1}{2})^{n-1}$ with ratio $r = \frac{1}{2}$ and first term $a = 1$:

$$1+\tfrac{1}{2}+\tfrac{1}{4}+\tfrac{1}{8}+\cdots$$

By Theorem 43.1(a), the series converges and has sum $\frac{1}{1-(\frac{1}{2})} = \frac{1}{\frac{1}{2}} = 2$. Thus, $\sum_{n=1}^{+\infty} (\frac{1}{2})^{n-1} = 2$.

We can multiply a series $\sum s_n$ by a constant c to obtain a new series $\sum cs_n$, and we can add two series $\sum s_n$ and $\sum t_n$ to obtain a new series $\sum (s_n + t_n)$.

Theorem 43.2: If $c \neq 0$, then $\sum cs_n$ converges if and only if $\sum s_n$ converges. Moreover, in the case of convergence,

$$\sum_{n=1}^{+\infty} cs_n = c\sum_{n=1}^{+\infty} s_n$$

To obtain this result, denote by $T_n = cs_1 + cs_2 + \cdots + cs_n$ the nth partial sum of the series $\sum cs_n$. Then $T_n = cS_n$, where S_n is the nth partial sum of $\sum s_n$. So, $\lim_{n\to+\infty} T_n$ exists if and only if $\lim_{n\to+\infty} S_n$ exists, and when the limits exist, $\lim_{n\to+\infty} T_n = c \lim_{n\to+\infty} S_n$. This yields Theorem 43.2.

Theorem 43.3: Assume that two series $\sum s_n$ and $\sum t_n$ both converge. Then their sum $\sum (s_n + t_n)$ also converges and

$$\sum_{n=1}^{+\infty} (s_n + t_n) = \sum_{n=1}^{+\infty} s_n + \sum_{n=1}^{+\infty} t_n$$

To see this, let S_n and T_n be the nth partial sums of $\sum s_n$ and $\sum t_n$, respectively. Then the nth partial sum U_n of $\sum (s_n + t_n)$ is easily seen to be $S_n + T_n$. So, $\lim_{n\to+\infty} U_n = \lim_{n\to+\infty} S_n + \lim_{n\to+\infty} T_n$. This yields Theorem 43.3.

Corollary 43.4: Assume that two series $\sum s_n$ and $\sum t_n$ both converge. Then their difference $\sum (s_n - t_n)$ also converges and

$$\sum_{n=1}^{+\infty} (s_n - t_n) = \sum_{n=1}^{+\infty} s_n - \sum_{n=1}^{+\infty} t_n$$

This follows directly from Theorems 43.2 and 43.3. Just note that $\sum (s_n - t_n)$ is the sum of $\sum s_n$ and the series $\sum (-1)t_n$.

Theorem 43.5: If $\sum s_n$ converges, then $\lim_{n\to+\infty} s_n = 0$.

To see this, assume that $\sum_{n=1}^{+\infty} s_n = S$. This means that $\lim_{n\to+\infty} S_n = S$, where, as usual, S_n is the nth partial sum of the series. We also have $\lim_{n\to+\infty} S_{n-1} = S$. But, $s_n = S_n - S_{n-1}$. So, $\lim_{n\to+\infty} s_n = \lim_{n\to+\infty} S_n - \lim_{n\to+\infty} S_{n-1} = S - S = 0$.

Corollary 43.6 (The Divergence Theorem): If $\lim_{n\to+\infty} s_n$ does not exist or $\lim_{n\to+\infty} s_n \neq 0$, then $\sum s_n$ diverges.

This is an immediate logical consequence of Theorem 43.5.

EXAMPLE 43.3: The series $\frac{1}{3}+\frac{2}{5}+\frac{3}{7}+\frac{4}{9}+\cdots$ diverges.

Here, $s_n = \frac{n}{2n+1}$. Since $\lim_{n\to+\infty} \frac{n}{2n+1} = \frac{1}{2} \neq 0$, the Divergence Theorem implies that the series diverges.

The converse of Theorem 43.5 is not valid: $\lim_{n\to+\infty} s_n = 0$ does not imply that $\sum s_n$ converges. This is shown by the following example.

EXAMPLE 43.4: Consider the so-called *harmonic series* $\sum \frac{1}{n} = 1+\frac{1}{2}+\frac{1}{3}+\frac{1}{4}+\frac{1}{5}+\cdots$. Let us look at the following partial sums of this series:

$$S_2 = 1+\frac{1}{2}$$

$$S_4 = 1+\frac{1}{2}+\frac{1}{3}+\frac{1}{4} > 1+\frac{1}{2}+\frac{1}{4}+\frac{1}{4} = 1+\frac{1}{2}+\frac{1}{2} = 1+\frac{2}{2}$$

$$S_8 = S_4+\frac{1}{5}+\frac{1}{6}+\frac{1}{7}+\frac{1}{8} > S_4+\frac{1}{8}+\frac{1}{8}+\frac{1}{8}+\frac{1}{8} = S_4+\frac{4}{8} = S_4+\frac{1}{2}$$

$$> 1+\frac{3}{2}$$

$$S_{16} = S_8+\frac{1}{9}+\frac{1}{10}+\frac{1}{11}+\frac{1}{12}+\frac{1}{13}+\frac{1}{14}+\frac{1}{15}+\frac{1}{16}$$

$$> S_8+\frac{1}{16}+\frac{1}{16}+\frac{1}{16}+\frac{1}{16}+\frac{1}{16}+\frac{1}{16}+\frac{1}{16}+\frac{1}{16} = S_8+\frac{1}{2}$$

$$> 1+\frac{4}{2}$$

Continuing in this manner, we would obtain $S_{32} > 1+\frac{5}{2}$, $S_{64} > 1+\frac{6}{2}$, and, in general, $S_{2^k} > 1+k/2$ when $k>1$. This implies that $\lim_{n\to+\infty} S_n = +\infty$ and, therefore, the harmonic series diverges. But notice that $\lim_{n\to+\infty} s_n = \lim_{n\to+\infty} 1/n = 0$.

Remark: Convergence or divergence is not affected by the addition or deletion of a finite number of terms at the beginning of a series. For example, if we delete the first k terms of a series and the sum of the deleted terms is c, then each new partial sum T_n has the form $S_{n+k} - c$. (For example, T_1 is $S_{k+1} - c$.) But $\lim_{n\to+\infty} (S_{n+k} - c)$ exists if and only if $\lim_{n\to+\infty} S_{n+k}$ exists, and $\lim_{n\to+\infty} S_{n+k}$ exists if and only if $\lim_{n\to+\infty} S_n$ exists.

Notation: It will often be useful to deal with series in which the terms of $\langle s_n \rangle$ are indexed by the nonnegative integers: $s_0, s_1, s_2, s_3, \ldots$. Then the partial sums S_n would also begin with $S_0 = s_0$, and the sum of a convergent series would be written as $\sum_{n=0}^{+\infty} s_n$.

SOLVED PROBLEMS

1. Examine the series $\frac{1}{5}+\frac{1}{5^2}+\frac{1}{5^3}+\cdots$ for convergence.

This is a geometric series with ratio $r=\frac{1}{5}$ and the first term $a=\frac{1}{5}$. Since $|r| = |\frac{1}{5}| < 1$, Theorem 43.1 (a) tells us that the series converges and that its sum is $\frac{a}{1-r} = \frac{1/5}{1-(1/5)} = \frac{1/5}{4/5} = \frac{1}{4}$.

2. Examine the series $\frac{1}{1\cdot 2}+\frac{1}{2\cdot 3}+\frac{1}{3\cdot 4}+\frac{1}{4\cdot 5}+\cdots$ for convergence.

The nth term is $\frac{1}{n\cdot(n+1)}$. This is equal to $\frac{1}{n}-\frac{1}{n+1}$. Hence, the nth partial sum

$$S_n = \frac{1}{1\cdot 2} + \frac{1}{2\cdot 3} + \frac{1}{3\cdot 4} + \frac{1}{4\cdot 5} + \cdots + \frac{1}{n\cdot(n+1)}$$
$$= \left(\frac{1}{1} - \frac{1}{2}\right) + \left(\frac{1}{2} - \frac{1}{3}\right) + \left(\frac{1}{3} - \frac{1}{4}\right) + \left(\frac{1}{4} - \frac{1}{5}\right) + \cdots + \left(\frac{1}{n} - \frac{1}{n+1}\right)$$
$$= 1 - \frac{1}{n+1}$$

Thus, $\lim_{n\to+\infty} S_n = \lim_{n\to+\infty}\left(1 - \frac{1}{n+1}\right) = 1 - 0 = 1$. Hence, the series converges and its sum is 1.

3. We know that the geometric series $1 + \frac{1}{2} + \frac{1}{4} + \frac{1}{8} + \frac{1}{16} + \cdots$ converges to $S = 2$. Examine the series that results when: (a) its first four terms are dropped; (b) the terms 3, 2, and 5 are added to the beginning of the series.

(a) The resulting series is a geometric series $\frac{1}{16} + \frac{1}{32} + \cdots$ with ratio $\frac{1}{2}$. It converges to $\frac{1/16}{1-(1/2)} = \frac{1/16}{1/2} = \frac{1}{8}$. Note that this is the same as $S - (1 + \frac{1}{2} + \frac{1}{4} + \frac{1}{8}) = 2 - (\frac{15}{8}) = \frac{1}{8}$.

(b) The new series is $3 + 2 + 5 + 1 + \frac{1}{2} + \frac{1}{4} + \frac{1}{8} + \frac{1}{16} + \cdots$. The new partial sums are the old ones plus $(3 + 2 + 5)$. Since the old partial sums converge to 2, the new ones converge to $2 + 10 = 12$. Thus, the new series is convergent and its sum is 12.

4. Show that the series $\frac{1}{2} + \frac{3}{4} + \frac{7}{8} + \frac{15}{16} + \cdots$ diverges.

Here, $s_n = \frac{2^n - 1}{2^n} = 1 - \frac{1}{2^n}$. Since $\lim_{n\to+\infty} \frac{1}{2^n} = 0$, it follows that $\lim_{n\to+\infty} s_n = 1 - 0 = 1 \neq 0$. So, by the Divergence Theorem, the series diverges.

5. Examine the series $9 - 12 + 16 - \frac{64}{3} + \frac{256}{9} - \cdots$ for convergence.

This is a geometric series with ratio $r = -\frac{4}{3}$. Since $|r| = \frac{4}{3} > 1$, Theorem 43.1(b) tells us that the series diverges.

6. Evaluate $\sum_{n=0}^{+\infty} \frac{(-1)^n}{2^n} = 1 - \frac{1}{2} + \frac{1}{4} - \frac{1}{8} - \frac{1}{16} - \cdots$.

This is a geometric series with ratio $r = -\frac{1}{2}$ and first term $a = 1$. Since $|r| = \frac{1}{2} < 1$, the series converges and its sum is $\frac{a}{1-r} = \frac{1}{1-(-1/2)} = \frac{1}{3/2} = \frac{2}{3}$.

7. Show that the infinite decimal 0.999... is equal to 1.

$0.999\cdots = \frac{9}{10} + \frac{9}{100} + \frac{9}{1000} + \cdots$. This is a geometric series with first term $a = \frac{9}{10}$ and ratio $r = \frac{1}{10}$.

Hence, it converges to the sum $\frac{a}{1-r} = \frac{9/10}{1-(1/10)} = \frac{9/10}{9/10} = 1$.

8. Examine the series $\frac{1}{1\cdot 3} + \frac{1}{3\cdot 5} + \frac{1}{5\cdot 7} + \frac{1}{7\cdot 9} + \cdots$.

Here, $s_n = \frac{1}{(2n-1)(2n+1)}$. Note that $\frac{1}{(2n-1)(2n+1)} = \frac{1}{2}\left(\frac{1}{2n-1} - \frac{1}{2n+1}\right)$. Hence, the nth partial sum S_n is

$$\frac{1}{2}\left(\frac{1}{1} - \frac{1}{3}\right) + \frac{1}{2}\left(\frac{1}{3} - \frac{1}{5}\right) + \frac{1}{2}\left(\frac{1}{5} - \frac{1}{7}\right) + \cdots + \frac{1}{2}\left(\frac{1}{2n-1} - \frac{1}{2n+1}\right) = \frac{1}{2}\left(1 - \frac{1}{2n+1}\right)$$

So, $\lim_{n\to+\infty} S_n = \frac{1}{2}$. Thus, the series converges to $\frac{1}{2}$.

9. Examine the series $3 + \sqrt{3} + \sqrt[3]{3} + \sqrt[4]{3} + \cdots$.

$s_n = \sqrt[n]{3} = 3^{1/n} = e^{(\ln 3)/n}$. Then $\lim_{n\to+\infty} s_n = e^0 = 1 \neq 0$. By the Divergence Theorem, the series diverges.

10. Examine the series $\frac{1}{10} + \frac{1}{11} + \frac{1}{12} + \frac{1}{13} + \cdots$.

This series is obtained from the harmonics series by deleting the first nine terms. Since the harmonic series diverges, so does this series.

11. **(Zeno's Paradox)** Achilles (A) and a tortoise (T) have a race. T gets a 1000 ft head start, but A runs at 10 ft/sec, whereas T only does 0.01 ft/sec. When A reaches T's starting point, T has moved a short distance ahead. When A reaches that point, T again has moved a short distance ahead, etc. Zeno claimed that A would never catch T. Show that this is not so.

When A reaches T's starting point, 100 seconds have passed and T has moved 0.01 (100) = 1 ft. A covers that additional 1 ft in 0.1 seconds, but T has moved 0.01(0.1) = 0.001 ft further. A needs 0.0001 seconds to cover that distance, but T meanwhile has moved 0.01(0.0001) = 0.000001 ft, etc. The limit of the distance between A and T approaches 0. The time involved is $100 + 0.1 + 0.0001 + 0.0000001 + \cdots$, which is a geometric series with first term $a = 100$ and ratio $r = 1/1000$. Its sum is

$$\frac{a}{1-r} = \frac{100}{1-(1/1000)} = \frac{100}{999/1000} = \frac{100000}{999}$$

which is a little more than 100 seconds. The seeming paradox arises from the artificial division of the event into infinitely many shorter and shorter steps.

SUPPLEMENTARY PROBLEMS

12. Examine each of the following geometric series. If the series converges, find its sum.

(a) $4 - 1 + \frac{1}{4} - \frac{1}{16} + \cdots$ *Ans.* $S = \frac{16}{5}$

(b) $1 + \frac{3}{2} + \frac{9}{4} + \frac{27}{8} + \cdots$ *Ans.* Diverges

(c) $1 - \frac{1}{3} + \frac{1}{9} - \frac{1}{27} + \cdots$ *Ans.* $S = \frac{3}{4}$

(d) $1 + e^{-1} + e^{-2} + e^{-3} + \cdots$ *Ans.* $S = \dfrac{e}{e-1}$

13. A rubber ball is dropped from a height of 10 ft. Whenever it hits the ground, it bounces straight up three-fourths of the previous height. What is the total distance traveled by the ball before it stops?

Ans. 70 ft

14. Examine the series $\sum \dfrac{1}{n(n+4)} = \dfrac{1}{1\cdot 5} + \dfrac{1}{2\cdot 6} + \dfrac{1}{3\cdot 7} + \cdots$.

Ans. $S = \frac{25}{48}$

15. Examine the series $\sum \dfrac{1}{n(n+1)(n+2)} = \dfrac{1}{1\cdot 2\cdot 3} + \dfrac{1}{2\cdot 3\cdot 4} + \dfrac{1}{3\cdot 4\cdot 5} + \cdots$

Ans. $S = \frac{1}{4}$

16. Evaluate $\sum_{n=1}^{+\infty} s_n$ when s_n is the following:

(a) 3^{-n} (b) $\dfrac{1}{n(n+2)}$ (c) $\dfrac{1}{n(n+3)}$ (d) $\dfrac{n}{(n+1)!}$

Ans. (a) $\frac{1}{2}$; (b) $\frac{3}{4}$; (c) $\frac{11}{18}$; (d) 1

17. Show that each of the following series diverges:

(a) $3 + \frac{5}{2} + \frac{7}{3} + \frac{9}{4} + \cdots$

(b) $2 + \sqrt{2} + \sqrt[3]{2} + \sqrt[4]{2} + \cdots$

(c) $\dfrac{1}{2} + \dfrac{1}{\sqrt{2}} + \dfrac{1}{\sqrt[3]{2}} + \dfrac{1}{\sqrt[4]{2}} + \cdots$

(d) $e + \frac{e^2}{8} + \frac{e^3}{27} + \frac{e^4}{64} + \cdots$

(e) $\sum \dfrac{1}{\sqrt{n} + \sqrt{n-1}}$

18. Evaluate the following

(a) $\sum_{n=0}^{+\infty} \left(\dfrac{1}{2^n} + \dfrac{1}{7^n}\right)$ (b) $\sum_{n=1}^{+\infty} \dfrac{1}{4n}$ (c) $\sum_{n=1}^{+\infty} \dfrac{2n+1}{n^2(n+1)^2}$

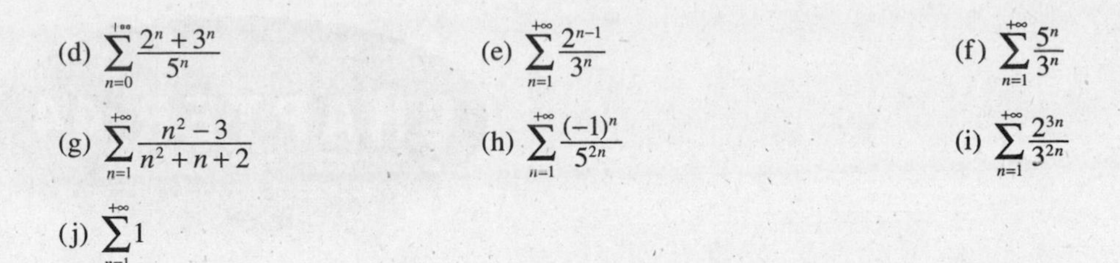

(d) $\sum_{n=0}^{+\infty} \frac{2^n + 3^n}{5^n}$ (e) $\sum_{n=1}^{+\infty} \frac{2^{n-1}}{3^n}$ (f) $\sum_{n=1}^{+\infty} \frac{5^n}{3^n}$

(g) $\sum_{n=1}^{+\infty} \frac{n^2 - 3}{n^2 + n + 2}$ (h) $\sum_{n=1}^{+\infty} \frac{(-1)^n}{5^{2n}}$ (i) $\sum_{n=1}^{+\infty} \frac{2^{3n}}{3^{2n}}$

(j) $\sum_{n=1}^{+\infty} 1$

Ans. (a) $\frac{19}{6}$; (b) $+\infty$; (c) 1; (d) $\frac{25}{6}$; (e) 1; (f) $+\infty$; (g) $+\infty$; (h) $-\frac{1}{26}$; (i) 8; (j) $+\infty$

19. (GC) In Problems 1 and 6, use a calculator to compute the first 10 partial sums and determine to how many decimal places the 10th partial sum is a correct estimate of the sum of the series.

20. (GC) (a) If $|x| < 1$, what function is represented by $\sum_{n=0}^{+\infty} x^n = 1 + x + x^2 + x^3 + \cdots$?

(b) Use a graphing calculator to graph $1 + x + x^2 + x^3 + \cdots + x^9$ on the interval $(-1, 1)$ and compare the graph with that of the function in (a).

Ans. (a) $\frac{1}{1-x}$

21. In each of the following, find those values of x for which the given series converges, and then find the function represented by the sum of the series for those values of x.

(a) $\sum_{n=0}^{+\infty} (3x)^n$ (b) $\sum_{n=0}^{+\infty} (x-2)^n$ (c) $\sum_{n=0}^{+\infty} \left(\frac{x}{2}\right)^n$ (d) $\sum_{n=0}^{+\infty} \left(\frac{x-1}{2}\right)^n$

Ans. (a) $|x| < \frac{1}{3}$, $\frac{1}{1-3x}$; (b) $1 < x < 3$, $\frac{1}{3-x}$; (c) $|x| < 2$, $\frac{2}{2-x}$; (d) $-1 < x < 3$, $\frac{2}{3-x}$

CHAPTER 44

Series with Positive Terms. The Integral Test. Comparison Tests

Series of Positive Terms

If all the terms of a series $\sum s_n$ are positive, then the series is called a *positive series*.

For a positive series $\sum s_n$, the sequence of partial sums $\langle S_n \rangle$ is an increasing sequence, since $S_{n+1} = S_n + s_{n+1} > S_n$. This yields the following useful result.

Theorem 44.1: A positive series $\sum s_n$ converges if and only if the sequence of partial sums $\langle s_n \rangle$ is bounded.

To see this, note first that, if $\sum s_n$ converges, then, by definition, $\langle S_n \rangle$ converges and, therefore, by Theorem 42.1, $\langle S_n \rangle$ is bounded. Conversely, if $\langle S_n \rangle$ is bounded, then, since $\langle S_n \rangle$ is increasing, Theorem 42.8 implies that $\langle S_n \rangle$ converges, that is, $\sum s_n$ converges.

Theorem 44.2 (Integral Test): Let $\sum s_n$ be a positive series and let $f(x)$ be a continuous, positive decreasing function on $[1, +\infty)$ such that $f(n) = s_n$ for all positive integers n. Then:

$$\sum s_n \text{ converges if and only if } \int_1^{+\infty} f(x)\,dx \text{ converges}$$

From Fig. 44-1 we see that $\int_1^n f(x)dx < s_1 + s_2 + \cdots + s_{n-1} = S_{n-1}$. If $\sum s_n$ converges, then $\langle S_n \rangle$ is bounded; so, $\int_1^u f(x)\,dx$ will be bounded for all $u \geq 1$ and, therefore, $\int_1^{+\infty} f(x)\,dx$ converges. Conversely, from Fig. 44-1 we have $s_2 + s_3 + \cdots + s_n < \int_1^n f(x)\,dx$ and, therefore, $S_n < \int_1^n f(x)\,dx + s_1$. Thus, if $\int_1^{+\infty} f(x)\,dx$ converges, then $S_n < \int_1^{+\infty} f(x)\,dx + s$, and so $\langle S_n \rangle$ will be bounded. Hence, by Theorem 44.1, $\sum s_n$ converges. This proves Theorem 44.2.

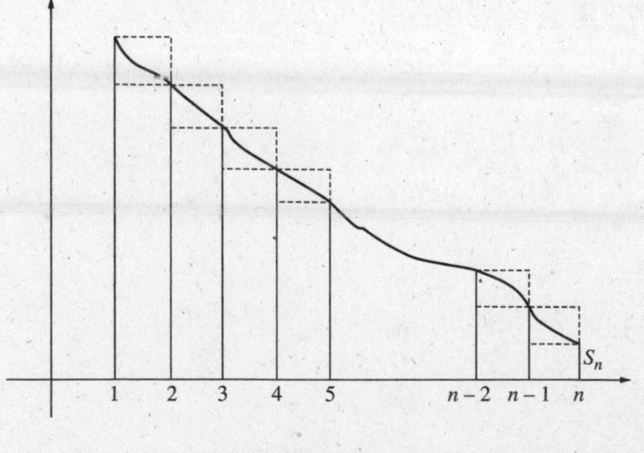

Fig. 44-1

EXAMPLE 44.1: $\sum \frac{\ln n}{n}$ diverges.

Let $f(x) = \frac{\ln x}{x}$. Now,

$$\int_1^{+\infty} \frac{\ln x}{x} dx = \lim_{u \to +\infty} \int_1^u \frac{\ln x}{x} dx = \lim_{u \to +\infty} \tfrac{1}{2}(\ln x)^2 \Big]_1^u = \lim_{u \to +\infty} \tfrac{1}{2}((\ln u)^2 - 0) = +\infty$$

Hence, by the integral test, $\sum \frac{\ln n}{n}$ diverges.

EXAMPLE 44.2: $\sum \frac{1}{n^2}$ converges.

Let $f(x) = \frac{1}{x^2}$. Now,

$$\int_1^{+\infty} \frac{1}{x^2} dx = \lim_{u \to +\infty} \int_1^u \frac{1}{x^2} dx = \lim_{u \to +\infty} -\frac{1}{x}\Big]_1^u = \lim_{u \to +\infty} -\left(\frac{1}{u} - 1\right) = 1$$

Hence, by the integral test, $\sum \frac{1}{n^2}$ converges.

Remark: The integral test can be easily be extended to the case where the lower limit of the integral is changed from 1 to any positive integer.

Theorem 44.3 (Comparison Test): Let $\sum a_n$ and $\sum b_n$ be two positive series such that there is a positive integer m for which $a_k \le b_k$ for all integers $k \ge m$. Then:

(1) If $\sum b_n$ converges, so does $\sum a_n$;
(2) If $\sum a_n$ diverges, so does $\sum b_n$.

We may assume in the derivation of Theorem 44.3 that $m = 1$, since convergence is not affected by deletion of a finite number of terms at the beginning of a series. Note also that (2) is a logical consequence of (1). To prove (1), assume that $\sum b_n$ converges. Let $B_n = b_1 + b_2 + \cdots + b_n$ be the nth partial sum for $\sum b_n$ and let $A_n = a_1 + a_2 + \cdots + a_n$ be the nth partial sum $\sum a_n$. Then $A_n \le B_n$, since $a_k \le b_k$ for all k. From the fact that $\sum b_n$ converges, it follows, by Theorem 44.1, that the sequence $\langle B_n \rangle$ is bounded. Since $A_n \le B_n$ for all n, it follows that the sequence $\langle A_n \rangle$ is bounded. So, by Theorem 44.1, $\sum a_n$ converges. This proves Theorem 44.3.

EXAMPLE 44.3: $\sum \frac{1}{n^2 + 5}$ converges.

Let $a_n = \frac{1}{n^2 + 5}$ and $b_n = \frac{1}{n^2}$. Then $a_n < b_n$ for all n. By Example 2, $\sum \frac{1}{n^2}$ converges. So, by the comparison test, $\sum \frac{1}{n^2 + 5}$ converges.

EXAMPLE 44.4: $\sum \frac{1}{3n + 5}$ diverges.

Let $a_n = \frac{1}{4n}$ and $b_n = \frac{1}{3n + 5}$. Now, $a_n \le b_n$ for $n \ge 5$. (To see this, observe that $\frac{1}{4n} \le \frac{1}{3n + 5}$ is equivalent to $3n + 5 \le 4n$, which is equivalent to $5 \le n$.) Recall that the harmonic series $\sum \frac{1}{n}$ diverges (by Chapter 43, Example 4). Hence, $\sum \frac{1}{4n}$ diverges by Theorem 43.2. The comparison test implies that $\sum \frac{1}{3n + 5}$ diverges.

Sometimes, as in Example 4, complicated maneuvers are needed in order to apply the comparison test. The following result offers a much more flexible tool.

Theorem 44.4 (Limit Comparison Test): Let $\sum a_n$ and $\sum b_n$ be two positive series such that $L = \lim_{n \to +\infty} \frac{a_n}{b_n}$ exists and $0 < L < +\infty$. Then $\sum a_n$ converges if and only if $\sum b_n$ converges.

Assume that $\sum b_n$ converges. Let c be a positive number such that $L < c$. Then there exists a positive integer m such that $a_n/b_n < c$ for all $n \geq m$. Hence, $a_n < cb_n$ for all $n \geq m$. But, since $\sum b_n$ converges, so does $\sum cb_n$. Therefore, by the comparison test, $\sum a_n$ converges. Conversely, if $\sum a_n$ converges, then $\sum b_n$ converges. (In fact, $\lim_{n\to+\infty} \frac{b_n}{a_n} = \frac{1}{L} > 0$ and we can use the same kind of argument that was just given.)

EXAMPLE 44.5: $\sum \frac{3n^2 - 5n + 4}{7n^3 + 2}$ diverges.

When dealing with quotients of polynomials, a good rule of thumb is to ignore everything except the leading terms. In this case, we have $\frac{3n^2}{7n^3} = \frac{3}{7}\frac{1}{n}$. Let us try a limit comparison with $\frac{1}{n}$. Now

$$\lim_{n\to+\infty}\left[\left(\frac{3n^2 - 5n + 4}{7n^3 + 2}\right)\bigg/\frac{1}{n}\right] = \lim_{n\to+\infty}\frac{3n^3 - 5n^2 + 4n}{7n^3 + 2} = \frac{3}{7}.$$

Since $\sum \frac{1}{n}$ diverges, the limit comparison test tells us that $\sum \frac{3n^2 - 5n + 4}{7n^3 + 2}$ diverges.

EXAMPLE 44.6: $\sum \frac{5n - 2}{\sqrt{n^6 - 4n^2 + 7}}$ converges.

Using the rule of thumb given in Example 5, we should look at $\frac{5n}{\sqrt{n^6}} = \frac{5n}{n^3} = \frac{5}{n^2}$. So, let us try a limit comparison with $\frac{1}{n^2}$:

$$\lim_{n\to+\infty}\left[\left(\frac{5n - 2}{\sqrt{n^6 - 4n^2 + 7}}\right)\bigg/\frac{1}{n^2}\right] = \lim_{n\to+\infty}\frac{5n^3 - 2n^2}{\sqrt{n^6 - 4n^2 + 7}}$$

Let us divide the numerator and denominator by n^3. Note that, in the denominator, we would get

$$\frac{1}{n^3}\sqrt{n^6 - 4n^2 + 7} = \frac{1}{\sqrt{n^6}}\sqrt{n^6 - 4n^2 + 7} = \sqrt{1 - \frac{4}{n^4} + \frac{6}{n^6}}$$

So, the result would be

$$\lim_{n\to+\infty}\frac{5 - \frac{2}{n}}{\sqrt{1 - \frac{4}{n^4} + \frac{7}{x^6}}} = \frac{5}{1} = 5$$

Hence, since we know, by Example 2, that $\sum \frac{1}{n^2}$ converges, the limit comparison test implies that $\sum \frac{5n - 2}{\sqrt{n^6 - 4n^2 + 7}}$ converges.

SOLVED PROBLEMS

1. Consider the series $\sum \frac{1}{n^p}$, where p is constant. This is called a *p-series*. Then:

   (a) If $p > 1$, the series $\sum \frac{1}{n^p}$ converges.

   (b) If $p \leq 1$, the series $\sum \frac{1}{n^p}$ diverges.

   We may assume that $p \neq 1$, since we already know that the harmonic series $\sum \frac{1}{n}$ diverges. We may also assume that $p > 0$; if $p \leq 0$, $\lim_{n\to+\infty} \frac{1}{n^p} \neq 0$ and the Divergence Theorem implies that the series diverges. Let us apply the integral test with $f(x) = 1/x^p$. ($f(x)$ is positive and decreasing in $[1, +\infty)$.) Now,

$$\int_1^{+\infty}\frac{1}{x^p}dx = \lim_{u\to+\infty}\int_1^u \frac{1}{x^p}dx = \lim_{u\to+\infty}\frac{x^{1-p}}{1-p}\bigg]_1^u$$

$$= \lim_{u\to+\infty}\left(\frac{u^{1-p}}{1-p} - \frac{1}{1-p}\right).$$

(a) $p > 1$. Then $p-1>0$ and $\lim_{u\to+\infty} u^{1-p} = \lim_{u\to+\infty} \frac{1}{u^{p-1}} = 0$. So, $\lim_{u\to+\infty}\left(\frac{u^{1-p}}{1-p} - \frac{1}{1-p}\right) = \frac{1}{p-1}$. By the integral test, $\sum \frac{1}{n^p}$ converges.

(b) $p<1$. Then $1-p>0$ and $\lim_{u\to+\infty} u^{1-p} = +\infty$. So, $\lim_{u\to+\infty}\left(\frac{u^{1-p}}{1-p} - \frac{1}{1-p}\right) = +\infty$ and, by the integral test, $\sum \frac{1}{n^p}$ diverges.

In Problems 2–7, examine the given series for convergence.

2. $1 + \frac{1}{\sqrt{3}} + \frac{1}{\sqrt{5}} + \frac{1}{\sqrt{7}} + \cdots$.

$s_n = \frac{1}{\sqrt{2n-1}}$. Let $f(x) = \frac{1}{\sqrt{2x-1}}$. On $[1,+\infty)$, $f(x)>0$ and f is decreasing.

$$\int_1^{+\infty} \frac{1}{\sqrt{2x-1}}\,dx = \lim_{u\to+\infty}\int_1^u \frac{dx}{\sqrt{2x-1}} = \lim_{u\to+\infty}\frac{1}{2}\int_1^u (2x-1)^{-1/2}(2)\,dx$$

$$= \lim_{u\to+\infty} \tfrac{1}{2}(2)(2x-1)^{1/2}\Big]_1^u = \lim_{u\to+\infty}((2u-1)^{1/2}-1) = +\infty$$

Hence, the series diverges by the integral test.

3. $\frac{1}{3} + \frac{1}{10} + \frac{1}{29} + \cdots + \frac{1}{n^3+2} + \cdots$.

$\frac{1}{n^3+2} < \frac{1}{n^3}$. $\sum \frac{1}{n^3}$ is convergent, since it is a p-series with $p = 3 > 1$. Thus, by the comparison test, $\sum \frac{1}{n^3+2}$ is convergent.

4. $1 + \frac{1}{2!} + \frac{1}{3!} + \frac{1}{4!} + \cdots$.

$s_n = \frac{1}{n!}$. Note that $\frac{1}{n!} = \frac{1}{n(n-1)\cdot\cdots\cdot 3\cdot 2} \le \frac{1}{2^{n-1}}$ for $n \ge 2$. Since $\sum \frac{1}{2^{n-1}}$ is a convergent geometric series (with ratio $\frac{1}{2}$), $\sum \frac{1}{n!}$ is convergent by the comparison test.

5. $2 + \frac{3}{2^3} + \frac{4}{3^3} + \frac{5}{4^3} + \cdots$.

$s_n = \frac{n+1}{n^3}$. Use limit comparison with $\frac{n}{n^3} = \frac{1}{n^2}$.

$$\lim_{n\to+\infty} \frac{n+1}{n^3}\Big/\frac{1}{n^2} = \lim_{n\to+\infty}\frac{n^3+n^2}{n^3} = 1$$

We know that $\sum \frac{1}{n^2}$ converges. So, by the limit comparison test, $\sum \frac{n+1}{n^3}$ converges.

6. $1 + \frac{1}{2^2} + \frac{1}{3^3} + \frac{1}{4^4} + \cdots$

$s_n = \frac{1}{n^n}$. Now, $\frac{1}{n^n} = \frac{1}{n\cdot n\cdot\cdots\cdot n} \le \frac{1}{2^{n-1}}$ and $\sum \frac{1}{2^{n-1}}$ is a convergent geometric series ($r = \frac{1}{2}$). So, by the comparison test, $\sum \frac{1}{n^n}$ converges.

7. $1 + \frac{2^2+1}{2^3+1} + \frac{3^2+1}{3^3+1} + \frac{4^2+1}{4^3+1} + \cdots$.

$s_n = \frac{n^2+1}{n^3+1}$. Use limit comparison with $\frac{n^2}{n^3} = \frac{1}{n}$:

$$\lim_{n\to+\infty}\left[\left(\frac{n^2+1}{n^3+1}\right)\Big/\frac{1}{n}\right] = \lim_{n\to+\infty}\frac{n^3+n}{n^3+1} = 1$$

We know that the harmonic series $\sum \frac{1}{n}$ diverges. So, by the limit comparison test, $\sum \frac{n^2+1}{n^3+1}$ diverges.

8. $\frac{1}{2\ln 2} + \frac{1}{3\ln 3} + \frac{1}{4\ln 4} + \cdots$.

$s_n = \frac{1}{n\ln n}$ is defined for $n \ge 2$.

$$\int_2^{+\infty} \frac{dx}{x\ln x} = \lim_{u\to+\infty} \int_2^u \frac{dx}{x\ln x} = \lim_{u\to+\infty} \ln(\ln u)\Big]_2^u = \lim_{u\to+\infty} (\ln(\ln u) - \ln(\ln 2)) = +\infty.$$

Hence, the series diverges by the integral test.

9. How many terms of $\sum \frac{1}{n^2}$ suffice to obtain two-decimal place accuracy (that is, an error $< 5/10^3$)?

If we use k terms, then we require that the error

$$\sum_{n=1}^{+\infty} \frac{1}{n^2} - \sum_{n=1}^{k} \frac{1}{n^2} = \sum_{n=k+1}^{+\infty} \frac{1}{n^2} \le \int_k^{+\infty} \frac{1}{x^2}\,dx = \lim_{u\to+\infty} \int_k^u \frac{1}{x^2}\,dx = \lim_{u\to+\infty} -\frac{1}{x}\Big]_k^u = \lim_{u\to+\infty} -\left(\frac{1}{u} - \frac{1}{k}\right)$$

$$= \frac{1}{k} < \frac{5}{10^3} = \frac{1}{200}$$

Hence, $200 < k$. Thus, it suffices to use 201 terms of the series. (The graphing calculator can be used to find $\sum_{n=1}^{201} \frac{1}{n^2} \approx 1.64$.)

10. Assume $\sum s_n$ converges by virtue of the integral test applied to $f(x)$ and, for each n, the error (or remainder) R_k after k terms is defined to be

$$\sum_{n=1}^{+\infty} s_n - \sum_{n=1}^{k} s_n. \quad \text{Then} \quad R_k = \sum_{n=k+1}^{+\infty} s_n < \int_k^{+\infty} f(x)\,dx.$$

Find a bound on the error when $\sum_{n=1}^{+\infty} \frac{1}{n^2}$ is approximated by the first five terms: $1 + \frac{1}{4} + \frac{1}{9} + \frac{1}{16} + \frac{1}{25} = \frac{5269}{3600} \approx 1.4636$.

The error $R_5 < \int_5^{+\infty} \frac{1}{x^2}\,dx = \frac{1}{5} = 0.2$.

11. Assume $\sum s_n$ and $\sum c_n$ are positive series, $\sum c_n$ converges, and $s_n \le c_n$ for all n. Then the error R_k after k terms is

$$\sum_{n=1}^{+\infty} s_n - \sum_{n=1}^{k} s_n = \sum_{n=k+1}^{+\infty} s_n \le \sum_{n=k+1}^{+\infty} c_n.$$

At least how many terms will suffice to estimate $\sum_{n=1}^{+\infty} \frac{1}{n^5+1}$ with an error < 0.00001?

In this case, $s_n = \frac{1}{n^5+1}$ and $c_n = \frac{1}{n^5}$. It suffices to have $\sum_{n=k+1}^{+\infty} \frac{1}{n^5} < 0.00001$. Now, $\sum_{n=k+1}^{+\infty} \frac{1}{n^5} < \int_k^{+\infty} \frac{1}{x^5}\,dx = \frac{1}{4k^4}$.

So, we need $\frac{1}{4k^4} < 0.00001 = \frac{1}{100{,}000}$. Equivalently, $100{,}000 < 4k^4$, $25{,}000 < k^4$, $k \ge 13$.

SUPPLEMENTARY PROBLEMS

For Problems 12–43, determine whether the series converges.

12. $\sum \frac{3}{n(n+1)}$ *Ans.* converges; comparison with $\sum \frac{3}{n^2}$

13. $\sum \frac{n}{(n+1)(n+2)}$ *Ans.* diverges; limit comparison with $\sum \frac{1}{n}$

14. $\sum \frac{n}{n^2+1}$ *Ans.* diverges; limit comparison with $\sum \frac{1}{n}$

15. $\sum \frac{n}{e^n}$ *Ans.* converges; integral test

16. $\sum \frac{2n}{(n+1)(n+2)(n+3)}$ *Ans.* converges; limit comparison with $\sum \frac{1}{n^2}$

17. $\sum \frac{1}{(2n+1)^2}$ *Ans.* converges; limit comparison with $\sum \frac{1}{n^2}$

18. $\sum \frac{1}{n^3-1}$ *Ans.* converges; limit comparison with $\sum \frac{1}{n^3}$

19. $\sum \frac{n-2}{n^3}$ *Ans.* converges; limit comparison with $\sum \frac{1}{n^2}$

20. $\sum \frac{\ln n}{n^2+2}$ *Ans.* converges; limit comparison with $\sum \frac{1}{n^{3/2}}$

21. $\sum n \sin\left(\frac{1}{n}\right)$ *Ans.* diverges; Divergence Theorem

22. $\sum \frac{1}{\sqrt[3]{n}}$ *Ans.* diverges; *p*-series, $p = \frac{1}{3} < 1$

23. $\sum \frac{1}{n^{n-1}}$ *Ans.* converges; comparison with $\sum \frac{1}{2^{n-1}}, n \geq 2$

24. $\sum \frac{\ln n}{\sqrt{n}}$ *Ans.* diverges; comparison with $\sum \frac{\ln n}{n}$

25. $\sum \frac{1}{1+\ln n}$ *Ans.* diverges; comparison with $\sum \frac{1}{n}$

26. $\sum \frac{n+1}{n\sqrt{3n-2}}$ *Ans.* diverges; limit comparison with $\sum \frac{1}{\sqrt{n}}$

27. $\sum \frac{1}{n \ln n \ln(\ln n)}$ (for $n \geq 3$) *Ans.* diverges; integral test

28. $\sum \frac{1}{n \ln n (\ln(\ln n))^2}$ (for $n \geq 3$) *Ans.* converges, integral test

29. $\frac{1}{4^2}+\frac{1}{7^2}+\frac{1}{10^2}+\frac{1}{13^2}+\cdots$.

Ans. $s_n=\frac{1}{(3n+1)^2}$; converges; limit comparison with $\sum\frac{1}{n^2}$

30. $3+\frac{3}{2^{1/3}}+\frac{3}{3^{1/3}}+\frac{3}{4^{1/3}}+\cdots$.

Ans. $s_n=\frac{3}{n^{1/3}}$; diverges; p-series, $p=\frac{1}{3}<1$

31. $1+\frac{1}{5}+\frac{1}{9}+\frac{1}{13}+\cdots$.

Ans. $s_n=\frac{1}{4n-3}$; diverges; limit comparison with $\sum\frac{1}{n}$

32. $\frac{1}{2}+\frac{1}{3\cdot 4}+\frac{1}{4\cdot 5\cdot 6}+\frac{1}{5\cdot 6\cdot 7\cdot 8}+\cdots$.

Ans. $s_n=\frac{1}{(n+1)(n+2)\cdots(2n)}$; converges; limit comparison with $\sum\frac{1}{n^2}$

33. $\frac{2}{3}+\frac{3}{2\cdot 3^2}+\frac{4}{3\cdot 3^3}+\frac{5}{4\cdot 3^4}+\cdots$.

Ans. $s_n=\frac{n+1}{n\cdot 3^n}$; converges; limit comparison with $\sum\frac{1}{3^n}$

34. $\frac{1}{2}+\frac{1}{2\cdot 2^2}+\frac{1}{3\cdot 2^3}+\frac{1}{4\cdot 2^4}+\cdots$.

Ans. $s_n=\frac{1}{n2^n}$; converges; comparison with $\sum\frac{1}{2^n}$

35. $\frac{2}{1\cdot 3}+\frac{3}{2\cdot 4}+\frac{4}{3\cdot 5}+\frac{5}{4\cdot 6}+\cdots$.

Ans. $s_n=\frac{n+1}{n(n+2)}$; diverges; limit comparison with $\sum\frac{1}{n}$

36. $\frac{1}{2}+\frac{2}{3^2}+\frac{3}{4^3}+\frac{4}{5^4}+\cdots$.

Ans. $s_n=\frac{n}{(n+1)^n}$; converges; comparison with $\sum\frac{1}{2^{n-1}}$

37. $1+\frac{1}{2^2}+\frac{1}{3^{5/2}}+\frac{1}{4^3}+\cdots$.

Ans. $s_n=\frac{1}{n^{(n+2)/2}}$; converges; comparison with $\sum\frac{1}{n^2}$

38. $1+\frac{3}{5}+\frac{4}{10}+\frac{5}{17}+\cdots$.

Ans. $s_n=\frac{n+1}{n^2+1}$; diverges; limit comparison with $\sum\frac{1}{n}$

39. $\frac{2}{5}+\frac{2\cdot 4}{5\cdot 8}+\frac{2\cdot 4\cdot 6}{5\cdot 8\cdot 11}+\frac{2\cdot 4\cdot 6\cdot 8}{5\cdot 8\cdot 11\cdot 14}+\cdots$.

Ans. $s_n=\frac{2\cdot 4\cdot\cdots\cdot(2n)}{5\cdot 8\cdots(2+3n)}$; converges; comparison with $\sum(\frac{2}{3})^n$

40. $\frac{3}{2}+\frac{5}{10}+\frac{7}{30}+\frac{9}{68}+\cdots$.

Ans. $s_n=\frac{2n+1}{n^3+n}$; converges, limit comparison with $\sum\frac{1}{n^2}$

41. $\frac{3}{2}+\frac{10}{24}+\frac{29}{108}+\frac{66}{320}+\cdots$.

Ans. $s_n=\frac{n^3+2}{n^4+n^3}$; diverges; limit comparison with $\sum\frac{1}{n}$

42. $\frac{1}{2^2-1}+\frac{2}{3^2-2}+\frac{3}{4^2-3}+\frac{4}{5^2-4}+\cdots$.

Ans. $s_n=\frac{n}{(n+1)^2-n}$; diverges; limit comparison with $\sum\frac{1}{n}$

43. $\frac{1}{2^3-1^2}+\frac{1}{3^3-2^2}+\frac{1}{4^3-3^2}+\frac{1}{5^3-4^2}+\cdots$.

Ans. $s_n=\frac{1}{(n+1)^3-n^2}$; converges; limit comparison with $\sum\frac{1}{n^3}$

44. (GC) Estimate the error when:

(a) $\sum_{n=1}^{+\infty}\frac{1}{3^n+1}$ is approximated by the sum of its first six terms.

(b) $\sum_{n=1}^{+\infty}\frac{1}{4^n+3}$ is approximated by the sum of its first six terms.

Ans. (a) 0.0007; (b) 0.00009

45. (GC) (a) Estimate the error when the geometric series $\sum\frac{3}{2^n}$ is approximated by the sum of its first six terms.

(b) How many terms suffice to compute the sum if the allowable error is 0.00005?

Ans. (a) 0.047; (b) 16

46. (GC) (a) How many terms suffice to approximate $\sum_{n=1}^{+\infty}\frac{1}{n^4}$ with an error < 0.001?

(b) Find a bound on the error if we approximate $\sum_{n=1}^{+\infty}\frac{1}{n^4}$ by the sixth partial sum.

(c) What is your approximation to $\sum_{n=1}^{+\infty}\frac{1}{n^4}$ by the sixth partial sum, correct to four decimal places?

Ans. (a) 7; (b) 0.0015; (c) 1.0811

47. (GC) Let S_n be the nth partial sum $1+\frac{1}{2}+\cdots+\frac{1}{n}$ of the divergent harmonic series.

(a) Prove $\ln(n+1)\le S_n\le 1+\ln n$.

(b) Let $E_n=S_n-\ln n$. Prove that $\langle E_n\rangle$ is bounded and decreasing.

(c) Prove that $\langle E_n\rangle$ converges. Its limit is denoted γ and is called *Euler's constant*.

(d) Use a graphing calculator to approximate E_{999} to eight decimal places.

Ans. (d) 0.57771608 (in fact, $\gamma\sim 0.57721566$.)

48. (Extension of the limit comparison test.) Assume $\sum s_n$ and $\sum t_n$ are positive series. Prove:

(a) If $\lim_{n\to+\infty}\frac{s_n}{t_n}=0$ and $\sum t_n$ converge, so does $\sum s_n$.

(b) If $\lim_{n\to+\infty}\frac{s_n}{t_n}=+\infty$ and $\sum t_n$ diverges, so does $\sum s_n$.

49. Use the extension of the limit comparison test to determine whether $\sum \frac{(\ln n)^4}{n^3}$ converges.

Ans. converges; use $\sum \frac{1}{n^2}$ and Problem 48(a)

50. Assume $\sum s_n$ is a positive series and $\lim_{n\to+\infty} ns_n$ exists and is positive. Prove that $\sum s_n$ diverges. (*Hint*: Limit comparison with $\sum (1/n)$.)

51. Assume $\sum s_n$ and $\sum t_n$ are convergent positive series. Prove that $\sum s_n t_n$ converges.

Alternating Series. Absolute and Conditional Convergence. The Ratio Test

Alternating Series

A series whose terms are alternately positive and negative is said to be an *alternating series*. It can be written in the form

$$\sum (-1)^{n+1} a_n = a_1 - a_2 + a_3 - a_4 + a_5 - \cdots$$

where a_n are all positive.

Theorem 45.1 (Alternating Series Theorem): Let $\sum (-1)^{n+1} a_n$ be an alternating series. Assume that: (1) the sequence $\langle a_n \rangle$ is decreasing; (2) $\lim_{n\to+\infty} a_n = 0$. Then:

(I) $\sum (-1)^{n+1} a_n$ converges to a sum A, and

(II) If A_n is the nth partial sum and $R_n = A - A_n$ is the corresponding error, then $|R_n| < a_{n+1}$ (that is, the error is less in magnitude than the first term omitted).

(I) Since $\langle a_n \rangle$ is decreasing, $a_{2n+1} > a_{2n+2}$ and, therefore, $a_{2n+1} - a_{2n+2} > 0$. Hence,

$$\begin{aligned} A_{2n+2} &= (a_1 - a_2) + (a_3 - a_4) + \cdots + (a_{2n-1} - a_{2n}) + (a_{2n+1} - a_{2n+2}) \\ &= A_{2n} + (a_{2n+1} - a_{2n+2}) > A_{2n} > 0 \end{aligned}$$

So, the sequence $\langle A_{2n} \rangle$ is increasing. Also,

$$A_{2n} = a_1 - (a_2 - a_3) - (a_4 - a_5) - \cdots - (a_{2n-2} - a_{2n-1}) - a_{2n} < a_1$$

Hence, $\langle A_{2n} \rangle$ is bounded. Therefore, by Theorem 42.8, $\langle A_{2n} \rangle$ converges to a limit L. Now, $A_{2n+1} = A_{2n} + a_{2n+1}$. Hence,

$$\lim_{n\to+\infty} A_{2n+1} = \lim_{n\to+\infty} A_{2n} + \lim_{n\to+\infty} a_{2n+1} = L + 0 = L$$

Thus, $\lim_{n\to+\infty} A_n = L$ and, therefore, $\sum (-1)^{n+1} a_n$ converges.

(II) $R_{2n} = (a_{2n+1} - a_{2n+2}) + (a_{2n+3} - a_{2n+4}) + \cdots > 0$, and $R_{2n} = a_{2n+1} - (a_{2n+2} - a_{2n+3}) - (a_{2n+4} - a_{2n+5}) - \cdots < a_{2n+1}$. Hence, $|R_{2n}| < a_{2n+1}$. For odd indices, $R_{2n+1} = -(a_{2n+2} - a_{2n+3}) - (a_{2n+4} - a_{2n+5}) - \cdots < 0$ and $R_{2n+1} = -a_{2n+2} + (a_{2n+3} - a_{2n+4}) + (a_{2n+5} - a_{2n+6}) + \cdots > -a_{2n+2}$. Hence, $|R_{2n+1}| < a_{2n+2}$. Thus, for all k, $|R_k| < a_{k+1}$.

EXAMPLE 45.1: The alternating harmonic series

$$1-\frac{1}{2}+\frac{1}{3}-\frac{1}{4}+\frac{1}{5}-\frac{1}{6}+\cdots$$

converges by virtue of the Alternating Series Theorem. By part (II) of that theorem, the magnitude $|R_n|$ of the error after n terms is less than $\frac{1}{n+1}$. If we want an error less than 0.1, then it suffices to take $\frac{1}{n+1} \le 0.1 = \frac{1}{10}$, which is equivalent to $10 \le n+1$. So, $n \ge 9$. Thus, we must use

$$A_9 = 1-\frac{1}{2}+\frac{1}{3}-\frac{1}{4}+\frac{1}{5}-\frac{1}{6}+\frac{1}{7}-\frac{1}{8}+\frac{1}{9} = \frac{1879}{2520} \sim 0.7456$$

Definition

Consider an arbitrary series $\sum s_n$.

$\sum s_n$ is said to be *absolutely convergent* if $\sum |s_n|$ is convergent.
$\sum s_n$ is said to be *conditionally convergent* if it is convergent but not absolutely convergent.

EXAMPLE 45.2: The alternating harmonic series $\sum (-1)^{n+1}\frac{1}{n}$ is conditionally convergent.

EXAMPLE 45.3: The series $\sum (-1)^{n+1}\frac{1}{n^2}$ is absolutely convergent.

We shall state without proof two significant results about absolute and conditional convergence. In what follows, by a rearrangement of a series we mean a series obtained from the given series by rearranging its terms (that is, by changing the order in which the terms occur).

(1) If $\sum s_n$ is absolutely convergent, then every rearrangement of $\sum s_n$ is convergent and has the same sum as $\sum s_n$.
(2) If $\sum s_n$ is conditionally convergent, then if c is any real number or $+\infty$ or $-\infty$, there is a rearrangement of $\sum s_n$ with sum c.

Theorem 45.2: If a series is absolutely convergent, then it is convergent.
For a proof, see Problem 1.

Note that a positive series is absolutely convergent if and only if it is convergent.
The following test is probably the most useful of all convergence tests.

Theorem 45.3 (The Ratio Test): Let $\sum s_n$ be any series.

(1) If $\lim_{n\to+\infty}\left|\frac{s_{n+1}}{s_n}\right| = r < 1$, then $\sum s_n$ is absolutely convergent.

(2) If $\lim_{n\to+\infty}\left|\frac{s_{n+1}}{s_n}\right| = r$ and ($r > 1$ or $r = +\infty$), then $\sum s_n$ diverges.

(3) If $\lim_{n\to+\infty}\left|\frac{s_{n+1}}{s_n}\right| = 1$, then we can draw no conclusion about the convergence or divergence of $\sum s_n$. For a proof, see Problem 14.

Theorem 45.4 (The Root Test): Let $\sum s_n$ be any series.

(1) If $\lim_{n\to+\infty}\sqrt[n]{|s_n|} = r < 1$, then $\sum s_n$ is absolutely convergent.
(2) If $\lim_{n\to+\infty}\sqrt[n]{|s_n|} = r$ and ($r > 1$ or $r = +\infty$), then $\sum s_n$ diverges.
(3) If $\lim_{n\to+\infty}\sqrt[n]{|s_n|} = 1$, then we can draw no conclusion about the convergence or divergence of $\sum s_n$.

For a proof, see Problem 15.

EXAMPLE 45.4: Consider the series $\sum \frac{2^{2n}}{n^n}$. Then $\lim_{n\to+\infty} \sqrt[n]{|s_n|} = \lim_{n\to+\infty} \frac{4}{n} = 0$. So, by the root test, the series converges absolutely.

SOLVED PROBLEMS

1. Show that, if $\sum s_n$ is absolutely convergent, then it is convergent.

$0 \le s_n + |s_n| \le 2|s_n|$. Since $\sum |s_n|$ converges, so does $\sum 2|s_n|$. Then, by the comparison test, $\sum (s_n + |s_n|)$ converges. Hence, $\sum s_n = \sum ((s_n + |s_n|) - |s_n|)$ converges by Corollary 43.4.

In Problems 2–13, determine whether the given series converges absolutely, conditionally, or not at all.

2. $\frac{1}{2} - \frac{1}{5} + \frac{1}{10} - \frac{1}{17} + \cdots$.

$s_n = (-1)^{n+1}\frac{1}{n^2+1}$. $\sum \frac{1}{n^2+1}$ converges by comparison with the convergent p-series $\sum \frac{1}{n^2}$. So, $\sum (-1)^{n+1}\frac{1}{n^2+1}$ is absolutely convergent.

3. $\frac{1}{e} - \frac{2}{e^2} + \frac{3}{e^3} - \frac{4}{e^4} + \cdots$.

$s_n = (-1)^{n+1}\frac{n}{e^n}$. The series $\sum \frac{n}{e^n}$ converges by the integral test $\left(\text{using } f(x) = \frac{x}{e^x}\right)$. Hence, $\sum (-1)^{n+1}\frac{n}{e^n}$ is absolutely convergent.

4. $1 - \frac{1}{\sqrt{2}} + \frac{1}{\sqrt{3}} - \frac{1}{\sqrt{4}} + \frac{1}{\sqrt{5}} - \cdots$.

$s_n = (-1)^{n+1}\frac{1}{\sqrt{n}}$. Since $\left\langle \frac{1}{\sqrt{n}} \right\rangle$ is a decreasing sequence, the series converges by virtue of the alternating series test. But $\sum \frac{1}{\sqrt{n}}$ is divergent, since it is a p-series with $p = \frac{1}{2} < 1$.

5. $1 - \frac{1}{2} + \frac{1}{4} - \frac{1}{8} + \cdots$.

The series $1 + \frac{1}{2} + \frac{1}{4} - \frac{1}{8} + \cdots$ is a geometric series with ratio $r = \frac{1}{2}$. Since $|r| < 1$, it converges and, therefore, the given series is absolutely convergent.

6. $1 - \frac{2}{3} + \frac{3}{3^2} - \frac{4}{3^3} + \cdots$.

$s_n = (-1)^{n+1}\frac{n}{3^{n-1}}$. Let us apply the ratio test:

$$\lim_{n\to+\infty}\left|\frac{s_{n+1}}{s_n}\right| = \frac{n+1}{3^n} \bigg/ \frac{n}{3^{n-1}} = \frac{n+1}{n}\frac{1}{3}. \qquad \text{So,} \qquad \left|\frac{s_{n+1}}{s_n}\right| = \frac{1}{3} < 1$$

Hence, the given series is absolutely convergent.

7. $\frac{1}{2} - \frac{2}{3}\frac{1}{2^3} + \frac{3}{4}\frac{1}{3^3} - \frac{4}{5}\frac{1}{4^3} + \cdots$.

$s_n = (-1)^{n+1}\frac{n}{n+1}\frac{1}{n^3}$. Look at $\sum |s_n|$. $|s_n| = \frac{n}{n+1}\frac{1}{n^3} < \frac{1}{n^3}$. So, $\sum |s_n|$ converges by comparison with the convergent p-series $\sum \frac{1}{n^3}$. Hence, the given series is absolutely convergent.

8. $\frac{2}{3} - \frac{3}{4}\frac{1}{2} + \frac{4}{5}\frac{1}{3} - \frac{5}{6}\frac{1}{4} + \cdots$.

$s_n = (-1)^{n+1}\frac{n+1}{n+2}\frac{1}{n}$. Note that $\left\langle \frac{n+1}{n+2}\frac{1}{n} \right\rangle$ is a decreasing sequence $\left(\text{since } D_x\left(\frac{x+1}{(x+2)x}\right) < 0\right)$. Hence, the given series is convergent by the Alternating Series Theorem. However, $|s_n| > \frac{1}{2}\frac{1}{n}$. So, $\sum |s_n|$ diverges by comparison with $\sum \frac{1}{n}$. Thus, the given series is conditionally convergent.

9. $2-\frac{2^3}{3!}+\frac{2^5}{5!}-\frac{2^7}{7!}+\cdots$.

$s_n=(-1)^{n+1}\frac{2^{2n-1}}{(2n-1)!}$. Apply the ratio test:

$$\left|\frac{s_{n+1}}{s_n}\right|=\frac{2^{2n+1}}{(2n+1)!}\bigg/\frac{2^{2n-1}}{(2n-1)!}=\frac{4}{(2n+1)(2n)}$$

Hence, $\lim_{n\to+\infty}\left|\frac{s_{n+1}}{s_n}\right|=0$ and, therefore, the series is absolutely convergent.

10. $\frac{1}{2}-\frac{4}{2^3+1}+\frac{9}{3^3+1}-\frac{16}{4^3+1}+\cdots$.

$s_n=(-1)^{n+1}\frac{n^2}{n^3+1}$. Since $\left\langle\frac{n^2}{n^3+1}\right\rangle$ is a decreasing sequence for $n\geq 2$, the given series converges by the Alternating Series Theorem. The series $\sum|s_n|$ is divergent by limit comparison with $\sum\frac{1}{n}$. Hence, the given series is conditionally convergent.

11. $\frac{1}{2}-\frac{2}{2^3+1}+\frac{3}{3^3+1}-\frac{4}{4^3+1}+\cdots$.

$s_n=(-1)^{n+1}\frac{n}{n^3+1}$. $\sum|s_n|$ is convergent by limit comparison with $\sum\frac{1}{n^2}$. Hence, the given series is absolutely convergent.

12. $\frac{1}{1\cdot 2}-\frac{1}{2\cdot 2^2}+\frac{1}{3\cdot 2^3}-\frac{1}{4\cdot 2^4}+\cdots$.

$s_n=(-1)^{n+1}\frac{1}{n2^n}$. Apply the ratio test:

$$\left|\frac{s_{n+1}}{s_n}\right|=\frac{1}{(n+1)2^{n+1}}\bigg/\frac{1}{n2^n}=\frac{n}{n+1}\frac{1}{2}$$

Thus, $\lim_{n\to+\infty}\left|\frac{s_{n+1}}{s_n}\right|=\frac{1}{2}<1$. So the given series is absolutely convergent.

13. $\sum(-1)^{n+1}\frac{n^3}{(n+1)!}$.

Apply the ratio test:

$$\left|\frac{s_{n+1}}{s_n}\right|=\frac{(n+1)^3}{(n+2)!}\bigg/\frac{n^3}{(n+1)!}=\left(\frac{n+1}{n}\right)^3\left(\frac{1}{n+2}\right)$$

So, $\lim_{n\to+\infty}\left|\frac{s_{n+1}}{s_n}\right|=0$. Hence, the given series is absolutely convergent.

14. Justify the ratio test (Theorem 45.3).

(a) Assume $\lim_{n\to+\infty}\left|\frac{s_{n+1}}{s_n}\right|=r<1$. Choose t such that $r<t<1$. Then there exists a positive integer m such that, if $n\geq m$, $\left|\frac{s_{n+1}}{s_n}\right|\leq t$. Hence,

$$|s_{m+1}|\leq t|s_m|,\qquad |s_{m+2}|\leq t|s_{m+1}|\leq t^2|s_m|,\qquad \ldots,\qquad |s_{m+k}|\leq t^k|s_m|$$

But, $\sum t^k|s_m|$ is a convergent geometric series (with ratio $t<1$). So, by the comparison test, $\sum|s_n|$ converges. Hence, $\sum s_n$ is absolutely convergent.

(b) Assume $\lim_{n\to+\infty}\left|\frac{s_{n+1}}{s_n}\right| = r$ and ($r > 1$ or $r = +\infty$). Choose t so that $1 < t < r$. There exists a positive integer m such that, if $n \geq m$, $\left|\frac{s_{n+1}}{s_n}\right| \geq t$. Hence,

$$|s_{m+1}| \geq t|s_m|, \qquad |s_{m+2}| \geq t|s_{m+1}| \geq t^2|s_m|, \qquad \cdots, \qquad |s_{m+k}| \geq t^k|s_m|$$

Therefore, $\lim_{n\to+\infty} s_n = \infty$ and, by the divergence theorem, $\sum s_n$ diverges.

(c) Consider $\sum\frac{1}{n}$. $\lim_{n\to+\infty}\left|\frac{s_{n+1}}{s_n}\right| = \lim_{n\to+\infty}\left[\left(\frac{1}{n+1}\right)\Big/\frac{1}{n}\right] = \lim_{n\to+\infty}\frac{n}{n+1} = 1$. In this case, the series diverges. Now consider $\sum\frac{1}{n^2}$:

$$\lim_{n\to+\infty}\left|\frac{s_{n+1}}{s_n}\right| = \lim_{n\to+\infty}\left(\frac{1}{(n+1)^2}\Big/\frac{1}{n^2}\right) = \lim_{n\to+\infty}\left(\frac{n}{n+1}\right)^2 = 1$$

In this case, the series converges.

15. Justify the root test (Theorem 45.4).

(a) Assume $\lim_{n\to+\infty}\sqrt[n]{|s_n|} = r < 1$. Choose t so that $r < t < 1$. Then, there exists a positive integer m such that $\sqrt[n]{|s_n|} \leq t$ for $n \geq m$. Hence, $|s_n| \leq t^n$ for $n \geq m$. Therefore, $\sum|s_n|$ converges by comparison with the convergent geometric series $\sum t^n$. So, $\sum s_n$ is absolutely convergent.

(b) Assume $\lim_{n\to+\infty}\sqrt[n]{|s_n|} = r$ and ($r > 1$ or $r = +\infty$). Choose t so that $1 < t <$ r. For some positive integer m, $\sqrt[n]{|s_n|} > t$ for $n \geq m$. Then $|s_n| \geq t^n$ for $n \geq m$. Since $\lim_{n\to+\infty} t^n = +\infty$, $\lim_{n\to+\infty} s_n = \infty$. So, by the Divergence Theorem, $\sum s_n$ diverges.

(c) Consider $\sum\frac{1}{n}$ and $\sum\frac{1}{n^2}$. In both cases, $\lim_{n\to+\infty}\sqrt[n]{|s_n|} = 1$. (Note that $\lim_{n\to+\infty} n^{-n} = \lim_{n\to+\infty} e^{-(\ln n)/n} = 1$.)

In Problems 16–22, use the ratio test to test the series for convergence.

16. $\frac{1}{3}+\frac{2}{3^2}+\frac{3}{3^3}+\frac{4}{3^4}+\cdots$.

$$\frac{s_{n+1}}{s_n} = \frac{n+1}{3^{n+1}}\Big/\frac{n}{3^n} = \frac{1}{3}\frac{n+1}{n}. \qquad \text{So,} \quad \lim_{n\to+\infty}\left|\frac{s_{n+1}}{s_n}\right| = \frac{1}{3} < 1$$

So, the series converges by the ratio test.

17. $\frac{1}{3}+\frac{2!}{3^2}+\frac{3!}{3^3}+\frac{4!}{3^4}+\cdots$.

$$s_n = \frac{n!}{3^n}. \qquad \text{So,} \quad \frac{s_{n+1}}{s_n} = \frac{(n+1)!}{3^{n+1}}\Big/\frac{n!}{3^n} = \frac{n+1}{3}$$

Hence, $\lim_{n\to+\infty}\left|\frac{s_{n+1}}{s_n}\right| = +\infty$ and the series diverges by the ratio test.

18. $1+\frac{1\cdot 2}{1\cdot 3}+\frac{1\cdot 2\cdot 3}{1\cdot 3\cdot 5}+\frac{1\cdot 2\cdot 3\cdot 4}{1\cdot 3\cdot 5\cdot 7}+\cdots$.

$$s_n = \frac{n!}{1\cdot 3\cdot 5\cdot\cdots\cdot(2n-1)}. \qquad \text{Then} \quad \frac{s_{n+1}}{s_n} = \frac{(n+1)!}{1\cdot 3\cdot 5\cdot\cdots\cdot(2n+1)}\Big/\frac{n!}{1\cdot 3\cdot 5\cdot\cdots\cdot(2n-1)} = \frac{n+1}{2n+1}.$$

So, $\lim_{n\to+\infty}\left|\frac{s_{n+1}}{s_n}\right| = \frac{1}{2} < 1$. Hence, the series converges by the ratio test.

19. $2+\frac{3}{2}\frac{1}{4}+\frac{4}{3}\frac{1}{4^2}+\frac{5}{4}\frac{1}{4^3}+\cdots$.

$$s_n = \frac{n+1}{n}\frac{1}{4^{n-1}}. \qquad \text{Then} \quad \frac{s_{n+1}}{s_n} = \left(\frac{n+2}{n+1}\frac{1}{4^n}\right)\Big/\left(\frac{n+1}{n}\frac{1}{4^{n-1}}\right) = \frac{1}{4}\frac{n(n+2)}{(n+1)^2}.$$

So, $\lim_{n\to+\infty}\left|\frac{s_{n+1}}{s_n}\right| = \frac{1}{4} < 1$. Hence, the series converges by the ratio test.

20. $1+\frac{2^2+1}{2^3+1}+\frac{3^2+1}{3^3+1}+\frac{4^2+1}{4^3+1}+\cdots$

$$s_n=\frac{n^2+1}{n^3+1}. \qquad \text{Then} \qquad \frac{s_{n+1}}{s_n}=\frac{(n+1)^2+1}{(n+1)^3+1}\Big/\frac{n^2+1}{n^3+1}=\frac{((n+1)^2+1)(n^3+1)}{((n+1)^3+1)(n^2+1)}.$$

Then $\lim_{n\to+\infty}\left|\frac{s_{n+1}}{s_n}\right|=1$. So the ratio test yields no conclusion. However, limit comparison with $\sum\frac{1}{n}$ shows that the series diverges.

21. $\sum\frac{n3^n}{(n+1)!}$.

$$\frac{s_{n+1}}{s_n}=\frac{(n+1)3^{n+1}}{(n+2)!}\Big/\frac{n3^n}{(n+1)!}=\frac{n+1}{n}\frac{3}{n+2}. \qquad \text{So,} \qquad \lim_{n\to+\infty}\left|\frac{s_{n+1}}{s_n}\right|=0$$

Hence, the series converges by the ratio test.

22. $\sum\frac{n^n}{n!}$.

$$\frac{s_{n+1}}{s_n}=\frac{(n+1)^{n+1}}{(n+1)!}\Big/\frac{n^n}{n!}=\left(\frac{n+1}{n}\right)^n=\left(1+\frac{1}{n}\right)^n. \qquad \text{So,} \qquad \lim_{n\to+\infty}\left|\frac{s_{n+1}}{s_n}\right|=e>1$$

Hence, the series diverges by the ratio test.

SUPPLEMENTARY PROBLEMS

In Problems 23–40, determine whether the given alternating series is absolutely convergent, conditionally convergent, or divergent.

23. $\sum(-1)^{n+1}\frac{1}{n!}$ *Ans.* absolutely convergent

24. $\sum(-1)^{n+1}\frac{1}{\ln n}$ *Ans.* conditionally convergent

25. $\sum(-1)^{n+1}\frac{n}{n+1}$ *Ans.* divergent

26. $\sum(-1)^{n+1}\frac{\ln n}{3n+1}$ *Ans.* conditionally convergent

27. $\sum(-1)^{n+1}\frac{1}{2n-1}$ *Ans.* conditionally convergent

28. $\sum(-1)^{n+1}\frac{1}{\sqrt[n]{3}}$ *Ans.* divergent

29. $\sum(-1)^{n+1}\frac{1}{(2n-1)^2}$ *Ans.* absolutely convergent

30. $\sum(-1)^{n+1}\frac{1}{\sqrt{n(n+1)}}$ *Ans.* conditionally convergent

31. $\sum(-1)^{n+1}\frac{1}{(n+1)^2}$ *Ans.* absolutely convergent

32. $\sum (-1)^{n+1} \frac{1}{n^2+2}$ *Ans.* absolutely convergent

33. $\sum (-1)^{n+1} \frac{1}{(n!)^2}$ *Ans.* absolutely convergent

34. $\sum (-1)^{n+1} \frac{n}{n^2+1}$ *Ans.* conditionally convergent

35. $\sum (-1)^{n+1} \frac{n^2}{n^4+2}$ *Ans.* absolutely convergent

36. $\sum (-1)^{n+1} n \left(\frac{3}{4}\right)^4$ *Ans.* absolutely convergent

37. $\sum (-1)^{n+1} \frac{n^2-3}{n^2+n+2}$ *Ans.* divergent

38. $\sum (-1)^{n+1} \frac{n+1}{2^n}$ *Ans.* absolutely convergent

39. $\sum (-1)^{n+1} \frac{n^3}{2^{n+2}}$ *Ans.* absolutely convergent

40. $\sum \frac{\cos \pi n}{n^2}$ *Ans.* absolutely convergent

41. (GC) How many terms of $\sum (-1)^{n+1} \frac{1}{n!}$ will suffice to get an approximation within 0.0005 of the actual sum? Find that approximation.

Ans. $n = 6; \frac{91}{144} \sim 0.632$

42. (GC) How many terms of $\sum (-1)^{n+1} \frac{1}{(2n-1)!}$ will suffice to get an approximation of the actual sum with an error < 0.001? Find that approximation.

Ans. $n = 3$; 0.842.

43. (GC) How many terms of $\sum (-1)^{n+1} \frac{1}{n}$ will suffice to get an approximation of the actual sum with an error < 0.001? Find that approximation.

Ans. $n = 1000$; 0.693

In Problems 44–49, determine whether the series converges.

44. $\sum \frac{(n!)^2}{(2n)!}$ *Ans.* convergent

45. $\sum \frac{(2n)!}{n^4}$ *Ans.* divergent

46. $\sum \frac{n^3}{(\ln 2)^n}$ *Ans.* divergent

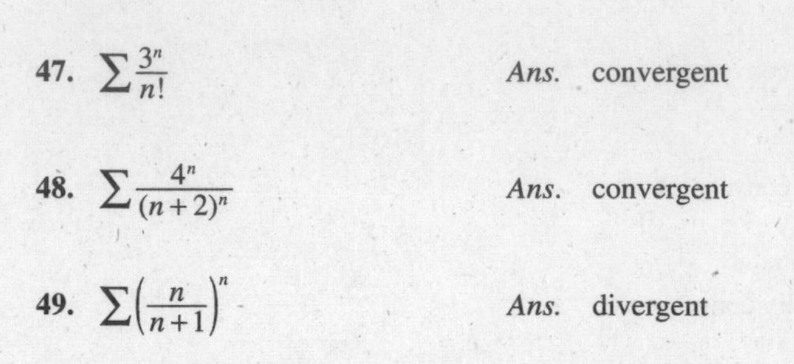

47. $\sum \frac{3^n}{n!}$ *Ans.* convergent

48. $\sum \frac{4^n}{(n+2)^n}$ *Ans.* convergent

49. $\sum \left(\frac{n}{n+1}\right)^n$ *Ans.* divergent

50. Determine whether $\sum (-1)^{n+1}(\sqrt{n+1}-\sqrt{n})$ is absolutely convergent, conditionally convergent, or divergent.

Ans. conditionally convergent

In Problems 51 and 52, find the number of terms that suffice to approximate the sum of the given series to four-decimal-place accuracy (that is, with an error $< 5/10^5$) and compute the approximation.

51. (GC) $\sum_{n=1}^{+\infty} (-1)^{n+1}\frac{1}{n^5}$ *Ans.* $n = 6$; 0.9721

52. (GC) $\sum_{n=1}^{+\infty} (-1)^{n+1}\frac{1}{(2n-1)!}$ *Ans.* $n = 4$; 0.8415

53. Let $|r| < 1$

(a) Prove that $\sum nr^n = r + 2r^2 + 3r^3 + 4r^4 + \cdots$ converges.

(b) Show that $\sum_{n=1}^{+\infty} nr^n = \frac{r}{(1-r)^2}$. (*Hint:* Let $S = r + 2r^2 + 3r^2 + 4r^4 + \cdots$, multiply this equation by r, and subtract the result from the original equation.)

(c) Show that $\sum_{n=1}^{+\infty} \frac{n}{2^n} = 2$.

CHAPTER 46

Power Series

Power Series

An infinite series

$$\sum_{n=0}^{+\infty} a_n (x-c)^n = a_0 + a_1(x-c) + a_2(x-c)^2 + \cdots \tag{46.1}$$

is called a *power series* in x about c with coefficients $\langle a_n \rangle$. An important special case

$$\sum_{n=0}^{+\infty} a_n x^n = a_0 + a_1 x + a_2 x^2 + \cdots \tag{46.2}$$

is a power series about 0.

For a given value of x, the series (46.1) either converges or diverges. Hence, (46.1) determines a function f whose domain is the set of all x for which (46.1) converges and whose corresponding value $f(x)$ is the sum of the series.

Note that (46.1) converges when $x = c$.

EXAMPLE 46.1: The power series about 0

$$\sum_{n=0}^{+\infty} x^n = 1 + x + x^2 + \cdots$$

is a geometric series with ratio x. Thus, it converges for $|x| < 1$, and its sum is $\frac{1}{1-x}$. So, the domain of the corresponding function is an interval around 0.

Theorem 46.1: Assume that the power series $\sum_{n=0}^{+\infty} a_n (x-c)^n$ converges for $x_0 \neq c$. Then it converges absolutely for all x such that $|x - c| < |x_0 - c|$ (that is, for all x that are closer to c than x_0).

For a proof, see Problem 4.

Theorem 46.2: For a power series $\sum_{n=0}^{+\infty} a_n (x-c)^n$, one of the following three cases holds:

(a) it converges for all x; or
(b) it converges for all x in an open interval $(c - R_1, c + R_1)$ around c, but not outside the closed interval $[c - R_1, c + R_1]$; or
(c) it converges only for $x = c$.

By the *interval of convergence* of $\sum_{n=0}^{+\infty} a_n (x-c)^n$ we mean:

In case (a): $(-\infty, +\infty)$
In case (b): $(c - R_1, c + R_1)$
In case (c): $\{c\}$

By the *radius of convergence* of $\sum_{n=0}^{+\infty} a_n(x-c)^n$ we mean:

In case (a): ∞

In case (b): R_1

In case (c): 0

Note: In case (b), whether the power series converges at neither endpoint of its interval of convergence or at one or both of those endpoints depends upon the given series.

For a sketch of a proof of Theorem 46.2, see Problem 5.

EXAMPLE 46.2: The power series

$$\sum_{n=1}^{+\infty} \frac{(x-2)^n}{n} = (x-2) + \frac{(x-2)^2}{2} + \frac{(x-2)^3}{3} + \cdots$$

is a power series about 2. Let us use the ratio test to find the interval of convergence.

$$\left|\frac{s_{n+1}}{s_n}\right| = \frac{|x-2|^{n+1}}{n+1} \Big/ \frac{|x-2|^n}{n} = \frac{n}{n+1}|x-2|. \quad \text{Thus,} \quad \lim_{n\to+\infty}\left|\frac{s_{n+1}}{s_n}\right| = |x-2|.$$

So, by the ratio test, the series converges absolutely for $|x-2| < 1$. The latter inequality is equivalent to $-1 < x-2 < 1$, which, in turn, is equivalent to $1 < x < 3$. Hence, the interval of convergence is (1, 3) and the radius of convergence is 1. At the endpoint $x = 1$, the series becomes $\sum_{n=1}^{+\infty}[(-1)^n/n]$, which converges by the Alternating Series Theorem. At the endpoint $x = 3$, the series becomes $\sum_{n=1}^{+\infty}(1/n)$, the divergent harmonic series. Thus, the power series converges for $1 \le x < 3$.

EXAMPLE 46.3: The power series

$$\sum_{n=0}^{+\infty} \frac{x^n}{n!} = 1 + x + \frac{x^2}{2!} + \frac{x^3}{3!} + \cdots$$

is a power series about 0. (Recall that $0! = 1$.) Let us use the ratio test:

$$\left|\frac{s_{n+1}}{s_n}\right| = \frac{|x|^{n+1}}{(n+1)!} \Big/ \frac{|x|^n}{n!} = \frac{|x|}{n+1}. \quad \text{So,} \quad \lim_{n\to+\infty}\left|\frac{s_{n+1}}{s_n}\right| = 0.$$

Hence, by the ratio test, the series converges (absolutely) for all x. Its interval of convergence is $(-\infty, +\infty)$ and its radius of convergence is ∞.

EXAMPLE 46.4: The power series

$$\sum_{n=0}^{+\infty} n!x^n = 1 + x + 2!x^2 + 3!x^3 + \cdots$$

is a power series about 0. Let us use the ratio test again:

$$\left|\frac{s_{n+1}}{s_n}\right| = \frac{(n+1)!\,|x|^{n+1}}{n!\,|x|^n} = (n+1)|x|. \quad \text{So,} \quad \lim_{n\to+\infty}\left|\frac{s_{n+1}}{s_n}\right| = +\infty.$$

except when $x = 0$. Thus, the series converges only for $x = 0$. Its (degenerate) "interval" of convergence is $\{0\}$ and its radius of convergence is 0.

Uniform Convergence

Let $\langle f_n \rangle$ be a sequence of functions, all defined on a set A. Let f be a function defined on A. Then $\langle f_n \rangle$ is said to *converge uniformly* to f on A if, for every $\epsilon > 0$, there exists a positive integer m such that, for each x in A and every $n \geq m$, $|f_n(x) - f(x)| < \epsilon$.

Theorem 46.3: If a power series $\sum_{n=0}^{+\infty} a_n(x-c)^n$ converges for $x_0 \neq c$ and $d < |x_0 - c|$, then the sequence of partial sums $\langle S_k(x) \rangle$, where $S_k(x) = \sum_{n=0}^{k} a_n(x-c)^n$, converges uniformly to $\sum_{n=0}^{+\infty} a_n(x-c)^n$ on the interval consisting of all x such that $|x - c| < d$. Hence, the convergence is uniform on any interval strictly inside the interval of convergence.

The reader is referred to more advanced books on analysis for a proof of this result.

Theorem 46.4: If $\langle f_n \rangle$ converges uniformly to f on a set A and each f_n is continuous on A, then f is continuous on A.

For a proof, see Problem 6.

Corollary 46.5: The function defined by a power series $\sum_{n=0}^{+\infty} a_n(x-c)^n$ is continuous at all points within its interval of convergence.

This follows from Theorems 46.3 and 46.4.

Theorem 46.6 (Integration of Power Series): Let f be the function defined by a power series $\sum_{n=0}^{+\infty} a_n(x-c)^n$ on its interval of convergence (with radius of convergence R_1). Then:

(a)
$$\int f(x)\,dx = \sum_{n=0}^{+\infty} a_n \frac{(x-c)^{n+1}}{n+1} + K \quad \text{for} \quad |x-c| < R_1 \tag{46.3}$$

where the interval of convergence of the power series on the right side of formula (46.3) is the same as that of the original series. K is an arbitrary constant of integration. Note that the antiderivative of f is obtained by term-by-term integration of the given power series.

(b) If a and b are in the interval of convergence, then:

$$\int_a^b f(x)\,dx = \sum_{n=0}^{+\infty} \left(a_n \frac{(x-c)^{n+1}}{n+1} \right) \Bigg]_a^b \tag{46.4}$$

Thus, $\int_a^b f(x)\,dx$ is obtained by term-by-term integration.

For a proof of Theorem 46.6, the reader should consult a more advanced book on analysis.

Theorem 46.7 (Differentiation of Power Series): Let f be the function defined by a power series $\sum_{n=0}^{+\infty} a_n(x-c)^n$ on its interval of convergence (with radius of convergence R_1). Then f is differentiable in that interval and

$$f'(x) = \sum_{n=0}^{+\infty} n a_n (x-c)^{n-1} \quad \text{for} \quad |x-c| < R_1 \tag{46.5}$$

Thus, the derivative f' is obtained by term-by-term differentiation of the power series. The interval of convergence of the power series on the right side of formula (46.5) will be the same as for the original power series.

For a proof, the reader is referred to more advanced texts in analysis.

EXAMPLE 46.5: We already know by Example 1 that, for $|x| < 1$,

$$\frac{1}{1-x} = \sum_{n=0}^{+\infty} x^n = 1 + x + x^2 + x^3 + \cdots + x^n + \cdots \tag{46.6}$$

Now, $D_x\left(\frac{1}{1-x}\right) = \frac{1}{(1-x)^2}$. So, by Theorem 46.7,

$$\frac{1}{(1-x)^2} = 1 + 2x + 3x^2 + \cdots + nx^{n-1} + \cdots \quad \text{for } |x| < 1$$

$$= \sum_{n=1}^{+\infty} nx^{n-1} = \sum_{n=0}^{+\infty} (n+1)x^n$$

EXAMPLE 46.6: We know already that

$$\frac{1}{1-x} = \sum_{n=0}^{+\infty} x^n = 1 + x + x^2 + x^3 + \cdots + x^n + \cdots \quad \text{for } |x| < 1$$

Replace x by $-x$. (This is permissible, since $|-x| = |x| < 1$.) The result is

$$\frac{1}{1+x} = \sum_{n=0}^{+\infty} (-x)^n = \sum_{n=0}^{+\infty} (-1)^n x^n = 1 - x + x^2 - x^3 + \cdots \tag{46.7}$$

By Theorem 46.6(a), we can integrate term by term:

$$\int \frac{dx}{1+x} = \sum_{n=0}^{+\infty} (-1)^n \frac{x^{n+1}}{n+1} + K = \sum_{n=1}^{+\infty} (-1)^{n-1} \frac{x^n}{n} + K \quad \text{for } |x| < 1$$

$$\ln|1+x| = \sum_{n=1}^{+\infty} (-1)^{n-1} \frac{x^n}{n} + K \quad \text{for } |x| < 1$$

Letting $x = 0$ and noting that $\ln 1 = 0$, we find that $K = 0$.

Note also that, for $|x| < 1$, we have $-1 < x < 1$, $0 < 1 + x < 2$, and, therefore, $|1 + x| = 1 + x$. Hence,

$$\ln(1+x) = \sum_{n=1}^{+\infty} (-1)^{n-1} \frac{x^n}{n} \quad \text{for } |x| < 1$$

$$= x - \frac{1}{2}x^2 + \frac{1}{3}x^3 - \frac{1}{4}x^4 + \cdots \tag{46.8}$$

The ratio test shows that this series converges.

If we replace x by $x - 1$, we obtain:

$$\ln x = \sum_{n=1}^{+\infty} (-1)^{n-1} \frac{(x-1)^n}{n} \quad \text{for } |x-1| < 1 \tag{46.9}$$

Note that $|x - 1| < 1$ is equivalent to $0 < x < 2$.

Thus, $\ln x$ is definable by a power series within $(0, 2)$.

Theorem 46.8 (Abel's Theorem): Assume that the power series $\sum_{n=0}^{+\infty} a_n(x-c)^n$ has a finite interval of convergence $|x - c| < R_1$ and let f be a function whose values in that interval are given by that power series. If the power series also converges at the right-hand endpoint $b = c + R_1$ of the interval of convergence, then $\lim_{x \to b^-} f(x)$ exists and is equal to the sum of the series at b. The analogous result holds at the left-hand endpoint $a = c - R_1$.

The reader is referred to advanced books on analysis for a proof.

EXAMPLE 46.7: This is a continuation of Example 6. By formula (46.8),

$$\ln(1+x) = \sum_{n=1}^{+\infty} (-1)^{n-1} \frac{x^n}{n} \quad \text{for } |x| < 1$$

At the right-hand endpoint $x = 1$ of the interval of convergence, the power series becomes the convergent alternating harmonic series

$$\sum_{n=1}^{+\infty} (-1)^{n-1} \frac{1}{n} = 1 - \tfrac{1}{2} + \tfrac{1}{3} + \tfrac{1}{4} + \cdots$$

By Abel's Theorem, this series is equal to $\lim\limits_{x \to 1^-} \ln(1+x) = \ln 2$. So,

$$\ln 2 = 1 - \tfrac{1}{2} + \tfrac{1}{3} - \tfrac{1}{4} + \cdots \tag{46.10}$$

EXAMPLE 46.8: Start again with

$$\frac{1}{1-x} = \sum_{n=0}^{+\infty} x^n = 1 + x + x^2 + x^3 + \cdots + x^n + \quad \text{for } |x| < 1$$

Replace x by $-x^2$, obtaining

$$\frac{1}{1+x^2} = \sum_{n=0}^{+\infty} (-1)^n x^{2n} = 1 - x^2 + x^4 - x^6 + \cdots \tag{46.11}$$

Since $|-x^2| < 1$ is equivalent to $|x| < 1$, (46.11) holds for $|x| < 1$.
Now, by Theorem 46.6(a), the antiderivative $\tan^{-1} x$ of $\frac{1}{1+x^2}$ can be obtained by term-by-term integration:

$$\begin{aligned} \tan^{-1} x &= \sum_{n=0}^{+\infty} (-1)^n \frac{x^{2n+1}}{2n+1} + K \quad \text{for } |x| < 1 \\ &= K + x - \tfrac{1}{3}x^3 + \tfrac{1}{5}x^5 - \tfrac{1}{7}x^7 + \cdots \end{aligned}$$

Here K is the constant of integration. If we let $x = 0$ and note that $\tan^{-1} 0 = 0$, it follows that $K = 0$. Hence,

$$\tan^{-1} x = \sum_{n=0}^{+\infty} (-1)^n \frac{x^{2n+1}}{2n+1} = x - \tfrac{1}{3}x^3 + \tfrac{1}{5}x^5 - \tfrac{1}{7}x^7 + \cdots \tag{46.12}$$

At the right-hand endpoint $x = 1$ of the interval of convergence, the series in (46.12) becomes

$$\sum_{n=0}^{+\infty} (-1)^n \frac{1}{2n+1} = 1 - \tfrac{1}{3} + \tfrac{1}{5} - \tfrac{1}{7} + \cdots$$

which converges by virtue of the Alternating Series Theorem. So, by Abel's Theorem,

$$1 - \tfrac{1}{3} + \tfrac{1}{5} - \tfrac{1}{7} + \cdots = \lim_{x \to 1^-} \tan^{-1}(x) = \tan^{-1} 1 = \frac{\pi}{4} \tag{46.13}$$

EXAMPLE 46.9: We know already, by Example 3, that $\sum_{n=0}^{+\infty} \frac{x^n}{n!}$ converges for all x. Let $f(x) = \sum_{n=0}^{+\infty} \frac{x^n}{n!}$ for all x. By term-by-term differentiation (Theorem 46.7),

$$f'(x) = \sum_{n=1}^{+\infty} \frac{x^{n-1}}{(n-1)!} = \sum_{n=0}^{+\infty} \frac{x^n}{n!} = f(x)$$

Note that $f(0) = 1$. Therefore, by formula (28.2), $f(x) = e^x$. Thus,

$$e^x = \sum_{n=0}^{+\infty} \frac{x^n}{n!} \quad \text{for all } x \tag{46.14}$$

SOLVED PROBLEMS

1. Find the interval of convergence of the power series

$$\sum_{n=1}^{+\infty} \frac{(x-2)^n}{n} = (x-2) + \frac{(x-2)^2}{2} + \frac{(x-2)^3}{3} + \cdots$$

and identify the function represented by this power series.

Use the ratio test:

$$\left|\frac{s_{n+1}}{s_n}\right| = \frac{|x-2|^{n+1}}{n+1} \bigg/ \frac{|x-2|^n}{n} = \frac{n}{n+1}|x-2|. \quad \text{So,} \quad \lim_{n\to+\infty}\left|\frac{s_{n+1}}{s_n}\right| = |x-2|$$

Hence, the interval of convergence is $|x-2| < 1$. (This is equivalent to $-1 < x-2 < 1$. which, in turn, is equivalent to $1 < x < 3$.) At the right endpoint $x = 3$, the series is the divergent harmonic series, and, at the left endpoint $x = 1$, the series is the negative of the convergent alternating harmonic series. So, the series converges for $1 \le x < 3$.

Let $h(x) = \sum_{n=1}^{+\infty} \frac{(x-2)^n}{n}$. By Theorem 46.7, $h'(x) = \sum_{n=1}^{+\infty} (x-2)^{n-1}$. This series is a geometric series with first term 1 and ratio $(x-2)$; so its sum is $\frac{1}{1-(x-2)} = \frac{1}{3-x}$. Thus, $h'(x) = \frac{1}{3-x}$. Hence, $h(x) = \int \frac{dx}{3-x} = -\ln|3-x| + C$. Now,

$$h(2) = \sum_{n=1}^{+\infty} \frac{(2-2)^n}{n} = 0 \quad \text{and} \quad -\ln|3-2| + C = 0. \quad \text{So,} \quad C = 0$$

Moreover, since $x < 3$ in the interval of convergence, $3 - x > 0$ and, therefore, $|3-x| = 3-x$. Thus, $h(x) = -\ln(3-x)$.

In Problems 2 and 3, find the interval of convergence of the given series and the behavior at the endpoints (if any).

2. $\sum_{n=1}^{+\infty} \frac{x^n}{n^2} = x + \frac{x^2}{4} + \frac{x^3}{9} + \cdots$.

Use the ratio test:

$$\left|\frac{s_{n+1}}{s_n}\right| = \frac{|x|^{n+1}}{(n+1)^2} \bigg/ \frac{|x|^n}{n^2} = \left(\frac{n}{n+1}\right)^2 |x|. \quad \text{Hence,} \quad \lim_{n\to+\infty}\left|\frac{s_{n+1}}{s_n}\right| = |x|.$$

Hence, the interval of convergence is $|x| < 1$. The radius of convergence is 1. At $x = 1$, we obtain the convergent p-series with $p = 2$. At $x = -1$, the series converges by the alternating series test. Thus, the series converges for $-1 \le x \le 1$.

3. $\sum_{n=1}^{+\infty} \frac{(x+1)^n}{\sqrt{n}} = (x+1) + \frac{(x+1)^2}{\sqrt{2}} + \frac{(x+1)^3}{\sqrt{3}} + \cdots.$

Use the ratio test:

$$\left|\frac{s_{n+1}}{s_n}\right| = \frac{|x+1|^{n+1}}{\sqrt{n+1}} \bigg/ \frac{|x+1|^n}{\sqrt{n}} = \sqrt{\frac{n}{n+1}}\,|x+1|. \quad \text{Hence,} \quad \lim_{n\to+\infty}\left|\frac{s_{n+1}}{s_n}\right| = |x+1|.$$

Hence, the interval of convergence is $|x+1| < 1$. This is equivalent to $-1 < x+1 < 1$, which, in turn, is equivalent to $-2 < x < 0$. The radius of convergence is 1. At the right endpoint $x = 0$, we get the divergent p-series $\sum_{n=1}^{+\infty} \frac{1}{\sqrt{n}}$ (with $p = \frac{1}{2}$). At the left endpoint $x = -2$, we get the alternating series $\sum_{n=1}^{+\infty} \frac{(-1)^n}{\sqrt{n}}$, which converges by the Alternating Series Theorem. Thus, the series converges for $-2 \le x < 0$.

4. Prove Theorem 46.1.

Since $\sum a_n(x_0 - c)^n$ converges, $\lim_{n\to+\infty} a_n(x_0 - c)^n = 0$ by Theorem 43.5. Hence, there is a positive number M such that $|a_n|\,|x_0 - c|^n < M$ for all n, by Theorem 42.1. Assume $|x - c| < |x_0 - c|$. Let

$$r = \frac{|x-c|}{|x_0 - c|} < 1. \quad \text{Then,} \quad |a_n||x-c|^n = |a_n||x_0 - c|^n\, r^n < Mr^n.$$

Therefore, $\sum |a_n(x-c)^n|$ is convergent by comparison with the convergent geometric series $\sum Mr^n$. Thus, $\sum a_n(x-c)^n$ is absolutely convergent.

5. Prove Theorem 46.2.

Only a very intuitive argument is possible here. Assume that neither case (a) nor case (c) holds. Since case (a) does not hold, the power series does not converge for some $x \neq c$. Since case (c) does not hold, the series does converge for some $x \neq c$. Theorem 46.1 implies that there is an interval $(c - K, c + K)$ around c in which the series converges. The interval of convergence is the maximal such interval. (Using Theorem 46.1, one takes the "least upper bound" R_1 of all K such that the series converges in $(c - K, c + K)$. Then, $(c - R_1, c + R_1)$ is the desired interval.)

6. Prove Theorem 46.4.

Assume x is in A and $\epsilon > 0$. Since $\langle f_n \rangle$ converges uniformly to f on A, there is a positive integer m such that, if $n \ge m$, then $|f_n(y) - f(y)| < \epsilon/3$ for all y in A. Since f_m is continuous at x, there exists $\delta > 0$ such that, for any x^* in A, if $|x^* - x| < \delta$, then $|f_m(x^*) - f_m(x)| < \epsilon/3$. Hence if $|x^* - x| < \delta$,

$$\begin{aligned} |f(x^*) - f(x)| &= |(f(x^*) - f_m(x^*)) + (f_m(x^*) - f_m(x)) + (f_m(x) - f(x))| \\ &\le |f(x^*) - f_m(x^*)| + |f_m(x^*) - f_m(x)| + |f_m(x) - f(x)| \\ &< \frac{\epsilon}{3} + \frac{\epsilon}{3} + \frac{\epsilon}{3} = \epsilon \end{aligned}$$

This proves the continuity of f at x.

7. If $\langle f_n \rangle$ converges uniformly to f on $[a, b]$ and each f_n is continuous on $[a, b]$, then $\int_a^b f(x)\,dx = \lim_{n\to+\infty} \int_a^b f_n(x)\,dx$.

Assume $\epsilon > 0$. There is a positive integer m such that, if $n \ge m$, then $|f_n(x) - f(x)| < \frac{\epsilon}{b-a}$ for all x in $[a, b]$. Therefore, $\int_a^b |f_n(x) - f(x)|\,dx < \epsilon$. Then

$$\left|\int_a^b f(x)\,dx - \int_a^b f_n(x)\,dx\right| = \left|\int_a^b (f_n(x) - f(x))\,dx\right| \le \int_a^b |f_n(x) - f(x)|\,dx < \epsilon \qquad \text{for } n \ge m$$

8. Prove that the function f defined by a power series is continuous within its interval of convergence (Corollary 46.5).

$f(x) = \lim_{n\to+\infty} S_n(x)$ and the convergence is uniform by Theorem 46.3. Each $S_n(x)$, being a polynomial, is continuous. Hence, f is continuous by Theorem 46.4.

9. Find a power series about 0 that represents the function $\frac{x}{1+x^2}$. In what interval is the representation valid?

By formula (46.11), $\frac{1}{1+x^2} = \sum_{n=0}^{+\infty} (-1)^n x^{2n}$ for $|x| < 1$. Hence

$$\frac{x}{1+x^2} = \sum_{n=0}^{+\infty} (-1)^n x^{2n+1} \quad \text{for } |x| < 1$$

The series diverges at both endpoints $x = 1$ and $x = -1$.

In Problems 10 and 11, use the ratio test to find the interval of convergence and indicate what happens at the endpoints (if any).

10. $\sum \frac{n}{10^n} x^n$.

$$\left|\frac{s_{n+1}}{s_n}\right| = \frac{(n+1)\,|x|^{n+1}}{10^{n+1}} \Big/ \frac{n\,|x|^n}{10^n} = \left(\frac{n+1}{n}\right)\frac{|x|}{10}. \quad \text{Hence,} \quad \lim_{n\to+\infty}\left|\frac{s_{n+1}}{s_n}\right| = \frac{|x|}{10}.$$

We get convergence when $|x|/10 < 1$, that is, when $|x| < 10$. That is the interval of convergence. The series diverges at both endpoints $\pm$ 10.

11. $\sum \frac{n}{3^n}(x-\pi)^n$.

$$\left|\frac{s_{n+1}}{s_n}\right| = \frac{(n+1)\,|x-\pi|^{n+1}}{3^{n+1}} \Big/ \frac{n\,|x-\pi|^n}{3^n} = \frac{n+1}{n}\,\frac{|x-\pi|}{3} \quad \text{Hence,} \quad \lim_{n\to+\infty}\left|\frac{s_{n+1}}{s_n}\right| = \frac{|x-\pi|}{3}.$$

So, the interval of convergence is $|x - \pi| < 3$. The series diverges at both endpoints.

12. Find the interval of convergence of $\sum \frac{(n!)^2}{(2n)!} x^n$.

Apply the ratio test:

$$\left|\frac{s_{n+1}}{s_n}\right| = \frac{((n+1)!)^2\,|x|^{n+1}}{(2n+2)!} \Big/ \frac{(n!)^2\,|x|^n}{(2n)!} = \frac{(n+1)^2}{(2n+2)(2n+1)}|x| \quad \text{Hence,} \quad \lim_{n\to+\infty}\left|\frac{s_{n+1}}{s_n}\right| = \frac{|x|}{4}.$$

So, the interval of convergence is $|x| < 4$.

13. Find a power series about 0 that represents $\frac{x}{1-x^3}$.

Start with $\frac{1}{1-x} = \sum_{n=0}^{+\infty} x^n$ for $|x| < 1$. Replace x by x^3:

$$\frac{1}{1-x^3} = \sum_{n=0}^{+\infty} x^{3n} \quad \text{for} \quad |x| < 1$$

(since $|x^3| < 1$ is equivalent to $|x| < 1$). Multiply by x:

$$\frac{x}{1-x^3} = \sum_{n=0}^{+\infty} x^{3n+1} \quad \text{for} \quad |x| < 1$$

In Problems 14–16, find simple formulas for the function $f(x)$ represented by the given power series.

14. $\frac{x}{2!}+\frac{x^2}{3!}+\frac{x^3}{4!}+\cdots$.

Let $f(x)=\sum_{n=1}^{+\infty}\frac{x^n}{(n+1)!}$.

$$xf(x)=\sum_{n=1}^{+\infty}\frac{x^{n+1}}{(n+1)!}=\sum_{n=0}^{+\infty}\frac{x^n}{n!}-1-x=e^x-1-x$$

Hence, $f(x)=\frac{e^x-1-x}{x}$.

15. $\frac{1}{3}x^3+\frac{1}{6}x^6+\frac{1}{9}x^9+\cdots$.

Let $f(x)=\sum_{n=1}^{+\infty}\frac{x^{3n}}{3n}$. Then $f'(x)=\sum_{n=1}^{+\infty}x^{3n-1}=x^2+x^5+x^8+\cdots$.

This is a geometric series with ratio x^3. So, it converges for $|x^3|<1$, which is equivalent to $|x|<1$. Hence, $f'(x)=\frac{x^2}{1-x^3}$ for $|x|<1$. Therefore, $f(x)=\int\frac{x^2}{1-x^3}dx=-\frac{1}{3}\ln|1-x^3|+C$. But $f(0)=0$. Hence, $C=0$. Also, $1-x^3>0$ for $|x|<1$. Therefore,

$$f(x)=-\tfrac{1}{3}\ln(1-x^3) \qquad \text{for} \qquad |x|<1.$$

16. $x+2x^3+3x^5+4x^7+\cdots$.

The ratio test shows that the series converges for $|x|<1$. Let

$$g(x)=x+2x^3+3x^5+4x^7+\cdots=\sum_{n=1}^{+\infty}nx^{2n-1}$$

Then $2g(x)=\sum_{n=1}^{+\infty}2nx^{2n-1}$ Hence, taking antiderivatives,

$$2\int g(x)dx=K+\sum_{n=1}^{+\infty}x^{2n}=K+\frac{x^2}{1-x^2} \qquad \left(\text{since } \sum_{n=1}^{+\infty}x^{2n} \text{ is a geometric series with ratio } x^2\right)$$

Now differentiate:

$$2g(x)=D_x\left(\frac{x^2}{1-x^2}\right)=\frac{2x}{(1-x^2)^2}, \qquad g(x)=\frac{x}{(1-x^2)^2} \qquad \text{for} \qquad |x|<1$$

17. (GC) Approximate $\int_0^{1/2}\frac{\ln(1+x)}{x}dx$ to two-decimal-place accuracy (that is, with an error $< 5/10^3$).

By formula (46.8), $\ln(1+x)=x-\frac{1}{2}x^2+\frac{1}{3}x^3-\frac{1}{4}x^4+\cdots$ for $|x|<1$, So

$$\frac{\ln(1+x)}{x}=1-\tfrac{1}{2}x+\tfrac{1}{3}x^2-\tfrac{1}{4}x^3+\cdots=\sum_{n=0}^{+\infty}\frac{(-1)^n x^n}{n+1}$$

By Theorem 46.6(b),

$$\int_0^{1/2}\frac{\ln(1+x)}{x}dx=\sum_{n=0}^{+\infty}\frac{(-1)^n}{n+1}\frac{x^{n+1}}{n+1}\Bigg]_0^{1/2}=\sum_{n=0}^{+\infty}\frac{(-1)^n}{(n+1)^2}\frac{1}{2^{n+1}}$$

which is a convergent alternating series.

In order to get an approximation with an error less than $5/10^3$, we must find n such that the first omitted term $\frac{1}{(n+1)^2}\frac{1}{2^{n+1}}$ is $\le\frac{5}{10^3}=\frac{1}{200}$. So, we must have $200\le(n+1)^2\,2^{n+1}$. Trial and error shows that $n\ge 3$. Hence, we can use the terms corresponding to $n=0, 1, 2$:

$$\frac{1}{2}-\frac{1}{16}+\frac{1}{72}=\frac{65}{144}\sim 0.45$$

This answer can be confirmed by a graphing calculator (which yields 0.44841421 as an approximation).

18. Find the function defined by $\sum_{n=0}^{+\infty} 2^n x^n$.

This is a geometric series with ratio $2x$ and first term 1. Hence, it converges for $|2x| < 1$, that is, for $|x| < \frac{1}{2}$, and its sum is $\frac{1}{1-2x}$.

19. Find the interval of convergence of $\sum_{n=1}^{+\infty} \frac{x^n}{\ln(n+1)}$.

Apply the ratio test:

$$\left|\frac{s_{n+1}}{s_n}\right| = \frac{|x|^{n+1}}{\ln(n+2)} \bigg/ \frac{|x|^n}{\ln(n+1)} = \frac{\ln(n+1)}{\ln(n+2)}|x|$$

By L'Hôpital's rule, $\lim_{n\to+\infty}\left|\frac{s_{n+1}}{s_n}\right| = |x|$. Hence, the interval of convergence is given by $|x| < 1$. (For $x = 1$, we get $\sum_{n=1}^{+\infty} \frac{1}{\ln(n+1)}$, which we know is divergent. For $x = -1$, we get the convergent alternating series $\sum_{n=1}^{+\infty} \frac{(-1)^n}{\ln(n+1)}$.)

20. Approximate $\frac{1}{e}$ with an error less than 0.0001.

By formula (46.14),

$$e^x = \sum_{n=0}^{+\infty} \frac{x^n}{n!} \quad \text{for all } x. \quad \text{Hence,} \quad \frac{1}{e} = e^{-1} = \sum_{n=0}^{+\infty} \frac{(-1)^n}{n!}$$

By the Alternating Series Theorem, we seek the least n such that $1/n! \le 0.0001 = 1/10{,}000$, that is, $10{,}000 \le n!$. Trial and error shows that $n \ge 8$. So, we must use the terms corresponding to $n = 0, 1, \ldots, 7$:

$$1 - 1 + \frac{1}{2} - \frac{1}{6} + \frac{1}{24} - \frac{1}{120} + \frac{1}{720} - \frac{1}{5040} = \frac{103}{280} \sim 0.3679$$

(A graphing calculator yields the answer 0.367 8794412, correct to 10 decimal places.)

21. Approximate $\int_0^1 e^{-x^2}\,dx$ to two-decimal-place accuracy, that is, with an error less than $5/10^3 = 0.005$.

By formula (46.14),

$$e^x = \sum_{n=0}^{+\infty} \frac{x^n}{n!} \quad \text{for all } x. \qquad \text{Hence,} \qquad e^{-x^2} = \sum_{n=0}^{+\infty} \frac{(-1)^n}{n!} x^{2n} \qquad \text{for all } x$$

By Theorem 46.6(b),

$$\int_0^1 e^{-x^2}\,dx = \sum_{n=0}^{+\infty} \frac{(-1)^n}{n!} \frac{x^{2n+1}}{2n+1}\Bigg]_0^1 = \sum_{n=0}^{+\infty} \frac{(-1)^n}{n!} \frac{1}{2n+1}$$

We can apply the Alternating Series Theorem. The magnitude of the first term omitted $\frac{1}{(2n+1)n!}$ should be $\le 0.005 = 1/200$. So, $200 \le (2n+1)n!$. Trial and error shows that $n \ge 4$. Hence, we should use the first four terms, that is, those corresponding to $n = 0, 1, 2, 3$:

$$1 - \frac{1}{3} + \frac{1}{10} - \frac{1}{42} = \frac{26}{35} \sim 0.743$$

(A graphing calculator yields the approximation 0.74682413, correct to eight decimal places.)

22. Find a power series expansion for $\frac{1}{x+3}$ about 0.

$\frac{1}{x+3} = \frac{1}{3}\frac{1}{(x/3)+1}$. By formula (46.7), $\frac{1}{1+x} = \sum_{n=0}^{+\infty} (-1)^n x^n = 1 - x + x^2 - x^3 + \cdots$ for $|x| < 1$.

Hence,

$$\frac{1}{(x/3)+1} = \sum_{n=0}^{+\infty}(-1)^n\left(\frac{x}{3}\right)^n = \sum_{n=0}^{+\infty}(-1)^n\frac{x^n}{3^n} \quad \text{for} \quad \left|\frac{x}{3}\right| < 1$$

Thus,

$$\frac{1}{x+3} = \sum_{n=0}^{+\infty}(-1)^n\frac{x^n}{3^{n+1}} \quad \text{for} \quad |x| < 3$$

The series diverges at $x = \pm 3$.

23. Find a power series expansion for $\frac{1}{x}$ about 1.

$\frac{1}{x} = \frac{1}{1+(x-1)}$. By formula (46.7), $\frac{1}{1+x} = \sum_{n=0}^{+\infty}(-1)^n x^n$ for $|x| < 1$. Hence,

$$\frac{1}{x} = \frac{1}{1+(x-1)} = \sum_{n=0}^{+\infty}(-1)^n(x-1)^n \quad \text{for} \quad |x-1| < 1$$

SUPPLEMENTARY PROBLEMS

In Problems 24–31, find the interval of convergence of the given power series.

24. $\sum nx^n$ *Ans.* $-1 < x < 1$

25. $\sum \frac{x^n}{n(n+1)}$ *Ans.* $-1 \le x \le 1$

26. $\sum \frac{x^n}{n5^n}$ *Ans.* $-5 \le x < 5$

27. $\sum \frac{x^{2n}}{n(n+1)(n+2)}$ *Ans.* $-1 \le x \le 1$

28. $\sum \frac{x^{n+1}}{(\ln(n+1))^2}$ *Ans.* $-1 \le x < 1$

29. $\sum \frac{x^n}{1+n^3}$ *Ans.* $-1 \le x \le 1$

30. $\sum \frac{(x-4)^n}{n^2}$ *Ans.* $3 \le x \le 5$

31. $\sum \frac{(3x-2)^n}{5^n}$ *Ans.* $-1 < x < \frac{7}{3}$

32. Express e^{-2x} as a power series about 0.

Ans. $\sum_{n=0}^{+\infty} \frac{(-1)^n 2^n}{n!} x^n$

33. Represent $e^{x/2}$ as a power series about 2.

Ans. $\sum_{n=0}^{+\infty} \frac{e}{2^n(n!)}(x-2)^n$

34. Represent ln x as a power series about 2.

Ans. $\ln 2+\sum_{n=1}^{+\infty}\frac{(-1)^{n+1}}{n2^n}(x-2)^n$

35. (GC) Find ln (0.97) with seven-decimal-place accuracy. (*Hint:* Use the power series for ln $(1-x)$ about 0.)

Ans. −0.0304592

36. How many terms in the power series for ln $(1+x)$ about 0 must be used to find ln 1.02 with an error ≤ 0.00000005?

Ans. Three

37. (GC) Use a power series to compute e^{-2} to four-decimal-place accuracy.

Ans. 0.1353

38. (GC) Evaluate $\int_0^{1/2}\frac{dx}{1+x^4}$ to four-decimal-place accuracy.

Ans. 0.4940

In Problems 39 and 40, find the interval of convergence of the given series.

39. $\sum_{n=1}^{+\infty}\frac{x^n}{n^n}$ *Ans.* $(-\infty,+\infty)$

40. $\sum_{n=0}^{+\infty}\frac{n!}{10^n}x^n$ *Ans.* $x=0$

41. Represent $\cosh x=\frac{e^x+e^{-x}}{2}$ as a power series about 0.

Ans. $\sum_{n=0}^{+\infty}\frac{x^{2n}}{(2n)!}$

42. Find a power series about 0 for the normal distribution function $\int_0^x e^{-t^2/2}dt$.

Ans. $\sum_{n=0}^{+\infty}\frac{(-1)^n}{n!(2^n)}\frac{x^{2n+1}}{2n+1}$

43. Find a power series expansion about 0 for $\ln\frac{1+x}{1-x}$.

Ans. $2\sum_{n=0}^{+\infty}\frac{x^{2n+1}}{2n+1}$

44. (GC) Approximate $\tan^{-1}\frac{1}{2}$ to two-decimal-place accuracy.

Ans. 0.46

45. Show that the converse of Abel's Theorem is not valid, that is, if $f(x)=\sum_{n=0}^{+\infty} a_n x^n$ for $|x|<r$, where r is the radius of convergence of the power series, and $\lim_{x\to r^-} f(x)$ exists, then $\sum_{n=0}^{+\infty} a_n r^n$ need not converge. (*Hint:* Look at $f(x)=\frac{1}{1+x}$.)

46. Find a simple formula for the function $f(x)$ represented by $\sum_{n=1}^{+\infty} n^2 x^n$.

Ans. $\frac{x(x+1)}{(1-x)^3}$

47. Find a simple formula for the function $f(x)$ represented by $\sum_{n=2}^{+\infty} \frac{x^n}{(n-1)n}$.

Ans. $x+(1-x)\ln(1-x)$

48. (a) Show that $\frac{x}{(1-x)^2}=\sum_{n=1}^{+\infty} nx^n$ for $|x|<1$. (*Hint:* Use Example 5.)

(b) Show that $\frac{2x^2}{(1-x)^3}=\sum_{n=2}^{+\infty} n(n-1)x^n$ for $|x|<1$. (*Hint*: First divide the series by x, integrate, factor out x, use part (a), and differentiate.)

(c) Show that $\frac{x(x+1)}{(1-x)^3}=\sum_{n=1}^{+\infty} n^2 x^n$ for $|x|<1$.

(d) Evaluate $\sum_{n=1}^{+\infty} \frac{n}{2^n}$ and $\sum_{n=1}^{+\infty} \frac{n^2}{2^n}$.

Ans. (d) 2 and 6

CHAPTER 47

Taylor and Maclaurin Series. Taylor's Formula with Remainder

Taylor and Maclaurin Series

Let f be a function that is infinitely differentiable at $x = c$, that is, the derivatives $f^{(n)}(c)$ exist for all positive integers n.

The *Taylor series for f about c* is the power series

$$\sum_{n=0}^{+\infty} a_n(x-c)^n = a_0 + a_1(x-c) + a_2(x-c)^2 + \cdots$$

where $a_n = \dfrac{f^{(n)}(c)}{n!}$ for all n. Note that $f^{(0)}$ is taken to mean the function f itself, so that $a_0 = f(c)$.

The *Maclaurin series for f* is the Taylor series for f about 0, that is, the power series

$$\sum_{n=0}^{+\infty} a_n x^n = a_0 + a_1 x + a_2 x^2 + \cdots$$

where $a_n = \dfrac{f^{(n)}(0)}{n!}$ for all n.

EXAMPLE 47.1: The Maclaurin series for sin *x*

Let $f(x) = \sin x$. Then

$$f'(x) = \cos x,$$

$$f''(x) = -\sin x,$$

$$f'''(x) = -\cos x,$$

Since $f^{(4)}(x) = \sin x$, further derivatives repeat this cycle of four functions. Since $\sin 0 = 0$ and $\cos 0 = 1$, $f^{(2k)}(0) = 0$ and $f^{(2k+1)}(0) = (-1)^k$. Hence, $a_{2k} = 0$ and $a_{2k+1} = \dfrac{(-1)^k}{(2k+1)!}$. So, the Maclaurin series for $\sin x$ is

$$\sum_{k=0}^{+\infty} \frac{(-1)^k}{(2k+1)!} x^{2k+1} = x - \frac{x^3}{3!} + \frac{x^5}{5!} - \frac{x^7}{7!} + \cdots$$

An application of the ratio test shows that this series converges for all x. We do not know that $\sin x$ is equal to its Maclaurin series. This will be proved later.

EXAMPLE 47.2: Let us find the Maclaurin series for $f(x) = \dfrac{1}{1-x}$.

$$f'(x) = \frac{1}{(1-x)^2}, \quad f''(x) = \frac{2}{(1-x)^3}, \quad f'''(x) = \frac{3\cdot 2}{(1-x)^4},$$

$$f^4(x) = \frac{4\cdot 3\cdot 2}{(1-x)^5}, \quad f^5(x) = \frac{5\cdot 4\cdot 3\cdot 2}{(1-x)^6}$$

We can see the pattern: $f^{(n)}(x)=\frac{n!}{(1-x)^{n+1}}$. Hence, $a_n=\frac{f^{(n)}(0)}{n!}=1$ for all n, and the Maclaurin series for $\frac{1}{1-x}$ is $\sum_{n=0}^{+\infty} x^n$. In this case, we already know that $\frac{1}{1-x}$ is equal to its Maclaurin series for $|x|<1$.

Theorem 47.1: If $f(x)=\sum_{n=0}^{+\infty} b_n(x-c)^n$ for some $x \neq c$, then this series is the Taylor series for f, that is, $b_n=\frac{f^{(n)}(c)}{n!}$ for all n. In particular, if $f(x)=\sum_{n=0}^{+\infty} b_n x^n$ for some $x \neq 0$, then this series is the Maclaurin series for f.

Assume $f(x)=\sum_{n=0}^{+\infty} b_n(x-c)^n$ for some $x \neq c$. Then $f(c)=b_0$. By term-by-term differentiation (Theorem 46.7), $f'(x)=\sum_{n=0}^{+\infty} nb_n(x-c)^{n-1}$ in the interval of convergence of $\sum_{n=0}^{+\infty} b_n(x-c)^n$. Hence $f'(c)=b_1$. Differentiating again, we get $f''(x)=\sum_{n=0}^{+\infty} n(n-1)b_n(x-c)^{n-2}$. So, $f''(c)=2b_2$ and, therefore, $b_2=\frac{f''(c)}{2!}$.

Differentiating again, we get $f'''(x)=\sum_{n=0}^{+\infty} n(n-1)(n-2)b_n(x-c)^{n-3}$. So, $f'''(c)=3!b_3$ and, therefore, $b_3=\frac{f'''(c)}{3!}$. Iterating this procedure, we obtain

$$b_n=\frac{f^{(n)}(c)}{n!} \quad \text{for all } n \geq 0$$

Thus, the series is the Taylor series for f.

EXAMPLE 47.3: We already know by formula (46.8) that

$$\ln(1+x)=\sum_{n=1}^{+\infty}(-1)^{n-1}\frac{x^n}{n} \quad \text{for } |x|<1$$

Hence, by Theorem 47.1, the series $\sum_{n=1}^{+\infty}(-1)^{n-1}\frac{x^n}{n}$ must be the Maclaurin series for ln $(1+x)$. It is not necessary to go through the laborious process of computing the Maclaurin series for In $(1+x)$ directly from the definition of Maclaurin series.

EXAMPLE 47.4: If $f(x)=\frac{1}{1-x}$, find $f^{(47)}(0)$.

We know that $\frac{1}{1-x}=\sum_{n=0}^{+\infty} x^n$ for $|x|<1$. Hence, by Theorem 47.1, the coefficient of x^n, namely 1, is equal to $\frac{f^{(n)}(0)}{n!}$. So, for $n=47$, $1=\frac{f^{(47)}(0)}{(47)!}$ and, therefore, $f^{(47)}(0)=(47)!$.

Theorem 47.2 (Taylor's Formula with Remainder): Let f by a function such that its $(n+1)$st derivative $f^{(n+1)}$ exists in (α, β). Assume also that c and x are in (α, β). Then there is some x^* between c and x such that

$$\begin{aligned} f(x) &= f(c)+f'(c)(x-c)+\frac{f''(c)}{2!}(x-c)^2+\cdots+\frac{f^{(n)}(c)}{n!}(x-c)^n+\frac{f^{(n+1)}(x^*)}{(n+1)!}(x-c)^{n+1} \\ &= \sum_{k=0}^{n}\frac{f^{(k)}(c)}{k!}(x-c)^k+R_n(x) \end{aligned} \tag{47.1}$$

Here, $R_n(x)=\frac{f^{(n+1)}(x^*)}{(n+1)!}(x-c)^{n+1}$ is called the *remainder term* or the *error*.

Theorem 47.2 can be derived from Theorem 13.6 (the Higher-Order Law of the Mean).

Applications of Taylor's Formula with Remainder

(I) Showing that certain functions are represented by their Taylor series by proving that $\lim_{n\to+\infty} R_n(x) = 0$

From Taylor's formula (47.1),

$$R_n(x) = f(x) - \sum_{k=0}^{n} \frac{f^{(k)}(c)}{k!}(x-c)^k$$

If $\lim_{n\to+\infty} R_n(x) = 0$ then

$$f(x) = \lim_{n\to+\infty} \sum_{k=0}^{n} \frac{f^{(k)}(c)}{k!}(x-c)^k = \sum_{k=0}^{+\infty} \frac{f^{(k)}(c)}{k!}(x-c)^k$$

that is, $f(x)$ is equal to its Taylor series.

Remark: $\lim_{n\to+\infty} \frac{d^n}{n!} = 0$ for any d. To see this, recall that $\sum_{n=0}^{+\infty} \frac{x^n}{n!}$ converges for all x. Hence, by Theorem 43.5, $\lim_{n\to+\infty} \frac{x^n}{n!} = 0$ for any x.

EXAMPLE 47.5: $\sin x$ is equal to its Maclaurin series.

When $f(x) = \sin x$, then every derivative $f^{(n)}(x)$ is either $\sin x$, $\cos x$, $-\sin x$, or $-\cos x$, and, therefore, $|f^{(n)}(x)| \le 1$. So,

$$|R_n(x)| = \left|\frac{f^{(n+1)}(x^*)}{(n+1)!}(x-c)^{n+1}\right| \le \frac{|(x-c)^{n+1}|}{(n+1)!}$$

By the Remark above, $\lim_{n\to+\infty} \frac{|(x-c)^{n+1}|}{(n+1)!} = 0$. Hence, $\lim_{n\to+\infty} R_n(x) = 0$. Therefore, $\sin x$ is equal to its Maclaurin series:

$$\sin x = \sum_{k=0}^{+\infty} \frac{(-1)^k}{(2k+1)!} x^{2k+1} = x - \frac{x^3}{3!} + \frac{x^5}{5!} - \frac{x^7}{7!} + \cdots \quad (47.2)$$

(II) Approximating values of functions or integrals

Use a bound on $R_n(x)$ to get a bound on the error when we approximate the sum of an infinite series by a partial sum.

EXAMPLE 47.6 Let us approximate e to four decimal places, that is, with an error less than 0.00005.

Preliminary result: $e < 3$. To see this, note that, since $e^x = \sum_{n=0}^{+\infty} \frac{x^n}{n!}$,

$$\begin{aligned} e = e^1 &= \sum_{n=0}^{+\infty} \frac{1}{n!} = 1 + 1 + \frac{1}{2!} + \frac{1}{3!} + \frac{1}{4!} + \frac{1}{5!} + \cdots \\ &< 1 + 1 + \frac{1}{2} + \frac{1}{2^2} + \frac{1}{2^3} + \frac{1}{2^4} + \cdots \\ &= 1 + \sum_{n=0}^{+\infty} \frac{1}{2^n} = 1 + \frac{1}{1-(1/2)} = 1 + 2 = 3 \end{aligned}$$

Now, for the function $f(x) = e^x$, we wish to make the magnitude of the error $R_n(1) < 0.00005$. By Taylor's formula with remainder, with $x = 1$,

$$|R_n(1)| = \left|\frac{f^{(n+1)}(x^*)}{(n+1)!}\right|, \quad \text{where } 0 < x^* < 1$$

Since $D_x(e^x) = e^x$, $f^{(n+1)}(x) = e^x$ for all x. So, $f^{(n+1)}(x^*) = e^{x^*}$. Since e^x is an increasing function, $e^{x^*} < e^1 = e < 3$. Thus, $|R_n(1)| < \frac{3}{(n+1)!}$. Since we wish to make the error <0.00005, it suffices to have

$$\frac{3}{(n+1)!} \leq 0.00005, \quad \text{that is,} \quad \frac{3}{(n+1)!} \leq \frac{1}{20{,}000}, \quad 60{,}000 \leq (n+1)!.$$

Trial and error shows that this holds for $n \geq 8$. So, we can use the partial sum $\sum_{n=0}^{8} \frac{1}{n!} \sim 1.7183$.

Theorem 47.3 (The Binomial Series): Assume $r \neq 0$. Then

$$(1+x)^r = 1 + \sum_{n=1}^{+\infty} \frac{r(r-1)(r-2)\cdots(r-n+1)}{n!} x^n \quad \text{for } |x| < 1$$

$$= 1 + rx + \frac{r(r-1)}{2!}x^2 + \frac{r(r-1)(r-2)}{3!}x^3 + \cdots \tag{47.3}$$

Apply the ratio test to the given series:

$$\left|\frac{s_{n+1}}{s_n}\right| = \left|\frac{r(r-1)(r-2)\cdots(r-n)x^{n+1}}{(n+1)!} \bigg/ \frac{r(r-1)(r-2)\cdots(r-n+1)x^n}{n!}\right|$$

So,

$$\lim_{n\to+\infty}\left|\frac{s_{n+1}}{s_n}\right| = \lim_{n\to+\infty}\left|\frac{(r-n)x}{n+1}\right| = |x|$$

Hence, the series converges for $|x| < 1$. For a sketch of the proof that this series is equal to $(1 + x)^r$, see Problem 31.

Note that, if r is a positive integer k, then the coefficients of x^n for $n > k$ are 0 and we get the binomial formula

$$(1+x)^k = \sum_{n=0}^{k} \frac{k!}{n!(k-n)!} x^n$$

EXAMPLE 47.7: Let us expand $\sqrt{1+x}$ as a power series about 0. This is the binomial series for $r = \frac{1}{2}$.

$$\sqrt{1+x} = 1 + \frac{1/2}{1!}x + \frac{(1/2)(-1/2)}{2!}x^2 + \frac{(1/2)(-1/2)(-3/2)}{3!}x^3$$

$$+ \frac{(1/2)(-1/2)(-3/2)(-5/2)}{4!}x^4 + \cdots$$

$$= 1 + \frac{1}{2}x - \frac{1}{8}x^2 + \frac{1}{16}x^3 - \frac{5}{128}x^4 + \cdots \tag{47.4}$$

EXAMPLE 47.8: Let us find a power series expansion about 0 for $\frac{1}{\sqrt{1-x}}$

Take the binomial series for $r = -\frac{1}{2}$, and then replace x by $-x$:

$$\frac{1}{\sqrt{1-x}} = 1 + \frac{-1/2}{1!}(-x) + \frac{(-1/2)(-3/2)}{2!}(-x)^2 + \frac{(-1/2)(-3/2)(-5/2)}{3!}(-x)^3 + \cdots$$

$$+ \frac{1\cdot 3\cdot 5\cdots(2n-1)}{n!2^n}x^n + \cdots$$

$$= 1 + \sum_{n=1}^{+\infty} \frac{1\cdot 3\cdot 5\cdots(2n-1)}{2\cdot 4\cdot 6\cdots(2n)} x^n \tag{47.5}$$

Theorem 47.4: If $f(x)=\sum_{n=0}^{+\infty} a_n x^n$ for $|x| < R_1$ and $g(x)=\sum_{n=0}^{+\infty} b_n x^n$ for $|x| < R_2$, then $f(x)g(x)=\sum_{n=0}^{+\infty} c_n x^n$ for $|x| <$ minimum (R_1, R_2), where $c_n = \sum_{k=0}^{n} a_k b_{n-k}$.

The reader is referred to more advanced treatments of analysis for a proof. Theorem 47.4 guarantees that, if f and g have power series expansions, then so does their product.

SOLVED PROBLEMS

1. Find a power series expansion about 0 for cos x.

We know by Example 5 that

$$\sin x = \sum_{k=0}^{+\infty} \frac{(-1)^k}{(2k+1)!} x^{2k+1} = x - \frac{x^3}{3!} + \frac{x^5}{5!} - \frac{x^7}{7!} + \cdots \quad \text{for all } x$$

Then, by Theorem 46.7, we can differentiate term by term:

$$\cos x = \sum_{k=0}^{+\infty} \frac{(-1)^k}{(2k)!} x^{2k} = 1 - \frac{x^2}{2!} + \frac{x^4}{4!} - \frac{x^6}{6!} + \cdots \quad \text{for all } x$$

2. Find a power series about $\frac{\pi}{2}$ for sin x.

Use the identity $\sin x = \cos\left(x - \frac{\pi}{2}\right)$. Then, by Problem 1,

$$\sin x = \sum_{k=0}^{+\infty} \frac{(-1)^k}{(2k)!}\left(x - \frac{\pi}{2}\right)^{2k} = 1 - \frac{1}{2!}\left(x - \frac{\pi}{2}\right)^2 + \frac{1}{4!}\left(x - \frac{\pi}{2}\right)^4 - \cdots$$

3. If $f(x) = \tan^{-1} x$, evaluate $f^{(38)}(0)$.

We know by formula (46.12) that

$$\tan^{-1} x = \sum_{n=0}^{+\infty} (-1)^n \frac{x^{2n+1}}{2n+1} = x - \tfrac{1}{3}x^3 + \tfrac{1}{5}x^5 - \tfrac{1}{7}x^7 + \cdots \quad \text{for } |x| < 1$$

Hence, by Theorem 47.1, the coefficient of x^{38} in this power series is equal to $\frac{f^{(38)}(0)}{(38)!}$. But the coefficient of x^{38} is 0. So, $f^{(38)}(0) = 0$.

4. Find power series expansions about 0 for the following functions:

(a) $\cos(x^2)$ (b) xe^{-2x} (c) $1/\sqrt[3]{1+x}$

(a) $\cos x = \sum_{k=0}^{+\infty} \frac{(-1)^k}{(2k)!} x^{2k}$ by Problem 1. Therefore, $\cos(x^2) = \sum_{k=0}^{+\infty} \frac{(-1)^k}{(2k)!} x^{4k}$.

(b) We know that $e^x = \sum_{k=0}^{+\infty} \frac{x^k}{k!}$. So, $e^{-2x} = \sum_{k=0}^{+\infty} \frac{(-1)^k 2^k}{k!} x^k$. Hence,

$$xe^{-2x} = \sum_{k=0}^{+\infty} \frac{(-1)^k 2^k}{k!} x^{k+1} = \sum_{n=1}^{+\infty} \frac{(-1)^{n-1} 2^{n-1}}{(n-1)!} x^n$$

(c) This is the binomial series for $r = -\frac{1}{3}$.

$$1/\sqrt[3]{1+x} = 1 - \frac{1}{3}x + \frac{(-1/3)(-4/3)}{2!}x^2 + \frac{(-1/3)(-4/3)(-7/3)}{3!}x^3 + \frac{(-1/3)(-4/3)(-7/3)(-10/3)}{4!}x^4 + \cdots$$

$$= 1 + \sum_{n=1}^{+\infty} \frac{(-1)^n (1 \cdot 4 \cdot 7 \cdots (3n-2))}{3^n n!} x^n$$

5. Find the first five terms of the Maclaurin series for $e^x(\sin x)$.
Method 1: Let $f(x) = e^x(\sin x)$. Then

$$f'(x) = e^x(\sin x + \cos x), \quad f''(x) = 2e^x(\cos x), \quad f'''(x) = 2e^x(\cos x - \sin x)$$

$$f^{(4)}(x) = -4e^x(\sin x), \quad \text{and} \quad f^{(5)}(x) = -4e^x(\sin x + \cos x)$$

Hence, since $a_n = \dfrac{f^{(n)}(0)}{n!}$, we get $a_0 = 0$, $a_1 = 1$, $a_2 = 1$, $a_3 = \frac{1}{3}$, $a_4 = 0$, and $a_5 = -\frac{1}{30}$. Thus

$$e^x(\sin x) = x + x^2 + \frac{x^3}{3} - \frac{x^5}{30} + \cdots$$

Method 2: $e^x(\sin x) = \left(1 + x + \dfrac{x^2}{2!} + \dfrac{x^3}{3!} + \cdots\right)\left(x - \dfrac{x^3}{3!} + \dfrac{x^5}{5!} - \cdots\right)$. If we multiply out according to the rule in Theorem 47.4, we get the same result as above. For example, $c_5 = \frac{1}{24} - \frac{1}{12} + \frac{1}{120} = -\frac{1}{30}$.

6. We know that $\sin x = x - \dfrac{x^3}{3!} + \dfrac{x^5}{5!} - \cdots$. For what values of x will approximating $\sin x$ by x produce an error < 0.005?

$|R_2(x)| = \left|\dfrac{f^{(3)}(x^*)}{3!}x^3\right| \le \dfrac{|x|^3}{6}$. (Here, $|f^{(3)}(x)| \le 1$ since $f^{(3)}$ is $-\cos x$.) So, we require $|x|^3/6 < 0.005$, which is equivalent to $|x|^3 < 0.03$. So, we want $|x| < \sqrt[3]{0.03} \sim 0.31$.

7. If we approximate $\sin x$ by $x - \dfrac{x^3}{3!}$ for $|x| < 0.5$, what is a bound on the error?
Since $\sin x$ is equal to an alternating series for any x, the error will be less than the magnitude of the first term omitted, in this case $|x|^5/5!$. When $|x| < 0.5$, the error will be less than $\dfrac{1}{120}(0.5)^5 \sim 0.00026$.

8. Approximate $\displaystyle\int_0^1 \frac{\sin x}{x}\,dx$ with an error less than 0.005.

$$\sin x = \sum_{k=0}^{+\infty} \frac{(-1)^k}{(2k+1)!}x^{2k+1} = x - \frac{x^3}{3!} + \frac{x^5}{5!} - \frac{x^7}{7!} + \cdots$$

Hence,

$$\frac{\sin x}{x} = \sum_{k=0}^{+\infty} \frac{(-1)^k}{(2k+1)!}x^{2k} = 1 - \frac{x^2}{3!} + \frac{x^4}{5!} - \frac{x^6}{7!} + \cdots$$

Therefore,

$$\int_0^1 \frac{\sin x}{x}\,dx = \sum_{k=0}^{+\infty} \frac{(-1)^k}{(2k+1)!}\int_0^1 x^{2k}\,dx = \sum_{k=0}^{+\infty} \frac{(-1)^k}{(2k+1)!}\left.\frac{x^{2k+1}}{2k+1}\right]_0^1$$

$$= \sum_{k=0}^{+\infty} \frac{(-1)^k}{(2k+1)!}\frac{1}{2k+1}$$

This is an alternating series. We must find k so that $\dfrac{1}{(2k+1)!}\dfrac{1}{2k+1} \le 0.005$, or, equivalently, $200 \le (2k+1)!(2k+1)$. It is true for $k \ge 2$. So, we need $1 - \frac{1}{18} = \frac{17}{18} \sim 0.944$.

9. Find a power series about 0 for $\sin^{-1} x$.
By formula (47.5),

$$\frac{1}{\sqrt{1-x}} = 1 + \sum_{n=1}^{+\infty} \frac{1\cdot 3\cdot 5\cdots(2n-1)}{2\cdot 4\cdot 6\cdots(2n)}x^n \quad \text{for} \quad |x| < 1$$

Replace x by t^2.

$$\frac{1}{\sqrt{1-t^2}} = 1 + \sum_{n=1}^{+\infty} \frac{1\cdot 3\cdot 5\cdots(2n-1)}{2\cdot 4\cdot 6\cdots(2n)}t^{2n} \quad \text{for} \quad |t| < 1$$

So, for $|x| < 1$,

$$\sin^{-1} x = \int_0^x \frac{1}{\sqrt{1-t^2}}\,dt = x + \sum_{n=1}^{+\infty} \frac{1\cdot 3\cdot 5\cdots(2n-1)}{2\cdot 4\cdot 6\cdots(2n)} \frac{x^{2n+1}}{2n+1}$$

10. Find Maclaurin series for the following functions: (a) $\sin(x^3)$; (b) $\sin^2 x$.

Recall that, if a function has a power series expansion in an interval about 0, then that power series is the Maclaurin series of the function.

(a) $\sin x = \sum_{k=0}^{+\infty} \frac{(-1)^k}{(2k+1)!} x^{2k+1}$ for all x. Hence, $\sin(x^3) = \sum_{k=0}^{+\infty} \frac{(-1)^k}{(2k+1)!} x^{6k+3}$ and this series is the Maclaurin series for $\sin(x^3)$.

(b) $\sin^2 x = \frac{1-\cos(2x)}{2} = \frac{1}{2}\left(1 - \sum_{k=0}^{+\infty} \frac{(-1)^k 2^{2k}}{(2k)!} x^{2k}\right) = \sum_{k=1}^{+\infty} \frac{(-1)^{k+1} 2^{2k-1}}{(2k)!} x^{2k}$ by Problem 1. So, the Maclaurin series for $\sin^2 x$ is $\sum_{k=1}^{+\infty} \frac{(-1)^{k+1} 2^{2k-1}}{(2k)!} x^{2k}$.

11. Find the first four nonzero terms of the Maclaurin series for $f(x) = \sec x$.

It would be very tedious to compute the successive derivatives. Instead, since $\sec x \cos x = 1$, we can proceed differently. We assume $\sec x = \sum_{n=0}^{+\infty} a_n x^n$. Then

$$\left(\sum_{n=0}^{+\infty} a_n x^n\right)\left(\sum_{k=0}^{+\infty} \frac{(-1)^k}{(2k)!} x^{2k}\right) = 1$$

$$(a_0 + a_1 x + a_2 x^2 + a_3 x^3 + \cdots)\left(1 - \frac{x^2}{2} + \frac{x^4}{24} - \frac{x^6}{720} + \cdots\right) = 1$$

We then "multiply out," compare coefficients on the two sides of the equation, and solve for the a_n.

$$a_0 = 1,\ a_1 = 0,\ a_2 = \tfrac{1}{2};\ a_3 = 0;\ a_4 = \tfrac{5}{24};\ a_5 = 0;\ a_6 = \tfrac{61}{720}$$

Thus,

$$\sec x = 1 + \frac{1}{2}x^2 + \frac{5}{24}x^4 + \frac{61}{270}x^6 + \cdots$$

An alternative method would be to carry out a "long division" of 1 by $1 - \frac{x^2}{2} + \frac{x^4}{24} - \frac{x^6}{720} + \cdots$

SUPPLEMENTARY PROBLEMS

12. Find the Maclaurin series for the following functions:

(a) $\sin(x^5)$; (b) $\frac{1}{1+x^5}$; (c) $\cos^2 x$.

Ans. (a) $\sum_{k=0}^{+\infty} \frac{(-1)^k}{(2k+1)!} x^{10k+5}$; (b) $\sum_{k=0}^{+\infty} (-1)^n x^{5n}$; (c) $1 + \sum_{k=1}^{+\infty} \frac{(-1)^k 2^{2k-1}}{(2k)!} x^{2k}$

13. Find the Taylor series for $\ln x$ about 2.

Ans. $\ln 2 + \sum_{n=1}^{+\infty} (-1)^{n-1} \frac{(x-2)^n}{n 2^n}$

14. Find the first three nonzero terms of the Maclaurin series for (a) $\frac{\sin x}{e^x}$; (b) $e^x \cos x$.

Ans. (a) $x - x^2 + \frac{1}{3}x^3 + \cdots$; (b) $1 + x - \frac{1}{3}x^3 + \cdots$

15. Compute the first three nonzero terms of the Maclaurin series for tan x.

Ans. $x+\frac{1}{3}x^3+\frac{2}{15}x^5+\cdots$

16. Compute the first three nonzero terms of the Maclaurin series for $\sin^{-1} x$.

Ans. $x+\frac{1}{6}x^3+\frac{3}{40}x^5+\cdots$

17. Find the Taylor series for cos x about $\frac{\pi}{3}$. [*Hint*: Use an identity for $\cos\left(\frac{\pi}{3}+\left(x-\frac{\pi}{3}\right)\right)$.]

Ans. $\frac{1}{2}\sum_{k=0}^{+\infty}\frac{(-1)^k}{(2k)!}\left(x-\frac{\pi}{3}\right)^{2k}-\frac{\sqrt{3}}{2}\sum_{k=0}^{+\infty}\frac{(-1)^k}{(2k+1)!}\left(x-\frac{\pi}{3}\right)^{2k+1}$

18. (GC) Use power series to approximate $\int_0^{1/2}\frac{\tan^{-1}x}{x}dx$

Ans. 0.4872

19. (GC) Use power series to approximate $\int_0^{1/2}\frac{\ln(1+x)}{x}dx$ correctly to four decimal places.

Ans. 0.4484

20. (GC) Use power series to approximate $\int_0^1\sqrt[3]{1+x^2}\,dx$ correctly to four decimal places.

Ans. 1.0948

21. (GC) What is a bound on the error if we approximate e^x by $1+x+\frac{1}{2}x^2$ for $|x|\le 0.05$? (You may use $e^{0.05}<1.06$.)

Ans. 0.0000221

22. (GC) What is a bound on the error if we approximate $\ln(1+x)$ by x for $|x|\le 0.05$?

Ans. 0.00125

23. (GC) Use the Taylor series for sin x about $\frac{\pi}{3}$ to approximate sin 62° correctly to five decimal places.

Ans. 0.88295

24. (GC) In what interval can you choose the angle if the values of cos x are to be computed using three terms of its Taylor series about $\frac{\pi}{3}$ and the error must not exceed 0.000 05?

Ans. $\left|x-\frac{\pi}{3}\right|\le 0.0669$

25. (GC) Use power series to compute to four-decimal-place accuracy: (a) e^{-2}; (b) sin 32°; (c) cos 36°.

Ans. (a) 0.1353; (b) 0.5299; (c) 0.8090

26. (GC) For what range of x can:

(a) e^x be replaced by $1+x+\frac{1}{2}x^2$ if the allowable error is 0.0005?
(b) sin x be replaced by $x-\frac{1}{6}x^3+\frac{1}{120}x^5$ if the allowable error is 0.00005?

Ans. (a) $|x|<0.1$; (b) $|x|<47°$

27. Use power series to evaluate: (a) $\lim_{x\to 0}\frac{e-e^{\sin x}}{x^3}$; (b) $\lim_{x\to 0}\frac{e-e^{\cos x}}{x^2}$.

Ans. (a) $\frac{1}{6}$; (b) $\frac{e}{2}$

28. (GC) Use power series to evaluate:

(a) $\int_0^{\pi/2}(1-\frac{1}{2}\sin^2 x)^{-1/2}\,dx$ (to three-decimal-place accuracy).

(b) $\int_0^1 \cos\sqrt{x}\,dx$ (to five-decimal-place accuracy).

(c) $\int_0^{1/2}\frac{dx}{1+x^4}$ (to four-decimal-place accuracy).

Ans. (a) 1.854; (b) 0.76355; (c) 0.4940

29. (GC) Use power series to approximate the length of the curve $y=\frac{1}{3}x^3$ from $x=0$ to $x=.5$, with four-decimal-place accuracy.

Ans. 0.5031

30. (GC) Use power series to approximate the area between the curve $y=\sin(x^2)$ and the x axis from $x=0$ to $x=1$, with four-decimal-place accuracy.

Ans. 0.3103

31. Prove that the binomial series expansion in Theorem 47.3 is correct.

Hint: Let $y=1+\sum_{n=1}^{+\infty}\frac{r(r-1)(r-2)\cdots(r-n+1)}{n!}x^n$. Use term-by-term differentiation to find the series for $\frac{dy}{dx}$ and show that $\frac{dy}{dx}=\frac{ry}{1+x}$. Then derive $y=(1+x)^r$. [Use "separation of variables"; $\int\frac{dy}{y}=\int\frac{r\,dx}{1+x}$.]

32. Expand the polynomial $f(x)=x^4-11x^3+43x^2-60x+14$ as a power series about 3, and find $\int_3^{3.2}f(x)\,dx$.

Ans. 1.185

CHAPTER 48

Partial Derivatives

Functions of Several Variables

If a real number z is assigned to each point (x, y) of a part of the xy plane, then z is said to be given as a function, $z = f(x, y)$, of the independent variables x and y. The set of all points (x, y, z) satisfying $z = f(x, y)$ is a surface in three-dimensional space. In a similar manner, functions $w = f(x, y, z, \ldots)$ of three or more independent variables may be defined, but no geometric picture is available.

There are a number of differences between the calculus of one and two variables. However, the calculus of functions of three or more variables differs only slightly from that of functions of two variables. The study here will be limited largely to functions of two variables.

Limits

By an *open disk* with center (a, b) we mean the set of points (x, y) within some fixed distance δ from (a, b), that is, such that $\sqrt{(x-a)^2+(y-b)^2} < \delta$. By a *deleted disk* around (a, b) we mean an open disk without its center (a, b).

Let f be a function of two variables and assume that there are points in the domain of f arbitrarily close to (a, b). To say that $f(x, y)$ has the limit L as (x, y) approaches (a, b) means intuitively that $f(x, y)$ can be made arbitrarily close to L when (x, y) is sufficiently close to (a, b). More precisely,

$$\lim_{(x,y)\to(a,b)} f(x,y) = L$$

if, for any $\epsilon > 0$, there exists $\delta > 0$ such that, for any (x, y) in the domain of f and in the deleted disk of radius δ around (a, b), $|f(x, y) - L| < \epsilon$. This is equivalent to saying that, for any $\epsilon > 0$, there exists $\delta > 0$ such that $0 < \sqrt{(x-a)^2+(y-b)^2} < \delta$ implies $|f(x, y) - L| < \epsilon$ for any (x, y) in the domain of f. Note that it is not assumed that $f(a, b)$ is defined.

Laws for limits analogous to those for functions of one variable (Theorems 7.1–7.6) also hold here and with similar proofs.

EXAMPLE 48.1: Using these standard laws for limits, we see that

$$\lim_{(x,y)\to(3,1)} \left(\frac{3xy^2}{7+y} + \frac{1}{2}xy\right) = \frac{3(3)(1)}{7+1} + \tfrac{1}{2}(3)(1) = \tfrac{9}{8} + \tfrac{3}{2} = \tfrac{21}{8}$$

EXAMPLE 48.2: In some cases, these standard laws do not suffice.

Let us show that $\lim_{(x,y)\to(0,0)} \dfrac{3xy^2}{x^2+y^2} = 0$. Our usual limit rules would yield $\dfrac{0}{0}$, which is indeterminate. So, we need a more involved argument. Assume $\epsilon > 0$. Now,

$$\left|\frac{3xy^2}{x^2+y^2} - 0\right| = \left|\frac{3xy^2}{x^2+y^2}\right| = 3|x|\left|\frac{y^2}{x^2+y^2}\right| \le 3|x| \le 3\sqrt{x^2+y^2} < 3\delta = \epsilon$$

if we choose $\delta = \epsilon/3$ and we assume that $0 < \sqrt{x^2+y^2} < \delta$.

EXAMPLE 48.3: Let us show that $\lim_{(x,y)\to(0,0)} \frac{x^2-y^2}{x^2+y^2}$ does not exist.

Let $(x, y)\to(0, 0)$ along the x axis, where $y = 0$. Then $\frac{x^2-y^2}{x^2+y^2} = \frac{x^2}{x^2} = 1$. So, the limit along the x axis is 1. Now let $(x, y)\to(0, 0)$ along the y axis, where $x = 0$. Then $\frac{x^2-y^2}{x^2+y^2} = -\frac{y^2}{y^2} = -1$. So, the limit along the y axis is -1. Hence, there can be no common limit as one approaches (0, 0), and the limit does not exist.

EXAMPLE 48.4: Let us show that $\lim_{(x,y)\to(0,0)} \left(\frac{x^2-y^2}{x^2+y^2}\right)^2$ does not exist.

Here, we cannot use the same argument as in Example 3, since $\left(\frac{x^2-y^2}{x^2+y^2}\right)^2$ approaches 1 as (x, y) approaches (0, 0) along both the x axis and the y axis. However, we can let (x, y) approach (0, 0) along the line $y = x$. Then $\left(\frac{x^2-y^2}{x^2+y^2}\right)^2 = \left(\frac{x^2-x^2}{x^2+x^2}\right)^2 = 0$. So, $\left(\frac{x^2-y^2}{x^2+y^2}\right)^2 \to 0$ along $y = x$. Since this is different from the limit 1 approached along the x axis, there is no limit as $(x, y)\to(0, 0)$.

Continuity

Let f be a function of two variables and assume that there are points in the domain of f arbitrarily close to (a, b). Then *f is continuous at (a, b)* if and only if f is defined at (a, b), $\lim_{(x,y)\to(a,b)} f(x, y)$ exists, and $\lim_{(x,y)\to(a,b)} f(x, y) = f(a, b)$

We say that f is continuous on a set A if f is continuous at each point of A.

This is a generalization to two variables of the definition of continuity for functions of one variable. The basic properties of continuous functions of one variable (Theorem 8.1) carry over easily to two variables. In addition, every polynomial in two variables, such as $7x^5 - 3xy^3 - y^4 + 2xy^2 + 5$, is continuous at all points. Every continuous function of one variable is also continuous as a function of two variables.

The notions of limit and continuity have obvious generalizations to functions of three or more variables.

Partial Derivatives

Let $z = f(x, y)$ be a function of two variables. If x varies while y is held fixed, z becomes a function of x. Then its derivative with respect to x

$$\lim_{\Delta x\to 0} \frac{f(x+\Delta x, y) - f(x, y)}{\Delta x}$$

is called the *(first) partial derivative* of f with respect to x and is denoted $f_x(x, y)$ or $\frac{\partial z}{\partial x}$ or $\frac{\partial f}{\partial x}$.

Similarly, if y varies while x is held fixed, the *(first) partial derivative* of f with respect to y is

$$f_y(x, y) = \frac{\partial z}{\partial y} = \frac{\partial f}{\partial y} = \lim_{\Delta y\to 0} \frac{f(x, y+\Delta y) - f(x, y)}{\Delta y}$$

EXAMPLE 48.5: Let $f(x, y) = x^2 \sin y$. Then $f_x(x, y) = 2x \sin y$ and $f_y(x, y) = x^2 \cos y$.

Note that, when f_x is computed, y is temporarily treated like a constant, and, when f_y is computed, x is temporarily treated like a constant.

The partial derivatives have simple geometric interpretations. Consider the surface $z = f(x, y)$ in Fig. 48-1. Through the point $P(x, y, z)$, there is a curve APB that is the intersection with the surface of the plane through P parallel to the xz plane (the plane determined by the x axis and the z axis). Similarly, CPD is the curve through P that is the intersection with the surface $z = f(x, y)$ of the plane through P that is parallel to the yz plane. As x varies while y is held fixed, P moves along the curve APB, and the value of $\frac{\partial z}{\partial x}$ at (x, y) is the slope of the tangent line to the curve APB at P. Similarly, as y varies while x is held fixed, P moves along the curve CPD, and the value of $\frac{\partial z}{\partial y}$ at (x, y) is the slope of the tangent line to the curve CPD at P.

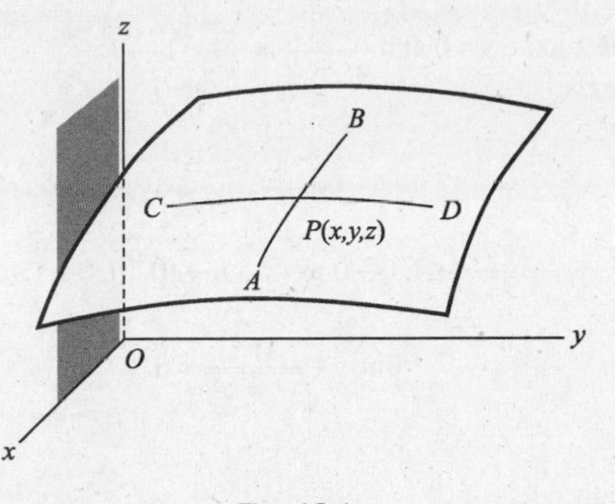

Fig. 48-1

Partial Derivatives of Higher Order

We can take the partial derivatives with respect to x and y of $\frac{\partial z}{\partial x}$, yielding

$$\frac{\partial^2 z}{\partial x^2} = f_{xx}(x, y) = \frac{\partial}{\partial x}\left(\frac{\partial z}{\partial x}\right) \text{ and } \frac{\partial^2 z}{\partial y \partial x} = f_{yx}(x, y) = \frac{\partial}{\partial y}\left(\frac{\partial z}{\partial x}\right)$$

Similarly, from $\frac{\partial z}{\partial y}$ we obtain

$$\frac{\partial^2 z}{\partial y^2} = f_{yy}(x, y) = \frac{\partial}{\partial y}\left(\frac{\partial z}{\partial y}\right) \text{ and } \frac{\partial^2 z}{\partial x \partial y} = f_{xy}(x, y) = \frac{\partial}{\partial x}\left(\frac{\partial z}{\partial y}\right)$$

Theorem 48.1: Assume that f_{xy} and f_{yx} exist and are continuous in an open disk. Then $f_{xy} = f_{yx}$ at every point of the disk.

For a proof, see Problem 30.

EXAMPLE 48.6: Let us verify Theorem 48.1 for $f(x, y) = x^2(\sin yx)$.

$$f_x(x,y) = x^2(\cos yx)(y) + 2x(\sin yx) = x[xy(\cos yx) + 2\sin yx]$$

$$f_y(x,y) = x^2(\cos yx)x + x^3(\cos yx)$$

$$f_{yx}(x,y) = x[x(y(-\sin yx)(x) + \cos yx) + 2(\cos yx)(x)]$$

$$= x^2[-xy \sin yx + 3\cos yx]$$

$$f_{xy}(x,y) = x^3(-\sin yx)(y) + 3x^2 \cos yx = x^2[-xy \sin yx + 3\cos yx]$$

Partial derivatives also can be defined for functions of three or more variables. An analogue of Theorem 48.1 holds for any two orderings of given subscripts.

Note that partial derivatives may fail to exist when the required limits do not exist.

SOLVED PROBLEMS

1. Evaluate: (a) $\lim_{(x,y)\to(3,2)} (2xy^4 - 7x^2y^2)$; (b) $\lim_{(x,y)\to(\pi,0)} x\cos\left(\frac{x-y}{4}\right)$.

Since the standard limit laws apply, the limits are:

(a) $2(3)(2)^4 - 7(3)^2(2)^2 = 96 - 252 = -156$; (b) $\pi\cos\frac{\pi}{4} = \frac{\pi\sqrt{2}}{2}$

2. Evaluate $\lim_{(x,y)\to(0,0)} \frac{x^2}{x^2 + y^2}$.

As $(x, y) \to (0, 0)$ along the y axis, $x = 0$ and $\frac{x^2}{x^2 + y^2} = 0 \to 0$.

As $(x, y) \to (0, 0)$ along the x axis, $y = 0$ and $\frac{x^2}{x^2+y^2} = \frac{x^2}{x^2} = 1 \to 1$.
Hence, the limit does not exist.

3. Evaluate $\lim_{(x,y)\to(0,0)} \frac{xy}{\sqrt{x^2+y^2}}$.

Since $|x| = \sqrt{x^2} \le \sqrt{x^2+y^2}$, $\left|\frac{xy}{\sqrt{x^2+y^2}}\right| \le |y| \to 0$ as $(x, y) \to (0, 0)$. So,

$$\lim_{(x,y)\to(0,0)} \frac{xy}{\sqrt{x^2+y^2}} = 0.$$

4. The function $f(x,y) = \frac{\sin(x+y)}{x+y}$ is continuous everywhere except at (0, 0) and on the line $y = -x$, where it is not defined. Can $f(0, 0)$ be defined so that the new function is continuous?

As $(x, y) \to (0, 0)$, $x + y \to 0$ and, therefore, $\frac{\sin(x+y)}{x+y} \to 1$, since $\lim_{u\to 0} \frac{\sin u}{u} = 1$. So, if we let $f(0, 0) = 1$, the new function will be continuous at (0, 0). Thus, the original discontinuity was removable.

In Problems 5–9, find the first partial derivatives.

5. $z = 2x^2 - 3xy + 4y^2$.

Treating y as a constant and differentiating with respect to x yields $\frac{\partial z}{\partial x} = 4x - 3y$.

Treating x as a constant and differentiating with respect to y yields $\frac{\partial z}{\partial y} = -3x + 8y$.

6. $z = \frac{x^2}{y} + \frac{y^2}{x}$.

Treating y as a constant and differentiating with respect to x yields $\frac{\partial z}{\partial x} = \frac{2x}{y} - \frac{y^2}{x^2}$.

Treating x as a constant and differentiating with respect to y yields $\frac{\partial z}{\partial y} = -\frac{x^2}{y^2} + \frac{2y}{x}$.

7. $z = \sin(2x + 3y)$.

$$\frac{\partial z}{\partial x} = 2\cos(2x+3y) \quad \text{and} \quad \frac{\partial z}{\partial y} = 3\cos(2x+3y)$$

8. $z = \tan^{-1}(x^2 y) + \tan^{-1}(xy^2)$.

$$\frac{\partial z}{\partial x} = \frac{2xy}{1+x^4y^2} + \frac{y^2}{1+x^2y^4} \quad \text{and} \quad \frac{\partial z}{\partial y} = \frac{x^2}{1+x^4y^2} + \frac{2xy}{1+x^2y^4}$$

9. $z = e^{x^2+xy}$

$$\frac{\partial z}{\partial x} = e^{x^2+xy}(2x+y) \quad \text{and} \quad \frac{\partial z}{\partial y} = xe^{x^2+xy}$$

10. The area of a triangle is given by $K = \frac{1}{2}ab\sin C$. When $a = 20$, $b = 30$, and $C = 30°$, find:

(a) The rate of change of K with respect to a, when b and C are constant.
(b) The rate of change of K with respect to C, when a and b are constant.
(c) The rate of change of b with respect to a, when K and C are constant.

(a) $\frac{\partial K}{\partial a} = \frac{1}{2}b\sin C = \frac{1}{2}(30)(\sin 30°) = \frac{15}{2}$

(b) $\frac{\partial K}{\partial C} = \frac{1}{2}ab\cos C = \frac{1}{2}(20)(30)(\cos 30°) = 150\sqrt{3}$

(c) $b = \frac{2K}{a\sin C}$ and $\frac{\partial b}{\partial a} = -\frac{2K}{a^2\sin C} = -\frac{2(\frac{1}{2}ab\sin C)}{a^2\sin C} = -\frac{b}{a} = -\frac{3}{2}$

In Problems 11–13, find the first partial derivatives of z with respect to the independent variables x and y.

11. $x^2 + y^2 + z^2 = 25$. [This is the equation of a sphere of radius 5 and center (0, 0, 0).]
Differentiate implicitly with respect to x, treating y as a constant, to obtain:

$$2x + 2z\frac{\partial z}{\partial x} = 0. \qquad \text{Hence,} \qquad \frac{\partial z}{\partial x} = -\frac{x}{z}$$

Differentiate implicitly with respect to y, treating x as a constant:

$$2y + 2z\frac{\partial z}{\partial y} = 0. \qquad \text{Hence,} \qquad \frac{\partial z}{\partial y} = -\frac{y}{z}$$

12. $x^2(2y + 3z) + y^2(3x - 4z) + z^2(x - 2y) = xyz$.
Differentiate implicitly with respect to x:

$$2x(2y+3z) + 3x^2\frac{\partial z}{\partial x} + 3y^2 - 4y^2\frac{\partial z}{\partial x} + 2z(x-2y)\frac{\partial z}{\partial x} + z^2 = yz + xy\frac{\partial z}{\partial x}$$

Solving for $\frac{\partial z}{\partial x}$ yields: $\frac{\partial z}{\partial x} = -\frac{4xy + 6xz + 3y^2 + z^2 - yz}{3x^2 - 4y^2 + 2xz - 4yz - xy}$.

Differentiate implicitly with respect to y:

$$2x^2 + 3x^2\frac{\partial z}{\partial y} + 2y(3x-4z) - 4y^2\frac{\partial z}{\partial y} + 2z(x-2y)\frac{\partial z}{\partial y} - 2z^2 = xz + xy\frac{\partial z}{\partial y}$$

Solving for $\frac{\partial z}{\partial y}$ yields: $\frac{\partial z}{\partial y} = -\frac{2x^2 + 6xy - 8yz - 2z^2 - xz}{3x^2 - 4y^2 + 2xz - 4yz - xy}$.

13. $xy + yz + zx = 1$.
Differentiating with respect to x yields $y + y\frac{\partial z}{\partial x} + x\frac{\partial z}{\partial x} + z = 0$, whence $\frac{\partial z}{\partial x} = -\frac{y+z}{x+y}$.

Differentiating with respect to y yields $x + y\frac{\partial z}{\partial y} + z + x\frac{\partial z}{\partial y} = 0$, whence $\frac{\partial z}{\partial y} = -\frac{x+z}{x+y}$.

14. Considering x and y as independent variables, find $\frac{\partial r}{\partial x}, \frac{\partial r}{\partial y}, \frac{\partial \theta}{\partial x}, \frac{\partial \theta}{\partial y}$ when $x = e^{2r}\cos\theta$, $y = e^{3r}\sin\theta$.
First differentiate the given relations with respect to x:

$$1 = 2e^{2r}\cos\theta\frac{\partial r}{\partial x} - e^{2r}\sin\theta\frac{\partial \theta}{\partial x} \qquad \text{and} \qquad 0 = 3e^{3r}\sin\theta\frac{\partial r}{\partial x} + e^{3r}\cos\theta\frac{\partial \theta}{\partial x}$$

Then solve simultaneously to obtain $\frac{\partial r}{\partial x} = \frac{\cos\theta}{e^{2r}(2+\sin^2\theta)}$ and $\frac{\partial \theta}{\partial x} = -\frac{3\sin\theta}{e^{2r}(2+\sin^2\theta)}$.

Now differentiate the given relations with respect to y:

$$0 = 2e^{2r}\cos\theta\frac{\partial r}{\partial y} - e^{2r}\sin\theta\frac{\partial \theta}{\partial y} \qquad \text{and} \qquad 1 = 3e^{3r}\sin\theta\frac{\partial r}{\partial y} + e^{3r}\cos\theta\frac{\partial \theta}{\partial y}$$

Then solve simultaneously to obtain $\frac{\partial r}{\partial y} = \frac{\sin\theta}{e^{3r}(2+\sin^2\theta)}$ and $\frac{\partial \theta}{\partial y} = \frac{2\cos\theta}{e^{3r}(2+\sin^2\theta)}$.

15. Find the slopes of the tangent lines to the curves cut from the surface $z = 3x^2 + 4y^2 - 6$ by planes through the point (1, 1, 1) and parallel to the xz and yz planes.

The plane $x = 1$, parallel to the yz plane, intersects the surface in the curve $z = 4y^2 - 3$, $x = 1$. Then $\frac{\partial z}{\partial y} = 8y = 8(1) = 8$ is the required slope.

The plane $y = 1$, parallel to the xz plane, intersects the surface in the curve $z = 3x^2 + 2$, $y = 1$. Then $\frac{\partial z}{\partial x} = 6x = 6$ is the required slope.

In Problems 16 and 17, find all second partial derivatives of z and verify Theorem 48.1.

16. $z = x^2 + 3xy + y^2$.

$$\frac{\partial z}{\partial x} = 2x + 3y, \qquad \frac{\partial^2 z}{\partial x^2} = \frac{\partial}{\partial x}\left(\frac{\partial z}{\partial x}\right) = 2, \qquad \frac{\partial^2 z}{\partial y\,\partial x} = \frac{\partial}{\partial y}\left(\frac{\partial z}{\partial x}\right) = 3$$

$$\frac{\partial z}{\partial y} = 3x + 2y, \qquad \frac{\partial^2 z}{\partial y^2} = \frac{\partial}{\partial y}\left(\frac{\partial z}{\partial y}\right) = 2, \qquad \frac{\partial^2 z}{\partial x\,\partial y} = \frac{\partial}{\partial x}\left(\frac{\partial z}{\partial y}\right) = 3$$

Note that $\dfrac{\partial^2 z}{\partial y\,\partial x} = \dfrac{\partial^2 z}{\partial x\,\partial y}$.

17. $z = x \cos y - y \cos x$.

$$\frac{\partial z}{\partial x} = \cos y + y \sin x, \qquad \frac{\partial^2 z}{\partial x^2} = \frac{\partial}{\partial x}\left(\frac{\partial z}{\partial x}\right) = y \cos x$$

$$\frac{\partial^2 z}{\partial y\,\partial x} = \frac{\partial}{\partial y}\left(\frac{\partial z}{\partial x}\right) = -\sin y + \sin x$$

$$\frac{\partial z}{\partial y} = -x \sin y - \cos x, \qquad \frac{\partial^2 z}{\partial y^2} = \frac{\partial}{\partial y}\left(\frac{\partial z}{\partial y}\right) = -x \cos y$$

$$\frac{\partial^2 z}{\partial x\,\partial y} = \frac{\partial}{\partial x}\left(\frac{\partial z}{\partial y}\right) = -\sin y + \sin x$$

Note that $\dfrac{\partial^2 z}{\partial y\,\partial x} = \dfrac{\partial^2 z}{\partial x\,\partial y}$.

18. Let $f(x, y, z) = x \cos (yz)$. Find all partial derivatives of the first, second, and third order.

$f_x = \cos (yz), \qquad f_{xx} = 0, \qquad f_{yx} = -z \sin (yz), \qquad f_{zx} = -y \sin (yz)$

$f_y = -xz \sin (yz), \qquad f_{yy} = -xz^2 \cos (yz), \qquad f_{xy} = -z \sin (yz)$

$f_{zy} = -x(zy \cos (yz) + \sin(yz))$

$f_z = -xy \sin (yz), \qquad f_{zz} = -xy^2 \cos (yz), \qquad f_{xz} = -y \sin (yz)$

$f_{yz} = -x(zy \cos (yz) + \sin (yz))$

Note that $f_{xy} = f_{yx}$ and $f_{xz} = f_{zx}$ and $f_{yz} = f_{zy}$.

$f_{xxx} = 0, \qquad f_{xxy} = f_{xyx} = 0, \qquad f_{xxz} = f_{xzx} = 0$

$f_{xyy} = -z^2 \cos (yz), \qquad f_{xyz} = f_{xzy} = -(zy \cos (yz) + \sin (yz))$

$f_{xzz} = -y^2 \cos (yz)$

$f_{yyy} = xz^3 \sin (yz), \qquad f_{yxx} = 0, \qquad f_{yxy} = f_{yyx} = -z^2 \cos (yz)$

$f_{yxz} = f_{yzx} = -(yz \cos (yz) + \sin (yz))$

$f_{yyz} = f_{yzy} = -x(-z^2y \sin (yz) + z \cos (yz) + z \cos (yz))$

$\quad = xz(zy \sin (yz) - 2 \cos (yz))$

$f_{yzz} = -x(-y^2z \sin (yz) + 2y \cos (yz))$

$\quad = xy(z \sin (yz) - 2 \cos (yz))$

$f_{zzz} = xy^3 \sin (yz), \qquad f_{zxx} = 0, \qquad f_{zxy} = f_{zyx} = -(zy \cos (yz) + \sin (yz))$

$f_{zxz} = f_{zzx} = -y^2 \cos (yz)$

$f_{zyy} = -x(-z^2y \sin (yz) + 2z \cos (yz)) = xz(zy \sin (yz) - 2 \cos (yz))$

$f_{zyz} = f_{zzy} = -x(-zy^2 \sin (yz) + y \cos (yz) + y \cos (yz))$

$\quad = xy(zy \sin (yz) - 2 \cos (yz))$

Note that, in the third order, any two rearrangements of subscripts will be equal. For example, $f_{xyz} = f_{xzy} = f_{yxz} = f_{yzx} = f_{zxy} = f_{zyx} = -(zy \cos (yz) + \sin (yz))$.

19. Determine whether the following functions are solutions of Laplace's equation $\frac{\partial^2 z}{\partial x^2} + \frac{\partial^2 z}{\partial y^2} = 0$:

(a) $z = e^x \cos y$ (b) $z = \frac{1}{2}(e^{x+y})$ (c) $z = x^2 - y^2$

(a) $\frac{\partial z}{\partial x} = e^x \cos y, \quad \frac{\partial^2 z}{\partial x^2} = e^x \cos y$

$\frac{\partial z}{\partial y} = -e^x \sin y, \quad \frac{\partial^2 z}{\partial y^2} = -e^x \cos y$

Then $\frac{\partial^2 z}{\partial x^2} + \frac{\partial^2 z}{\partial y^2} = 0$.

(b) $\frac{\partial z}{\partial x} = \frac{1}{2}(e^{x+y}), \quad \frac{\partial^2 z}{\partial x^2} = \frac{1}{2}(e^{x+y})$

$\frac{\partial z}{\partial y} = \frac{1}{2}(e^{x+y}), \quad \frac{\partial^2 z}{\partial y^2} = \frac{1}{2}(e^{x+y})$

So, $\frac{\partial^2 z}{\partial x^2} + \frac{\partial^2 z}{\partial y^2} = e^{x+y} \neq 0$.

(c) $\frac{\partial z}{\partial x} = 2x, \quad \frac{\partial^2 z}{\partial x^2} = 2$

$\frac{\partial z}{\partial y} = -2y, \quad \frac{\partial^2 z}{\partial y^2} = -2$

So, $\frac{\partial^2 z}{\partial x^2} + \frac{\partial^2 z}{\partial y^2} = 0$.

SUPPLEMENTARY PROBLEMS

In Problems 20–24, evaluate the given limit.

20. $\lim_{(x,y)\to(-1,2)} \frac{x - 2y}{x^2 + y}$ *Ans.* $-\frac{5}{3}$

21. $\lim_{(x,y)\to(0,0)} \frac{x - y}{x^2 + y^2}$ *Ans.* no limit

22. $\lim_{(x,y)\to(0,0)} \frac{3xy}{2x^2 + y^2}$ *Ans.* no limit

23. $\lim_{(x,y)\to(0,0)} \frac{xy^2}{x^2 + y^4}$ *Ans.* no limit

24. $\lim_{(x,y)\to(0,0)} \frac{x^2 + y^2}{\sqrt{x^2 + y^2 + 4} - 2}$ *Ans.* 4

25. Determine whether each of the following functions can be defined at (0, 0) so as to be continuous:

(a) $\frac{y^2}{x^2 + y^2}$ (b) $\frac{x - y}{x + y}$ (c) $\frac{x^3 + y^3}{x^2 + y^2}$ (d) $\frac{x + y}{x^2 + y^2}$

Ans. (a) no; (b) no; (c) yes; (d) no

26. For each of the following functions z, find $\frac{\partial z}{\partial x}$ and $\frac{\partial z}{\partial y}$.

(a) $z = x^2 + 3xy + y^2$ *Ans.* $\frac{\partial z}{\partial x} = 2x + 3y;\ \frac{\partial z}{\partial y} = 3x + 2y$

(b) $z = \frac{x}{y^2} - \frac{y}{x^2}$ *Ans.* $\frac{\partial z}{\partial x} = \frac{1}{y^2} + \frac{2y}{x^3};\ \frac{\partial z}{\partial y} = -\frac{2x}{y^3} - \frac{1}{x^2}$

(c) $z = \sin 3x \cos 4y$ *Ans.* $\frac{\partial z}{\partial x} = 3\cos 3x \cos 4y;\ \frac{\partial z}{\partial y} = -4\sin 3x \sin 4y$

(d) $z = \tan^{-1}\left(\frac{y}{x}\right)$ *Ans.* $\frac{\partial z}{\partial x} = \frac{-y}{x^2 + y^2};\ \frac{\partial z}{\partial y} = \frac{x}{x^2 + y^2}$

(e) $x^2 - 4y^2 + 9z^2 = 36$ *Ans.* $\frac{\partial z}{\partial x} = -\frac{x}{9z};\ \frac{\partial z}{\partial y} = \frac{4y}{9z}$

(f) $z^3 - 3x^2y + 6xyz = 0$ *Ans.* $\frac{\partial z}{\partial x} = \frac{2y(x-z)}{z^2 + 2xy};\ \frac{\partial z}{\partial y} = \frac{x(x-2z)}{z^2 + 2xy}$

(g) $yz + xz + xy = 0$ *Ans.* $\frac{\partial z}{\partial x} = -\frac{y+z}{x+y};\ \frac{\partial z}{\partial y} = -\frac{x+z}{x+y}$.

27. For each of the following functions z, find $\frac{\partial^2 z}{\partial x^2}, \frac{\partial^2 z}{\partial y \partial x}, \frac{\partial^2 z}{\partial x \partial y}$, and $\frac{\partial^2 z}{\partial y^2}$.

(a) $z = 2x^2 - 5xy + y^2$ *Ans.* $\frac{\partial^2 z}{\partial x^2} = 4;\ \frac{\partial^2 z}{\partial x \partial y} = \frac{\partial^2 z}{\partial y \partial x} = -5;\ \frac{\partial^2 z}{\partial y^2} = 2$

(b) $z = \frac{x}{y^2} - \frac{y}{x^2}$ *Ans.* $\frac{\partial^2 z}{\partial x^2} = -\frac{6y}{x^4};\ \frac{\partial^2 z}{\partial x \partial y} = \frac{\partial^2 z}{\partial y \partial x} = 2\left(\frac{1}{x^3} - \frac{1}{y^3}\right);\ \frac{\partial^2 z}{\partial y^2} = \frac{6x}{y^4}$

(c) $z = \sin 3x \cos 4y$ *Ans.* $\frac{\partial^2 z}{\partial x^2} = -9z;\ \frac{\partial^2 z}{\partial x \partial y} = \frac{\partial^2 z}{\partial y \partial x} = -12\cos 3x \sin 4y;\ \frac{\partial^2 z}{\partial y^2} = -16z$

(d) $z = \tan^{-1}\left(\frac{y}{x}\right)$ *Ans.* $\frac{\partial^2 z}{\partial x^2} = -\frac{\partial^2 z}{\partial y^2} = \frac{2xy}{(x^2 + y^2)^2};\ \frac{\partial^2 z}{\partial x \partial y} = \frac{\partial^2 z}{\partial y \partial x} = \frac{y^2 - x^2}{(x^2 + y^2)^2}$

28. (a) If $z = \frac{xy}{x-y}$, show that $x^2 \frac{\partial^2 z}{\partial x^2} + 2xy \frac{\partial^2 z}{\partial x \partial y} + y^2 \frac{\partial^2 z}{\partial y^2} = 0$.

(b) If $z = e^{\alpha x} \cos \beta y$ and $\beta = \pm\alpha$, show that $\frac{\partial^2 z}{\partial x^2} + \frac{\partial^2 z}{\partial y^2} = 0$.

(c) If $z = e^{-t}(\sin x + \cos y)$, show that $\frac{\partial^2 z}{\partial x^2} + \frac{\partial^2 z}{\partial y^2} = \frac{\partial z}{\partial t}$.

(d) If $z = \sin ax \sin by \sin kt\sqrt{a^2 + b^2}$, show that $\frac{\partial^2 z}{\partial t^2} = k^2\left(\frac{\partial^2 z}{\partial x^2} + \frac{\partial^2 z}{\partial y^2}\right)$.

29. For the gas formula $\left(p + \frac{a}{v^2}\right)(v - b) = ct$, where a, b, and c are constants, show that

$$\frac{\partial p}{\partial v} = \frac{2a(v-b) - (p + a/v^2)v^3}{v^3(v-b)}, \quad \frac{\partial v}{\partial t} = \frac{cv^3}{(p + a/v^2)v^3 - 2a(v-b)}$$

$$\frac{\partial t}{\partial p} = \frac{v-b}{c}, \quad \frac{\partial p}{\partial v}\frac{\partial v}{\partial t}\frac{\partial t}{\partial p} = -1$$

30. Fill in the following sketch of a proof of Theorem 48.1. Assume that f_{xy} and f_{yx} exist and are continuous in an open disk. We must prove that $f_{xy}(a, b) = f_{yx}(a, b)$ at every point (a, b) of the disk. Let $\Delta_h = (f(a+h, b+h) - f(a+h, b)) - (f(a, b+h) - f(a, b))$ for h sufficiently small and $\neq 0$. Let $F(x) = f(x, b+h) - f(x, b)$. Then $\Delta_h = F(a+h) - F(a)$. Apply the Mean-Value Theorem to get a^* between a and $a+h$ so that $F(a+h) - F(a) = F'(a^*)h = [f_x(a^*, b+h) - f_x(a^*, b)]h$, and apply the Mean-Value Theorem to get b^* between b and $b+h$ so that $f_x(a^*, b+h) - f_x(a^*, b) = f_{xy}(a^*, b^*)h$. Then,

$$\Delta_h = h^2 f_{xy}(a^*, b^*) \text{ and } \lim_{h \to 0} \frac{\Delta_h}{h^2} = \lim_{(a^*, b^*) \to (a, b)} f_{xy}(a^*, b^*) = f_{xy}(a, b)$$

By a similar argument using $\Delta_h = (f(a+h, b+h) - f(a, b+h)) - (f(a+h, b) - f(a, b))$ and the Mean-Value Theorem, we get

$$\lim_{h \to 0} \frac{\Delta_h}{h^2} = f_{yx}(a, b)$$

31. Show that Theorem 48.1 no longer holds if the continuity assumption for f_{xy} and f_{yx} is dropped. Use the following function:

$$f(x, y) = \begin{cases} \dfrac{xy(x^2 - y^2)}{x^2 + y^2} & \text{if } (x, y) \neq (0, 0) \\ 0 & \text{if } (x, y) = (0, 0) \end{cases}$$

[Find formulas for $f_x(x, y)$ and $f_y(x, y)$ for $(x, y) \neq (0, 0)$; evaluate $f_x(0, 0)$ and $f_y(0, 0)$, and then $f_{xy}(0, 0)$ and $f_{yx}(0, 0)$.]

CHAPTER 49

Total Differential. Differentiability. Chain Rules

Total Differential

Let $z = f(x, y)$. Let Δx and Δy be any numbers. Δx and Δy are called *increments of x and y*, respectively. For these increments of x and y, the corresponding *change in z*, denoted Δz, is defined by

$$\Delta z = f(x+\Delta x, y+\Delta y) - f(x,y) \tag{49.1}$$

The *total differential dz* is defined by:

$$dz = \frac{\partial z}{\partial x}\Delta x + \frac{\partial z}{\partial y}\Delta y = f_x(x,y)\,\Delta x + f_y(x,y)\,\Delta y \tag{49.2}$$

Note that, if $z = f(x, y) = x$, then $\frac{\partial z}{\partial x} = 1$ and $\frac{\partial z}{\partial y} = 0$, and, therefore, $dz = \Delta x$. So, $dx = \Delta x$. Similarly, $dy = \Delta y$. Hence, equation (49.2) becomes

$$dz = \frac{\partial z}{\partial x}dx + \frac{\partial z}{\partial y}dy = f_x(x,y)\,dx + f_y(x,y)\,dy \tag{49.3}$$

Notation: dz is also denoted df.

These definitions can be extended to functions of three or more variables. For example, if $u = f(x, y, z)$, then we get:

$$du = \frac{\partial u}{\partial x}dx + \frac{\partial u}{\partial y}dy + \frac{\partial u}{\partial z}dz$$

$$= f_x(x,y,z)\,dx + f_y(x,y,z)\,dy + f_z(x,y,z)\,dz$$

EXAMPLE 49.1: Let $z = x\cos y - 2x^2 + 3$. Then $\frac{\partial z}{\partial x} = \cos y - 4x$ and $\frac{\partial z}{\partial y} = -x\sin y$. Then the total differential for z is $dz = (\cos y - 4x)\,dx - (x\sin y)\,dy$.

In the case of a function of one variable $y = f(x)$, we used the approximation principle $\Delta y \sim f'(x)\,\Delta x = dy$ to estimate values of f. However, in the case of a function $z = f(x, y)$ of two variables, the function f has to satisfy a special condition in order to make good approximations possible.

Differentiability

A function $z = f(x, y)$ is said to be *differentiable* at (a, b) if functions ϵ_1 and ϵ_2 exist such that

$$\Delta z = f_x(a, b)\,\Delta x + f_y(a, b)\,\Delta y + \epsilon_1\,\Delta x + \epsilon_2\,\Delta y \tag{49.4}$$

and $\lim_{(\Delta x,\Delta y)\to(0,0)} \epsilon_1 = \lim_{(\Delta x,\Delta y)\to(0,0)} \epsilon_2 = 0$

Note that formula (49.4) can be written as

$$\Delta z = dz + \epsilon_1\,\Delta x + \epsilon_2\,\Delta y \tag{49.5}$$

We say that $z = f(x, y)$ is differentiable on a set A if it is differentiable at each point of A.

As in the case of one variable, differentiability implies continuity. (See Problem 23.)

EXAMPLE 49.2: Let us see that $z = f(x, y) = x + 2y^2$ is differentiable at every point (a, b). Note that $f_x(x, y) = 1$ and $f_y(x, y) = 4y$. Then

$$\begin{aligned}\Delta z &= f(a + \Delta x, b + \Delta y) - f(a, b) = a + \Delta x + 2(b + \Delta y)^2 - a - 2b^2 \\ &= \Delta x + 4b\,\Delta y + 2(\Delta y)^2 = f_x(a, b)\,\Delta x + f_y(a, b)\,\Delta y + (2\,\Delta y)\,\Delta y\end{aligned}$$

Let $\epsilon_1 = 0$ and $\epsilon_2 = 2\,\Delta y$.

Definition: By an *open set* in a plane, we mean a set A of points in the plane such that every point of A belongs to an open disk that is included in A.

Examples of open sets are an open disk and the interior of a rectangle.

Theorem 49.1: Assume that $f(x, y)$ is such that f_x and f_y are continuous in an open set A. Then f is differentiable in A.

For the proof, see Problem 43.

EXAMPLE 49.3: Let $z = f(x, y) = \sqrt{9 - x^2 - y^2}$. Then $f_x = \dfrac{-x}{\sqrt{9 - x^2 - y^2}}$ and $f_y = \dfrac{-y}{\sqrt{9 - x^2 - y^2}}$. So, by Theorem 49.1, f is differentiable in the open disk of radius 3 and center at the origin (0, 0) (where the denominators of f_x and f_y exist and are continuous). In that disk, $x^2 + y^2 < 9$. Take the point $(a, b) = (1, 2)$ and let us evaluate the change Δz as we move from (1, 2) to (1.03, 2.01). So, $\Delta x = 0.03$ and $\Delta y = 0.01$. Let us approximate Δz by

$$dz = f_x(1, 2)\,\Delta x + f_y(1, 2)\,\Delta y = \frac{-1}{2}(0.03) + \frac{-2}{2}(0.01) = -0.025$$

The actual difference Δz is $\sqrt{9 - (1.03)^2 - (2.01)^2} - \sqrt{9 - 1 - 4} \sim 1.9746 - 2 = -0.0254$.

Chain Rules

Chain Rule $(2 \to 1)$

Let $z = f(x, y)$, where f is differentiable, and let $x = g(t)$ and $y = h(t)$, where g and h are differentiable functions of one variable. Then $z = f(g(t), h(t))$ is a differentiable function of one variable, and

$$\frac{dz}{dt} = \frac{\partial z}{\partial x}\frac{dx}{dt} + \frac{\partial z}{\partial y}\frac{dy}{dt} \tag{49.6}$$

Warning: Note the double meaning of z, x, and y in (49.6). In $\frac{dz}{dt}$, z means $f(g(t), h(t))$, whereas, in $\frac{\partial z}{\partial x}$ and $\frac{\partial z}{\partial y}$, z means $f(x, y)$. In $\frac{\partial z}{\partial x}$, x is an independent variable, whereas, in $\frac{dx}{dt}$, x means $g(t)$. Likewise, y has two different meanings.

To prove (49.6), note first that, by (49.4),

$$\Delta z = \frac{\partial z}{\partial x}\Delta x + \frac{\partial z}{\partial y}\Delta y + \epsilon_1 \Delta x + \epsilon_2 \Delta y$$

Then $$\frac{\Delta z}{\Delta t} = \frac{\partial z}{\partial x}\frac{\Delta x}{\Delta t} + \frac{\partial z}{\partial y}\frac{\Delta y}{\Delta t} + \epsilon_1 \frac{\Delta x}{\Delta t} + \epsilon_2 \frac{\Delta y}{\Delta t}.$$

Letting $\Delta t \to 0$, we obtain

$$\frac{dz}{dt} = \frac{\partial z}{\partial x}\frac{dx}{dt} + \frac{\partial z}{\partial y}\frac{dy}{dt} + 0(\Delta x) + 0(\Delta y) = \frac{\partial z}{\partial x}\frac{dx}{dt} + \frac{\partial z}{\partial y}\frac{dy}{dt}$$

(Note that, since g and h are differentiable, they are continuous. Hence, as $\Delta t \to 0$, $\Delta x \to 0$ and $\Delta y \to 0$ and, therefore, $\epsilon_1 \to 0$ and $\epsilon_2 \to 0$.)

EXAMPLE 49.4: Let $z = xy + \sin x$, and let $x = t^2$ and $y = \cos t$. Note that $\frac{\partial z}{\partial x} = y + \cos x$ and $\frac{\partial z}{\partial y} = x$. Moreover, $\frac{dx}{dt} = 2t$ and $\frac{dy}{dt} = -\sin t$. Now, as a function of t, $z = t^2 \cos t + \sin(t^2)$.

By formula (49.6),

$$\frac{dz}{dt} = (y + \cos x)2t + x(-\sin t) = (\cos t + \cos(t^2))2t - t^2 \sin t$$

In this particular example, the reader can check the result by computing $D_t(t^2 \cos t + \sin(t^2))$.

Chain Rule (2 → 2)

Let $z = f(x, y)$, where f is differentiable, and let $x = g(t, s)$ and $y = h(t, s)$, where g and h are differentiable functions. Then $z = f(g(t, s), h(t, s))$ is a differentiable function, and

$$\frac{\partial z}{\partial t} = \frac{\partial z}{\partial x}\frac{\partial x}{\partial t} + \frac{\partial z}{\partial y}\frac{\partial y}{\partial t} \quad \text{and} \quad \frac{\partial z}{\partial s} = \frac{\partial z}{\partial x}\frac{\partial x}{\partial s} + \frac{\partial z}{\partial y}\frac{\partial y}{\partial s} \tag{49.7}$$

Here again, as in the previous chain rule, the symbols z, x, and y have obvious double meanings.

This chain rule can be considered a special case of the chain rule (2 → 1). For example, the partial derivative $\frac{\partial z}{\partial t}$ can be thought of as an ordinary derivative $\frac{dz}{dt}$, because s is treated as a constant. Then the formula for $\frac{\partial z}{\partial t}$ in (49.7) is the same as the formula for $\frac{dz}{dt}$ in (49.6).

EXAMPLE 49.5: Let $z = e^x \sin y$ and $x = ts^2$ and $y = t + 2s$. Now, $\frac{\partial z}{\partial x} = e^x \sin y$, $\frac{\partial x}{\partial t} = s^2$, $\frac{\partial z}{\partial y} = e^x \cos y$, and $\frac{\partial y}{\partial t} = 1$. Hence, by (49.7),

$$\frac{\partial z}{\partial t} = (e^x \sin y)s^2 + (e^x \cos y) = e^x(s^2 \sin y + \cos y) = e^{ts^2}(s^2 \sin(t + 2s) + \cos(t + 2s))$$

Similarly,

$$\frac{\partial z}{\partial s} = 2(e^x \sin y)ts + 2(e^x \cos y) = 2e^x(ts \sin y + \cos y) = 2e^{ts^2}(ts \sin(t + 2s) + \cos(t + 2s))$$

Generalizations of the chain rule (49.47) hold for cases ($m \to n$), where $z = f(x, y, \ldots)$ is a function of m variables and each of those variables is a function of a given set of n variables.

Implicit Differentiation

Assume that the equation $F(x, y, z) = 0$ defines z implicitly as a function of x and y. Then, by the chain rule ($3 \to 2$), if we differentiate both sides of the equation with respect to x, we get

$$\frac{\partial F}{\partial x}\frac{\partial x}{\partial x} + \frac{\partial F}{\partial y}\frac{\partial y}{\partial x} + \frac{\partial F}{\partial z}\frac{\partial z}{\partial x} = 0$$

Since $\dfrac{\partial x}{\partial x} = 1$ and $\dfrac{\partial y}{\partial x} = 0$, $\dfrac{\partial F}{\partial x} + \dfrac{\partial F}{\partial z}\dfrac{\partial z}{\partial x} = 0$

Similarly, $\dfrac{\partial F}{\partial y} + \dfrac{\partial F}{\partial z}\dfrac{\partial z}{\partial y} = 0$. So, if $\dfrac{\partial F}{\partial z} \neq 0$,

$$\frac{\partial z}{\partial x} = -\frac{\partial F/\partial x}{\partial F/\partial z} \quad \text{and} \quad \frac{\partial z}{\partial y} = -\frac{\partial F/\partial y}{\partial F/\partial z} \tag{49.8}$$

This also can be written as $\dfrac{\partial z}{\partial x} = -\dfrac{F_x}{F_z}$ and $\dfrac{\partial z}{\partial y} = -\dfrac{F_y}{F_z}$.

EXAMPLE 49.6: The equation $xy + yz^3 + xz = 0$ determines z as a function of x and y. Let $F(x, y, z) = xy + yz^3 + xz$. Since $F_z = x + 3yz^2$, $F_x = y + z$, and $F_y = x + z^3$, (49.8) implies

$$\frac{\partial z}{\partial x} = -\frac{y+z}{x+3yz^2} \quad \text{and} \quad \frac{\partial z}{\partial y} = -\frac{x+z^3}{x+3yz^2}$$

SOLVED PROBLEMS

In Problems 1 and 2, find the total differential.

1. $z = x^3y + x^2y^2 + xy^3$

We have $$\frac{\partial z}{\partial x} = 3x^2y + 2xy^2 + y^3 \quad \text{and} \quad \frac{\partial z}{\partial y} = x^3 + 2x^2y + 3xy^2$$

Then $$dz = \frac{\partial z}{\partial x}dx + \frac{\partial z}{\partial y}dy = (3x^2y + 2xy^2 + y^3)\,dx + (x^3 + 2x^2y + 3xy^2)\,dy$$

2. $z = x\sin y - y\sin x$

We have $$\frac{\partial z}{\partial x} = \sin y - y\cos x \quad \text{and} \quad \frac{\partial z}{\partial y} = x\cos y - \sin x$$

Then $$dz = \frac{\partial z}{\partial x}dx + \frac{\partial z}{\partial y}dy = (\sin y - y\cos x)\,dx + (x\cos y - \sin x)\,dy$$

3. Compare dz and Δz, given $z = x^2 + 2xy - 3y^2$.

$\dfrac{\partial z}{\partial x} = 2x + 2y$ and $\dfrac{\partial z}{\partial y} = 2x - 6y$. So $dz = 2(x+y)\,dx + 2(x-3y)\,dy$

Also, $$\begin{aligned}\Delta z &= \left[(x+dx)^2 + 2(x+dx)(y+dy) - 3(y+dy)^2\right] - (x^2 + 2xy - 3y^2) \\ &= 2(x+y)\,dx + 2(x-3y)\,dy + (dx)^2 + 2\,dx\,dy - 3(dy)^2\end{aligned}$$

Thus dz and Δz differ by $(dx)^2 + 2\,dx\,dy - 3(dy)^2$.

4. Approximate the area of a rectangle of dimensions 35.02 by 24.97 units.

For dimensions x by y, the area is $A = xy$ so that $dA = \frac{\partial A}{\partial x}dx + \frac{\partial A}{\partial y}dy = y\,dx + x\,dy$. With $x = 35$, $dx = 0.02$, $y = 25$, and $dy = -0.03$, we have $A = 35(25) = 875$ and $dA = 25(0.02) + 35(-0.03) = -0.55$. The area is approximately $A + dA = 874.45$ square units. The actual area is 874. 4494.

5. Approximate the change in the hypotenuse of a right triangle of legs 6 and 8 inches when the shorter leg is lengthened by $\frac{1}{4}$ inch and the longer leg is shortened by $\frac{1}{8}$ inch.

Let x, y, and z be the shorter leg, the longer leg, and the hypotenuse of the triangle. Then

$$z = \sqrt{x^2 + y^2}, \quad \frac{\partial z}{\partial x} = \frac{x}{\sqrt{x^2 + y^2}}, \quad \frac{\partial z}{\partial y} = \frac{y}{\sqrt{x^2 + y^2}}, \quad dz = \frac{\partial z}{\partial x}dx + \frac{\partial z}{\partial y}dy = \frac{x\,dx + y\,dy}{\sqrt{x^2 + y^2}}$$

When $x = 6$, $y = 8$, $dx = \frac{1}{4}$, and $dy = -\frac{1}{8}$, then $dz = \frac{6(\frac{1}{4}) + 8(-\frac{1}{8})}{\sqrt{6^2 + 8^2}} = \frac{1}{20}$ inch. Thus the hypotenuse is lengthened by approximately $\frac{1}{20}$ inch.

6. The power consumed in an electrical resistor is given by $P = E^2/R$ (in watts). If $E = 200$ volts and $R = 8$ ohms, by how much does the power change if E is decreased by 5 volts and R is decreased by 0.2 ohm?

We have

$$\frac{\partial P}{\partial E} = \frac{2E}{R}, \quad \frac{\partial P}{\partial R} = -\frac{E^2}{R^2}, \quad dP = \frac{2E}{R}dE - \frac{E^2}{R^2}dR$$

When $E = 200$, $R = 8$, $dE = -5$, and $dR = -0.2$, then

$$dP = \frac{2(200)}{8}(-5) - \left(\frac{200}{8}\right)^2(-0.2) = -250 + 125 = -125$$

The power is reduced by approximately 125 watts.

7. The dimensions of a rectangular block of wood were found to be 10, 12, and 20 inches, with a possible error of 0.05 in each of the measurements. Find (approximately) the greatest error in the surface area of the block and the percentage error in the area caused by the errors in the individual measurements.

The surface area is $S = 2(xy + yz + zx)$; then

$$dS = \frac{\partial S}{\partial x}dx + \frac{\partial S}{\partial y}dy + \frac{\partial S}{\partial z}dz = 2(y + z)\,dx + 2(x + z)\,dy + 2(y + x)\,dz$$

The greatest error in S occurs when the errors in the lengths are of the same sign, say positive. Then

$$dS = 2(12 + 20)(0.05) + 2(10 + 20)(0.05) + 2(12 + 10)(0.05) = 8.4\,\text{in}^2$$

The percentage error is (error/area)(100) = (8.4/1120)(100) = 0.75%.

8. For the formula $R = E/C$, find the maximum error and the percentage error if $C = 20$ with a possible error of 0.1 and $E = 120$ with a possible error of 0.05.

Here

$$dR = \frac{\partial R}{\partial E}dE + \frac{\partial R}{\partial C}dC = \frac{1}{C}dE - \frac{E}{C^2}dC$$

The maximum error will occur when $dE = 0.05$ and $dC = -0.1$; then $dR = \frac{0.05}{20} - \frac{120}{400}(-0.1) = 0.0325$ is the approximate maximum error. The percentage error is $\frac{dR}{R}(100) = \frac{0.0325}{8}(100) = 0.40625 = 0.41\%$.

9. Two sides of a triangle were measured as 150 and 200 ft, and the included angle is 60°. If the possible errors are 0.2 ft in measuring the sides and 1° in the angle, what is the greatest possible error in the computed area?

Here

$$A = \frac{1}{2}xy\sin\theta, \quad \frac{\partial A}{\partial x} = \frac{1}{2}y\sin\theta, \quad \frac{\partial A}{\partial y} = \frac{1}{2}x\sin\theta, \quad \frac{\partial A}{\partial \theta} = \frac{1}{2}xy\cos\theta$$

and

$$dA = \tfrac{1}{2}y\sin\theta\,dx + \tfrac{1}{2}x\sin\theta\,dy + \tfrac{1}{2}xy\cos\theta\,d\theta$$

When $x = 150$, $y = 200$, $\theta = 60°$, $dx = 0.2$, $dy = 0.2$, and $d\theta = 1° = \pi/180$, then

$$dA = \tfrac{1}{2}(200)(\sin 60°)(0.2) + \tfrac{1}{2}(150)(\sin 60°)(0.2) + \tfrac{1}{2}(250)(200)(\cos 60°)(\pi/180) = 161.21\,\text{ft}^2$$

10. Find dz/dt, given $z = x^2 + 3xy + 5y^2$; $x = \sin t$, $y = \cos t$.

Since

$$\frac{\partial z}{\partial x} = 2x + 3y, \quad \frac{\partial z}{\partial y} = 3x + 10y, \quad \frac{dx}{dt} = \cos t, \quad \frac{dy}{dt} = -\sin t$$

we have

$$\frac{dz}{dt} = \frac{\partial z}{\partial x}\frac{dx}{dt} + \frac{\partial z}{\partial y}\frac{dy}{dt} = (2x + 3y)\cos t - (3x + 10y)\sin t$$

11. Find dz/dt, given $z = \ln(x^2 + y^2)$; $x = e^{-t}$, $y = e^t$.

Since

$$\frac{\partial z}{\partial x} = \frac{2x}{x^2 + y^2}, \quad \frac{\partial z}{\partial y} = \frac{2y}{x^2 + y^2}, \quad \frac{dx}{dt} = -e^{-t}, \quad \frac{dy}{dt} = e^t$$

we have

$$\frac{dz}{dt} = \frac{\partial z}{\partial x}\frac{dx}{dt} + \frac{\partial z}{\partial y}\frac{dy}{dt} = \frac{2x}{x^2 + y^2}(-e^{-t}) + \frac{2y}{x^2 - y^2}e^t = 2\frac{ye^t - xe^{-t}}{x^2 + y^2}$$

12. Find $\frac{dz}{dx}$, given $z = f(x, y) = x^2 + 2xy + 4y^2$, $y = e^{ax}$.

$$\frac{dz}{dx} = f_x + f_y\frac{dy}{dx} = (2x + 2y) + (2x + 8)ae^{ax} = 2(x + y) + 2a(x + 4y)e^{ax}$$

13. Find (a) $\frac{dz}{dx}$ and (b) $\frac{dz}{dy}$, given $z = f(x, y) = xy^2 + yx^2$, $y = \ln x$.

(a) Here x is the independent variable:

$$\frac{dz}{dx} = \frac{\partial f}{\partial x} + \frac{\partial f}{\partial y}\frac{dy}{dx} = (y^2 + 2xy) + (2xy + x^2)\frac{1}{x} = y^2 + 2xy + 2y + x$$

(b) Here y is the independent variable:

$$\frac{dz}{dy} = \frac{\partial f}{\partial x}\frac{dx}{dy} + \frac{\partial f}{\partial y} = (y^2 + 2xy)x + (2xy + x^2) = xy^2 + 2x^2y + 2xy + x^2$$

14. The altitude of a right circular cone is 15 inches and is increasing at 0.2 in/min. The radius of the base is 10 inches and is decreasing at 0.3 in/min. How fast is the volume changing?

Let x be the radius, and y the altitude of the cone (Fig. 49-1). From $V = \frac{1}{3}\pi x^2 y$, considering x and y as functions of time t, we have

$$\frac{dV}{dt} = \frac{\partial V}{\partial x}\frac{dx}{dt} + \frac{\partial V}{\partial y}\frac{dy}{dt} = \frac{\pi}{3}\left(2xy\frac{dx}{dt} + x^2\frac{dy}{dt}\right) = \frac{\pi}{3}[2(10)(15)(-0.3) + 10^2(0.2)] = -\frac{70\pi}{3}\text{ in}^3/\text{min}$$

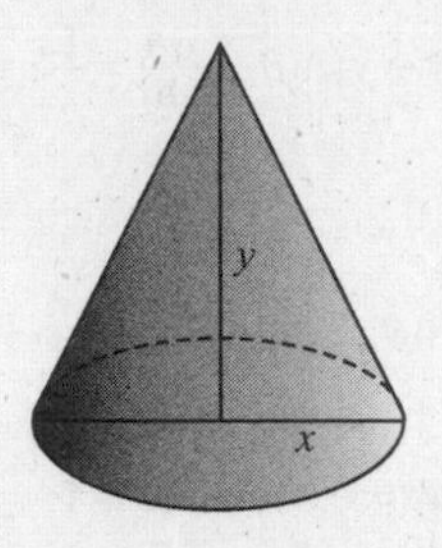

Fig. 49-1

15. A point P is moving along the curve that is the intersection of the surfaces $\frac{x^2}{16} - \frac{y^2}{9} = z$ and $x^2 + y^2 = 5$, with x, y, and z expressed in inches. If x is increasing at the rate of 0.2 inches per minute, how fast is z changing when $x = 2$?

From $z = \frac{x^2}{16} - \frac{y^2}{9}$, we obtain $\frac{dz}{dt} = \frac{\partial z}{\partial x}\frac{dx}{dt} + \frac{\partial z}{\partial y}\frac{dy}{dt} = \frac{x}{8}\frac{dx}{dt} - \frac{2y}{9}\frac{dy}{dt}$. Since $x^2 + y^2 = 5$, $y = \pm 1$ when $x = 2$; also, differentiation yields $x\frac{dx}{dt} + y\frac{dy}{dt} = 0$.

When $y = 1$, $\frac{dy}{dt} = -\frac{x}{y}\frac{dx}{dt} = -\frac{2}{1}(0.2) = -0.4$ and $\frac{dz}{dt} = \frac{2}{8}(0.2) - \frac{2}{9}(-0.4) = \frac{5}{36}$ in/min.

When $y = -1$, $\frac{dy}{dt} = -\frac{x}{y}\frac{dx}{dt} = 0.4$ and $\frac{dz}{dt} = \frac{2}{8}(0.2) - \frac{2}{9}(-1)(0.4) = \frac{5}{36}$ in/min.

16. Find $\frac{\partial z}{\partial r}$ and $\frac{\partial z}{\partial s}$, given $z = x^2 + xy + y^2$; $x = 2r + s$, $y = r - 2s$.

Here

$$\frac{\partial z}{\partial x} = 2x + y,\quad \frac{\partial z}{\partial y} = x + 2y,\quad \frac{\partial x}{\partial r} = 2,\quad \frac{\partial x}{\partial s} = 1,\quad \frac{\partial y}{\partial r} = 1,\quad \frac{\partial y}{\partial s} = -2$$

Then

$$\frac{\partial z}{\partial r} = \frac{\partial z}{\partial x}\frac{\partial x}{\partial r} + \frac{\partial z}{\partial y}\frac{\partial y}{\partial r} = (2x + y)(2) + (x + 2y)(1) = 5x + 4y$$

and

$$\frac{\partial z}{\partial s} = \frac{\partial z}{\partial x}\frac{\partial x}{\partial s} + \frac{\partial z}{\partial y}\frac{\partial y}{\partial s} = (2x + y)(1) + (x + 2y)(-2) = -3y$$

17. Find $\frac{\partial u}{\partial \rho}$, $\frac{\partial u}{\partial \beta}$, and $\frac{\partial u}{\partial \theta}$, given $u = x^2 + 2y^2 + 2z^2$; $x = \rho \sin \beta \cos \theta$, $y = \rho \sin \beta \sin \theta$, $z = \rho \cos \beta$.

$$\frac{\partial u}{\partial \rho} = \frac{\partial u}{\partial x}\frac{\partial x}{\partial \rho} + \frac{\partial u}{\partial y}\frac{\partial y}{\partial \rho} + \frac{\partial u}{\partial z}\frac{\partial z}{\partial \rho} = 2x\sin\beta\cos\theta + 4y\sin\beta\sin + 4z\cos\beta$$

$$\frac{\partial u}{\partial \beta} = \frac{\partial u}{\partial x}\frac{\partial x}{\partial \beta} + \frac{\partial u}{\partial y}\frac{\partial y}{\partial \beta} + \frac{\partial u}{\partial z}\frac{\partial z}{\partial \beta} = 2x\rho\cos\beta\cos\theta + 4y\rho\cos\beta\sin\theta - 4z\rho\sin\beta$$

$$\frac{\partial u}{\partial \theta} = \frac{\partial u}{\partial x}\frac{\partial x}{\partial \theta} + \frac{\partial u}{\partial y}\frac{\partial y}{\partial \theta} + \frac{\partial u}{\partial z}\frac{\partial z}{\partial \theta} = -2x\rho\sin\beta\sin\theta + 4y\rho\sin\beta\cos\theta$$

18. Find $\frac{du}{dx}$, given $u = f(x, y, z) = xy + yz + zx$; $y = \frac{1}{x}$, $z = x^2$.

$$\frac{du}{dx} = \frac{\partial f}{\partial x} + \frac{\partial f}{\partial y}\frac{dy}{dx} + \frac{\partial f}{\partial z}\frac{dz}{dx} = (y+z) + (x+z)\left(-\frac{1}{x^2}\right) + (y+x)2x = y + z + 2x(x+y) - \frac{x+z}{x^2}$$

19. Use implicit differentiation (formula (49.8)) to find $\frac{\partial z}{\partial x}$ and $\frac{\partial z}{\partial y}$, given $F(x, y, z) = x^2 + 3xy - 2y^2 + 3xz + z^2 = 0$.

$$\frac{\partial z}{\partial x} = -\frac{F_x}{F_z} = -\frac{2x+3y+3z}{3x+2z} \quad \text{and} \quad \frac{\partial z}{\partial x} = -\frac{F_y}{F_z} = -\frac{3x-4y}{3x+2z}$$

20. Use implicit differentiation (formula (49.8)) to find $\frac{\partial z}{\partial x}$ and $\frac{\partial z}{\partial y}$, given $\sin xy + \sin yz + \sin zx = 1$.

Set $F(x, y, z) = \sin xy + \sin yz + \sin zx - 1$; then

$$\frac{\partial F}{\partial x} = y\cos xy + z\cos zx, \quad \frac{\partial F}{\partial y} = x\cos xy + z\cos yz, \quad \frac{\partial F}{\partial z} = y\cos yz + x\cos zx$$

and
$$\frac{\partial z}{\partial x} = -\frac{\partial F/\partial x}{\partial F/\partial z} = -\frac{y\cos xy + z\cos zx}{y\cos yz + x\cos zx}, \quad \frac{\partial z}{\partial y} = -\frac{\partial F/\partial y}{\partial F/\partial z} = -\frac{x\cos xy + z\cos yz}{y\cos yz + x\cos zx}$$

21. If u and v are defined as functions of x and y by the equations $f(x, y, u, v) = x + y^2 + 2uv = 0$ and $g(x, y, u, v) = x^2 - xy + y^2 + u^2 + v^2 = 0$, find (a) $\frac{\partial u}{\partial x}$ and $\frac{\partial v}{\partial x}$; (b) $\frac{\partial u}{\partial y}$ and $\frac{\partial v}{\partial y}$.

(a) Differentiating f and g partially with respect to x, we obtain

$$1 + 2v\frac{\partial u}{\partial x} + 2u\frac{\partial v}{\partial x} = 0 \quad \text{and} \quad 2x - y + 2u\frac{\partial u}{\partial x} + 2v\frac{\partial v}{\partial x} = 0$$

Solving these relations simultaneously for $\partial u/\partial x$ and $\partial v/\partial x$, we find

$$\frac{\partial u}{\partial x} = \frac{v + u(y-2x)}{2(u^2 - v^2)} \quad \text{and} \quad \frac{\partial v}{\partial x} = \frac{v(2x-y) - u}{2(u^2 - v^2)}$$

(b) Differentiating f and g partially with respect to y, we obtain

$$2y + 2v\frac{\partial u}{\partial y} + 2u\frac{\partial v}{\partial y} = 0 \quad \text{and} \quad -x + 2y + 2u\frac{\partial u}{\partial y} + 2v\frac{\partial v}{\partial y} = 0$$

Then
$$\frac{\partial u}{\partial y} = \frac{u(x-2y) + 2vy}{2(u^2 - v^2)} \quad \text{and} \quad \frac{\partial v}{\partial y} = \frac{v(2y-x) - 2uy}{2(u^2 - v^2)}$$

22. Given $u^2 - v^2 + 2x + 3y = 0$ and $uv + x - y = 0$, find (a) $\frac{\partial u}{\partial x}, \frac{\partial v}{\partial x}, \frac{\partial u}{\partial y}, \frac{\partial v}{\partial y}$ and (b) $\frac{\partial x}{\partial u}, \frac{\partial y}{\partial u}, \frac{\partial x}{\partial v}, \frac{\partial y}{\partial v}$.

(a) Here x and y are to be considered as independent variables. Differentiate the given equations partially with respect to x, obtaining

$$2u\frac{\partial u}{\partial x} - 2v\frac{\partial v}{\partial x} + 2 = 0 \quad \text{and} \quad v\frac{\partial u}{\partial x} + u\frac{\partial v}{\partial x} + 1 = 0$$

Solve these relations simultaneously to obtain $\frac{\partial u}{\partial x} = -\frac{u+v}{u^2+v^2}$ and $\frac{\partial v}{\partial x} = \frac{v-u}{u^2+v^2}$.

Differentiate the given equations partially with respect to y, obtaining

$$2u\frac{\partial u}{\partial y} - 2v\frac{\partial v}{\partial y} + 3 = 0 \quad \text{and} \quad v\frac{\partial u}{\partial y} + u\frac{\partial v}{\partial y} - 1 = 0$$

Solve simultaneously to obtain $\frac{\partial u}{\partial y} = \frac{2v - 3u}{2(u^2+v^2)}$ and $\frac{\partial v}{\partial y} = \frac{2u + 3v}{2(u^2+v^2)}$.

(b) Here u and v are to be considered as independent variables. Differentiate the given equations partially with respect to u, obtaining

$$2u + 2\frac{\partial x}{\partial u} + 3\frac{\partial y}{\partial u} = 0 \quad \text{and} \quad v + \frac{\partial x}{\partial u} - \frac{\partial y}{\partial u} = 0$$

Then $$\frac{\partial x}{\partial u} = -\frac{2u + 3v}{5} \quad \text{and} \quad \frac{\partial y}{\partial u} = \frac{2(v-u)}{5}.$$

Differentiate the given equations with respect to v, obtaining

$$-2v + 2\frac{\partial x}{\partial v} + 3\frac{\partial y}{\partial v} = 0 \quad \text{and} \quad u + \frac{\partial x}{\partial v} - \frac{\partial y}{\partial v} = 0$$

Then $$\frac{\partial x}{\partial v} = \frac{2v - 3u}{5} \quad \text{and} \quad \frac{\partial y}{\partial v} = \frac{2u(u+v)}{5}.$$

23. Show that differentiability of $z = f(x, y)$ at (a, b) implies that f is continuous at (a, b).

From (49.4), $\Delta z = (f_x(a, b) + \epsilon_1)\,\Delta x + (f_y(a, b) + \epsilon_2)\,\Delta y$, where $\lim_{(\Delta x, \Delta y) \to (0,0)} \epsilon_1 = \lim_{(\Delta x, \Delta y) \to (0,0)} \epsilon_2 = 0$. Hence, $\Delta z \to 0$ as $(\Delta x, \Delta y) \to (0, 0)$, which implies that f is continuous at (a, b).

SUPPLEMENTARY PROBLEMS

24. Find the total differential of the following functions:

(a) $z = xy^3 + 2xy^3$ *Ans.* $dz = (3x^2 + 2y^2)\,dx + (x^2 + 6y^2)\,dy$

(b) $\theta = \tan^{-1}\left(\frac{y}{x}\right)$ *Ans.* $d\theta = \frac{x\,dy - y\,dx}{x^2 + y^2}$

(c) $z = e^{x^2 - y^2}$ *Ans.* $dz = 2z(x\,dx - y\,dy)$

(d) $z = x(x^2 + y^2)^{-1/2}$ *Ans.* $dz = \frac{y(y\,dx - x\,dy)}{(x^2 + y^2)^{3/2}}$

25. Use differentials to approximate (a) the volume of a box with square base of side 8.005 and height 9.996 ft; (b) the diagonal of a rectangular box of dimensions 3.03 by 5.98 by 6.01 ft.

Ans. (a) 640.544 ft^3; (b) 9.003 ft

26. Approximate the maximum possible error and the percentage of error when z is computed by the given formula.

(a) $z = \pi r^2 h$; $r = 5 \pm 0.05$, $h = 12 \pm 0.1$ *Ans.* 8.5π; 2.8%

(b) $1/z = 1/f + 1/g$; $f = 4 \pm 0.01$, $g = 8 \pm 0.02$ *Ans.* 0.0067; 0.25%

(c) $z = y/x$; $x = 1.8 \pm 0.1$, $y = 2.4 \pm 0.1$ *Ans.* 0.13; 10%

27. Find the approximate maximum percentage of error in:

(a) $\omega = \sqrt[3]{g/b}$ if there is a possible 1% error in measuring g and a possible $\frac{1}{2}$% error in measuring b.

(*Hint:* $\ln \omega = \frac{1}{3}(\ln g - \ln b)$; $\frac{d\omega}{\omega} = \frac{1}{3}\left(\frac{dg}{g} - \frac{db}{b}\right)$; $\left|\frac{dg}{g}\right| = 0.01$; $\left|\frac{db}{b}\right| = 0.005$)

Ans. 0.005

(b) $g = 2s/t^2$ if there is a possible 1% error in measuring s and $\frac{1}{4}$% error in measuring t.

Ans. 0.015

28. Find du/dt, given:

(a) $u = x^2y^3$; $x = 2t^3$, $y = 3t^2$

Ans. $6xy^2t(2yt + 3x)$

(b) $u = x \cos y + y \sin x$; $x = \sin 2t$, $y = \cos 2t$

Ans. $2(\cos y + y \cos x) \cos 2t - 2(-x \sin y + \sin x) \sin 2t$

(c) $u = xy + yz + zx$; $x = e^t$, $y = e^{-t}$, $z = e^t + e^{-t}$

Ans. $(x + 2y + z)e^t - (2x + y + z)e^{-t}$

29. At a certain instant, the radius of a right circular cylinder is 6 inches and is increasing at the rate 0.2 in/sec, while the altitude is 8 inches and is decreasing at the rate 0.4 in/sec. Find the time rate of change (a) of the volume and (b) of the surface at that instant.

Ans. (a) 4.8π in^3/sec; (b) 3.2π in^2/sec

30. A particle moves in a plane so that at any time t its abscissa and ordinate are given by $x = 2 + 3t$, $y = t^2 + 4$ with x and y in feet and t in minutes. How is the distance of the particle from the origin changing when $t = 1$?

Ans. $5/\sqrt{2}$ ft/min

31. A point is moving along the curve of intersection of $x^2 + 3xy + 3y^2 = z^2$ and the plane $x - 2y + 4 = 0$. When $x = 2$ and is increasing at 3 units/sec, find (a) how y is changing, (b) how z is changing, and (c) the speed of the point.

Ans. (a) increasing 3/2 units/sec; (b) increasing 75/14 units/sec at (2, 3, 7) and decreasing 75/14 units/sec at (2, 3, –7); (c) 6.3 units/sec

32. Find $\partial z/\partial s$ and $\partial z/\partial t$, given:

(a) $z = x^2 - 2y^2$; $x = 3s + 2t$, $y = 3s - 2t$ *Ans.* $6(x - 2y)$; $4(x + 2y)$

(b) $z = x^2 + 3xy + y^2$; $x = \sin s + \cos t$, $y = \sin s - \cos t$ *Ans.* $5(x + y) \cos s$; $(x - y) \sin t$

(c) $z = x^2 + 2y^2$; $x = e^s - e^t$, $y = e^s + e^t$ *Ans.* $2(x + 2y)e^s$; $2(2y - x)e^t$

(d) $z = \sin(4x + 5y)$; $x = s + t$, $y = s - t$ *Ans.* $9 \cos(4x + 5y)$; $-\cos(4x + 5y)$

(e) $z = e^{xy}$; $x = s^2 + 2st$, $y = 2st + t^2$ *Ans.* $2e^{xy}[tx + (s + t)y]$; $2e^{xy}[(s + t)x + sy]$

33. (a) If $u = f(x, y)$ and $x = r \cos\theta$, $y = r \sin\theta$, show that

$$\left(\frac{\partial u}{\partial x}\right)^2 + \left(\frac{\partial u}{\partial y}\right)^2 = \left(\frac{\partial u}{\partial r}\right)^2 + \frac{1}{r^2}\left(\frac{\partial u}{\partial \theta}\right)^2$$

(b) If $u = f(x, y)$ and $x = r \cosh s$, $y = r \sinh s$, show that

$$\left(\frac{\partial u}{\partial x}\right)^2 - \left(\frac{\partial u}{\partial y}\right)^2 = \left(\frac{\partial u}{\partial r}\right)^2 - \frac{1}{s^2}\left(\frac{\partial u}{\partial s}\right)^2$$

34. (a) If $z = f(x + \alpha y) + g(x - \alpha y)$, show that $\frac{\partial^2 z}{\partial x^2} = \frac{1}{\alpha^2}\frac{\partial^2 z}{\partial y^2}$. (*Hint:* Write $z = f(u) + g(v)$, $u = x + \alpha y$, $v = x - \alpha y$.)

(b) If $z = x^n f(y/x)$, show that $x\partial z/\partial x + y\partial z/\partial y = nz$.

(c) If $z = f(x, y)$ and $x = g(t)$, $y = h(t)$, show that, subject to continuity conditions,

$$\frac{d^2z}{dt^2} = f_{xx}(g')^2 + 2f_{xy}g'h' + f_{yy}(h')^2 + f_x g'' + f_y h''$$

(d) If $z = f(x, y)$; $x = g(r, s)$, $y = h(r, s)$, show that, subject to continuity conditions,

$$\frac{\partial^2 z}{\partial r^2} = f_{xx}(g_r)^2 + 2f_{xy}g_r h_r + f_{yy}(h_r)^2 + f_x g_{rr} + f_y h_{rr}$$

$$\frac{\partial^2 z}{\partial r \partial s} = f_{xx}g_r g_s + f_{xy}(g_r h_s + g_s h_r) + f_{yy}h_r h_s + f_x g_{rs} + f_y h_{rs}$$

$$\frac{\partial^2 z}{\partial s^2} = f_{xx}(g_s)^2 + 2f_{xy}g_s h_s + f_{yy}(h_s)^2 + f_x g_{ss} + f_y h_{ss}$$

35. A function $f(x, y)$ is called *homogeneous of order n* if $f(tx, ty) = t^n f(x, y)$. (For example, $f(x, y) = x^2 + 2xy + 3y^2$ is homogeneous of order 2; $f(x, y) = x \sin(y/x) + y \cos(y/x)$ is homogeneous of order 1.) Differentiate $f(tx, ty) = t^n f(x, y)$ with respect to t and replace t by 1 to show that $xf_x + yf_y = nf$. Verify this formula using the two given examples. See also Problem 34(*b*).

36. If $z = \phi(u, v)$, where $u = f(x, y)$ and $v = g(x, y)$, and if $\frac{\partial u}{\partial x} = \frac{\partial v}{\partial y}$ and $\frac{\partial u}{\partial y} = -\frac{\partial v}{\partial x}$, show that

(a) $\frac{\partial^2 u}{\partial x^2} + \frac{\partial^2 u}{\partial y^2} = \frac{\partial^2 v}{\partial x^2} + \frac{\partial^2 v}{\partial y^2} = 0$ (b) $\frac{\partial^2 \phi}{\partial x^2} + \frac{\partial^2 \phi}{\partial y^2} = \left\{\left(\frac{\partial u}{\partial x}\right)^2 + \left(\frac{\partial v}{\partial x}\right)^2\right\}\left(\frac{\partial^2 \phi}{\partial u^2} + \frac{\partial^2 \phi}{\partial v^2}\right)$

37. Find $\frac{\partial z}{\partial x}$ and $\frac{\partial z}{\partial y}$, given

(a) $3x^2 + 4y^2 - 5z^2 = 60$ *Ans.* $\frac{\partial z}{\partial x} = \frac{3x}{5z}$; $\frac{\partial z}{\partial y} = \frac{4y}{5z}$

(b) $x^2 + y^2 + z^2 + 2xy + 4yz + 8zx = 20$ *Ans.* $\frac{\partial z}{\partial x} = -\frac{x + y + 4z}{4x + 2y + z}$; $\frac{\partial z}{\partial y} = -\frac{x + y + 2z}{4x + 2y + z}$

(c) $x + 3y + 2z = \ln z$ *Ans.* $\frac{\partial z}{\partial x} = \frac{z}{1 - 2z}$; $\frac{\partial z}{\partial y} = \frac{3z}{1 - 2z}$

(d) $z = e^x \cos(y + z)$ *Ans.* $\frac{\partial z}{\partial x} = \frac{z}{1 + e^x \sin(y + z)}$; $\frac{\partial z}{\partial y} = \frac{-e^x \sin(y + z)}{1 + e^x \sin(y + z)}$

(e) $\sin(x + y) + \sin(y + z) + \sin(z + x) = 1$

Ans. $\frac{\partial z}{\partial x} = -\frac{\cos(x + y) + \cos(z + x)}{\cos(y + z) + \cos(z + x)}$; $\frac{\partial z}{\partial y} = -\frac{\cos(x + y) + \cos(y + z)}{\cos(y + z) + \cos(z + x)}$

38. Find all the first and second partial derivatives of z, given $x^2 + 2yz + 2zx = 1$.

Ans. $\frac{\partial z}{\partial x} = -\frac{x + z}{x + y}$, $\frac{\partial z}{\partial y} = -\frac{z}{x + y}$; $\frac{\partial^2 z}{\partial x^2} = \frac{x - y + 2z}{(x + y)^2}$; $\frac{\partial^2 z}{\partial x \partial y} = \frac{x + 2z}{(x + y)^2}$; $\frac{\partial^2 z}{\partial y^2} = \frac{2z}{(x + y)^2}$

39. If $F(x, y, z) = 0$, show that $\frac{\partial x}{\partial y}\frac{\partial y}{\partial z}\frac{\partial z}{\partial x} = -1$.

40. If $f(x, y) = 0$ and $g(z, x) = 0$, show that $\frac{\partial f}{\partial y}\frac{\partial g}{\partial x}\frac{\partial y}{\partial z} = \frac{\partial f}{\partial x}\frac{\partial g}{\partial z}$.

41. Find the first partial derivatives of u and v with respect to x and y and the first partial derivatives of x and y with respect to u and v, given $2u - v + x^2 + xy = 0$, $u + 2v + xy - y^2 = 0$.

Ans. $\frac{\partial u}{\partial x} = -\frac{1}{5}(4x + 3y)$; $\frac{\partial v}{\partial x} = \frac{1}{5}(2x - y)$; $\frac{\partial u}{\partial y} = \frac{1}{5}(2y - 3x)$; $\frac{\partial v}{\partial y} = \frac{4y - x}{5}$; $\frac{\partial x}{\partial u} = \frac{4y - x}{2(x^2 - 2xy - y^2)}$;

$\frac{\partial y}{\partial u} = \frac{y - 2x}{2(x^2 - 2xy - y^2)}$; $\frac{\partial x}{\partial v} = \frac{3x - 2y}{2(x^2 - 2xy - y^2)}$; $\frac{\partial y}{\partial v} = \frac{-4x - 3y}{2(x^2 - 2xy - y^2)}$

42. If $u = x + y + z$, $v = x^2 + y^2 + z^2$, and $w = x^3 + y^3 + z^3$, show that

$$\frac{\partial x}{\partial u} = \frac{yz}{(x-y)(x-z)}, \quad \frac{\partial y}{\partial v} = \frac{x+z}{2(x-y)(y-z)}, \quad \frac{\partial z}{\partial w} = \frac{1}{3(x-z)(y-z)}$$

43. Fill in the gaps in the following sketch of a proof of Theorem 49.1. Assume that $f(x, y)$ is such that f_x and f_y are continuous in an open set A. We must prove that f is differentiable in A.

There exists x^* between a and $a + \Delta x$ such that

$$f(a + \Delta x, b) - f(a, b) = f_x(x^*, b)\,\Delta x$$

and there exists y^* between b and $b + \Delta y$ such that

$$f(a + \Delta x, b + \Delta y) - f(a + \Delta x, b) = f_y(a + \Delta x, y^*)\,\Delta y.$$

Then

$$\begin{aligned}\Delta z &= f(a + \Delta x, b + \Delta y) - f(a, b)\\ &= [f(a + \Delta x, b) - f(a, b)] + [f(a + \Delta x, b + \Delta y) - f(a + \Delta x, b)]\\ &= f_x(x^*, b)\Delta x + f_y(a + \Delta x, y^*)\,\Delta y\end{aligned}$$

Let $\epsilon_1 = f_x(x^*, y) - f_x(a, b)$ and $\epsilon_2 = f_y(a + \Delta x, y^*) - f_y(a, b)$. Then

$$\Delta z = f_x(a, b)\,\Delta x + f_y(a, b)\,\Delta y + \epsilon_1\,\Delta x + \epsilon_2\,\Delta y$$

To show that $\epsilon_1 \to 0$ and $\epsilon_2 \to 0$, use the continuity of f_x and f_y.

44. Show that continuity of $f(x, y)$ does not imply differentiability, even when f_x and f_y both exist. Use the function

$$f(x, y) = \begin{cases} \dfrac{xy}{x^2 + y^2} & \text{if } (x, y) \neq (0, 0) \\ 0 & \text{if } (x, y) = (0, 0) \end{cases}$$

[*Hint:* Show that f is not continuous at (0, 0) and, therefore, not differentiable. Show the existence of $f_x(0, 0)$ and $f_y(0, 0)$ by a direct computation.]

45. Find a function $f(x, y)$ such that $f_x(0, 0) = f_y(0, 0) = 0$, and f is not continuous at (0, 0). This shows that existence of the first partial derivatives does not imply continuity. [*Hint*: Define $f(x, y) = \dfrac{xy}{x^2 + y^2}$ for $(x, y) \neq (0, 0)$ and $f(0, 0) = 0$.]

CHAPTER 50

Space Vectors

Vectors in Space

As in the plane (see Chapter 39), a vector in space is a quantity that has both magnitude and direction. Three vectors **a**, **b**, and **c**, not in the same plane and no two parallel, issuing from a common point are said to form a *right-handed system* or *triad* if **c** has the direction in which the right-threaded screw would move when rotated through the smaller angle in the direction from **a** to **b**, as in Fig. 50-1. Note that, as seen from a point on **c**, the rotation through the smaller angle from **a** to **b** is counterclockwise.

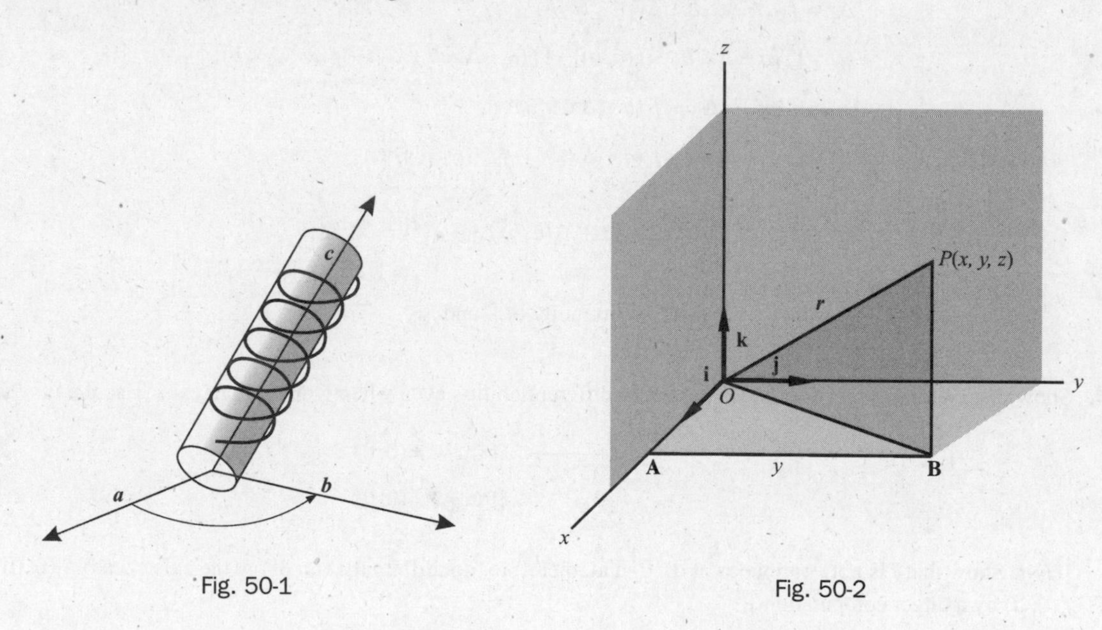

Fig. 50-1

Fig. 50-2

We choose a right-handed rectangular coordinate system in space and let **i**, **j**, and **k** be unit vectors along the positive x, y and z axes, respectively, as in Fig. 50-2. The coordinate axes divide space into eight parts, called *octants*. The *first octant*, for example, consists of all points (x, y, z) for which $x > 0, y > 0, z > 0$.

As in Chapter 39, any vector **a** may be written as

$$\mathbf{a} = a_1\mathbf{i} + a_2\mathbf{j} + a_3\mathbf{k}$$

If $P(x, y, z)$ is a point in space (Fig. 50-2), the vector **r** from the origin O to P is called the *position vector* of P and may be written as

$$\mathbf{r} = \mathbf{OP} = \mathbf{OB} + \mathbf{BP} = \mathbf{OA} + \mathbf{AB} + \mathbf{BP} = x\mathbf{i} + y\mathbf{j} + z\mathbf{k} \tag{50.1}$$

The algebra of vectors developed in Chapter 39 holds here with only such changes as the difference in dimensions requires. For example, if $\mathbf{a}=a_1\mathbf{i}+a_2\mathbf{j}+a_3\mathbf{k}$ and $\mathbf{b}=b_1\mathbf{i}+b_2\mathbf{j}+b_3\mathbf{k}$. then

$$k\mathbf{a}=ka_1\mathbf{i}+ka_2\mathbf{j}+ka_3\mathbf{k} \text{ for } k \text{ any scalar}$$

$$\mathbf{a}=\mathbf{b} \text{ if and only if } a_1=b_1, a_2=b_2, and\ a_3=b_3$$

$$\mathbf{a}\pm\mathbf{b}=(a_1\pm b_1)\mathbf{i}+(a_2\pm b_2)\mathbf{j}+(a_3\pm b_3)\mathbf{k}$$

$$\mathbf{a}\cdot\mathbf{b}=|\mathbf{a}||\mathbf{b}|\cos\theta, \text{ where } \theta \text{ is the smaller angle between } \mathbf{a} \text{ and } \mathbf{b}$$

$$\mathbf{i}\cdot\mathbf{i}=\mathbf{j}\cdot\mathbf{j}=\mathbf{k}\cdot\mathbf{k}=1 \text{ and } \mathbf{i}\cdot\mathbf{j}=\mathbf{j}\cdot\mathbf{k}=\mathbf{k}\cdot\mathbf{i}=0$$

$$|\mathbf{a}|=\sqrt{\mathbf{a}\cdot\mathbf{a}}=\sqrt{a_1^2+a_2^2+a_3^2}$$

$$\mathbf{a}\cdot\mathbf{b}=0 \text{ if and only if } \mathbf{a}=\mathbf{0}, \text{ or } \mathbf{b}=\mathbf{0}, \text{ or } \mathbf{a} \text{ and } \mathbf{b} \text{ are perpendicular}$$

From (50.1), we have

$$|\mathbf{r}|=\sqrt{\mathbf{r}\,.\,\mathbf{r}}=\sqrt{x^2+y^2+z^2} \tag{50.2}$$

as the distance of the point $P(x, y, z)$ from the origin. Also, if $P_1(x_1, y_1, z_1)$ and $P_2(x_2, y_2, z_2)$ are any two points (see Fig. 50-3), then

$$\mathbf{P_1P_2}=\mathbf{P_1B}+\mathbf{BP_2}=\mathbf{P_1A}+\mathbf{AB}+\mathbf{BP_2}=(x_2-x_1)\mathbf{i}+(y_2-y_1)\mathbf{j}+(z_2-z_1)\mathbf{k}$$

and

$$|\mathbf{P_1P_2}|=\sqrt{(x_2-x_1)^2+(y_2-y_1)^2+(z_2-z_1)^2} \tag{50.3}$$

is the familiar formula for the distance between two points. (See Problems 1–3.)

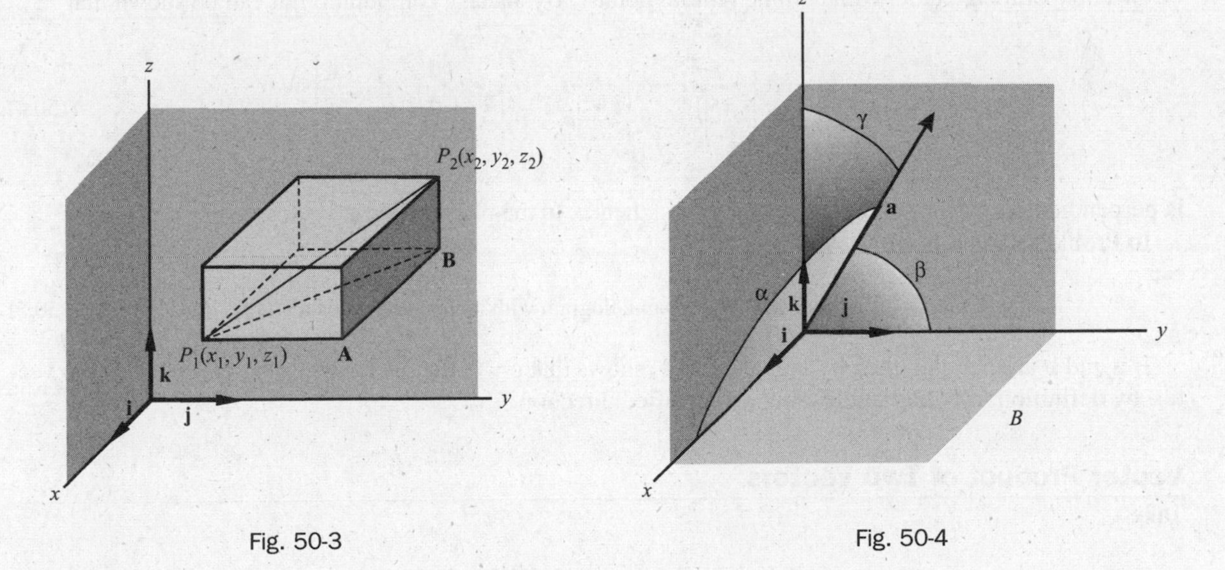

Fig. 50-3

Fig. 50-4

Direction Cosines of a Vector

Let $\mathbf{a}=a_1\mathbf{i}+a_2\mathbf{j}+a_3\mathbf{k}$ make angles α, β, and γ, respectively, with the positive x, y, and z axes, as in Fig. 50-4. From

$$\mathbf{i}\cdot\mathbf{a}=|\mathbf{i}||\mathbf{a}|\cos\alpha=|\mathbf{a}|\cos\alpha, \qquad \mathbf{j}\cdot\mathbf{a}=|\mathbf{a}|\cos\beta, \quad \mathbf{k}\cdot\mathbf{a}=|\mathbf{a}|\cos\gamma$$

we have

$$\cos\alpha = \frac{\mathbf{i}\,.\,\mathbf{a}}{|\mathbf{a}|} = \frac{a_1}{|\mathbf{a}|}, \qquad \cos\beta = \frac{\mathbf{j}\,.\,\mathbf{a}}{|\mathbf{a}|} = \frac{a_2}{|\mathbf{a}|}, \qquad \cos\gamma = \frac{\mathbf{k}\,.\,\mathbf{a}}{|\mathbf{a}|} = \frac{a_3}{|\mathbf{a}|}$$

These are the *direction cosines* of **a**. Since

$$\cos^2\alpha + \cos^2\beta + \cos^2\gamma = \frac{a_1^2 + a_2^2 + a_3^2}{|\mathbf{a}|^2} = 1$$

the vector $\mathbf{u} = \mathbf{i}\cos\alpha + \mathbf{j}\cos\beta + \mathbf{k}\cos\gamma$ is a unit vector parallel to **a**.

Determinants

We shall assume familiarity with 2×2 and 3×3 determinants. In particular,

$$\begin{vmatrix} a & b \\ c & d \end{vmatrix} = ad - bc \quad \text{and} \quad \begin{vmatrix} a & b & c \\ d & e & f \\ g & h & i \end{vmatrix} = a\begin{vmatrix} e & f \\ h & i \end{vmatrix} - b\begin{vmatrix} d & f \\ g & i \end{vmatrix} + c\begin{vmatrix} d & e \\ g & h \end{vmatrix}$$

That expansion of the 3×3 determinant is said to be "along the first row." It is equal to suitable expansions along the other rows and down the columns.

Vector Perpendicular to Two Vectors

Let

$$\mathbf{a} = a_1\mathbf{i} + a_2\mathbf{j} + a_3\mathbf{k} \quad \text{and} \quad \mathbf{b} = b_1\mathbf{i} + b_2\mathbf{j} + b_3\mathbf{k}$$

be two nonparallel vectors with common initial point *P*. By an easy computation, it can be shown that

$$\mathbf{c} = \begin{vmatrix} a_2 & a_3 \\ b_2 & b_3 \end{vmatrix}\mathbf{i} + \begin{vmatrix} a_3 & a_1 \\ b_3 & b_1 \end{vmatrix}\mathbf{j} + \begin{vmatrix} a_1 & a_2 \\ b_1 & b_2 \end{vmatrix}\mathbf{k} = \begin{vmatrix} \mathbf{i} & \mathbf{j} & \mathbf{k} \\ a_1 & a_2 & a_3 \\ b_1 & b_2 & b_3 \end{vmatrix} \tag{50.4}$$

is perpendicular to (normal to) both **a** and **b** and, hence, to the plane of these vectors.

In Problems 5 and 6, we show that

$$|\mathbf{c}| = |\mathbf{a}||\mathbf{b}|\sin\theta = \text{area of a parallelogram with nonparallel sides } \mathbf{a} \text{ and } \mathbf{b} \tag{50.5}$$

If **a** and **b** are parallel, then $\mathbf{b} = k\mathbf{a}$, and (50.4) shows that $\mathbf{c} = \mathbf{0}$; that is, **c** is the zero vector. The zero vector, by definition, has magnitude 0 but no specified direction.

Vector Product of Two Vectors

Take

$$\mathbf{a} = a_1\mathbf{i} + a_2\mathbf{j} + a_3\mathbf{k} \quad \text{and} \quad \mathbf{b} = b_1\mathbf{i} + b_2\mathbf{j} + b_3\mathbf{k}$$

with initial point *P* and denote by **n** the unit vector normal to the plane of **a** and **b**, so directed that **a**, **b**, and **n** (in that order) form a right-handed triad at *P*, as in Fig. 50-5. The *vector product* or *cross product* of **a** and **b** is defined as

$$\mathbf{a} \times \mathbf{b} = |\mathbf{a}||\mathbf{b}|\sin\theta\,\mathbf{n} \tag{50.6}$$

where θ is again the smaller angle between **a** and **b**. Thus, $\mathbf{a} \times \mathbf{b}$ is a vector perpendicular to both **a** and **b**.

We show in Problem 6 that $|\mathbf{a} \times \mathbf{b}| = |\mathbf{a}|\,|\mathbf{b}| \sin\theta$ is the area of the parallelogram having **a** and **b** as non-parallel sides.

If **a** and **b** are parallel, then $\theta = 0$ or π and $\mathbf{a} \times \mathbf{b} = \mathbf{0}$. Thus,

$$\mathbf{i}\times\mathbf{i} = \mathbf{j}\times\mathbf{j} = \mathbf{k}\times\mathbf{k} = \mathbf{0} \tag{50.7}$$

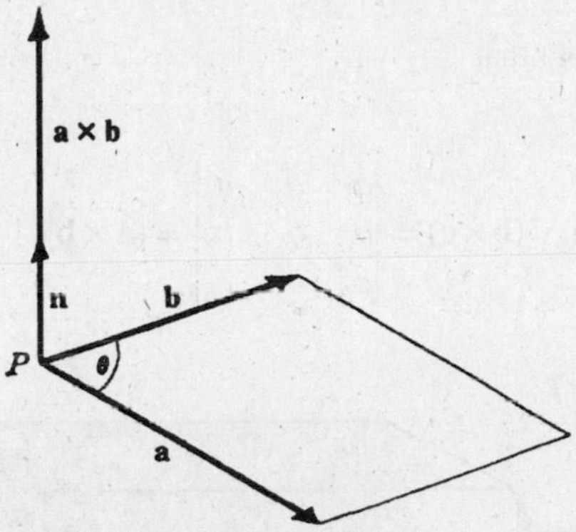

Fig. 50-5

In (50.6), if the order of **a** and **b** is reversed, then **n** must be replaced by $-\mathbf{n}$; hence,

$$\mathbf{b}\times\mathbf{a} = -(\mathbf{a}\times\mathbf{b}) \tag{50.8}$$

Since the coordinate axes were chosen as a right-handed system, it follows that

$$\begin{array}{lll} \mathbf{i}\times\mathbf{j} = \mathbf{k}, & \mathbf{j}\times\mathbf{k} = \mathbf{i}, & \mathbf{k}\times\mathbf{i} = \mathbf{j} \\ \mathbf{j}\times\mathbf{i} = -\mathbf{k}, & \mathbf{k}\times\mathbf{j} = -\mathbf{i}, & \mathbf{i}\times\mathbf{k} = -\mathbf{j} \end{array} \tag{50.9}$$

In Problem 8, we prove for any vectors **a**, **b**, and **c**, the distributive law

$$(\mathbf{a}+\mathbf{b})\times\mathbf{c} = (\mathbf{a}\times\mathbf{c}) + (\mathbf{b}\times\mathbf{c}) \tag{50.10}$$

Multiplying (50.10) by -1 and using (50.8), we have the companion distributive law

$$\mathbf{c}\times(\mathbf{a}+\mathbf{b}) = (\mathbf{c}\times\mathbf{a}) + (\mathbf{c}\times\mathbf{b}) \tag{50.11}$$

Then, also,

$$(\mathbf{a}+\mathbf{b})\times(\mathbf{c}+\mathbf{d}) = \mathbf{a}\times\mathbf{c}+\mathbf{a}\times\mathbf{d}+\mathbf{b}\times\mathbf{c}+\mathbf{b}\times\mathbf{d} \tag{50.12}$$

and

$$\mathbf{a}\times\mathbf{b} = \begin{vmatrix} \mathbf{i} & \mathbf{j} & \mathbf{k} \\ a_1 & a_2 & a_3 \\ b_1 & b_2 & b_3 \end{vmatrix} \tag{50.13}$$

(See Problems 9 and 10.)

Triple Scalar Product

In Fig. 50-6, let θ be the smaller angle between **b** and **c** and let ϕ be the smaller angle between **a** and $\mathbf{b} \times \mathbf{c}$. Let h denote the height and A the area of the base of the parallelepiped. Then the triple scalar product is by definition

$$\mathbf{a} \cdot (\mathbf{b} \times \mathbf{c}) = \mathbf{a} \cdot |\mathbf{b}|\,|\mathbf{c}| \sin\theta \; \mathbf{n} = |\mathbf{a}|\,|\mathbf{b}|\,|\mathbf{c}| \sin\theta \cos\phi = (|\mathbf{a}| \cos\phi)(|\mathbf{b}|\,|\mathbf{c}| \sin\theta) = hA$$

$$= \text{volume of parallelepiped}$$

It may be shown (see Problem 11) that

$$\mathbf{a} \cdot (\mathbf{b} \times \mathbf{c}) = \begin{vmatrix} a_1 & a_2 & a_3 \\ b_1 & b_2 & b_3 \\ c_1 & c_2 & c_3 \end{vmatrix} = (\mathbf{a} \times \mathbf{b}) \cdot \mathbf{c} \tag{50.14}$$

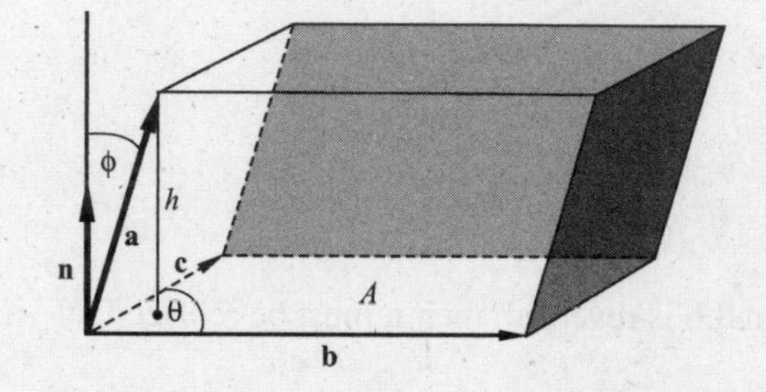

Fig. 50-6

Also
$$\mathbf{c} \cdot (\mathbf{a} \times \mathbf{b}) = \begin{vmatrix} c_1 & c_2 & c_3 \\ a_1 & a_2 & a_3 \\ b_1 & b_2 & b_3 \end{vmatrix} = \begin{vmatrix} a_1 & a_2 & a_3 \\ b_1 & b_2 & b_3 \\ c_1 & c_2 & c_3 \end{vmatrix} = \mathbf{a} \cdot (\mathbf{b} \times \mathbf{c})$$

whereas
$$\mathbf{b} \cdot (\mathbf{a} \times \mathbf{c}) = \begin{vmatrix} b_1 & b_2 & b_3 \\ a_1 & a_2 & a_3 \\ c_1 & c_2 & c_3 \end{vmatrix} = -\begin{vmatrix} a_1 & a_2 & a_3 \\ b_1 & b_2 & b_3 \\ c_1 & c_2 & c_3 \end{vmatrix} = -\mathbf{a} \cdot (\mathbf{b} \times \mathbf{c})$$

Similarly, we have

$$\mathbf{a} \cdot (\mathbf{b} \times \mathbf{c}) = \mathbf{c} \cdot (\mathbf{a} \times \mathbf{b}) = \mathbf{b} \cdot (\mathbf{c} \times \mathbf{a}) \tag{50.15}$$

and

$$\mathbf{a} \cdot (\mathbf{b} \times \mathbf{c}) = -\mathbf{b} \cdot (\mathbf{a} \times \mathbf{c}) = -\mathbf{c} \cdot (\mathbf{b} \times \mathbf{a}) = -\,\mathbf{a} \cdot (\mathbf{c} \times \mathbf{b}) \tag{50.16}$$

From the definition of $\mathbf{a} \cdot (\mathbf{b} \times \mathbf{c})$ as a volume, it follows that if **a**, **b**, and **c** are coplanar, then $\mathbf{a} \cdot (\mathbf{b} \times \mathbf{c}) = 0$, and conversely.

The parentheses in $\mathbf{a} \cdot (\mathbf{b} \times \mathbf{c})$ and $(\mathbf{a} \times \mathbf{b}) \cdot \mathbf{c}$ are not necessary. For example, $\mathbf{a} \cdot \mathbf{b} \times \mathbf{c}$ can be interpreted only as $\mathbf{a} \cdot (\mathbf{b} \times \mathbf{c})$ or $(\mathbf{a} \cdot \mathbf{b}) \times \mathbf{c}$. But $\mathbf{a} \cdot \mathbf{b}$ is a scalar, so $(\mathbf{a} \cdot \mathbf{b}) \times \mathbf{c}$ is without meaning. (See Problem 12.)

Triple Vector Product

In Problem 13, we show that

$$\mathbf{a} \times (\mathbf{b} \times \mathbf{c}) = (\mathbf{a} \cdot \mathbf{c})\mathbf{b} - (\mathbf{a} \cdot \mathbf{b})\mathbf{c} \tag{50.17}$$

Similarly,

$$(\mathbf{a} \times \mathbf{b}) \times \mathbf{c} = (\mathbf{a} \cdot \mathbf{c})\mathbf{b} - (\mathbf{b} \cdot \mathbf{c})\mathbf{a} \tag{50.18}$$

Thus, except when **b** is perpendicular to both **a** and **c**, $\mathbf{a} \times (\mathbf{b} \times \mathbf{c}) \neq (\mathbf{a} \times \mathbf{b}) \times \mathbf{c}$ and the use of parentheses is necessary.

The Straight Line

A line in space through a given point $P_0(x_0, y_0, z_0)$ may be defined as the locus of all points $P(x, y, z)$ such that P_0P is parallel to a given direction $\mathbf{a} = a_1\mathbf{i} + a_2\mathbf{j} + a_3\mathbf{k}$. Let $\mathbf{r}_0$ and $\mathbf{r}$ be the position vectors of P_0 and P (Fig. 50-7). Then

$$\mathbf{r} - \mathbf{r}_0 = k\mathbf{a} \quad \text{where } k \text{ is a scalar variable} \tag{50.19}$$

is the vector equation of line PP_0. Writing (50.19) as

$$(x - x_0)\mathbf{i} + (y - y_0)\mathbf{j} + (z - z_0)\mathbf{k} = k(a_1\mathbf{i} + a_2\mathbf{j} + a_3\mathbf{k})$$

then separating components to obtain

$$x - x_0 = ka_1, \quad y - y_0 = ka_2, \quad z - z_0 = ka_3$$

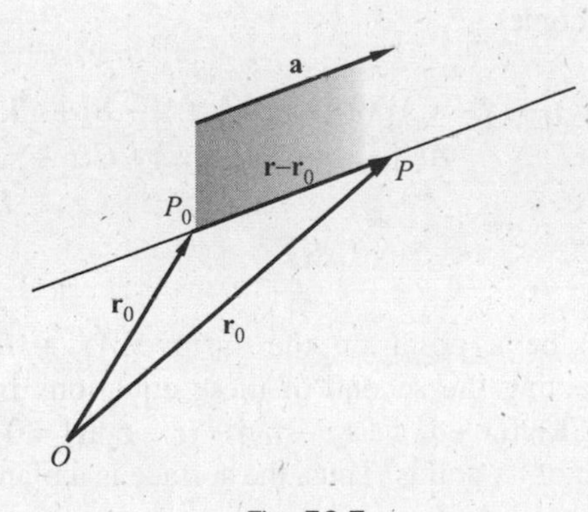

Fig. 50-7

and eliminating k, we have

$$\frac{x - x_0}{a_1} = \frac{y - y_0}{a_2} = \frac{z - z_0}{a_3} \tag{50.20}$$

as the equations of the line in rectangular coordinates. Here, $[a_1, a_2, a_3]$ is a set of *direction numbers* for the line and $\left[\frac{a_1}{|\mathbf{a}|}, \frac{a_2}{|\mathbf{a}|}, \frac{a_3}{|\mathbf{a}|}\right]$ is a set of *direction cosines* of the line.

If any one of the numbers a_1, a_2, or a_3 is zero, the corresponding numerator in (50.20) must be zero. For example, if $a_1 = 0$ but $a_2, a_3 \neq 0$, the equations of the line are

$$x - x_0 = 0 \quad \text{and} \quad \frac{y - y_0}{a_2} = \frac{z - z_0}{a_3}$$

The Plane

A plane in space through a given point $P_0(x_0, y_0, z_0)$ can be defined as the locus of all lines through P_0 and perpendicular (normal) to a given line (direction) $\mathbf{a} = A\mathbf{i} + B\mathbf{j} + C\mathbf{k}$ (Fig. 50-8). Let $P(x, y, z)$ be any other point in the plane. Then $\mathbf{r} - \mathbf{r}_0 = \mathbf{P_0P}$ is perpendicular to **a**, and the equation of the plane is

$$(\mathbf{r} - \mathbf{r}_0) \cdot \mathbf{a} = 0 \tag{50.21}$$

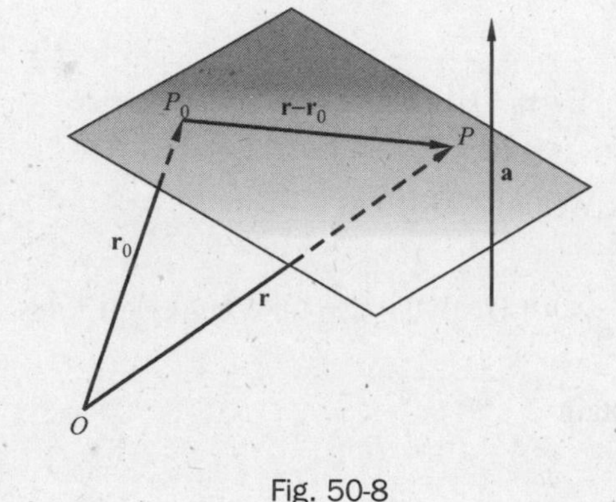

Fig. 50-8

In rectangular coordinates, this becomes

$$[(x - x_0)\mathbf{i} + (y - y_0)\mathbf{j} + (z - z_0)\mathbf{k}] \cdot (A\mathbf{i} + B\mathbf{j} + C\mathbf{k}) = 0$$

or

$$A(x - x_0) + B(y - y_0) + C(z - z_0) = 0$$

or

$$Ax + By + Cz + D = 0 \tag{50.22}$$

where $D = -(Ax_0 + By_0 + Cz_0)$.

Conversely, let $P_0(x_0, y_0, z_0)$ be a point on the surface $Ax + By + Cz + D = 0$. Then also $Ax_0 + By_0 + Cz_0 + D = 0$. Subtracting the second of these equations from the first yields $A(x - x_0) + B(y - y_0) + C(z - z_0) = (A\mathbf{i} + B\mathbf{j} + C\mathbf{k}) \cdot [(x - x_0)\mathbf{i} + (y - y_0)\mathbf{j} + (z - z_0)\mathbf{k}] = 0$ and the constant vector $A\mathbf{i} + B\mathbf{j} + C\mathbf{k}$ is normal to the surface at each of its points. Thus, the surface is a plane.

SOLVED PROBLEMS

1. Find the distance of the point $P_1(1, 2, 3)$ from (a) the origin, (b) the x axis, (c) the z axis, (d) the xy plane, and (e) the point $P_2(3, -1, 5)$.

In Fig. 50-9,

(a) $r = \mathbf{OP}_1 = \mathbf{i} + 2\mathbf{j} + 3\mathbf{k}$; hence, $|\mathbf{r}| = \sqrt{1^2 + 2^2 + 3^2} = \sqrt{14}$.

(b) $\mathbf{AP}_1 = \mathbf{AB} + \mathbf{BP}_1 = 2\mathbf{j} + 3\mathbf{k}$; hence, $|\mathbf{AP}_1| = \sqrt{4 + 9} = \sqrt{13}$.

(c) $\mathbf{DP}_1 = \mathbf{DE} + \mathbf{EP}_1 = 2\mathbf{j} + \mathbf{i}$; hence, $|\mathbf{DP}_1| = \sqrt{5}$.

(d) $\mathbf{BP}_1 = 3\mathbf{k}$, so $|\mathbf{BP}_1| = 3$.

(e) $\mathbf{P_1P_2} = (3 - 1)\mathbf{i} + (-1 - 2)\mathbf{j} + (5 - 3)\mathbf{k} = 2\mathbf{i} - 3\mathbf{j} + 2\mathbf{k}$; hence, $|\mathbf{P_1P_2}| = \sqrt{4 + 9 + 4} = \sqrt{17}$.

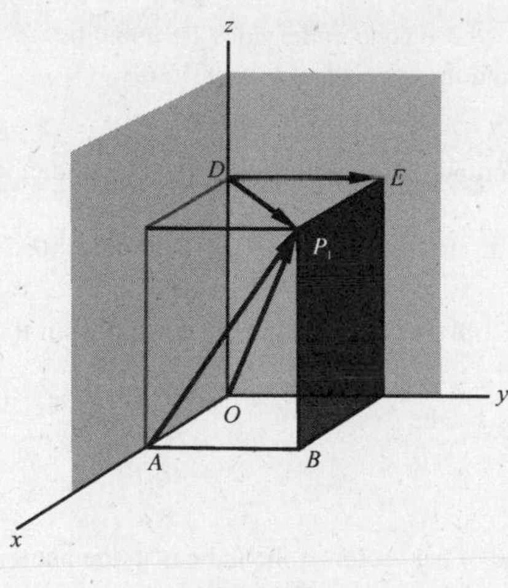

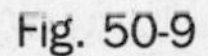

Fig. 50-9

2. Find the angle θ between the vectors joining O to $P_1(1, 2, 3)$ and $P_2(2, -3, -1)$.

Let $\mathbf{r}_1 = \mathbf{OP}_1 = \mathbf{i} + 2\mathbf{j} + 3\mathbf{k}$ and $\mathbf{r}_2 = \mathbf{OP}_2 = 2\mathbf{i} - 3\mathbf{j} - \mathbf{k}$. Then

$$\cos\theta = \frac{\mathbf{r}_1 \cdot \mathbf{r}_2}{|\mathbf{r}_1||\mathbf{r}_2|} = \frac{1(2) + 2(-3) + 3(-1)}{\sqrt{14}\sqrt{14}} = -\frac{1}{2} \quad \text{and} \quad \theta = 120^\circ.$$

3. Find the angle $\alpha = \angle BAC$ of the triangle ABC (Fig. 50-10) whose vertices are $A(1, 0, 1)$, $B(2, -1, 1)$, $C(-2, 1, 0)$.

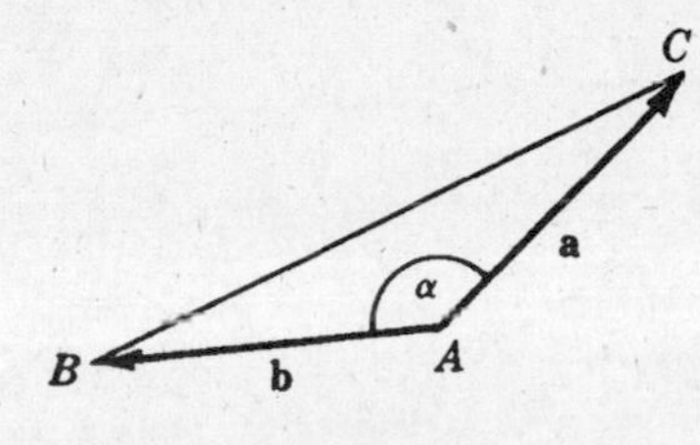

Fig. 50-10

Let $\mathbf{a} = \mathbf{AC} = -3\mathbf{i} + \mathbf{j} - \mathbf{k}$ and $\mathbf{b} = \mathbf{AB} = \mathbf{i} - \mathbf{j}$. Then

$$\cos\alpha = \frac{\mathbf{a} \cdot \mathbf{b}}{|\mathbf{a}||\mathbf{b}|} = \frac{-3-1}{\sqrt{22}} \sim -0.85280 \quad \text{and} \quad \alpha \sim 148^\circ 31'.$$

4. Find the direction cosines of $\mathbf{a} = 3\mathbf{i} + 12\mathbf{j} + 4\mathbf{k}$.

The direction cosines are $\cos\alpha = \frac{\mathbf{i} \cdot \mathbf{a}}{|\mathbf{a}|} = \frac{3}{13}$, $\cos\beta = \frac{\mathbf{j} \cdot \mathbf{a}}{|\mathbf{a}|} = \frac{12}{13}$, $\cos\gamma = \frac{\mathbf{k} \cdot \mathbf{a}}{|\mathbf{a}|} = \frac{4}{13}$.

5. If $\mathbf{a} = a_1\mathbf{i} + a_2\mathbf{j} + a_3\mathbf{k}$ and $\mathbf{b} = b_1\mathbf{i} + b_2\mathbf{j} + b_3\mathbf{k}$ are two vectors issuing from a point P and if

$$\mathbf{c} = \begin{vmatrix} a_2 & a_3 \\ b_2 & b_3 \end{vmatrix}\mathbf{i} + \begin{vmatrix} a_1 & a_3 \\ b_1 & b_3 \end{vmatrix}\mathbf{j} + \begin{vmatrix} a_1 & a_2 \\ b_1 & b_2 \end{vmatrix}\mathbf{k},$$

show that $|\mathbf{c}| = |\mathbf{a}||\mathbf{b}| \sin\theta$, where θ is the smaller angle between $\mathbf{a}$ and $\mathbf{b}$.

We have $\cos\theta = \frac{\mathbf{a} \cdot \mathbf{b}}{|\mathbf{a}||\mathbf{b}|}$ and

$$\sin\theta = \sqrt{1 - \left(\frac{\mathbf{a} \cdot \mathbf{b}}{|\mathbf{a}||\mathbf{b}|}\right)^2} = \frac{\sqrt{(a_1^2 + a_2^2 + a_3^2)(b_1^2 + b_2^2 + b_3^2) - (a_1b_1 + a_2b_2 + a_3b_3)^2}}{|\mathbf{a}||\mathbf{b}|} = \frac{|\mathbf{c}|}{|\mathbf{a}||\mathbf{b}|}$$

Hence, $|\mathbf{c}| = |\mathbf{a}|\,|\mathbf{b}| \sin\theta$ as required.

6. Find the area of the parallelogram whose nonparallel sides are **a** and **b**.

From Fig. 50-11, $h = |\mathbf{b}| \sin \theta$ and the area is $h|\mathbf{a}| = |\mathbf{a}||\mathbf{b}| \sin \theta$.

7. Let $\mathbf{a}_1$ and $\mathbf{a}_2$, respectively, be the components of **a** parallel and perpendicular to **b**, as in Fig. 50-12. Show that $\mathbf{a}_2 \times \mathbf{b} = \mathbf{a} \times \mathbf{b}$ and $\mathbf{a}_1 \times \mathbf{b} = 0$.

If θ is the angle between **a** and **b**, then $|\mathbf{a}_1| = |\mathbf{a}| \cos\theta$ and $|\mathbf{a}_2| = |\mathbf{a}| \sin\theta$. Since **a**, $\mathbf{a}_2$, and **b** are coplanar,

$$\mathbf{a}_2 \times \mathbf{b} = |\mathbf{a}_2||\mathbf{b}| \sin \phi\, \mathbf{n} = |\mathbf{a}| \sin \theta\, |\mathbf{b}|\, \mathbf{n} = |\mathbf{a}||\mathbf{b}| \sin \theta\, \mathbf{n} = \mathbf{a} \times \mathbf{b}$$

Since $\mathbf{a}_1$ and **b** are parallel, $\mathbf{a}_1 \times \mathbf{b} = \mathbf{0}$.

8. Prove: $(\mathbf{a} + \mathbf{b}) \times \mathbf{c} = (\mathbf{a} \times \mathbf{c}) + (\mathbf{b} \times \mathbf{c})$.

In Fig. 50-13, the initial point P of the vectors **a**, **b**, and **c** is in the plane of the paper, while their endpoints are above this plane, The vectors $\mathbf{a}_1$ and $\mathbf{b}_1$ are, respectively, the components of **a** and **b** perpendicular to **c**. Then $\mathbf{a}_1$, $\mathbf{b}_1$, $\mathbf{a}_1 + \mathbf{b}_1$, $\mathbf{a}_1 \times \mathbf{c}$, $\mathbf{b}_1 \times \mathbf{c}$, and $(\mathbf{a}_1 + \mathbf{b}_1) \times \mathbf{c}$ all lie in the plane of the paper.

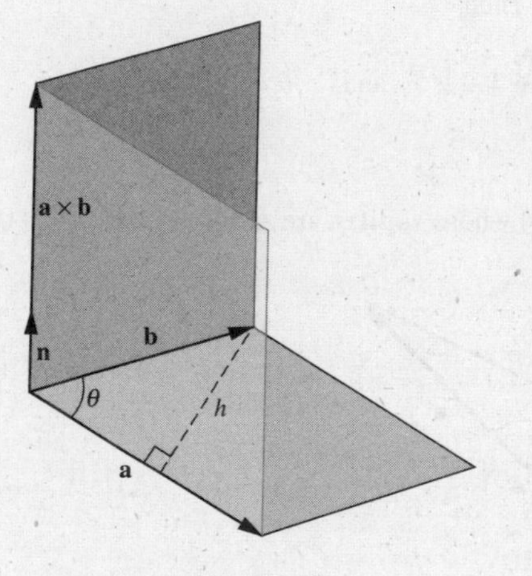

Fig. 50-11

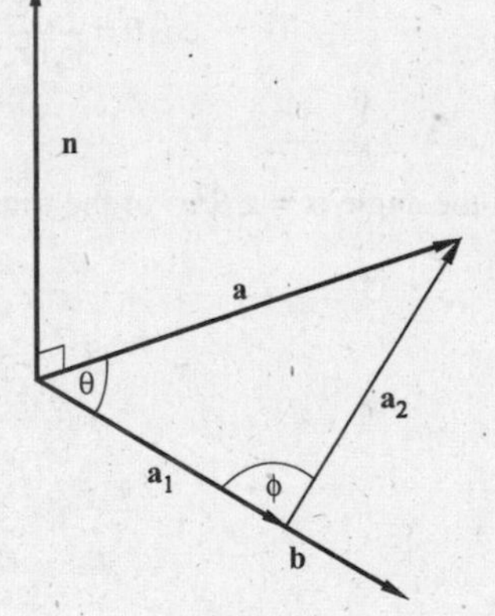

Fig. 50-12

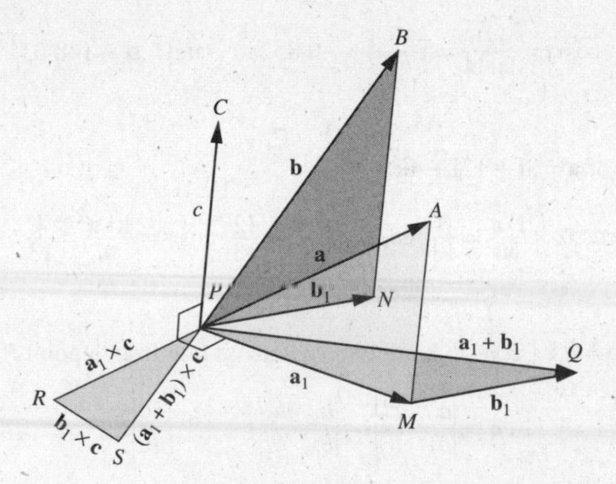

Fig. 50-13

In triangles PRS and PMQ,

$$\frac{RS}{PR} = \frac{|\mathbf{b}_1 \times \mathbf{c}|}{|\mathbf{a}_1 \times \mathbf{c}|} = \frac{|\mathbf{b}_1||\mathbf{c}|}{|\mathbf{a}_1||\mathbf{c}|} = \frac{|\mathbf{b}_1|}{|\mathbf{a}_1|} = \frac{MQ}{PM}$$

Thus, *PRS* and *PMQ* are similar. Now *PR* is perpendicular to *PM*, and *RS* is perpendicular to *MQ*; hence *PS* is perpendicular to *PQ* and $\mathbf{PS} = \mathbf{PQ} \times \mathbf{c}$. Then, since $\mathbf{PS} = \mathbf{PQ} \times \mathbf{c} = \mathbf{PR} + \mathbf{RS}$, we have

$$(\mathbf{a}_1 + \mathbf{b}_1) \times \mathbf{c} = (\mathbf{a}_1 \times \mathbf{c}) + (\mathbf{b}_1 \times \mathbf{c})$$

By Problem 7, $\mathbf{a}_1$ and $\mathbf{b}_1$ may be replaced by **a** and **b**, respectively, to yield the required result.

9. When $\mathbf{a} = a_1\mathbf{i} + a_2\mathbf{j} + a_3\mathbf{k}$ and $\mathbf{b} = b_1\mathbf{i} + b_2\mathbf{j} + b_3\mathbf{k}$, show that $\mathbf{a} \times \mathbf{b} = \begin{vmatrix} \mathbf{i} & \mathbf{j} & \mathbf{k} \\ a_1 & a_2 & a_3 \\ b_1 & b_2 & b_3 \end{vmatrix}$.

We have, by the distributive law,

$$\begin{aligned}
\mathbf{a} \times \mathbf{b} &= (a_1\mathbf{i} + a_2\mathbf{j} + a_3\mathbf{k}) \times (b_1\mathbf{i} + b_2\mathbf{j} + b_3\mathbf{k}) \\
&= a_1\mathbf{i} \times (b_1\mathbf{i} + b_2\mathbf{j} + b_3\mathbf{k}) + a_2\mathbf{j} \times (b_1\mathbf{i} + b_2\mathbf{j} + b_3\mathbf{k}) + a_3\mathbf{k} \times (b_1\mathbf{i} + b_2\mathbf{j} + b_3\mathbf{k}) \\
&= (a_1b_2\mathbf{k} - a_1b_3\mathbf{j}) + (-a_2b_1\mathbf{k} + a_2b_3\mathbf{i}) + (a_3b_1\mathbf{j} - a_3b_2\mathbf{i}) \\
&= (a_2b_3 - a_3b_2)\mathbf{i} - (a_1b_3 - a_3b_1)\mathbf{j} + (a_1b_2 - a_2b_1)\mathbf{k} \\
&= \begin{vmatrix} \mathbf{i} & \mathbf{j} & \mathbf{k} \\ a_1 & a_2 & a_3 \\ b_1 & b_2 & b_3 \end{vmatrix}
\end{aligned}$$

10. Derive the law of sines of plane trigonometry.

Consider the triangle *ABC*, whose sides **a**, **b**, **c** are of magnitudes *a*, *b*, *c*, respectively, and whose interior angles are α, β, γ. We have

$$\mathbf{a} + \mathbf{b} + \mathbf{c} = 0$$

Then $\quad \mathbf{a} \times (\mathbf{a} + \mathbf{b} + \mathbf{c}) = \mathbf{a} \times \mathbf{b} + \mathbf{a} \times \mathbf{c} = 0 \quad$ or $\quad \mathbf{a} \times \mathbf{b} = \mathbf{c} \times \mathbf{a}$

and $\quad \mathbf{b} \times (\mathbf{a} + \mathbf{b} + \mathbf{c}) = \mathbf{b} \times \mathbf{a} + \mathbf{b} \times \mathbf{c} = 0 \quad$ or $\quad \mathbf{b} \times \mathbf{c} = \mathbf{a} \times \mathbf{b}$

Thus, $\quad \mathbf{a} \times \mathbf{b} = \mathbf{b} \times \mathbf{c} = \mathbf{c} \times \mathbf{a}$

so that $\quad |\mathbf{a}||\mathbf{b}| \sin \gamma = |\mathbf{b}||\mathbf{c}| \sin \alpha = |\mathbf{c}||\mathbf{a}| \sin \beta$

or $\quad ab \sin \gamma = bc \sin \alpha = ca \sin \beta$

and $\quad \dfrac{\sin \gamma}{c} = \dfrac{\sin \alpha}{a} = \dfrac{\sin \beta}{b}$

11. If $\mathbf{a} = a_1\mathbf{i} + a_2\mathbf{j} + a_3\mathbf{k}$, $\mathbf{b} = b_1\mathbf{i} + b_2\mathbf{j} + b_3\mathbf{k}$, and $\mathbf{c} = c_1\mathbf{i} + c_2\mathbf{j} + c_3\mathbf{k}$, show that

$$\mathbf{a} \cdot (\mathbf{b} \times \mathbf{c}) = \begin{vmatrix} a_1 & a_2 & a_3 \\ b_1 & b_2 & b_3 \\ c_1 & c_2 & c_3 \end{vmatrix}$$

By (50.13),

$$\begin{aligned}
\mathbf{a} \cdot (\mathbf{b} \times \mathbf{c}) &= (a_1\mathbf{i} + a_2\mathbf{j} + a_3\mathbf{k}) \cdot \begin{vmatrix} \mathbf{i} & \mathbf{j} & \mathbf{k} \\ b_1 & b_2 & b_3 \\ c_1 & c_2 & c_3 \end{vmatrix} \\
&= (a_1\mathbf{i} + a_2\mathbf{j} + a_3\mathbf{k}) \cdot [(b_2c_3 - b_3c_2)\mathbf{i} + (b_3c_1 - b_1c_3)\mathbf{j} + (b_1c_2 - b_2c_1)\mathbf{k}] \\
&= a_1(b_2c_3 - b_3c_2) + a_2(b_3c_1 - b_1c_3) + a_3(b_1c_2 - b_2c_1) = \begin{vmatrix} a_1 & a_2 & a_3 \\ b_1 & b_2 & b_3 \\ c_1 & c_2 & c_3 \end{vmatrix}
\end{aligned}$$

12. Show that $\mathbf{a} \cdot (\mathbf{a} \times \mathbf{c}) = 0$.

By (50.14), $\mathbf{a} \cdot (\mathbf{a} \times \mathbf{c}) = (\mathbf{a} \times \mathbf{a}) \cdot \mathbf{c} = 0$.

13. For the vectors **a**, **b**, and **c** of Problem 11, show that $\mathbf{a} \times (\mathbf{b} \times \mathbf{c}) = (\mathbf{a} \cdot \mathbf{c})\mathbf{b} - (\mathbf{a} \cdot \mathbf{b})\mathbf{c}$.

Here

$$\mathbf{a} \times (\mathbf{b} \times \mathbf{c}) = (a_1\mathbf{i} + a_2\mathbf{j} + a_3\mathbf{k}) \times \begin{vmatrix} \mathbf{i} & \mathbf{j} & \mathbf{k} \\ b_1 & b_2 & b_3 \\ c_1 & c_2 & c_3 \end{vmatrix}$$

$$= (a_1\mathbf{i} + a_2\mathbf{j} + a_3\mathbf{k}) \times [(b_2c_3 - b_3c_2)\mathbf{i} + (b_3c_1 - b_1c_3)\mathbf{j} + (b_1c_2 - b_2c_1)\mathbf{k}]$$

$$= \begin{vmatrix} \mathbf{i} & \mathbf{j} & \mathbf{k} \\ a_1 & a_2 & a_3 \\ b_2c_3 - b_3c_2 & b_3c_1 - b_1c_3 & b_1c_2 - b_2c_1 \end{vmatrix}$$

$$= \mathbf{i}(a_2b_1c_2 - a_2b_2c_1 - a_3b_3c_1 + a_3b_1c_3) + \mathbf{j}(a_3b_2c_3 - a_3b_3c_2 - a_1b_1c_2 + a_1b_2c_1)$$
$$+ \mathbf{k}(a_1b_3c_1 - a_1b_1c_3 - a_2b_2c_3 + a_2b_3c_2)$$

$$= \mathbf{i}b_1(a_1c_1 + a_2c_2 + a_3c_3) + \mathbf{j}b_2(a_1c_1 + a_2c_2 + a_3c_3) + \mathbf{k}b_3(a_1c_1 + a_2c_2 + a_3c_3)$$
$$- [\mathbf{i}c_1(a_1b_1 + a_2b_2 + a_3b_3) + \mathbf{j}c_2(a_1b_1 + a_2b_2 + a_3b_3) + \mathbf{k}c_3(a_1b_1 + a_2b_2 + a_3b_3)]$$

$$= (b_1\mathbf{i} + b_2\mathbf{j} + b_3\mathbf{k})(\mathbf{a} \cdot \mathbf{c}) - (c_1\mathbf{i} + c_2\mathbf{j} + c_3\mathbf{k})(\mathbf{a} \cdot \mathbf{b})$$

$$= \mathbf{b}(\mathbf{a} \cdot \mathbf{c}) - \mathbf{c}(\mathbf{a} \cdot \mathbf{b}) = (\mathbf{a} \cdot \mathbf{c})\mathbf{b} - (\mathbf{a} \cdot \mathbf{b})\mathbf{c}$$

14. If l_1 and l_2 are two nonintersecting lines in space, show that the shortest distance d between them is the distance from any point on l_1 to the plane through l_2 and parallel to l_1; that is, show that if P_1 is a point on l_1 and P_2 is a point on l_2 then, apart from sign, d is the scalar projection of $\mathbf{P}_1\mathbf{P}_2$ on a common perpendicular to l_1 and l_2.

Let l_1 pass through $P_1(x_1, y_1, z_1)$ in the direction $\mathbf{a} = a_1\mathbf{i} + a_2\mathbf{j} + a_3\mathbf{k}$, and let l_2 pass through $P_2(x_2, y_2, z_2)$ in the direction $\mathbf{b} = b_1\mathbf{i} + b_2\mathbf{j} + b_3\mathbf{k}$.

Then $\mathbf{P}_1\mathbf{P}_2 = (x_2 - x_1)\mathbf{i} + (y_2 - y_1)\mathbf{j} + (z_2 - z_1)\mathbf{k}$, and the vector $\mathbf{a} \times \mathbf{b}$ is perpendicular to both l_1 and l_2. Thus,

$$d = \left| \frac{\mathbf{P}_1\mathbf{P}_2 \cdot (\mathbf{a} \times \mathbf{b})}{|\mathbf{a} \times \mathbf{b}|} \right| = \left| \frac{(\mathbf{r}_2 - \mathbf{r}) \cdot (\mathbf{a} \times \mathbf{b})}{|\mathbf{a} \times \mathbf{b}|} \right|$$

15. Write the equation of the line passing through $P_0(1, 2, 3)$ and parallel to $\mathbf{a} = 2\mathbf{i} - \mathbf{j} - 4\mathbf{k}$. Which of the points $A(3, 1, -1)$, $B(\frac{1}{2}, \frac{9}{4}, 4)$, $C(2, 0, 1)$ are on this line?

From (50.19), the vector equation is

$$(x\mathbf{i} + y\mathbf{j} + z\mathbf{k}) - (\mathbf{i} + 2\mathbf{j} + 3\mathbf{k}) = k(2\mathbf{i} - \mathbf{j} - 4\mathbf{k})$$

or

$$(x - 1)\mathbf{i} + (y - 2)\mathbf{j} + (z - 3)\mathbf{k} = k(2\mathbf{i} - \mathbf{j} - 4\mathbf{k}) \tag{1}$$

The rectangular equations are

$$\frac{x-1}{2} = \frac{y-2}{-1} = \frac{z-3}{-4} \tag{2}$$

Using (2), it is readily found that A and B are on the line while C is not.

In the vector equation (1), a point $P(x, y, z)$ on the line is found by giving k a value and comparing components. The point A is on the line because

$$(3 - 1)\mathbf{i} + (1 - 2)\mathbf{j} + (-1 - 3)\mathbf{k} = k(2\mathbf{i} - \mathbf{j} - 4\mathbf{k})$$

when $k = 1$. Similarly B is on the line because

$$-\tfrac{1}{2}\mathbf{i}+\tfrac{1}{4}\mathbf{j}+\mathbf{k}=k(2\mathbf{i}-\mathbf{j}-4\mathbf{k})$$

when $k = -\frac{1}{4}$. The point C is not on the line because

$$\mathbf{i}-2\mathbf{j}-2\mathbf{k}=k(2\mathbf{i}-\mathbf{j}-4\mathbf{k})$$

for no value of k.

16. Write the equation of the plane:
(a) Passing through $P_0(1, 2, 3)$ and parallel to $3x - 2y + 4z - 5 = 0$.
(b) Passing through $P_0(1, 2, 3)$ and $P_1(3, -2, 1)$, and perpendicular to the plane $3x - 2y + 4z - 5 = 0$.
(c) Through $P_0(1, 2, 3)$, $P_1(3, -2, 1)$ and $P_2(5, 0, -4)$.
Let $P(x, y, z)$ be a general point in the required plane.
(a) Here $\mathbf{a} = 3\mathbf{i} - 2\mathbf{j} + 4\mathbf{k}$ is normal to the given plane and to the required plane. The vector equation of the latter is $(\mathbf{r} - \mathbf{r}_0)\cdot \mathbf{a} = \mathbf{0}$ and the rectangular equation is

$$3(x-1)-2(y-2)+4(z-3)=0$$

or

$$3x-2y+4z-11=0$$

(b) Here $\mathbf{r}_1 - \mathbf{r}_0 = 2\mathbf{i} - 4\mathbf{j} - 2\mathbf{k}$ and $\mathbf{a} = 3\mathbf{i} - 2\mathbf{j} + 4\mathbf{k}$ are parallel to the required plane; thus, $(\mathbf{r}_1 - \mathbf{r}_0) \times \mathbf{a}$ is normal to this plane. Its vector equation is $(\mathbf{r} - \mathbf{r}_0)\cdot[(\mathbf{r}_1 - \mathbf{r}_0) \times \mathbf{a}] = 0$. The rectangular equation is

$$(\mathbf{r}-\mathbf{r}_0)\cdot\begin{vmatrix}\mathbf{i} & \mathbf{j} & \mathbf{k}\\ 2 & -4 & -2\\ 3 & -2 & 4\end{vmatrix} = [(x-1)\mathbf{i}+(y-2)\mathbf{j}+(z-3)\mathbf{k}]\cdot[-20\mathbf{i}-14\mathbf{j}+8\mathbf{k}]$$

$$= -20(x-1)-14(y-2)+8(z-3)=0$$

or $20x + 14y - 8z - 24 = 0$.
(c) Here $\mathbf{r}_1 - \mathbf{r}_0 = 2\mathbf{i} - 4\mathbf{j} - 2\mathbf{k}$ and $\mathbf{r}_2 - \mathbf{r}_0 = 4\mathbf{i} = 2\mathbf{j} - 7\mathbf{k}$ are parallel to the required plane, so that $(\mathbf{r}_1 - \mathbf{r}_0) \times (\mathbf{r}_2 - \mathbf{r}_0)$ is normal to it. The vector equation is $(\mathbf{r} - \mathbf{r}_0)\cdot[(\mathbf{r}_1 - \mathbf{r}_0) \times (\mathbf{r}_2 - \mathbf{r}_0)] = \mathbf{0}$ and the rectangular equation is

$$(\mathbf{r}-\mathbf{r}_0)\cdot\begin{vmatrix}\mathbf{i} & \mathbf{j} & \mathbf{k}\\ 2 & -4 & -2\\ 4 & -2 & -7\end{vmatrix} = [(x-1)\mathbf{i}+(y-2)\mathbf{j}+(z-3)\mathbf{k}]\cdot[-24\mathbf{i}+6\mathbf{j}+12\mathbf{k}]$$

$$= 24(x-1)+6(y-2)+12(z-3)=0$$

or $4x + y + 2z - 12 = 0$.

17. Find the shortest distance d between the point $P_0(1, 2, 3)$ and the plane Π given by the equation $3x - 2y + 5z - 10 = 0$.

A normal to the plane is $\mathbf{a} = 3\mathbf{i} - 2\mathbf{j} + 5\mathbf{k}$. Take $P_1(2, 3, 2)$ as a convenient point in Π. Then, apart from sign, d is the scalar projection of $\mathbf{P_0P_1}$ on $\mathbf{a}$. Hence,

$$d=\left|\frac{(\mathbf{r}_1-\mathbf{r}_0)\cdot\mathbf{a}}{|\mathbf{a}|}\right|=\left|\frac{(\mathbf{i}+\mathbf{j}-\mathbf{k})\cdot(3\mathbf{i}-2\mathbf{j}+5\mathbf{k})}{\sqrt{38}}\right|=\frac{2}{19}\sqrt{38}$$

SUPPLEMENTARY PROBLEMS

18. Find the length of (a) the vector $\mathbf{a} = 2\mathbf{i} + 3\mathbf{j} + \mathbf{k}$; (b) the vector $\mathbf{b} = 3\mathbf{i} - 5\mathbf{j} + 9\mathbf{k}$; and (c) the vector $\mathbf{c}$, joining $P_1(3, 4, 5)$ to $P_2(1, -2, 3)$.

Ans. (a) $\sqrt{14}$; (b) $\sqrt{115}$; (c) $2\sqrt{11}$

19. For the vectors of Problem 18:

(a) Show that **a** and **b** are perpendicular.
(b) Find the smaller angle between **a** and **c**, and that between **b** and **c**.
(c) Find the angles that **b** makes with the coordinate axes.

Ans. (b) 165°14′, 85°10′; (c) 73°45′, 117°47′, 32°56′

20. Prove: $\mathbf{i}\cdot\mathbf{i} = \mathbf{j}\cdot\mathbf{j} = \mathbf{k}\cdot\mathbf{k} = 1$ and $\mathbf{i}\cdot\mathbf{j} = \mathbf{j}\cdot\mathbf{k} = \mathbf{k}\cdot\mathbf{i} = 0$.

21. Write a unit vector in the direction of **a** and a unit vector in the direction of **b** of Problem 18.

Ans. (a) $\dfrac{\sqrt{14}}{7}\mathbf{i} + \dfrac{3\sqrt{14}}{14}\mathbf{j} + \dfrac{\sqrt{14}}{14}\mathbf{k}$; (b) $\dfrac{3}{\sqrt{115}}\mathbf{i} - \dfrac{5}{\sqrt{115}}\mathbf{j} + \dfrac{9}{\sqrt{115}}\mathbf{k}$

22. Find the interior angles β and γ of the triangle of Problem 3.

Ans. $\beta = 22°12′$; $\gamma = 9°16′$

23. For the unit cube in Fig. 50-14, find (a) the angle between its diagonal and an edge, and (b) the angle between its diagonal and a diagonal of a face.

Ans. (a) 54°44′; (b) 35°16′

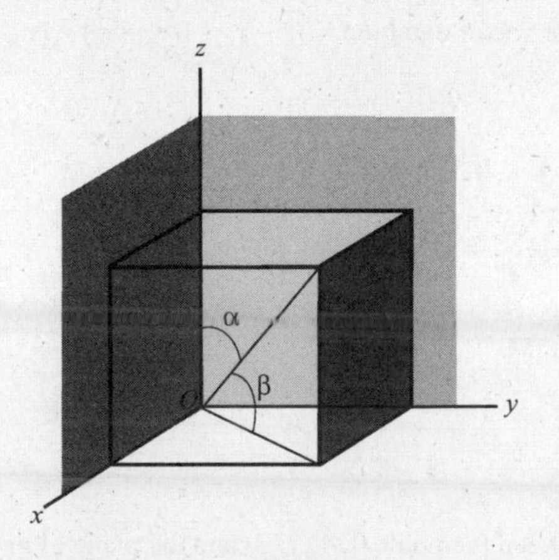

Fig. 50-14

24. Show that the scalar projection of **b** onto **a** is given by $\dfrac{\mathbf{a}\cdot\mathbf{b}}{|\mathbf{a}|}$.

25. Show that the vector **c** of (50.4) is perpendicular to both **a** and **b**.

26. Given $\mathbf{a} = \mathbf{i} + \mathbf{j}$, $\mathbf{b} = \mathbf{i} - 2\mathbf{k}$, and $\mathbf{c} = 2\mathbf{i} + 3\mathbf{j} + 4\mathbf{k}$, confirm the following equations:

(a) $\mathbf{a} \times \mathbf{b} = -2\mathbf{i} + 2\mathbf{j} - \mathbf{k}$
(b) $\mathbf{b} \times \mathbf{c} = 6\mathbf{i} - 8\mathbf{j} + 3\mathbf{k}$
(c) $\mathbf{c} \times \mathbf{a} = -4\mathbf{i} + 4\mathbf{j} - \mathbf{k}$
(d) $(\mathbf{a} + \mathbf{b}) \times (\mathbf{a} - \mathbf{b}) = 4\mathbf{i} - 4\mathbf{j} + 2\mathbf{k}$
(e) $\mathbf{a} \cdot (\mathbf{a} \times \mathbf{b}) = 0$
(f) $\mathbf{a} \cdot (\mathbf{b} \times \mathbf{c}) = -2$
(g) $\mathbf{a} \times (\mathbf{b} \times \mathbf{c}) = 3\mathbf{i} - 3\mathbf{j} - 14\mathbf{k}$
(h) $\mathbf{c} \times (\mathbf{a} \times \mathbf{b}) = -11\mathbf{i} - 6\mathbf{j} + 10\mathbf{k}$

27. Find the area of the triangle whose vertices are $A(1, 2, 3)$, $B(2, -1, 1)$, and $C(-2, 1, -1)$. (*Hint*: $|\mathbf{AB} \times \mathbf{AC}|$ = twice the area.)

Ans. $5\sqrt{3}$

28. Find the volume of the parallelepiped whose edges are *OA*, *OB*, and *OC*, for $A(1, 2, 3)$, $B(1, 1, 2)$, and $C(2, 1, 1)$.

Ans. 2

29. If $\mathbf{u} = \mathbf{a} \times \mathbf{b}$, $\mathbf{v} = \mathbf{b} \times \mathbf{c}$, $\mathbf{w} = \mathbf{c} \times \mathbf{a}$, show that:

(a) $\mathbf{u} \cdot \mathbf{c} = \mathbf{v} \cdot \mathbf{a} = \mathbf{w} \cdot \mathbf{b}$
(b) $\mathbf{a} \cdot \mathbf{u} = \mathbf{b} \cdot \mathbf{u} = 0$, $\mathbf{b} \cdot \mathbf{v} = \mathbf{c} \cdot \mathbf{v} = 0$, $\mathbf{c} \cdot \mathbf{w} = \mathbf{a} \cdot \mathbf{w} = 0$
(c) $\mathbf{u} \cdot (\mathbf{v} \times \mathbf{w}) = [\mathbf{a} \cdot (\mathbf{b} \times \mathbf{c})]^2$

30. Show that $(\mathbf{a} + \mathbf{b}) \cdot [(\mathbf{b} + \mathbf{c}) \times (\mathbf{c} + \mathbf{a})] = 2\mathbf{a} \cdot (\mathbf{b} \times \mathbf{c})$.

31. Find the smaller angle of intersection of the planes $5x - 14y + 2z - 8 = 0$ and $10x - 11y + 2z + 15 = 0$. (*Hint:* Find the angle between their normals.)

Ans. 22°25′

32. Write the vector equation of the line of intersection of the planes $x + y - z - 5 = 0$ and $4x - y - z + 2 = 0$.

Ans. $(x - 1)\mathbf{i} + (y - 5)\mathbf{j} + (z - 1)\mathbf{k} = k(-2\mathbf{i} - 3\mathbf{j} - 5\mathbf{k})$, where $P_0(1, 5, 1)$ is a point on the line.

33. Find the shortest distance between the line through $A(2, -1, -1)$ and $B(6, -8, 0)$ and the line through $C(2, 1, 2)$ and $D(0, 2, -1)$.

Ans. $\sqrt{6}/6$

34. Define a line through $P_0(x_0, y_0, z_0)$ as the locus of all points $P(x, y, z)$ such that $\mathbf{P_0P}$ and $\mathbf{OP_0}$ are perpendicular. Show that its vector equation is $(\mathbf{r} - \mathbf{r}_0) \cdot \mathbf{r}_0 = 0$.

35. Find the rectangular equations of the line through $P_0(2, -3, 5)$ and

(a) Perpendicular to $7x - 4y + 2z - 8 = 0$.
(b) Parallel to the line $x - y + 2z + 4 = 0$, $2x + 3y + 6z - 12 = 0$.
(c) Through $P_1(3, 6, -2)$.

Ans. (a) $\frac{x-2}{7} = \frac{y+3}{-4} = \frac{z-5}{2}$; (b) $\frac{x-2}{12} = \frac{y+3}{2} = \frac{z-5}{-5}$; (c) $\frac{x-2}{1} = \frac{y+3}{9} = \frac{z-5}{-7}$

36. Find the equation of the plane:

(a) Through $P_0(1, 2, 3)$ and parallel to $\mathbf{a} = 2\mathbf{i} + \mathbf{j} - \mathbf{k}$ and $\mathbf{b} = 3\mathbf{i} + 6\mathbf{j} - 2\mathbf{k}$.
(b) Through $P_0(2, -3, 2)$ and the line $6x + 4y + 3z + 5 = 0$, $2x + y + z - 2 = 0$.
(c) Through $P_0(2, -1, -1)$ and $P_1(1, 2, 3)$ and perpendicular to $2x + 3y - 5z - 6 = 0$.

Ans. (a) $4x + y + 9z - 33 = 0$; (b) $16x + 7y + 8z - 27 = 0$; (c) $9x - y + 3z - 16 = 0$

37. If $\mathbf{r}_0 = \mathbf{i} + \mathbf{j} + \mathbf{k}$, $\mathbf{r}_1 = 2\mathbf{i} + 3\mathbf{j} + 4\mathbf{k}$, and $\mathbf{r}_2 = 3\mathbf{i} + 5\mathbf{j} + 7\mathbf{k}$ are three position vectors, show that $\mathbf{r}_0 \times \mathbf{r}_1 + \mathbf{r}_1 \times \mathbf{r}_2 + \mathbf{r}_2 \times \mathbf{r}_0 = \mathbf{0}$. What can be said of the terminal points of these vectors?

Ans. They are collinear.

38. If P_0, P_1, and P_2 are three noncollinear points and $\mathbf{r}_0$, $\mathbf{r}_1$, and $\mathbf{r}_2$ are their position vectors, what is the position of $\mathbf{r}_0 \times \mathbf{r}_1 + \mathbf{r}_1 \times \mathbf{r}_2 + \mathbf{r}_2 \times \mathbf{r}_0$ with respect to the plane $P_0P_1P_2$?

Ans. normal

39. Prove: (a) $\mathbf{a} \times (\mathbf{b} \times \mathbf{c}) + \mathbf{b} \times (\mathbf{c} \times \mathbf{a}) + \mathbf{c} \times (\mathbf{a} \times \mathbf{b}) = \mathbf{0}$; (b) $(\mathbf{a} \times \mathbf{b}) \cdot (\mathbf{c} \times \mathbf{d}) = (\mathbf{a} \cdot \mathbf{c})(\mathbf{b} \cdot \mathbf{d}) - (\mathbf{a} \cdot \mathbf{d})(\mathbf{b} \cdot \mathbf{c})$.

40. Prove: (a) The perpendiculars erected at the midpoints of the sides of a triangle meet in a point; (b) the perpendiculars dropped from the vertices to the opposite sides (produced if necessary) of a triangle meet in a point.

41. Let $A(1, 2, 3)$, $B(2, -1, 5)$, and $C(4, 1, 3)$ be three vertices of the parallelogram $ABCD$. Find (a) the coordinates of D; (b) the area of $ABCD$; and (c) the area of the orthogonal projection of $ABCD$ on each of the coordinate planes.

Ans. (a) $D(3, 4, 1)$; (b) $2\sqrt{26}$; (c) 8, 6, 2

42. Prove that the area of a parallogram in space is the square root of the sum of the squares of the areas of projections of the parallelogram on the coordinate planes.

Surfaces and Curves in Space

Planes

We already know (formula (50.22)) that the equation of a plane has the form $Ax + By + Cz + D = 0$, where $A\mathbf{i} + B\mathbf{j} + C\mathbf{k}$ is a nonzero vector perpendicular to the plane. The plane passes through the origin (0, 0, 0) when and only when $D = 0$.

Spheres

From the distance formula (50.3), we see that an equation of the sphere with radius r and center (a, b, c) is

$$(x-a)^2 + (y-b)^2 + (z-c)^2 = r^2$$

So a sphere with center at the origin (0, 0, 0) and radius r has the equation

$$x^2 + y^2 + z^2 = r^2$$

Cylindrical Surfaces

An equation $F(x, y) = 0$ ordinarily defines a curve $\mathscr{C}$ in the xy plane. Now, if a point (x, y) satisfies this equation, then, for any z, the point (x, y, z) in space also satisfies the equation. So, the equation $F(x, y) = 0$ determines the cylindrical *surface* obtained by moving the curve $\mathscr{C}$ parallel to the z axis. For example, the equation $x^2 + y^2 = 4$ determines a circle in the xy plane with radius 2 and center at the origin. If we move this circle parallel to the z axis, we obtain a right circular cylinder. Thus, what we ordinarily call a cylinder is a special case of a cylindrical surface.

Similarly, an equation $F(y, z) = 0$ determines the cylindrical surface obtained by moving the curve in the yz plane defined by $F(y, z) = 0$ parallel to the x axis. An equation $F(x, z) = 0$ determines the cylindrical surface obtained by moving the curve in the xz plane defined by $F(x, z) = 0$ parallel to the y axis.

More precisely, the cylindrical surfaces defined above are called right cylindrical surfaces. Other cylindrical surfaces can be obtained by moving the given curve parallel to a line that is not perpendicular to the plane of the curve.

EXAMPLE 51.1: The equation $z = x^2$ determines a cylindrical surface generated by moving the parabola $z = x^2$ lying in the xz plane parallel to the y axis.

Now we shall look at examples of surfaces determined by equations of the second degree in x, y, and z. Such surfaces are called *quadric* surfaces. Imagining what they look like is often helped by describing their intersections with planes parallel to the coordinate planes. Such intersections are called *traces*.

Ellipsoid

$$x^2 + \frac{y^2}{9} + \frac{z^2}{4} = 1$$

The nontrivial traces are ellipses. See Fig. 51-1. In general, the equation of an ellipsoid has the form

$$\frac{x^2}{a^2} + \frac{y^2}{b^2} + \frac{z^2}{c^2} = 1 \qquad (a > 0, b > 0, c > 0)$$

When $a = b = c$, we obtain a sphere.

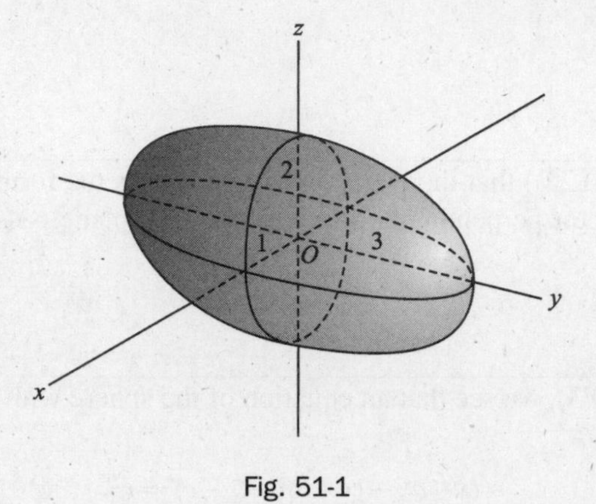

Fig. 51-1

Elliptic Paraboloid

$$z = x^2 + y^2$$

The surface lies on or above the *xy* plane. The traces parallel to the *xy* plane (for a fixed positive *z*) are circles. The traces parallel to the *xz* or *yz* plane are parabolas. See Fig. 51-2. In general, the equation of an elliptic paraboloid has the form

$$\frac{z}{c} = \frac{x^2}{a^2} + \frac{y^2}{b^2} \quad (a > 0, b > 0, c > 0)$$

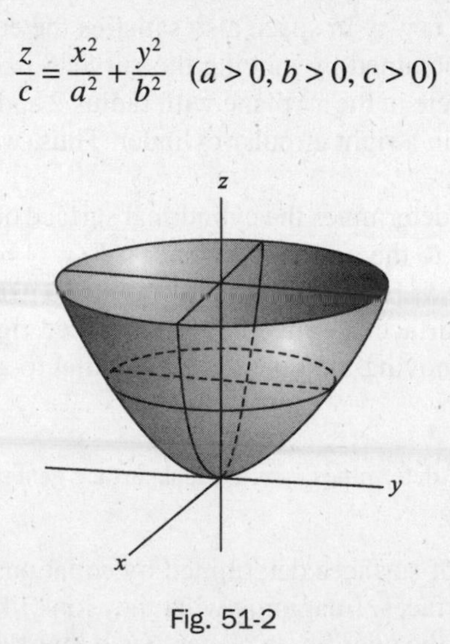

Fig. 51-2

and the traces parallel to the *xy* plane are ellipses. When $a = b$, we obtain a circular paraboloid, as in the given example.

Elliptic Cone

$$z^2 = x^2 + y^2$$

See Fig. 51-3. This is a pair of ordinary cones, meeting at the origin. The traces parallel to the xy plane are circles. The traces parallel to the xz or yz plane are hyperbolas. In general, the equation of an elliptic cone has the form

$$\frac{z^2}{c^2} = \frac{x^2}{a^2} + \frac{y^2}{b^2} \qquad (a > 0, b > 0, c > 0)$$

and the traces parallel to the xy plane are ellipses. When $a = b$, we obtain a right circular cone, as in the given example.

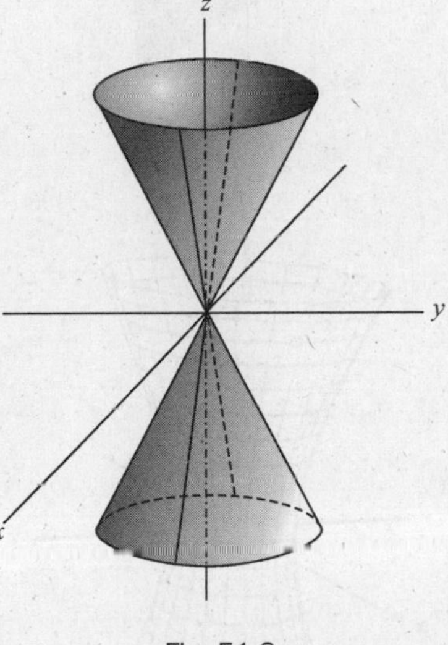

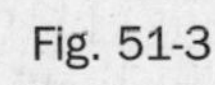

Fig. 51-3

Hyperbolic Paraboloid

$$z = 2y^2 - x^2$$

See Fig. 51-4. The surface resembles a saddle. The traces parallel to the xy plane are hyperbolas. The other traces are parabolas. In general, the equation of a hyperbolic paraboloid has the form

$$\frac{z}{c} = \frac{y^2}{b^2} + \frac{x^2}{a^2} \qquad (a > 0, b > 0, c \neq 0)$$

In the given example, $c = 1$, $a = 1$, and $b = 1/\sqrt{2}$.

Hyperboloid of One Sheet

$$x^2 + y^2 - \frac{z^2}{9} = 1$$

See Fig. 51-5. The traces parallel to the xy plane are circles and the other traces are hyperbolas. In general, a hyperboloid of one sheet has an equation of the form

$$\frac{x^2}{a^2}+\frac{y^2}{b^2}-\frac{z^2}{c^2}=1$$

and the traces parallel to the xy plane are ellipses.

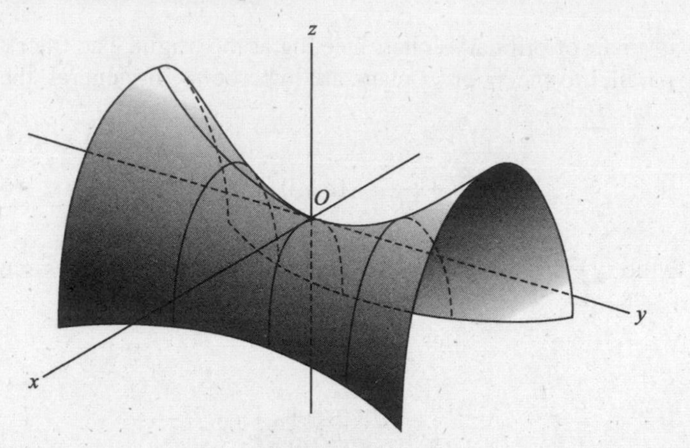

Fig. 51-4

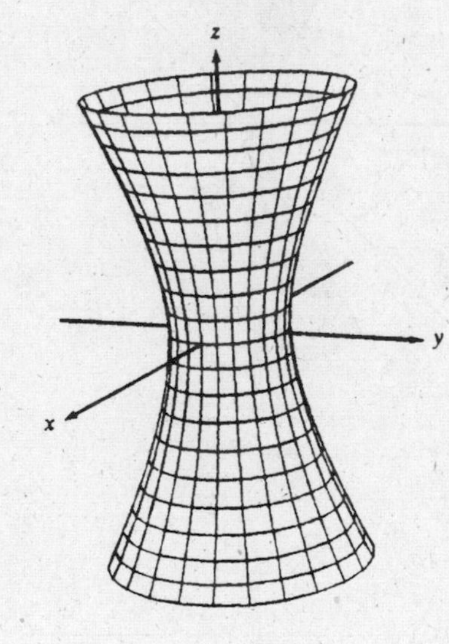

Fig. 51-5

Hyperboloid of Two Sheets

$$\frac{z^2}{4}-\frac{x^2}{9}-\frac{y^2}{9}=1$$

See Fig. 51-6. The traces parallel to the xy plane are circles, and the other traces are hyperbolas. In general, a hyperboloid of two sheets has an equation of the form

$$\frac{z^2}{c^2}-\frac{x^2}{a^2}-\frac{y^2}{b^2}=1 \quad (a>0, b>0, c>0)$$

and the traces parallel to the xy plane are ellipses.

In general equations given above for various quadric surfaces, permutation of the variables x, y, z is understood to produce quadric surfaces of the same type. For example, $\frac{y^2}{c^2}-\frac{z^2}{a^2}-\frac{x^2}{b^2}=1$ also determines a hyperboloid of two sheets.

Tangent Line and Normal Plane to a Space Curve

A space curve may be defined parametrically by the equations

$$x = f(t), \quad y = g(t), \quad z = h(t) \tag{51.1}$$

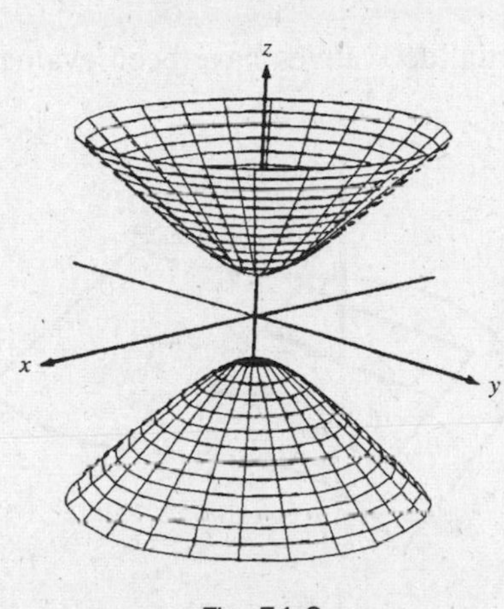

Fig. 51-6

At the point $P_0(x_0, y_0, z_0)$ of the curve (determined by $t = t_0$), the equations of the tangent line are

$$\frac{x - x_0}{dx/dt} = \frac{y - y_0}{dy/dt} = \frac{z - z_0}{dz/dt} \tag{51.2}$$

and the equations of the normal plane (the plane through P_0 perpendicular to the tangent line there) are

$$\frac{dx}{dt}(x - x_0) + \frac{dy}{dt}(y - y_0) + \frac{dz}{dt}(z - z_0) = 0 \tag{51.3}$$

See Fig. 51-7. In both (51.2) and (51.3), it is understood that the derivative has been evaluated at the point P_0. (See Problems 1 and 2.)

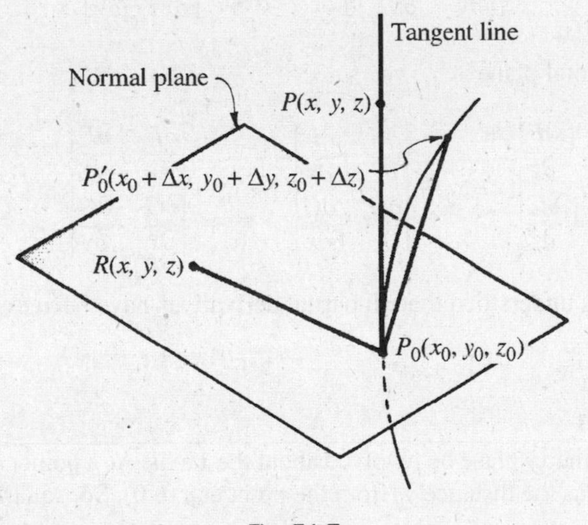

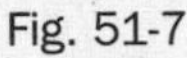
Fig. 51-7

Tangent Plane and Normal Line to a Surface

The equation of the tangent plane to the surface $F(x, y, z) = 0$ at one of its points $P_0(x_0, y_0, z_0)$ is

$$\frac{\partial F}{\partial x}(x - x_0) + \frac{\partial F}{\partial y}(y - y_0) + \frac{\partial F}{\partial z}(z - z_0) = 0 \tag{51.4}$$

and the equations of the normal line at P_0 are

$$\frac{x - x_0}{\partial F / \partial x} = \frac{y - y_0}{\partial F / \partial y} = \frac{z - z_0}{\partial F / \partial z} \tag{51.5}$$

with the understanding that the partial derivatives have been evaluated at the point P_0. See Fig. 51-8. (See Problems 3–9.)

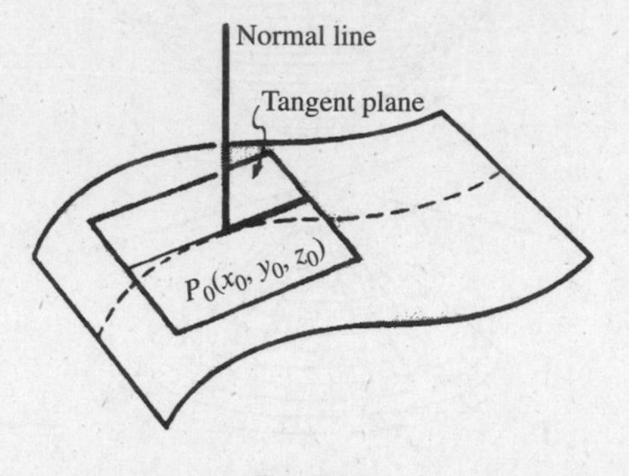

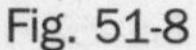
Fig. 51-8

A space curve may also be defined by a pair of equations

$$F(x, y, z) = 0, \quad G(x, y, z) = 0 \tag{51.6}$$

At the point $P_0(x_0, y_0, z_0)$ of the curve, the equations of the tangent line are

$$\frac{x - x_0}{\begin{vmatrix} \dfrac{\partial F}{\partial y} & \dfrac{\partial F}{\partial z} \\ \dfrac{\partial G}{\partial y} & \dfrac{\partial G}{\partial z} \end{vmatrix}} = \frac{y - y_0}{\begin{vmatrix} \dfrac{\partial F}{\partial z} & \dfrac{\partial F}{\partial x} \\ \dfrac{\partial G}{\partial z} & \dfrac{\partial G}{\partial x} \end{vmatrix}} = \frac{z - z_0}{\begin{vmatrix} \dfrac{\partial F}{\partial x} & \dfrac{\partial F}{\partial y} \\ \dfrac{\partial G}{\partial x} & \dfrac{\partial G}{\partial y} \end{vmatrix}} \tag{51.7}$$

and the equation of the normal plane is

$$\begin{vmatrix} \dfrac{\partial F}{\partial y} & \dfrac{\partial F}{\partial z} \\ \dfrac{\partial G}{\partial y} & \dfrac{\partial G}{\partial z} \end{vmatrix}(x - x_0) + \begin{vmatrix} \dfrac{\partial F}{\partial z} & \dfrac{\partial F}{\partial x} \\ \dfrac{\partial G}{\partial z} & \dfrac{\partial G}{\partial x} \end{vmatrix}(y - y_0) + \begin{vmatrix} \dfrac{\partial F}{\partial x} & \dfrac{\partial F}{\partial y} \\ \dfrac{\partial G}{\partial x} & \dfrac{\partial G}{\partial y} \end{vmatrix}(z - z_0) = 0 \tag{51.8}$$

In (51.7) and (51.8), it is understood that all partial derivatives have been evaluated at the point P_0. (See Problems 10 and 11.)

Surface of Revolution

Let the graph of $y = f(x)$ in the xy plane be revolved about the x axis. As a point (x_0, y_0) on the graph revolves, a resulting point (x_0, y, z) has the distance y_0 from the point $(x_0, 0, 0)$. So, squaring that distance, we get

$$(x_0 - x_0)^2 + y^2 + z^2 = (y_0)^2 = (f(x_0))^2 \qquad \text{and, therefore,} \qquad y^2 + z^2 = (f(x_0))^2$$

Then, the equation of the surface of revolution is

$$y^2 + z^2 = (f(x))^2 \tag{51.9}$$

SOLVED PROBLEMS

1. Derive (51.2) and (51.3) for the tangent line and normal plane to the space curve $x = f(t)$, $y = g(t)$, $z = h(t)$ at the point $P_0(x_0, y_0, z_0)$ determined by the value $t = t_0$. Refer to Fig. 51-7.

Let $P'_0(x_0 + \Delta x, y_0 + \Delta y, z_0 + \Delta z)$, determined by $t = t_0 + \Delta t$, be another point on the curve. As $P_0 \to P_0$ along the curve, the chord $P_0P'_0$ approaches the tangent line to the curve at P_0 as the limiting position.

A simple set of direction numbers for the chord $P_0P'_0$ is $[\Delta x, \Delta y, \Delta z]$, but we shall use $\left[\frac{\Delta x}{\Delta t}, \frac{\Delta y}{\Delta t}, \frac{\Delta z}{\Delta t}\right]$. Then as $P_0 \to P_0$, $\Delta t \to 0$ and $\left[\frac{\Delta x}{\Delta t}, \frac{\Delta y}{\Delta t}, \frac{\Delta z}{\Delta t}\right] \to \left[\frac{dx}{dt}, \frac{dy}{dt}, \frac{dz}{dt}\right]$, a set of direction numbers of the tangent line at P_0. Now if $P(x, y, z)$ is an arbitrary point on this tangent line, then $[x - x_0, y - y_0, z - z_0]$ is a set of direction numbers of P_0P. Thus, since the sets of direction numbers are proportional, the equations of the tangent line at P_0 are

$$\frac{x - x_0}{dx/dt} = \frac{y - y_0}{dy/dt} = \frac{z - z_0}{dz/dt}$$

If $R(x, y, z)$ is an arbitrary point in the normal plane at P_0, then, since P_0R and P_0P are perpendicular, the equation of the normal plane at P_0 is

$$(x - x_0)\frac{dx}{dt} + (y - y_0)\frac{dy}{dt} + (z - z_0)\frac{dz}{dt} = 0$$

2. Find the equations of the tangent line and normal plane to:

(a) The curve $x = t$, $y = t^2$, $z = t^3$ at the point $t = 1$.
(b) The curve $x = t - 2$, $y = 3t^2 + 1$, $z = 2t^3$ at the point where it pierces the yz plane.

(a) At the point $t = 1$ or $(1, 1, 1)$, $dx/dt = 1$, $dy/dt = 2t = 2$, and $dz/dt = 3t^2 = 3$. Using (51.2) yields, for the equations of the tangent line, $\frac{x-1}{1} = \frac{y-1}{2} = \frac{z-1}{3}$; using (51.3) gives the equation of the normal plane as $(x - 1) + 2(y - 1) + 3(z - 1) = x + 2y + 3z - 6 = 0$.

(b) The given curve pierces the yz plane at the point where $x = t - 2 = 0$, that is, at the point $t = 2$ or $(0, 13, 16)$. At this point, $dx/dt = 1$, $dy/dt = 6t = 12$, and $dz/dt = 6t^2 = 24$. The equations of the tangent line are $\frac{x}{1} = \frac{y-13}{12} = \frac{z-16}{24}$, and the equation of the normal plane is $x + 12(y - 13) + 24(z - 16) = x + 12y + 24z - 540 = 0$.

3. Derive (51.4) and (51.5) for the tangent plane to the surface $F(x, y, z) = 0$ at the point $P_0(x_0, y_0, z_0)$. Refer to Fig. 51-8.

Let $x = f(t)$, $y = g(t)$, $z = h(t)$ be the parametric equations of any curve on the surface $F(x, y, z) = 0$ and passing through the point P_0. Then, at P_0,

$$\frac{\partial F}{\partial x}\frac{dx}{dt} + \frac{\partial F}{\partial y}\frac{dy}{dt} + \frac{\partial F}{\partial z}\frac{dz}{dt} = 0$$

with the understanding that all derivatives have been evaluated at P_0.

This relation expresses the fact that the line through P_0 with direction numbers $\left[\frac{dx}{dt}, \frac{dy}{dt}, \frac{dz}{dt}\right]$ is perpendicular to the line through P_0 having direction numbers $\left[\frac{\partial F}{\partial x}, \frac{\partial F}{\partial y}, \frac{\partial F}{\partial z}\right]$. The first set of direction numbers belongs to the tangent to the curve which lies in the tangent plane of the surface. The second set defines the normal line to the surface at P_0. The equations of this normal are

$$\frac{x - x_0}{\partial F/\partial x} = \frac{y - y_0}{\partial F/\partial y} = \frac{z - z_0}{\partial F/\partial z}$$

and the equation of the tangent plane at P_0 is

$$\frac{\partial F}{\partial x}(x-x_0)+\frac{\partial F}{\partial y}(y-y_0)+\frac{\partial F}{\partial z}(z-z_0)=0$$

In Problems 4 and 5, find the equations of the tangent plane and normal line to the given surface at the given point.

4. $z = 3x^2 + 2y^2 - 11$; (2, 1, 3).

Put $F(x, y, z) = 3x^2 + 2y^2 - z - 11 = 0$. At (2, 1, 3), $\frac{\partial F}{\partial x} = 6x = 12, \frac{\partial F}{\partial y} = 4y = 4$, and $\frac{\partial F}{\partial z} = -1$. The equation of the tangent plane is $12(x-2) + 4(y-1) - (z-3) = 0$ or $12x + 4y - z = 25$.

The equations of the normal line are $\frac{x-2}{12} = \frac{y-1}{4} = \frac{z-3}{-1}$.

5. $F(x, y, z) = x^2 + 3y^2 - 4z^2 + 3xy - 10yz + 4x - 5z - 22 = 0$; (1, –2, 1).

At (1, –2, 1), $\frac{\partial F}{\partial x} = 2x + 3y + 4 = 0$, $\frac{\partial F}{\partial y} = 6y + 3x - 10z = -19$, and $\frac{\partial F}{\partial z} = -8z - 10y - 5 = 7$. The equation of the tangent plane is $0(x-1) - 19(y+2) + 7(z-1) = 0$ or $19y - 7z + 45 = 0$.

The equations of the normal line are $x - 1 = 0$ and $\frac{y+2}{-19} = \frac{z-1}{7}$ or $x = 1$, $7y + 19z - 5 = 0$.

6. Show that the equation of the tangent plane to the surface $\frac{x^2}{a^2} - \frac{y^2}{b^2} - \frac{z^2}{c^2} = 1$ at the point $P_0(x_0, y_0, z_0)$ is $\frac{xx_0}{a^2} - \frac{yy_0}{b^2} - \frac{zz_0}{c^2} = 1$.

At P_0, $\frac{\partial F}{\partial x} = \frac{2x_0}{a^2}, \frac{\partial F}{\partial y} = -\frac{2y_0}{b^2}$, and $\frac{\partial F}{\partial z} = -\frac{2z_0}{c^2}$. The equation of the tangent plane is

$\frac{2x_0}{a^2}(x-x_0) - \frac{2y_0}{b^2}(y-y_0) - \frac{2z_0}{c^2}(z-z_0) = 0$.

This becomes $\frac{xx_0}{a^2} - \frac{yy_0}{b^2} - \frac{zz_0}{c^2} = \frac{x_0^2}{a^2} - \frac{y_0^2}{b^2} - \frac{z_0^2}{c^2} = 1$, since P_0 is on the surface.

7. Show that the surfaces $F(x, y, z) = x^2 + 4y^2 - 4z^2 - 4 = 0$ and $G(x, y, z) = x^2 + y^2 + z^2 - 6x - 6y + 2z + 10 = 0$ are tangent at the point (2, 1, 1).

It is to be shown that the two surfaces have the same tangent plane at the given point. At (2, 1, 1),

$$\frac{\partial F}{\partial x} = 2x - 4, \qquad \frac{\partial F}{\partial y} = 8y = 8, \qquad \frac{\partial F}{\partial z} = -8z = -8$$

$$\text{and} \qquad \frac{\partial G}{\partial x} = 2x - 6 = -2, \qquad \frac{\partial G}{\partial y} = 2y - 6 = -4, \qquad \frac{\partial G}{\partial z} = 2z + 2 = 4$$

Since the sets of direction numbers [4, 8, –8] and [–2, –4, 4] of the normal lines of the two surfaces are proportional, the surfaces have the common tangent plane

$$1(x-2) + 2(y-1) - 2(z-1) = 0 \quad \text{or} \quad x + 2y - 2z = 2$$

8. Show that the surfaces $F(x, y, z) = xy + yz - 4zx = 0$ and $G(x, y, z) = 3z^2 - 5x + y = 0$ intersect at right angles at the point (1, 2, 1).

It is to be shown that the tangent planes to the surfaces at the point are perpendicular or, what is the same, that the normal lines at the point are perpendicular. At (1, 2, 1),

$$\frac{\partial F}{\partial x} = y - 4z = -2, \qquad \frac{\partial F}{\partial y} = x + z = 2, \qquad \frac{\partial F}{\partial z} = y - 4x = -2$$

A set of direction numbers for the normal line to $F(x, y, z) = 0$ is $[l_1, m_1, n_1] = [1, -1, 1]$. At the same point,

$$\frac{\partial G}{\partial x} = -5, \qquad \frac{\partial G}{\partial y} = 1, \qquad \frac{\partial G}{\partial z} = 6z = 6$$

A set of direction numbers for the normal line to $G(x, y, z) = 0$ is $[l_2, m_2, n_2] = [-5, 1, 6]$. Since $l_1l_2 + m_1m_2 + n_1n_2 = 1(-5) + (-1)1 + 1(6) = 0$, these directions are perpendicular.

9. Show that the surfaces $F(x, y, z) = 3x^2 + 4y^2 + 8z^2 - 36 = 0$ and $G(x, y, z) = x^2 + 2y^2 - 4z^2 - 6 = 0$ intersect at right angles.

At any point $P_0(x_0, y_0, z_0)$ on the two surfaces, $\frac{\partial F}{\partial x} = 6x_0, \frac{\partial F}{\partial y} = 8y_0$, and $\frac{\partial F}{\partial z} = 16z_0$; hence $[3x_0, 4y_0, 8z_0]$ is a set of direction numbers for the normal to the surface $F(x, y, z) = 0$ at P_0. Similarly, $[x_0, 2y_0, -4z_0]$ is a set of direction numbers for the normal line to $G(x, y, z) = 0$ at P_0. Now, since

$$6(x_0^2 + 2y_0^2 - 4z_0^2) - (3x_0^2 + 4y_0^2 + 8z_0^2) = 6(6) - 36 = 0,$$

these directions are perpendicular.

10. Derive (51.7) and (51.8) for the tangent line and normal plane to the space curve C: $F(x, y, z) = 0$, $G(x, y, z) = 0$ at one of its points $P_0(x_0, y_0, z_0)$.

At P_0, the directions $\left[\frac{\partial F}{\partial x}, \frac{\partial F}{\partial y}, \frac{\partial F}{\partial z}\right]$ and $\left[\frac{\partial G}{\partial x}, \frac{\partial G}{\partial y}, \frac{\partial G}{\partial z}\right]$ are normal, respectively, to the tangent planes of the surfaces $F(x, y, z) = 0$ and $G(x, y, z) = 0$. Now the direction

$$\left[\begin{vmatrix} \partial F/\partial y & \partial F/\partial z \\ \partial G/\partial y & \partial G/\partial z \end{vmatrix}, \begin{vmatrix} \partial F/\partial z & \partial F/\partial x \\ \partial G/\partial z & \partial G/\partial x \end{vmatrix}, \begin{vmatrix} \partial F/\partial x & \partial F/\partial y \\ \partial G/\partial x & \partial G/\partial y \end{vmatrix}\right]$$

being perpendicular to each of these directions, is that of the tangent line to C at P_0. Hence, the equations of the tangent line are

$$\frac{x - x_0}{\begin{vmatrix} \partial F/\partial y & \partial F/\partial z \\ \partial G/\partial y & \partial G/\partial z \end{vmatrix}} = \frac{y - y_0}{\begin{vmatrix} \partial F/\partial z & \partial F/\partial x \\ \partial G/\partial z & \partial G/\partial x \end{vmatrix}} = \frac{z - z_0}{\begin{vmatrix} \partial F/\partial x & \partial F/\partial y \\ \partial G/\partial x & \partial G/\partial y \end{vmatrix}}$$

and the equation of the normal plane is

$$\begin{vmatrix} \partial F/\partial y & \partial F/\partial z \\ \partial G/\partial y & \partial G/\partial z \end{vmatrix}(x - x_0) + \begin{vmatrix} \partial F/\partial z & \partial F/\partial x \\ \partial G/\partial z & \partial G/\partial x \end{vmatrix}(y - y_0) + \begin{vmatrix} \partial F/\partial x & \partial F/\partial y \\ \partial G/\partial x & \partial G/\partial y \end{vmatrix}(z - z_0) = 0$$

11. Find the equations of the tangent line and the normal plane to the curve $x^2 + y^2 + z^2 = 14$, $x + y + z = 6$ at the point $(1, 2, 3)$.

Set $F(x, y, z) = x^2 + y^2 + z^2 - 14 = 0$ and $G(x, t, z) = x + y + z - 6 = 0$. At $(1, 2, 3)$,

$$\begin{vmatrix} \partial F/\partial y & \partial F/\partial z \\ \partial G/\partial y & \partial G/\partial z \end{vmatrix} = \begin{vmatrix} 2y & 2z \\ 1 & 1 \end{vmatrix} = \begin{vmatrix} 4 & 6 \\ 1 & 1 \end{vmatrix} = -2$$

$$\begin{vmatrix} \partial F/\partial z & \partial F/\partial x \\ \partial G/\partial z & \partial G/\partial x \end{vmatrix} = \begin{vmatrix} 6 & 2 \\ 1 & 1 \end{vmatrix} = 4, \qquad \begin{vmatrix} \partial F/\partial x & \partial F/\partial y \\ \partial G/\partial x & \partial G/\partial y \end{vmatrix} = \begin{vmatrix} 2 & 4 \\ 1 & 1 \end{vmatrix} = -2$$

With $[1, -2, 1]$ as a set of direction numbers of the tangent, its equations are $\frac{x - 1}{1} = \frac{y - 2}{-2} = \frac{z - 3}{1}$. The equation of the normal plane is $(x - 1) - 2(y - 2) + (z - 3) = x - 2y + z = 0$.

12. Find equations of the surfaces of revolution generated by revolving the given curve about the given axis: (a) $y = x^2$ about the x axis; (b) $y = \frac{1}{x}$ about the y axis; (c) $z = 4y$ about the y axis.

In each case, we use an appropriate form of (51.9): (a) $y^2 + z^2 = x^4$; (b) $x^2 + z^2 = \frac{1}{y^2}$; (c) $x^2 + z^2 = 16y^2$.

13. Identify the locus of all points (x, y, z) that are equidistant from the point $(0, -1, 0)$ and the plane $y = 1$.

Squaring the distances, we get $x^2 + (y + 1^2) + z^2 = (y - 1)^2$, whence $x^2 + z^2 = -4y$, a circular paraboloid.

14. Identify the surface $4x^2 - y^2 + z^2 - 8x + 2y + 2z + 3 = 0$ by completing the squares.

We have

$$4(x^2 - 2x) - (y^2 - 2y) + (z^2 + 2z) + 3 = 0$$

$$4(x - 1)^2 - (y - 1)^2 + (z + 1)^2 + 3 = 4$$

$$4(x - 1)^2 - (y - 1)^2 + (z + 1)^2 = 1$$

This is a hyperboloid of one sheet, centered at $(1, 1, -1)$.

SUPPLEMENTARY PROBLEMS

15. Find the equations of the tangent line and the normal plane to the given curve at the given point:

(a) $x = 2t,\ y = t^2,\ z = t^3;\ t = 1$ *Ans.* $\frac{x-2}{2} = \frac{y-1}{2} = \frac{z-1}{3};\ 2x + 2y + 3z - 9 = 0$

(b) $x = te^t,\ y = e^t,\ z = t;\ t = 0$ *Ans.* $\frac{x}{1} = \frac{y-1}{1} = \frac{z}{1};\ x + y + z - 1 = 0$

(c) $x = t\cos t,\ y = t\sin t,\ z = t;\ t = 0$ *Ans.* $x = z,\ y = 0;\ x + z = 0$

16. Show that the curves (a) $x = 2 - t,\ y = -1/t,\ z = 2t^2$ and (b) $x = 1 + \theta,\ y = \sin\theta - 1,\ z = 2\cos\theta$ intersect at right angles at $P(1, -1, 2)$. Obtain the equations of the tangent line and normal plane of each curve at P.

Ans. (a) $\frac{x-1}{-1} = \frac{y+1}{1} = \frac{z-2}{4};\ x - y - 4z + 6 = 0$; (b) $x - y = 2,\ z = 2;\ x + y = 0$

17. Show that the tangent lines to the helix $x = a\cos t,\ y = a\sin t,\ z = bt$ meet the xy plane at the same angle.

18. Show that the length of the curve (51.1) from the point $t = t_0$ to the point $t = t_1$ is given by

$$\int_{t_0}^{t_1} \sqrt{\left(\frac{dx}{dt}\right)^2 + \left(\frac{dy}{dt}\right)^2 + \left(\frac{dz}{dt}\right)^2}\, dt$$

Find the length of the helix of Problem 17 from $t = 0$ to $t = t_1$.

Ans. $\sqrt{a^2 + b^2}\, t_1$

19. Find the equations of the tangent line and the normal plane to the given curve at the given point:

(a) $x^2 + 2y^2 + 2z^2 = 5,\ 3x - 2y - z = 0;\ (1, 1, 1)$.

(b) $9x^2 + 4y^2 - 36z = 0,\ 3x + y + z - z^2 - 1 = 0;\ (2, -3, 2)$.

(c) $4z^2 = xy,\ x^2 + y^2 = 8z;\ (2, 2, 1)$.

Ans. (a) $\frac{x-1}{2}=\frac{y-1}{7}=\frac{z-1}{-8}$; $2x+7y-8z-1=0$; (b) $\frac{x-2}{1}=\frac{y-2}{1}$, $y+3=0$; $x+z-4=0$;
(c) $\frac{x-2}{1}=\frac{y-2}{-1}$, $z-1=0$; $x-y=0$

20. Find the equations of the tangent plane and normal line to the given surface at the given point:

(a) $x^2+y^2+z^2=14$; $(1, -2, 3)$ *Ans.* $x-2y+3z=14$; $\frac{x-1}{1}=\frac{y+2}{-2}=\frac{z-3}{3}$

(b) $x^2+y^2+z^2=r^2$; (x_1, y_1, z_1) *Ans.* $x_1x+y_1y+z_1z=r^2$; $\frac{x-x_1}{x_1}=\frac{y+y_1}{y_1}=\frac{z-z_1}{z_1}$

(c) $x^2+2z^3+3y^2$; $(2, -2, -2)$ *Ans.* $x+3y-2z=0$; $\frac{x-2}{1}=\frac{y+2}{3}=\frac{z+2}{-2}$

(d) $2x^2+2xy+y^2+z+1=0$; $(1, -2, -3)$ *Ans.* $z-2y=1$; $x-1=0$, $\frac{y+2}{2}=\frac{z+3}{-1}$

(e) $z=xy$; $(3, -4, -12)$ *Ans.* $4x-3y+z=12$; $\frac{x-3}{4}=\frac{y+4}{-3}=\frac{z+12}{1}$

21. (a) Show that the sum of the intercepts of the plane tangent to the surface $x^{1/2}+y^{1/2}+z^{1/2}=a^{1/2}$ at any of its points is a.

(b) Show that the square root of the sum of the squares of the intercepts of the plane tangent to the surface $x^{2/3}+y^{2/3}+z^{2/3}=a^{2/3}$ at any of its points is a.

22. Show that each pair of surfaces are tangent at the given point:

(a) $x^2+y^2+z^2=18$, $xy=9$; $(3, 3, 0)$.
(b) $x^2+y^2+z^2-8x-8y-6z+24=0$, $x^2+3y^2+2z^2=9$; $(2, 1, 1)$.

23. Show that each pair of surfaces are perpendicular at the given point:

(a) $x^2+2y^2-4z^2=8$, $4x^2-y^2+2z^2=14$; $(2, 2, 1)$.
(b) $x^2+y^2+z^2-50$, $x^2+y^2-10z+25=0$; $(3, 4, 5)$.

24. Show that each of the surfaces (a) $14x^2+11y^2+8z^2=66$, (b) $3z^2-5x+y=0$, and (c) $xy+yz-4zx=0$ is perpendicular to the other two at the point $(1, 2, 1)$.

25. Identify the following surfaces.

(a) $36y^2-x^2+36z^2=9$.
(b) $5y=-z^2+x^2$.
(c) $x^2+4y^2-4z^2-6x-16y-16z+5=0$.

Ans. (a) hyperboloid of one sheet (around the x axis); (b) hyperbolic paraboloid; (c) hyperboloid of one sheet, centered at $(3, 2, -2)$

26. Find an equation of a curve that, when revolved about a suitable axis, yields the paraboloid $y^2+z^2-2x=0$.

Ans. $y=\sqrt{2x}$ or $z=\sqrt{2x}$, about the x axis

27. Find an equation of the surface obtained by revolving the given curve about the given axis. Identify the type of surface: (a) $x=y^2$ about the x axis; (b) $x=2y$ about the x axis.

Ans. (a) $x=y^2+z^2$ (circular paraboloid); (b) $y^2+z^2=\frac{x^2}{4}$ (right circular cone)

CHAPTER 52

Directional Derivatives. Maximum and Minimum Values

Directional Derivatives

Let $P(x, y, z)$ be a point on a surface $z = f(x, y)$. Through P, pass planes parallel to the xz and yz planes, cutting the surface in the arcs PR and PS, and cutting the xy plane in the lines P^*M and P^*N, as shown in Fig. 52-1. Note that P^* is the foot of the perpendicular from P to the xy plane. The partial derivatives $\partial z/\partial x$ and $\partial z/\partial y$, evaluated at $P^*(x, y)$, give, respectively, the rates of change of $z = P^*P$ when y is held fixed and when x is held fixed. In other words, they give the rates of change of z in directions parallel to the x and y axes. These rates of change are the slopes of the tangent lines of the curves PR and PS at P.

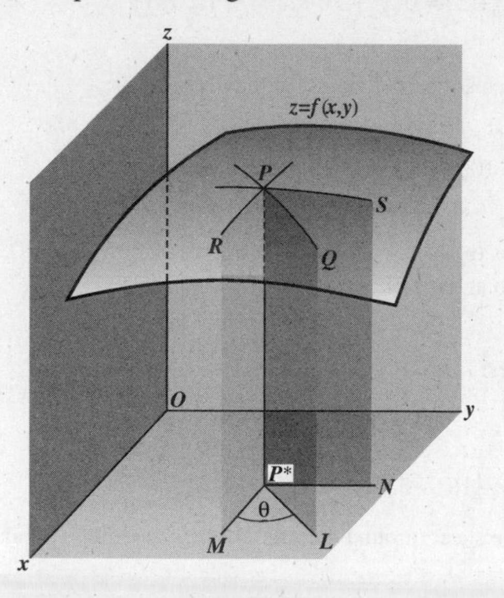

Fig. 52-1

Consider next a plane through P perpendicular to the xy plane and making an angle θ with the x axis. Let it cut the surface in the curve PQ and the xy plane in the line P^*L. The *directional derivative* of $f(x, y)$ at P^* in the direction θ is given by

$$\frac{dz}{ds} = \frac{\partial z}{\partial x}\cos\theta + \frac{\partial z}{\partial y}\sin\theta \tag{52.1}$$

The direction θ is the direction of the vector $(\cos\theta)\mathbf{i} + (\sin\theta)\mathbf{j}$.

The directional derivative gives the rate of change of $z = P^*P$ in the direction of P^*L; it is equal to the slope of the tangent line of the curve PQ at P. (See Problem 1.)

The directional derivative at a point P^* is a function of θ. We shall see that there is a direction, determined by a vector called the *gradient* of f at P^* (see Chapter 53), for which the directional derivative at P^* has a maximum value. That maximum value is the slope of the steepest tangent line that can be drawn to the surface at P.

For a function $w = F(x, y, z)$, the directional derivative at $P(x, y, z)$ in the direction determined by the angles α, β, γ is given by

$$\frac{dF}{ds} = \frac{\partial F}{\partial x}\cos\alpha + \frac{\partial F}{\partial y}\cos\beta + \frac{\partial F}{\partial z}\cos\gamma$$

By the direction determined by α, β, and γ, we mean the direction of the vector $(\cos\alpha)\mathbf{i} + (\cos\beta)\mathbf{j} + (\cos\gamma)\mathbf{k}$.

Relative Maximum and Minimum Values

Assume that $z = f(x, y)$ has a relative maximum (or minimum) value at $P_0(x_0, y_0, z_0)$. Any plane through P_0 perpendicular to the xy plane will cut the surface in a curve having a relative maximum (or minimum) point at P_0. Thus, the directional derivative $\frac{\partial f}{\partial x}\cos\theta + \frac{\partial f}{\partial y}\sin\theta$ of $z = f(x, y)$ must equal zero at P_0. In particular, when $\theta = 0$, $\sin\theta = 0$ and $\cos\theta = 1$, so that $\frac{\partial f}{\partial x} = 0$. When $\theta = \frac{\pi}{2}$, $\sin\theta = 1$ and $\cos\theta = 0$, so that $\frac{\partial f}{\partial y} = 0$. Hence, we obtain the following theorem.

Theorem 52.1: If $z = f(x, y)$ has a relative extremum at $P_0(x_0, y_0, z_0)$ and $\frac{\partial f}{\partial x}$ and $\frac{\partial f}{\partial y}$ exist at (x_0, y_0), then $\frac{\partial f}{\partial x} = 0$ and $\frac{\partial f}{\partial y} = 0$ at (x_0, y_0).

We shall cite without proof the following sufficient conditions for the existence of a relative maximum or minimum.

Theorem 52.2: Let $z = f(x, y)$ have first and second partial derivatives in an open set including a point (x_0, y_0) at which $\frac{\partial f}{\partial x} = 0$ and $\frac{\partial f}{\partial y} = 0$. Define $\Delta = \left(\frac{\partial^2 f}{\partial x\,\partial y}\right)^2 - \left(\frac{\partial^2 f}{\partial x^2}\right)\left(\frac{\partial^2 f}{\partial y^2}\right)$. Assume $\Delta < 0$ at (x_0, y_0). Then:

$$z = f(x, y) \text{ has } \begin{cases} \text{a relative } \textbf{min}\text{imum at } (x_0, y_0) & \text{if } \dfrac{\partial^2 f}{\partial x^2} + \dfrac{\partial^2 f}{\partial y^2} > 0 \\ \text{a relative } \textbf{max}\text{imum at } (x_0, y_0) & \text{if } \dfrac{\partial^2 f}{\partial x^2} + \dfrac{\partial^2 f}{\partial y^2} < 0 \end{cases}$$

If $\Delta > 0$, there is neither a relative maximum nor a relative minimum at (x_0, y_0).
If $\Delta = 0$, we have no information.

Absolute Maximum and Minimum Values

Let A be a set of points in the xy plane. We say that A is *bounded* if A is included in some disk. By the *complement* of A in the xy plane, we mean the set of all points in the xy plane that are not in A. A is said to be *closed* if the complement of A is an open set.

Example 1: The following are instances of closed and bounded sets.

(a) Any closed disk D, that is, the set of all points whose distance from a fixed point is less than or equal to some fixed positive number r. (Note that the complement of D is open because any point not in D can be surrounded by an open disk having no points in D.)

(b) The inside and boundary of any rectangle. More generally, the inside and boundary of any "simple closed curve," that is, a curve that does not interset itself except at its initial and terminal point.

Theorem 52.3: Let $f(x, y)$ be a function that is continuous on a closed, bounded set A. Then f has an absolute maximum and an absolute minimum value in A.

The reader is referred to more advanced texts for a proof of Theorem 52.3. For three or more variables, an analogous result can be derived.

SOLVED PROBLEMS

1. Derive formula (52.1).

In Fig. 52-1, let $P^{**}(x + \Delta x, y + \Delta y)$ be a second point on P^*L and denote by Δs the distance P^*P^{**}. Assuming that $z = f(x, y)$ possesses continuous first partial derivatives, we have, by Theorem 49.1,

$$\Delta z = \frac{\partial z}{\partial x}\,\Delta x + \frac{\partial z}{\partial y}\,\Delta y + \epsilon_1 \Delta x + \epsilon_2 \Delta y$$

where ϵ_1 and $\epsilon_2 \to 0$ as Δx and $\Delta y \to 0$. The average rate of change between points P^* and P^{**} is

$$\frac{\Delta z}{\Delta s} = \frac{\partial z}{\partial x}\frac{\Delta x}{\Delta s} + \frac{\partial z}{\partial y}\frac{\Delta y}{\Delta s} + \epsilon_1 \frac{\Delta x}{\Delta s} + \epsilon_2 \frac{\Delta y}{\Delta s}$$

$$= \frac{\partial z}{\partial x}\cos\theta + \frac{\partial z}{\partial y}\sin\theta + \epsilon_1 \cos\theta + \epsilon_2 \sin\theta$$

where θ is the angle that the line P^*P^{**} makes with the x axis. Now let $P^{**} \to P^*$ along P^*L. The directional derivative at P^*, that is, the instantaneous rate of change of z, is then

$$\frac{dz}{ds} = \frac{\partial z}{\partial x}\cos\theta + \frac{\partial z}{\partial y}\sin\theta$$

2. Find the directional derivative of $z = x^2 - 6y^2$ at $P^*(7, 2)$ in the direction: (a) $\theta = 45°$; (b) $\theta = 135°$.

The directional derivative at any point $P^*(x, y)$ in the direction θ is

$$\frac{dz}{ds} = \frac{\partial z}{\partial x}\cos\theta + \frac{\partial z}{\partial y}\sin\theta = 2x\cos\theta - 12y\sin\theta$$

(a) At $P^*(7, 2)$ in the direction $\theta = 45°$,

$$\frac{dz}{ds} = 2(7)(\tfrac{1}{2}\sqrt{2}) - 12(2)(\tfrac{1}{2}\sqrt{2}) = -5\sqrt{2}$$

(b) At $P^*(7, 2)$ in the direction $\theta = 135°$,

$$\frac{dz}{ds} = 2(7)(-\tfrac{1}{2}\sqrt{2}) - 12(2)(\tfrac{1}{2}\sqrt{2}) = -19\sqrt{2}$$

3. Find the directional derivative of $z = ye^x$ at $P^*(0, 3)$ in the direction (a) $\theta = 30°$; (b) $\theta = 120°$.

Here, $dz/ds = ye^x \cos\theta + e^x \sin\theta$.

(a) At (0, 3) in the direction $\theta = 30°$, $dz/ds = 3(1)(\tfrac{1}{2}\sqrt{3}) + \tfrac{1}{2} = \tfrac{1}{2}(3\sqrt{3} + 1)$.

(b) At (0, 3) in the direction $\theta = 120°$, $dz/ds = 3(1)(-\tfrac{1}{2}) + \tfrac{1}{2}\sqrt{3} = \tfrac{1}{2}(-3 + \sqrt{3})$.

4. The temperature T of a heated circular plate at any of its points (x, y) is given by $T = \dfrac{64}{x^2 + y^2 + 2}$, the origin being at the center of the plate. At the point (1, 2), find the rate of change of T in the direction $\theta = \pi/3$.

We have

$$\frac{dT}{ds} = -\frac{64(2x)}{(x^2+y^2+2)^2}\cos\theta - \frac{64(2y)}{(x^2+y^2+2)^2}\sin\theta$$

At (1, 2) in the direction $\theta = \frac{\pi}{3}$, $\frac{dT}{ds} = -\frac{128}{49}\frac{1}{2} - \frac{256}{49}\frac{\sqrt{3}}{2} = -\frac{64}{49}(1+2\sqrt{3})$.

5. The electrical potential V at any point (x, y) is given by $V = \ln\sqrt{x^2+y^2}$. Find the rate of change of V at the point (3, 4) in the direction toward the point (2, 6).

Here,

$$\frac{dV}{ds} = \frac{x}{x^2+y^2}\cos\theta + \frac{y}{x^2+y^2}\sin\theta$$

Since θ is a second-quadrant angle and $\tan\theta = (6-4)/(2-3) = -2$, $\cos\theta = -1/\sqrt{5}$ and $\sin\theta = 2/\sqrt{5}$.

Hence, at (3, 4) in the indicated direction, $\frac{dV}{ds} = \frac{3}{25}\left(-\frac{1}{\sqrt{5}}\right) + \frac{4}{25}\frac{2}{\sqrt{5}} = \frac{\sqrt{5}}{25}$.

6. Find the maximum directional derivative for the surface and point of Problem 2.

At $P^*(7, 2)$ in the direction θ, $dz/ds = 14\cos\theta - 24\sin\theta$.

To find the value of θ for which $\frac{dz}{ds}$ is a maximum, set $\frac{d}{d\theta}\left(\frac{dz}{ds}\right) = -14\sin\theta - 24\cos\theta = 0$. Then $\tan\theta = -\frac{24}{14} = -\frac{12}{7}$ and θ is either a second- or fourth-quadrant angle. For the second-quadrant angle, $\sin\theta = 12/\sqrt{193}$ and $\cos = -7/\sqrt{193}$. For the fourth-quadrant angle, $\sin\theta = -12/\sqrt{193}$ and $\cos\theta = 7/\sqrt{193}$.

Since $\frac{d^2}{d\theta^2}\left(\frac{dz}{ds}\right) = \frac{d}{d\theta}(-14\sin\theta - 24\cos\theta) = -14\cos\theta + 24\sin\theta$ is negative for the fourth-quadrant angle, the maximum directional derivative is $\frac{dz}{dz} = 14\left(\frac{7}{\sqrt{193}}\right) - 24\left(-\frac{12}{\sqrt{193}}\right) = 2\sqrt{193}$, and the direction is $\theta = 300°15'$.

7. Find the maximum directional derivative for the function and point of Problem 3.

At $P^*(0, 3)$ in the direction θ, $dz/ds = 3\cos\theta + \sin\theta$.

To find the value of θ for which $\frac{dz}{ds}$ is a maximum, set $\frac{d}{d\theta}\left(\frac{dz}{ds}\right) = -3\sin\theta + \cos\theta = 0$. Then $\tan\theta = \frac{1}{3}$ and θ is either a first- or third-quadrant angle.

Since $\frac{d^2}{d\theta^2}\left(\frac{dz}{ds}\right) = \frac{d}{d\theta}(-3\sin\theta + \cos\theta) = -3\cos\theta - \sin\theta$ is negative for the first-quadrant angle, the maximum directional derivative is $\frac{dz}{ds} = 3\frac{3}{\sqrt{10}} + \frac{1}{\sqrt{10}} = \sqrt{10}$, and the direction is $\theta = 18°26'$.

8. In Problem 5, show that V changes most rapidly along the set of radial lines through the origin.

At any point (x_1, y_1) in the direction θ, $\frac{dV}{ds} = \frac{x_1}{x_1^2+y_1^2}\cos\theta + \frac{y_1}{x_1^2+y_1^2}\sin\theta$. Now V changes most rapidly when $\frac{d}{d\theta}\left(\frac{dV}{ds}\right) = -\frac{x_1}{x_1^2+y_1^2}\sin\theta + \frac{y_1}{x_1^2+y_1^2}\cos\theta = 0$, and then $\tan\theta = \frac{y_1/(x_1^2+y_1^2)}{x_1/(x_1^2+y_1^2)} = \frac{y_1}{x_1}$. Thus, θ is the angle of inclination of the line joining the origin and the point (x_1, y_1).

9. Find the directional derivative of $F(x, y, z) = xy + 2xz - y^2 + z^2$ at the point (1, −2, 1) along the curve $x = t$, $y = t - 3$, $z = t^2$ in the direction of increasing z.

A set of direction numbers of the tangent to the curve at (1, −2, 1) is [1, 1, 2]; the direction cosines are $[1/\sqrt{6}, 1/\sqrt{6}, 2/\sqrt{6}]$. The directional derivative is

$$\frac{\partial F}{\partial x}\cos\alpha + \frac{\partial F}{\partial y}\cos\beta + \frac{\partial F}{\partial z}\cos\gamma = 0\frac{1}{\sqrt{6}} + 5\frac{1}{\sqrt{6}} + 4\frac{2}{\sqrt{6}} = \frac{13\sqrt{6}}{6}$$

10. Examine $f(x, y) = x^2 + y^2 - 4x + 6y + 25$ for maximum and minimum values.

The conditions $\frac{\partial f}{\partial x} = 2x - 4 = 0$ and $\frac{\partial f}{\partial y} = 2y + 6 = 0$ are satisfied when $x = 2$, $y = -3$. Since

$$f(x, y) = (x^2 - 4x + 4) + (y^2 + 6y + 9) + 25 - 4 - 9 = (x - 2)^2 + (y + 3)^2 + 12$$

it is evident that $f(2, -3) = 12$ is the absolute minimum value of the function. Geometrically, $(2, -3, 12)$ is the lowest point on the surface $z = x^2 + y^2 - 4x + 6y + 25$. Clearly, $f(x, y)$ has no absolute maximum value.

11. Examine $f(x,y) = x^3 + y^3 + 3xy$ for maximum and minimum values.

We shall use Theorem 52.2. The conditions $\frac{\partial f}{\partial x} = 3(x^2 + y) = 0$ and $\frac{\partial f}{\partial y} = 3(y^2 + x) = 0$ are satisfied when $x = 0$, $y = 0$ and when $x = -1$, $y = -1$.

At $(0, 0)$, $\frac{\partial^2 f}{\partial x^2} = 6x = 0$, $\frac{\partial^2 f}{\partial x\, \partial y} = 3$, and $\frac{\partial^2 f}{\partial y^2} = 6y = 0$. Then

$$\left(\frac{\partial^2 f}{\partial x\, \partial y}\right)^2 - \left(\frac{\partial^2 f}{\partial x^2}\right)\left(\frac{\partial^2 f}{\partial y^2}\right) = 9 > 0$$

and $(0, 0)$ yields neither a relative maximum nor minimum.

At $(-1, -1)$, $\frac{\partial^2 f}{\partial x^2} = -6$, $\frac{\partial^2 f}{\partial x\, \partial y} = 3$, and $\frac{\partial^2 f}{\partial y^2} = -6$. Then

$$\left(\frac{\partial^2 f}{\partial x\, \partial y}\right)^2 - \left(\frac{\partial^2 f}{\partial x^2}\right)\left(\frac{\partial^2 f}{\partial y^2}\right) = -27 < 0 \quad \text{and} \quad \frac{\partial^2 f}{\partial x^2} + \frac{\partial^2 f}{\partial y^2} < 0$$

Hence, $f(-1, -1) = 1$ is a relative maximum value of the function.

Clearly, there are no absolute maximum or minimum values. (When $y = 0$, $f(x, y) = x^3$ can be made arbitrarily large or small.)

12. Divide 120 into three nonnegative parts such that the sum of their products taken two at a time is a maximum.

Let x, y, and $120 - (x + y)$ be the three parts. The function to be maximized is $S = xy + (x + y)(120 - x - y)$. Since $0 \le x + y \le 120$, the domain of the function consists of the solid triangle shown in Fig. 52-2. Theorem 52.3 guarantees an absolute maximum.

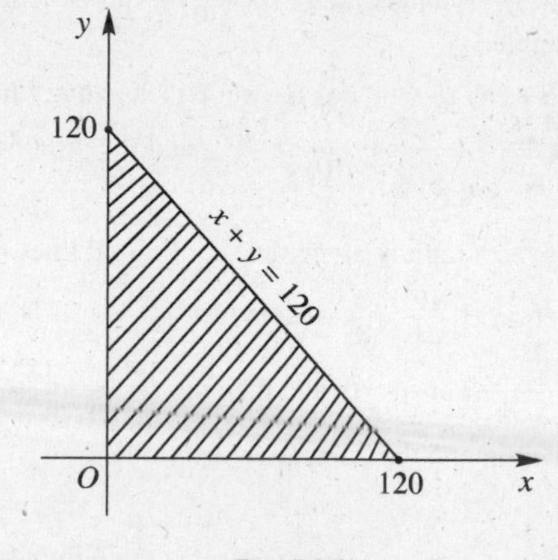

Fig. 52-2

Now,

$$\frac{\partial S}{\partial x} = y + (120 - x - y) - (x + y) = 120 - 2x - y$$

and

$$\frac{\partial S}{\partial y} = x + (120 - x - y) - (x + y) = 120 - x - 2y$$

Setting $\partial S/\partial x = \partial S/\partial y = 0$ yields $2x + y = 120$ and $x + 2y = 120$.

Simultaneous solution gives $x = 40$, $y = 40$, and $120 - (x + 4) = 40$ as the three parts, and $S = 3(40^2) = 4800$. So, if the absolute maximum occurs in the interior of the triangle, Theorem 52.1 tells us we have found it. It is still necessary to check the boundary of the triangle. When $y = 0$, $S = x(120 - x)$. Then $dS/dx = 120 - 2x$, and the critical number is $x = 60$. The corresponding maximum value of S is $60(60) = 3600$, which is < 4800. A similar result holds when $x = 0$. Finally, on the hypotenuse, where $y = 120 - x$, $S = x(120 - x)$ and we again obtain a maximum of 3600. Thus, the absolute maximum is 4800, and $x = y = z = 40$.

13. Find the point in the plane $2x - y + 2z = 16$ nearest the origin.

Let (x, y, z) be the required point; then the square of its distance from the origin is $D = x^2 + y^2 + z^2$. Since also $2x - y + 2z = 16$, we have $y = 2x + 2z - 16$ and $D = x^2 + (2x + 2z - 16)^2 + z^2$.

Then the conditions $\partial D/\partial x = 2x + 4(2x + 2z - 16) = 0$ and $\partial D/\partial z = 4(2x + 2z - 16) + 2z = 0$ are equivalent to $5x + 4z = 32$ and $4x + 5z = 32$, and $x = z = \frac{32}{9}$. Since it is known that a point for which D is a minimum exists, $(\frac{32}{9}, -\frac{16}{9}, \frac{32}{9})$ is that point.

14. Show that a rectangular parallelepiped of maximum volume V with constant surface area S is a cube.

Let the dimensions be x, y, and z. Then $V = xyz$ and $S = 2(xy + yz + zx)$.

The second relation may be solved for z and substituted in the first, to express V as a function of x and y. We prefer to avoid this step by simply treating z as a function of x and y. Then

$$\frac{\partial V}{\partial x} = yz + xy\frac{\partial z}{\partial x}, \qquad \frac{\partial V}{\partial y} = xz + xy\frac{\partial z}{\partial y}$$

$$\frac{\partial S}{\partial x} = 0 = 2\left(y + z + x\frac{\partial z}{\partial y} + y\frac{\partial z}{\partial x}\right), \qquad \frac{\partial S}{\partial y} = 0 = 2\left(x + z + x\frac{\partial z}{\partial y} + y\frac{\partial z}{\partial y}\right)$$

From the latter two equations, $\frac{\partial z}{\partial x} = -\frac{y+z}{x+y}$ and $\frac{\partial z}{\partial y} = -\frac{x+z}{x+y}$. Substituting in the first two yields the conditions $\frac{\partial V}{\partial x} = yz - \frac{xy(y+z)}{x+y} = 0$ and $\frac{\partial V}{\partial y} = xz - \frac{xy(x+z)}{x+y} = 0$, which reduce to $y^2(z - x) = 0$ and $x^2(z - y) = 0$. Thus $x = y = z$, as required.

15. Find the volume V of the largest rectangular parallelepiped that can be inscribed in the ellipsoid $\frac{x^2}{a^2} + \frac{y^2}{b^2} + \frac{z^2}{c^2} = 1$.

Let $P(x, y, z)$ be the vertex in the first octant. Then $V = 8xyz$. Consider z to be defined as a function of the independent variables x and y by the equation of the ellipsoid. The necessary conditions for a maximum are

$$\frac{\partial V}{\partial x} = 8\left(yz + xy\frac{\partial z}{\partial x}\right) = 0 \quad \text{and} \quad \frac{\partial V}{\partial y} = 8\left(xz + xy\frac{\partial z}{\partial y}\right) = 0 \tag{1}$$

From the equation of the ellipsoid, obtain $\frac{2x}{a^2} + \frac{2z}{c^2}\frac{\partial z}{\partial x} = 0$ and $\frac{2y}{b^2} + \frac{2z}{c^2}\frac{\partial z}{\partial y} = 0$. Eliminate $\partial z/\partial x$ and $\partial z/\partial y$ between these relations and (1) to obtain

$$\frac{\partial V}{\partial x} = 8\left(yz - \frac{c^2x^2y}{a^2z}\right) = 0 \quad \text{and} \quad \frac{\partial V}{\partial y} = 8\left(xz - \frac{c^2xy^2}{b^2z}\right) = 0$$

and, finally,

$$\frac{x^2}{a^2} = \frac{z^2}{c^2} = \frac{y^2}{b^2} \tag{2}$$

Combine (2) with the equation of the ellipsoid to get $x = a\sqrt{3}/3$, $y = b\sqrt{3}/3$, and $z = c\sqrt{3}/3$. Then $V = 8xyz = (8\sqrt{3}/9)abc$ cubic units.

SUPPLEMENTARY PROBLEMS

16. Find the directional derivatives of the given function at the given point in the indicated direction.

(a) $z = x^2 + xy + y^2$, (3, 1), $\theta = \frac{\pi}{3}$.
(b) $z = x^3 - 3xy + y^3$, (2, 1), $\theta = \tan^{-1}(\frac{2}{3})$.
(c) $z = y + x\cos xy$, (0, 0), $\theta = \frac{\pi}{3}$.
(d) $z = 2x^2 + 3xy - y^2$, (1, −1), toward (2, 1).

Ans. (a) $\frac{1}{2}(7+5\sqrt{3})$; (b) $21\sqrt{13}/13$; (c) $\frac{1}{2}(1+\sqrt{3})$; (d) $11\sqrt{5}/5$

17. Find the maximum directional derivative for each of the functions of Problem 16 at the given point.

Ans. (a) $\sqrt{74}$; (b) $3\sqrt{10}$; (c) $\sqrt{2}$; (d) $\sqrt{26}$

18. Show that the maximal directional derivative of $V = \ln\sqrt{x^2+y^2}$ of Problem 8 is constant along any circle $x^2 + y^2 = r^2$.

19. On a hill represented by $z = 8 - 4x^2 - 2y^2$, find (a) the direction of the steepest grade at (1, 1, 2) and (b) the direction of the contour line (the direction for which z = constant). Note that the directions are mutually perpendicular.

Ans. (a) $\tan^{-1}(\frac{1}{2})$, third quadrant; (b) $\tan^{-1}(-2)$

20. Show that the sum of the squares of the directional derivatives of $z = f(x, y)$ at any of its points is constant for any two mutually perpendicular directions and is equal to the square of the maximum directional derivative.

21. Given $z = f(x, y)$ and $w = g(x, y)$ such that $\partial z/\partial x = \partial w/\partial y$ and $\partial z/\partial y = -\partial w/\partial x$. If θ_1 and θ_2 are two mutually perpendicular directions, show that at any point $P(x, y)$, $\partial z/\partial s_1 = \partial w/\partial s_2$ and $\partial z/\partial s_2 = -\partial w/\partial s_1$.

22. Find the directional derivative of the given function at the given point in the indicated direction:

(a) xy^2z, (2, 1, 3), [1, −2, 2].
(b) $x^2 + y^2 + z^2$, (1, 1, 1), toward (2, 3, 4).
(c) $x^2 + y^2 - 2xz$, (1, 3, 2), along $x^2 + y^2 - 2xz = 6$, $3x^2 - y^2 + 3z = 0$ in the direction of increasing z.

Ans. (a) $-\frac{17}{3}$; (b) $6\sqrt{14}/7$; (c) 0

23. Examine each of the following functions for relative maximum and minimum values.

(a) $z = 2x + 4y - x^2 - y^2 - 3$ *Ans.* maximum = 2 when $x = 1$, $y = 2$
(b) $z = x^3 + y^3 - 3xy$ *Ans.* minimum = −1 when $x = 1$, $y = 1$
(c) $z = x^2 + 2xy + 2y^2$ *Ans.* minimum = 0 when $x = 0$, $y = 0$
(d) $z = (x - y)(1 - xy)$ *Ans.* neither maximum nor minimum
(e) $z = 2x^2 + y^2 + 6xy + 10x - 6y + 5$ *Ans.* neither maximum nor minimum
(f) $z = 3x - 3y - 2x^3 - xy^2 + 2x^2y + y^3$ *Ans.* minimum = $-\sqrt{6}$ when $x = -\sqrt{6}/6$, $y = \sqrt{6}/3$; maximum $\sqrt{6}$ when $x = \sqrt{6}/6$, $y = -\sqrt{6}/3$
(g) $z = xy(2x + 4y + 1)$ *Ans.* maximum $\frac{1}{216}$ when $x = -\frac{1}{6}$, $y = -\frac{1}{12}$

24. Find positive numbers x, y, z such that

(a) $x + y + z = 18$ and xyz is a maximum
(b) $xyz = 27$ and $x + y + z$ is a minimum
(c) $x + y + z = 20$ and xyz^2 is a maximum
(d) $x + y + z = 12$ and xy^2z^3 is a maximum

Ans. (a) $x = y = z = 6$; (b) $x = y = z = 3$; (c) $x = y = 5$, $z = 10$; (d) $x = 2$, $y = 4$, $z = 6$

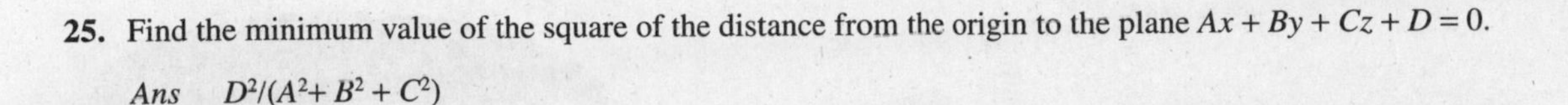

25. Find the minimum value of the square of the distance from the origin to the plane $Ax + By + Cz + D = 0$.

Ans $D^2/(A^2 + B^2 + C^2)$

26. (a) The surface area of a rectangular box without a top is to be 108 ft². Find the greatest possible volume.
(b) The volume of a rectangular box without a top is to be 500 ft³. Find the minimum surface area.

Ans. (a) 108 ft³; (b) 300 ft²

27. Find the point on $z = xy - 1$ nearest the origin.

Ans. $(0, 0, -1)$

28. Find the equation of the plane through (1, 1, 2) that cuts off the least volume in the first octant.

Ans. $2x + 2y + z = 6$

29. Determine the values of p and q so that the sum S of the squares of the vertical distances of the points (0, 2), (1, 3), and (2, 5) from the line $y = px + q$ is a minimum. (*Hint:* $S = (q - 2)^2 + (p + q - 3)^2 + (2p + q - 5)^2$.)

Ans. $p = \frac{3}{2}$; $q = \frac{11}{6}$

CHAPTER 53

Vector Differentiation and Integration

Vector Differentiation

Let

$$\mathbf{r} = \mathbf{i}f_1(t) + \mathbf{j}f_2(t) + \mathbf{k}f_3(t) = \mathbf{i}f_1 + \mathbf{j}f_2 + \mathbf{k}f_3$$
$$\mathbf{s} = \mathbf{i}g_1(t) + \mathbf{j}g_2(t) + \mathbf{k}g_3(t) = \mathbf{i}g_1 + \mathbf{j}g_2 + \mathbf{k}g_3$$
$$\mathbf{u} = \mathbf{i}h_1(t) + \mathbf{j}h_2(t) + \mathbf{k}h_3(t) = \mathbf{i}h_1 + \mathbf{j}h_2 + \mathbf{k}h_3$$

be vectors whose components are functions of a single scalar variable t having continuous first and second derivatives.

We can show, as in Chapter 39 for plane vectors, that

$$\frac{d}{dt}(\mathbf{r} \cdot \mathbf{s}) = \frac{d\mathbf{r}}{dt} \cdot s + \mathbf{r} \cdot \frac{d\mathbf{s}}{dt} \tag{53.1}$$

Also, from the properties of determinants whose entries are functions of a single variable, we have

$$\frac{d}{dt}(\mathbf{r} \times \mathbf{s}) = \frac{d}{dt}\begin{vmatrix} \mathbf{i} & \mathbf{j} & \mathbf{k} \\ f_1 & f_2 & f_3 \\ g_1 & g_2 & g_3 \end{vmatrix} = \begin{vmatrix} \mathbf{i} & \mathbf{j} & \mathbf{k} \\ f_1' & f_2' & f_3' \\ g_1 & g_2 & g_3 \end{vmatrix} + \begin{vmatrix} \mathbf{i} & \mathbf{j} & \mathbf{k} \\ f_1 & f_2 & f_3 \\ g_1' & g_2' & g_3' \end{vmatrix}$$

$$= \frac{d\mathbf{r}}{dt} \times \mathbf{s} + \mathbf{r} \times \frac{d\mathbf{s}}{dt} \tag{53.2}$$

and $$\frac{d}{dt}[\mathbf{r} \cdot (\mathbf{s} \times \mathbf{u})] = \frac{d\mathbf{r}}{dt} \cdot (\mathbf{s} \times \mathbf{u}) + \mathbf{r} \cdot \left(\frac{d\mathbf{s}}{dt} \times \mathbf{u}\right) + \mathbf{r} \cdot \left(\mathbf{s} \times \frac{d\mathbf{u}}{dt}\right) \tag{53.3}$$

These formulas may also be established by expanding the products before differentiating.

From (53.2) follows

$$\frac{d}{dt}[\mathbf{r} \times (\mathbf{s} \times \mathbf{u})] = \frac{d\mathbf{r}}{dt} \times (\mathbf{s} \times \mathbf{u}) + \mathbf{r} \times \frac{d}{dt}(\mathbf{s} \times \mathbf{u})$$
$$= \frac{d\mathbf{r}}{dt} \times (\mathbf{s} \times \mathbf{u}) + \mathbf{r} \times \left(\frac{d\mathbf{s}}{dt} \times \mathbf{u}\right) + \mathbf{r} \times \left(\mathbf{s} \times \frac{d\mathbf{u}}{dt}\right) \tag{53.4}$$

Space Curves

Consider the space curve

$$x = f(t), \qquad y = g(t), \qquad z = h(t) \tag{53.5}$$

where $f(t)$, $g(t)$, and $h(t)$ have continuous first and second derivatives. Let the position vector of a general variable point $P(x, y, z)$ of the curve be given by

$$\mathbf{r} = x\mathbf{i} + y\mathbf{j} + z\mathbf{k}$$

As in Chapter 39, $\mathbf{t} = d\mathbf{r}/ds$ is the unit tangent vector to the curve. If $\mathbf{R}$ is the position vector of a point (X, Y, Z) on the tangent line at P, the vector equation of this line is (see Chapter 50)

$$\mathbf{R} - \mathbf{r} = k\mathbf{t} \quad \text{for } k \text{ a scalar variable} \tag{53.6}$$

and the equations in rectangular coordinates are

$$\frac{X - x}{dx/ds} = \frac{Y - y}{dy/ds} = \frac{Z - z}{dz/ds}$$

where $\left[\frac{dx}{ds}, \frac{dy}{ds}, \frac{dz}{ds}\right]$ is a set of direction cosines of the line. In the corresponding equation (51.2), a set of direction numbers $\left[\frac{dx}{dt}, \frac{dy}{dt}, \frac{dz}{dt}\right]$ was used.

The vector equation of the normal plane to the curve at P is given by

$$(\mathbf{R} - \mathbf{r}) \cdot \mathbf{t} = 0 \tag{53.7}$$

where $\mathbf{R}$ is the position vector of a general point of the plane.

Again, as in Chapter 39, $d\mathbf{t}/ds$ is a vector perpendicular to $\mathbf{t}$. If $\mathbf{n}$ is a unit vector having the direction of $d\mathbf{t}/ds$, then

$$\frac{d\mathbf{t}}{ds} = |K|\mathbf{n}$$

where $|K|$ is the magnitude of the curvature at P. The unit vector

$$\mathbf{n} = \frac{1}{|K|}\frac{d\mathbf{t}}{ds} \tag{53.8}$$

is called the *principal normal* to the curve at P.

The unit vector $\mathbf{b}$ at P, defined by

$$\mathbf{b} = \mathbf{t} \times \mathbf{n} \tag{53.9}$$

is called the *binormal* at P. The three vectors $\mathbf{t}$, $\mathbf{n}$, $\mathbf{b}$ form at P a right-handed triad of mutually orthogonal vectors. (See Problems 1 and 2.)

At a general point P of a space curve (Fig. 53-1), the vectors $\mathbf{t}$, $\mathbf{n}$, $\mathbf{b}$ determine three mutually perpendicular planes:

1. The *osculating plane*, containing $\mathbf{t}$ and $\mathbf{n}$, having the equation $(\mathbf{R} - \mathbf{r}) \cdot \mathbf{b} = 0$
2. The *normal plane*, containing $\mathbf{n}$ and $\mathbf{b}$, having the equation $(\mathbf{R} - \mathbf{r}) \cdot \mathbf{t} = 0$
3. The *rectifying plane*, containing $\mathbf{t}$ and $\mathbf{b}$, having the equation $(\mathbf{R} - \mathbf{r}) \cdot \mathbf{n} = 0$

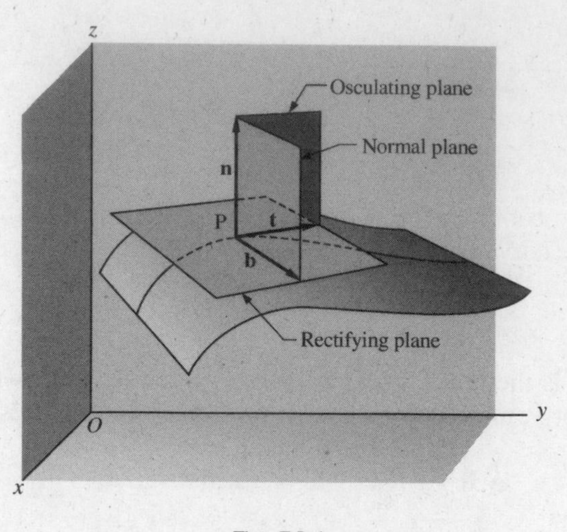

Fig. 53-1

In each equation, **R** is the position vector of a general point in the particular plane.

Surfaces

Let $F(x, y, z) = 0$ be the equation of a surface. (See Chapter 51.) A parametric representation results when x, y, and z are written as functions of two independent variables or parameters u and v, for example, as

$$x = f_1(u, v), \qquad y = f_2(u, v), \qquad z = f_3(u, v) \tag{53.10}$$

When u is replaced with u_0, a constant, (53.10) becomes

$$x = f_1(u_0, v), \qquad y = f_2(u_0, v), \qquad z = f_3(u_0, v) \tag{53.11}$$

the equation of a space curve (u curve) lying on the surface. Similarly, when v is replaced with v_0, a constant, (53.10) becomes

$$x = f_1(u, v_0), \qquad y = f_2(u, v_0), \qquad z = f_3(u, v_0) \tag{53.12}$$

the equation of another space curve (v curve) on the surface. The two curves intersect in a point of the surface obtained by setting $u = u_0$ and $v = v_0$ simultaneously in (53.10).

The position vector of a general point P on the surface is given by

$$\mathbf{r} = x\mathbf{i} + y\mathbf{j} + z\mathbf{k} = \mathbf{i}f_1(u, v) + \mathbf{j}f_2(u, v) + \mathbf{k}f_3(u, v) \tag{53.13}$$

Suppose (53.11) and (53.12) are the u and v curves through P. Then, at P,

$$\frac{\partial \mathbf{r}}{\partial v} = \mathbf{i}\frac{\partial}{\partial v} f_1(u_0, v) + \mathbf{j}\frac{\partial}{\partial v} f_2(u_0, v) + \mathbf{k}\frac{\partial}{\partial v} f_3(u_0, v)$$

is a vector tangent to the u curve, and

$$\frac{\partial \mathbf{r}}{\partial u} = \mathbf{i}\frac{\partial}{\partial u} f_1(u, v_0) + \mathbf{j}\frac{\partial}{\partial u} f_2(u, v_0) + \mathbf{k}\frac{\partial}{\partial u} f_3(u, v_0)$$

is a vector tangent to the v curve. The two tangents determine a plane that is the tangent plane to the surface at P (Fig. 53-2). Clearly, a normal to this plane is given by $\frac{\partial \mathbf{r}}{\partial u} \times \frac{\partial \mathbf{r}}{\partial v}$. The *unit normal* to the surface at P is defined by

$$\mathbf{n} = \frac{\frac{\partial \mathbf{r}}{\partial u} \times \frac{\partial \mathbf{r}}{\partial v}}{\left|\frac{\partial \mathbf{r}}{\partial u} \times \frac{\partial \mathbf{r}}{\partial v}\right|} \tag{53.14}$$

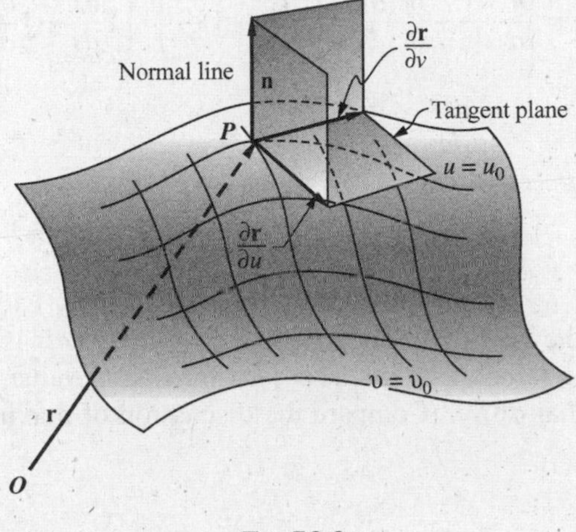

Fig. 53-2

If $\mathbf{R}$ is the position vector of a general point on the normal to the surface at P, its vector equation is

$$(\mathbf{R} - \mathbf{r}) = k\left(\frac{\partial \mathbf{r}}{\partial u} \times \frac{\partial \mathbf{r}}{\partial v}\right) \tag{53.15}$$

If $\mathbf{R}$ is the position vector of a general point on the tangent plane to the surface at P, its vector equation is

$$(\mathbf{R} - \mathbf{r}) \cdot \left(\frac{\partial \mathbf{r}}{\partial u} \times \frac{\partial \mathbf{r}}{\partial v}\right) = 0 \tag{53.16}$$

(See Problem 3.)

The Operation ∇

In Chapter 52, the directional derivative of $z = f(x, y)$ at an arbitrary point (x, y) and in a direction making an angle θ with the positive x axis is given as

$$\frac{dz}{ds} = \frac{\partial f}{\partial x}\cos\theta + \frac{\partial f}{\partial y}\sin\theta$$

Let us write

$$\frac{\partial f}{\partial x}\cos\theta + \frac{\partial f}{\partial y}\sin\theta = \left(\mathbf{i}\frac{\partial f}{\partial x} + \mathbf{j}\frac{\partial f}{\partial y}\right) \cdot (\mathbf{i}\cos\theta + \mathbf{j}\sin\theta) \tag{53.17}$$

Now $\mathbf{a} = \mathbf{i}\cos\theta + \mathbf{j}\sin\theta$ is a unit vector whose direction makes the angle θ with the positive x axis. The other factor on the right of (53.17), when written as $\left(\mathbf{i}\frac{\partial}{\partial x} + \mathbf{j}\frac{\partial}{\partial y}\right)f$, suggests the definition of a vector differential operator ∇ (del), defined by

$$\nabla = \mathbf{i}\frac{\partial}{\partial x} + \mathbf{j}\frac{\partial}{\partial y} \tag{53.18}$$

In vector analysis, $\nabla f = \mathbf{i}\frac{\partial f}{\partial x} + \mathbf{j}\frac{\partial f}{\partial y}$ is called the *gradient* of f or *grad* f. From (53.17), we see that the component of ∇f in the direction of a *unit vector* $\mathbf{a}$ is the directional derivative of f in the direction of $\mathbf{a}$.

Let $\mathbf{r} = x\mathbf{i} + y\mathbf{j}$ be the position vector to $P(x, y)$. Since

$$\frac{df}{ds} = \frac{\partial f}{\partial x}\frac{dx}{ds} + \frac{\partial f}{\partial y}\frac{dy}{ds} = \left(\mathbf{i}\frac{\partial f}{\partial x} + \mathbf{j}\frac{\partial f}{\partial y}\right) \cdot \left(\mathbf{i}\frac{dx}{ds} + \mathbf{j}\frac{\partial y}{\partial s}\right)$$

$$= \nabla f \cdot \frac{d\mathbf{r}}{ds}$$

and

$$\left|\frac{df}{ds}\right| = |\nabla f| \cos \phi$$

where ϕ is the angle between the vectors ∇f and $d\mathbf{r}/ds$, we see that df/ds is maximal when $\cos \phi = 1$, that is, when ∇f and $d\mathbf{r}/ds$ have the same direction. Thus, the maximum value of the directional derivative at P is $|\nabla f|$; and its direction is that of ∇f. (Compare the discussion of maximum directional derivatives in Chapter 52.) (See Problem 4.)

For $w = F(x, y, z)$, we define

$$\nabla F = \mathbf{i}\frac{\partial F}{\partial x} + \mathbf{j}\frac{\partial F}{\partial y} + \mathbf{k}\frac{\partial F}{\partial z}$$

and the directional derivative of $F(x, y, z)$ at an arbitrary point $P(x, y, z)$ in the direction $\mathbf{a} = \mathbf{a}_1\mathbf{i} + a_2\mathbf{j} + a_2\mathbf{k}$ is

$$\frac{dF}{ds} = \nabla F \cdot \mathbf{a} \tag{53.19}$$

As in the case of functions of two variables, $|\nabla F|$ is the maximum value of the directional derivative of $F(x, y, z)$ at $P(x, y, z)$, and its direction is that of ∇F. (See Problem 5.)

Consider now the surface $F(x, y, z) = 0$. The equation of the tangent plane to the surface at one of its points $P_0(x_0, y_0, z_0)$ is given by

$$(x - x_0)\frac{\partial F}{\partial x} + (y - y_0)\frac{\partial F}{\partial y} + (z - z_0)\frac{\partial F}{\partial z}$$

$$= [(x - x_0)\mathbf{i} + (y - y_0)\mathbf{j} + (z - z_0)\mathbf{k}] \cdot \left[\mathbf{i}\frac{\partial F}{\partial x} + \mathbf{j}\frac{\partial F}{\partial y} + \mathbf{k}\frac{\partial F}{\partial z}\right] = 0 \tag{53.20}$$

with the understanding that the partial derivatives are evaluated at P_0. The first factor is an arbitrary vector through P_0 in the tangent plane; hence the second factor ∇F, evaluated at P_0, is normal to the tangent plane, that is, is normal to the surface at P_0. (See Problems 6 and 7.)

Divergence and Curl

The *divergence* of a vector function $\mathbf{F} = \mathbf{i}f_1(x, y, z) + \mathbf{j}f_2(x, y, z) + \mathbf{k}f_3(x, y, z)$, sometimes called *del dot* $\mathbf{F}$, is defined by

$$\text{div } \mathbf{F} = \nabla \cdot \mathbf{F} = \frac{\partial}{\partial x}f_1 + \frac{\partial}{\partial y}f_2 + \frac{\partial}{\partial z}f_3 \tag{53.21}$$

The *curl* of the vector function **F**, or *del cross* **F**, is defined by

$$\text{curl } \mathbf{F} = \nabla \times \mathbf{F} = \begin{vmatrix} \mathbf{i} & \mathbf{j} & \mathbf{k} \\ \dfrac{\partial}{\partial x} & \dfrac{\partial}{\partial y} & \dfrac{\partial}{\partial z} \\ f_1 & f_2 & f_3 \end{vmatrix}$$

$$= \left(\frac{\partial}{\partial y} f_3 - \frac{\partial}{\partial z} f_2\right)\mathbf{i} + \left(\frac{\partial}{\partial z} f_1 - \frac{\partial}{\partial x} f_3\right)\mathbf{j} + \left(\frac{\partial}{\partial x} f_2 - \frac{\partial}{\partial y} f_1\right)\mathbf{k} \tag{53.22}$$

(See Problem 8.)

Integration

Our discussion of integration here will be limited to ordinary integration of vectors and to so-called "line integrals." As an example of the former, let

$$\mathbf{F}(u) = \mathbf{i} \cos u + \mathbf{j} \sin u + au\mathbf{k}$$

be a vector depending upon the scalar variable u. Then

$$\mathbf{F}'(u) = -\mathbf{i} \sin u + \mathbf{j} \cos u + a\mathbf{k}$$

and

$$\int \mathbf{F}'(u)du = \int (-\mathbf{i}\sin u + \mathbf{j}\cos u + a\mathbf{k})\, du$$

$$= \mathbf{i}\int -\sin u\, du + \mathbf{j}\int \cos u\, du + \mathbf{k}\int a\, du$$

$$= \mathbf{i}\cos u + \mathbf{j}\sin u + au\mathbf{k} + \mathbf{c}$$

$$= \mathbf{F}(u) + \mathbf{c}$$

where **c** is an arbitrary constant vector independent of u. Moreover,

$$\int_{u=a}^{u=b} \mathbf{F}'(u)\, du = [\mathbf{F}(u) + c]_{u=a}^{u=b} = \mathbf{F}(b) - \mathbf{F}(a)$$

(See Problems 9 and 10.)

Line Integrals

Consider two points P_0 and P_1 in space, joined by an arc C. The arc may be a segment of a straight line or a portion of a space curve $x = g_1(t)$, $y = g_2(t)$, $z = g_3(t)$, or it may consist of several subarcs of curves. In any case, C is assumed to be continuous at each of its points and not to intersect itself. Consider further a vector function

$$\mathbf{F} = \mathbf{F}(x, y, z) = \mathbf{i}f_1(x, y, z) + \mathbf{j}f_2(x, y, z) + \mathbf{k}f_3(x, y, z)$$

which at every point in a region about C and, in particular, at every point of C, defines a vector of known magnitude and direction. Denote by

$$\mathbf{r} = x\mathbf{i} + y\mathbf{j} + z\mathbf{k} \tag{53.23}$$

the position vector of $P(x, y, z)$ on C. The integral

$$\int_{\substack{P_0 \\ C}}^{P_1} \left(\mathbf{F} \cdot \frac{d\mathbf{r}}{ds}\right) ds = \int_{\substack{P_0 \\ C}}^{P_1} \mathbf{F} \cdot d\mathbf{r} \tag{53.24}$$

is called a line integral, that is, an integral along a given path C.

As an example, let $\mathbf{F}$ denote a force. The work done by it in moving a particle over $d\mathbf{r}$ is given by (see Problem 16 of Chapter 39)

$$|\mathbf{F}||d\mathbf{r}|\cos\theta = \mathbf{F} \cdot d\mathbf{r}$$

and the work done in moving the particle from P_0 to P_1 along the arc C is given by

$$\int_{\substack{P_0 \\ C}}^{P_1} \mathbf{F} \cdot d\mathbf{r}$$

From (53.23),

$$d\mathbf{r} = \mathbf{i}\, dx + \mathbf{j}\, dy + \mathbf{k}\, dz$$

and (53.24) becomes

$$\int_{\substack{P_0 \\ C}}^{P_1} \mathbf{F} \cdot d\mathbf{r} = \int_{\substack{P_0 \\ C}}^{P_1} (f_1\, dx + f_2\, dy + f_3\, dz) \tag{53.25}$$

(See Problem 11.)

SOLVED PROBLEMS

1. A particle moves along the curve $x = 4\cos t$, $y = 4\sin t$, $z = 6t$. Find the magnitude of its velocity and acceleration at times $t = 0$ and $t = \frac{1}{2}\pi$.

Let $P(x, y, z)$ be a point on the curve, and

$$\mathbf{r} = x\mathbf{i} + y\mathbf{j} + z\mathbf{k} = 4\mathbf{i}\cos t + 4\mathbf{j}\sin t + 6\mathbf{k}t$$

be its position vector. Then

$$\mathbf{v} = \frac{d\mathbf{r}}{dt} = -4\mathbf{i}\sin t + 4\mathbf{j}\cos t + 6\mathbf{k} \quad \text{and} \quad \mathbf{a} = \frac{d^2\mathbf{r}}{dt^2} = -4\mathbf{i}\cos t - 4\mathbf{j}\sin t$$

At $t = 0$: $\mathbf{v} = 4\mathbf{j} + 6\mathbf{k}$ $\quad |\mathbf{v}| = \sqrt{16+36} = 2\sqrt{13}$

$\mathbf{a} = -4\mathbf{i}$ $\quad |\mathbf{a}| = 4$

At $t = \frac{1}{2}\pi$: $\mathbf{v} = -4\mathbf{i} + 6\mathbf{k}$ $\quad |\mathbf{v}| = \sqrt{16+36} = 2\sqrt{13}$

$\mathbf{a} = -4\mathbf{j}$ $\quad |\mathbf{a}| = 4$

2. At the point (1, 1, 1) or $t = 1$ of the space curve $x = t,\ y = t^2,\ z = t^3$, find:

(a) The equations of the tangent line and normal plane.
(b) The unit tangent, principal normal, and binormal.
(c) The equations of the principal normal and binormal.

We have

$$\mathbf{r} = t\mathbf{i} + t^2\mathbf{j} + t^3\mathbf{k}$$
$$\frac{d\mathbf{r}}{dt} = \mathbf{i} + 2t\mathbf{j} + 3t^2\mathbf{k}$$
$$\frac{ds}{dt} = \left|\frac{d\mathbf{r}}{dt}\right| = \sqrt{1+4t^2+9t^4}$$
$$\mathbf{t} = \frac{d\mathbf{r}}{ds} = \frac{d\mathbf{r}}{dt}\frac{dt}{ds} = \frac{\mathbf{i}+2t\mathbf{j}+3t^2\mathbf{k}}{\sqrt{1+4t^2+9t^4}}$$

At $t = 1$, $\mathbf{r} = \mathbf{i} + \mathbf{j} + \mathbf{k}$ and $\mathbf{t} = \frac{1}{\sqrt{14}}(\mathbf{i} + 2\mathbf{j} + 3\mathbf{k})$.

(a) If $\mathbf{R}$ is the position vector of a general point (X, Y, Z) on the tangent line, its vector equation is $\mathbf{R} - \mathbf{r} = k\mathbf{t}$ or

$$(X-1)\mathbf{i} + (Y-1)\mathbf{j} + (Z-1)\mathbf{k} = \frac{k}{\sqrt{14}}(\mathbf{i} + 2\mathbf{j} + 3\mathbf{k})$$

and its rectangular equations are

$$\frac{X-1}{1} = \frac{Y-1}{2} = \frac{Z-1}{3}$$

If $\mathbf{R}$ is the position vector of a general point (X, Y, Z) on the normal plane, its vector equation is $(\mathbf{R} - \mathbf{r}) \cdot \mathbf{t} = 0$ or

$$[(X-1)\mathbf{i} + (Y-1)\mathbf{j} + (Z-1)\mathbf{k}] \cdot \frac{1}{\sqrt{14}}(\mathbf{i} + 2\mathbf{j} + 3\mathbf{k}) = 0$$

and its rectangular equation is

$$(X-1) + 2(Y-1) + 3(Z-1) = X + 2Y + 3Z - 6 = 0$$

(See Problem 2(a) of Chapter 51.)

(b) $$\frac{d\mathbf{t}}{ds} = \frac{d\mathbf{t}}{dt}\frac{dt}{ds} = \frac{(-4t-18t^3)\mathbf{i} + (2-18t^4)\mathbf{j} + (6t+12t^3)\mathbf{k}}{(1+4t^2+9t^4)^2}$$

At $t = 1$, $$\frac{d\mathbf{t}}{ds} = \frac{-11\mathbf{i}-8\mathbf{j}+9\mathbf{k}}{98} \quad \text{and} \quad \left|\frac{d\mathbf{t}}{ds}\right| = \frac{1}{7}\sqrt{\frac{19}{14}} = |K|.$$

Then $$\mathbf{n} = \frac{1}{|K|}\frac{d\mathbf{t}}{ds} = \frac{-11\mathbf{i}-8\mathbf{j}+9\mathbf{k}}{\sqrt{266}}$$

and $$\mathbf{b} = \mathbf{t} \times \mathbf{n} = \frac{1}{\sqrt{14}\sqrt{266}}\begin{vmatrix} \mathbf{i} & \mathbf{j} & \mathbf{k} \\ 1 & 2 & 3 \\ -11 & -8 & 9 \end{vmatrix} = \frac{1}{\sqrt{19}}(3\mathbf{i} - 3\mathbf{j} + \mathbf{k})$$

(c) If $\mathbf{R}$ is the position vector of a general point (X, Y, Z) on the principal normal, its vector equation is $\mathbf{R} - \mathbf{r} = k\mathbf{n}$ or

$$(X-1)\mathbf{i} + (Y-1)\mathbf{j} + (Z-1)\mathbf{k} = k\frac{-11\mathbf{i}-8\mathbf{j}+9\mathbf{k}}{\sqrt{266}}$$

and the equations in rectangular coordinates are

$$\frac{X-1}{-11}=\frac{Y-1}{-8}=\frac{Z-1}{9}$$

If **R** is the position vector of a general point (X, Y, Z) on the binormal, its vector equation is $\mathbf{R}-\mathbf{r}=k\mathbf{b}$ or

$$(X-1)\mathbf{i}+(Y-1)\mathbf{j}+(Z-1)\mathbf{k}=k\frac{3\mathbf{i}-3\mathbf{j}+\mathbf{k}}{\sqrt{19}}$$

and the equations in rectangular coordinates are

$$\frac{X-1}{3}=\frac{Y-1}{-3}=\frac{Z-1}{1}$$

3. Find the equations of the tangent plane and normal line to the surface $x=2(u+v)$, $y=3(u-v)$, $z=uv$ at the point $P(u=2, v=1)$.

Here

$$\mathbf{r}=2(u+v)\mathbf{i}+3(u-v)\mathbf{j}+uv\mathbf{k}, \qquad \frac{\partial \mathbf{r}}{\partial u}=2\mathbf{i}+3\mathbf{j}+v\mathbf{k}, \qquad \frac{\partial \mathbf{r}}{\partial v}=2\mathbf{i}-3\mathbf{j}+u\mathbf{k}$$

and at the point P,

$$\mathbf{r}=6\mathbf{i}+3\mathbf{j}+2\mathbf{k}, \qquad \frac{\partial \mathbf{r}}{\partial u}=2\mathbf{i}+3\mathbf{j}+\mathbf{k}, \qquad \frac{\partial \mathbf{r}}{\partial v}=2\mathbf{i}-3\mathbf{j}+2\mathbf{k}$$

and
$$\frac{\partial \mathbf{r}}{\partial u}\times\frac{\partial \mathbf{r}}{\partial v}=9\mathbf{i}-2\mathbf{j}-12\mathbf{k}$$

The vector and rectangular equations of the normal line are

$$\mathbf{R}-\mathbf{r}=k\frac{\partial \mathbf{r}}{\partial u}\times\frac{\partial \mathbf{r}}{\partial v}$$

or
$$(X-6)\mathbf{i}+(Y-3)\mathbf{j}+(Z-2)\mathbf{k}=k(9\mathbf{i}-2\mathbf{j}-12\mathbf{k})$$

and
$$\frac{X-6}{9}+\frac{Y-3}{-2}=\frac{Z-2}{-12}$$

The vector and rectangular equations of the tangent plane are

$$(\mathbf{R}-\mathbf{r})\cdot\left(\frac{\partial \mathbf{r}}{\partial u}\times\frac{\partial \mathbf{r}}{\partial v}\right)=0$$

or
$$[(X-6)\mathbf{i}+(Y-3)\mathbf{j}+(Z-2)\mathbf{k}]\cdot[9\mathbf{i}-2\mathbf{j}-12\mathbf{k}]=0$$

and
$$9X-2Y-12Z-24=0$$

4. (a) Find the directional derivative of $f(x, y)=x^2-6y^2$ at the point $(7, 2)$ in the direction $\theta=\frac{1}{4}\pi$.
(b) Find the maximum value of the directional derivative at $(7, 2)$.

(a) $\nabla f=\left(\mathbf{i}\frac{\partial}{\partial x}+\mathbf{j}\frac{\partial}{\partial y}\right)(x^2-6y^2)=\mathbf{i}\frac{\partial}{\partial x}(x^2-6y^2)+\mathbf{j}\frac{\partial}{\partial y}(x^2-6y^2)=2x\mathbf{i}-12y\mathbf{j}$

and

$$\mathbf{a}=\mathbf{i}\cos\theta+\mathbf{j}\sin\theta=\frac{1}{\sqrt{2}}\mathbf{i}+\frac{1}{\sqrt{2}}\mathbf{j}$$

At (7, 2), $\nabla f = 14\mathbf{i} - 24\mathbf{j}$, and

$$\nabla f \cdot \mathbf{a} = (14\mathbf{i} - 24\mathbf{j}) \cdot \left(\frac{1}{\sqrt{2}}\mathbf{i} + \frac{1}{\sqrt{2}}\mathbf{j}\right) = 7\sqrt{2} - 12\sqrt{2} = -5\sqrt{2}$$

is the directional derivative.

(b) At (7, 2), with $\nabla f = 14\mathbf{i} - 24\mathbf{j}$, $|\nabla f| = \sqrt{14^2 + 24^2} = 2\sqrt{193}$ is the maximum directional derivative. Since

$$\frac{\nabla f}{|\nabla f|} = \frac{7}{\sqrt{193}}\mathbf{i} - \frac{12}{\sqrt{193}}\mathbf{j} = \mathbf{i}\cos\theta + \mathbf{j}\sin\theta$$

the direction is $\theta = 300°15'$. (See Problems 2 and 6 of Chapter 52.)

5. (a) Find the directional derivative of $F(x, y, z) = x^2 - 2y^2 + 4z^2$ at $P(1, 1, -1)$ in the direction $\mathbf{a} = 2\mathbf{i} + \mathbf{j} - \mathbf{k}$.
(b) Find the maximum value of the directional derivative at P.

Here

$$\nabla F = \left(\mathbf{i}\frac{\partial}{\partial x} + \mathbf{j}\frac{\partial}{\partial y} + \mathbf{k}\frac{\partial}{\partial z}\right)(x^2 - 2y^2 + 4z^2) = 2x\mathbf{i} - 4y\mathbf{j} + 8z\mathbf{k}$$

and at (1, 1, −1), $\nabla F = 2\mathbf{i} - 4\mathbf{j} - 8\mathbf{k}$.

(a) $\nabla F \cdot \mathbf{a} = (2\mathbf{i} - 4\mathbf{j} - 8\mathbf{k}) \cdot (2\mathbf{i} + \mathbf{j} - \mathbf{k}) = 8$
(b) At P, $|\nabla F| = \sqrt{84} = 2\sqrt{21}$. The direction is $\mathbf{a} = 2\mathbf{i} - 4\mathbf{j} - 8\mathbf{k}$.

6. Given the surface $F(x, y, z) = x^3 + 3xyz + 2y^3 - z^3 - 5 = 0$ and one of its points $P_0(1, 1, 1)$, find (a) a unit normal to the surface at P_0; (b) the equations of the normal line at P_0, and (c) the equation of the tangent plane at P_0.

Here

$$\nabla F = (3x^2 + 3yz)\mathbf{i} + (3xz + 6y^2)\mathbf{j} + (3xy - 3z^2)\mathbf{k}$$

and at $P_0(1, 1, 1)$, $\nabla F = 6\mathbf{i} + 9\mathbf{j}$.

(a) $\dfrac{\nabla F}{|\nabla F|} = \dfrac{2}{\sqrt{13}}\mathbf{i} + \dfrac{3}{\sqrt{13}}\mathbf{j}$ is a unit normal at P_0; the other $-\dfrac{2}{\sqrt{13}}\mathbf{i} - \dfrac{3}{\sqrt{13}}\mathbf{j}$.
(b) The equations of the normal line are $\dfrac{X-1}{2} = \dfrac{Y-1}{3}$, $Z = 1$.
(c) The equation of the tangent plane is $2(X - 1) + 3(Y - 1) = 2X + 3Y - 5 = 0$.

7. Find the angle of intersection of the surfaces

$$F_1 = x^2 + y^2 + z^2 - 9 = 0 \quad \text{and} \quad F_2 = x^2 + 2y^2 - z - 8 = 0$$

at the point (2, 1, −2).

We have

$$\nabla F_1 = \nabla(x^2 + y^2 + z^2 - 9) = 2x\mathbf{i} + 2y\mathbf{j} + 2z\mathbf{k}$$

and

$$\nabla F_2 = \nabla(x^2 + 2y^2 - z - 8) = 2x\mathbf{i} + 4y\mathbf{j} - \mathbf{k}$$

At (2, 1, −2), $\nabla F_1 = 4\mathbf{i} + 2\mathbf{j} - 4\mathbf{k}$ and $\nabla F_2 = 4\mathbf{i} + 4\mathbf{j} - \mathbf{k}$.

Now $\nabla F_1 \cdot \nabla F_2 = |\nabla F_1||\nabla F_2| \cos\theta$, where θ is the required angle. Thus,

$$(4\mathbf{i} + 2\mathbf{j} - 4\mathbf{k}) \cdot (4\mathbf{i} + 4\mathbf{j} - \mathbf{k}) = |4\mathbf{i} + 2\mathbf{j} - 4\mathbf{k}||4\mathbf{i} + 4\mathbf{j} - \mathbf{k}| \cos\theta$$

from which $\cos\theta = \frac{14}{99}\sqrt{33} = 0.81236$, and $\theta = 35°40'$.

8. When $\mathbf{B} = xy^2\mathbf{i} + 2x^2yz\mathbf{j} - 3yz^2\mathbf{k}$, find (a) div **B** and (b) curl **B**.

(a)
$$\text{div}\,\mathbf{B} = \nabla \cdot \mathbf{B} = \left(\frac{\partial}{\partial x}\mathbf{i} + \frac{\partial}{\partial y}\mathbf{j} + \frac{\partial}{\partial z}\mathbf{k}\right) \cdot (xy^2\mathbf{i} + 2x^2yz\mathbf{j} - 3yz^2\mathbf{k})$$
$$= \frac{\partial}{\partial x}(xy^2) + \frac{\partial}{\partial y}(2x^2yz) + \frac{\partial}{\partial z}(-3yz^2)$$
$$= y^2 + 2x^2z - 6yz$$

(b)
$$\text{curl}\,\mathbf{B} = \nabla \times \mathbf{B} = \begin{vmatrix} \mathbf{i} & \mathbf{j} & \mathbf{k} \\ \frac{\partial}{\partial x} & \frac{\partial}{\partial y} & \frac{\partial}{\partial z} \\ xy^2 & 2x^2yz & -3yz^2 \end{vmatrix}$$
$$= \left[\frac{\partial}{\partial y}(-3yz^2) - \frac{\partial}{\partial z}(2x^2yz)\right]\mathbf{i} + \left[\frac{\partial}{\partial z}(xy^2) - \frac{\partial}{\partial x}(-3yz^2)\right]\mathbf{j}$$
$$+ \left[\frac{\partial}{\partial x}(2x^2yz) - \frac{\partial}{\partial y}(xy^2)\right]\mathbf{k}$$
$$= -(3z^2 + 2x^2y)\mathbf{i} + (4xyz - 2xy)\mathbf{k}$$

9. Given $\mathbf{F}(u) = u\mathbf{i} + (u^2 - 2u)\mathbf{j} + (3u^2 + u^3)\mathbf{k}$, find (a) $\int \mathbf{F}(u)\,du$ and (b) $\int_0^1 \mathbf{F}(u)\,du$.

(a)
$$\int \mathbf{F}(u)\,du = \int [u\mathbf{i} + (u^2 - 2u)\mathbf{j} + (3\mathbf{u}^2 + u^3)\mathbf{k}]\,du$$
$$= \mathbf{i}\int u\,du + \mathbf{j}\int (u^2 - 2u)\,du + \mathbf{k}\int (3u^2 + u^3)\,du$$
$$= \frac{u^2}{2}\mathbf{i} + \left(\frac{u^3}{3} - u^2\right)\mathbf{j} + \left(u^3 + \frac{u^4}{4}\right)\mathbf{k} + \mathbf{c}$$

where $\mathbf{c} = c_1\mathbf{i} + c_2\mathbf{j} + c_3\mathbf{k}$ with c_1, c_2, c_3 arbitrary scalars.

(b)
$$\int_0^1 \mathbf{F}(u)\,du = \left[\frac{u^2}{2}\mathbf{i} + \left(\frac{u^3}{3} - u^2\right)\mathbf{j} + \left(u^3 + \frac{u^4}{4}\right)\mathbf{k}\right]_0^1 = \frac{1}{2}\mathbf{i} - \frac{2}{3}\mathbf{j} + \frac{5}{4}\mathbf{k}$$

10. The acceleration of a particle at any time $t \geq 0$ is given by $\mathbf{a} = d\mathbf{v}/dt = e^t\mathbf{i} + e^{2t}\mathbf{j} + \mathbf{k}$. If at $t = 0$, the displacement is $\mathbf{r} = 0$ and the velocity is $\mathbf{v} = \mathbf{i} + \mathbf{j}$, find **r** and **v** at any time *t*.

Here

$$\mathbf{v} = \int \mathbf{a}\,dt = \mathbf{i}\int e^t\,dt + \mathbf{j}\int e^{2t}\,dt + \mathbf{k}\int dt$$
$$= e^t\mathbf{i} + \tfrac{1}{2}e^{2t}\mathbf{j} + t\mathbf{k} + \mathrm{c}_1$$

At $t = 0$, we have $\mathbf{v} = \mathbf{i} + \frac{1}{2}\mathbf{j} + \mathbf{c}_1 = \mathrm{i} + \mathrm{j}$, from which $\mathbf{c}_1 = \frac{1}{2}\mathbf{j}$. Then

$$\mathbf{v} = e^t\mathbf{i} + \tfrac{1}{2}(e^{2t} + 1)\mathbf{j} + t\mathbf{k}$$

and
$$\mathbf{r} = \int \mathbf{v}\,dt = e^t\mathbf{i} + \left(\tfrac{1}{4}e^{2t} + \tfrac{1}{2}t\right)\mathbf{j} + \tfrac{1}{2}t^2\mathbf{k} + \mathbf{c}_2$$

At $t = 0$, $\mathbf{r} = \mathbf{i} + \frac{1}{4}\mathbf{j} + \mathbf{c}_2 = 0$, from which $\mathbf{c}_2 = -\mathbf{i} - \frac{1}{4}\mathbf{j}$. Thus,

$$\mathbf{r} = (e^t - 1)\mathbf{i} + \left(\tfrac{1}{4}e^{2t} + \tfrac{1}{2}t - \tfrac{1}{4}\right)\mathbf{j} + \tfrac{1}{2}t^2\mathbf{k}$$

11. Find the work done by a force $\mathbf{F} = (x + yz)\mathbf{i} + (y + xz)\mathbf{j} + (z + xy)\mathbf{k}$ in moving a particle from the origin O to $C(1, 1, 1)$, (a) along the straight line OC; (b) along the curve $x = t$, $y = t^2$, $z = t^3$; and (c) along the straight lines from O to $A(1, 0, 0)$, A to $B(1, 1, 0)$, and B to C.

$$\mathbf{F} \cdot d\mathbf{r} = [(x + yz)\mathbf{i} + (y + xy)\mathbf{j} + (z + xy)\mathbf{k}] \cdot [\mathbf{i}\, dx + \mathbf{j}\, dy + \mathbf{k}\, dz]$$
$$= (x + yz)\, dx + (y + xz)\, dy + (z + xy)\, dz$$

(a) Along the line OC, $x = y = z$ and $dx = dy = dz$. The integral to be evaluated becomes

$$W = \int_{C\,(0,0,0)}^{(1,1,1)} \mathbf{F} \cdot d\mathbf{r} = 3\int_0^1 (x + x^2)dx = \left[\left(\tfrac{3}{2}x^2 + x^3\right)\right]_0^1 = \tfrac{5}{2}$$

(b) Along the given curve, $x = t$ and $dx = dt$; $y = t^2$ and $dy = 2t\, dt$; $z = t^3$ and $dz = 3t^2\, dt$. At O, $t = 0$; at C, $t = 1$. Then

$$W = \int_0^1 (t + t^5)\, dt + (t^2 + t^4)2t\, dt + (t^3 + t^3)3t^2\, dt$$

$$= \int_0^1 (t + 2t^3 + 9t^5)\, dt = \left[\tfrac{1}{2}t^2 + \tfrac{1}{2}t^4 + \tfrac{3}{2}t^6\right]_0^1 = \tfrac{5}{2}$$

(c) From O to A: $y = z = 0$ and $dy = dz = 0$, and x varies from 0 to 1.
From A to B: $x = 1$, $z = 0$, $dx = dz = 0$, and y varies from 0 to 1.
From B to C: $x = y = 1$ and $dx = dy = 0$, and z varies from 0 to 1.

Now, for the distance from O to A, $W_1 = = \int_0^1 x dx = \frac{1}{2}$; for the distance from A to B, $W_2 = \int_0^1 y dy = \frac{1}{2}$; and for the distance from B to C, $W_3 = \int_0^1 (z + 1)\, dz = \frac{3}{2}$. Thus, $W = W_1 + W_2 + W_3 = \frac{5}{2}$.

In general, the value of a line integral depends upon the path of integration. Here is an example of one which does not, that is, one which is independent of the path. It can be shown that a line integral $\int_C (f_1 dx + f_2 dy + f_3 dz)$ is independent of the path if there exists a function $\phi(x, y, z)$ such that $d\phi = f_1 dx + f_2\, dy + f_3\, dz$. In this problem, the integrand is

$$(x + yz)\, dx + (y + xz)\, dy + (z + xy)\, dz = d\left[\tfrac{1}{2}(x^2 + y^2 + z^2) + xyz\right]$$

SUPPLEMENTARY PROBLEMS

12. Find $d\mathbf{s}/dt$ and $d^2\mathbf{s}/dt^2$, given (a) $\mathbf{s} = (t + 1)\mathbf{i} + (t^2 + t + 1)\mathbf{j} + (t^3 + t^2 + t + 1)\mathbf{k}$ and (b) $\mathbf{s} = \mathbf{i}e^t \cos 2t + \mathbf{j}e^t \sin 2t + t^2\mathbf{k}$.

Ans. (a) $\mathbf{i} + (2t + 1)\mathbf{j} + (3t^2 + 2t + 1)\mathbf{k}$, $2\mathbf{j} + (6t + 2)\mathbf{k}$;
(b) $e^t(\cos 2t - 2\sin 2t)\mathbf{i} + e^t(\sin 2t + 2\cos 2t)\mathbf{j} + 2t\mathbf{k}$,
$e^t(-4\sin 2t - 3\cos 2t)\mathbf{i} + e^t(-3\sin 2t + 4\cos 2t)\mathbf{j} + 2\mathbf{k}$

13. Given $\mathbf{a} = u\mathbf{i} + u^2\mathbf{j} + u^3\mathbf{k}$, $\mathbf{b} = \mathbf{i}\cos u + \mathbf{j}\sin u$, and $\mathbf{c} = 3u^2\mathbf{i} - 4u\mathbf{k}$. First compute $\mathbf{a}\cdot\mathbf{b}$, $\mathbf{a}\times\mathbf{b}$, $\mathbf{a}\cdot(\mathbf{b}\times\mathbf{c})$, and $\mathbf{a}\times(\mathbf{b}\times\mathbf{c})$, and find the derivative of each. Then find the derivatives using the formulas.

14. A particle moves along the curve $x = 3t^2$, $y = t^2 - 2t$, $z = t^3$, where t is time. Find (a) the magnitudes of its velocity and acceleration at time $t = 1$; (b) the components of velocity and acceleration at time $t = 1$ in the direction $\mathbf{a} = 4\mathbf{i} - 2\mathbf{j} + 4\mathbf{k}$.

Ans. (a) $|\mathbf{v}| = 3\sqrt{5}$, $|\mathbf{a}| = 2\sqrt{19}$; (b) 6, $\frac{22}{3}$

15. Using vector methods, find the equations of the tangent line and normal plane to the curves of Problem 15 of Chapter 51.

16. Solve Problem 16 of Chapter 51 using vector methods.

17. Show that the surfaces $x = u$, $y = 5u - 3v^2$, $z = v$ and $x = u$, $y = v$, $z = \dfrac{uv}{4u - v}$ are perpendicular at $P(1, 2, 1)$.

18. Using vector methods, find the equations of the tangent plane and normal line to the surface:

(a) $x = u$, $y = v$, $z = uv$ at the point $(u, v) = (3, -4)$.
(b) $x = u$, $y = v$, $z = u^2 - v^2$ at the point $(u, v) = (2, 1)$.

Ans. (a) $4X - 3Y + Z - 12 = 0$, $\dfrac{X-3}{-4} = \dfrac{Y+4}{3} = \dfrac{Z+12}{-1}$

(b) $4X - 2Y - Z - 3 = 0$, $\dfrac{X-2}{-4} = \dfrac{Y-1}{2} = \dfrac{Z-3}{1}$

19. (a) Find the equations of the osculating and rectifying planes to the curve of Problem 2 at the given point.
(b) Find the equations of the osculating, normal, and rectifying planes to $x = 2t - t^2$, $y = t^2$, $z = 2t + t^2$ at $t = 1$.

Ans. (a) $3X - 3Y + Z - 1 = 0$, $11X + 8Y - 9Z - 10 = 0$

(b) $X + 2Y - Z = 0$, $Y + 2Z - 7 = 0$, $5X - 2Y + Z - 6 = 0$

20. Show that the equation of the osculating plane to a space curve at P is given by

$$(\mathbf{R} - \mathbf{r})\cdot\left(\frac{d\mathbf{r}}{dt}\times\frac{d^2\mathbf{r}}{dt^2}\right) = 0$$

21. Solve Problems 16 and 17 of Chapter 52, using vector methods.

22. Find $\int_a^b \mathbf{F}(u)\,du$, given

(a) $\mathbf{F}(u) = u^3\mathbf{i} + (3u^2 - 2u)\mathbf{j} + 3\mathbf{k}$; $a = 0$, $b = 2$; (b) $\mathbf{F}(u) = e^u\mathbf{i} + e^{-2u}\mathbf{j} + u\mathbf{k}$; $a = 0$, $b = 1$

Ans. (a) $4\mathbf{i} + 4\mathbf{j} + 6\mathbf{k}$; (b) $(e - 1)\mathbf{i} + \frac{1}{2}(1 - e^{-2})\mathbf{j} + \frac{1}{2}\mathbf{k}$

23. The acceleration of a particle at any time t is given by $\mathbf{a} = d\mathbf{v}/dt = (t+1)\mathbf{i} + t^2\mathbf{j} + (t^2 - 2)\mathbf{k}$. If at $t = 0$, the displacement is $\mathbf{r} = 0$ and the velocity is $\mathbf{v} = \mathbf{i} - \mathbf{k}$, find $\mathbf{v}$ and $\mathbf{r}$ at any time t.

Ans. $\mathbf{v} = (\frac{1}{2}t^2 + t + 1)\mathbf{i} + \frac{1}{3}t^3\mathbf{j} + (\frac{1}{3}t^3 - 2t - 1)\mathbf{k}$; $\mathbf{r} = (\frac{1}{6}t^3 + \frac{1}{2}t^2 + t)\mathbf{i} + \frac{1}{12}t^4\mathbf{j} + (\frac{1}{12}t^4 - t^2 - t)\mathbf{k}$

24. In each of the following, find the work done by the given force $\mathbf{F}$ in moving a particle from $O(0, 0, 0)$ to $C(1, 1, 1)$ along (1) the straight line $x = y = z$, (2) the curve $x = t$, $y = t^2$, $z = t^3$, and (3) the straight lines from O to $A(1, 0, 0)$, A to $B(1, 1, 0)$, and B to C.

(a) $\mathbf{F} = x\mathbf{i} + 2y\mathbf{j} + 3x\mathbf{k}$.
(b) $\mathbf{F} = (y + z)\mathbf{i} + (x + z)\mathbf{j} + (x + y)\mathbf{k}$.
(c) $\mathbf{F} = (x + xyz)\mathbf{i} + (y + x^2z)\mathbf{j} + (z + x^2y)\mathbf{k}$.

Ans. (a) 3; (b) 3; (c) $\frac{9}{4}, \frac{33}{14}, \frac{5}{2}$

25. If $\mathbf{r} = x\mathbf{i} + y\mathbf{j} + z\mathbf{k}$, show that (a) div $\mathbf{r} = 3$ and (b) curl $\mathbf{r} = 0$.

26. If $f = f(x, y, z)$ has partial derivatives of order at least two, show that (a) $\nabla \times \nabla f = 0$; (b) $\nabla \cdot (\nabla \times f) = 0$; (c) $\nabla \cdot \nabla f = \left(\frac{\partial^2}{\partial x^2} + \frac{\partial^2}{\partial y^2} + \frac{\partial^2}{\partial z^2}\right) f$.

CHAPTER 54

Double and Iterated Integrals

The Double Integral

Consider a function $z = f(x, y)$ that is continuous on a bounded region R of the xy plane. Define a partition $\mathcal{P}$ of R by drawing a grid of horizontal and vertical lines. This divides the region into n subregions $R_1, R_2, \ldots, R_n$ of areas $\Delta_1 A, \Delta_2 A, \ldots, \Delta_n A$, respectively. (See Fig. 54-1.) In each subregion, R_k, choose a point $P_k(x_k, y_k)$ and form the sum

$$\sum_{k=1}^{n} f(x_k, y_k)\Delta_k A = f(x_1, y_1)\Delta_1 A + \cdots + f(x_n, y_n)\Delta_n A \tag{54.1}$$

Define the diameter of a subregion to be the greatest distance between any two points within or on its boundary, and denote by $d_{\mathcal{P}}$ the maximum diameter of the subregions. Suppose that we select partitions so that $d_{\mathcal{P}} \to 0$ and $n \to +\infty$. (In other words, we choose more and more subregions and we make their diameters smaller and smaller.) Then the *double integral* of $f(x, y)$ over R is defined as

$$\iint_R f(x, y)\,dA = \lim_{n \to +\infty} \sum_{k=1}^{n} f(x_k, y_k)\Delta_k A \tag{54.2}$$

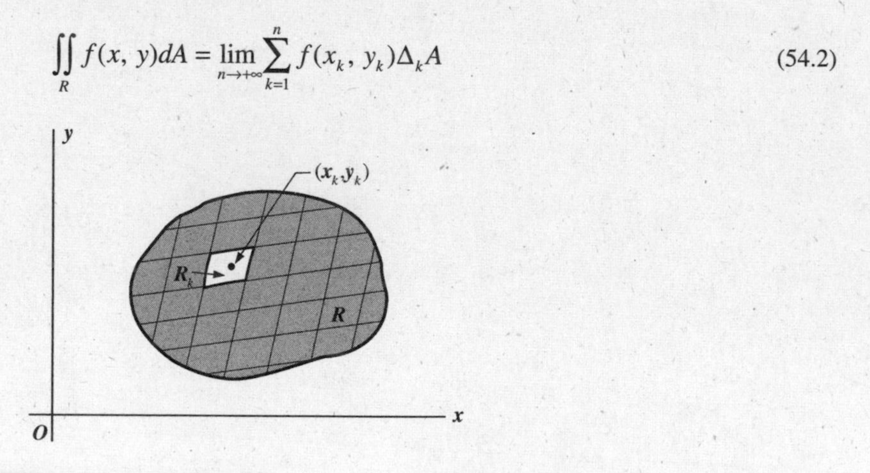

Fig. 54-1

This is not a genuine limit statement. What (54.2) really says is that $\iint_R f(x,y)\,dA$ is a number such that, for any $\epsilon > 0$, there exists a positive integer n_0 such that, for any $n \geq n_0$ and any partition with $d_{\mathcal{P}} < 1/n_0$, and any corresponding approximating sum $\sum_{k=1}^{n} f(x_k, y_k)\Delta_k A$, we have

$$\left| \sum_{k=1}^{n} f(x_k, y_k)\Delta_k A - \iint_R f(x, y)\,dA \right| < \epsilon$$

When $z = f(x, y)$ is nonnegative on the region R, as in Fig. 54-2, the double integral (54.2) may be interpreted as a volume. Any term $f(x_k, y_k)\,\Delta_k A$ of (54.1) gives the volume of a vertical column whose base is of area $\Delta_k A$ and whose altitude is the distance $z_k = f(x_k, y_k)$ measured along the vertical from the selected point $P_k(x_k, y_k)$ to the surface $z = f(x, y)$. This, in turn, may be taken as an approximation of the volume of the vertical column whose lower base is the subregion R_k and whose upper base is the projection of R_k on the surface. Thus, (54.1) is an approximation of the volume "under the surface" (that is, the volume with lower base R and upper base the surface cut off by moving a line parallel to the z axis along the boundary of R). It is intuitively clear that (54.2) is the measure of this volume.

The evaluation of even the simplest double integral by direct summation is usually very difficult.

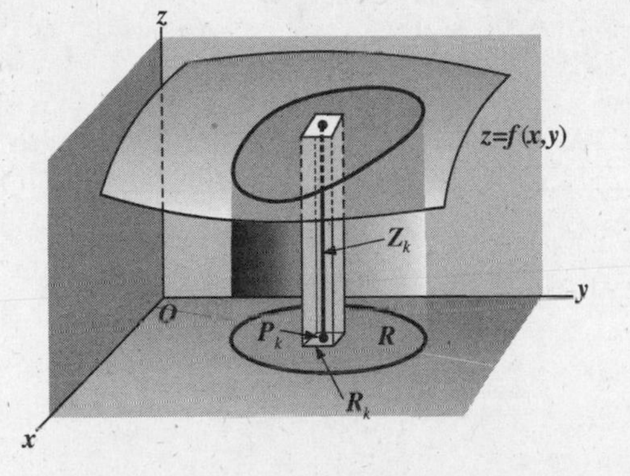

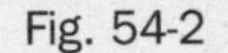
Fig. 54-2

The Iterated Integral

Consider a volume defined as above, and assume that the boundary of R is such that no line parallel to the x axis or to the y axis cuts it in more than two points. Draw the tangent lines $x = a$ and $x = b$ to the boundary with points of tangency K and L, and the tangent lines $y = c$ and $y = d$ with points of tangency M and N. (See Fig. 54-3.) Let the equation of the plane arc LMK be $y = g_1(x)$, and that of the plane arc LNK be $y = g_2(x)$.

Divide the interval $a \le x \le b$ into m subintervals $h_1, h_2, \ldots, h_m$ of respective lengths $\Delta_1 x, \Delta_2 x, \ldots \Delta_m x$ by the insertion of points $\xi_1, \xi_2, \ldots, \xi_{m-1}$ so that $a = \xi_0 < \xi_1 < \xi_2 < \ldots < \xi_{m-1} < \xi_m = b$. Similarly, divide the interval $c \le y \le d$ into n subintervals $k_1, k_2, \ldots, k_n$ of respective lengths $\Delta_1 y, \Delta_2 y, \ldots, \Delta_n y$ by the insertion points $\eta_1, \eta_2, \ldots, \eta_{n-1}$ so that $c = \eta_0 < \eta_1 < \eta_2 < \ldots < \eta_{n-1} < \eta_n = d$. Let λ_m be the greatest $\Delta_i x$ and let μ_n be the greatest $\Delta_j y$. Draw the parallel lines $x = \xi_1, x = \xi_2, \ldots, x = \xi_{m-1}$ and the parallel lines $y = \eta_1, y = \eta_2, \ldots,$ $y = \eta_{n-1}$, thus dividing the region R into a set of rectangles R_{ij} of areas $\Delta_i x\, \Delta_j y$, plus a set of nonrectangles along the boundary (whose areas will be small enough to be safely ignored). In each subinterval h_i select a point $x = x_i$ and, in each subinterval k_j select a point $y = y_j$, thereby determining in each subregion R_{ij} a point $P_{ij}(x_i, y_j)$. With each subregion R_{ij} associate, by means of the equation of the surface, a number $z_{ij} = f(x_i, y_j)$, and form the sum

$$\sum_{\substack{i=1,2,\ldots,m \\ j=1,2,\ldots,n}} f(x_i, y_j)\Delta_i x \Delta_j y \qquad (54.3)$$

Now, (54.3) is merely a special case of (54.1). So, if the number of rectangles is indefinitely increased in such a manner that both $\lambda_m \to 0$ and $\mu_n \to 0$, the limit of (54.3) should be equal to the double integral (54.2).

In effecting this limit, let us first choose one of the subintervals, say h_i, and form the sum

$$\left[\sum_{j=1}^{n} f(x_i, y_j)\Delta_j y\right]\Delta_i x \quad (i \text{ fixed})$$

of the contributions of all rectangles having h_i as one dimension, that is, the contributions of all rectangles lying on the ith column. When $n \to +\infty$, $\mu_n \to 0$,

$$\lim_{n\to+\infty}\left[\sum_{j=1}^{n} f(x_i, y_j)\Delta_j y\right]\Delta_i x = \left[\int_{g_1(x_i)}^{g_2(x_i)} f(x_i, y)\,dy\right]\Delta_i x = \phi(x_i)\Delta_i x$$

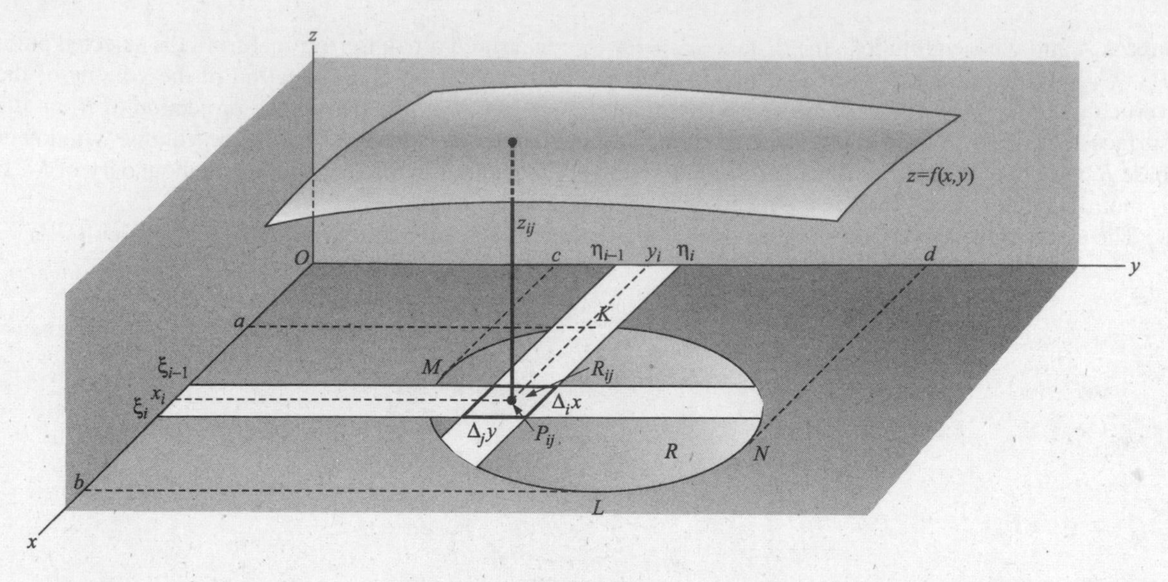

Fig. 54-3

Now summing over the m columns and letting $m \to +\infty$, we have

$$\lim_{m \to +\infty} \sum_{i=1}^{m} \phi(x_i)\Delta_i x = \int_a^b \phi(x)\,dx = \int_a^b \left[\int_{g_1(x_i)}^{g_2(x_i)} f(x,y)\,dy\right] dx$$

$$= \int_a^b \int_{g_1(x_i)}^{g_2(x_i)} f(x,y)\,dy\,dx \tag{54.4}$$

Although we shall not use the brackets hereafter, it must be clearly understood that (54.4) calls for the evaluation of two simple definite integrals in a prescribed order: first, the integral of $f(x, y)$ with respect to y (considering x as a constant) from $y = g_1(x)$, the lower boundary of R, to $y = g_2(x)$, the upper boundary of R, and then the integral of this result with respect to x from the abscissa $x = a$ of the leftmost point of R to the abscissa $x = b$ of the rightmost point of R. The integral (54.4) is called an *iterated* or *repeated integral*.

It will be left as an exercise to sum first for the contributions of the rectangles lying in each row and then over all the rows to obtain the equivalent iterated integral

$$\int_c^d \int_{h_1(y)}^{h_2(y)} f(x,y)\,dx\,dy \tag{54.5}$$

where $x = h_1(y)$ and $x = h_2(y)$ are the equations of the plane arcs MKN and MLN, respectively.

In Problem 1, it is shown by a different procedure that the iterated integral (54.4) measures the volume under discussion. For the evaluation of iterated integrals, see Problems 2 to 6.

The principal difficulty in setting up the iterated integrals of the next several chapters will be that of inserting the limits of integration to cover the region R. The discussion here assumed the simplest of regions; more complex regions are considered in Problems 7 to 9.

SOLVED PROBLEMS

1. Let $z = f(x, y)$ be nonnegative and continuous over the region R of the xy plane whose boundary consists of the arcs of two curves $y = g_1(x)$ and $y = g_2(x)$ intersecting at the points K and L, as in Fig. 54-4. Find a formula for the volume V under the surface $z = f(x, y)$.

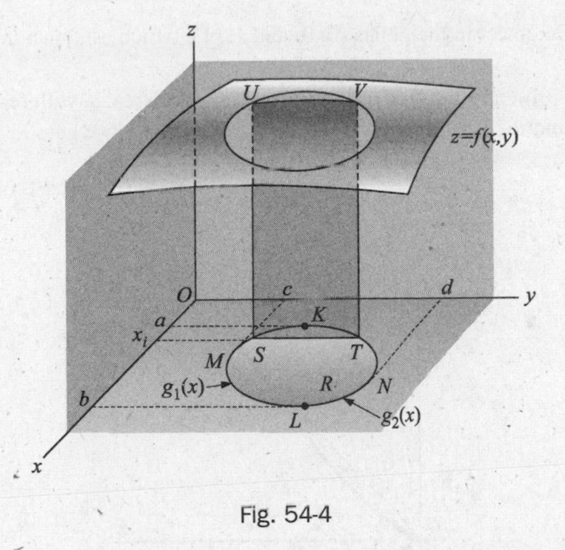

Fig. 54-4

Let the section of this volume cut by a plane $x = x_i$, where $a < x_i < b$, meet the boundary of R at the points $S(x_i, g_1(x_i))$ and $T(x_i, g_2(x_i))$, and let it meet the surface $z = f(x, y)$ in the arc UV along which $z = f(x_i, y)$. The area of this section $STUV$ is given by

$$A(x_i) = \int_{g_1(x_i)}^{g_2(x_i)} f(x_i, y)\,dy$$

Thus, the areas of cross sections of the volume cut by planes parallel to the yz plane are known functions $A(x) = \int_{g_1(x_i)}^{g_2(x_i)} f(x,y)\,dy$ of x, where x is the distance of the sectioning plane from the origin. By the cross-section formula of Chapter 30, the required volume is given by

$$V = \int_a^b A(x)dx = \int_a^b \left[\int_{g_1(x_i)}^{g_2(x_i)} f(x,y)\,dy\right]dx$$

This is the iterated integral of (54.4).

In Problems 2–6, evaluate the integral on the left.

2. $\int_0^1 \int_{x^2}^{x} dy\,dx = \int_0^1 [y]_{x^2}^{x}\,dx = \int_0^1 (x - x^2)dx = \left[\frac{x^2}{2} - \frac{x^3}{3}\right] = \frac{1}{6}$

3. $\int_1^2 \int_y^{3y} (x + y)dx\,dy = \int_1^2 [\frac{1}{2}x^2 + xy]_y^{3y}\,dy = \int_1^2 6y^2 dy = [2y^3]_1^2 = 14$

4. $\int_{-1}^2 \int_{2x^2-2}^{x^2+x} x\,dy\,dx = \int_{-1}^2 [xy]_{2x^2-2}^{x^2+x}\,dx = \int_{-1}^2 (x^3 + x^2 - 2x^3 + 2x)\,dx = \frac{9}{4}$

5. $\int_0^\pi \int_0^{\cos\theta} \rho\sin\theta\,d\rho\,d\theta = \int_0^\pi [\frac{1}{2}\rho^2 \sin\theta]_0^{\cos\theta}\,d\theta = \frac{1}{2}\int_0^\pi \cos^2\theta\sin\theta\,d\theta = [-\frac{1}{6}\cos^3\theta]_0^\pi = \frac{1}{3}$

6. $\int_0^{\pi/2} \int_2^{4\cos\theta} \rho^3 d\rho\,d\theta = \int_0^{\pi/2} \left[\frac{1}{4}\rho^4\right]_2^{4\cos\theta} d\theta = \int_0^{\pi/2} (64\cos^4\theta - 4)\,d\theta$

$$= \left[64\left(\frac{3\theta}{8} + \frac{\sin\theta}{4} + \frac{\sin 4\theta}{32}\right) - 40\right]_0^{\pi/2} = 10\pi$$

7. Evaluate $\iint_R dA$, where R is the region in the first quadrant bounded by the semicubical parabola $y^2 = x^3$ and the line $y = x$.

The line and parabola intersect in the points (0, 0) and (1, 1), which establish the extreme values of x and y on the region R.

Solution 1 (Fig. 54-5): Integrating first over a horizontal strip, that is, with respect to x from $x = y$ (the line) to $x = y^{2/3}$ (the parabola), and then with respect to y from $y = 0$ to $y = 1$, we get

$$\iint_R dA = \int_0^1 \int_y^{y^{2/3}} dx\,dy = \int_0^1 (y^{2/3} - y)\,dy = [\tfrac{3}{5}y^{5/3} - \tfrac{1}{2}y^2]_0^1 = \tfrac{1}{10}$$

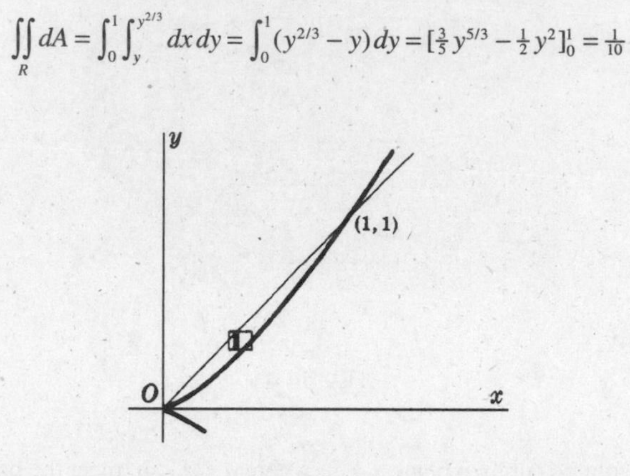

Fig. 54-5

Solution 2 (Fig. 54-6): Integrating first over a vertical strip, that is, with respect to y from $y = x^{3/2}$ (the parabola) to $y = x$ (the line), and then with respect to x from $x = 0$ to $x = 1$, we obtain

$$\iint_R dA = \int_0^1 \int_{x^{3/2}}^{x} dy\,dx = \int_0^1 (x - x^{3/2})\,dx = [\tfrac{1}{2}x^2 - \tfrac{2}{5}x^{5/2}]_0^1 = \tfrac{1}{10}$$

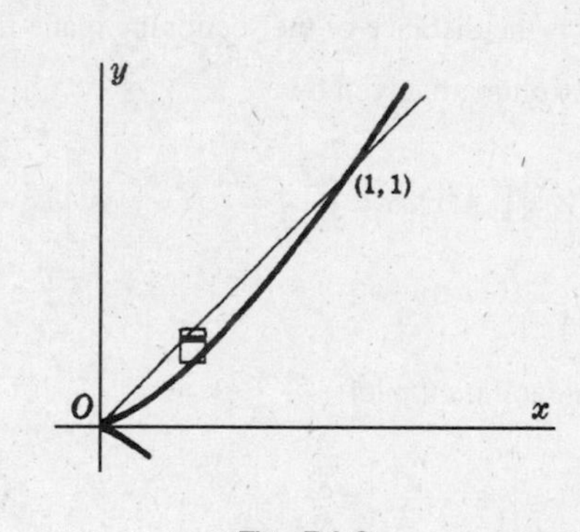

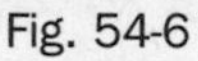

Fig. 54-6

8. Evaluate $\iint_R dA$ where R is the region between $y = 2x$ and $y = x^2$ lying to the left of $x = 1$.

Integrating first over the vertical strip (see Fig. 54-7), we have

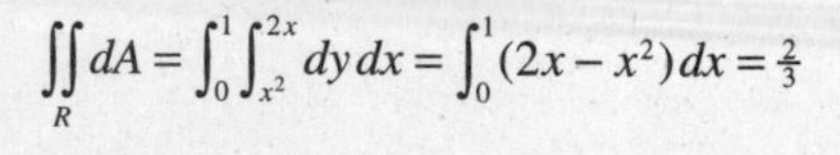

$$\iint_R dA = \int_0^1 \int_{x^2}^{2x} dy\,dx = \int_0^1 (2x - x^2)\,dx = \tfrac{2}{3}$$

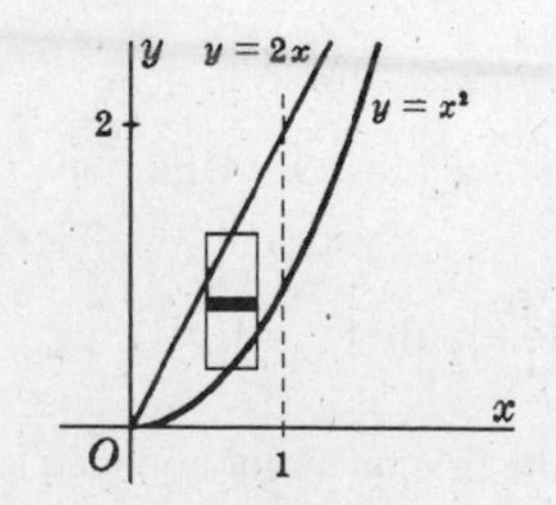

Fig. 54-7

When horizontal strips are used (see Fig. 54-8), two iterated integrals are necessary. Let R_1 denote the part of R lying below AB, and R_2 the part above AB. Then

$$\iint_R dA = \iint_{R_1} dA + \iint_{R_2} dA = \int_0^1 \int_{y/2}^{\sqrt{y}} dx\,dy + \int_1^2 \int_{y/2}^1 dx\,dy = \tfrac{5}{12} + \tfrac{1}{4} = \tfrac{2}{3}$$

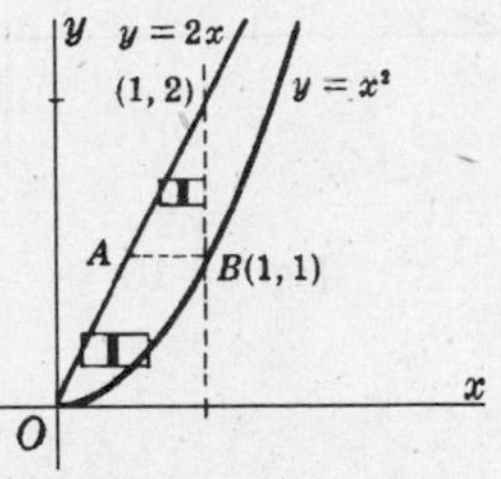

Fig. 54-8

9. Evaluate $\iint_R x^2 dA$ where R is the region in the first quadrant bounded by the hyperbola $xy = 16$ and the lines $y = x$, $y = 0$, and $x = 8$. (See Fig. 54-9.)

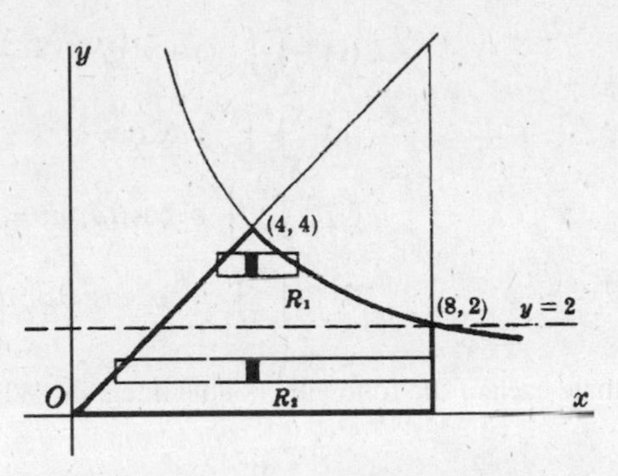

Fig. 54-9

It is evident from Fig. 54-9 that R must be separated into two regions, and an iterated integral evaluated for each. Let R_1 denote the part of R lying above the line $y = 2$, and R_2 the part below that line. Then

$$\iint_R x^2 dA = \iint_{R_1} x^2 dA + \iint_{R_2} x^2 dA = \int_2^4 \int_y^{16/y} x^2 dx\,dy + \int_0^2 \int_y^8 x^2 dx\,dy$$

$$= \frac{1}{3}\int_2^4 \left(\frac{16^3}{y^3} - y^3\right) dy + \frac{1}{3}\int_0^2 (8^3 - y^3)\,dy = 448$$

As an exercise, you might separate R with the line $x = 4$ and obtain

$$\iint_R x^2 dA = \int_0^4 \int_0^x x^2 dy\,dx + \int_4^8 \int_0^{16/x} x^2 dy\,dx$$

10. Evaluate $\int_0^1 \int_{3y}^3 e^{x^2} dx\,dy$ by first reversing the order of integration.

The given integral cannot be evaluated directly, since $\int e^{x^2} dx$ is not an elementary function. The region R of integration (see Fig. 54-10) is bounded by the lines $x = 3y$, $x = 3$, and $y = 0$. To reverse the order of integration, first integrate with respect to y from $y = 0$ to $y = x/3$, and then with respect to x from $x = 0$ to $x = 3$. Thus,

$$\int_0^1 \int_{3y}^3 e^{x^2} dx\,dy = \int_0^3 \int_0^{x/3} e^{x^2} dy\,dx = \int_0^3 [e^{x^2} y]_0^{x/3} dx$$

$$= \frac{1}{3}\int_0^3 e^{x^2} x\,dx = [\tfrac{1}{6} e^{x^2}]_0^3 = \tfrac{1}{6}(e^9 - 1)$$

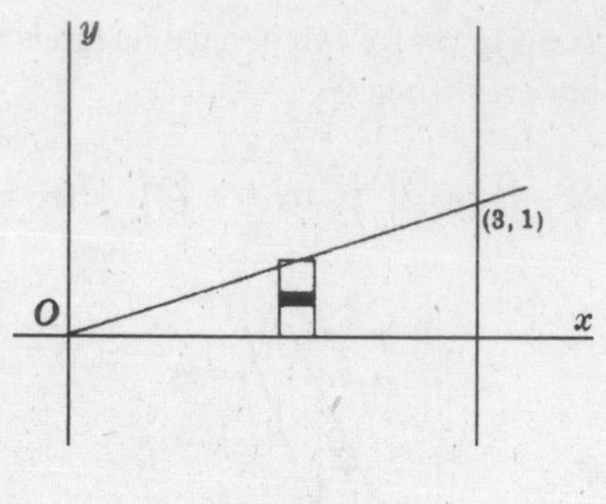

Fig. 54-10

SUPPLEMENTARY PROBLEMS

11. Evaluate the iterated integral at the left:

(a) $\int_0^1\int_1^2 dx\,dy = 1$ (b) $\int_0^2\int_0^3 (x+y)\,dx\,dy = 9$

(c) $\int_2^4\int_1^2 (x^2+y^2)\,dy\,dx = \frac{70}{3}$ (d) $\int_0^1\int_{x^2}^{x} xy^2\,dy\,dx = \frac{1}{40}$

(e) $\int_1^2\int_0^{y^{3/2}} x/y^2\,dx\,dy = \frac{3}{4}$ (f) $\int_0^1\int_x^{\sqrt{x}} (x+y^3)\,dy\,dx = \frac{7}{60}$

(g) $\int_0^1\int_0^{x^2} xe^y\,dy\,dx = \frac{1}{2}e - 1$ (h) $\int_2^4\int_y^{8-y} y\,dx\,dy = \frac{32}{3}$

(i) $\int_0^{\tan^{-1}(3/2)}\int_0^{2\sec\theta} \rho\,d\rho\,d\theta = 3$ (j) $\int_0^{\pi/2}\int_0^{2} \rho^2\cos\theta\,d\rho\,d\theta = \frac{8}{3}$

(k) $\int_0^{\pi/4}\int_0^{\tan\theta\sec\theta} \rho^3\cos^2\theta\,d\rho\,d\theta = \frac{1}{20}$ (l) $\int_0^{2\pi}\int_0^{1-\cos\theta} \rho^3\cos^2\theta\,d\rho\,d\theta = \frac{49}{32}\pi$

12. Using an iterated integral, evaluate each of the following double integrals. When feasible, evaluate the iterated integrals in both orders.

(a) x over the region bounded by $y = x^2$ and $y = x^3$ *Ans.* $\frac{1}{20}$
(b) y over the region of part (a) *Ans.* $\frac{1}{35}$
(c) x^2 over the region bounded by $y = x$, $y = 2x$, and $x = 2$ *Ans.* 4
(d) 1 over each first-quadrant region bounded by $2y = x^2$, $y = 3x$, and $x + y = 4$ *Ans.* $\frac{8}{3}$; $\frac{46}{3}$
(e) y over the region above $y = 0$ bounded by $y^2 = 4x$ and $y^2 = 5 - x$ *Ans.* 5
(f) $\dfrac{1}{\sqrt{2y-y^2}}$ over the region in the first quadrant bounded by $x^2 = 4 - 2y$ *Ans.* 4

13. In Problems 11(a) to (h), reverse the order of integration and evaluate the resulting iterated integral.

CHAPTER 55

Centroids and Moments of Inertia of Plane Areas

Plane Area by Double Integration

If $f(x, y) = 1$, the double integral of Chapter 54 becomes $\iint_R dA$. In cubic units, this measures the volume of a cylinder of unit height; in square units, it measures the area A of the region R.

In polar coordinates

$$A = \iint_R dA = \int_\alpha^\beta \int_{\rho_1(\theta)}^{\rho_2(\theta)} \rho\, d\rho\, d\theta$$

where $\theta = \alpha$, $\theta = \beta$, $\rho = \rho_1(\theta)$, and $\rho = \rho_2(\theta)$ are chosen as boundaries of the region R.

Centroids

The centroid $(\bar{x}, \bar{y})$ of a plane region R is intuitively thought of in the following way. If R is supposed to have a uniform unit density, and if R is supported from below at the point $(\bar{x}, \bar{y})$, then R will balance (that is, R will not rotate at all).

To locate $(\bar{x}, \bar{y})$, first consider the vertical line $x = \bar{x}$. If we divide R into subregions $R_1, \ldots, R_n$, of areas $\Delta_1 A, \ldots, \Delta_n A$ as in Chapter 54, and if we select points (x_k, y_k) in each R_k, then the moment (rotational force) of R_k about the line $x = \bar{x}$ is approximately $(x_k - \bar{x})\Delta_k A$. So, the moment of R about $x = \bar{x}$ is approximately $\sum_{k=1}^{n}(x_k - \bar{x})\Delta_k A$. Making the partition of R finer and finer, we get $\iint_R (x - \bar{x})\, dA$ as the moment of R about $x = \bar{x}$. In order to have no rotation about $x = \bar{x}$, we must have $\iint_R (x - \bar{x})\, dA = 0$. But

$$\iint_R (x - \bar{x})\, dA = \iint_R x\, dA - \iint_R \bar{x}\, dA = \iint_R x\, dA - \bar{x}\iint_R dA$$

Hence, we must have $\iint_R x\, dA = \bar{x}\iint_R dA$. Similarly, we get $\iint_R y\, dA = \bar{y}\iint_R dA$. So, the centroid is determined by the equations

$$\iint_R x\, dA = \bar{x}\iint_R dA \quad \text{and} \quad \iint_R y\, dA = \bar{y}\iint_R dA$$

Note that $\iint_R dA$ is equal to the area A of the region R.

Moments of Inertia

The moments of inertia of a plane region R with respect to the coordinate axes are given by

$$I_x = \iint_R y^2\, dA \quad \text{and} \quad I_y = \iint_R x^2 dA$$

The polar moment of inertia (the moment of inertia with respect to a line through the origin and perpendicular to the plane of the area) of a plane region R is given by

$$I_0 = I_x + I_y = \iint_R (x^2 + y^2)\, dA$$

SOLVED PROBLEMS

1. Find the area bounded by the parabola $y = x^2$ and the line $y = 2x + 3$.

Using vertical strips (see Fig. 55-1), we have

$$A = \int_{-1}^{3}\int_{x^2}^{2x+3} dy\, dx = \int_{-1}^{3}(2x + 3 - x^2)dx = 32/3 \text{ square units}$$

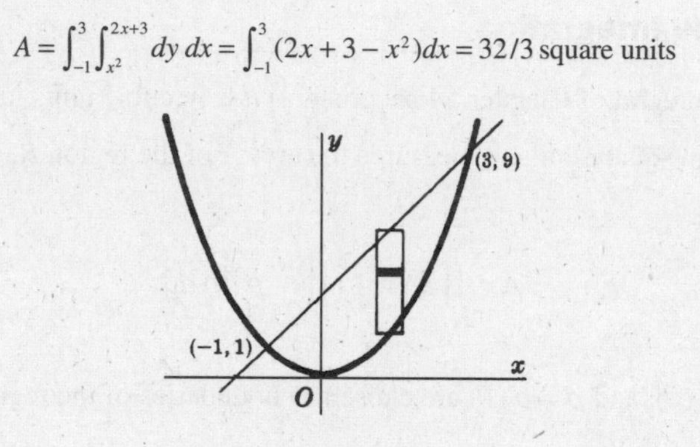

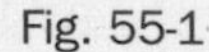
Fig. 55-1

2. Find the area bounded by the parabolas $y^2 = 4 - x$ and $y^2 = 4 - 4x$.

Using horizontal strips (Fig. 55-2) and taking advantage of symmetry, we have

$$A = 2\int_0^2\int_{1-y^2/4}^{4-y^2} dx\, dy = 2\int_0^2[(4 - y^2) - (1 - \tfrac{1}{4}y^2)]dy$$

$$= 6\int_0^2(1 - \tfrac{1}{4}y^2)\, dy = 8 \text{ square units}$$

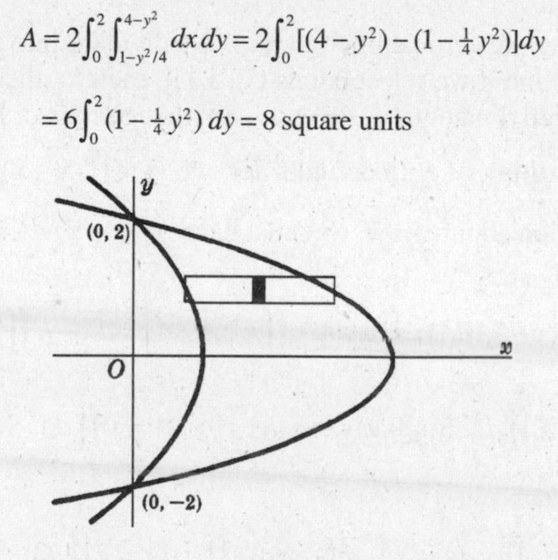

Fig. 55-2

3. Find the area outside the circle $\rho = 2$ and inside the cardiod $\rho = 2(1 + \cos\theta)$.

Owing to symmetry (see Fig. 55-3), the required area is twice that swept over as θ varies from $\theta = 0$ to $\theta = \frac{1}{2}\pi$. Thus,

$$A = 2\int_0^{\pi/2}\int_2^{2(1+\cos\theta)} \rho\, d\rho\, d\theta = 2\int_0^{\pi/2} [\tfrac{1}{2}\rho^2]_2^{2(1+\cos\theta)} d\theta = 4\int_0^{\pi/2} (2\cos\theta + \cos^2\theta) d\theta$$

$$= 4[\sin\theta + \tfrac{1}{2}\theta + \tfrac{1}{4}\sin 2\theta]_0^{\pi/2} = (\pi + 8) \text{ square units}$$

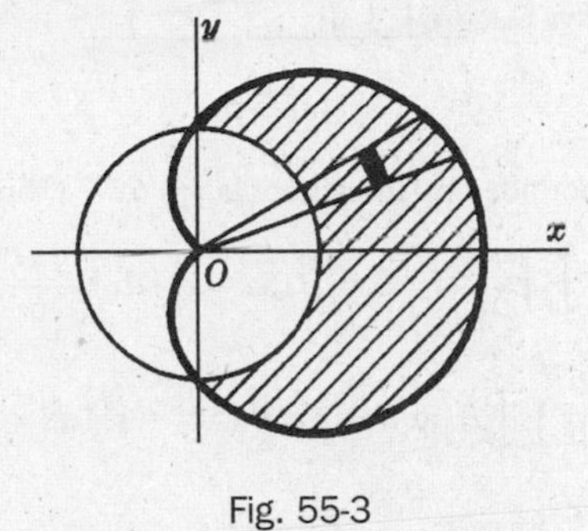

Fig. 55-3

4. Find the area inside the circle $\rho = 4 \sin\theta$ and outside the lemniscate $\rho^2 = 8 \cos 2\theta$.

The required area is twice that in the first quadrant bounded by the two curves and the line $\theta = \frac{1}{2}\pi$. Note in Fig. 55-4 that the arc AO of the lemniscate is described as θ varies from $\theta = \pi/6$ to $\theta = \pi/4$, while the arc AB of the circle is described as θ varies from $\theta = \pi/6$ to $\theta = \pi/2$. This area must then be considered as two regions, one below and one above the line $\theta = \pi/4$. Thus,

$$A = 2\int_{\pi/6}^{\pi/4}\int_{2\sqrt{2\cos 2\theta}}^{4\sin\theta} \rho\, d\rho\, d\theta + 2\int_{\pi/4}^{\pi/2}\int_0^{4\sin\theta} \rho\, d\rho\, d\theta$$

$$= \int_{\pi/6}^{\pi/4} (16\sin^2\theta - 8\cos 2\theta) d\theta + \int_{\pi/4}^{\pi/2} 16\sin^2\theta\, d\theta$$

$$= (\tfrac{8}{3}\pi + 4\sqrt{3} - 4) \text{ square units}$$

Fig. 55-4

5. Evaluate $N = \int_0^{+\infty} e^{-x^2} dx$. (See Fig. 55-5.)

Since $\int_0^{+\infty} e^{-x^2} dx = \int_0^{+\infty} e^{-y^2} dy$, we have

$$N^2 = \int_0^{+\infty} e^{-x^2} dx \int_0^{+\infty} e^{-y^2} dy = \int_0^{+\infty}\int_0^{+\infty} e^{-(x^2+y^2)} dx\, dy = \iint_R e^{-(x^2+y^2)} dA$$

Changing to polar coordinates, $(x^2 + y^2) = \rho^2$, $dA = \rho\, d\rho\, d\rho$ yields

$$N^2 = \int_0^{\pi/2}\int_0^{+\infty} e^{-\rho^2} \rho\, d\rho\, d\theta = \int_0^{\pi/2} \lim_{a\to+\infty}\left[-\frac{1}{2}e^{-\rho^2}\right]_0^a d\theta = \frac{1}{2}\int_0^{\pi/2} d\theta = \frac{\pi}{4}$$

and $N = \sqrt{\pi}/2$.

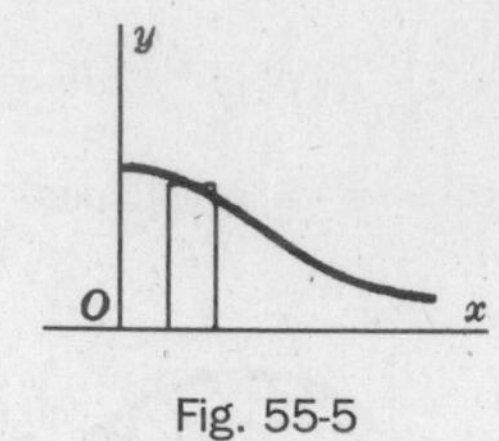

Fig. 55-5

6. Find the centroid of the plane area bounded by the parabola $y = 6x - x^2$ and the line $y = x$. (See Fig. 55-6.)

$$A = \iint_R dA = \int_0^5 \int_x^{6x-x^2} dy\, dx = \int_0^5 (5x - x^2)\, dx = \frac{125}{6}$$

$$M_y = \iint_R x\, dA = \int_0^5 \int_x^{6x-x^2} x\, dy\, dx = \int_0^5 (5x^2 - x^3)\, dx = \frac{625}{12}$$

$$M_x = \iint_R y\, dA = \int_0^5 \int_x^{6x-x^2} y\, dy\, dx = \frac{1}{2}\int_0^5 [(6x - x^2)^2 - x^2]\, dx = \frac{625}{6}$$

Hence, $\bar{x} = M_y/A = \frac{5}{2}$, $\bar{y} = M_x/A = 5$, and the coordinates of the centroid are $(\frac{5}{2}, 5)$.

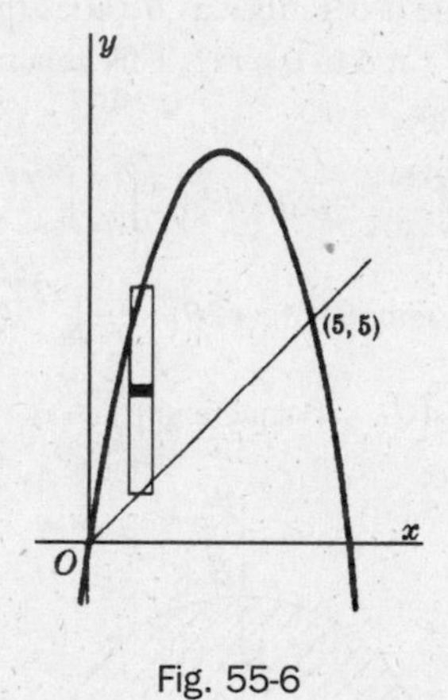

Fig. 55-6

7. Find the centroid of the plane area bounded by the parabolas $y = 2x - x^2$ and $y = 3x^2 - 6x$. (See Fig. 55-7.)

$$A = \iint_R dA = \int_0^2 \int_{3x^2-6x}^{2x-x^2} dy\, dx = \int_0^2 (8x - 4x^2)\, dx = \frac{16}{3}$$

$$M_y = \iint_R x\, dA = \int_0^2 \int_{3x^2-6x}^{2x-x^2} x\, dy\, dx = \int_0^2 (8x^2 - 4x^3)\, dx = \frac{16}{3}$$

$$M_x = \iint_R y\, dA = \int_0^2 \int_{3x^2-6x}^{2x-x^2} y\, dy\, dx = \frac{1}{2}\int_0^2 [(2x - x^2)^2 - (3x^2 - 6x)^2]\, dx = -\frac{64}{15}$$

Hence, $\bar{x} = M_y/A = 1$, $\bar{y} = M_x/A = -\frac{4}{5}$, and the centroid is $(1, -\frac{4}{5})$.

8. Find the centroid of the plane area outside the circle $\rho = 1$ and inside the cardiod $\rho = 1 + \cos\theta$.

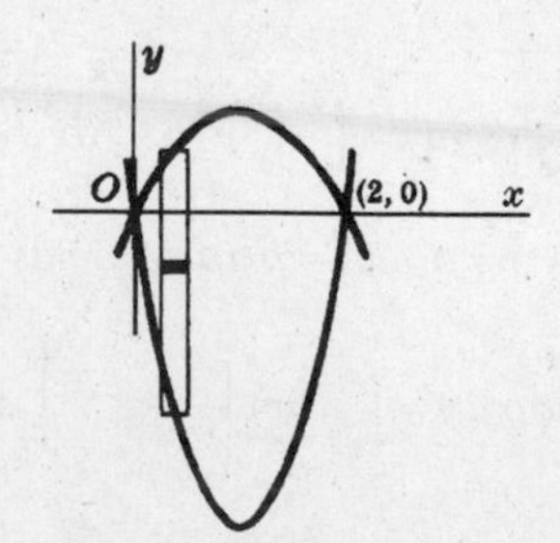

Fig. 55-7

From Fig. 55-8 it is evident that $\bar{y} = 0$ and that $\bar{x}$ is the same whether computed for the given area or for the half lying above the polar axis. For the latter area,

$$A = \iint_R dA = \int_0^{\pi/2}\int_1^{1+\cos\theta} \rho\, d\rho\, d\theta = \frac{1}{2}\int_0^{\pi/2}[(1+\cos\theta)^2 - 1^2]d\theta = \frac{\pi+8}{8}$$

$$M_y = \iint_R x\, dA = \int_0^{\pi/2}\int_1^{1+\cos\theta}(\rho\cos\theta)\rho\, d\rho\, d\theta = \frac{1}{3}\int_0^{\pi/2}(3\cos^2\theta + 3\cos^3\theta + \cos^4\theta)\, d\theta$$

$$= \frac{1}{3}\left[\frac{3}{2}\theta + \frac{3}{4}\sin 2\theta + 3\sin\theta - \sin^3\theta + \frac{3}{8}\theta + \frac{1}{4}\sin 2\theta + \frac{1}{32}\sin 4\theta\right]_0^{\pi/2} = \frac{15\pi+32}{48}$$

The coordinates of the centroid are $\left(\frac{15\pi+32}{6(\pi+8)}, 0\right)$.

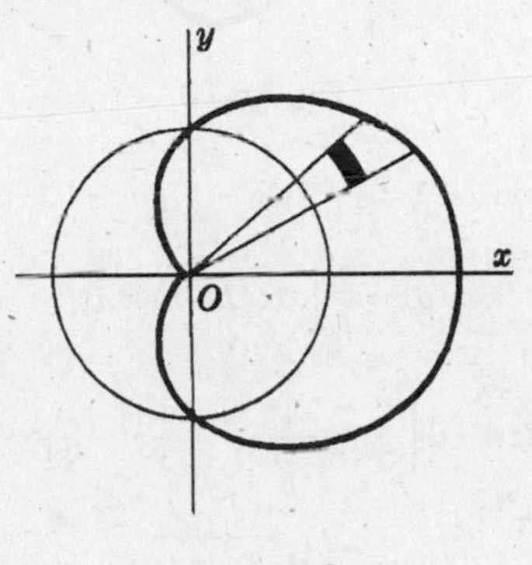

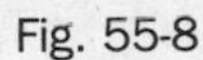
Fig. 55-8

9. Find the centroid of the area inside $\rho = \sin\theta$ and outside $\rho = 1 - \cos\theta$. (See Fig. 55-9.)

$$A = \iint_R dA = \int_0^{\pi/2}\int_{1-\cos\theta}^{\sin\theta} \rho\, d\rho\, d\theta = \frac{1}{2}\int_0^{\pi/2}(2\cos\theta - 1 - \cos 2\theta)d\theta = \frac{4-\pi}{4}$$

$$M_y = \iint_R x\, dA = \int_0^{\pi/2}\int_{1-\cos\theta}^{\sin\theta}(\rho\cos\theta)\rho\, d\rho\, d\theta$$

$$= \frac{1}{3}\int_0^{\pi/2}(\sin^3\theta - 1 + 3\cos\theta - 3\cos^2\theta + \cos^3\theta)\cos\theta\, d\theta = \frac{15\pi - 44}{48}$$

$$M_x = \iint_R y\, dA = \int_0^{\pi/2}\int_{1-\cos\theta}^{\sin\theta}(\rho\sin\theta)\rho\, d\rho\, d\theta$$

$$= \frac{1}{3}\int_0^{\pi/2}(\sin^3\theta - 1 + 3\cos\theta - 3\cos^2\theta + \cos^3\theta)\sin\theta\, d\theta = \frac{3\pi - 4}{48}$$

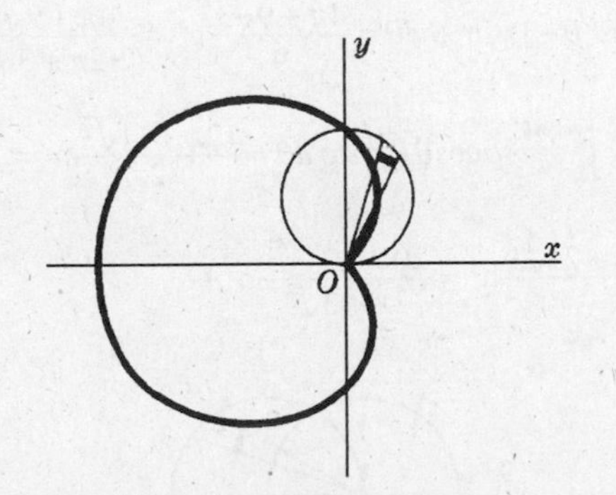

Fig. 55-9

The coordinates of the centroid are $\left(\frac{15\pi - 44}{12(4-\pi)}, \frac{3\pi - 4}{12(4-\pi)}\right)$.

10. Find I_x, I_y, and I_0 for the area enclosed by the loop of $y^2 = x^2(2-x)$. (See Fig. 55-10.)

$$A = \iint_R dA = 2\int_0^2 \int_0^{x\sqrt{2-x}} dy\,dx = 2\int_0^2 x\sqrt{2-x}\,dx$$

$$= -4\int_{\sqrt{2}}^0 (2z^2 - z^4)dz = -4\left[\frac{2}{3}z^3 - \frac{1}{5}z^5\right]_{\sqrt{2}}^0 = \frac{32\sqrt{2}}{15}$$

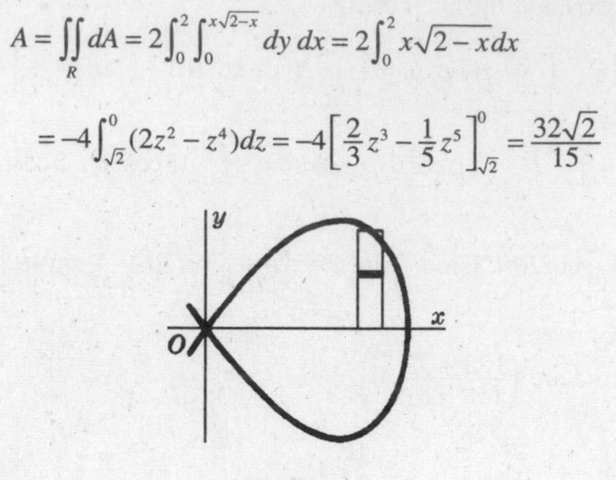

Fig. 55-10

where we have used the transformation $2 - x = z^2$. Then

$$I_x = \iint_R y^2 dA = 2\int_0^2 \int_0^{x\sqrt{2-x}} y^2 dy\,dx = \frac{2}{3}\int_0^2 x^3(2-x)^{3/2}dx$$

$$= -\frac{4}{3}\int_{\sqrt{2}}^0 (2-z^2)^3 z^4 dz = -\frac{4}{3}\left[\frac{8}{5}z^5 - \frac{12}{7}z^7 + \frac{2}{3}z^9 - \frac{1}{11}z^{11}\right]_{\sqrt{2}}^0 = \frac{2048\sqrt{2}}{3465} = \frac{64}{231}A$$

$$I_y = \iint_R x^2 dA = 2\int_0^2 \int_0^{x\sqrt{2-x}} x^2 dy\,dx = 2\int_0^2 x^3\sqrt{2-x}\,dx$$

$$= -4\int_{\sqrt{2}}^0 (2-z^2)^3 z^2 dz = -4\left[\frac{8}{3}z^3 - \frac{12}{5}z^5 + \frac{6}{7}z^7 - \frac{1}{9}z^9\right]_{\sqrt{2}}^0 = \frac{1024\sqrt{2}}{315} = \frac{32}{21}A$$

$$I_0 = I_x + I_y = \frac{13{,}312\sqrt{2}}{3465} = \frac{416}{231}A$$

11. Find I_x, I_y, and I_0 for the first-quadrant area outside the circle $\rho = 2a$ and inside the circle $\rho = 4a\cos\theta$. (See Fig. 55-11.)

$$A = \iint_R dA = \int_0^{\pi/3}\int_2^{4a\cos\theta} \rho\,d\rho\,d\theta = \frac{1}{2}\int_0^{\pi/3}[(4a\cos\theta)^2 - (2a)^2]d\theta = \frac{2\pi + 3\sqrt{3}}{3}a^2$$

$$I_x = \iint_R y^2 dA = \int_0^{\pi/3}\int_{2a}^{4a\cos\theta} (\rho\sin\theta)^2 \rho\,d\rho\,d\theta = \frac{1}{4}\int_0^{\pi/3}[(4a\cos\theta)^4 - (2a)^4]\sin^2\theta\,d\theta$$

$$= 4a^4\int_0^{\pi/3}(16\cos^4\theta - 1)\sin^2\theta\,d\theta = \frac{4\pi + 9\sqrt{3}}{6}a^4 = \frac{4\pi + 9\sqrt{3}}{2(2\pi + 3\sqrt{3})}a^2 A$$

$$I_y = \iint_R x^2 dA = \int_0^{\pi/3}\int_{2a}^{4a\cos\theta} (\rho\cos\theta)^2 \rho\,d\rho\,d\theta = \frac{12\pi + 11\sqrt{3}}{2}a^4 = \frac{3(12\pi + 11\sqrt{3})}{2(2\pi + 3\sqrt{3})}a^2 A$$

$$I_0 = I_x + I_y = \frac{20\pi + 21\sqrt{3}}{3}a^4 = \frac{20\pi + 21\sqrt{3}}{2\pi + 3\sqrt{3}}a^2 A$$

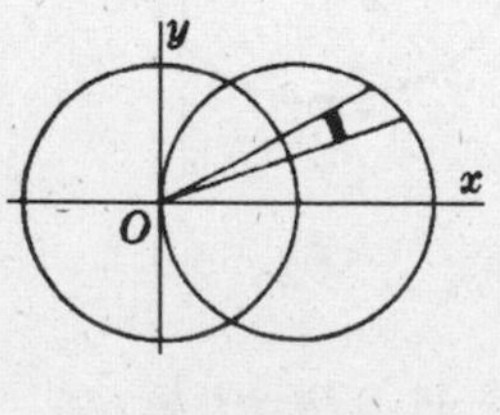

Fig. 55-11

12. Find I_x, I_y, and I_0 for the area of the circle $\rho = 2(\sin\theta + \cos\theta)$. (See Fig. 55-12.)

Since $x^2+y^2 = \rho^2$,

$$I_0 = \iint_R (x^2+y^2)\,dA = \int_{-\pi/4}^{3\pi/4}\int_0^{2(\sin\theta+\cos\theta)} \rho^2\rho\,d\rho\,d\theta = 4\int_{-\pi/4}^{3\pi/4}(\sin\theta+\cos\theta)^4\,d\theta$$

$$= 4[\tfrac{3}{2}\theta - \cos 2\theta - \tfrac{1}{8}\sin 4\theta]_{-1/4\pi}^{3/4\pi} = 6\pi = 3A$$

It is evident from Fig. 55-12 that $I_x = I_y$. Hence, $I_x = I_y = \frac{1}{2}I_0 = \frac{3}{2}A$.

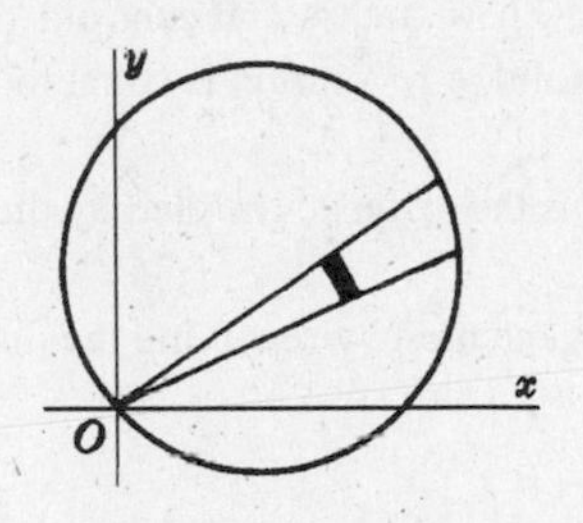

Fig. 55-12

SUPPLEMENTARY PROBLEMS

13. Use double integration to find the area.

(a) Bounded by $3x + 4y = 24$, $x = 0$, $y = 0$ — *Ans.* 24 square units
(b) Bounded by $x + y = 2$, $2y = x + 4$, $y = 0$ — *Ans.* 6 square units
(c) Bounded by $x^2 = 4y$, $8y = x^2 + 16$ — *Ans.* $\frac{32}{3}$ square units
(d) Within $\rho = 2(1 - \cos\theta)$ — *Ans.* 6π square units
(e) Bounded by $\rho = \tan\theta\sec\theta$ and $\theta = \pi/3$ — *Ans.* $\frac{1}{2}\sqrt{3}$ square units
(f) Outside $\rho = 4$ and inside $\theta = 8\cos\theta$ — *Ans.* $8(\frac{2}{3}\pi + \sqrt{3})$ square units

14. Locate the centroid of each of the following areas.

(a) The area of Problem 13(a) — *Ans.* $(\frac{8}{3}, 2)$
(b) The first-quadrant area of Problem 13(c) — *Ans.* $(\frac{3}{2}, \frac{8}{5})$
(c) The first-quadrant area bounded by $y^2 = 6x$, $y = 0$, $x = 6$ — *Ans.* $(\frac{18}{5}, \frac{9}{4})$
(d) The area bounded by $y^2 = 4x$, $x^2 = 5 - 2y$, $x = 0$ — *Ans.* $(\frac{13}{40}, \frac{26}{15})$
(e) The first-quadrant area bounded by $x^2 - 8y + 4 = 0$, $x^2 = 4y$, $x = 0$ — *Ans.* $(\frac{3}{4}, \frac{2}{5})$
(f) The area of Problem 13(e) — *Ans.* $(\frac{1}{2}\sqrt{3}, \frac{6}{5})$
(g) The first-quadrant area of Problem 13(f) — *Ans.* $\left(\dfrac{16\pi + 6\sqrt{3}}{2\pi + 3\sqrt{3}}, \dfrac{22}{2\pi + 3\sqrt{3}}\right)$

15. Verify that $\frac{1}{2}\int_\alpha^\beta [g_2^2(\theta) - g_1^2(\theta)]d\theta = \int_\alpha^\beta\int_{g_1(\theta)}^{g_2(\theta)} \rho\,d\rho\,d\theta = \iint_R dA$; then infer that

$$\iint_R f(x,y)dA = \iint_R f(\rho\cos\theta, \rho\sin\theta)\rho\,d\rho\,d\theta$$

16. Find I_x and I_y for each of the following areas:

(a) The area of Problem 13(a) — *Ans.* $I_x = 6A$; $I_y = \frac{32}{3}A$
(b) The area cut from $y^2 = 8x$ by its latus rectum — *Ans.* $I_x = \frac{16}{5}A$; $I_y = \frac{12}{7}A$
(c) The area bounded by $y = x^2$ and $y = x$ — *Ans.* $I_x = \frac{3}{14}A$; $I_y = \frac{3}{10}A$
(d) The area bounded by $y = 4x - x^2$ and $y = x$ — *Ans.* $I_x = \frac{459}{70}A$; $I_y = \frac{27}{10}A$

17. Find I_x and I_y for one loop of $\rho^2 = \cos 2\,\theta$.

Ans. $I_x = \left(\frac{\pi}{16} - \frac{1}{6}\right)A$; $I_y = \left(\frac{\pi}{16} + \frac{1}{6}\right)A$

18. Find I_0 for (a) the loop of $\theta = \sin 2\,\theta$ and (b) the area enclosed by $\theta = 1 + \cos\,\theta$.

Ans. (a) $\frac{3}{8}A$; (b) $\frac{35}{24}A$

19. (a) Let the region R shown in Fig. 55-13 have area A and centroid $(\bar{x}, \bar{y})$. If R is revolved about the x axis, show that the volume V of the resulting solid of revolution is equal to $2\pi\bar{x}A$. (*Hint:* Use the method of cylindrical shells.)

(b) Prove the Theorem of Pappus: If d is the distance traveled by the centroid during the revolution (of part (a)), show that $V = Ad$.

(c) Prove that the volume of the torus generated by revolving the disk shown in Fig. 55-14 about the x axis is $2\pi^2a^2b$. (It is assumed that $0 < a < b$.)

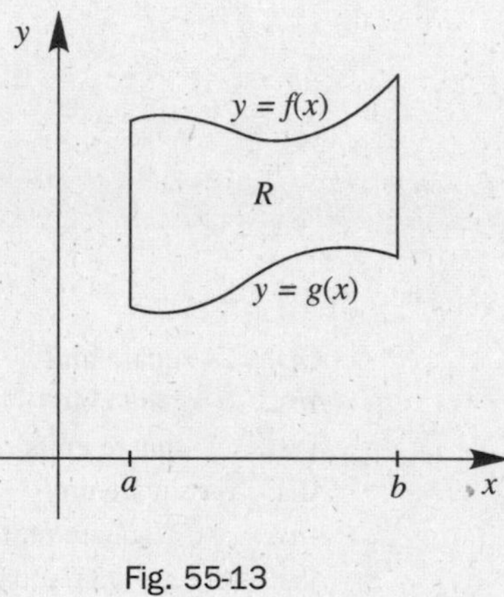

Fig. 55-13

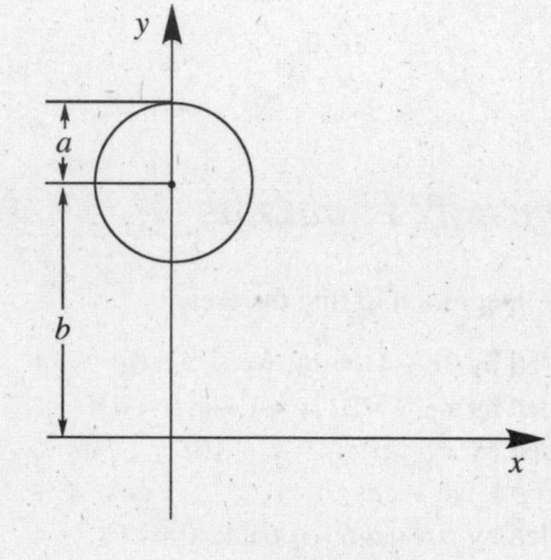

Fig. 55-14

CHAPTER 56

Double Integration Applied to Volume Under a Surface and the Area of a Curved Surface

Let $z = f(x, y)$ or $z = f(\rho, \theta)$ define a surface.

The volume V under the surface, that is, the volume of a vertical column whose upper base is in the surface and whose lower base is in the xy plane, is given by the double integral

$$V = \iint_R z \, dA \tag{56.1}$$

where R is the region forming the lower base.

The area S of the portion R^* of the surface lying above the region R is given by the double integral

$$S = \iint_R \sqrt{1 + \left(\frac{\partial z}{\partial x}\right)^2 + \left(\frac{\partial z}{\partial y}\right)^2} \, dA \tag{56.2}$$

If the surface is given by $x = f(y, z)$ and the region R lies in the yz plane, then

$$S = \iint_R \sqrt{1 + \left(\frac{\partial x}{\partial y}\right)^2 + \left(\frac{\partial x}{\partial z}\right)^2} \, dA \tag{56.3}$$

If the surface is given by $y = f(x, z)$ and the region R lies in the xz plane, then

$$S = \iint_R \sqrt{1 + \left(\frac{\partial y}{\partial x}\right)^2 + \left(\frac{\partial y}{\partial z}\right)^2} \, dA \tag{56.4}$$

SOLVED PROBLEMS

1. Find the volume in the first octant between the planes $z = 0$ and $z = x + y + 2$, and inside the cylinder $x^2 + y^2 = 16$.

From Fig. 56-1, it is evident that $z = x + y + 2$ is to be integrated over a quadrant of the circle $x^2 + y^2 = 16$ in the xy plane. Hence,

$$V = \iint_R z\,dA = \int_0^4 \int_0^{\sqrt{16-x^2}} (x + y + 2)\,dy\,dx = \int_0^4 \left(x\sqrt{16 - x^2} + 8 - \frac{1}{2}x^2 + 2\sqrt{16 - x^2}\right)dx$$

$$= \left[-\frac{1}{3}(16 - x^2)^{3/2} + 8x - \frac{x^3}{6} + x\sqrt{16 - x^2} + 16\sin^{-1}\frac{1}{4}x\right]_0^4 = \left(\frac{128}{3} + 8\pi\right) \text{ cubic units}$$

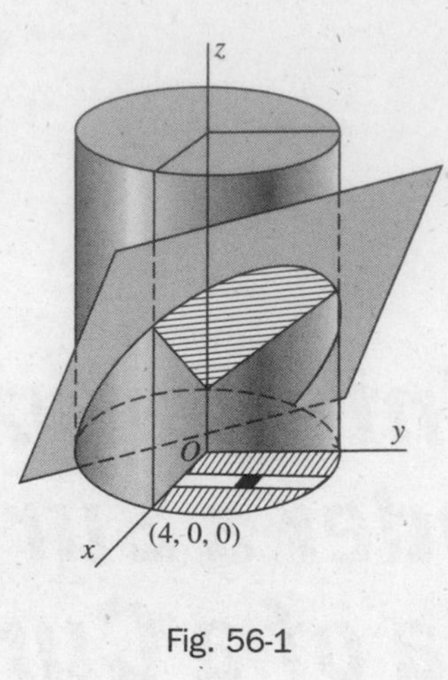

Fig. 56-1

2. Find the volume bounded by the cylinder $x^2 + y^2 = 4$ and the planes $y + z = 4$ and $z = 0$.

From Fig. 56-2, it is evident that $z = 4 - y$ is to be integrated over the circle $x^2 + y^2 = 4$ in the xy plane. Hence,

$$V = \int_{-2}^{2}\int_{-\sqrt{4-y^2}}^{\sqrt{4-y^2}} (4 - y)dx\,dy = 2\int_{-2}^{2}\int_{0}^{\sqrt{4-y^2}} (4 - y)dx\,dy = 16\pi \text{ cubic units}$$

3. Find the volume bounded above by the paraboloid $x^2 + 4y^2 = z$, below by the plane $z = 0$, and laterally by the cylinders $y^2 = x$ and $x^2 = y$. (See Fig. 56-3.)

The required volume is obtained by integrating $z = x^2 + 4y^2$ over the region R common to the parabolas $y^2 = x$ and $x^2 = y$ in the xy plane. Hence,

$$V = \int_{0}^{1}\int_{x^2}^{\sqrt{x}} (x^2 + 4y^2)dy\,dx = \int_{0}^{1}\left[x^2y + \frac{4}{3}y^3\right]_{x^2}^{\sqrt{x}} dx = \tfrac{3}{7} \text{ cubic units}$$

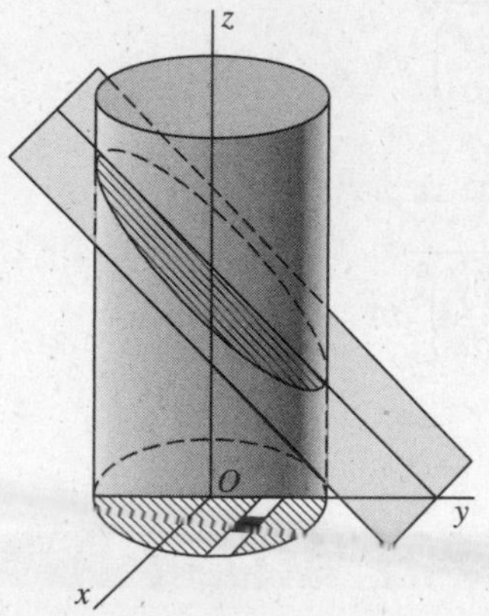

Fig. 56-2

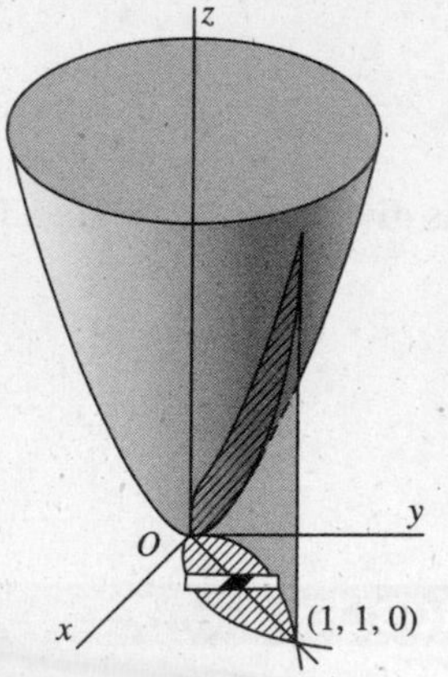

Fig. 56-3

4. Find the volume of one of the wedges cut from the cylinder $4x^2 + y^2 = a^2$ by the planes $z = 0$ and $z = my$. (See Fig. 56-4.)

The volume is obtained by integrating $z = my$ over half the ellipse $4x^2 + y^2 = a^2$. Hence,

$$V = 2\int_{0}^{a/2}\int_{0}^{\sqrt{a^2-4x^2}} my\,dy\,dx = m\int_{0}^{a/2} [y^2]_{0}^{\sqrt{a^2-4x^2}}\,dx = \frac{ma^3}{3} \text{ cubic units}$$

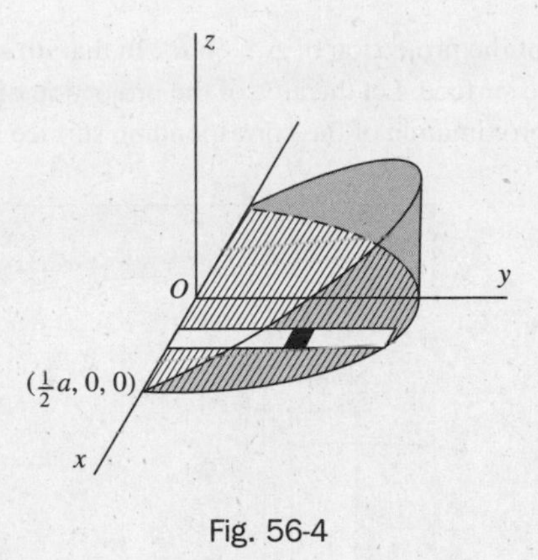

Fig. 56-4

5. Find the volume bounded by the paraboloid $x^2 + y^2 = 4z$, the cylinder $x^2 + y^2 = 8y$, and the plane $z = 0$. (See Fig. 56-5.)

The required volume is obtained by integrating $z = \frac{1}{4}(x^2 + y^2)$ over the circle $x^2 + y^2 = 8y$. Using cylindrical coordinates (see Chapter 57), the volume is obtained by integrating $z = \frac{1}{4}\rho^2$ over the circle $\rho = 8 \sin \theta$. Then,

$$V = \iint_R z\,dA = \int_0^{\pi}\int_0^{8\sin\theta} z\rho\,d\rho\,d\theta = \frac{1}{4}\int_0^{\pi}\int_0^{8\sin\theta} \rho^3 d\rho\,d\theta$$

$$= \frac{1}{16}\int_0^{\pi} [\rho^4]_0^{8\sin\theta}\,d\theta = 256\int_0^{\pi} \sin^4\theta\,d\theta = 96\pi \text{ cubic units}$$

6. Find the volume removed when a hole of radius a is bored through a sphere of radius $2a$, the axis of the hole being a diameter of the sphere. (See Fig. 56-6.)

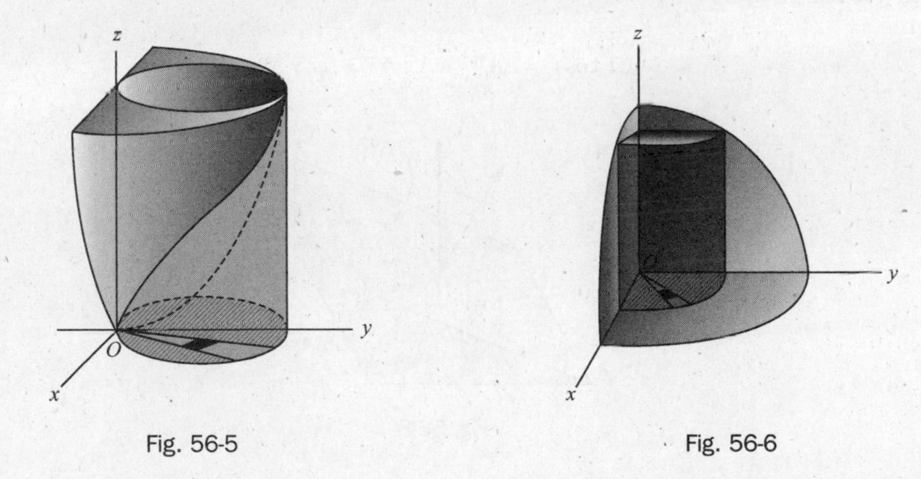

Fig. 56-5 Fig. 56-6

From the figure, it is obvious that the required volume is eight times the volume in the first octant bounded by the cylinder $\rho^2 = a^2$, the sphere $\rho^2 + z^2 = 4a^2$, and the plane $z = 0$. The latter volume is obtained by integrating $z = \sqrt{4a^2 - \rho^2}$ over a quadrant of the circle $\rho = a$. Hence,

$$V = 8\int_0^{\pi/2}\int_0^{a} \sqrt{4a^2 - \rho^2}\,\rho\,d\rho\,d\theta = \frac{8}{3}\int_0^{\pi/2} (8a^3 - 3\sqrt{3}a^3)d\theta = \tfrac{4}{3}(8 - 3\sqrt{3})a^3\pi \text{ cubic units}$$

7. Derive formula (56.2).

Consider a region R^* of area S on the surface $z = f(x, y)$. Through the boundary of R^* pass a vertical cylinder (see Fig. 56-7) cutting the xy plane in the region R. Now divide R into n subregions $R_1, \ldots, R_n$ of areas $\Delta A_1, \ldots,$

ΔA_n, and denote by ΔS_i, the area of the projection of ΔA_i on R^*. In that ith subregion of R^*, choose a point P_i and draw there the tangent plane to the surface. Let the area of the projection of R_i on this tangent plane be denoted by ΔT_i. We shall use ΔT_i as an approximation of the corresponding surface area ΔS_i.

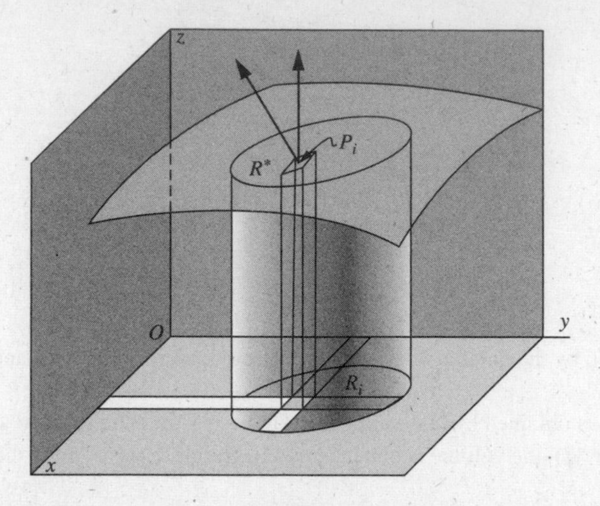

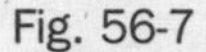
Fig. 56-7

Now the angle between the xy plane and the tangent plane at P_i is the angle γ_i between the z axis with direction numbers [0, 0, 1] and the normal, $\left[-\frac{\partial f}{\partial x}, -\frac{\partial f}{\partial y}, 1\right] = \left[-\frac{\partial z}{\partial x}, -\frac{\partial z}{\partial y}, 1\right]$, to the surface at P_i. Thus,

$$\cos \gamma_i = \frac{1}{\sqrt{\left(\frac{\partial z}{\partial x}\right)^2 + \left(\frac{\partial z}{\partial y}\right)^2 + 1}}$$

Then (see Fig. 56-8)

$$\Delta T_i \cos \gamma_i = \Delta A_i \quad \text{and} \quad \Delta T_i = \sec \gamma_i \Delta A_i$$

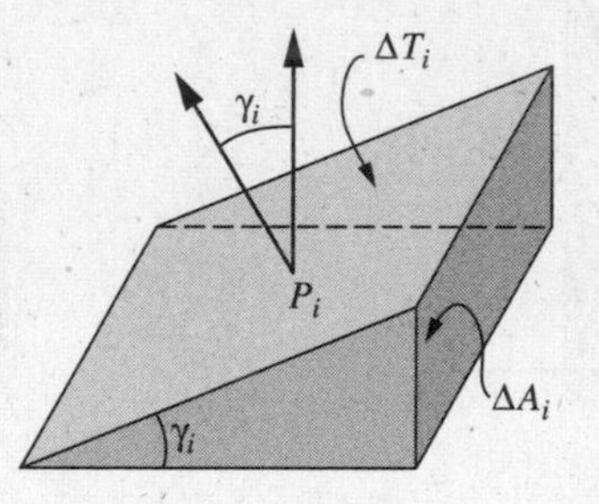

Fig. 56-8

Hence, an approximation of S is $\sum_{i=1}^{n} \Delta T_i = \sum_{i=1}^{n} \sec \gamma_i \Delta A_i$, and

$$S = \lim_{n \to +\infty} \sum_{i=1}^{n} \sec \gamma_i \Delta A_i = \iint_R \sec \gamma \, dA = \iint_R \sqrt{\left(\frac{\partial z}{\partial x}\right)^2 + \left(\frac{\partial z}{\partial y}\right)^2 + 1} \, dA$$

8. Find the area of the portion of the cone $x^2 + y^2 = 3z^2$ lying above the xy plane and inside the cylinder $x^2 + y^2 = 4y$.

Solution 1: Refer to Fig. 56-9. The projection of the required area on the xy plane is the region R enclosed by the circle $x^2 + y^2 = 4y$. For the cone,

$$\frac{\partial z}{\partial x} = \frac{1}{3}\frac{x}{z} \text{ and } \frac{\partial z}{\partial y} = \frac{1}{3}\frac{y}{z}. \quad \text{So} \quad 1 + \left(\frac{\partial z}{\partial x}\right)^2 + \left(\frac{\partial z}{\partial y}\right)^2 = \frac{9z^2 + x^2 + y^2}{9z^2} = \frac{12z^2}{9z^2} = \frac{4}{3}$$

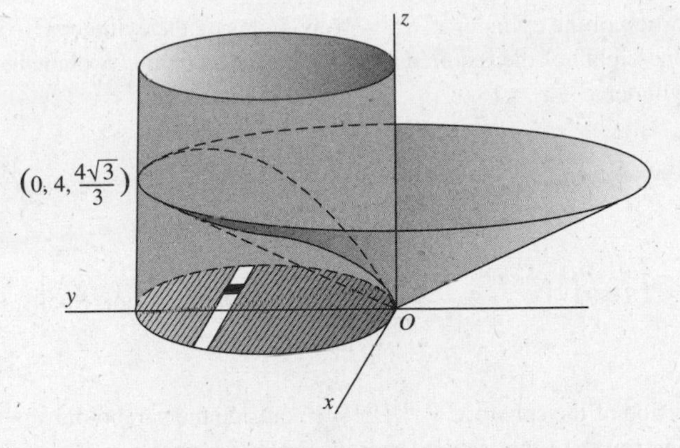

Fig. 56-9

Then $$S = \iint_R \sqrt{1+\left(\frac{\partial z}{\partial x}\right)^2+\left(\frac{\partial z}{\partial y}\right)^2}\, dA = \int_0^4 \int_{-\sqrt{4y-y^2}}^{\sqrt{4y-y^2}} \frac{2}{\sqrt{3}}\, dx\, dy = 2\frac{2}{\sqrt{3}}\int_0^4 \int_0^{\sqrt{4y-y^2}} dx\, dy$$

$$= \frac{4}{\sqrt{3}}\int_0^4 \sqrt{4y-y^2}\, dy = \frac{8\sqrt{3}}{3}\pi \text{ square units}$$

Solution 2: Refer to Fig. 56-10. The projection of one-half the required area on the yz plane is the region R bounded by the line $y = \sqrt{3}z$ and the parabola $y = \frac{3}{4}z^2$, the latter having been obtained by eliminating x from the equations of the two surfaces. For the cone,

$$\frac{\partial x}{\partial y} = -\frac{y}{x} \quad \text{and} \quad \frac{\partial x}{\partial z} = \frac{3z}{x}. \quad \text{So} \quad 1+\left(\frac{\partial x}{\partial y}\right)^2+\left(\frac{\partial x}{\partial z}\right)^2 = \frac{x^2+y^2+9z^2}{x^2} = \frac{12z^2}{x^2} = \frac{12z^2}{3z^2-y^2}.$$

Then $$S = 2\int_0^4 \int_{y/\sqrt{3}}^{2\sqrt{y}/\sqrt{3}} \frac{2\sqrt{3}z}{\sqrt{3z^2-y^2}}\, dz\, dy = \frac{4\sqrt{3}}{3}\int_0^4 [\sqrt{3z^2-y^2}]_{y/\sqrt{3}}^{2\sqrt{y}/\sqrt{3}}\, dy = \frac{4\sqrt{3}}{3}\int_0^4 \sqrt{4y-y^2}\, dy.$$

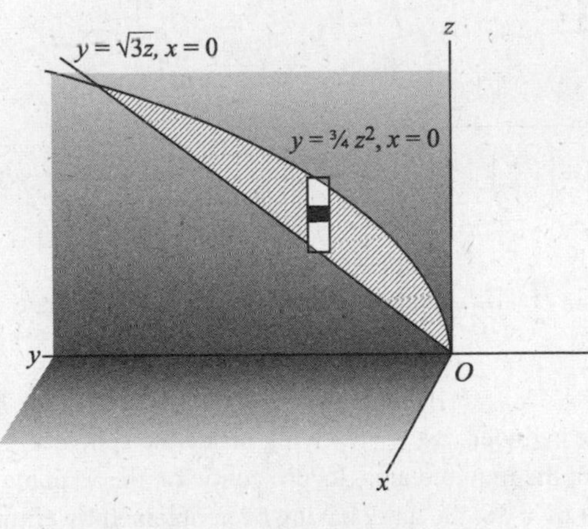

Fig. 56-10

Solution 3: Using polar coordinates in solution 1, we must integrate $\sqrt{1+\left(\frac{\partial z}{\partial x}\right)^2+\left(\frac{\partial z}{\partial y}\right)^2} = \frac{2}{\sqrt{3}}$ over the region R enclosed by the circle $\rho = 4 \sin \theta$. Then,

$$S = \iint_R \frac{2}{\sqrt{3}}\, dA = \int_0^\pi \int_0^{4\sin\theta} \frac{2}{\sqrt{3}}\rho\, d\rho\, d\theta = \frac{1}{\sqrt{3}}\int_0^\pi [\rho^2]_0^{4\sin\theta}\, d\theta$$

$$= \frac{16}{\sqrt{3}}\int_0^\pi \sin^2\theta\, d\theta = \frac{8\sqrt{3}}{3}\pi \text{ square units}$$

9. Find the area of the portion of the cylinder $x^2 + z^2 = 16$ lying inside the cylinder $x^2 + y^2 = 16$.

Fig. 56-11 shows one-eighth of the required area, its projection on the xy plane being a quadrant of the circle $x^2 + y^2 = 16$. For the cylinder $x^2 + z^2 = 16$,

$$\frac{\partial z}{\partial x} = -\frac{x}{z} \quad \text{and} \quad \frac{\partial z}{\partial y} = 0. \quad \text{So} \quad 1 + \left(\frac{\partial z}{\partial x}\right)^2 + \left(\frac{\partial z}{\partial y}\right)^2 = \frac{x^2 + z^2}{z^2} = \frac{16}{16 - x^2}.$$

Then

$$S = 8\int_0^4 \int_0^{\sqrt{16-x^2}} \frac{4}{\sqrt{16 - x^2}}\, dy\, dx = 32\int_0^4 dx = 128 \text{ square units}$$

10. Find the area of the portion of the sphere $x^2 + y^2 + z^2 = 16$ outside the paraboloid $x^2 + y^2 + z = 16$.

Fig. 56-12 shows one-fourth of the required area, its projection on the yz plane being the region R bounded by the circle $y^2 + z^2 = 16$, the y and z axes, and the line $z = 1$. For the sphere,

$$\frac{\partial x}{\partial y} = -\frac{y}{x} \quad \text{and} \quad \frac{\partial x}{\partial z} = -\frac{z}{x}. \quad \text{So} \quad 1 + \left(\frac{\partial x}{\partial y}\right)^2 + \left(\frac{\partial x}{\partial z}\right)^2 = \frac{x^2 + y^2 + z^2}{x^2} = \frac{16}{16 - y^2 - z^2}.$$

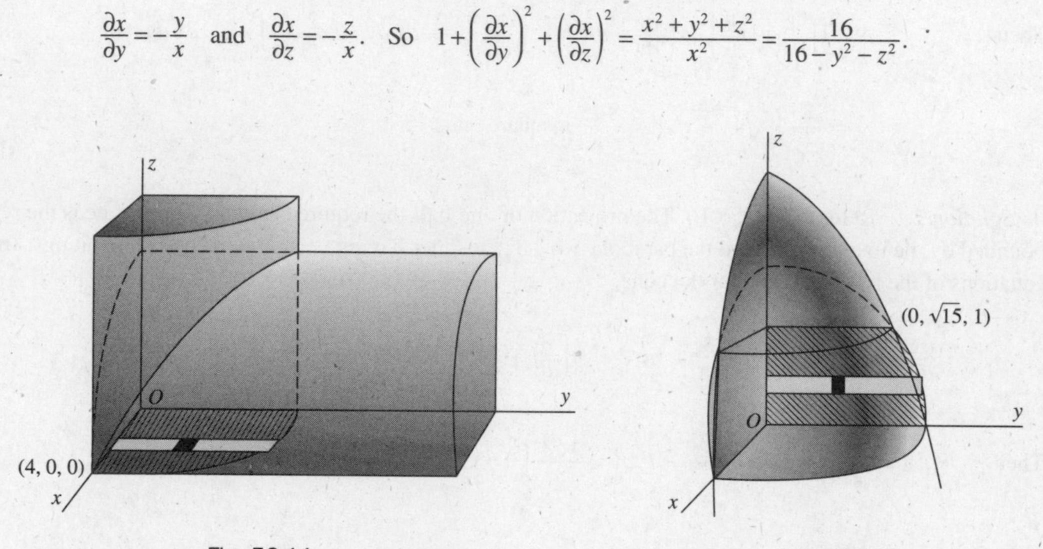

Fig. 56-11

Fig. 56-12

Then

$$S = 4\iint_R \sqrt{1 + \left(\frac{\partial x}{\partial y}\right)^2 + \left(\frac{\partial x}{\partial z}\right)^2}\, dA = 4\int_0^1 \int_0^{\sqrt{16-z^2}} \frac{4}{\sqrt{16 - y^2 - z^2}}\, dy\, dz$$

$$= 16\int_0^1 \left[\sin^{-1}\left(\frac{y}{\sqrt{16 - z^2}}\right)\right]_0^{\sqrt{16-z^2}} dz = 16\int_0^1 \frac{\pi}{2}\, dz = 8\pi \text{ square units}$$

11. Find the area of the portion of the cylinder $x^2 + y^2 = 6y$ lying inside the sphere $x^2 + y^2 + z^2 = 36$.

Fig. 56-13 shows one-fourth of the required area. Its projection on the yz plane is the region R bounded by the z and y axes and the parabola $z^2 + 6y = 36$, the latter having been obtained by eliminating x from the equations of the two surfaces. For the cylinder,

$$\frac{\partial x}{\partial y} = \frac{3 - y}{x} \quad \text{and} \quad \frac{\partial x}{\partial z} = 0. \quad \text{So} \quad 1 + \left(\frac{\partial x}{\partial y}\right)^2 + \left(\frac{\partial x}{\partial z}\right)^2 = \frac{x^2 + 9 - 6y + y^2}{x^2} = \frac{9}{6y - y^2}$$

Then

$$S = 4\int_0^6 \int_0^{\sqrt{36-6y}} \frac{3}{\sqrt{6y - y^2}}\, dz\, dy = 12\int_0^6 \frac{\sqrt{6}}{\sqrt{y}}\, dy = 144 \text{ square units}$$

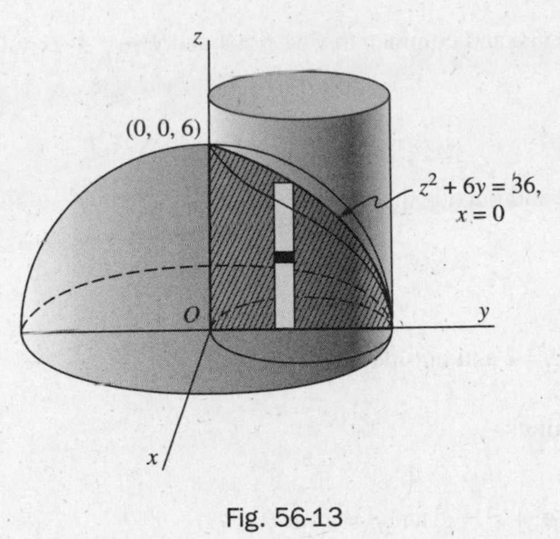

Fig. 56-13

SUPPLEMENTARY PROBLEMS

12. Find the volume cut from $9x^2 + 4y^2 + 36z = 36$ by the plane $z = 0$.

Ans. 3π cubic units

13. Find the volume under $z = 3x$ and above the first-quadrant area bounded by $x = 0$, $y = 0$, $x = 4$, and $x^2 + y^2 = 25$.

Ans. 98 cubic units

14. Find the volume in the first octant bounded by $x^2 + z = 9$, $3x + 4y = 24$, $x = 0$, $y = 0$, and $z = 0$.

Ans. 1485/16 cubic units

15. Find the volume in the first octant bounded by $xy = 4z$, $y = x$, and $x = 4$.

Ans. 8 cubic units

16. Find the volume in the first octant bounded by $x^2 + y^2 = 25$ and $z = y$.

Ans. 125/3 cubic units

17. Find the volume common to the cylinders $x^2 + y^2 = 16$ and $x^2 + z^2 = 16$.

Ans. 1024/3 cubic units

18. Find the volume in the first octant inside $y^2 + z^2 = 9$ and outside $y^2 = 3x$.

Ans. $27\pi/16$ cubic units

19. Find the volume in the first octant bounded by $x^2 + z^2 = 16$ and $x - y = 0$.

Ans. 64/3 cubic units

20. Find the volume in front of $x = 0$ and common to $y^2 + z^2 = 4$ and $y^2 + z^2 + 2x = 16$.

Ans. 28π cubic units

21. Find the volume inside $\rho = 2$ and outside the cone $z^2 = \rho^2$.

Ans. $32\pi/3$ cubic units

22. Find the volume inside $y^2 + z^2 = 2$ and outside $x^2 - y^2 - z^2 = 2$.

Ans. $8\pi(4 - \sqrt{2})/3$ cubic units

23. Find the volume common to $\rho^2 + z^2 = a^2$ and $\rho = a \sin \theta$.

Ans. $2(3\pi - 4)a^2/9$ cubic units

24. Find the volume inside $x^2 + y^2 = 9$, bounded below by $x^2 + y^2 + 4z = 16$ and above by $z = 4$.

Ans. $81\pi/8$ cubic units

25. Find the volume cut from the paraboloid $4x^2 + y^2 = 4z$ by the plane $z - y = 2$.

Ans. 9π cubic units

26. Find the volume generated by revolving the cardiod $\rho = 2(1 - \cos \theta)$ about the polar axis.

Ans. $V = 2\pi \iint y\rho \, d\rho \, d\theta = 64\pi/3$ cubic units

27. Find the volume generated by revolving a petal of $\rho = \sin 2\theta$ about either axis.

Ans. $32\pi/105$ cubic units

28. Find the area of the portion of the cone $x^2 + y^2 = z^2$ inside the vertical prism whose base is the triangle bounded by the lines $y = x$, $x = 0$, and $y = 1$ in the xy plane.

Ans. $\frac{1}{2}\sqrt{2}$ square units

29. Find the area of the portion of the plane $x + y + z = 6$ inside the cylinder $x^2 + y^2 = 4$.

Ans. $4\sqrt{3}\pi$ square units

30. Find the area of the portion of the sphere $x^2 + y^2 + z^2 = 36$ inside the cylinder $x^2 + y^2 = 6y$.

Ans. $72(\pi - 2)$ square units

31. Find the area of the portion of the sphere $x^2 + y^2 + z^2 = 4z$ inside the paraboloid $x^2 + y^2 = z$.

Ans. 4π square units

32. Find the area of the portion of the sphere $x^2 + y^2 + z^2 = 25$ between the planes $z = 2$ and $z = 4$.

Ans. 20π square units

33. Find the area of the portion of the surface $z = xy$ inside the cylinder $x^2 + y^2 = 1$.

Ans. $2\pi(2\sqrt{2} - 1)/3$ square units

34. Find the area of the surface of the cone $x^2 + y^2 - 9z^2 = 0$ above the plane $z = 0$ and inside the cylinder $x^2 + y^2 = 6y$.

Ans. $3\sqrt{10}\pi$ square units

35. Find the area of that part of the sphere $x^2 + y^2 + z^2 = 25$ that is within the elliptic cylinder $2x^2 + y^2 = 25$.

Ans. 50π square units

36. Find the area of the surface of $x^2 + y^2 - az = 0$ which lies directly above the lemniscate $4\rho^2 = a^2 \cos 2\theta$.

Ans. $S = \frac{4}{a}\iint \sqrt{4\rho^2 + a^2}\,\rho\, d\rho\, d\theta = \frac{a^2}{3}\left(\frac{5}{3} - \frac{\pi}{4}\right)$ square units

37. Find the area of the surface of $x^2 + y^2 + z^2 = 4$ which lies directly above the cardioid $\rho = 1 - \cos\theta$.

Ans. $8[\pi - \sqrt{2} - \ln(\sqrt{2} + 1)]$ square units

CHAPTER 57

Triple Integrals

Cylindrical and Spherical Coordinates

Assume that a point P has coordinates (x, y, z) in a right-handed rectangular coordinate system. The corresponding *cylindrical coordinates* of P are (r, θ, z), where (r, θ) are polar coordinates for the point (x, y) in the xy plane. (Note the notational change here from (ρ, θ) to (r, θ) for the polar coordinates of (x, y); see Fig. 57-1.) Hence, we have the relations

$$x = r\cos\theta, \qquad y = r\sin\theta, \qquad r^2 = x^2 + y^2, \qquad \tan\theta = \frac{y}{x}$$

In cylindrical coordinates, an equation $r = c$ represents a right cylinder of radius c with the z axis as its axis of symmetry. An equation $\theta = c$ represents a plane through the z axis.

A point P with rectangular coordinates (x, y, z) has the *spherical coordinates* (ρ, θ, ϕ), where $\rho = |OP|$, θ is the same as in cylindrical coordinates, and ϕ is the directed angle from the positive z axis to the vector **OP**. (See Fig. 57-2.) In spherical coordinates, an equation $\rho = c$ represents a sphere of radius c with center at the origin. An equation $\phi = c$ represents a cone with vertex at the origin and the z axis as its axis of symmetry.

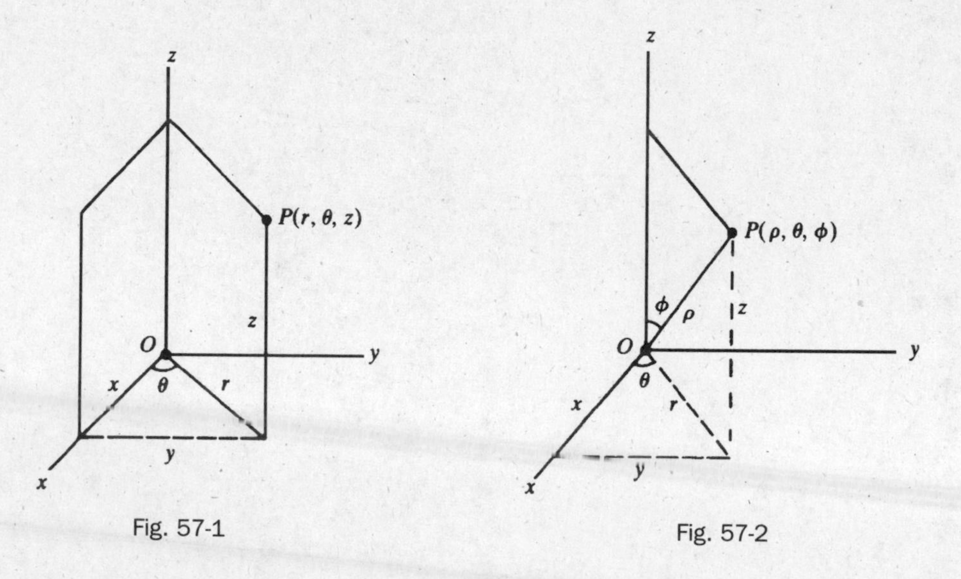

Fig. 57-1

Fig. 57-2

The following additional relations, easily deduced from Fig. 57-2 and the equations above, hold among spherical, cylindrical, and rectangular coordinates:

$$r = \rho\sin\phi, \qquad \rho^2 = x^2 + y^2 + z^2$$

$$x = \rho\sin\phi\cos\theta, \qquad y = \rho\sin\phi\sin\theta, \qquad z = \rho\cos\phi$$

The Triple Integral

Let $f(x, y, z)$ be a continuous function on a three-dimensional region R. The definition of the double integral can be extended in an obvious way to obtain the definition of the triple integral $\iiint_R f(x, y, z)\,dV$.

If $f(x, y, z) = 1$, then $\iiint_R f(x, y, z)\,dV$ may be interpreted as measuring the volume of the region R.

Evaluation of Triple Integrals

As in the case of double integrals, a triple integral can be evaluated in terms of iterated integrals.

In rectangular coordinates,

$$\iiint_R f(x, y, z)\,dV = \int_a^b \int_{y_1(x)}^{y_2(x)} \int_{z_1(x,y)}^{z_2(x,y)} f(x, y, z)\,dz\,dy\,dx$$

$$= \int_c^d \int_{x_1(y)}^{x_2(y)} \int_{z_1(x,y)}^{z_2(x,y)} f(x, y, z)\,dz\,dx\,dy, \text{ etc.}$$

where the limits of integration are chosen to cover the region R.

In cylindrical coordinates,

$$\iiint_R f(r, \theta, z)\,dV = \int_\alpha^\beta \int_{r_1(\theta)}^{r_2(\theta)} \int_{z_1(r,\theta)}^{z_2(r,\theta)} f(r, \theta, z) r\,dz\,dr\,d\theta$$

where the limits of integration are chosen to cover the region R. (See Problem 23.)

In spherical coordinates,

$$\iiint_R f(\rho, \phi, \theta)\,dV = \int_\alpha^\beta \int_{\phi_1(\theta)}^{\phi_2(\theta)} \int_{\rho_1(\phi,\theta)}^{\rho_2(\phi,\theta)} f(\rho, \phi, \theta)\rho^2 \sin\phi\,d\rho\,d\phi\,d\theta$$

where the limits of integration are chosen to cover the region R. (See Problem 24.)

Discussion of the definitions: Consider the function $f(x, y, z)$, continuous over a region R of ordinary space. After slicing R with planes $x = \xi_i$ and $y = \eta_j$ as in Chapter 54, let these subregions be further sliced by planes $z = \zeta_k$. The region R has now been separated into a number of rectangular parallelepipeds of volume $\Delta V_{ijk} = \Delta x_i \Delta y_j \Delta z_k$ and a number of partial parallelepipeds which we shall ignore. In each complete parallelepiped, select a point $P_{ijk}(x_i, y_j, z_k)$; then compute $f(x_i, y_j, z_k)$ and form the sum

$$\sum_{\substack{i=1,\dots,m\\ j=1,\dots,n\\ k=1,\dots,p}} f(x_i, y_j, z_k)\Delta V_{ijk} = \sum_{\substack{i=1,\dots,m\\ j=1,\dots,n\\ k=1,\dots,p}} f(x_i, y_j, z_k)\Delta x_i \Delta y_j \Delta z_k \tag{57.1}$$

The triple integral of $f(x, y, z)$ over the region R is defined to be the limit of (57.1) as the number of parallelepipeds is indefinitely increased in such a manner that all dimensions of each go to zero.

In evaluating this limit, we may sum first each set of parallelepipeds having $\Delta_i x$ and $\Delta_j y$, for fixed i and j, as two dimensions and consider the limit as each $\Delta_k z \to 0$. We have

$$\lim_{p\to+\infty} \sum_{k=1}^{p} f(x_i, y_i, z_k)\Delta_k z \Delta_i x \Delta_j y = \int_{z_1}^{z_2} f(x_i, y_i, z)\,dz\,\Delta_i x \Delta_j y$$

Now these are the columns, the basic subregions, of Chapter 54; hence,

$$\lim_{\substack{m\to+\infty\\ n\to+\infty\\ p\to+\infty}} \sum_{\substack{i=1,\dots,m\\ j=1,\dots,n\\ k=1,\dots,p}} f(x_i, y_j, z_k)\Delta V_{ijk} = \iiint_R f(x, y, z)\,dz\,dx\,dy = \iiint_R f(x, y, z)\,dz\,dy\,dx$$

Centroids and Moments of Inertia

The coordinates $(\overline{x}, \overline{y}, \overline{z})$ of the *centroid of a volume* satisfy the relations

$$\overline{x}\iiint_R dV = \iiint_R x\,dV, \quad \overline{y}\iiint_R dV = \iiint_R y\,dV, \quad \overline{z}\iiint_R dV = \iiint_R z\,dV$$

The *moments of inertia of a volume* with respect to the coordinate axes are given by

$$I_x = \iiint_R (y^2+z^2)dV, \quad I_y = \iiint_R (z^2+x^2)dV, \quad I_z = \iiint_R (x^2+y^2)dV$$

SOLVED PROBLEMS

1. Evaluate the given triple integrals:

(a) $\int_0^1\int_0^{1-x}\int_0^{2-x} xyz\,dz\,dy\,dx$

$$= \int_0^1\left[\int_0^{1-x}\left(\int_0^{2-x} xyz\,dz\right)dy\right]dx$$

$$= \int_0^1\left[\int_0^{1-x}\left(\frac{xyz^2}{2}\Bigg|_{z=0}^{z=2-x}\right)dy\right]dx = \int_0^1\left[\int_0^{1-x}\frac{xy(2-x)^2}{2}dy\right]dx$$

$$= \int_0^1\left[\frac{xy^2(2-x)^2}{4}\right]_{y=0}^{y=1-x}dx = \frac{1}{4}\int_0^1(4x-12x^2+13x^3-6x^4+x^5)\,dx = \frac{13}{240}$$

(b) $\int_0^{\pi/2}\int_0^1\int_0^2 zr^2\sin\theta\,dz\,dr\,d\theta$

$$= \int_0^{\pi/2}\int_0^1\left[\frac{z^2}{2}\right]_0^2 r^2\sin\theta\,dr\,d\theta = 2\int_0^{\pi/2}\int_0^1 r^2\sin\theta\,dr\,d\theta$$

$$= \frac{2}{3}\int_0^{\pi/2}[r^3]_0^1\sin\theta\,d\theta = -\frac{2}{3}[\cos\theta]_0^{\pi/2} = \frac{2}{3}$$

(c) $\int_0^{\pi}\int_0^{\pi/4}\int_0^{\sec\phi}\sin 2\phi\,d\rho\,d\phi\,d\theta$

$$= 2\int_0^{\pi}\int_0^{\pi/4}\sin\phi\,d\phi\,d\theta = 2\int_0^{\pi}(1-\tfrac{1}{2}\sqrt{2})\,d\theta = (2-\sqrt{2})\pi$$

2. Compute the triple integral of $F(x, y, z) = z$ over the region R in the first octant bounded by the planes $y = 0$, $z = 0$, $x + y = 2$, $2y + x = 6$, and the cylinder $y^2 + z^2 = 4$. (See Fig. 57-3.)

Integrate first with respect to z from $z = 0$ (the xy plane) to $z = \sqrt{4-y^2}$ (the cylinder), then with respect to x from $x = 2 - y$ to $x = 6 - 2y$, and finally with respect to y from $y = 0$ to $y = 2$. This yields

$$\iiint_R z\,dV = \int_0^2\int_{2-y}^{6-2y}\int_0^{\sqrt{4-y^2}} z\,dz\,dx\,dy = \int_0^2\int_{2-y}^{6-2y}[\tfrac{1}{2}z^2]_0^{\sqrt{4-y^2}}\,dx\,dy$$

$$= \frac{1}{2}\int_0^2\int_{2-y}^{6-2y}(4-y^2)\,dx\,dy = \frac{1}{2}\int_0^2[(4-y^2)x]_{2-y}^{6-2y}\,dy = \frac{26}{3}$$

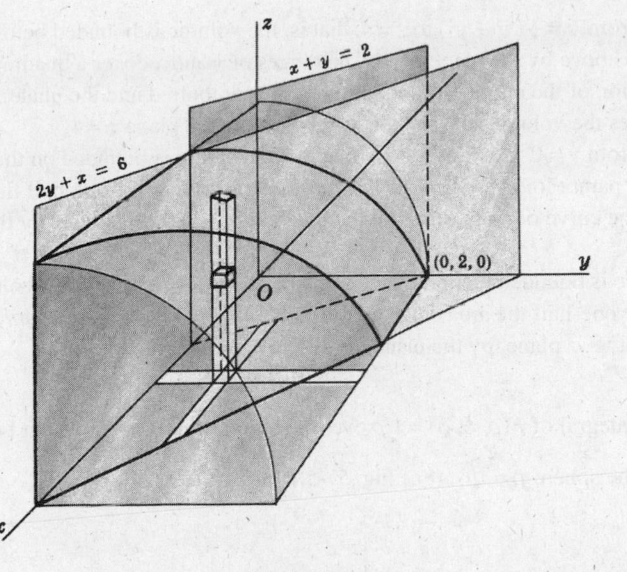

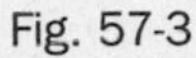
Fig. 57-3

3. Compute the triple integral of $f(r, \theta, z) = r^2$ over the region R bounded by the paraboloid $r^2 = 9 - z$ and the plane $z = 0$. (See Fig. 57-4.)

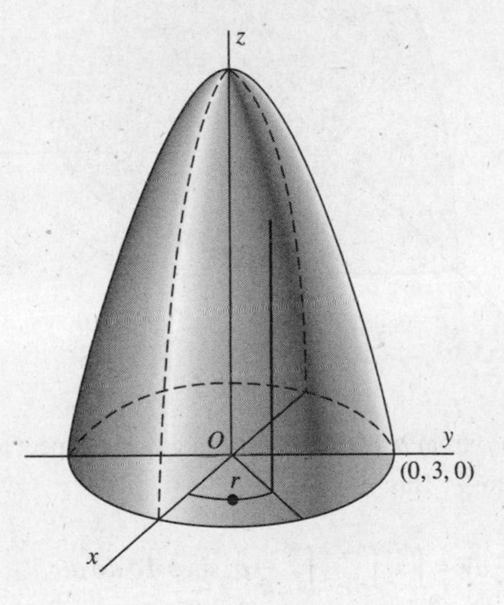

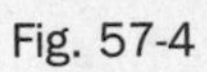
Fig. 57-4

Integrate first with respect to z from $z = 0$ to $z = 9 - r^2$, then with respect to r from $r = 0$ to $r = 3$, and finally with respect to θ from $\theta = 0$ to $\theta = 2\pi$. This yields

$$\iiint_R r^2 dV = \int_0^{2\pi}\int_0^3\int_0^{9-r^2} r^2(r\,dz\,dr\,d\theta) = \int_0^{2\pi}\int_0^3 r^3(9-r^2)dr\,d\theta$$

$$= \int_0^{2\pi}\left[\frac{9}{4}r^4 - \frac{1}{6}r^6\right]_0^3 d\theta = \int_0^{2\pi}\frac{243}{4}d\theta = \frac{243}{2}\pi$$

4. Show that the following integrals give the same volume: (a) $4\int_0^4\int_0^{\sqrt{16-x^2}}\int_{(x^2+y^2)/4}^4 dz\,dy\,dx$, (b) $\int_0^4\int_0^{2\sqrt{z}}\int_0^{\sqrt{4z-x^2}} dy\,dxdz$; and (c) $4\int_0^4\int_{y^2/4}^4\int_0^{\sqrt{4z-y^2}} dx\,dz\,dy$.

(a) Here z ranges from $z=\frac{1}{4}(x^2+y^2)$ to $z=4$, that is, the volume is bounded below by the paraboloid $4z=x^2+y^2$ and above by the plane $z=4$. The ranges of y and x cover a quadrant of the circle $x^2+y^2=16$, $z=0$, the projection of the curve of intersection of the paraboloid and the plane $z=4$ on the xy plane. Thus, the integral gives the volume cut from the paraboloid by the plane $z=4$.

(b) Here y ranges from $y=0$ to $y=\sqrt{4z-x^2}$, that is, the volume is bounded on the left by the xz plane and on the right by the paraboloid $y^2=4z-x^2$. The ranges of x and z cover one-half the area cut from the parabola $x^2=4z$, $y=0$, the curve of intersection of the paraboloid and the xz plane, by the plane $z=4$. The region R is that of (a).

(c) Here the volume is bounded behind by the yz plane and in front by the paraboloid $4z=x^2+y^2$. The ranges of z and y cover one-half the area cut from the parabola $y^2=4z$, $x=0$, the curve of intersection of the paraboloid and the yz plane, by the plane $z=4$. The region R is that of (a).

5. Compute the triple integral of $F(\rho,\theta,\phi)=1/\rho$ over the region R in the first octant bounded by the cones $\phi=\frac{\pi}{4}$ and $\phi=\tan^{-1}2$ and the sphere $\rho=\sqrt{6}$. (See Fig. 57-5.)

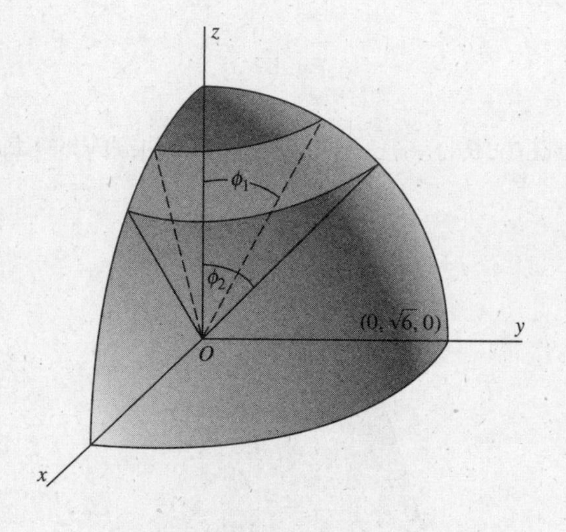

Fig. 57-5

Integrate first with respect to ρ from $\rho=0$ to $\rho=\sqrt{6}$, then with respect to ϕ from $\phi=\frac{\pi}{4}$ to $\phi=\tan^{-1}2$, and finally with respect to θ from 0 to $\frac{\pi}{2}$. This yields

$$\iiint_R \frac{1}{\rho}\,dV=\int_0^{\pi/2}\int_{\pi/4}^{\tan^{-1}2}\int_0^{\sqrt{6}}\frac{1}{\rho}\rho^2\sin\phi\,d\rho\,d\phi\,d\theta$$

$$=3\int_0^{\pi/2}\int_{\pi/4}^{\tan^{-1}2}\sin\phi\,d\phi\,d\theta$$

$$=-3\int_0^{\pi/2}\left(\frac{1}{\sqrt{5}}-\frac{1}{\sqrt{2}}\right)d\theta=\frac{3\pi}{2}\left(\frac{1}{\sqrt{2}}-\frac{1}{\sqrt{5}}\right)$$

6. Find the volume bounded by the paraboloid $z=2x^2+y^2$ and the cylinder $z=4-y^2$. (See Fig. 57-6.)

Integrate first with respect to z from $z=2x^2+y^2$ to $z=4-y^2$, then with respect to y from $y=0$ to $y=\sqrt{2-x^2}$ (obtain $x^2+y^2=2$ by eliminating x between the equations of the two surfaces), and finally with respect to x from $x=0$ to $x=\sqrt{2}$ (obtained by setting $y=0$ in $x^2+y^2=2$) to obtain one-fourth of the required volume. Thus,

$$V=4\int_0^{\sqrt{2}}\int_0^{\sqrt{2-x^2}}\int_{2x^2+y^2}^{4-y^2}dz\,dy\,dx=4\int_0^{\sqrt{2}}\int_0^{\sqrt{2-x^2}}[(4-y^2)+(2x^2+y^2)]dy\,dx$$

$$=4\int_0^{\sqrt{2}}\left[4y-2x^2y-\frac{2y^3}{3}\right]_0^{\sqrt{2-x^2}}dx=\frac{16}{3}\int_0^{\sqrt{2}}(2-x^2)^{3/2}dx=4\pi\text{ cubic units}$$

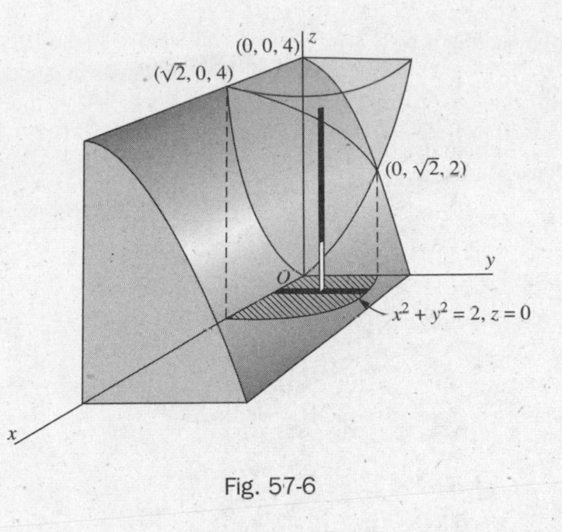

Fig. 57-6

7. Find the volume within the cylinder $r = 4\cos\theta$ bounded above by the sphere $r^2 + z^2 = 16$ and below by the plane $z = 0$. (See Fig. 57-7.)

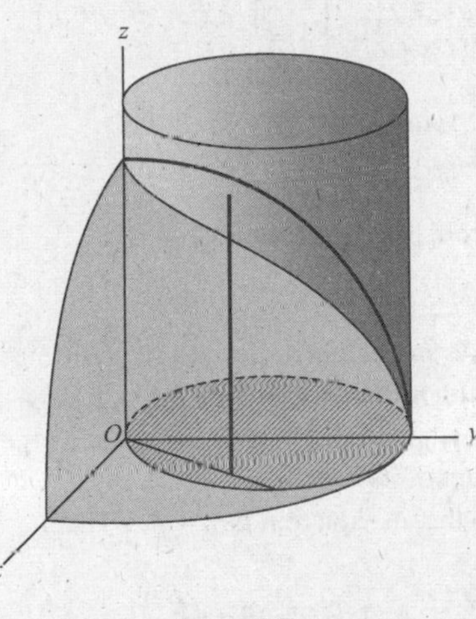

Fig. 57-7

Integrate first with respect to z from $z = 0$ to $z = \sqrt{16 - r^2}$, then with respect to r from $r = 0$ to $r = 4\cos\theta$, and finally with respect to θ from $\theta = 0$ to $\theta = \pi$ to obtain the required volume. Thus,

$$V = \int_0^{\pi}\int_0^{4\cos\theta}\int_0^{\sqrt{16-r^2}} r\,dz\,dy\,d\theta = \int_0^{\pi}\int_0^{4\cos\theta} r\sqrt{16-r^2}\,dr\,d\theta$$

$$= -\frac{64}{3}\int_0^{\pi}(\sin^3\theta - 1)\,d\theta = \frac{64}{9}(3\pi - 4)\text{ cubic units}$$

8. Find the coordinates of the centroid of the volume within the cylinder $r = 2\cos\theta$, bounded above by the paraboloid $z = r^2$ and below by the plane $z = 0$. (See Fig. 57-8.)

$$V = 2\int_0^{\pi/2}\int_0^{2\cos\theta}\int_0^{r^2} r\,dz\,dr\,d\theta = 2\int_0^{\pi/2}\int_0^{2\cos\theta} r^2\,dr\,d\theta$$

$$= \frac{1}{2}\int_0^{\pi/2}[r^4]_0^{2\cos\theta}\,d\theta = 8\int_0^{\pi/2}\cos^4\theta\,d\theta = \tfrac{3}{2}\pi$$

$$M_{yz} = \iiint_R x\,dV = 2\int_0^{\pi/2}\int_0^{2\cos\theta}\int_0^{r^2}(r\cos\theta)r\,dz\,dr\,d\theta$$

$$= 2\int_0^{\pi/2}\int_0^{2\cos\theta} r^4\cos\theta\,dr\,d\theta = \frac{64}{5}\int_0^{\pi/2}\cos^6\theta\,d\theta = 2\pi$$

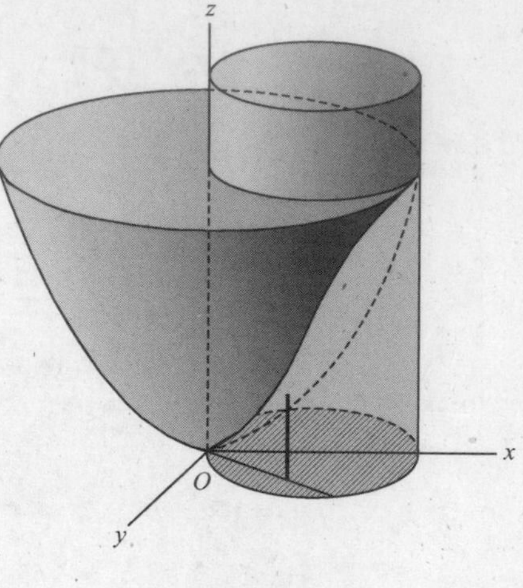

Fig. 57-8

Then $\bar{x} = M_{yz}/V = \frac{4}{3}$. By symmetry, $\bar{y} = 0$. Also,

$$M_{xy} = \iiint_R z\,dV = 2\int_0^{\pi/2}\int_0^{2\cos\theta}\int_0^{r^2} zr\,dz\,dr\,d\theta = \int_0^{\pi/2}\int_0^{2\cos\theta} r^5\,dr\,d\theta$$

$$= \frac{32}{3}\int_0^{\pi/2} \cos^6\theta\,d\theta = \tfrac{5}{3}\pi$$

and $\bar{z} = M_{xy}/V = \frac{10}{9}$. Thus, the centroid has coordinates $(\frac{4}{3}, 0, \frac{10}{9})$.

9. For the right circular cone of radius a and height h, find (a) the centroid; (b) the moment of inertia with respect to its axis; (c) the moment of inertia with respect to any line through its vertex and perpendicular to its axis; (d) the moment of inertia with respect to any line through its centroid and perpendicular to its axis; and (e) the moment of inertia with respect to any diameter of its base.

Take the cone as in Fig. 57-9, so that its equation is $r = \frac{a}{h}z$. Then:

$$V = 4\int_0^{\pi/2}\int_0^{a}\int_{hr/a}^{h} r\,dz\,dr\,d\theta = 4\int_0^{\pi/2}\int_0^{a}\left(hr - \frac{h}{a}r^2\right)dr\,d\theta$$

$$= \frac{2}{3}ha^2\int_0^{\pi/2} d\theta = \frac{1}{3}\pi ha^2$$

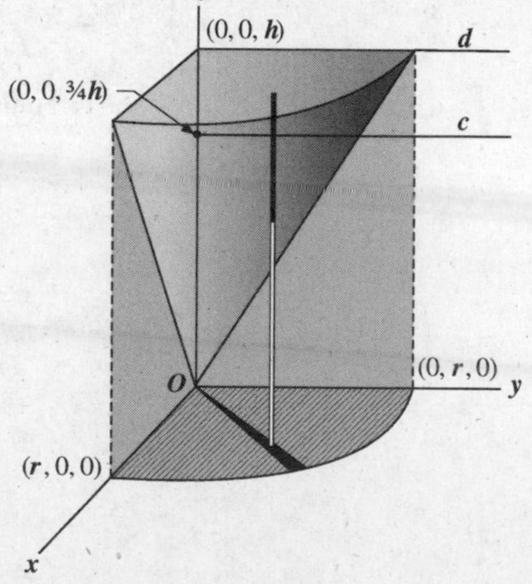

Fig. 57-9

(a) The centroid lies on the z axis, and we have

$$M_{xy} = \iiint_R z\,dV = 4\int_0^{\pi/2}\int_0^a\int_{hr/a}^h zr\;dz\;dr\;d\theta$$

$$= 2\int_0^{\pi/2}\int_0^a\left(h^2r - \frac{h^2}{a^2}r^3\right)dr\;d\theta - \frac{1}{2}h^2a^2\int_0^{\pi/2}d\theta = \frac{1}{4}\pi h^2a^2$$

Then $\bar{z} = M_{xy}/V = \frac{3}{4}h$, and the centroid has coordinates $(0, 0, \frac{3}{4}h)$.

(b) $I_z = \iiint_R (x^2+y^2)\,dV = 4\int_0^{\pi/2}\int_0^a\int_{hr/a}^h (r^2)r\;dz\;dr\;d\theta = \frac{1}{10}\pi ha^4 = \frac{3}{10}a^2V$

(c) Take the line as the y axis. Then

$$I_y = \iiint_R (x^2+z^2)\,dV = 4\int_0^{\pi/2}\int_0^a\int_{hr/a}^h (r^2\cos^2\theta + z^2)r\;dz\;dr\;d\theta$$

$$= 4\int_0^{\pi/2}\int_0^a\left[\left(hr^3 - \frac{h}{a}r^4\right)\cos^2\theta + \frac{1}{3}\left(h^3r - \frac{h^3}{a^3}r^4\right)\right]dr\;d\theta$$

$$= \frac{1}{5}\pi ha^2\left(h^2 + \frac{1}{4}a^2\right) = \frac{3}{5}\left(h^2 + \frac{1}{4}a^2\right)V$$

(d) Let the line c through the centroid be parallel to the y axis.

$$I_y = I_c + V(\tfrac{3}{4}h)^2 \quad\text{and}\quad I_c = \tfrac{3}{5}(h^2 + \tfrac{1}{4}a^2)V - \tfrac{9}{16}h^2V = \tfrac{3}{80}(h^2 + 4a^2)V$$

(e) Let d denote the diameter of the base of the cone parallel to the y axis. Then

$$I_d = I_c + V(\tfrac{1}{4}h)^2 = \tfrac{3}{80}(h^2 + 4a^2)V + \tfrac{1}{16}h^2V = \tfrac{1}{20}(2h^2 + 3a^2)V$$

10. Find the volume cut from the cone $\phi = \frac{1}{4}\pi$ by the sphere $\rho = 2a\cos\phi$. (See Fig. 57-10.)

$$V = 4\iiint_R dV = 4\int_0^{\pi/2}\int_0^{\pi/4}\int_0^{2a\cos\phi}\rho^2\sin\phi\;d\rho\;d\phi\;d\theta$$

$$= \frac{32a^3}{3}\int_0^{\pi/2}\int_0^{\pi/4}\cos^3\phi\sin\phi\;d\phi\;d\theta = 2a^3\int_0^{\pi/2}d\theta = \pi a^3 \text{ cubic units}$$

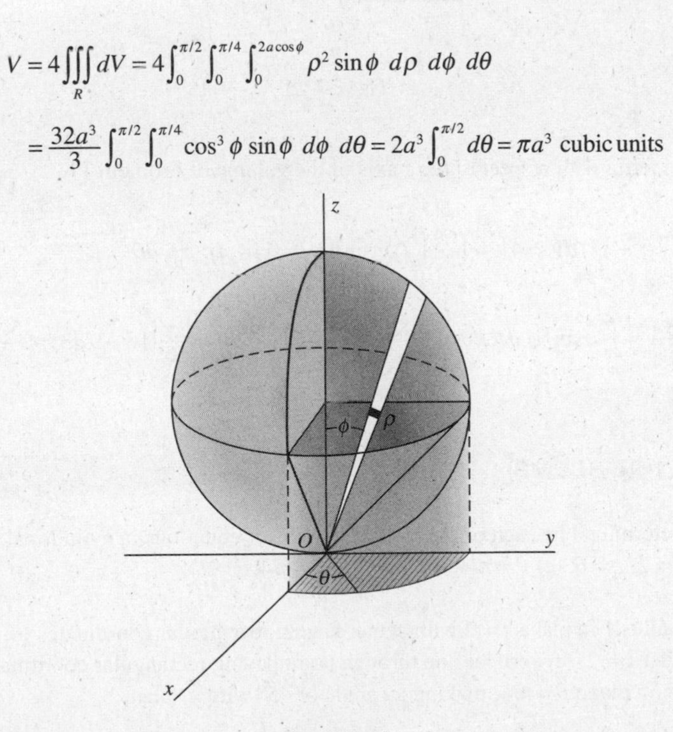

Fig. 57-10

11. Locate the centroid of the volume cut from one nappe of a cone of vertex angle 60° by a sphere of radius 2 whose center is at the vertex of the cone.

Take the surface as in Fig. 57-11, so that $x = y = 0$. In spherical coordinates, the equation of the cone is $\phi = \pi/6$, and the equation of the sphere is $\rho = 2$. Then

$$V = \iiint_R dV = 4\int_0^{\pi/2}\int_0^{\pi/6}\int_0^2 \rho^2 \sin\phi \, d\rho \, d\phi \, d\theta = \frac{32}{3}\int_0^{\pi/2}\int_0^{\pi/6} \sin\phi \, d\phi \, d\theta$$

$$= -\frac{32}{3}\left(\frac{\sqrt{3}}{2} - 1\right)\int_0^{\pi/2} d\theta = \frac{8\pi}{3}(2 - \sqrt{3})$$

$$M_{xy} = \iiint_R z \, dV = 4\int_0^{\pi/2}\int_0^{\pi/6}\int_0^2 (\rho\cos\phi)\rho^2 \sin\phi \, d\rho \, d\phi \, d\theta$$

$$= 8\int_0^{\pi/2}\int_0^{\pi/6} \sin 2\phi \, d\phi \, d\theta = \pi$$

and $\bar{z} = M_{xy}/V = \frac{3}{8}(2 + \sqrt{3})$.

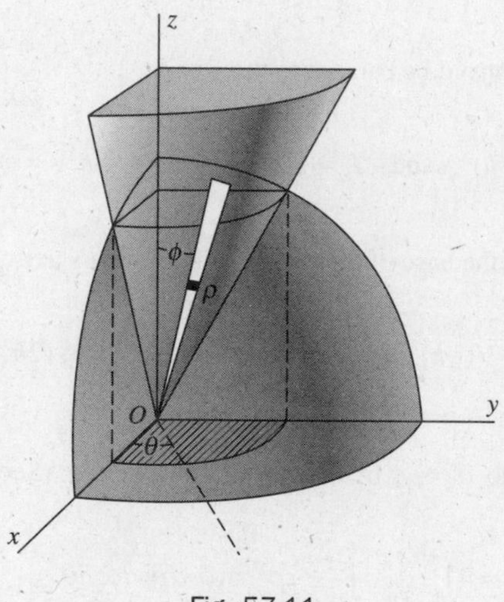

Fig. 57-11

12. Find the moment of inertia with respect to the z axis of the volume of Problem 11.

$$I_z = \iiint_R (x^2 + y^2) dV = 4\int_0^{\pi/2}\int_0^{\pi/6}\int_0^2 (\rho^2 \sin^2\phi)\rho^2 \sin\phi \, d\rho \, d\phi \, d\theta$$

$$= \frac{128}{5}\int_0^{\pi/2}\int_0^{\pi/6} \sin^3\phi \, d\phi \, d\theta = \frac{128}{5}\left(\frac{2}{3} - \frac{3}{8}\sqrt{3}\right)\int_0^{\pi/2} d\theta = \frac{8\pi}{15}(16 - 9\sqrt{3}) = \frac{5 - 2\sqrt{3}}{5}V$$

SUPPLEMENTARY PROBLEMS

13. Describe the curve determined by each of the following pairs of equations in cylindrical coordinates: (a) $r = 1, z = 2$; (b) $r = 2, z = \theta$; (c) $\theta = \pi/4, r = \sqrt{2}$; (d) $\theta = \pi/4, z = r$.

Ans. (a) circle of radius 1 in plane $z = 2$ with center having rectangular coordinates (0, 0, 2); (b) helix on right circular cylinder $r = 2$; (c) vertical line through point having rectangular coordinates (1, 1, 0); (d) line through origin in plane $\theta = \pi/4$, making an angle of 45° with xy plane

14. Describe the curve determined by each of the following pairs of equations in spherical coordinates: (a) $\rho = 1$, $\theta = \pi$; (b) $\theta = \frac{\pi}{4}$, $\phi = \frac{\pi}{6}$; (c) $\rho = 2$, $\phi = \frac{\pi}{4}$.

Ans. (a) circle of radius 1 in xz plane with center at origin; (b) halfline on intersection of plane $\theta = \pi/4$ and cone $\phi = \pi/6$; (c) circle of radius $\sqrt{2}$ in plane $z = \sqrt{2}$ with center on z axis

15. Transform each of the following equations in either rectangular, cylindrical, or spherical coordinates into equivalent equations in the two other coordinate systems:
(a) $\rho = 5$; (b) $z^2 = r^2$; (c) $x^2 + y^2 + (z-1)^2 = 1$

Ans. (a) $x^2 + y^2 + z^2 = 25$, $r^2 + z^2 = 25$; (b) $z^2 = x^2 + y^2$, $\cos^2\phi = \frac{1}{2}$ (that is, $\phi = \pi/4$ or $\phi = 3\pi/4$); (c) $r^2 + z^2 = 2z$, $\rho = 2\cos\phi$

16. Evaluate the triple integral on the left in each of the following:

(a) $\int_0^1 \int_1^2 \int_2^3 dz\, dx\, dy = 1$

(b) $\int_0^1 \int_{x^2}^{x} \int_0^{xy} dz\, dy\, dx = \frac{1}{24}$

(c) $\int_0^6 \int_0^{12-2y} \int_0^{4-2y/3-x/3} x\, dz\, dx\, dy = 144 \quad \left[= \int_0^{12} \int_0^{6-x/2} \int_0^{4-2y/3-x/3} x\, dz\, dy\, dx \right]$

(d) $\int_0^{\pi/2} \int_0^4 \int_0^{\sqrt{16-z^2}} (16 - r^2)^{1/2} rz\, dr\, d\theta = \frac{256}{5}\pi$

(e) $\int_0^{2\pi} \int_0^{\pi} \int_0^5 \rho^4 \sin\phi\, d\rho\, d\phi\, d\theta = 2500\pi$

17. Evaluate the integral of Problem 16(b) after changing the order to $dz\, dx\, dy$.

18. Evalute the integral of Problem 16(c), changing the order to $dx\, dy\, dz$ and to $dy\, dz\, dx$.

19. Find the following volumes, using integrals in rectangular coordinates:
(a) Inside $x^2 + y^2 = 9$, above $z = 0$, and below $x + z = 4$ *Ans.* 36π cubic units
(b) Bounded by the coordinate planes and $6x + 4y + 3z = 12$ *Ans.* 4 cubic units
(c) Inside $x^2 + y^2 = 4x$, above $z = 0$, and below $x^2 + y^2 = 4z$ *Ans.* 6π cubic units

20. Find the following volumes, using triple integrals in cylindrical coordinates:
(a) The volume of Problem 4.
(b) The volume of Problem 19(c).
(c) That inside $r^2 = 16$, above $z = 0$, and below $2z = y$ *Ans.* 64/3 cubic units

21. Find the centroid of each of the following volumes:
(a) Under $z^2 = xy$ and above the triangle $y = x$, $y = 0$, $x = 4$ in the plane $z = 0$ *Ans.* $(3, \frac{9}{5}, \frac{9}{8})$
(b) That of Problem 19(b) *Ans.* $(\frac{1}{2}, \frac{3}{4}, 1)$
(c) The first-octant volume of Problem 19(a) *Ans.* $\left(\frac{64 - 9\pi}{16(\pi - 1)}, \frac{23}{8(\pi - 1)}, \frac{73\pi - 128}{32(\pi - 1)}\right)$
(d) That of Problem 19(c) *Ans.* $(\frac{8}{3}, 0, \frac{10}{9})$
(e) That of Problem 20(c) *Ans.* $(0, 3\pi/4, 3\pi/16)$

22. Find the moments of inertia I_x, I_y, I_z of the following volumes:

(a) That of Problem 4 *Ans.* $I_x = I_y = \frac{32}{3}V;\ I_z = \frac{16}{3}V$

(b) That of Problem 19(b) *Ans.* $I_x = \frac{5}{2}V;\ I_y = 2V;\ I_z = \frac{13}{10}V$

(c) That of Problem 19(c) *Ans.* $I_x = \frac{55}{18}V;\ I_y = \frac{175}{18}V;\ I_z = \frac{80}{9}V$

(d) That cut from $z = r^2$ by the plane $z = 2$ *Ans.* $I_x = I_y = \frac{7}{3}V;\ I_z = \frac{2}{3}V$

23. Show that, in cylindrical coordinates, the triple integral of a function $f(r, \theta, z)$ over a region R may be represented by

$$\int_{\alpha}^{\beta}\int_{r_1(\theta)}^{r_2(\theta)}\int_{z_1(r,\theta)}^{z_2(r,\theta)} f(r, \theta, z) r\, dz\, dr\, d\theta$$

[*Hint*: Consider, in Fig. 57-12, a representative subregion of R bounded by two cylinders having the z axis as axis and of radii r and $r + \Delta r$, respectively, cut by two horizontal planes through $(0, 0, z)$ and $(0, 0, z + \Delta z)$, respectively, and by two vertical planes through the z axis making angles θ and $\theta + \Delta\theta$, respectively, with the xz plane. Take $\Delta V = (r\,\Delta\theta)\,\Delta r\,\Delta z$ as an approximation of its volume.]

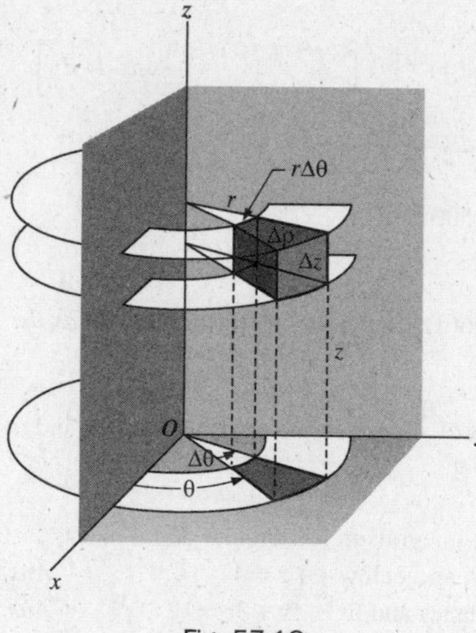

Fig. 57-12

24. Show that, in spherical coordinates, the triple integral of a function $f(\rho, \phi, \theta)$ over a region R may be represented by

$$\int_{\alpha}^{\beta}\int_{\phi_1(\theta)}^{\phi_2(\theta)}\int_{\rho_1(\phi,\theta)}^{\rho_2(\phi,\theta)} f(\rho, \phi, \theta)\rho^2 \sin\phi\, d\rho\, d\phi\, d\theta$$

[*Hint*: Consider, in Fig. 57-13, a representative subregion of R bounded by two spheres centered at O, of radii ρ and $\rho + \Delta\rho$, respectively, by two cones having O as vertex, the z axis as axis, and semivertical angles ϕ and $\phi + \Delta\phi$, respectively, and by two vertical planes through the z axis making angles θ and $\theta + \Delta\theta$, respectively, with the yz plane. Take $\Delta V = (\rho\Delta\phi)(\rho\sin\phi\,\Delta\theta)(\Delta\rho) = \rho^2\sin\phi\,\Delta\rho\,\Delta\phi\,\Delta\theta$ as an approximation of its volume.]

25. Change the following points from rectangular to cylindrical coordinates: (a) $(1, 0, 0)$; (b) $(\sqrt{2}, \sqrt{2}, 2)$; (c) $(-\sqrt{3}, 1, 5)$.

Ans. (a) $(1, 0, 0)$; (b) $\left(2, \frac{\pi}{4}, 1\right)$; (c) $\left(2, \frac{5\pi}{6}, 5\right)$

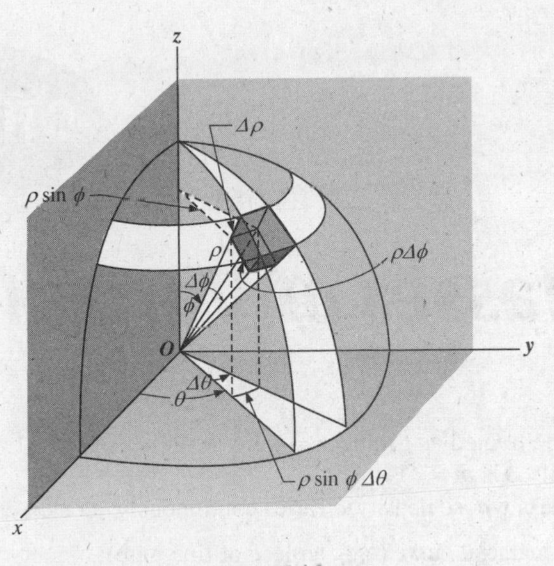

Fig. 57-13

26. Change the following points from cylindrical to rectangular coordinates: (a) $\left(5, \frac{\pi}{3}, 1\right)$; (b) $\left(2, -\frac{\pi}{6}, 0\right)$; (c) (0, 7, 1).

Ans. (a) $\left(\frac{5}{2}, \frac{5\sqrt{3}}{2}, 1\right)$; (b) $(\sqrt{3}, -1, 0)$; (c) (0, 0, 1)

27. Change the following points from rectangular to spherical coordinates: (a) (1, 0, 0); (b) $(\sqrt{2}, \sqrt{2}, 2)$; (c) $(1, -1, -\sqrt{2})$.

Ans. (a) $\left(1, 0, \frac{\pi}{2}\right)$; (b) $\left(2\sqrt{2}, \frac{\pi}{4}, \frac{\pi}{4}\right)$; (c) $\left(2, \frac{7\pi}{4}, \frac{3\pi}{4}\right)$

28. Change the following points from spherical to rectangular coordinates: (a) (1, 0, 0); (b) (2, 0, π); (c) $\left(4, \frac{\pi}{4}, \frac{\pi}{6}\right)$.

Ans. (a) (0, 0, 1); (b) (0, 0, –2); (c) $(\sqrt{2}, \sqrt{2}, 2\sqrt{3})$

29. Describe the surfaces determined by the following equations:

(a) $z = r^2$; (b) $r = 4 \cos \theta$; (c) $\rho \cos \phi = 4$; (d) $\rho \sin \phi = 4$; (e) $\phi = \frac{\pi}{2}$; (f) $\theta = \frac{\pi}{4}$; (g) $\rho = 2 \sin \phi$

Ans. (a) circular paraboloid; (b) right circular cylinder $(x-2)^2 + y^2 = 4$; (c) plane $z = 4$; (d) right circular cylinder $x^2 + y^2 = 16$; (e) the xy plane; (f) right circular cone with the z axis as its axis; (g) right circular cylinder $x^2 + y^2 = 4$

CHAPTER 58

Masses of Variable Density

Homogeneous masses can be treated as geometric figures with density $\delta = 1$. The mass of a homogeneous body of volume V and density δ is $m = \delta V$.

For a nonhomogenous mass whose density δ varies continuously, an element of mass dm is given by

(1) $\delta(x, y)\, ds$ for a planar material curve (e.g., a piece of fine wire);
(2) $\delta(x, y)\, dA$ for a material two-dimensional plate (e.g., a thin sheet of metal);
(3) $\delta(x, y, z)\, dV$ for a material body.

The center of mass $(\bar{x}, \bar{y})$ of a planar plate that is distributed over a region R with density $\delta(x, y)$ is determined by the equations

$$m\bar{x} = M_y \quad \text{and} \quad m\bar{y} = M_x, \quad \text{where} \quad M_y = \iint_R \delta(x,y)x\, dA \quad \text{and} \quad M_x = \iint_R \delta(x,y)y\, dA$$

An analogous result holds for the center of mass of a three-dimensional body. The reasoning is similar to that for centroids in Chapter 55.

The moments of inertia of a planar mass with respect to the x axis and the y axis are $I_x = \iint_R \delta(x, y)y^2\, dA$ and $I_y = \iint_R \delta(x,y)x^2\, dA$. Similar formulas with triple integrals hold for three-dimensional bodies. (For example, $I_x = \iiint_R \delta(x, y, z)(y^2 + z^2)\, dA$.)

SOLVED PROBLEMS

1. Find the mass of a semicircular wire whose density varies as the distance from the diameter joining the ends.

Take the wire as in Fig. 58-1, so that $\delta(x, y) = ky$. Then, from $x^2 + y^2 = r^2$.

$$ds = \sqrt{1+\left(\frac{dy}{dx}\right)^2}\, dx = \frac{r}{y}\, dx$$

and

$$m = \int \delta(x,y)\, ds = \int_{-r}^{r} ky\frac{r}{y}\, dx = kr\int_{-r}^{r} dx = 2kr^2 \text{ units}$$

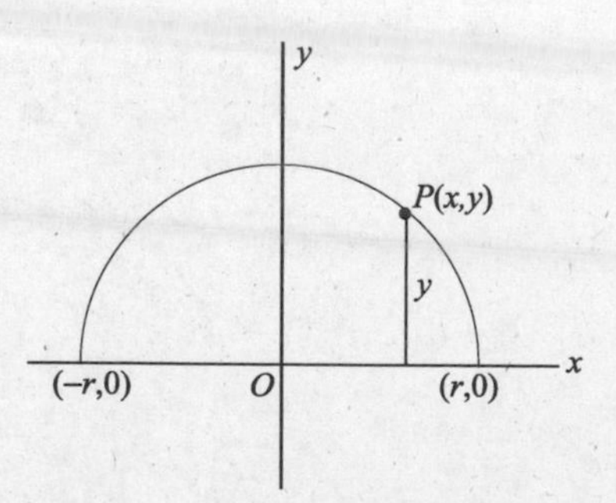

Fig. 58-1

2. Find the mass of a square plate of side a if the density varies as the square of the distance from a vertex.

Take the square as in Fig. 58-2, and let the vertex from which distances are measured be at the origin. Then $\delta(x, y) = k(x^2 + y^2)$ and

$$m = \iint_R \delta(x, y)\, dA = \int_0^a \int_0^a k(x^2 + y^2)\, dx\, dy = k\int_0^a (\tfrac{1}{3}a^3 + ay^2)\, dy = \tfrac{2}{3}ka^4 \text{ units}$$

Fig. 58-2

3. Find the mass of a circular plate of radius r if the density varies as the square of the distance from a point on the circumference.

Take the circle as in Fig. 58-3 and let $A(r, 0)$ be the fixed point on the circumference. Then $\delta(x, y) = k[(x - r)^2 + y^2]$ and

$$m = \iint_R \delta(x, y)\, dA = 2\int_{-r}^{r} \int_0^{\sqrt{r^2 - x^2}} k[(x - r)^2 + y^2]\, dy\, dx = \tfrac{3}{2}k\pi r^4 \text{ units}$$

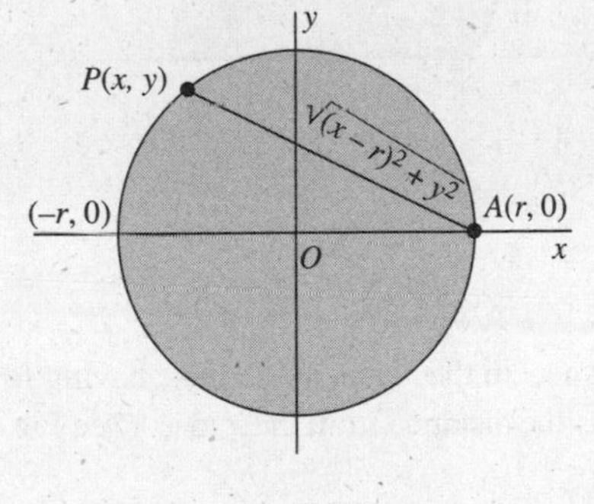

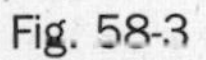
Fig. 58-3

4. Find the center of mass of a plate in the form of the segments cut from the parabola $y^2 = 8x$ by its latus rectum $x = 2$ if the density varies as the distance from the latus rectum. (See Fig. 58-4.)

Here, $\delta(x, y) = 2 - x$ and, by symmetry, $\bar{y} = 0$. For the upper half of the plate,

$$m = \iint_R \delta(x, y)\, dA = \int_0^4 \int_{y^2/8}^2 k(2 - x)\, dx\, dy = k\int_0^4 \left(2 - \frac{y^2}{4} + \frac{y^4}{128}\right) dy = \frac{64}{15}k$$

$$M_y = \iint_R \delta(x, y)x\, dA = \int_0^4 \int_{y^2/8}^2 k(2 - x)x\, dx\, dy = k\int_0^4 \left[\frac{4}{3} - \frac{y^4}{64} + \frac{y^6}{(24)(64)}\right] dy = \frac{128}{35}k$$

and $\bar{x} = M_y / m = \frac{6}{7}$. The center of mass has coordinates $(\frac{6}{7}, 0)$.

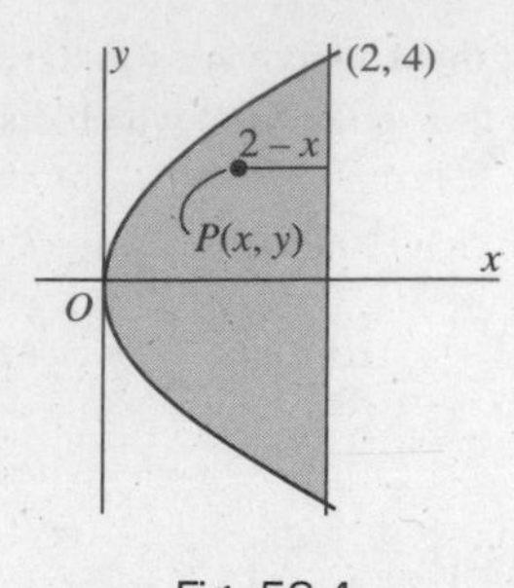

Fig. 58-4

5. Find the center of mass of a plate in the form of the upper half of the cardioid $r = 2(1 + \cos\theta)$ if the density varies as the distance from the pole. (See Fig. 58-5.)

$$m = \iint_R \delta(r,\theta)\,dA = \int_0^{\pi}\int_0^{2(1+\cos\theta)} (kr)r\,dr\,d\theta = \tfrac{8}{3}k\int_0^{\pi}(1+\cos\theta)^3\,d\theta = \tfrac{20}{3}k\pi$$

$$M_x = \iint_R \delta(r,\theta)y\,dA = \int_0^{\pi}\int_0^{2(1+\cos\theta)} (kr)(r\sin\theta)r\,dr\,d\theta$$

$$= 4k\int_0^{\pi}(1+\cos\theta)^4\sin\theta\,d\theta = \tfrac{128}{5}k$$

$$M_y = \iint_R \delta(r,\theta)x\,dA = \int_0^{\pi}\int_0^{2(1+\cos\theta)} (kr)(r\cos\theta)r\,dr\,d\theta = 14k\pi$$

Then $\bar{x} = \dfrac{M_y}{m} = \dfrac{21}{10}$, $\bar{y} = \dfrac{M_x}{m} = \dfrac{96}{25\pi}$, and the center of mass has coordinates $\left(\dfrac{21}{10}, \dfrac{96}{25\pi}\right)$.

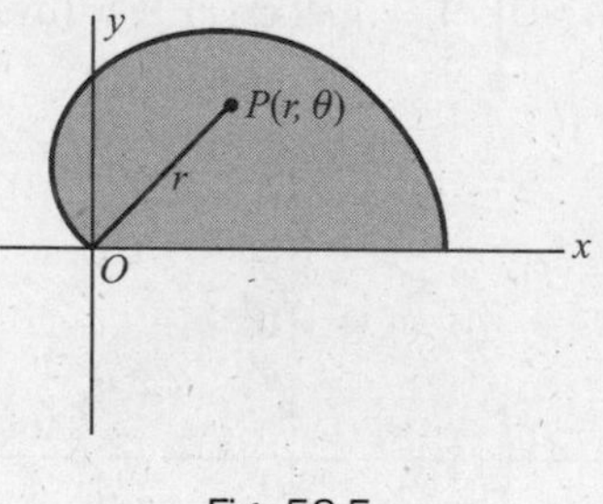

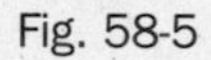

Fig. 58-5

6. Find the moment of inertia with respect to the x axis of the plate having for edges one arch of the curve $y = \sin x$ and the x axis if its density varies as the distance from the x axis. (See Fig. 58-6.)

$$m = \iint_R \delta(x,y)\,dA = \int_0^{\pi}\int_0^{\sin x} ky\,dy\,dx = \tfrac{1}{2}k\int_0^{\pi}\sin^2 x\,dx = \tfrac{1}{4}k\pi$$

$$I_x = \iint_R \delta(x,y)y^2\,dA = \int_0^{\pi}\int_0^{\sin x}(ky)(y^2)\,dy\,dx = \tfrac{1}{4}k\int_0^{\pi}\sin^4 x\,dx = \tfrac{3}{32}k\pi = \tfrac{3}{8}m$$

Fig. 58-6

7. Find the mass of a sphere of radius a if the density varies inversely as the square of the distance from the center.

Take the sphere as in Fig. 58-7. Then $\delta(x,y,z) = \dfrac{k}{x^2+y^2+z^2} = \dfrac{k}{\rho^2}$ and

$$m = \iiint_R \delta(x,y,z)\,dV = 8\int_0^{\pi/2}\int_0^{\pi/2}\int_0^a \frac{k}{\rho^2}\rho^2 \sin\phi\, d\rho\, d\phi\, d\theta$$

$$= 8ka\int_0^{\pi/2}\int_0^{\pi/2} \sin\phi\, d\phi\, d\theta = 8ka\int_0^{\pi/2} d\theta = 4k\pi a \text{ units}$$

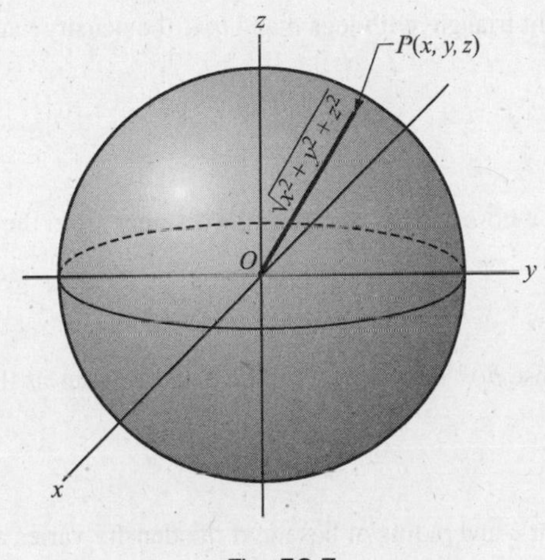

Fig. 58-7

8. Find the center of mass of a right circular cylinder of radius a and height h if the density varies as the distance from the base.

Take the cylinder as in Fig. 58-8, so that its equation is $r = a$ and the volume in question is that part of the cylinder between the planes $z = 0$ and $z = h$. Clearly, the center of mass lies on the z axis. Then

$$m = \iiint_R \delta(z,r,\theta)\,dV = 4\int_0^{\pi/2}\int_0^a\int_0^h (kz)r\,dz\,dr\,d\theta = 2kh^2\int_0^{\pi/2}\int_0^a r\,dr\,d\theta$$

$$= kh^2a^2\int_0^{\pi/2} d\theta = \tfrac{1}{2}k\pi h^2a^2$$

$$M_{xy} = \iiint_R \delta(z,r,\theta)z\,dV = 4\int_0^{\pi/2}\int_0^a\int_0^h (kz^2)r\,dz\,dr\,d\theta = \tfrac{4}{3}kh^3\int_0^{\pi/2}\int_0^a r\,dr\,d\theta$$

$$= \tfrac{2}{3}kh^3a^2\int_0^{\pi/2} d\theta = \tfrac{1}{3}k\pi h^3a^2$$

and $\bar{z} = M_{xy}/m = \frac{2}{3}h$. Thus the center of mass has coordinates $(0, 0, \frac{2}{3}h)$.

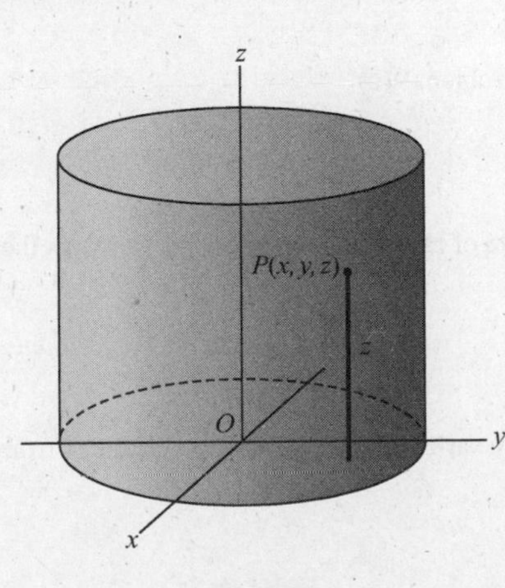

Fig. 58-8

SUPPLEMENTARY PROBLEMS

9. Find the mass of

(a) A straight rod of length a whose density varies as the square of the distance from one end

Ans. $\frac{1}{3}ka^3$ units

(b) A plate in the form of a right triangle with legs a and b, if the density varies as the sum of the distance from the legs

Ans. $\frac{1}{6}kab\ (a+b)$ units

(c) A circular plate of radius a whose density varies as the distance from the center

Ans. $\frac{2}{3}ka^3\pi$ units

(d) A plate in the form of an ellipse $b^2x^2 + a^2y^2 = a^2b^2$, if the density varies as the sum of the distances from its axes

Ans. $\frac{4}{3}kab(a+b)$ units

(e) A circular cylinder of height b and radius of base a, if the density varies as the square of the distance from its axis

Ans. $\frac{1}{2}ka^4b\pi$ units

(f) A sphere of radius a whose density varies as the distance from a fixed diametral plane

Ans. $\frac{1}{2}ka^4\pi$ units

(g) A circular cone of height b and radius of base a whose density varies as the distance from its axis

Ans. $\frac{1}{6}ka^3b\pi$ units

(h) A spherical surface whose density varies as the distance from a fixed diametral plane

Ans. $2ka^3\pi$ units

10. Find the center of mass of

(a) One quadrant of the plate of Problem 9(c)

Ans. $(3a/2\pi, 3a/2\pi)$

(b) One quadrant of a circular plate of radius a, if the density varies as the distance from a bounding radius (the x axis)

Ans. $(3a/8, 3a\pi/16)$

(c) A cube of edge a, if the density varies as the sum of the distances from three adjacent edges (on the coordinate axes)

Ans. $(5a/9, 5a/9, 5a/9)$

(d) An octant of a sphere of radius a, if the density varies as the distance from one of the plane faces

Ans. $(16a/15\pi, 16a/15\pi, 8a/15)$

(e) A right circular cone of height b and radius of base a, if the density varies as the distance from its base

Ans. $(0, 0, 2b/5)$

11. Find the moment of inertia of:

(a) A square plate of side a with respect to a side, if the density varies as the square of the distance from an extremity of that side

Ans. $\frac{7}{15}a^2m$

(b) A plate in the form of a circle of radius a with respect to its center, if the density varies as the square of the distance from the center

Ans. $\frac{2}{3}a^2m$

(c) A cube of edge a with respect to an edge, if the density varies as the square of the distance from one extremity of that edge

Ans. $\frac{38}{45}a^2m$

(d) A right circular cone of height b and radius of base a with respect to its axis, if the density varies as the distance from the axis

Ans. $\frac{2}{5}a^2m$

(e) The cone of (d), if the density varies as the distance from the base

Ans. $\frac{1}{5}a^2m$

CHAPTER 59

Differential Equations of First and Second Order

A differential equation is an equation that involves a function, say y, of one variable, say x, and derivatives of y or differentials of x and y. Examples are $\frac{d^2y}{dx^2}+2\frac{dy}{dx}+3y-7\sin x+4x=0$ and $dy=(x+2y)\,dx$. The first equation also can be written as $y''+2y'+3y-7\sin x+4x=0$.

The *order* of a differential equation is the order of the derivative of highest order appearing in it. The first of the above equations is of order two, and the second is of order one.

A *solution* of a differential equation is a function y that satisfies the equation. A *general solution* of an equation is a formula that describes all solutions of the equation. It turns out that a general solution of a differential equation of order n will contain n arbitrary constants.

Separable Differential Equations

A separable differential equation is a first-order equation that can be put in the form

$$f(x)\,dx+g(y)\,dy=0, \qquad \text{which is equivalent to } \frac{dy}{dx}=-\frac{f(x)}{g(y)}$$

A separable equation can be solved by taking antiderivatives

$$\int f(x)\,dx+\int g(y)\,dy=C$$

The result is an equation involving x and y that determines y as a function of x. (See Problems 4–6, and for justification, see Problem 61.)

Homogeneous Functions

A function $f(x, y)$ is said to be *homogeneous of degree n* if $f(\lambda x, \lambda y)=\lambda^n f(x, y)$. The equation $M(x, y)\,dx + N(x, y)\,dy=0$ is said to be *homogeneous* if $M(x, y)$ and $N(x, y)$ are homogeneous of the same degree. It is easy to verify that the substitution

$$y=vx, \quad dy=v\,dx+x\,dv$$

will transform a homogeneous equation into a separable equation in the variables x and v.

Integrating Factors

Certain differential equations may be solved after multiplication by a suitable function of x and y produces an integrable combination of terms. Such a function is called an *integrating factor* of the equations. In looking for integrable combinations, note that:

(i) $d(xy)=x\,dy+y\,dx$ (ii) $d(y/x)=\frac{x\,dy-y\,dx}{x^2}$

(iii) $d(\ln xy)=\frac{x\,dy+y\,dx}{xy}$ (iv) $d(\frac{1}{k+1}u^{k+1})=u^k\,du$

Moreover, $d(F) + d(G) + \cdots = 0$ yields $F + G + . = \text{constant}$. (See Problems 10–14.)

The so-called *linear differential equations of the first order*, $\frac{dy}{dx} + Py = Q$, where P and Q are functions of x alone, have the function $\xi(x) = e^{\int P\,dx}$ as integrating factor. (See Problems 15–17.)

An equation of the form $\frac{dy}{dx} + Py = Qy^n$, where $n \neq 0, 1$ and where P and Q are functions of x alone, can be reduced to the linear form by the substitution

$$y^{1-n} = z, \qquad y^{-n}\frac{dy}{dx} = \frac{1}{1-n}\frac{dz}{dx}$$

(See Problems 18–19).

Second-Order Equations

The second-order equations that will be solved in this chapter are of the following types:

$\frac{d^2y}{dx^2} = f(x)$ (See Problem 23.)

$\frac{d^2y}{dx^2} = f\left(x, \frac{dy}{dx}\right)$ (See Problems 24 and 25.)

$\frac{d^2y}{dx^2} = f(y)$ (See Problems 26 and 27.)

$\frac{d^2y}{dx^2} + P\frac{dy}{dx} + Qy = R$, where P and Q are constants and R is a constant or function of x only.

(See Problems 28–33.)

If the equation $m^2 + Pm + Q = 0$ has two distinct roots m_1 and m_2, then $y = C_1e^{m_1x} + C_2e^{m_2x}$ is the general solution of the equation $\frac{d^2y}{dx^2} + P\frac{dy}{dx} + Qy = 0$. If the two roots are identical so that $m_1 = m_2 = m$, then

$$y = C_1e^{mx} + C_2xe^{mx} = e^{mx}(C_1 + C_2x)$$

is the general solution.

The general solution of $\frac{d^2y}{dx^2} + P\frac{dy}{dx} + Qy = 0$ is called the *complementary function* of the equation

$$\frac{d^2y}{dx^2} + P\frac{dy}{dx} + Qy = R(x) \tag{59.1}$$

If $f(x)$ satisfies (59.1), then the general solution of (59.1) is

$$y = \text{complementary function} + f(x)$$

The function $f(x)$ is called a *particular solution* of (59.1).

SOLVED PROBLEMS

1. Show that (a) $y = 2e^x$, (b) $y = 3x$, and (c) $y = C_1e^x + C_2x$, where C_1 and C_2 are arbitrary constants, are solutions of the differential equation $y''(1-x) + y'x - y = 0$.

(a) Differentiate $y = 2e^x$ twice to obtain $y' = 2e^x$ and $y'' = 2e^x$. Substitute in the differential equation to obtain the identity $2e^x(1-x) + 2e^x x - 2e^x = 0$.

(b) Differentiate $y = 3x$ twice to obtain $y' = 3$ and $y'' = 0$. Substitute in the differential equation to obtain the identity $0(1-x) + 3x - 3x = 0$.

(c) Differentiate $y = C_1e^x + C_2x$ twice to obtain $y' = C_1e^x + C_2$ and $y'' = C_1e^x$. Substitute in the differential equation to obtain the identity $C_1e^x(1-x) + (C_1e^x + C_2)x - (C_1e^x + C_2x) = 0$.

Solution (c) is the *general solution* of the differential equation because it satisfies the equation and contains the proper number of essential arbitrary constants. Solutions (a) and (b) are called *particular solutions* because each may be obtained by assigning particular values to the arbitrary constants of the general solution.

2. Form the differential equation whose general solution is:

(a) $y = Cx^2 - x$; (b) $y = C_1x^3 + C_2x + C_3$.

(a) Differentiate $y = Cx^2 - x$ once to obtain $y' = 2Cx - 1$. Solve for $C = \frac{1}{2}\left(\frac{y'+1}{x}\right)$ and substitute in the given relation (general solution) to obtain $y = \frac{1}{2}\left(\frac{y'+1}{x}\right)x^2 - x$ or $y'x = 2y + x$.

(b) Differentiate $y = C_1x^3 + C_2x + C_3$ three times to obtain $y' = 3C_1x^2 + C_2$, $y'' = 6C_1x$, $y''' = 6C_1$. Then $y'' = xy'''$ is the required equation. Note that the given relation is a solution of the equation $y^{(4)} = 0$ but is not the general solution, since it contains only three arbitrary constants.

3. Form the second-order differential equation of all parabolas with principal axis along the x axis.

The system of parabolas has equation $y^2 = Ax + B$, where A and B are arbitrary constants. Differentiate twice to obtain $2yy' = A$ and $2yy'' + 2(y')^2 = 0$. The latter is the required equation.

4. Solve $\frac{dy}{dx} + \frac{1+y^3}{xy^2(1+x^2)} = 0$.

Here $xy^2(1 + x^2)dy + (1 + y^3)dx = 0$, or $\frac{y^2}{1+y^3}dy + \frac{1}{x(1+x^2)}dx = 0$ with the variables separated. Then the partial-fraction decomposition yields

$$\frac{y^2dy}{1+y^3} + \frac{dx}{x} - \frac{x\,dx}{1+x^2} = 0,$$

and integration yields

$$\tfrac{1}{3}\ln|1+y^3| + \ln|x| - \tfrac{1}{2}\ln(1+x^2) = c$$

or

$$2\ln|1+y^3| + 6\ln|x| - 3\ln(1+x^2) = 6c$$

from which

$$\ln\frac{x^6(1+y^3)^2}{(1+x^2)^3} = 6c \quad \text{and} \quad \frac{x^6(1+y^3)^2}{(1+x^2)^3} = e^{6c} = C$$

5. Solve $\frac{dy}{dx} = \frac{1+y^2}{1+x^2}$.

Separate the variables: $\frac{dy}{1+y^2} = \frac{dx}{1+x^2}$. Integration yields $\tan^{-1} y = \tan^{-1} x + \tan^{-1} C$, and then

$$y = \tan(\tan^{-1} x + \tan^{-1} C) = \frac{x+C}{1-Cx}$$

6. Solve $\frac{dy}{dx} = \frac{\cos^2 y}{\sin^2 x}$.

The variables are easily separated to yield $\frac{dy}{\cos^2 y} = \frac{dx}{\sin^2 x}$.

Hence, $\sec^2 y\,dy = \csc^2 x\,dx$ and integration yields $\tan y = -\cot x + C$.

7. Solve $2xy\,dy = (x^2 - y^2)\,dx$.

The equation is homogeneous of degree two. The transformation $y = vx$, $dy = v\,dx + x\,dv$ yields $(2x)(vx)(v\,dx + x\,dv) = (x^2 - v^2x)\,dx$ or $\frac{2v\,dv}{1-3v^2} = \frac{dx}{x}$. Then integration yields

$$-\tfrac{1}{3}\ln|1-3v^2| = \ln|x| + \ln c$$

from which $\ln|1 - 3v^2| + 3\ln|x| + \ln C' = 0$ or $C''|x^3(1 - 3v^2)| = 1$.

Now $\pm C'x^3(1 - 3v^2) = Cx^3(1 - 3v^2) = 1$, and using $v = y/x$ produces $C(x^3 - 3xy^2) = 1$.

8. Solve $x\sin\frac{y}{x}(y\,dx + x\,dy) + \cos\frac{y}{x}(x\,dy - y\,dx) = 0.$

The equation is homogeneous of degree two. The transformation $y = vx$, $dy = v\,dx + x\,dv$ yields

$$x\sin v(vx\,dx + x^2 dv + vx\,dx) + vx\cos v(x^2\,dv + vx\,dx - vx\,dx) = 0$$

or
$$\sin v(2v\,dx + x\,dv) + xv\cos v\,dv = 0$$

or
$$\frac{\sin v + v\cos v}{v\sin v}dv + 2\frac{dx}{x} = 0$$

Then $\ln|v\sin v| + 2\ln|x| = \ln C'$, so that $x^2 v\sin v = C$ and $xy\sin\frac{y}{x} = C.$

9. Solve $(x^2 - 2y^2)\,dy + 2xy\,dx = 0.$

The equation is homogeneous of degree two, and the standard transformation yields

$$(1 - 2v^2)(v\,dx + x\,dv) + 2v\,dx = 0$$

or
$$\frac{1 - 2v^2}{v(3 - 2v^2)}dv + \frac{dx}{x} = 0$$

or
$$\frac{dv}{3v} - \frac{4v\,dv}{3(3 - 2v^2)} + \frac{dx}{x} = 0$$

Integration yields $\frac{1}{3}\ln|v| + \frac{1}{3}\ln|3 - 2v^2| + \ln|x| = \ln c$, which we may write as $\ln|v| + \ln|3 - 2v^2| + 3\ln|x| = \ln C'$. Then $x^3(3 - 2vv^2) = C$ and $y(3x^2 - 2y^2) = C.$

10. Solve $(x^2 + y)\,dx + (y^3 + x)\,dy = 0.$

Integrate $x^2\,dx + (y\,dx + x\,dy) + y^3\,dy = 0$, term by term, to obtain

$$\frac{x^3}{3} + xy + \frac{y^4}{4} = C$$

11. Solve $(x + e^{-x}\sin y)\,dx - (y + e^{-x}\cos y)\,dy = 0.$

Integrate $x\,dx - y\,dy - (e^{-x}\cos y\,dy - e^{-x}\sin y\,dx) = 0$, term by term, to obtain

$$\tfrac{1}{2}x^2 - \tfrac{1}{2}y^2 - e^{-x}\sin y = C$$

12. Solve $x\,dy - y\,dx = 2x^3\,dx.$

The combination $x\,dy - y\,dx$ suggests $d\left(\frac{y}{x}\right) = \frac{x\,dy - y\,dx}{x^2}$. Hence, multiplying the given equation by $\xi(x) = \frac{1}{x^2}$, we obtain $\frac{x\,dy - y\,dx}{x^2} = 2x\,dx$, from which

$$\frac{y}{x} = x^2 + C \quad \text{or} \quad y = x^3 + Cx$$

13. Solve $x\,dy + y\,dx = 2x^2y\,dx.$

The combination $x\,dy + y\,dx$ suggests $d(\ln xy) = \frac{x\,dy + y\,dx}{xy}$. Hence, multiplying the given equation by $\xi(x, y) = \frac{1}{xy}$, we obtain $\frac{x\,dy + y\,dx}{xy} = 2x\,dx$, from which In $|xy| = x^2 + C.$

14. Solve $x\,dy + (3y - e^x)\,dx = 0$.

Multiply the equation by $\xi(x) = x^2$ to obtain $x^3\,dy + 3x^2y\,dx = x^2e^x\,dx$. This yields

$$x^3y = \int x^2e^x\,dx = x^2e^x - 2xe^x + 2e^x + C$$

15. $\dfrac{dy}{dx} + \dfrac{2}{x}y = 6x^3$.

Here $P(x) = \frac{2}{x}$, $\int P(x) = \ln x^2$, and an integrating factor is $\xi(x) = e^{\ln x^2} = x^2$. We multiply the given equation by $\xi(x) = x^2$ to obtain $x^2\,dy + 2xy\,dx = 6x^5\,dx$. Then integration yields $x^2y = x^6 + C$.

Note 1: After multiplication by the integrating factor, the terms on the left side of the resulting equation are an *integrable combination*.

Note 2: The integrating factor for a given equation is not unique. In this problem, x^2, $3x^2$, $\frac{1}{2}x^2$, etc., are all integrating factors. Hence, we write the simplest particular integral of $P(x)\,dx$ rather than the general integral, $\ln x^2 + \ln C = \ln Cx^2$.

16. Solve $\tan x\dfrac{dy}{dx} + y = \sec x$.

Since $\dfrac{dy}{dx} + y\cot x = \csc x$, we have $\int P(x)dx = \int \cot x\,dx = \ln|\sin x|$, and $\xi(x) = e^{\ln|\sin x|} = |\sin x|$. Then multiplication by $\xi(x)$ yields

$$\sin x\left(\frac{dy}{dx} + y\cot x\right) = \sin x\csc x \quad \text{or} \quad \sin x\,dy + y\cos x\,dx = dx$$

and integration gives

$$y\sin x = x + C$$

17. Solve $\dfrac{dy}{dx} - xy = x$.

Here $P(x) = -x$, $\int P(x)dx = -\frac{1}{2}x^2$, and $\xi(x) = e^{-\frac{1}{2}x^2}$. This produces

$$e^{-\frac{1}{2}x^2}dy - xye^{-\frac{1}{2}x^2}dx = xe^{-\frac{1}{2}x^2}dx$$

and integration yields

$$ye^{-\frac{1}{2}x^2} = -e^{-\frac{1}{2}x^2} + C, \quad \text{or} \quad y = Ce^{\frac{1}{2}x^2} - 1$$

18. Solve $\dfrac{dy}{dx} + y = xy^2$.

The equation is of the form $\dfrac{dy}{dx} + Py - Qy^n$, with $n = 2$. Hence we use the substitution $y^{1-n} = y^{-1} = z$, $y^{-2}\dfrac{dy}{dx} = -\dfrac{dz}{dx}$. For convenience, we write the original equation in the form $y^{-2}\dfrac{dy}{dx} + y^{-1} = x$, obtaining $-\dfrac{dz}{dx} + z = x$, or $\dfrac{dz}{dx} - z = -x$.

The integrating factor is $\xi(x) = e^{\int P\,dx} = e^{-\int dx} = e^{-x}$. It gives us $e^{-x}\,dx - ze^{-x}\,dx = -xe^{-x}\,dx$, from which $ze^{-x} = xe^{-x} + e^{-x} + C$. Finally, since $z = y^{-1}$, we have

$$\frac{1}{y} = x + 1 + Ce^x.$$

19. Solve $\frac{dy}{dx} + y\tan x = y^3 \sec x$.

Write the equation in the form $y^{-3}\frac{dy}{dx} + y^{-2}\tan x = \sec x$. Then use the substitution $y^{-2} = z$, $y^{-3}\frac{dy}{dx} = -\frac{1}{2}\frac{dz}{dx}$ to obtain $\frac{dz}{dx} - 2z\tan x = -2\sec x$.

The integrating factor is $\xi(x) = e^{-2\int \tan x\, dx} = \cos^2 x$. It gives $\cos^2 x\, dz - 2z\cos x \sin x\, dx = -2\cos x\, dx$, from which

$$z\cos^2 x = -2\sin x + C, \quad \text{or} \quad \frac{\cos^2 x}{y^2} = -2\sin x + C$$

20. When a bullet is fired into a sand bank, its retardation is assumed equal to the square root of its velocity on entering. For how long will it travel if its velocity on entering the bank is 144 ft/sec?

Let v represent the bullet's velocity t seconds after striking the bank. Then the retardation is $-\frac{dv}{dt} = \sqrt{v}$, so $\frac{dv}{\sqrt{v}} = -dt$ and $2\sqrt{v} = -t + C$.

When $t = 0$, $v = 144$ and $C = 2\sqrt{144} = 24$. Thus, $2\sqrt{v} = -t + 24$ is the law governing the motion of the bullet. When $v = 0$, $t = 24$; the bullet will travel for 24 seconds before coming to rest.

21. A tank contains 100 gal of brine holding 200 lb of salt in solution. Water containing 1 lb of salt per gallon flows into the tank at the rate of 3 gal/min, and the mixture, kept uniform by stirring, flows out at the same rate. Find the amount of salt at the end of 90 min.

Let q denote the number of pounds of salt in the tank at the end of t minutes. Then $\frac{dq}{dt}$ is the rate of change of the amount of salt at time t.

Three pounds of salt enter the tank each minute, and $0.03q$ pounds are removed. Thus, $\frac{dq}{dt} = 3 - 0.03q$.

Rearranged, this becomes $\frac{dq}{3 - 0.03q} = dt$, and integration yields

$$\frac{\ln(0.03q - 3)}{0.03} = -t + C.$$

When $t = 0$, $q = 200$ and $C = \frac{\ln 3}{0.03}$ so that $\ln(0.03q - 3) = -0.03t + \ln 3$. Then $0.01q - 1 = e^{-0.03t}$, and $q = 100 + 100e^{-0.03t}$. When $t = 90$, $q = 100 + 100e^{-2.7} \sim 106.72$ lb.

22. Under certain conditions, cane sugar in water is converted into dextrose at a rate proportional to the amount that is unconverted at any time. If, of 75 grams at time $t = 0$, 8 grams are converted during the first 30 min, find the amount converted in $1\frac{1}{2}$ hours.

Let q denote the amount converted in t minutes. Then $\frac{dq}{dt} = k(75 - q)$, from which $\frac{dq}{75 - q} = k\,dt$, and integration gives $\ln(75 - q) = -kt + C$.

When $t = 0$, $q = 0$ and $C = \ln 75$, so that $\ln(75 - q) = -kt + \ln 75$.

When $t = 30$ and $q = 8$, we have $30k = \ln 75 - \ln 67$; hence, $k = 0.0038$, and $q = 75(1 - e^{-0.0038t})$.

When $t = 90$, $q = 75(1 - e^{-0.34}) \sim 21.6$ grams.

23. Solve $\frac{d^2y}{dx^2} = xe^x + \cos x$.

Here $\frac{d}{dx}\left(\frac{dy}{dx}\right) = xe^x + \cos x$. Hence, $\frac{dy}{dx} = \int (xe^x + \cos x)dx = xe^x - e^x + \sin x + C_1$, and another integration yields $y = xe^x - 2e^x - \cos x + C_1x + C_2$.

24. Solve $x^2\frac{d^2y}{dx^2} + x\frac{dy}{dx} = a$.

Let $p = \frac{dy}{dx}$; then $\frac{d^2y}{dx^2} = \frac{dp}{dx}$ and the given equation becomes $x^2\frac{dp}{dx} + xp = a$ or $x\,dp + p\,dx = \frac{a}{x}dx$. Then integration yields $xp = a\ln|x| + C_1$, or $x\frac{dy}{dx} = a\ln|x| + C$. When this is written as $dy = a\ln|x|\frac{dx}{x} + C_1\frac{dx}{x}$, integration gives $y = \frac{1}{2}a\ln^2|x| + C_1\ln|x| + C_2$.

25. Solve $xy'' + y' + x = 0$.

Let $p = \frac{dy}{dx}$. Then $\frac{d^2y}{dx^2} = \frac{dp}{dx}$ and the given equation becomes $x\frac{dp}{dx} + p + x = 0$ or $x\,dp + p\,dx = -x\,dx$. Integration gives $xp = -\frac{1}{2}x^2 + C_1$, substitution for p gives $\frac{dy}{dx} = -\frac{1}{2}x + \frac{C_1}{x}$, and another integration yields $y = \frac{1}{4}x^2 + C_2 \ln|x| + C_2$.

26. Solve $\frac{d^2y}{dx^2} - 2y = 0$.

Since $\frac{d}{dx}[(y')^2] = 2y'y''$, we can multiply the given equation by $2y'$ to obtain $2y'y'' = 4yy'$, and integrate to obtain $(y')^2 = 4\int yy'dx = 4\int y\,dy = 2y^2 + C_1$.

Then $\frac{dy}{dx} = \sqrt{2y^2 + C_1}$, so that $\frac{dy}{\sqrt{2y^2 + C_1}} = dx$ and $\ln|\sqrt{2}y + \sqrt{2y^2 + C_1}| = \sqrt{2}x + \ln C_2$. The last equation yields $\sqrt{2}y + \sqrt{2y^2 + C_1} = C_2 e^{\sqrt{2}x}$.

27. Solve $y'' = -1/y^3$.

Multiply by $2y'$ to obtain $2y'y'' = -\frac{2y'}{y^3}$. Then integration yields

$$(y')^2 = \frac{1}{y^2} + C_1 \qquad \text{so that} \qquad \frac{dy}{dx} = \frac{\sqrt{1 + C_1 y^2}}{y} \qquad \text{or} \qquad \frac{y\,dy}{\sqrt{1 + C_1 y^2}} = dx$$

Another integration gives $\sqrt{1 + C_1 y^2} = C_1 x + C_2$ or $(C_1 x + C_2)^2 - C_1 y^2 = 1$.

28. Solve $\frac{d^2y}{dx^2} + 3\frac{dy}{dx} - 4y = 0$.

Here we have $m^2 + 3m - 4 = 0$, from which $m = 1, -4$. The general solution is $y = C_1 e^x + C_2 e^{-4x}$.

29. Solve $\frac{d^2y}{dx^2} + 3\frac{dy}{dx} = 0$.

Here $m^2 + 3m = 0$, from which $m = 0, -3$. The general solution is $y = C_1 + C_2 e^{-3x}$.

30. Solve $\frac{d^2y}{dx^2} - 4\frac{dy}{dx} + 13y = 0$.

Here $m^2 - 4m + 13 = 0$, with roots $m_1 = 2 + 3i$ and $m_2 = 2 - 3i$. The general solution is

$$y = C_1 e^{(2+3i)x} + C_2 e^{(2-3i)x} = e^{2x}(C_1 e^{3ix} + C_2 e^{-3ix})$$

Since $e^{iax} = \cos ax + i \sin ax$, we have $e^{3ix} = \cos 3x + i \sin 3x$ and $e^{-3ix} = \cos 3x - i \sin 3x$. Hence, the solution may be put in the form

$$\begin{aligned} y &= e^{2x}[C_1(\cos 3x + i \sin 3x) + C_2(\cos 3x - i \sin 3x)] \\ &= e^{2x}[(C_1 + C_2)\cos 3x + i(C_1 - C_2)\sin 3x)] \\ &= e^{2x}(A \cos 3x + B \sin 3x) \end{aligned}$$

31. Solve $\frac{d^2y}{dx^2} - 4\frac{dy}{dx} + 4y = 0$.

Here $m^2 - 4m + 4 = 0$, with roots $m = 2, 2$. The general solution is $y = C_1 e^{2x} + C_2 x e^{2x}$.

32. Solve $\frac{d^2y}{dx^2} + 3\frac{dy}{dx} - 4y = x^2$.

From Problem 6, the complementary function is $y = C_1 e^x + C_2 e^{-4x}$.

To find a particular solution of the equation, we note that the right-hand member is $R(x) = x^2$. This suggests that the particular solution will contain a term in x^2 and perhaps other terms obtained by successive differentiation. We assume it to be of the form $y = Ax^2 + Bx + C$, where the constants A, B, C are to be determined. Hence we substitute $y = Ax^2 + Bx + C$, $y' = 2Ax + B$, and $y'' = 2A$ in the differential equation to obtain

$$2A + 3(2Ax + B) - 4(Ax^2 + Bx + C) = x^2 \quad \text{or} \quad -4Ax^2 + (6A - 4B)x + (2A + 3B - 4C) = x^2$$

Since this latter equation is an identity in x, we have $-4A = 1$, $6A - 4B = 0$, and $2A + 3B - 4C = 0$. These yield $A = -\frac{1}{4}$, $B = -\frac{3}{8}$, $C = -\frac{13}{32}$, and $y = -\frac{1}{4}x^2 - \frac{3}{8}x - \frac{13}{32}$ is a particular solution. Thus, the general solution is $y = C_1e^x + C_2e^{-4x} - \frac{1}{4}x^2 - \frac{3}{8}x - \frac{13}{32}$.

33. Solve $\frac{d^2y}{dx^2} - 2\frac{dy}{dx} - 3y = \cos x$.

Here $m^2 - 2m - 3 = 0$, from which $m = -1, 3$; the complementary function is $y = C_1e^{-x} + C_2e^{3x}$. The right-hand member of the differential equation suggests that a particular solution is of the form $A \cos x + B \sin x$. Hence, we substitute $y = A \cos x + B \sin x$, $y' = B \cos x - A \sin x$, and $y'' = -A \cos x - B \sin x$ in the differential equation to obtain

$$(-A\cos x - B\sin x) - 2(B\cos x - A\sin x) - 3(A\cos x + B\sin x) = \cos x$$

or

$$-2(2A + B)\cos x + 2(A - 2B)\sin x = \cos x$$

The latter equation yields $-2(2A + B) = 1$ and $A - 2B = 0$, from which $A = -\frac{1}{5}$, $B = -\frac{1}{10}$. The general solution is $C_1e^{-x} + C_2e^{3x} - \frac{1}{5}\cos x - \frac{1}{10}\sin x$.

34. A weight attached to a spring moves up and down so that the equation of motion is $\frac{d^2s}{dt^2} + 16s = 0$, where s is the stretch of the spring at time t. If $s = 2$ and $\frac{ds}{dt} = 1$ when $t = 0$, find s in terms of t.

Here $m^2 + 16 = 0$ yields $m = \pm 4i$, and the general solution is $s = A \cos 4t + B \sin 4t$. Now when $t = 0$, $s = 2 = A$, so that $s = 2 \cos 4t + B \sin 4t$.

Also when $t = 0$, $ds/dt = 1 = -8 \sin 4t + 4B \cos 4t = 4B$, so that $B = \frac{1}{4}$. Thus, the required equation is $s = 2\cos 4t + \frac{1}{4}\sin 4t$.

35. The electric current in a certain circuit is given by $\frac{d^2I}{dt^2} + 4\frac{dI}{dt} + 2504I = 110$. If $I = 0$ and $\frac{dI}{dt} = 0$ when $t = 0$, find I in terms of t.

Here $m^2 + 4m + 2504 = 0$ yields $m = -2 + 50i, -2 - 50i$; the complementary function is $e^{-2t}(A \cos 50t + B \sin 50t)$. Because the right-hand member is a constant, we find that the particular solution is $I = 110/2504 = 0.044$. Thus, the general solution is $I = e^{-2t}(A \cos 50t + B \sin 50t) + 0.044$.

Also when $t = 0$, $dI/dt = 0 = e^{-2t}[(-2A + 50B) \cos 50t - (2B + 50A) \sin 50t] = -2A + 50B$. Then $B = -0.0018$, and the required relation is $I = -e^{-2t}(0.044 \cos 50t + 0.0018 \sin 50t) + 0.044$.

36. A chain 4 ft long starts to slide off a flat roof with 1 ft hanging over the edge. Discounting friction, find (a) the velocity with which it slides off and (b) the time required to slide off.

Let s denote the length of the chain hanging over the edge of the roof at time t.

(a) The force F causing the chain to slide off the roof is the weight of the part hanging over the edge. That weight is $mgs/4$. Hence,

$$F = \text{mass} \times \text{acceleration} = ms'' = \tfrac{1}{4}mgs \quad \text{or} \quad s'' = \tfrac{1}{4}gs$$

Multiplying by $2s'$ yields $2s's'' = \frac{1}{2}gss'$ and integrating once gives $(s')^2 = \frac{1}{4}gs^2 + C_1$. When $t = 0$, $s = 1$ and $s' = 0$. Hence, $C_1 = -\frac{1}{4}g$ and $s' = \frac{1}{2}\sqrt{g}\sqrt{s^2 - 1}$. When $s = 4$, $s' = \frac{1}{2}\sqrt{15g}$ ft/sec.

(b) Since $\frac{ds}{\sqrt{s^2-1}} = \frac{1}{2}\sqrt{g}\,dt$, integration yields $\ln\left|s+\sqrt{s^2-1}\right| = \frac{1}{2}\sqrt{g}t + C_2$. When $t=0$, $s=1$. Then $C_2 = 0$ and $\ln(s+\sqrt{s^2-1}) = \frac{1}{2}\sqrt{g}t$.

When $s=4$, $t = \frac{2}{\sqrt{g}}\ln(4+\sqrt{15})$seconds.

37. A boat of mass 1600 lb has a speed of 20 ft/sec when its engine is suddenly stopped (at $t=0$). The resistance of the water is proportional to the speed of the boat and is 200 lb when $t=0$. How far will the boat have moved when its speed is reduced to 5 ft/sec?

Let s denote the distance traveled by the boat t seconds after the engine is stopped. Then the force F on the boat is

$$F = ms'' = -Ks' \text{ from which } s'' = -ks'$$

To determine k, we note that at $t=0$, $s'=20$ and $s'' = \frac{\text{force}}{\text{mass}} = -\frac{200g}{1600} = -4$. Then $k = -s''/s' = \frac{1}{5}$. Now $s'' = \frac{dv}{dt} = -\frac{v}{5}$, and integration gives $\ln v = -\frac{1}{5}t + C_1$, or $v = C_1e^{-t/5}$.

When $t=0$, $v=20$. Then $C_1 = 20$ and $v = \frac{ds}{dt} = 20e^{-t/5}$. Another integration yields $s = -100e^{-t/5} + C_2$.

When $t=0$, $s=0$; then $C_2 = 100$ and $s = 100(1-e^{-t/5})$. We require the value of s when $v = 5 = 20e^{-t/5}$, that is, when $e^{-t/5} = \frac{1}{4}$. Then $s = 100(1-\frac{1}{4}) = 75$ ft.

SUPPLEMENTARY PROBLEMS

38. Form the differential equation whose general solution is:

(a) $y = Cx^2 + 1$
(b) $y = C^2x + C$
(c) $y = Cx^2 + C^2$
(d) $xy = x^3 - C$
(e) $y = C_1 + C_2x + C_3x^2$
(f) $y = C_1e^x + C_2e^{2x}$
(g) $y = C_1 \sin x + C_2 \cos x$
(h) $y = C_1e^x \cos(3x + C_2)$

Ans. (a) $xy' = 2(y-1)$; (b) $y' = (y - xy')^2$; (c) $4x^2y = 2x^3y' + (y')^2$; (d) $xy' + y = 3x^2$; (e) $y''' = 0$; (f) $y'' - 3y' + 2y = 0$; (g) $y'' + y = 0$; (h) $y'' - 2y' + 10y = 0$

39. Solve:

(a) $y\,dy - 4x\,dx = 0$ — *Ans.* $y^2 = 4x^2 + C$
(b) $y^2\,dy - 3x^5\,dx = 0$ — *Ans.* $2y^3 = 3x^6 + C$
(c) $x^3y' = y^2(x-4)$ — *Ans.* $x^2 - xy + 2y = Cx^2y$
(d) $(x-2y)\,dy + (y+4x)\,dx = 0$ — *Ans.* $xy - y^2 + 2x^2 = C$
(e) $(2y^2+1)y' = 3x^2y$ — *Ans.* $y^2 + \ln|y| = x^3 + C$
(f) $xy'(2y-1) = y(1-x)$ — *Ans.* $\ln|xy = x + 2y + C$
(g) $(x^2+y^2)\,dx = 2xy\,dy$ — *Ans.* $x^2 - y^2 - Cx$
(h) $(x+y)\,dy = (x-y)\,dx$ — *Ans.* $x^2 - 2xy - y^2 = C$
(i) $x(x+y)\,dy - y^2\,dx = 0$ — *Ans.* $y = Ce^{-y/x}$
(j) $x\,dy - y\,dx + xe^{-y/x}\,dx = 0$ — *Ans.* $e^{y/x} + \ln|Cx| = 0$
(k) $dy = (3y + e^{2x})\,dx$ — *Ans.* $y = (Ce^x - 1)e^{2x}$
(l) $x^2y^2\,dy = (1 - xy^3)\,dx$ — *Ans.* $2x^3y^3 = 3x^2 + C$

40. The tangent and normal to a curve at a point $P(x, y)$ meet the x axis in T and N, respectively, and the y axis in S and M, respectively. Determine the family of curves satisfying the conditions:

(a) $TP = PS$; (b) $NM = MP$; (c) $TP = OP$; (d) $NP = OP$

Ans. (a) $xy = C$; (b) $2x^2 + y^2 = C$; (c) $xy = C$, $y = Cx$; (d) $x^2 \pm y^2 = C$

41. Solve Problem 21, assuming that pure water flows into the tank at the rate of 3 gal/min and the mixture flows out at the same rate.

Ans. 13.44 lb

42. Solve Problem 41 assuming that the mixture flows out at the rate 4 gal/min. (*Hint:* $dq = -\frac{4q}{100-t}dt$).

Ans. 0.02 lb

In Problems 43–59, solve the given equation.

43. $\frac{d^2y}{dx^2} = 3x + 2$ *Ans.* $y = \frac{1}{2}x^3 + x^2 + C_1x + C_2$

44. $e^{2x}\frac{d^2y}{dx^2} = 4(e^{4x} + 1)$ *Ans.* $y = e^{2x} + e^{-2x} + C_1x + C_2$

45. $\frac{d^2y}{dx^2} = -9\sin 3x$ *Ans.* $y = \sin 3x + C_1x + C_2$

46. $x\frac{d^2y}{dx^2} - 3\frac{dy}{dx} + 4x = 0$ *Ans.* $y = x^2 + C_1x^4 + C_2$

47. $\frac{d^2y}{dx^2} - \frac{dy}{dx} = 2x - x^2$ *Ans.* $y = \frac{x^3}{3} + C_1e^x + C_2$

48. $x\frac{d^2y}{dx^2} - \frac{dy}{dx} = 8x^3$ *Ans.* $y = x^4 + C_1x^2 + C_2$

49. $\frac{d^2y}{dx^2} - 3\frac{dy}{dx} + 2y = 0$ *Ans.* $y = C_1e^x + C_2e^{2x}$

50. $\frac{d^2y}{dx^2} + 5\frac{dy}{dx} + 6y = 0$ *Ans.* $y = C_1e^{-2x} + C_2e^{-3x}$

51. $\frac{d^2y}{dx^2} - \frac{dy}{dx} = 0$ *Ans.* $y = C_1 + C_2e^x$

52. $\frac{d^2y}{dx^2} - 2\frac{dy}{dx} + y = 0$ *Ans.* $y = C_2xe^x + C_2e^x$

53. $\frac{d^2y}{dx^2} + 9y = 0$ *Ans.* $y = C_1\cos 3x + C_2\sin 3x$

54. $\frac{d^2y}{dx^2} - 2\frac{dy}{dx} + 5y = 0$ *Ans.* $y = e^x(C_1\cos 2x + C_2\sin 2x)$

55. $\frac{d^2y}{dx^2} - 4\frac{dy}{dx} + 5y = 0$ *Ans.* $y = e^{2x}(C_1\cos x + C_2\sin x)$

56. $\frac{d^2y}{dx^2} + 4\frac{dy}{dx} + 3y = 6x + 23$ *Ans.* $y = C_1e^{-x} + C_2e^{-3x} + 2x + 5$

57. $\dfrac{d^2y}{dx^2}+4y=e^{3x}$ *Ans.* $y=C_1\sin 2x+C_2\cos 2x+\dfrac{e^{3x}}{13}$

58. $\dfrac{d^2y}{dx^2}-6\dfrac{dy}{dx}+9y=x+e^{2x}$ *Ans.* $y=C_1e^{3x}+C_2xe^{3x}+e^{2x}+\dfrac{x}{9}+\dfrac{2}{27}$

59. $\dfrac{d^2y}{dx^2}-y=\cos 2x-2\sin 2x$ *Ans.* $y=C_1e^x+C_2e^{-x}-\frac{1}{5}\cos 2x+\frac{2}{5}\sin 2x$

60. A particle of mass m, moving in a medium that offers a resistance proportional to the velocity, is subject to an attracting force proportional to the displacement. Find the equation of motion of the particle if at time $t=0$, $s=0$ and $s'=v_0$. (*Hint*: Here $m\dfrac{d^2s}{dt^2}=-k_1\dfrac{ds}{dt}-k_2s$ or $\dfrac{d^2s}{dt^2}+2b\dfrac{ds}{dt}+c^2s=0, b>0$.)

Ans. If $b^2=c^2$, $s=v_0te^{-bt}$; if $b^2<c^2$, $s=\dfrac{v_0}{\sqrt{c^2-b^2}}e^{-bt}\sin\sqrt{c^2-b^2}\,t$; if $b^2>c^2$,

$$s=\frac{v_0}{2\sqrt{b^2-c^2}}\left(e^{(-b+\sqrt{b^2-c^2})t}-e^{(-b-\sqrt{b^2-c^2})t}\right)$$

61. Justify our method for solving a separable differential equation $\dfrac{dy}{dx}=-\dfrac{f(x)}{g(y)}$ by integration, that is, $\int f(x)\,dx+\int g(y)\,dy=C$.

Ans. Differentiate both sides of $\int f(x)\,dx+\int g(y)\,dy=C$ with respect to x, obtaining $f(x)+g(y)\dfrac{dy}{dx}=0$. Hence, $\dfrac{dy}{dx}=-\dfrac{f(x)}{g(y)}$ and the solution y satisfies the given equation.

Trigonometric Formulas

$\cos^2\theta + \sin^2\theta = 1$

$\cos(\theta + 2\pi) = \cos\theta$

$\sin(\theta + 2\pi) = \sin\theta$

$\cos(-\theta) = \cos\theta$

$\sin(-\theta) = -\sin\theta$

$\cos(u + v) = \cos u \cos v - \sin u \sin v$

$\cos(u - v) = \cos u \cos v + \sin u \sin v$

$\sin(u + v) = \sin u \cos v + \cos u \sin v$

$\sin(u - v) = \sin u \cos v - \cos u \sin v$

$\sin(2\theta) = 2\sin\theta\cos\theta$

$\cos 2\theta = \cos^2\theta - \sin^2\theta$

$\quad = 2\cos^2\theta - 1 = 1 - 2\sin^2\theta$

$\cos^2\dfrac{\theta}{2} = \dfrac{1 + \cos\theta}{2}$

$\sin^2\dfrac{\theta}{2} = \dfrac{1 - \cos\theta}{2}$

$\tan x = \dfrac{\sin x}{\cos x} = \dfrac{1}{\cot x}$

$\cot x = \dfrac{\cos x}{\sin x} = \dfrac{1}{\tan x}$

$\sec x = \dfrac{1}{\cos x}$

$\csc x = \dfrac{1}{\sin x}$

$\tan(-x) = -\tan x$

$\tan(x + \pi) = \tan x$

$1 + \tan^2 x = \sec^2 x$

$1 + \cot^2 x = \csc^2 x$

$\tan(u + v) = \dfrac{\tan u + \tan v}{1 - \tan u \tan v}$

$\tan(u - v) = \dfrac{\tan u - \tan v}{1 + \tan u \tan v}$

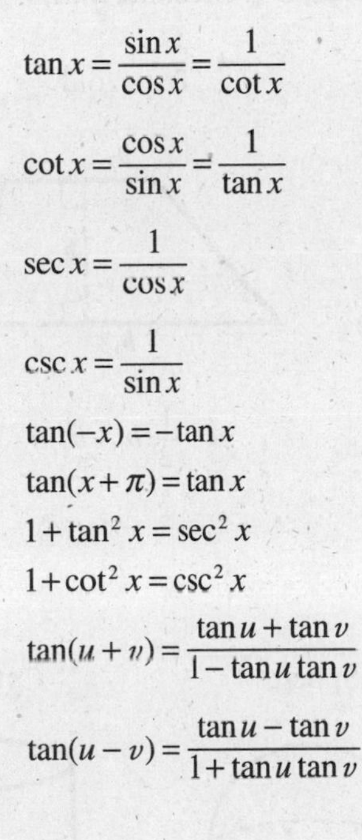

$\cos\left(\dfrac{\pi}{2} - \theta\right) = \sin\theta$; $\sin(\pi - \theta) = \sin\theta$; $\sin(\theta + \pi) = -\sin\theta$

$\sin\left(\dfrac{\pi}{2} - \theta\right) = \cos\theta$; $\cos(\pi - \theta) = -\cos\theta$; $\cos(\theta + \pi) = -\cos\theta$

Law of cosines: $c^2 = a^2 + b^2 - 2ab\cos\theta$

Law of sines: $\dfrac{\sin A}{a} = \dfrac{\sin B}{b} = \dfrac{\sin C}{c}$

Geometric Formulas

(A = area, C = circumference, V = volume, S = lateral surface area)

Triangle

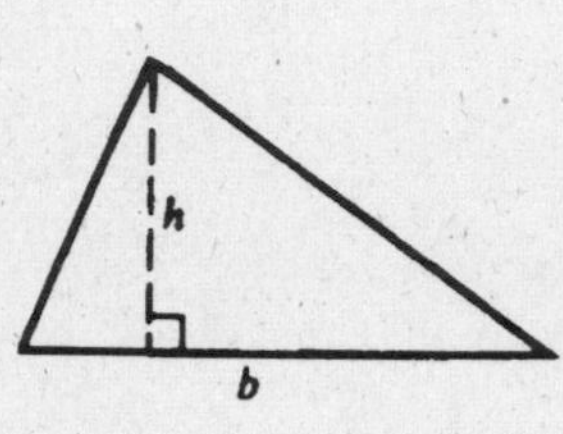

$A = \frac{1}{2}bh$

Trapezoid

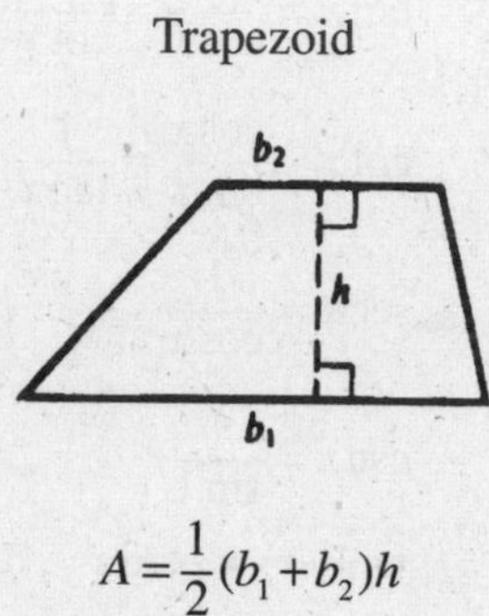

$A = \frac{1}{2}(b_1 + b_2)h$

Parallelogram

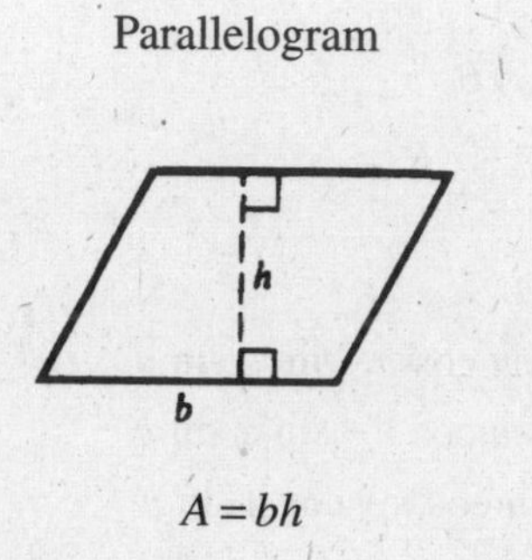

$A = bh$

Circle

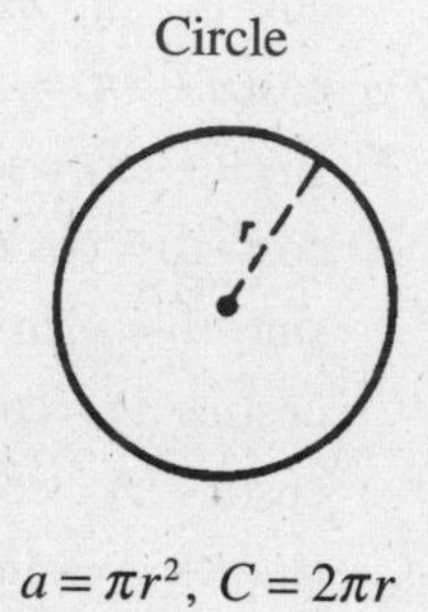

$a = \pi r^2,\ C = 2\pi r$

Sphere

Cylinder

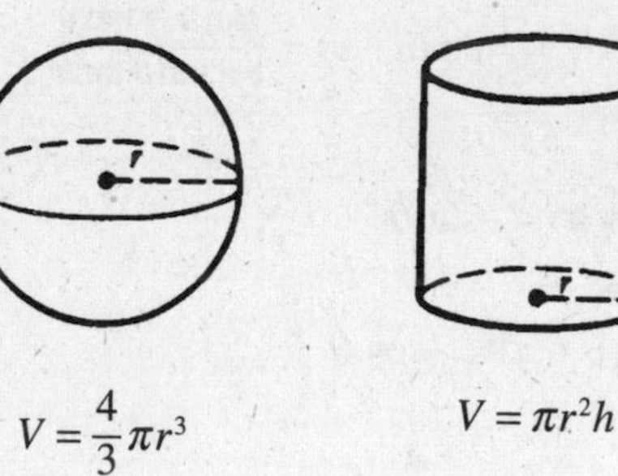

Sphere: $V = \frac{4}{3}\pi r^3$

$S = 4\pi r^2$

Cylinder: $V = \pi r^2 h$

$S = 2\pi rh$

Cone

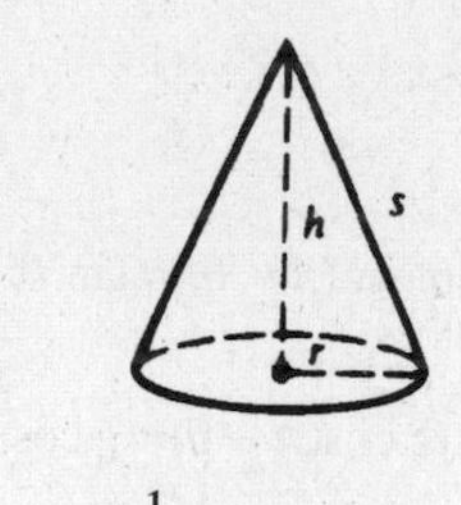

$V = \frac{1}{3}\pi r^2 h$

$S = \pi rs = \pi r\sqrt{r^2 + h^2}$

Index